SIGNALS, SYSTEMS, AND TRANSFORMS

THIRD EDITION

Linear $\quad a_1 x(t) + a_2 x_2(t) = y_1(t) + y_2(t)$

LST 1610
Lecture 11970

① The system impulse response signal is the inverse laplace xform of the system's x-fer function. alternatively, the system x-fer function is the Laplace xform of its impulse response signal.

② for the given transfer function in 's'domain the impulse response is the inverse of Laplace

$$\mathcal{L}\{\ \}\quad T(s) = \frac{s+9}{s^2-4s+2}$$

$$\mathcal{L}^{-1}\left\{\frac{s+7}{s^2-4s+2}\right\} = \frac{\frac{1}{2}-\frac{9}{4}\sqrt{2}}{s-2+2} + \frac{\frac{1}{2}+\frac{9}{4}\sqrt{2}}{s-2-\sqrt{2}}$$

$$\mathcal{L}^{-1}\{T(s)\} = \left(\frac{1}{2}-\frac{9}{4}\sqrt{2}\right)e^{(2-\sqrt{2})t} + \left(\frac{1}{2}+\frac{9}{4}\sqrt{2}\right)e^{(2+\sqrt{2})t}\ ]\ y(t)$$

· Stable if poles
left sid pole → stable
Right or both side pole → unstable

Time varying
$y_1(t) = y(t)\big|_{x(t-t_0)}$
$= y(t)\big|_{t-t_0}$

TI

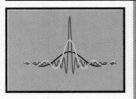

SIGNALS, SYSTEMS, AND TRANSFORMS

THIRD EDITION

CHARLES L. PHILLIPS
Emeritus
Auburn University
Auburn, Alabama

JOHN M. PARR
University of Evansville
Evansville, Indiana

EVE A. RISKIN
University of Washington
Seattle, Washington

Pearson Education, Inc.
Upper Saddle River, NJ 07458

Library of Congress Cataloging-in-Publication Data on file

Vice President and Editorial Director, ECS: *Marcia J. Horton*
Publisher: *Tom Robbins*
Aquisitions Editor: *Alice Dworkin*
Vice President and Director of Production and Manufacturing, ESM: *David W. Riccardi*
Executive Managing Editor: *Vince O'Brien*
Managing Editor: *David A. George*
Production Editor: *Scott Disanno*
Director of Creative Services: *Paul Belfanti*
Creative Director: *Carole Anson*
Art Director: *Jayne Conte*
Cover Designer: *Bruce Kenselaar*
Art Editor: *Greg Dulles*
Manufacturing Manager: *Trudy Pisciotti*
Manufacturing Buyer: *Lisa McDowell*
Marketing Manager: *Holly Stark*

© 2003 by Pearson Education, Inc.
Pearson Education, Inc.
Upper Saddle River, NJ 07458

MATLAB is a registered trademark of the Math Works, Inc.

The MathWorks, Inc., 3 Apple Hill Drive, Natick, MA 01760-2098.

Printed in the United States of America
10 9 8 7 6 5 4 3

ISBN 0-13-041207-4

Pearson Education Ltd., *London*
Pearson Education Australia Pty. Ltd., *Sydney*
Pearson Education Singapore, Pte. Ltd.
Pearson Education North Asia Ltd., *Hong Kong*
Pearson Education Canada, Inc., *Toronto*
Pearson Educación de Mexico, S.A. de C.V.
Pearson Education—Japan, *Tokyo*
Pearson Education Malaysia, Pte. Ltd.
Pearson Education, Inc., *Upper Saddle River, New Jersey*

To

Taylor, Justin, Jackson, Rebecca, and Alex

Judith, Dara, Johna, and Duncan

Gary, Noah, and Aden

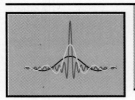

CONTENTS

3   CONTINUOUS-TIME LINEAR TIME-INVARIANT SYSTEMS   89

11 THE z-TRANSFORM   547

12 FOURIER TRANSFORMS OF DISCRETE-TIME SIGNALS   599

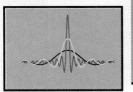

The basic structure and philosophy of the previous editions of *Signals, System and Transforms* are retained in the third edition. New examples have been added and some examples have been revised to demonstrate key concepts more clearly. New figures have been added to better illustrate concepts such as aliasing, orthogonality of exponentials, data reconstruction, etc. The wording of many passages throughout the text has been revised to ease reading and improve clarity. In particular, we have greatly simplified the development of convolution, the Fourier Transform, and the Discrete Fourier Transform. Further, we use sidebars in Sections 2.1 and 2.7 to demonstrate real-world applications of the material.

Chapters 5, 6, and 12 have been reorganized to consolidate the presentation on sampling and data construction and to reduce redundancy. Many end-of-chapter problems have been revised and numerous new problems are provided. Several of these new problems illustrate real-world concepts in digital communications, filtering, and control theory. In addition, in response to requests from students at our universities, we have included answers to selected problems in Appendix H. We hope that this will enable the student to obtain immediate feedback about his/her understanding of new material and concepts.

All MATLAB examples have been updated to ensure compatibility with Student Version Release 12. Several new MATLAB examples have been added.

New to this edition is a third co-author, Professor Eve Riskin from the University of Washington. Professor Riskin has contributed many ideas for the text including a companion web site at *http://www.ee.washington.edu/class/SST_textbook/textbook.html.*

This web site contains sample laboratories, lecture notes for Chapters 1–7 and Chapters 9–12, and the MATLAB files listed in the textbook as well as several additional MATLAB files. It also contains a link to a second web site at http://www.ee.washington.edu/class/235dl/, which contains interactive versions of the lecture notes for Chapters 1–7. Here, students and professors can find worked-out solutions to all the examples in the lecture notes, as well as animated demonstrations of various concepts including transformations of continuous-time signals, properties of continuous-time systems (including numerous examples on time-invariance), convolution, sampling, and aliasing. Additional examples for discrete-time material will be added as they are developed.

In addition to the website listed above, the Department of Electrical Engineering, University of Washington, maintains an electronic mail list server for your use. For information on how to subscribe and unsubscribe, simply send a plain text E-mail message with the word HELP as the message body (and nothing else) to *sst_textbook-request@ee.washington.edu.* This list server will be used to communicate any typos found in the book or solution manual as well as point out new updates to the above-mentioned web pages.

This book is intended to be used primarily as a text for junior-level students in engineering curricula and for self-study by practicing engineers. It is assumed that the reader has had some introduction to signal models, system models, and differential equations (as in, for example, circuits courses and courses in mathematics), and some laboratory work with physical systems.

The authors have attempted to consistently differentiate between signal and system models and physical signals and systems. Although a true understanding of this difference can be acquired only through experience, readers should understand that there are usually significant differences in performance between physical systems and their mathematical models.

We have attempted to relate the mathematical results to physical systems that are familiar to the readers (for example, the simple pendulum) or physical systems that students can visualize (for example, a picture in a picture for television). The descriptions of these physical systems, given in Chapter 1, are not complete in any sense of the word; these systems are introduced simply to illustrate practical applications of the mathematical procedures presented.

Generally, practicing engineers must in some manner validate their work. To introduce the topic of validation, the results of examples are verified using different procedures where practical. Many homework problems require verification of the results. Hence, students become familiar with the process of validating their own work.

The software tool MATLAB is integrated into the text in two ways. First, in appropriate examples, MATLAB programs are provided that will verify the computations. Then, in appropriate homework problems, the student is asked to verify the calculations using MATLAB. This verification should not be difficult because MATLAB programs given in examples similar to the problems are applicable. Hence, another procedure for verification is given. The MATLAB programs given in the examples may be downloaded from *http://www.ee.washington.edu/class/SST_textbook/textbook.html.* Students can alter data statements in these programs to apply them to the end-of-chapter problems. This should minimize programming errors. Hence, another procedure for verification is given. However, all references to MATLAB may be omitted, if the instructor or reader so desires.

Laplace transforms are covered in Chapter 7 and z-transforms are covered in Chapter 11. At many universities, one or both transforms are introduced prior to the signals and systems courses. Chapters 7 and 11 are written such that the material can be covered anywhere in the signals and systems course, or it can be omitted entirely, except for required references.

The more advanced material has been placed toward the end of the chapters wherever possible. Hence, this material may be omitted if desired. For example, sections 3.7, 3.8, 4.6, 5.5, 7.9, 10.7, 12.6, 12.7, and 12.8 could be omitted by instructors without loss of continuity in teaching. Further, Chapters 8 and 13 can be skipped if a professor does not wish to cover state-space material at the undergraduate level.

The material of this book is organized into two principal areas: continuous-time signals and systems, and discrete-time signals and systems. Some professors prefer to cover first one of these topics, followed by the second. Other professors prefer to cover continuous-time material and discrete-time material simultaneously. The authors have taken the first approach, with the continuous-time material covered in Chapters 2–8, and the discrete-time material covered in Chapters 9–13. The material on discrete-time concepts is essentially independent of the material on continuous-time concepts so that a professor or reader who desires to study the discrete-time material first could cover Chapters 9–11 and 13 before Chapters 2–8. The material may also be arranged such that basic continuous-time material and discrete-time material are intermixed. For example, Chapters 2 and 9 may be covered simultaneously and Chapters 3 and 10 may also be covered simultaneously.

In Chapter 1, we present a brief introduction to signals and systems, followed by short descriptions of several physical continuous-time and discrete-time systems. In addition, some of the signals that appear in these systems are described. Then a very brief introduction to MATLAB is given.

In Chapter 2, we present general material basic to continuous-time signals and systems; the same material for discrete-time signals and systems is presented in Chapter 9. However, as stated above, Chapter 9 can be covered before Chapter 2 or simultaneously with Chapter 2. Chapter 3 extends this basic material to continuous-time linear time-invariant systems, while Chapter 10 does the same for discrete-time linear time-invariant systems.

Presented in Chapters 4, 5, and 6 are the Fourier series and the Fourier transform for continuous-time signals and systems. The Laplace transform is then developed in Chapter 7. State variables for continuous-time systems are covered in Chapter 8; this development utilizes the Laplace transform.

The z-transform is developed in Chapter 11, with the discrete-time Fourier transform and the discrete Fourier transform presented in Chapter 12. However, Chapter 12 may be covered prior to Chapter 11. The development of the discrete-time Fourier transform and discrete Fourier transform in Chapter 12 assumes that the reader is familiar with the Fourier transform. State variables for discrete-time systems are given in Chapter 13. This material is independent of the state variables for continuous-time systems of Chapter 8.

In Appendix A, we give some useful integrals and trigonometric identities. In general, the table of integrals is used in the book, rather than taking the longer approach of integration by parts. Leibnitz's rule for the differentiation of an integral and L'Hôpital's rule for indeterminate forms are given in Appendix B and are referenced in the text where needed. Appendix C covers the closed forms for certain geometric series; this material is useful in discrete-time signals and systems. In Appendix D, we review complex numbers and introduce Euler's relation, in Appendix E the solution of linear differential equations with constant coefficients, and in Appendix F partial-fraction expansions. Matrices are reviewed in Appendix G; this appendix is required for the state-variable coverage of Chapters 8 and 13. As each matrix operation is defined, MATLAB statements that perform the operation are given.

This book may be covered in its entirety in two 3-semester-hour courses, or in quarter courses of approximately the equivalent of 6 semester hours. With the omission of appropriate material, the remaining parts of the book may be covered with fewer credits. For example, most of the material of Chapters 2, 3, 4, 5, 6, 8, 9, 10, 11 and 12 has been covered in one 4-semester-hour course. The students were already familiar with some linear-system analysis and the Laplace transform.

We wish to acknowledge the many colleagues and students at Auburn University, the University of Evansville, and the University of Washington who have contributed to the development of this book. In particular, the first author wishes to express thanks to Professors Charles M. Gross, Martial A. Honnell, and Charles L. Rogers of Auburn University for many stimulating discussions on the topics in this book, and to Professor Roger Webb, director of the School of Electrical Engineering at the Georgia Institute of Technology, for the opportunity to teach the signal and system courses at Georgia Tech. The second author wishes to thank Professors Dick Blandford and William Thayer for their encouragement and support for this effort, and Professor David Mitchell for his enthusiastic discussions of the subject matter. The third author wishes to thank the professors and many students in EE235 and EE341 at the University of Washington who contributed comments to this book and interactive web site, in particular Professors Mari Ostendorf and Mani Soma, Eddy Ferré, Wai Shan Lau, Bee Ngo, Sanaz Namdar, and Jessica Tsao. The interactive web site was developed under a grant from the Fund for the Improvement of Postsecondary Education (FIPSE), U.S. Department of Education.

CHARLES L. PHILLIPS
Auburn University

JOHN M. PARR
University of Evansville

EVE A. RISKIN
University of Washington

1 INTRODUCTION

In this book, we consider the topics of signals and systems as related to engineering. These topics involve the *modeling* of physical signals by mathematical functions, the *modeling* of physical systems by mathematical equations, and the solutions of the equations when excited by the functions.

1.1 MODELING

Engineers must model two distinct physical phenomena. First, *physical systems* are modeled by *mathematical equations*. For systems that contain no sampling (*continuous-time*, or *analog, systems*), we prefer to use ordinary differential equations with constant coefficients; a wealth of information is available for the analysis and the design of systems of this type. Of course, the equation must accurately model the physical systems. An example of the model of a physical system is a linear electric-circuit model of Figure 1.1:

$$L\frac{di(t)}{dt} + Ri(t) + \frac{1}{C}\int_{-\infty}^{t} i(\tau)d\tau = v(t). \tag{1.1}$$

Another example is Newton's second law,

$$f(t) = M\frac{d^2x(t)}{dt^2}, \tag{1.2}$$

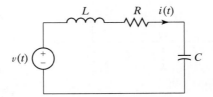

Figure 1.1 Example circuit.

where $f(t)$ is the force applied to the mass M and $x(t)$ is the resulting displacement of the mass.

A second physical phenomenon to be modeled is called *signals*. *Physical signals* are modeled by *mathematical functions*. One example of a physical signal is the voltage that is applied to the speaker in a radio. Another example is the temperature at a designated point in a particular room. This signal is a function of time because the temperature varies with time. We can express this temperature as

$$\text{temperature at a point} = \theta(t), \tag{1.3}$$

where $\theta(t)$ has the units of, for example, degrees Celsius.

Consider again Newton's second law. Equation (1.2) is the *model* of a physical system, and $f(t)$ and $x(t)$ are *models* of physical signals. Given the signal (function) $f(t)$, we solve the model (equation) (1.2) for the signal (function) $x(t)$. In analyzing physical systems, we apply mathematics to the *models* of systems and signals, not to the physical systems and signals. The usefulness of the results depends on the accuracy of the models.

In this book, we usually limit signals to having one independent variable. We choose this independent variable to be *time, t*, without loss of generality. Signals are divided into two natural categories. The first category to be considered is *continuous-time*, or simply *continuous, signals*. A signal of this type is defined for all values of time. A continuous-time signal is also called an *analog signal*. A continuous-time signal is illustrated in Figure 1.2(a).

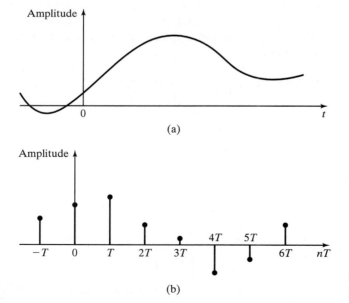

(a)

(b)

Figure 1.2 (a) Continuous-time signal; (b) discrete-time signal.

The second category for signals is *discrete-time*, or simply *discrete, signals*. A discrete signal is defined at only certain instants of time. For example, suppose that a signal $f(t)$ is to be processed by a digital computer. [This operation is called *digital signal processing* (DSP).] Because a computer can operate only on numbers and not on a continuum, the continuous signal must be converted into a sequence of numbers by sampling. If a signal $f(t)$ is sampled every T seconds, the number sequence $f(nT), n = \ldots, -2, -1, 0, 1, 2, \ldots,$ is available to the computer. This sequence of numbers is called a *discrete-time signal*. Insofar as the computer is concerned, $f(nT)$ with n a noninteger does not exist (is not available). A discrete-time signal is illustrated in Figure 1.2(b).

We define a *continuous-time system* as one in which all signals are continuous time. We define a *discrete-time system* as one in which all signals are discrete time. Both continuous-time and discrete-time signals appear in some physical systems; we call these systems *hybrid systems* or *sampled-data systems*. An example of a sampled-data system is an automatic aircraft-landing system, in which the control functions are implemented on a digital computer. We do not consider hybrid systems in this book.

The mathematical analysis of physical systems can be represented as in Figure 1.3 [1]. We first develop mathematical models of the physical systems and signals involved. One procedure for finding the model of a physical system is to use the laws of physics, as, for example, in (1.1). Once a model is developed, the equations are solved for typical excitation functions. This solution is compared to the response of the physical system with the same excitations. If the two responses are approximately equal, we can then use the model in analysis and design. If not, we must improve the model.

Improving the mathematical model of a system usually involves making the models more complex and is not a simple step. Several iterations of the process illustrated in Figure 1.3 may be necessary before a model of adequate accuracy results. For some simple systems, the modeling may be completed in hours; for very complex systems, the modeling may take years. An example of a complex model is that of NASA's shuttle; this model relates the position and attitude of the shuttle to the engine thrust, the wind, the positions of the control surfaces (e.g., the rudder),

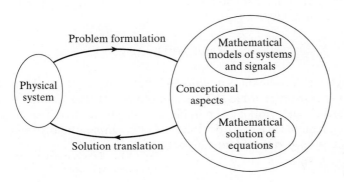

Figure 1.3 Mathematical solutions of physical problems.

and so on. As an additional point, for complex models of this type, the equations can be solved only by computer.

This book contains two main topics: (1) continuous-time signals and systems and (2) discrete-time signals and systems. Chapters 2 through 8 cover continuous-time signals and systems, while Chapters 9 through 13 cover discrete-time signals and systems. The material may be covered in the order of the chapters, in which continuous-time topics and discrete-time topics are covered separately. The basic material of the two topics may also be intermixed, with Chapters 2 and 9 covered simultaneously, followed by Chapters 3 and 10 covered simultaneously.

1.2   CONTINUOUS-TIME PHYSICAL SYSTEMS

In this section, we discuss several continuous-time physical systems. The descriptions are simplified; references are given that contain more complete descriptions. The systems described in this and the following section are used in examples in the remainder of the book.

We have already given the model of a rigid mass M in a frictionless environment

[eq(1.2)]
$$f(t) = M \frac{d^2x(t)}{dt^2}, \quad = M \times f(t)''$$

where $f(t)$ is the force applied to the mass and $x(t)$ is the displacement of the mass that results from the force applied. This model is a *second-order linear differential equation with constant coefficients*.

The terms *linear equation* and *nonlinear equation* are defined in Section 2.7. As we will see, an equation (or system) is *linear* if the principle of superposition applies. Otherwise, the equation is *nonlinear*.

Next we discuss several physical systems.

Electric Circuits

In this section, we give models for some electric-circuit elements [2]. We begin with the model for *resistance*, given by

$$v(t) = Ri(t), \tag{1.4}$$

where the voltage $v(t)$ has the units of volts (V), the current $i(t)$ has the units of amperes (A), and the resistance R has the units of ohms (Ω). This model is represented by the standard circuit symbol given in Figure 1.4. The dashed lines in this figure indicate that the elements are parts of circuits. For example, the resistance must be a part of a circuit, or else $v(t)$ is identically zero.

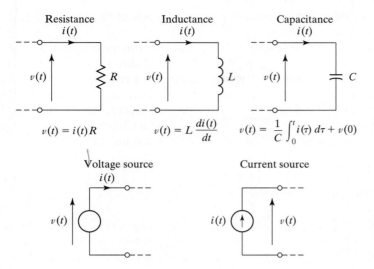

Figure 1.4 Electric-circuit elements. (From C. L. Phillips and R. D. Harbor, *Feedback Control Systems*, 3rd ed., Prentice Hall, Upper Saddle River, NJ, 1995.)

The model for *inductance* is given by

$$v(t) = L\frac{di(t)}{dt},\qquad\qquad(1.5)$$

where $v(t)$ and $i(t)$ are as defined earlier and L is the inductance in henrys. The model for *capacitance* is given by

$$v(t) = \frac{1}{C}\int_{-\infty}^{t} i(\tau)d\tau = \frac{1}{C}\int_{0}^{t} i(\tau)d\tau + v(0),\qquad\qquad(1.6)$$

where C is the capacitance in farads. The symbols for inductance and capacitance are also given in Figure 1.4.

For the ideal voltage source in Figure 1.4, the voltage at the terminals of the source is $v(t)$, independent of the circuit connected to these terminals. The current $i(t)$ that flows through the voltage source is determined by the circuit connected to the source. For the ideal current source, the current that flows through the current source is $i(t)$, independent of the circuit connected to the source. The voltage $v(t)$ that appears at the terminals of the current source is determined by the circuit connected to these terminals.

Consider now a circuit that is an interconnection of the elements shown in Figure 1.4. The circuit equations are written using the models given in the figure along with Kirchhoff's voltage and current laws. Kirchhoff's voltage law may be stated as follows:

The algebraic sum of voltages around any closed loop in an electric circuit is zero.

Kirchhoff's current law may be stated as follows:

The algebraic sum of currents into any junction in an electric circuit is zero.

Operational
amplifier

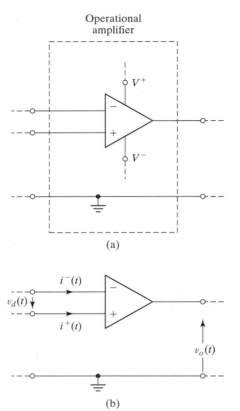

(a)

(b)   **Figure 1.5** Operational amplifier.

Operational Amplifier Circuits

A device called an *operational amplifier* (op amp) [3] is commonly used in circuits
for processing analog electrical signals. We do not investigate the internal structure
of this amplifier, but instead present only its terminal characteristics.

We denote an operational amplifier by the circuit symbol of Figure 1.5(a). The
circles indicate amplifier terminals, and the dashed lines indicate connections exter-
nal to the amplifier. The signal-input terminals are labeled with a minus sign for the
inverting input and a plus sign for the *noninverting input*. The power-supply terminals are labeled V^+ for the positive dc voltage and V^- for the negative dc voltage.
The op amp is normally shown as in Figure 1.5(b), with the power-supply terminals
omitted. In this circuit, $v_d(t)$ is the input voltage to be amplified and the amplified
voltage output is $v_o(t)$.

The operational amplifier is designed and constructed such that the input impedance is very high, resulting in the input currents $i^-(t)$ and $i^+(t)$ in Figure 1.5(b)
being very small. Additionally, the amplifier gain [the ratio $v_o(t)/v_d(t)$] is very large
(on the order of 10^5 or larger). This large gain results in a very small allowable input
voltage if the amplifier is to operate in its linear range (not saturated).

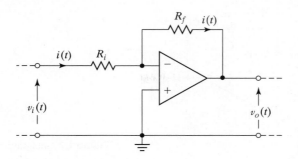

Figure 1.6 Practical voltage amplifier.

For this discussion, we assume that the amplifier is ideal, which is sufficiently accurate for most purposes. The ideal op amp has zero input currents $[i^-(t) = i^+(t) = 0]$. Additionally, the ideal amplifier operates in its linear range with infinite gain, resulting in the input voltage $v_d(t)$ being zero.

Because the op amp is a very high gain device, feedback is usually added for stabilization. The feedback is connected from the output terminal to the inverting input terminal (the minus terminal). This connection results in negative, or stabilizing, feedback and tends to prevent saturation of the op amp.

An example of a practical op-amp circuit is given in Figure 1.6. In this circuit, $v_i(t)$ is the circuit input voltage and $v_o(t)$ the circuit output voltage. Because $v_d(t)$ in Figure 1.5(b) is assumed to be zero, the equation for the input loop in Figure 1.6 is given by

$$v_i(t) - i(t)R_i = 0 \Rightarrow i(t) = \frac{v_i(t)}{R_i}. \tag{1.7}$$

Also, because $i^-(t)$ in Figure 1.5(b) is zero, the current through R_f in Figure 1.6 is equal to that through R_i. The equation for the outer loop is then

$$v_i(t) - i(t)R_i - i(t)R_f - v_o(t) = 0.$$

Using (1.7), we express this equation as

$$v_i(t) - v_i(t) - \frac{v_i(t)}{R_i}R_f - v_o(t) = 0 \Rightarrow \frac{v_o(t)}{v_i(t)} = -\frac{R_f}{R_i}. \tag{1.8}$$

This circuit is then a *voltage amplifier*. The ratio R_f/R_i is a real number; hence, the amplifier voltage gain $v_o(t)/v_i(t)$ is a negative real number. The model (1.8) is a *linear algebraic equation.*

A second practical op-amp circuit is given in Figure 1.7. We use the preceding procedure to analyze this circuit. Because the input loop is unchanged, (1.7) applies, with $R_i = R$. The equation of the outer loop is given by

$$v_i(t) - i(t)R - \frac{1}{C}\int_{-\infty}^{t} i(\tau)d\tau - v_o(t) = 0. \tag{1.9}$$

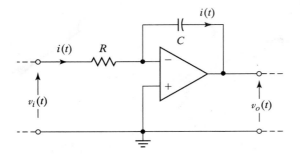

Figure 1.7 Integrating amplifier.

Substitution of (1.7) into (1.9) yields

$$v_i(t) - v_i(t) - \frac{1}{RC} \int_{-\infty}^{t} v_i(\tau)d\tau - v_o(t) = 0. \tag{1.10}$$

Thus, the equation describing this circuit is given by

$$v_0(t) = -\frac{1}{RC} \int_{-\infty}^{t} v_i(\tau)d\tau. \tag{1.11}$$

This circuit is called an *integrator* or an *integrating amplifier*; the output voltage is the integral of the input voltage multiplied by a negative constant $(-1/RC)$. This integrator is a commonly used circuit in analog signal processing and is used in several examples in this book.

If the positions of the resistance and the capacitance in Figure 1.7 are interchanged, the op-amp circuit of Figure 1.8 results. We state without proof that the equation of this circuit is given by

$$v_o(t) = -RC\frac{dv_i(t)}{dt}. \tag{1.12}$$

(The reader can show this using the previous procedure.) This circuit is called a *differentiator* or a *differentiating amplifier*; the output voltage is the derivative of the input voltage multiplied by a negative constant $(-RC)$. The differentiator has limited

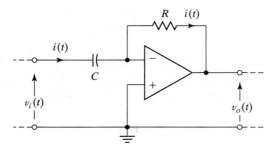

Figure 1.8 Differentiating amplifier.

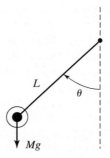

Figure 1.9 Simple pendulum.

use in analog signal processing, because the derivative of a signal that changes rapidly is large. Hence, the differentiator amplifies any high-frequency noise in $v_i(t)$. However, some practical applications require the use of a differentiator. For these applications, usually some type of high-frequency filtering is required before the differentiation, to reduce high-frequency noise.

Simple Pendulum

We now consider a differential-equation model of the simple pendulum, which is illustrated in Figure 1.9. The angle of the pendulum is denoted as θ, the mass of the pendulum bob is M, and the length of the (weightless) arm from the axis of rotation to the center of the bob is L.

The force acting on the bob of the pendulum is then Mg, where g is the gravitational acceleration, as shown in Figure 1.9. From physics we recall the equation of motion of the simple pendulum:

$$ML \frac{d^2\theta(t)}{dt^2} = -Mg \sin \theta(t). \tag{1.13}$$

This model is a *second-order nonlinear differential equation*; the term $\sin \theta(t)$ is nonlinear. (Superposition does not apply.)

We have great difficulty in solving nonlinear differential equations; however, we can *linearize* (1.13). The power-series expansion for $\sin \theta$ is given (from Appendix D) by

$$\sin \theta = \theta - \frac{\theta^3}{3!} + \frac{\theta^5}{5!} - \cdots. \tag{1.14}$$

For θ small, we can ignore all terms except the first one, resulting in $\sin \theta \approx \theta$. The error in this approximation is less than 1% for $\theta = 20°$ and decreases as θ becomes smaller. We then express the model of the pendulum as, from (1.13) and (1.14),

$$\frac{d^2\theta(t)}{dt^2} + \frac{g}{L}\theta(t) = 0 \tag{1.15}$$

for θ small. This model is a *second-order linear differential equation with constant coefficients.*

This derivation illustrates both a linear model (1.15) and a nonlinear model (1.13) and one procedure for linearizing a nonlinear model. Models (1.13) and (1.15) have unusual characteristics because friction has been ignored. Energy given to the system by displacing the bob and releasing it cannot be dissipated. Hence, we expect the bob to remain in motion for all time once it has been set in motion. Note that these comments relate to a model of a pendulum, not to the physical device. If we want to model a physical pendulum more accurately, we must as a minimum include a term in (1.13) and (1.15) for friction.

DC Power Supplies

Power supplies that convert an ac voltage (sinusoidal voltage) into a dc voltage (constant voltage) [3] are required in almost all electronic equipment. Shown in Figure 1.10 are voltages that appear in certain dc power supplies in which the ac voltage is converted to a nonnegative voltage.

The voltage in Figure 1.10(a) is called a *half-wave rectified signal*. This signal is generated from a sinusoidal signal by replacing the negative half cycles of the sinusoid with a value of zero. The positive half cycles are unchanged. In this figure, T_0 is the period of the waveform (the time of one cycle).

The signal in Figure 1.10(b) is called a *full-wave rectified signal*. This signal is generated from a sinusoidal signal by the amplitude reversal of each negative half cycle. The positive half cycles are unchanged. Note that the period T_0 of this signal is one-half that of the sinusoid and, hence, one-half that of the half-wave rectified signal.

Usually, these waveforms are generated using diodes. The circuit symbol for a diode is given in Figure 1.11(a). An ideal diode has the voltage–current characteristic shown by the heavy line in Figure 1.11(b). The diode allows current to flow

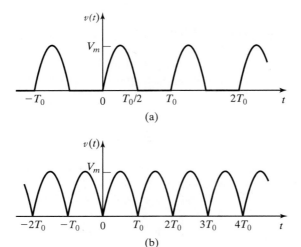

(a)

(b)

Figure 1.10 Rectified signals: (a) half wave; (b) full wave.

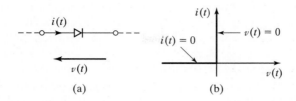

Figure 1.11 (a) Diode; (b) ideal diode characteristic.

unimpeded in the direction of the arrowhead in its symbol and blocks current flow in the opposite direction. Hence, the ideal diode is a short circuit for current flow in the direction of the arrowhead [when $v(t)$ tends to be positive] and an open circuit for current flow in the opposite direction [when $v(t)$ is negative]. The diode is a non-linear device; therefore, any circuit that contains a diode is a nonlinear circuit.

One circuit for a power supply is given in Figure 1.12(a). The power-supply load is represented by the resistance R_L, and the voltage across this load is the half-wave rectified signal of Figure 1.12(b). The load current $i_L(t)$ is the load voltage $v_L(t)$ divided by R_L; $i_L(t)$ is also shown in Figure 1.12(b). We see then that the voltage across the load is unidirectional; however, this voltage is not constant.

A practical dc power supply is illustrated in Figure 1.13. The inductor–capacitor (LC) circuit forms a low-pass filter and is added to the circuit to filter out the voltage variations such that the load voltage $v_L(t)$ is approximately constant.

A circuit that uses four diodes to generate a full-wave rectified signal is given in Figure 1.14. The diodes A and D conduct when the source voltage is positive, and the diodes B and C conduct when the source voltage is negative. However, the current

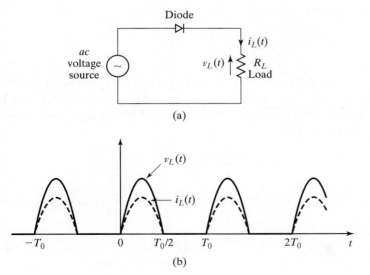

Figure 1.12 Half-wave rectifier.

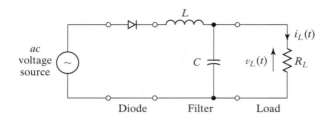

Figure 1.13 Practical dc power supply.

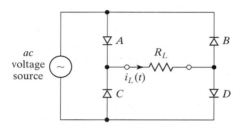

Figure 1.14 Full-wave rectifier.

through the load resistance R_L is unidirectional, in the direction shown. Hence, the voltage across the load is a full-wave rectified signal, as shown in Figure 1.10(b). As in the half-wave rectified case in Figure 1.13, a filter is usually added to convert the load voltage to (approximately) dc.

Analogous Systems

We introduce analogous systems with two examples. The model of a rigid mass M in a frictionless environment is given in (1.2):

$$M \frac{d^2 x(t)}{dt^2} = f(t), \tag{1.16}$$

where $f(t)$ is the force applied to the mass and $x(t)$ is the displacement of the mass that results from the applied force. We can represent this system with Figure 1.15(a).

Consider next the circuit of Figure 1.15(b), in which a voltage $v(t)$ is applied to an inductance. The loop equation is given by

$$L \frac{di(t)}{dt} = v(t). \tag{1.17}$$

Recall that $i(t) = dq(t)/dt$, where $q(t)$ is charge. Hence, we can express the loop equation (1.17) as

$$L \frac{d^2 q(t)}{dt^2} = v(t). \tag{1.18}$$

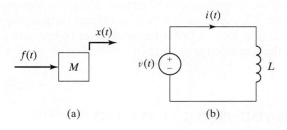

(a)   (b)   **Figure 1.15** Analogous systems.

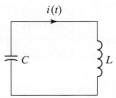

Figure 1.16 LC circuit.

We see that the model for the mass in (1.16) and for the circuit in (1.18) are of the same mathematical form; these two systems are called *analogous systems*. We define analogous systems as systems that are modeled by equations of the same mathematical form.

As a second example, consider the LC circuit in Figure 1.16, which is excited by initial conditions. The loop equation for this circuit is given by

$$L\frac{di(t)}{dt} + \frac{1}{C}\int_{-\infty}^{t} i(\tau)d\tau = 0. \tag{1.19}$$

Expressing this equation as a function of charge $q(t)$ yields

$$\frac{d^2q(t)}{dt^2} + \frac{1}{LC}q(t) = 0. \tag{1.20}$$

Recall the linearized equation for a simple pendulum:

[eq(1.15)] $$\frac{d^2\theta(t)}{dt^2} + \frac{g}{L}\theta(t) = 0.$$

Comparing the last two equations, we see that the pendulum and the LC circuit are analogous systems.

In the two preceding examples, analogous electrical circuits are found for two mechanical systems. We can also find analogous thermal systems, analogous fluidic systems, and so on. Suppose that we know the characteristics of the LC circuit; we then know the characteristics of the simple pendulum. We can transfer our knowledge of the characteristics of circuits to the understanding of other types of physical systems. This process can be generalized further; for example, we study the characteristics of a second-order linear differential equation with constant coefficients, with the knowledge that many different physical systems have these characteristics.

1.3   SAMPLERS AND DISCRETE-TIME PHYSICAL SYSTEMS

We now describe a physical sampler and some discrete-time physical systems. In many applications, we wish to apply a continuous-time signal to a discrete-time system. This operation requires the sampling of the continuous-time signal; we consider first an analog-to-digital converter, which is one type of physical sampler. This device is used extensively in the application of continuous-time physical signals to digital computers either for processing or for data storage.

Analog-to-Digital Converter

We begin with a description of a *digital-to-analog converter* (D/A or DAC), since this device is usually a part of an *analog-to-digital converter* (A/D or ADC). We assume that the D/A receives a binary number every T seconds, usually from a digital computer. The D/A converts the binary number to a constant voltage equal to the value of that number and outputs this voltage until the next binary number appears at the D/A input. The D/A is represented in block diagram as in Figure 1.17(a), and a typical response is depicted in Figure 1.17(b). We do not investigate the internal operation of the D/A.

Next we describe a comparator, which is also a part of an A/D. A comparator and its characteristics are depicted in Figure 1.18. The input voltage $v_i(t)$ is compared with a reference voltage $v_r(t)$. If $v_i(t)$ is greater than $v_r(t)$, the comparator outputs

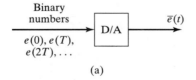

(a)

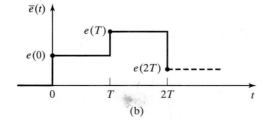

(b)

Figure 1.17 Digital-to-analog converter.

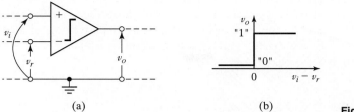

(a)   (b)

Figure 1.18 Comparator.

logic 1; for example, logic 1 is approximately 5 V for TTL (transistor-to-transistor logic). If $v_i(t)$ is less than $v_r(t)$, the comparator outputs logic zero, which is less than 1 V for TTL. The comparator is normally shown with the signal ground of Figure 1.18 omitted; however, all voltages are defined relative to the signal ground.

Several different circuits are used to implement analog-to-digital converters, with each circuit having different characteristics. We now describe the internal operation of a particular circuit. The *counter-ramp A/D* is depicted in Figure 1.19(a), with the device signals illustrated in Figure 1.19(b) [4]. The n-bit counter begins the count at value zero when the start-of-conversion (SOC) pulse arrives from the controlling

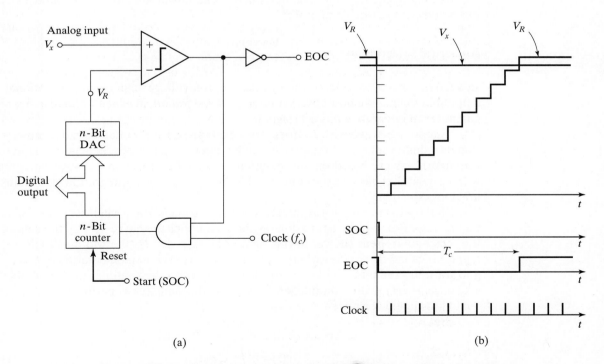

(a)   (b)

Figure 1.19 Counter-ramp analog-to-digital converter. (From L. Phillips and H. T. Nagle, *Digital Control Systems Analysis and Design*, 3rd ed., Prentice Hall, Upper Saddle River, NJ, 1995.)

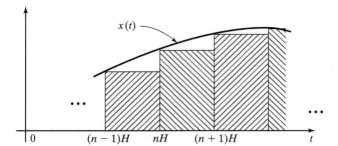

Figure 1.20 Euler's rule.

device (usually, a digital computer). The count increases by one with the arrival of each clock pulse. The n-bit D/A converts the count to a voltage V_R.

The analog input voltage V_x, which is to be converted to binary, is compared to V_R. When V_R becomes greater than V_x, the comparator outputs a logic 0, which halts the clock input through an AND gate. The end-of-conversion (EOC) pulse then signals the controlling device that the conversion is complete. At this time, the controlling device reads the counter output, which is a binary number that is approximately equal to the analog input V_x. The advantages and disadvantages of this converter are discussed in Ref. 4.

Numerical Integration

In Section 1.2, we considered the integration of a voltage signal using operational-amplifier circuits. We now consider *numerical integration*, in which we used a digital computer to integrate a physical signal.

Suppose that we wish to integrate a voltage signal $x(t)$ numerically. Integration by a digital computer requires use of a numerical algorithm. In general, numerical algorithms are based on approximating a signal that has an unknown integral with a signal that has a known integral. Hence, all numerical integration algorithms are approximate in nature.

We illustrate numerical integration using Euler's rule, which is depicted in Figure 1.20. Euler's rule approximates the area under the curve $x(t)$ by the sum of the rectangular areas shown. In this figure, the step size H (the width of each rectangle) is called the *numerical-integration increment*. The implementation of this algorithm requires that $x(t)$ be sampled every H seconds, resulting in the number sequence $x(nH)$, with n an integer. Generally, the sampling is performed using an analog-to-digital converter.

Let

$$y(t) = \int_0^t x(\tau)d\tau. \tag{1.21}$$

The integral of $x(t)$ from $t = 0$ to $t = nH$ in Figure 1.20 can be expressed as the integral from $t = 0$ to $t = (n - 1)H$ plus the integral from $(n - 1)H$ to nH. Thus, in (1.21),

$$y(t)|_{t=nH} = y(nH) = \int_0^{nH} x(\tau)d\tau$$

$$= \int_0^{(n-1)H} x(\tau)d\tau + \int_{(n-1)H}^{nH} x(\tau)d\tau \qquad (1.22)$$

$$\approx y[(n - 1)H] + Hx[(n - 1)H].$$

Ignoring the approximations involved, we expressed this equation as

$$y(nH) = y[(n - 1)H] + Hx[(n - 1)H]. \qquad (1.23)$$

However, $y(nH)$ is only an approximation to the integral of $x(t)$ at $t = nH$. Equation (1.23) is called a *first-order linear difference equation with constant coefficients*. Usually, the factor H that multiplies the independent variable n in (1.23) is omitted, resulting in the equation

$$y[n] - y[n - 1] = Hx[n - 1]. \qquad (1.24)$$

We can consider the numerical integrator to be a system with the input $x[n]$ and output $y[n]$ and the difference-equation model (1.24). A system described by a difference equation is called a *discrete-time system*.

Many algorithms are available for numerical integration [5]. Most of these algorithms have difference equations of the type (1.23). Others are more complex and cannot be expressed as a single difference equation. Euler's rule is seldom used in practice, because faster or more accurate algorithms are available. Euler's rule is presented here because of its simplicity.

Picture in a Picture

We now consider a television system that produces a *picture in a picture* [6]. This system is used in television to show two frames simultaneously, where a smaller picture is superimposed on a larger picture. Consider Figure 1.21, where a TV picture is depicted as having six lines (the actual number is greater than 500). Suppose that the picture is to be reduced in size by a factor of 3 and inserted into the upper right corner of a second picture.

First the lines of the picture are digitized (sampled). In Figure 1.21, each line produces six samples (the actual number can be more than 2000), which are called *picture elements* (or *pels*). The samples are also called *pixels*. Both the number of lines and the number of samples per line must be reduced by a factor of 3 to reduce

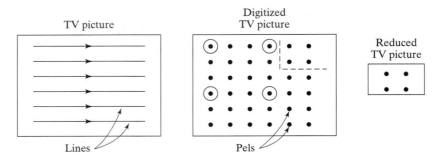

Figure 1.21 Television picture within a picture.

the size of the picture. Assume that the samples retained for the reduced picture are the four circled in Figure 1.21. In practical cases, the total number of pixels retained may be greater than 100,000.

Now let the digitized full picture in Figure 1.21 represent a different picture; the four pixels of the reduced picture then replace the four pixels in the upper right corner of the full picture. The inserted picture is outlined by the dashed lines.

The generation of a picture in a picture is more complex than as described here, but we do see the necessity to reduce the number of samples in a given line. Note that the information content of the signal is reduced, since information (samples) is discarded. We can investigate the reduction of information only after mathematical models are developed for the process of converting a continuous-time signal to a discrete-time signal by sampling. This reduction in the number of samples as indicated in Figure 1.21, called *time scaling*, is discussed in Section 9.2. The effects of sampling on the information content of a signal are discussed in Chapters 6 and 12.

Compact Disks

We next discuss compact disks (CDs). These disks store large amounts of data in a sampled form, and hence are a good example of practical sampling. We initially consider the audio compact disk (CD) [7]. The audio CD is a good example of a practical system in which the emphasis in sampling is to maintain the quality of the audio signal. A continuous-time signal to be stored on a CD is first passed through an analog antialiasing bandpass filter with a bandwidth of 5 to 20,000 Hz. This is because humans typically can only hear frequencies up to about 20 kHz. Any frequency component above 20 kHz would be inaudible to most people and, hence, can be removed from the signal without noticeable degradation in the quality of the music. The filtered signal is then sampled by an analog-to-digital converter at the rate of $f_s = 44,100$ Hz; hence, $f_s/2 = 22,050$ Hz. The data format used in storage of the samples includes error-correcting bits and is not discussed here.

The audio CD stores data for up to 74 minutes of playing time. For stereo music, two channels (signals) must be stored. The disk stores 650 megabytes of data, with the data stored on a continuous track that spirals outward. So that data from the CD may be read at a constant rate, the angular velocity of the motor driving the disk must decrease in time as the radius of the track increases in order to maintain a constant linear velocity. The speed of the motor varies from 200 rpm down to 50 rpm.

As a comparison, computer hard disks store data in circular tracks. These disks rotate at a constant speed, which is commonly 3600 rpm; hence, the data rate varies according to the radius of the track being read. High-speed CD-ROM drives use constant angular velocities as high as 12,000 rpm.

Because the data stored on the CD contain error-correcting bits, the CD player must process these data to reproduce the original samples. This processing is rather complex and is not discussed here.

The audio CD player contains three servos (closed-loop control systems). One servo controls the speed of the motor that rotates the disk, such that the data are read off the disk at a rate of 44,100 samples per second. The second servo points the laser beam at the required position on the CD, and the third servo keeps the laser beam focused on this position.

The Digital Video Disc (DVD) is a popular new medium for viewing movies and television programs. It stores sampled video and audio and allows for high-quality playback. Over two hours of video can be stored on one DVD. The DVD uses *video compression* to be able to fit this much data on one disk. (We will discuss the mathematics used in some video-compression algorithms in Chapter 12.)

Sampling in Telephone Systems

In this section, we consider the sampling of telephone signals [8]. The emphasis in these sampling systems is to reduce the number of samples required, even though the quality of the audio is degraded. Telephone signals are usually sampled at $f_s = 8000$ Hz. This sampling allows the transmission of a number of telephone signals simultaneously over a single pair of wires (or in a single communications channel), as described next.

A telephone signal is passed through an analog antialiasing filter with a passband of 200 to 3200 Hz, before sampling. Frequencies of less than 200 Hz are attenuated by this filter, to reduce the 60-Hz noise coupled into telephones circuits from ac power systems. The 3200-Hz cutoff frequency of the filter ensures that no significant frequency aliasing occurs because $f_s/2 = 4000$ Hz. However, the bandpass filter severely reduces the quality of the audio in telephone conversations; this reduction in quality is evident in telephone conversations.

The sampling of a telephone signal is illustrated in Figure 1.22. The pulses are of constant width; the amplitude of each pulse is equal to the value of the telephone signal at that instant. This process is called *pulse-amplitude modulation*; the information is carried in the amplitudes of the pulses.

Several pulse-amplitude–modulated signals can be transmitted simultaneously over a single channel as illustrated in Figure 1.23. In this figure, the numeral 1 denotes

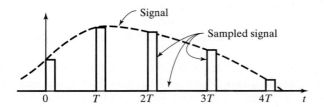

Figure 1.22 Pulse-amplitude modulation.

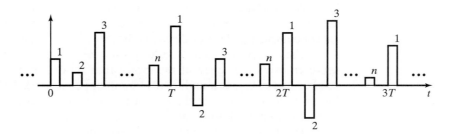

Figure 1.23 Time-division multiplexing.

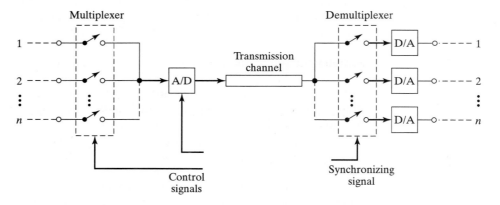

Figure 1.24 Multiplexed pulse-code–modulated telephone transmission.

the samples of the first telephone signal, 2 denotes the samples of the second one, and so on. The process depicted in Figure 1.23 is called *time-division multiplexing* because the signals are separated in time. This figure is simplified somewhat; information must be included with the signal that allows the receiving circuits to assign the pulses to the correct signals. This process is called *synchronization*.

If the sampling is performed by an analog-to-digital converter, the sample values are in binary and the process is called *pulse-code modulation*. Figure 1.24 illustrates the hardware of a time-division–multiplexed telephone system. The control signals switch each telephone signal to the analog-to-digital converter in order and then command the A/D to sample. The receiving circuits separate out the synchronizing information and switch the pulses, in order, to the correct digital-to-analog converter.

In *pulse-width modulation*, the sampling process produces a rectangular pulse of constant amplitude and variable width, with the pulse width proportional to the signal amplitude. In *delta modulation*, the output of the sampling process is the difference in the present sample and the previous sample. All four sampling processes are used in telephone communications.

Data-Acquisition System

Data-acquisition systems are instrumentation systems in which the taking of measurements is controlled automatically by a computer. A typical system is illustrated in Figure 1.25. In this figure, we have assumed that measurements from n different sensors are to be recorded in computer memory. The computer controls the multiplexer and, hence, determines which measurement is recorded at a given time. In some large data-acquisition systems, the number of sensors is greater than 1000.

When switching occurs in the input multiplexer in Figure 1.25, the transient circuit is often modeled as in Figure 1.26(a). The sensor is modeled as a Thévenin equivalent circuit [9], with (constant) source voltage V_s and source resistance R_s. The voltage $v_{ad}(t)$ is the voltage internal to the analog-to-digital converter that is converted to a binary number. (See Figure 1.19.) Resistance R_a is the equivalent resistance of the circuit from the input terminals to the voltage $v_{ad}(t)$, and C_a represents the stray capacitance in the circuit. Generally, the resistance of the remainder of the circuit at the voltage $v_{ad}(t)$ is sufficiently large that it can be ignored.

The voltage $v_{ad}(t)$ exponentially approaches the constant value V_s of the sensor, as shown in Figure 1.26(b). The initial value of $v_{ad}(t)$ (that value at the instant of switching) is in general unknown; a value is given in Figure 1.26(b). The transient term in $v_{ad}(t)$ is of the form $Ve^{-t/\tau}$, with V constant and the time constant $\tau = (R_s + R_a)C_a$. (Time constants are discussed in Section 2.3.) Theoretically, this term never goes to zero and the system never reaches steady state. We see then that

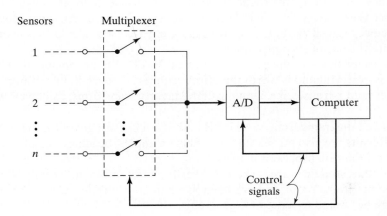

Figure 1.25 Data-acquisition system.

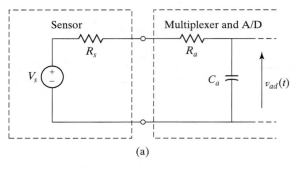

(a)

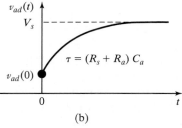

(b)

Figure 1.26 Transient circuit.

the accuracy of the measurement is affected by the length of elapsed time between the switching in the multiplexer and conversion by the A/D. To increase the accuracy of the recorded measurements, in some systems the conversion is delayed nine time constants (9τ) after the circuit is switched. Hence, the transient term has decayed to $Ve^{-9} = 0.000123V$, or to approximately 0.01% of its initial value.

1.4   MATLAB AND SIMULINK

The mathematical computer program MATLAB [10] and the related simulation computer program SIMULINK [11] are used throughout this book to illustrate computer programs available for calculations and simulations. No instructions are given for the use of these programs; that is beyond the scope of this book. If the reader is interested in these programs, instructions books and on-line help are available. Much experience is necessary for one to become proficient in these programs. However, this proficiency is well worth the effort because any realistic calculations in engineering require the use of such programs.

Programs using the symbolic mathematics of MATLAB are included. The symbolic math is powerful and is worth the effort required to learn it. For example, Laplace, Fourier, and z-transforms, along with the inverse transforms, can be calculated.

Programs and diagrams from MATLAB and SIMULINK are given for many of the examples. Results are also presented for certain examples, but these results may not be in the form that the programs display.

Generally the students are asked to verify most problem results with MATLAB. The programs required have been given in appropriate examples; usually, students are not asked to write new programs.

1.5 SIGNALS AND SYSTEMS REFERENCES

In general, references for particular topics are given throughout this book. Many good books in the general area of signals and systems are available. An incomplete list of these books is given as Refs. 12 through 20. Any omission of books from this list is inadvertent.

REFERENCES

1. W. A. Gardner, *Introduction to Random Processes*. New York: Macmillan, 1986.
2. C. L. Phillips and R. D. Harbor, *Feedback Control Systems*, 4th ed. Upper Saddle River, NJ: Prentice Hall, 1999.
3. J. Millman, *Microelectronics*, 2d ed. New York: McGraw–Hill, 1999.
4. C. L. Phillips and H. T. Nagle, *Digital Control System Analysis and Design*, 3d ed. Upper Saddle River, NJ: Prentice-Hall, 1996.
5. S. D. Conte and C. deBoor, *Elementary Numerical Analysis: An Algorithmic Approach*. New York: McGraw-Hill, 1982.
6. M. Burkert et al., "IC Set for a Picture-in-Picture System with On-Chip Memory," *IEEE Transactions on Consumer Electronics*, February 1990.
7. L. Buddine and E. Young, *The Brady Guide to CD-ROM*. Upper Saddle River, NJ: Prentice Hall, 1988.
8. B. E. Keiser and E. Strange, *Digital Telephony and Network Integration*. New York: Van Nostrand Reinhold, 1995.
9. J. D. Irwin, *Basic Engineering Circuit Analysis*, 6th ed. New York: Macmillan, 1999.
10. Learning MATLAB 6, Natick, MA: The Mathworks, Inc., 2001.
11. Learning SIMULINK 4, Natick, MA: The Mathworks, Inc., 2001.
12. R. A. Gabel and R. A. Roberts, *Signals and Linear Systems*. New York: Wiley, 1987.
13. L. B. Jackson, *Signals, Systems, and Transforms*. Reading, MA: Addison–Wesley, 1991.
14. B. P. Lathi, *Linear Systems and Signals*, New York: Berkeley-Cambridge, 1992.
15. R. J. Mayhan, *Discrete-Time and Continuous-Time Linear Systems*, 2d ed. Reading, MA: Addison–Wesley, 1998.
16. C. D. McGillem and G. R. Cooper, *Continuous and Discrete Signal and System Analysis*, 3d ed. New York: Holt, Rinehart and Winston, 1995.
17. M. O'Flynn and E. Moriarty, *Linear Systems Time Domain and Transform Analysis*. New York: Harper & Row, 1987.
18. A. V. Oppenheim and A. S. Willsky, *Signals and Systems*, 2d ed. Upper Saddle River, NJ: Prentice-Hall, 1996.
19. S. S. Soliman and M. D. Srinath, *Continuous and Discrete Signals and Systems*, 2d ed. Upper Saddle River, NJ: Prentice-Hall, 1997.
20. R. E. Ziemer, W. H. Tranter, and S. R. Fannin, *Signals and Systems Continuous and Discrete*, 4th ed. New York: Macmillan, 1998.

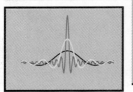

2

CONTINUOUS-TIME
SIGNALS AND SYSTEMS

As stated in Chapter 1, engineers must model two distinct physical phenomena. The first is physical *systems*, which can be modeled by mathematical *equations*. For example, continuous-time, or analog, systems (systems that contain no sampling) can be modeled by ordinary differential equations with constant coefficients. An example of such a system is a linear electrical circuit. Of course, it is important to remember that the accuracy of these mathematical models may vary.

A second physical phenomenon to be modeled is called a *signal*. *Physical signals* are modeled by *mathematical functions*. One example of a physical signal is the voltage that is applied to the speaker in a radio. Another example is the temperature at a designated point in a particular room. This signal is a function of time, since the temperature varies with time. We can express this temperature as

$$\text{temperature at a point} = \theta(t), \tag{2.1}$$

where $\theta(t)$ has the units of, for example, degrees Celsius. To be more precise in this example, the temperature in a room is a function of time and of space. We may designate a point in a room in the rectangular coordinates x, y, and z. Equation (2.1) then becomes

$$\text{temperature in a room} = \theta(x, y, z, t), \tag{2.2}$$

where the point in a room is identified by the three space coordinates x, y, and z. The signal in (2.1) is a function of one independent variable, whereas the signal in (2.2) is a function of four independent variables.

In this book, we limit signals to having one independent variable (except in Section 12.7, where we briefly discuss images which are functions of two dimensions). In general, this independent variable will be time t. Signals are divided into two natural categories. The first category to be considered is *continuous-time*, or simply *continuous, signals*. A signal of this type is defined for all values of time. A continuous-time signal is also called an *analog signal*. A *continuous-time system* is a system in which only continuous-time signals appear.

There are two types of continuous time signals. A continuous-time signal $x(t)$ can be a *continuous-amplitude signal*, for which the time-varying amplitude can assume any value. A continuous-time signal may also be a *discrete-amplitude signal*, which can assume only certain defined amplitudes. An example of a discrete-amplitude continuous-time signal is the output of a digital-to-analog converter. (See Figure 1.17.) For example, if the binary signal into the digital-to-analog converter is represented by eight bits, the output-signal amplitude can assume only $2^8 = 256$ different values.

The second category for signals is *discrete-time*, or simply *discrete*, signals. A discrete signal is defined at only certain instants of time. For example, suppose that a signal $f(t)$ is to be processed by a digital computer [this operation is called *digital signal processing* (DSP)]. Since a computer can operate on only a number and not a continuum, the continuous signal must be converted into a sequence of numbers by sampling. This sequence of numbers is called a *discrete-time signal*. Like continuous-time signals, discrete-time signals can be either continuous amplitude or discrete amplitude. These signals are described in detail in Chapter 9. A discrete-time system is a system in which only discrete-time signals appear.

In some physical systems, both continuous-time and discrete-time signals appear. We call these systems hybrid or sampled-data systems. An example is an automatic aircraft-landing system, in which the signals that control the aircraft are calculated by a digital computer. We do not consider hybrid systems in this book. Continuous-time signals and systems are introduced in this chapter, with discrete-time signals and systems introduced in Chapter 9.

2.1   TRANSFORMATIONS OF CONTINUOUS-TIME SIGNALS

In this chapter, we consider functions of the independent variable time t, such as $x(t)$; the variable time can assume all values $-\infty < t < \infty$. Later we consider the case that $x(t)$ can be complex, but time t is always real.

We begin by considering six transformations on a real function of the real variable, denoted as $x(t)$. The first three are transformations in time, and the second three are transformations in amplitude. These transformations are especially useful in the applications of Fourier series in Chapter 4 and in applications of the Fourier transforms, the z-transform, and the Laplace transform in later chapters.

Time Transformations

We first consider the transformation of *time reversal*.

Time Reversal
In time reversal, we create a new, transformed signal $y(t)$ by replacing t with $-t$ in the original signal, $x(t)$. Hence,

$$y(t) = x(-t), \tag{2.3}$$

where $y(t)$ denotes the transformed signal. The result of the time-reversal transformation is that for any particular value of time, $t = t_0$, $y(t_0) = x(-t_0)$, and $y(-t_0) = x(t_0)$.

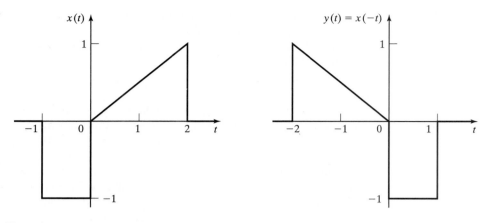

Figure 2.1   Time reversal of a signal.

An example of time reversal is given in Figure 2.1 in which the time-reversed signal, $y(t)$, is the mirror image of the original signal, $x(t)$, reflected about the vertical axis. We shall see later, when we study convolution in Chapter 3 that one application of time reversal is in calculating the responses of systems to input signals.

> A real-life example of time reversal is playing music on a CD backwards. Many people play popular music recordings in reverse. A number of musicians have included backwards tracks in their music including the Beatles, Pink Floyd, and Frank Zappa.

In general, we must verify any analysis or design in engineering. For the procedures of this section, we can assign a value of $t = t_0$ and find $y(t_0)$, where t_0 is a chosen fixed value of time. Next we let $t = -t_0$ and find $x(-t_0)$. If the time reversal is correct then $y(t_0)$ must equal $x(-t_0)$. Testing values in Figure 2.1 verifies the results of the example.

Next we consider time scaling.

> **Time Scaling**
> Given a signal $x(t)$, a time-scaled version of this signal is
>
> $$y(t) = x(at), \qquad (2.4)$$
>
> where a is a real constant.

Figure 2.2(a) shows a signal $x(t)$. As examples of time scaling, we plot the signals $y_1(t) = x(2t)$ and $y_2(t) = x(0.1t)$.

Figure 2.2(b) shows the transformed signal $y_1(t)$. A comparison of the plots of $x(t)$ and $y_1(t)$ shows that for any particular value of time $t = t_0$, $y_1(t_0) = x(2t_0)$ and $x(t_0) = y_1(t_0/2)$. It is seen that $y_1(t)$ is a time-compressed (sped up) version of $x(t)$.

Figure 2.2(c) shows the second transformed signal $y_2(t) = x(0.1t)$. A comparison of the plots of $x(t)$ and $y_2(t)$ shows that for any particular value of time $t = t_0$, $y_2(t_0) = x(0.1t_0)$ and $x(t_0) = y_2(10t_0)$. It is seen that $y_2(t)$ is a time-stretched (slowed down) version of $x(t)$.

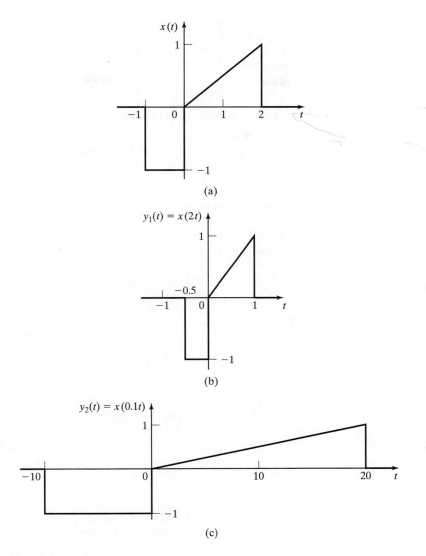

Figure 2.2   Time-scaled signals.

From the development just described, we see that the time-scaled signal $x(at)$ is compressed in time for $|a| > 1$ and is expanded in time for $|a| < 1$. One application of independent-variable scaling is the design of certain filters, as we will see later when we study filter design in Chapter 6.

The wavelet transform, a focus of much current research in signal processing, uses time-scaling to analyze signals simultaneously in both the time and the frequency domains. The wavelet transform is the basis of JPEG2000, the new standard for image compression. See *www.jpeg.org* for more information.

A real-life example of time scaling is listening to an answering-machine message on fast forward. This speeds up the signal in time and increases the pitch or frequency content of the speaker's voice. A second such example would be playing a forty-five-revolutions-per-minute (45-rpm) analog recording at 33 rpm. This would both slow down the signal in time and decrease the pitch of the signal. Similarly, playing a 33-rpm recording at 45 rpm would speed up the signal and increase the pitch of the voice.

Next we consider time shifting.

Time Shifting
Given a signal $x(t)$, a time-shifted version of this signal is

$$y(t) = x(t - t_0), \tag{2.5}$$

where t_0 is a constant.

Note that $y(t_0) = x(0)$. Hence, if t_0 is positive, the shifted signal $y(t)$ is delayed in time [shifted to the right relative to $x(t)$]. If t_0 is negative, $y(t)$ is advanced in time (shifted to the left). Consider the signal in Figure 2.3(a). We wish to plot the time-shifted signals $y_1(t) = x(t - 2)$ and $y_2(t) = x(t + 1)$. The transformed signals are plotted in Figure 2.3(b) and (c).

As a second example of time shifting, for the signal

$$x(t) = e^{-t} \cos(3t - \pi/2),$$

the time-shifted signal is given by

$$y(t) = x(t - t_0) = e^{-(t-t_0)} \cos(3(t - t_0) - \pi/2) = e^{t_0} e^{-t} \cos(3t - 3t_0 - \pi/2).$$

As will be shown later, one application of time shifting is in calculating the responses of systems to input signals.

We now develop a general approach to independent-variable transformations. The three transformations in time just considered are of the general form

$$y(t) = x(at + b). \tag{2.6}$$

In this equation, a and b are real constants. For clarity, we let τ denote time in the original signal. The transformed t-axis equation is found from

$$\tau = at + b \Rightarrow t = \frac{\tau}{a} - \frac{b}{a}. \tag{2.7}$$

For example,

$$y(t) = x(-2t + 3).$$

The value $a = -2$ yields time reversal (the minus sign) and time scaling ($|a| = 2$). The value $b = 3$ yields a time shift. The transformed t-axis equation for this example is found from (2.7)

$$\tau = -2t + 3 \Rightarrow t = -\frac{\tau}{2} + \frac{3}{2}.$$

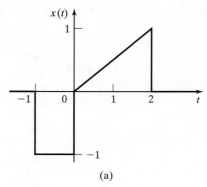

(a)

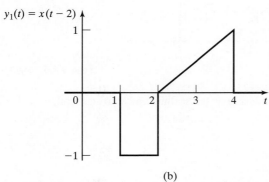

(b)

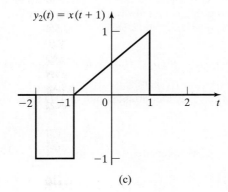

(c)

Figure 2.3   Time-shifted signals.

An example of the transformation in (2.6) and (2.7) is now given.

EXAMPLE 2.1   **Time transformation of a signal**

Consider the signal $x(\tau)$ in Figure 2.4(a). We wish to plot the transformed signal

$$y(t) = x\left(1 - \frac{t}{2}\right).$$

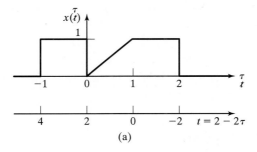

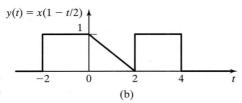

Figure 2.4 Signals for Example 2.1.

This transformation has reversal, scaling, and shifting. First, we solve the transformation for the variable t:

$$\tau = 1 - \frac{t}{2} \Rightarrow t = 2 - 2\tau.$$

The t-axis is shown below the time axis in Figure 2.4(a), and Figure 2.4(b) gives the desired plot of the transformed signal. As always, we should check our work. For any particular value of time, $t = t_0$, we can write, from (2.7),

$$y(t_0) = x(at_0 + b) \text{ and } x(t_0) = y\left(\frac{t_0}{a} - \frac{b}{a}\right).$$

Choosing $t_0 = 1$ as an easily identifiable point in $x(t)$ from Figure 2.4(a), we calculate

$$x(1) = y(2 - 2(1)) = y(0).$$

Again choosing $t_0 = 1$, we calculate

$$y(1) = x\left(1 - \frac{1}{2}\right) = x\left(\frac{1}{2}\right).$$

Both calculated points confirm the correct transformation.   ■

  A general approach for plotting transformations of the independent variable is as follows:

1. On the plot of the original signal, replace t with τ.

2. Given the time transformation $\tau = at + b$, solve for $t = \dfrac{\tau}{a} - \dfrac{b}{a}$.

3. Draw the transformed t-axis directly below the τ-axis.

4. Plot $y(t)$ on the t-axis.

Three transformations of time (the independent variable) have been described. Three equivalent transformations of the amplitude of a signal (the dependent variable) are now defined.

Amplitude Transformations

We now consider signal-amplitude transformations. One application of these transformations is in the amplification of signals by physical amplifiers. Some amplifiers not only amplify signals, but also add (or remove) a constant, or dc, value. A second use of amplitude transformations is given in Chapter 4, in applications of Fourier series. Amplitude transformations follow the same rules as time transformations.

The three transformations in amplitude are of the general form

$$y(t) = Ax(t) + B, \tag{2.8}$$

where A and B are constants. For example, consider

$$y(t) = -3x(t) - 5.$$

The value $A = -3$ yields amplitude reversal (the minus sign) and amplitude scaling ($|A| = 3$), and the value $B = -5$ shifts the amplitude of the signal. Many physical amplifiers invert the input signal in addition to amplifying the signal. (The gain is then a negative number.) An example of amplitude scaling is now given.

EXAMPLE 2.2   **Amplitude transformation of a signal**

Consider the signal of Example 2.1, which is shown again in Figure 2.5(a). Suppose that this signal is applied to an amplifier that has a gain of 3 and introduces a bias (a dc value) of -1, as shown in Figure 2.5(b). We wish to plot the amplifier output signal

$$y(t) = 3x(t) - 1.$$

We first plot the transformed amplitude axis, as shown in Figure 2.5(a). For example, when $x(t) = 1$, $y(t) = 2$. Figure 2.5(c) shows the desired plot of the transformed signal $y(t)$. ∎

EXAMPLE 2.3   **Time and amplitude transformation of a signal**

Next we consider the signal

$$y(t) = 3x\left(1 - \frac{t}{2}\right) - 1,$$

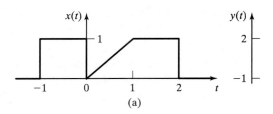

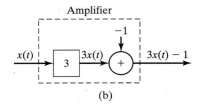

(b)

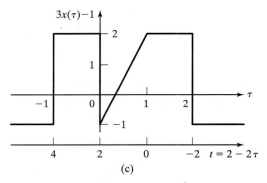

(c)

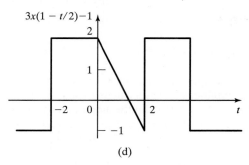

(d)

Figure 2.5   Signals for Examples 2.2 and 2.3.

which has (1) the time transformation of Example 2.1 and Figure 2.4 and (2) the amplitude transformation of Example 2.2 and Figure 2.5(c). To plot this transformed signal, we first transform the amplitude axis as shown in Figure 2.5(c). The t-axis of Figure 2.4(a) is redrawn in Figure 2.5(c), to facilitate the time transformation.

$$\tau = 1 - \frac{t}{2} \Rightarrow t = 2 - 2\tau.$$

The signal is then plotted on the t-axis, as shown in Figure 2.5(d). To verify one point, let $t = 1$. From Figure 2.5(d), $y(1) = 0.5$. From Figure 2.5(a),

$$3x(1 - t/2) - 1 \Big|_{t=1} = 3(0.5) - 1 = 0.5$$

and the point is verified.   ■

In summary, the six transformations defined in this section are (1) reversal, scaling, and shifting with respect to-time and (2) reversal, scaling, and shifting with respect to amplitude. These transformations are listed in Table 2.1. All six transformations have applications in signal and system analysis.

2.2   SIGNAL CHARACTERISTICS

In this section, certain characteristics of continuous-time signals are defined. These characteristics are needed for later derivations in signal and system analysis.

Even and Odd Signals

We first define the signal characteristics of *even symmetry* and *odd symmetry*. By definition, the function (signal) is *even* if

$$x_e(t) = x_e(-t). \tag{2.9}$$

An even function has symmetry with respect to the vertical axis; the signal for $t < 0$ is the mirror image of the signal for $t > 0$. The function $x(t) = \cos \omega t$ is even because $\cos \omega t = \cos(-\omega t)$. Another example of an even function is given in Figure 2.6.

By definition, a function is *odd* if

$$x_o(t) = -x_o(-t). \tag{2.10}$$

An odd function has symmetry with respect to the origin. The function $x(t) = \sin \omega t$ is odd because $\sin \omega t = -\sin(-\omega t)$. Another example of an odd function is given in Figure 2.7.

TABLE 2.1   Transformations of Signals

Name	$y(t)$
Time reversal	$x(-t)$
Time scaling	$x(at)$
Time shifting	$x(t - t_0)$
Amplitude reversal	$-x(t)$
Amplitude scaling	$Ax(t)$
Amplitude shifting	$x(t) + B$

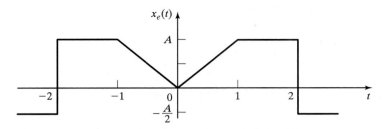

Figure 2.6   Even signal.

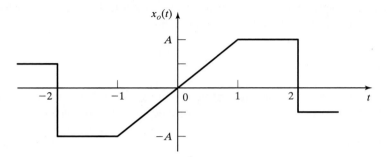

Figure 2.7   Odd signal.

Any signal can be expressed as the sum of an even part and an odd part; that is,

$$x(t) = x_e(t) + x_o(t), \tag{2.11}$$

where $x_e(t)$ is even and $x_o(t)$ is odd. Replacing t with $-t$ in this equation yields

$$x(-t) = x_e(-t) + x_o(-t) = x_e(t) - x_o(t) \tag{2.12}$$

from (2.9) and (2.10). Adding (2.11) and (2.12) and dividing by 2 yields

$$x_e(t) = \tfrac{1}{2}[x(t) + x(-t)]. \tag{2.13}$$

This equation is used to solve for the even part of a function $x(t)$. Subtracting (2.12) from (2.11) and dividing by 2 yields

$$x_o(t) = \tfrac{1}{2}[x(t) - x(-t)], \tag{2.14}$$

which is used to find the odd part of a function.

The *average value* A_x of a signal $x(t)$ is defined as

$$A_x = \lim_{T \to \infty} \frac{1}{2T} \int_{-T}^{T} x(t)\, dt.$$

The average value of a signal is contained in its even function, since the average value of an odd function is zero. (See Problem 2.9.)

Even and odd functions have the following properties:

1. The sum of two even functions is even.
2. The sum of two odd functions is odd.
3. The sum of an even function and an odd function is neither even nor odd.
4. The product of two even functions is even.
5. The product of two odd functions is even.
6. The product of an even function and an odd function is odd.

These properties are easily proved. (See Problem 2.10.) An example is given to illustrate some of the relations developed thus far in this section.

EXAMPLE 2.4   **Even and odd signals**

We consider the signal $x(t)$ of Example 2.1. This signal is given again in Figure 2.8(a). The time-reversed signal $x(-t)$ is also given. The two signals are added and scaled in amplitude by 0.5 to yield the even signal $x_e(t)$ of (2.13). This even signal is plotted in Figure 2.8(b). Next, $x(-t)$ is subtracted from $x(t)$, and the result is amplitude scaled by 0.5 to yield the odd signal $x_o(t)$ of (2.14). This odd signal is plotted in Figure 2.8(c). For verification, we see that adding $x_e(t)$ and $x_o(t)$ yields the signal $x(t)$.   ■

As will be shown in Chapter 5 and later chapters, even-function and odd-function properties aid us in understanding and applying the Fourier transform to system and signal analysis. These properties are useful in both the continuous-time Fourier transform and the discrete-time Fourier transform. In addition, these properties are useful in both the development and the applications of the Fourier series, as shown in Chapter 4.

Periodic Signals

Next we consider the important topic of periodic functions. By definition, a continuous-time signal $x(t)$ *is periodic* if

$$x(t) = x(t + T), \quad T > 0 \tag{2.15}$$

for all t, where the constant T is the period. A signal that is not periodic is said to be *aperiodic*. In (2.15), we replace t with $(t + T)$, resulting in

$$x(t + T) = x(t + 2T).$$

This equation is also equal to $x(t)$ from (2.15). By repeating this substitution, we see that a periodic function satisfies the equation

$$x(t) = x(t + nT),$$

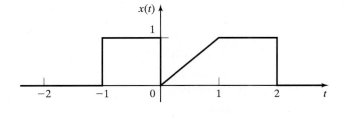

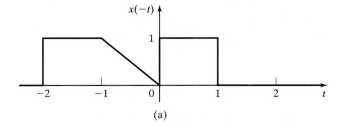

(a)

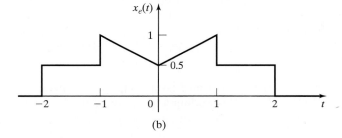

(b)

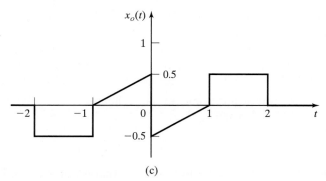

(c)

Figure 2.8 Signals for Example 2.4.

where n is any integer. Hence, a periodic signal with period $T > 0$ is also periodic with period nT.

    The minimum value of the period $T > 0$ that satisfies the definition $x(t) = x(t + T)$ is called the *fundamental period* of the signal and is denoted as T_0.

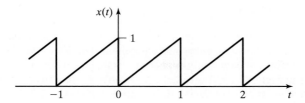

Figure 2.9   Sawtooth waveform.

With T_0 in seconds, the fundamental frequency in hertz (the number of periods per second) and the fundamental frequency in rad/s are given by

$$f_0 = \frac{1}{T_0} \text{ Hz}, \quad \omega_0 = 2\pi f_0 = \frac{2\pi}{T_0} \text{ rad/s}, \tag{2.16}$$

respectively.

Examples of periodic signals are the sinusoids $x_c(t) = \cos \omega t$ and $x_s(t) = \sin \omega t$. The movement of a clock pendulum is periodic, as is the voltage used for ac electric-power distribution. Both signals are usually *modeled* as sinusoids, even though neither is exactly sinusoidal. The movement of the earth is periodic with periods of one day and of one year (approximately). Most physical signals that are dependent on the earth's movement, such as the time of sunrise and average temperatures, are usually modeled as sinusoids.

A third example of a periodic signal is given in Figure 2.9. This signal, called a *sawtooth wave*, is useful in sweeping a beam of electrons across the face of a cathode ray tube (CRT). If a CRT uses an electric field to sweep the beam of electrons, the signal of Figure 2.9 is a voltage; if a magnetic field is used, the signal is a current.

A special case of a periodic function is that of $x(t)$ equal to a constant. A constant satisfies the definition $x(t) = x(t + T)$ for any value of T. Because there is no smallest value of T, the fundamental period of a constant signal is not defined. However, it is sometimes convenient to consider a constant signal A to be the limiting case of the sinusoid $x(t) = A \cos \omega t$, with ω approaching zero. For this case, the period T is unbounded.

Two examples concerning periodic functions will now be given.

EXAMPLE 2.5   **Periodic signals**

In this example, we test two functions for periodicity. The function $x(t) = e^{\sin t}$ is periodic because

$$x(t + T) = e^{\sin(t+T)} = e^{\sin t} = x(t)$$

with $\sin(t + T) = \sin t$ for $T = 2\pi$. The function $x(t) = te^{\sin t}$ is not periodic, since

$$x(t + T) = (t + T)e^{\sin(t+T)} = (t + T)e^{\sin t} \neq x(t)$$

for $T = 2\pi$. A factor in the last function is periodic, but the function itself is aperiodic. ∎

| EXAMPLE 2.6 | **Power-supply periodic signals** |

Power supplies that convert an ac voltage (sinusoidal voltage) into a dc voltage (constant voltage) are required in almost all electronic equipment that doesn't use batteries. Shown in Figure 2.10 are voltages that commonly appear in certain power supplies. (See Section 1.2.)

The voltage in Figure 2.10(a) is called a *half-wave rectified signal*. This signal is generated from a sinusoidal signal by replacing the negative half cycles of the sinusoid with a voltage of zero. The positive half cycles are unchanged.

The signal in Figure 2.10(b) is called *full-wave rectified signal*. This signal is generated from a sinusoidal signal by the amplitude reversal of each negative half cycle. The positive half cycles are unchanged. Note that the period of this signal is one-half that of the sinusoid and, hence, one-half that of the half-wave rectified signal.

It is necessary in the analysis and design of these power supplies to express these signals as mathematical functions. These mathematical functions will be written after the definitions of some additional signals.

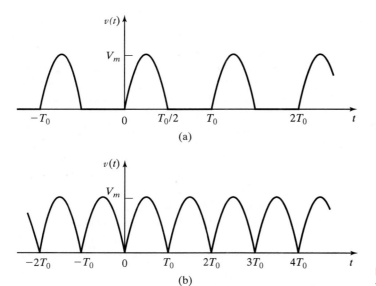

Figure 2.10   (a) Half-wave and (b) full-wave rectified signals.   ■

The sum of continuous-time periodic signals is periodic if and only if the ratios of the periods of the individual signals are ratios of integers. If a sum of N periodic signals is periodic, the fundamental period can be found as follows:

1. Convert each period ratio, T_{01}/T_{0i}, $2 \le i \le N$, to a ratio of integers, where T_{01} is the period of the first signal considered and T_{0i} is the period of one of the other N-1 signals. If one or more of these ratios is not rational, then the sum of signals is not periodic.

2. Eliminate common factors from the numerator and denominator of each ratio of integers.

3.   The fundamental period of the sum of signals is $T_0 = k_0 T_{01}$, where k_0 is the least common multiple of the denominators of the individual ratios of integers.

EXAMPLE 2.7   **Sum of periodical signals**

Three periodic signals $[x_1(t) = \cos(3.5t)$, $x_2(t) = \sin(2t)$, and $x_3(t) = 2\cos(\frac{7t}{6})]$ are summed to form $v(t)$. The signal $v(t)$ is shown in Figure 2.11(a). To determine if $v(t)$ is periodic, we must see if the ratios of the periods of $x_1(t)$, $x_2(t)$, and $x_3(t)$ are ratios of integers:

$$T_{01} = \frac{2\pi}{\omega_1} = \frac{2\pi}{3.5}, \quad T_{02} = \frac{2\pi}{\omega_2} = \frac{2\pi}{2}, \text{ and } T_{03} = \frac{2\pi}{\omega_3} = \frac{2\pi}{\frac{7}{6}}.$$

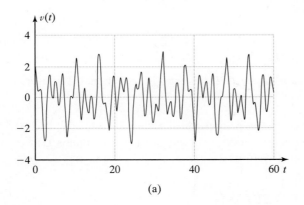

(a)

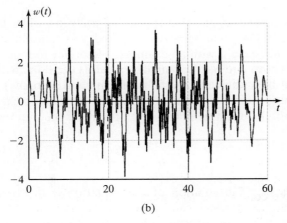

(b)

Figure 2.11   The sums of (a) three periodic signals and (b) four periodic signals.

The ratios of periods are

$$\frac{T_{01}}{T_{02}} = \frac{2\pi/3.5}{2\pi/2} = \frac{2}{3.5} = \frac{4}{7} \text{ and } \frac{T_{01}}{T_{03}} = \frac{2\pi/3.5}{2\pi/(7/6)} = \frac{7/6}{3.5} = \frac{7}{21}.$$

Both are ratios of integers; therefore, $v(t)$ is periodic.

The period of $v(t)$ is determined by first eliminating any common factors from the numerator and denominator of each ratio of periods. In this case, we have no common factor in the ratio T_{01}/T_{02}, but the ratio T_{01}/T_{03} has a common factor of 7. After eliminating the common factor, we have $T_{01}/T_{03} = 1/3$.

The next step is to find the least common multiple of the denominators of the ratios. In this case, the least common multiple is $n_1 = 3 \times 7 = 21$.

The fundamental period of $v(t)$ is $T_0 = n_1 T_{01} = 21 \times \dfrac{2\pi}{3.5} = 12\pi$ (s). This period can be confirmed by examination of Figure 2.11(a).

Another signal, $x_4(t) = 3\sin(5\pi t)$, is added to $v(t)$ to give $w(t) = x_1(t) + x_2(t) + x_3(t) + x_4(t)$.

This sum of four periodic signals is shown in Figure 2.11(b). The periodicity of this new sum of signals must be determined as was done previously for $v(t)$. The addition of $x_4(t)$ gives the ratio of periods,

$$\frac{T_{01}}{T_{04}} = \frac{2\pi/3.5}{2\pi/5\pi} = \frac{5\pi}{3.5}.$$

Since π is an irrational number, this ratio is not rational. Therefore the signal $w(t)$ is not periodic. ∎

In this section, the properties of evenness, oddness, and periodicity have been defined. In the next section, we consider certain signals that commonly appear in models of physical systems.

2.3 COMMON SIGNALS IN ENGINEERING

In this section, models of signals that appear naturally in a wide class of physical systems are presented. One such signal, the sinusoid, was mentioned in Section 2.2.

We begin this section with an example. We prefer to model continuous-time physical systems with ordinary linear differential equations with constant coefficients when possible (when this model is of sufficient accuracy). A signal that appears often in these models is one whose time rate of change is directly proportional to the signal itself. An example of this type of signal is the differential equation

$$\frac{dx(t)}{dt} = ax(t), \tag{2.17}$$

where a is constant. The solution of this equation is the exponential function $x(t) = x(0)e^{at}$ for $t \geq 0$, which can be verified by direct substitution in (2.17). An example is the current in the resistance-inductance (RL) circuit of Figure 2.12:

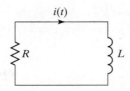

Figure 2.12 *RL circuit.*

$$L\frac{di(t)}{dt} + Ri(t) = 0 \Rightarrow \frac{di(t)}{dt} = -\frac{R}{L}i(t).$$

For this circuit, in (2.17) $a = -R/L$. The current is then $i(t) = i(0)e^{-Rt/L}$, where $i(0)$ is the initial current.

For the *RL* circuit, $i(0)$, R, and L are all real. It is sometimes convenient in the exponential function

$$x(t) = Ce^{at} \tag{2.18}$$

to consider the case that parameters C and a can be complex. Complex signals cannot appear in physical systems. However, the solutions of many differential equations are simplified by assuming that complex signals can appear both as excitations and in the solutions. Then, in translating the results back to physical systems, only the real or imaginary parts of the solutions apply.

An important relation that is utilized often in analysis using complex exponential functions is *Euler's relation*, given by

$$e^{j\theta} = \cos\theta + j\sin\theta. \tag{2.19}$$

(See Appendix D.) Replacing θ in (2.19) with $-\theta$ yields

$$e^{-j\theta} = \cos(-\theta) + j\sin(-\theta) = \cos\theta - j\sin\theta, \tag{2.20}$$

since the cosine function is even and the sine function is odd. The sum of (2.19) and (2.20) can be expressed as

$$\cos\theta = \frac{e^{j\theta} + e^{-j\theta}}{2}, \tag{2.21}$$

and the difference of (2.19) and (2.20) can be expressed as

$$\sin\theta = \frac{e^{j\theta} - e^{-j\theta}}{2j}. \tag{2.22}$$

The four relations (2.19), (2.20), (2.21), and (2.22) are so useful in signal and system analysis that they should be memorized.

The complex exponential in (2.19) can also be expressed in polar form as

$$e^{j\theta} = 1\underline{/\theta}, \tag{2.23}$$

where the notation $R\underline{/\theta}$ signifies the complex function of magnitude R at the angle θ. To prove (2.23), consider, from (2.19),

$$|e^{j\theta}| = [\cos^2\theta + \sin^2\theta]^{1/2} = 1$$

and

$$\arg e^{j\theta} = \tan^{-1}\left[\frac{\sin\theta}{\cos\theta}\right] = \theta, \tag{2.24}$$

where $\arg(\cdot)$ denotes the angle of $(\cdot)$. Three cases for exponential functions will now be investigated.

CASE 1

C and a Real

For the first case, both C and a are real for the exponential $x(t) = Ce^{at}$. The product (at) is unitless; hence, the units of a are the reciprocal of those of t. The units of C are the same as those of $x(t)$.

The signal $x(t) = Ce^{at}$ is plotted in Figure 2.13 for $C > 0$ with $a > 0$, $a < 0$, and $a = 0$. For $a > 0$, the signal magnitude increases monotonically without limit with increasing time. For $a < 0$, the signal magnitude decreases monotonically toward zero as time increases. For $a = 0$, the signal is constant.

For $a < 0$, the signal decays toward zero, but does not reach zero in finite time. To aid us in differentiating between exponentials that decay at different rates, we express the exponential as, for $a < 0$,

$$x(t) = Ce^{at} = Ce^{-t/\tau}, \quad \tau > 0. \tag{2.25}$$

The constant parameter τ is called the *time constant* of the exponential.

For example, for $x(t) = Ce^{-2t}$ with t in seconds, the time constant is $\tau = 0.5$ s. As a second example, the absorption of certain drugs into the human body is modeled as being exponential with time constants that can be in hours or even days. For a third example, voltages and currents in microcircuits can have time constants in nanoseconds.

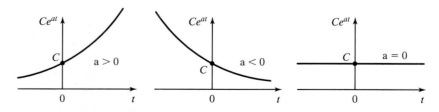

Figure 2.13   Exponential signals.

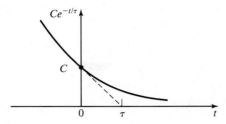

Figure 2.14 Signal illustrating the time constant.

The time constant of an exponential signal is illustrated in Figure 2.14. The derivative of $x(t)$ in (2.25) at $t = 0$ is given by

$$\frac{dx(t)}{dt}\bigg|_{t=0} = -\frac{C}{\tau} e^{-t/\tau}\bigg|_{t=0} = -\frac{C}{\tau}.$$

If the signal continued to decay from $t = 0$ at this rate, it would be zero at $t = \tau$. Actually, the value of the signal at $t = \tau$ is equal to $0.368C$; that is, the signal has decayed to 36.8% of its amplitude after τ seconds. This result is general; the signal $x(t) = Ce^{-t/\tau}$ at time $t_1 + \tau$ is equal to $0.368Ce^{-t_1/\tau}$, or $0.368x(t_1)$. The interested reader can show this.

Table 2.2 illustrates the decay of an exponential as a function of the time constant τ. While infinite time is required for an exponential to decay to zero, the exponential decays to less than 2% of its amplitude in 4τ units of time and to less than 1% in 5τ units of time.

In most applications, in a practical sense the exponential signal can be ignored after four or five time constants. Recall that the models of physical phenomena are never exact. Hence, in many circumstances high accuracy is unnecessary in either the parameters of a system model or in the amplitudes of signals.

CASE 2

C Complex, *a* Imaginary

Next we consider the case that C is complex and a is imaginary, namely,

$$x(t) = Ce^{at}; \quad C = Ae^{j\phi} = A\underline{/\phi}, \quad a = j\omega_0, \tag{2.26}$$

TABLE 2.2 Exponential Decay

t	$e^{-t/\tau}$
0	1.0
τ	0.3679
2τ	0.1353
3τ	0.0498
4τ	0.0183
5τ	0.0067

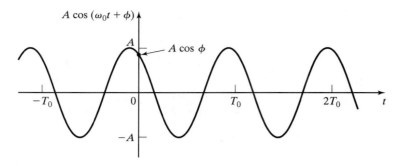

Figure 2.15   Sinusoidal signal.

where A, ϕ, and ω_0 are real and constant. The complex exponential signal $x(t)$ can be expressed as

$$x(t) = Ae^{j\phi}e^{j\omega_0 t} = Ae^{j(\omega_0 t + \phi)}$$
$$= A\cos(\omega_0 t + \phi) + jA\sin(\omega_0 t + \phi) \tag{2.27}$$

from Euler's relation in (2.19). In (2.27), all factors are real, except of course $j = \sqrt{-1}$. The sinusoids are periodic, with frequency ω_0 and period $T_0 = 2\pi/\omega_0$. Hence, the complex exponential is also periodic. (See Problem 2.12.) A plot of the real part of (2.27) is given in Figure 2.15.

With respect to (2.27), we make the following definition.

> **Harmonically Related Complex Exponentials**
> Harmonically related complex exponentials are a set of functions with frequencies related by integers, of the form
>
> $$x_k(t) = A_k e^{jk\omega_0 t}, \quad k = \pm 1, \pm 2, \ldots. \tag{2.28}$$

We will make extensive use of harmonically related complex exponentials later when we study the Fourier series representation of periodic signals.

CASE 3

Both C and a Complex

For this case, the complex exponential $x(t) = Ce^{at}$ has the parameters

$$x(t) = Ce^{at}; \quad C = Ae^{j\phi}; \quad a = \sigma_0 + j\omega_0, \tag{2.29}$$

where A, ϕ, σ, and ω_0 are real and constant. The complex exponential signal can then be expressed as

$$x(t) = Ae^{j\phi}e^{(\sigma_0 + j\omega_0)t} = Ae^{\sigma_0 t}e^{j(\omega_0 t + \phi)}$$
$$= Ae^{\sigma_0 t}\cos(\omega_0 t + \phi) + jAe^{\sigma_0 t}\sin(\omega_0 t + \phi) \tag{2.30}$$
$$= x_r(t) + jx_i(t).$$

In this expression, both $x_r(t) = \text{Re}[x(t)]$ and $x_i(t) = \text{Im}[x(t)]$ are real. The notation $\text{Re}[\cdot]$ denotes the real part of the expression, and $\text{Im}[\cdot]$ denotes the imaginary part. Plots of the

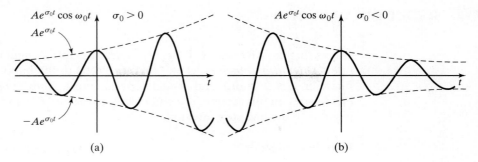

Figure 2.16   Real part of a complex exponential.

real part of (2.30) are given in Figure 2.16 for $\phi = 0$. Figure 2.16(a) shows the case that $\sigma_0 > 0$. Figure 2.16(b) shows the case that $\sigma_0 < 0$; this signal is called an *underdamped sinusoid*. For $\sigma_0 = 0$ as in Figure 2.15, the signal is called an *undamped sinusoid*.

In Figure 2.16(a), by definition the *envelope* of the signal is $\pm Ae^{\sigma_0 t}$. Because both the cosine function and the sine function have magnitudes that are less than or equal to unity, in (2.30),

$$-Ae^{\sigma_0 t} \leqq x_r(t) \leqq Ae^{\sigma_0 t}, \quad -Ae^{\sigma_0 t} \leqq x_i(t) \leqq Ae^{\sigma_0 t}. \tag{2.31}$$

For the case that the sinusoid is damped ($\sigma_0 < 0$), the envelope can be expressed as $\pm Ae^{-t/\tau}$; we say that this damped sinusoid has a time constant of τ seconds.

The signals defined in this section appear in the responses of a wide class of physical systems. In terms of circuit analysis, the real exponential of Case 1 appears in the transient, or natural, response of RL and RC circuits. The undamped sinusoid of Case 2 appears in the transient response of LC circuits (no resistance), and the underdamped sinusoid of Case 3 can appear in the transient response of RLC circuits.

A physical circuit always has resistance. Hence, an undamped sinusoid can appear in the natural response of a physical circuit only if the circuit is designed to replace the energy, in each cycle, that is dissipated in the resistance in that cycle. This operation also appears in a pendulum clock, where the mainspring exactly replaces the energy lost to friction in each cycle of the pendulum's swing.

EXAMPLE 2.8   **Time constants and a data-acquisition system**

Data-acquisition systems are instrumentation systems in which the taking of measurements is controlled automatically by a computer. (See Section 1.3.) In these systems, it is often necessary to switch the sensor circuits on command from the computer, before the taking of measurements. When switching occurs, the circuits used in the measurement process must be allowed to settle to steady state before the reading is taken. Usually, the transient terms in these circuits are exponential signals of the form $Ce^{-t/\tau}$, with C and τ real. To increase the accuracy of the measurements, in some systems the measurement is delayed nine time constants (9τ) after the circuit is switched. Hence, the transient term has decayed to $Ce^{-9} = 0.000123C$ or to approximately 0.01% of its initial value. ■

2.4 SINGULARITY FUNCTIONS

In this section, we consider a class of functions called singularity functions. We define a *singularity function* as one that is related to the impulse function (to be defined in this section) and associated functions. Two singularity functions are emphasized in this section: the unit step function and the unit impulse function. We begin with the unit step function.

Unit Step Function

The unit step function, denoted as $u(t)$, is usually employed to switch other signals on or off. The *unit step function* is defined as

$$u(\tau) = \begin{cases} 1, & \tau > 0 \\ 0, & \tau < 0 \end{cases}, \tag{2.32}$$

where the independent variable is denoted as τ. In the study of signals, we choose the independent variable to be a linear function of time. For example, if $\tau = (t - 5)$, the unit step is expressed as

$$u(t - 5) = \begin{cases} 1, & t - 5 > 0 \Rightarrow t > 5 \\ 0, & t - 5 < 0 \Rightarrow t < 5 \end{cases}.$$

This unit step function has a value of unity for $t > 5$ and a value of zero for $t < 5$. The general unit step is written as $u(t - t_0)$, with

$$u(t - t_0) = \begin{cases} 1, & t > t_0 \\ 0, & t < t_0 \end{cases}.$$

A plot of $u(t - t_0)$ is given in Figure 2.17 for a value of $t_0 > 0$.

The unit step function has the property

$$u(t - t_0) = [u(t - t_0)]^2 = [u(t - t_0)]^k, \tag{2.33}$$

with k any positive integer. This property is based on the relations $(0)^k = 0$ and $(1)^k = 1, k = 1, 2, \ldots$. A second property is related to time scaling:

$$u(at - t_0) = u(t - t_0/a), a \neq 0. \tag{2.34}$$

(See Problem 2.21.)

Note that we have not defined the value of the unit step function at the point that the step occurs. Unfortunately, no standard definition exists for this value. As is sometimes done, we leave this value undefined; some authors define the value as zero, while others define the value as unity.

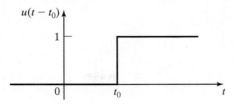

Figure 2.17 Unit step function.

As previously stated, the unit step is often used to switch functions. An example is given by

$$\cos \omega t \, u(t) = \begin{cases} \cos \omega t, & t > 0 \\ 0, & t < 0 \end{cases}.$$

The unit step allows us mathematically to switch this sinusoidal function on at $t = 0$. Another example is $v(t) = 12u(t)$ volts; this function is equal to 0 volts for $t < 0$ and to 12 V for $t > 0$. In this case, the unit step function is used to switch a 12-V source.

Another useful switching function is the unit rectangular pulse, $\text{rect}(t/T)$, which is *defined* as

$$\text{rect}(t/T) = \begin{cases} 1, & -T/2 < t < T/2 \\ 0, & \text{otherwise} \end{cases}.$$

This function is plotted in Figure 2.18(a). It can be expressed as three different functions of unit step signals:

$$\text{rect}(t/T) = \begin{cases} u(t + T/2) - u(t - T/2) \\ u(T/2 - t) - u(-T/2 - t). \\ u(t + T/2)u(T/2 - t) \end{cases} \tag{2.35}$$

These functions are plotted in Figure 2.18(b), (c), and (d).

The time-shifted rectangular pulse function is given by

$$\text{rect}[(t - t_0)/T] = \begin{cases} 1, & t_0 - T/2 < t < t_0 + T/2 \\ 0, & \text{otherwise} \end{cases}. \tag{2.36}$$

This function is plotted in Figure 2.19. Notice that in both (2.35) and (2.36) the rectangular pulse has a duration of T seconds.

The unit rectangular pulse is useful in extracting part of a signal. For example, the signal $x(t) = \cos t$ has a period $T_0 = 2\pi/\omega = 2\pi$. Consider a signal composed of one period of this cosine function beginning at $t = 0$, and zero for all other time. This signal can be expressed as

$$x(t) = (\cos t)[u(t) - u(t - 2\pi)] = \begin{cases} \cos t, & 0 < t < 2\pi \\ 0, & \text{otherwise} \end{cases}.$$

The rectangular-pulse notation allows us to write

$$x(t) = \cos t \, \text{rect}[(t - \pi)/2\pi].$$

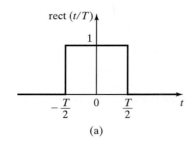

(a)

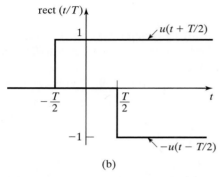

(b)

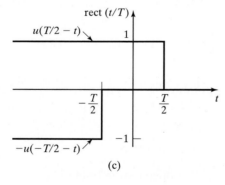

(c)

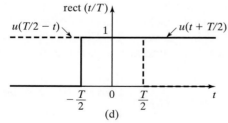

(d)

Figure 2.18 Unit rectangular pulse.

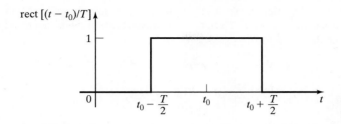

Figure 2.19 Time-shifted rectangular function.

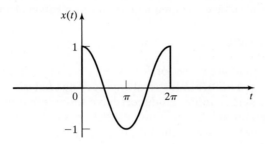

Figure 2.20 The function $x(t) = \cos t \; \text{rect}[(t - \pi)/2\pi]$.

This sinusoidal pulse is plotted in Figure 2.20. Another example of writing the equation of a signal using unit step functions will now be given.

EXAMPLE 2.9   **Equations for a half-wave rectified signal**

Consider again the half-wave rectified signal described in Section 1.2 and Example 2.6, and shown again as $v(t)$ in Figure 2.21. We assume that $v(t)$ is zero for $t < 0$ in this example. If a system containing this signal is to be analyzed or designed, the signal must be expressed as a mathematical function. In Figure 2.21, the signal for $0 \leq t \leq T_0$ can be written as

$$v_1(t) = (V_m \sin \omega_0 t)[u(t) - u(t - T_0/2)]$$
$$= V_m \sin(\omega_0 t) \, \text{rect}[(t - T_0/4)/(T_0/2)],$$

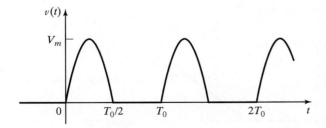

Figure 2.21 Half-wave rectified signal.

where $T_0 = 2\pi/\omega_0$. This signal, $v_1(t)$, is equal to the half-wave rectified signal for $0 \leq t \leq T_0$ and is zero elsewhere. Thus, the half-wave rectified signal can be expressed as a sum of shifted signals,

$$v(t) = v_1(t) + v_1(t - T_0) + v_1(t - 2T_0) + \cdots$$
$$= \sum_{k=0}^{\infty} v_1(t - kT_0), \tag{2.37}$$

since $v_1(t - T_0)$ is $v_1(t)$ delayed by one period, $v_1(t - 2T_0)$ is $v_1(t)$ delayed by two periods, and so on. If the half-wave rectified signal is specified as periodic for all time, the lower limit in (2.37) is changed to negative infinity. As indicated in this example, expressing a periodic signal as a mathematical function often requires the summation of an infinity of terms. ∎

Unit Impulse Function

Engineers have found great use for $j = \sqrt{-1}$ even though this is not a real number and cannot appear in nature. Electrical engineering analysis and design utilizes j extensively. In the same manner, engineers have found great use for the *unit impulse function* $\delta(t)$, even though this function cannot appear in nature. In fact, the impulse function is not a mathematical function in the usual sense [2].

To introduce the impulse function, we begin with the integral of the unit step function; this integral yields the unit ramp function

$$f(t) = \int_0^t u(\tau - t_0)d\tau = \int_{t_0}^t d\tau = \tau \Big|_{t_0}^t = [t - t_0]u(t - t_0), \tag{2.38}$$

where $[t - t_0]u(t - t_0)$ by definition is the *unit ramp function*. In (2.38), the factor $u(t - t_0)$ in the result is necessary, since the value of the integral is zero for $t < t_0$. The unit step function and the unit ramp function are illustrated in Figure 2.22.

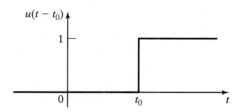

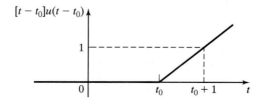

Figure 2.22   Integral of the unit step function.

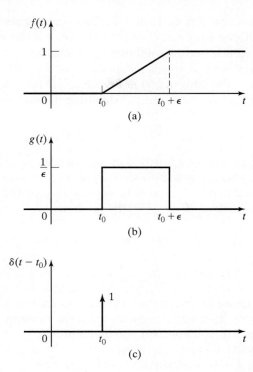

Figure 2.23 Generation of an impulse function.

Note that in Figure 2.22 and in (2.38), the unit step function is the derivative of the unit ramp function. We have no mathematical problems in (2.38) or in the derivative of (2.38). However, problems do occur if we attempt to take the second derivative of (2.38). We now consider this derivative.

The result of differentiating the unit step function $u(t - t_0)$ is not a function in the usual mathematical sense. The derivative is undefined at the only point, $t = t_0$, where it is not zero. (See Figure 2.22.) However, this derivative has been shown, using the rigorous mathematical theory of distributions [2–6], to be very useful in the modeling and analysis of systems. We now consider this derivative as the limit of a derivative that does exist.

No signal can change instantaneously in a physical system, since this change in general represents an instantaneous transfer of energy. Hence, we can consider the function $f(t)$ in Figure 2.23(a) to be a more accurate model of a physical step function. Note that we can differentiate this function and that this derivative is the rectangular pulse $g(t)$ of Figure 2.23(b); that is,

$$g(t) = \frac{df(t)}{dt}.$$

The practical function $f(t)$ in Figure 2.23(a) approaches the unit step function $u(t - t_0)$ if we allow ϵ to approach zero. For this case, the width of $g(t)$ approaches zero and the amplitude becomes unbounded. However, the area under $g(t)$ remains constant at unity, since this area is independent of the pulse width ϵ.

We often call the limit of $g(t)$ in Figure 2.23(b) as ϵ approaches zero the *unit impulse function*. Hence, with $\delta(t - t_0)$ denoting the unit impulse function,

$$\delta(t - t_0) = \lim_{\epsilon \to 0} g(t). \tag{2.39}$$

The impulse function is not a function in the ordinary sense, since it is zero at every point except t_0, where it is unbounded. However, the area under a unit impulse function is well defined and is equal to unity. Based on these properties, we *define* the unit impulse function $\delta(t - t_0)$ by the relations

$$\delta(t - t_0) = 0, \quad t \neq t_0;$$
$$\int_{-\infty}^{\infty} \delta(t - t_0)dt = 1. \tag{2.40}$$

We depict the impulse function as a vertical arrow as shown in Figure 2.23(c), where the number written beside the arrow denotes the multiplying constant of the unit impulse function. For example, for the function $5\delta(t - t_0)$, that number is 5. This multiplying constant is called the *weight* of the impulse. The *amplitude* of the impulse function at $t = t_0$ is unbounded, while the multiplying factor (the weight) is the *area* under the impulse function.

The definition of the impulse function (2.40) is not mathematically rigorous [2]; we now give the definition that is. For any function $f(t)$ that is continuous at $t = t_0$, $\delta(t - t_0)$ is *defined* by the integral

$$\int_{-\infty}^{\infty} f(t)\delta(t - t_0)dt = f(t_0). \tag{2.41}$$

The impulse function is defined by its *properties* rather than by its *values*. The two definitions of the impulse function, (2.40) and (2.41), are not exactly equivalent; use of the rectangular pulse in defining the impulse function is *not* mathematically rigorous and must be used with caution [4]. However, (2.40) allows us to derive in a simple nonrigorous manner some of the properties of the impulse function (2.41). In addition, (2.40) is useful when applying the impulse function in signal and system analysis.

We say that the impulse function $\delta(t - t_0)$ "occurs" at $t = t_0$ because this concept is useful. The quotation marks are used since the impulse function (1) is not an ordinary function and (2) is defined rigorously only under the integral in (2.41). The operation in (2.41) is often taken one step further; if $f(t)$ is continuous at $t = t_0$, then

$$f(t)\delta(t - t_0) = f(t_0)\delta(t - t_0). \tag{2.42}$$

TABLE 2.3   Properties of the Unit Impulse Function

1. $\displaystyle\int_{-\infty}^{\infty} f(t)\delta(t - t_0)dt = f(t_0),\ f(t)$ continuous at $t = t_0$

2. $\displaystyle\int_{-\infty}^{\infty} f(t - t_0)\delta(t)dt = f(-t_0),\ f(t)$ continuous at $t = -t_0$

3. $f(t)\delta(t - t_0) = f(t_0)\delta(t - t_0),\ f(t)$ continuous at $t = t_0$

4. $\delta(t - t_0) = \dfrac{d}{dt}u(t - t_0)$

5. $u(t - t_0) = \displaystyle\int_{-\infty}^{t} \delta(\tau - t_0)d\tau = \begin{cases} 1, & t > t_0 \\ 0, & t < t_0 \end{cases}$

6. $\displaystyle\int_{-\infty}^{\infty} \delta(at - t_0)dt = \dfrac{1}{|a|}\int_{-\infty}^{\infty}\delta\left(t - \dfrac{t_0}{a}\right)dt$

7. $\delta(-t) = \delta(t)$

The product of a continuous-time function $f(t)$ and $\delta(t - t_0)$ is an impulse with its weight equal to $f(t)$ evaluated at time t_0, the time that the impulse occurs. Equations (2.41) and (2.42) are sometimes called the *sifting* property of the impulse function.

One practical use of the impulse function, (2.41), is in modeling sampling operations, since the result of sampling is the selection of a value of the function at a particular instant of time. The sampling of a time signal by an analog-to-digital converter (a hardware device described in Section 1.2) such that samples of the signal can be either processed by a digital computer or stored in the memory of a digital computer is often modeled as in (2.41). If the model of sampling is based on the impulse function, the sampling is said to be *ideal*, because an impulse function cannot appear in a physical system. However, as we will see later, ideal sampling can *accurately* model physical sampling in many applications.

Table 2.3 lists the definition and several properties of the unit impulse function. See Refs. 2 through 6 for rigorous proofs of these properties. The properties listed in Table 2.3 are very useful in the signal and system analysis to be covered later.

EXAMPLE 2.10   **Integral evaluations for impulse functions**

This example illustrates the evaluation of some integrals containing impulse functions, using Table 2.3, for $f(t)$ given in Figure 2.24(a). First, from Property 1 in Table 2.3 with $t_0 = 0$,

$$\int_{-\infty}^{\infty} f(t)\delta(t)dt = f(0) = 2,$$

and the value of the integral is equal to the value of $f(t)$ at the point at which the impulse function occurs. Next, for Figure 2.24(b), from Property 2 in Table 2.3,

$$\int_{-\infty}^{\infty} f(t - 1)\delta(t)dt = f(-1) = 3.$$

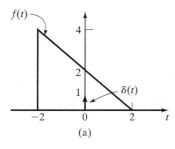

(a)

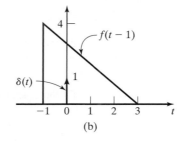

(b)

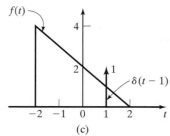

(c)

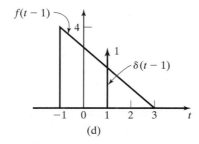

(d)

Figure 2.24   Signals for Example 2.10.

As a third example, for Figure 2.24(c),

$$\int_{-\infty}^{\infty} f(t)\delta(t-1)dt = f(1) = 1,$$

from Property 1 in Table 2.3. For Figure 2.24(d),

$$\int_{-\infty}^{\infty} f(t-1)\delta(t-1)dt = f(0) = 2,$$

from Property 1 in Table 2.3. We have considered all possible combinations of delaying the functions. In each case, the value of the integral is that value of $f(t)$ at which the impulse function occurs.

As a final example, consider the effects of time scaling the impulse function,

$$\int_{-\infty}^{\infty} f(t)\delta(4t)dt = \frac{1}{4}\int_{-\infty}^{\infty} f(t)\delta(t)dt = \frac{f(0)}{4} = \frac{1}{2},$$

from Property 6 in Table 2.3.   ∎

Property 6 is considered further in Problem 2.20.

In this section, we have defined two signals, the step and impulse functions, that are used extensively in signal and system analysis. These signals belong to a class called *singularity functions*; singularity functions are the set of functions obtained by integrating and differentiating the impulse function (or the step function).

2.5   MATHEMATICAL FUNCTIONS FOR SIGNALS

In Example 2.9, we wrote the equation of a half-wave rectified signal using unit step functions. This topic is considered further in this section. First, we consider an example.

EXAMPLE 2.11   **Plotting signal waveforms**

In this example, we consider a signal given in mathematical form:

$$f(t) = 3u(t) + tu(t) - [t - 1]u(t - 1) - 5u(t - 2). \qquad (2.43)$$

The terms of $f(t)$ are plotted in Figure 2.25(a), and $f(t)$ is plotted in Figure 2.25(b). We now verify these plots. The four terms of $f(t)$ are evaluated as

$$3u(t) = \begin{cases} 3, & t > 0 \\ 0, & t < 0 \end{cases};$$

$$tu(t) = \begin{cases} t, & t > 0 \\ 0, & t < 0 \end{cases};$$

$$(t - 1)u(t - 1) = \begin{cases} t - 1, & t > 1 \\ 0, & t < 1 \end{cases};$$

$$5u(t - 2) = \begin{cases} 5, & t > 2 \\ 0, & t < 2 \end{cases}.$$

Using these functions and (2.43), we can write the equations for $f(t)$ (as the sum of four terms) over each different range:

$$\begin{aligned}
t < 0, \quad & f(t) = 0 + 0 - 0 - 0 = 0 \,; \\
0 < t < 1, \quad & f(t) = 3 + t - 0 - 0 = 3 + t \,; \\
1 < t < 2, \quad & f(t) = 3 + t - (t - 1) - 0 = 4 \,; \\
2 < t, \quad & f(t) = 3 + t - (t - 1) - 5 = -1 \,.
\end{aligned}$$

and the graph of $f(t)$ given in Figure 2.25(b) is correct.   ■

Example 2.11 illustrates the construction of waveforms composed only of straight-line segments. Recall the general equation of a straight line:

$$y - y_0 = m[x - x_0] \,. \qquad (2.44)$$

In this equation, y is the ordinate (vertical axis), x is the abscissa (horizontal axis), m is the slope and is equal to dy/dx, and (x_0, y_0) is any point on the line.

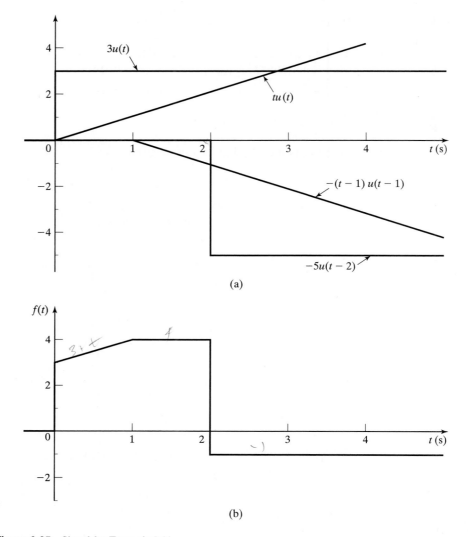

Figure 2.25   Signal for Example 2.11.

A technique is now developed for writing the equations for functions composed of straight-line segments. An example is given in Figure 2.26. The slopes of the segments are denoted as m_i. The signal is zero for $t < t_0$. For $t < t_1$, the equation of the signal, denoted as $x_0(t)$, is given by

$$x_0(t) = m_0[t - t_0]u(t - t_0); \quad t < t_1, \tag{2.45}$$

where $x_0(t) = x(t)$ for $t < t_1$.

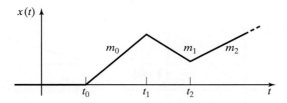

Figure 2.26   Signal.

To write the equation of the signal for $t < t_2$, first we set the slope to zero by subtracting the slope m_0:

$$x_1(t) = x_0(t) - m_0[t - t_1]u(t - t_1). \tag{2.46}$$

Next we add the term required to give the slope m_1,

$$x_2(t) = x_1(t) + m_1[t - t_1]u(t - t_1), \tag{2.47}$$

with $x_2(t) = x(t)$ for $t < t_2$. Then, from the last three equations, for $t < t_2$,

$$x_2(t) = m_0[t - t_0]u(t - t_0) - m_0[t - t_1]u(t - t_1) + m_1[t - t_1]u(t - t_1)$$

$$= m_0[t - t_0]u(t - t_0) + [m_1 - m_0][t - t_1]u(t - t_1); \quad t < t_2. \tag{2.48}$$

This result is general. When the slope of a signal changes, a ramp function is added at that point, with the slope of this ramp function equal to the new slope minus the previous slope $(m_1 - m_0)$. At any point that a step occurs in the signal, a step function is added. An example using this procedure is now given.

EXAMPLE 2.12   **Equations for straight-line-segments signal**

The equation for the signal in Figure 2.27 will be written. The slope of the signal changes from 0 to 3 for a change in slope of 3, beginning at $t = -2$. The slope changes from 3 to -3 at $t = -1$, for a change in slope of -6. At $t = 1$, the slope becomes 0 for a change in slope of 3. The function steps from -3 to 0 at $t = 3$, for a change in amplitude of 3. Hence, the equation for $x(t)$ is given by

$$x(t) = 3[t + 2]u(t + 2) - 6[t + 1]u(t + 1) + 3[t - 1]u(t - 1) + 3u(t - 3).$$

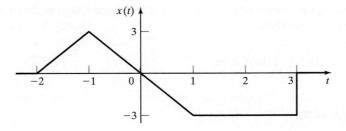

Figure 2.27   Signal for Example 2.12.

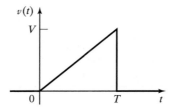

Figure 2.28   Linear sweep voltage.

To verify this result, we write the equation for each segment of the signal, using the procedure illustrated in Example 2.11.

$$t < -2, \quad f(t) = 0 - 0 + 0 + 0 = 0;$$
$$-2 < t < -1, \quad f(t) = 3[t + 2] - 0 + 0 + 0 = 3[t + 2];$$
$$-1 < t < 1, \quad f(t) = 3[t + 2] - 6[t + 1] + 0 + 0 = -3t;$$
$$1 < t < 3, \quad f(t) = [-3t] + 3[t - 1] + 0 = -3;$$
$$3 < t, \quad f(t) = [-3] + 3 = 0.$$

We see that the equation is correct. ■

Next, an example of constructing the equation for a practical periodic waveform will be given. A problem in the design of circuits to drive a cathode ray tube (CRT), such as are used in oscilloscopes, computer monitors, and television sets, is the generation of a constant lateral velocity for the electron beam as the beam sweeps across the face of the CRT. This problem was mentioned in Section 2.2. Control of the electron beam may be either by an electric field (set up by a voltage) or a magnetic field (set up by a current). In either case, a linear waveform, as shown in Figure 2.28, must be generated. For electric-field deflection, the waveform of Figure 2.28 is a voltage, and this is the case for the next example.

EXAMPLE 2.13   **The equation for a CRT sweep signal**

We wish to write the equation of the sweep voltage for a CRT, as shown in Figure 2.28. The value T is the time required to sweep the electron beam from the left edge to the right edge of the face of the CRT, facing the CRT. From (2.44),

$$v(t) = \frac{V}{T}t, \quad 0 < t < T.$$

We can multiply this pulse by the unit rectangular pulse to force $v(t)$ to be zero for all other values of time, with the final result

$$v(t) = \frac{V}{T}t[u(t) - u(t - T)] = \frac{V}{T}t \, \text{rect}\left[\left(t - \frac{T}{2}\right)\Big/ T\right]. \tag{2.49}$$

Recall from (2.35) that the rectangular pulse can also be expressed as other functions of unit step signals. ■

EXAMPLE 2.14   **The equation for a sawtooth waveform**

This example is a continuation of Example 2.13. The results of that example are used to write the equation of the periodic waveform of Figure 2.29. This waveform, called a *sawtooth wave* because of its appearance, is used to sweep an electron beam repeatedly across the face of the CRT. From Example 2.13, the equation of the sawtooth pulse for $0 < t < T$ is

[eq(2.49)]
$$v_1(t) = \frac{V}{T} t[u(t) - u(t - T)] = \frac{V}{T} t \, \text{rect}\left[\left(t - \frac{T}{2}\right)\Big/ T\right].$$

Note that this pulse has been denoted as $v_1(t)$. Hence, as in the case of the half-wave rectified signal of Example 2.9 and (2.37), the pulse from $T < t < 2T$ is $v_1(t - T)$:

$$v_1(t - T) = \frac{V}{T}[t - T][u(t - T) - u(t - 2T)]$$

$$= \frac{V}{T}[t - T] \, \text{rect}\left[\left(t - \frac{3T}{2}\right)\Big/ T\right].$$

The interested reader can plot this function to show its correctness. In a similar manner, the pulse from $kT < t < (k + 1)T$ is $v_1(t - kT)$:

$$v_1(t - kT) = \frac{V}{T}[t - kT][u(t - kT) - u(t - kT - T)]. \qquad (2.50)$$

This equation applies for k either positive or negative. Hence, the equation for the sawtooth wave of Figure 2.29 is

$$v(t) = \sum_{k=-\infty}^{\infty} v_1(t - kT),$$

where $v_1(t - kT)$ is defined in (2.50).

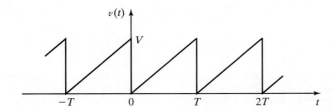

Figure 2.29   Sawtooth waveform.   ■

This section presents a procedure for writing equations for signals, with emphasis on signals composed only of straight-line segments. This procedure is practical, since many physical signals are modeled in this manner.

2.6   CONTINUOUS-TIME SYSTEMS

In this section, we present some of the properties of continuous-time systems. Recall that we define a continuous-time system as one in which no sampled signals (no discrete-time signals) appear.

We begin with a definition of a system that is suitable for this book.

> **System**
> A system is a process for which cause-and-effect relations exist.

For our purposes, the cause is the system input signal, the effect is the system output signal, and the relations are expressed as equations (the system model). We often refer to the input signal and the output signal as simply the input and the output, respectively.

An example of a physical system is an electric heater; for example, electric heaters have wide applications in the chemical processing industry. The input signal is the ac voltage, $v(t)$, applied to the heater. We consider the output signal to be the temperature, $\theta(t)$, of a certain point in space that is close to the heater. One representation of this system is the *block diagram* shown in Figure 2.30. The units of the input are volts, and the units of the output are degrees Celsius. The input signal is sinusoidal. If the system has settled to steady state, the output signal (the temperature) is (approximately) constant.

A second example of a physical system is a voltage amplifier such as that used in public-address systems. The amplifier input signal is a voltage that represents speech, music, and so on. We prefer that the output signal be of exactly the same form, but amplitude scaled to a higher voltage (and higher energy level). However, any physical system will change (distort) a signal. If the distortion is insignificant, we assume that the output signal of the amplifier is simply the input signal multiplied by a constant. We call an amplifier that introduces no distortion an *ideal amplifier*. The block diagram of Figure 2.31 represents an ideal amplifier with a gain of 10.

One representation of a general system is by a block diagram as shown in Figure 2.32. The input signal is $x(t)$, and the output signal is $y(t)$. The system may be denoted by the equation

$$y(t) = T[x(t)], \tag{2.51}$$

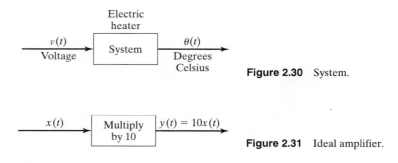

Figure 2.30   System.

Figure 2.31   Ideal amplifier.

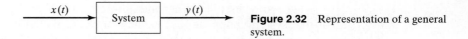

Figure 2.32   Representation of a general system.

where the notation $T[\,\cdot\,]$ indicates a *transformation*. This notation does not indicate a function; that is, $T[x(t)]$ is not a mathematical function into which we substitute $x(t)$ and directly calculate $y(t)$. The explicit set of equations relating the input $x(t)$ and the output $y(t)$ is called the *mathematical model*, or simply the model, of the system. Given the input $x(t)$, this set of equations must be solved to obtain $y(t)$. For continuous-time systems, the model is usually a set of *differential equations*.

Often we are careless in speaking of systems. Generally, when we use the word *system*, we are referring to the mathematical model of a system, not the physical system itself. This is common usage and will be followed in this book. If we are referring to a physical system, we call it a physical system if any confusion can occur. An example of a system will now be given.

EXAMPLE 2.15   **Transformation notation for a circuit**

Consider the circuit of Figure 2.33. We define the system input as the voltage source $v(t)$ and the system output as the inductor voltage $v_L(t)$. The *transformation notation* for the system is

$$v_L(t) = T[v(t)]. \qquad (2.52)$$

The equations that model the system are given by

$$L\frac{di(t)}{dt} + Ri(t) = v(t)$$

and

$$v_L(t) = L\frac{di(t)}{dt}. \qquad (2.53)$$

Hence, the transformation notation of (2.52) represents the explicit equations of (2.53). This model, (2.53), is two equations, with the first a first-order linear differential equation with constant coefficients. (L and R are constants.)

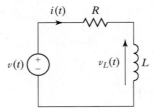

Figure 2.33   *RL circuit.*   ∎

We now discuss further the difference between a physical system and a model of that system. Suppose that we have a physical circuit with an inductor, a resistor, and some type of voltage source connected in series. Equation (2.53) may or may not be an accurate model for this physical system. Developing accurate models for physical systems can be one of the most difficult and time-consuming tasks for engineers. For example, the model that related the thrust from the engines to the attitude (pitch angle) of the Saturn V booster stage was a 27th-order differential equation.

As a final point, consider again the circuit diagram of Figure 2.33. We (engineers) are very careless in diagrams of this type. This diagram may represent either

1. the physical interconnections of a power supply, a coil, and a resistor or
2. a circuit model of a physical system that contains *any number* of physical devices (not necessarily three).

Hence, often, we do not differentiate between drawing a wiring diagram for physical elements and drawing a circuit model. This carelessness is frequently a source of confusion, even for experienced engineers.

Interconnecting Systems

In this section, the system-transformation notation of (2.51) will be used to specify the interconnection of systems. First, we define three block-diagram elements. The first element is a block as shown in Figure 2.34(a); this block is a graphical representation of a system described by (2.51), as shown in the figure. The second element is a circle that represents a summing junction as shown in Figure 2.34(b). The

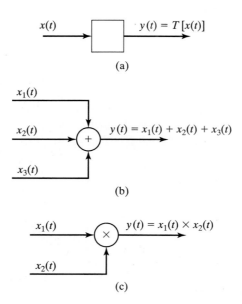

Figure 2.34   Block-diagram elements.

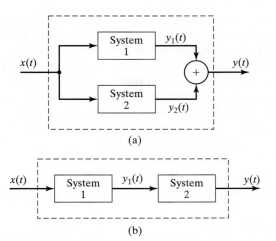

(a)

(b)

Figure 2.35   Basic connections of systems.

output signal of the junction is defined to be the sum of the input signals. The third element is a circle that represents a product junction, as shown in Figure 2.34(c). The output signal of the junction is defined to be the product of the input signals.

We next define two basic connections for systems. The first is the *parallel* connection and is illustrated in Figure 2.35(a). Let the output of System 1 be $y_1(t)$ and that of System 2 be $y_2(t)$. The output signal of the total system is then given by

$$y(t) = y_1(t) + y_2(t) = T_1[x(t)] + T_2[x(t)] = T[x(t)]. \qquad (2.54)$$

The notation for the total system is then $y(t) = T[x(t)]$.

The second basic connection for systems is illustrated in Figure 2.35(b). This connection is called the *series*, or *cascade, connection*. In this figure, the output signal of the first system is $y_1(t) = T_1[x(t)]$, and the total-system output signal is

$$y(t) = T_2[y_1(t)] = T_2(T_1[x(t)]) = T[x(t)]. \qquad (2.55)$$

The system equations of (2.54) and (2.55) cannot be simplified further until the mathematical models for the two systems are known.

The preceding analysis is based on the assumption that the interconnection of systems does not change the characteristics of any of the systems. This point can be illustrated with the cascade connection of two *RL* circuits in Figure 2.36. The equation

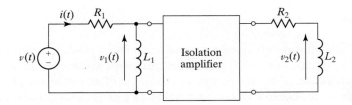

Figure 2.36   Cascade connection of two circuits.

of the current $i(t)$ in the first circuit is independent of the second circuit, provided that the isolation amplifier has a very high input impedance (usually, a good approximation). For this case, the current into the input terminals of the amplifier is negligible compared to $i(t)$, and the system characteristics of the first circuit are not affected by the presence of the second circuit. If the isolation amplifier is removed, R_2 and L_2 affect the voltage across L_1, and the system model for $v_1(t)$ as a function of $v(t)$ changes. When we draw an interconnection of systems, such as in Figures 2.34 and 2.35, the implicit assumption is made that the characteristics of all systems are unaffected by the presence of the other systems.

An example illustrating the interconnection of systems will now be given.

EXAMPLE 2.16   **Interconnections for a system**

Consider the system of Figure 2.37. Each block represents a system, with a number given to identify the system. The circle with the symbol $\times$ denotes the multiplication of the two input signals. We can write the following equations for the system:

$$y_3(t) = y_1(t) + y_2(t) = T_1[x(t)] + T_2[x(t)]$$

and

$$y_4(t) = T_3[y_3(t)] = T_3(T_1[x(t)] + T_2[x(t)]).$$

Thus,

$$y(t) = y_4(t) \times y_5(t) = [T_3(T_1[x(t)] + T_2[x(t)])]T_4[x(t)]. \qquad (2.56)$$

This equation denotes only the interconnection of the systems. The mathematical model of the total system depends on the mathematical models of the individual systems.

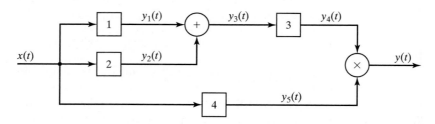

Figure 2.37   System for Example 2.16.   ■

Feedback System

We now consider an important system connection called a *feedback-control system* that is used in automatic control. By the term *automatic control* we mean control without the intervention of human beings. An example of an automatic control system is the temperature-control system in a home or office. A second example is a

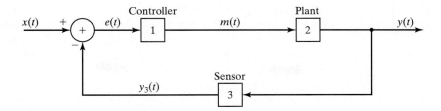

Figure 2.38   Feedback-control system.

system that lands aircraft automatically (without intervention of the pilot) at airports.

The basic configuration of a feedback-control system is given in Figure 2.38. The *plant* is the physical system to be controlled. The *controller* is a physical system inserted by the design engineers to give the total system certain desired characteristics. The *sensor* measures the signal to be controlled, and the input signal represents the desired output. The error signal $e(t)$ is a measure of the difference between the desired output, modeled as $x(t)$, and the measurement of the output $y(t)$. We write the equations of this system as

$$e(t) = x(t) - T_3[y(t)]$$

and

$$y(t) = T_2[m(t)] = T_2(T_1[e(t)]). \tag{2.57}$$

Hence, we can express the output signal as

$$y(t) = T_2[T_1(x(t) - T_3[y(t)])]. \tag{2.58}$$

The system output is expressed as a function of both the system input and the system output, as is always the case for feedback systems. Relationship (2.58) cannot be simplified further without knowing the mathematical models of the three subsystems. A simple example of the model of a feedback control system is now given.

EXAMPLE 2.17   **Interconnections for a feedback system**

In the feedback-control system of Figure 2.38, suppose that the controller and the sensor can be modeled as simple gains (amplitude scaling). These models are adequate in some physical control systems. Thus,

$$m(t) = T_1[e(t)] = K_1 e(t)$$

and

$$y_3(t) = T_3[y(t)] = K_3 y(t),$$

where K_1 and K_3 are real constants. Now,

$$e(t) = x(t) - K_3 y(t),$$

and thus,

$$m(t) = K_1 e(t) = K_1 x(t) - K_1 K_3 y(t).$$

Suppose also that the plant is modeled as a first-order differentiator such that

$$y(t) = T_2[m(t)] = \frac{dm(t)}{dt}.$$

Hence,

$$y(t) = \frac{d}{dt}[K_1 x(t) - K_1 K_3 y(t)] = K_1 \frac{dx(t)}{dt} - K_1 K_3 \frac{dy(t)}{dt}.$$

This equation is the system model and can be expressed as

$$K_1 K_3 \frac{dy(t)}{dt} + y(t) = K_1 \frac{dx(t)}{dt}.$$

This system is modeled by a first-order linear differential equation with constant coefficients. ∎

This section covers a general procedure for representing the interconnection of systems. In addition, two mathematical models are developed.

In Example 2.17 the term *linear* was used without being defined. Linearity is one of the most important properties that a system can have. This property, along with several other system properties, is defined in the next section.

2.7   PROPERTIES OF CONTINUOUS-TIME SYSTEMS

In Section 2.6, continuous-time systems were introduced. In this section, we define some of the properties, or characteristics, of these systems. These definitions allow us to test the mathematical representation of a system to determine its properties.

When testing for the existence of a property, it is often much easier to establish that a system does not exhibit the property in question. To prove that a system does not have a particular property, one only needs to show one counter example. To prove that a system does have the property one must present an analytical argument that is valid for an arbitrary input.

In the following relation, $x(t)$ denotes the input signal and $y(t)$ denotes the output signal of a system.

$$x(t) \rightarrow y(t). \tag{2.59}$$

We read this notation as "$x(t)$ produces $y(t)$"; it has the same meaning as the block diagram of Figure 2.32 and the transformation notation

[eq(2.51)] $$y(t) = T[x(t)].$$

The following are the six properties of continuous-time systems.

Memory
A system has memory if its output at time t_0, $y(t_0)$, depends on input values other than $x(t_0)$. Otherwise, the system is memoryless.

A system with memory is also called a *dynamic system*. An example of a system with memory is an *integrating amplifier*, described by

$$y(t) = K \int_{-\infty}^{t} x(\tau)d\tau. \tag{2.60}$$

(See Section 1.2.) The output voltage $y(t)$ depends on all past values of the input voltage $x(t)$, as one can see by examining the limits of integration. A capacitor also has memory if its current is defined to be the input and its voltage the output:

$$v(t) = \frac{1}{C} \int_{-\infty}^{t} i(\tau)d\tau.$$

The voltage across the capacitor at time t_0 depends on the current $i(t)$ for all time before t_0. Thus, the system has memory.

A memoryless system is also called a *static system*. An example of a memoryless system is the ideal amplifier defined earlier. With $x(t)$ as its input and $y(t)$ as its output, the model of an ideal amplifier with (constant) gain K is given by

$$y(t) = Kx(t)$$

for all t. A second example is resistance, for which $v(t) = Ri(t)$. A third example is a squaring circuit, such that

$$y(t) = x^2(t). \tag{2.61}$$

Clearly, a system $y_1(t) = 5x(t)$ would be memoryless whereas a second system $y_2(t) = x(t + 5)$ has memory, because $y_2(t_0)$ depends on the value of $x(t_0 + 5)$, which is five units of time ahead of t_0.

Invertibility
A system is said to be invertible if distinct inputs result in distinct outputs.

For an invertible system, the system input can be determined uniquely from its output. As an example, consider the squaring circuit mentioned earlier, which is described by

$$y(t) = x^2(t) \Rightarrow x(t) = \pm\sqrt{y(t)}. \tag{2.62}$$

Suppose that the output of this circuit is constant at 4 V. The input could be either $+2$ V or -2 V. Hence, this system is not invertible. An example of an invertible system is an ideal amplifier of gain K:

$$y(t) = Kx(t) \Rightarrow x(t) = \frac{1}{K}y(t). \tag{2.63}$$

A definition related to invertibility is the *inverse* of a system. Before giving this definition, we define the *identity system* to be that system for which the output is equal to its input. An example of an identity system is an ideal amplifier with a gain of unity. We now define the inverse of a system.

Inverse of a System
The inverse of a system (denoted by T) is a second system (denoted by T_i) that, when cascaded with the system T, yields the identity system.

The notation for an inverse transformation is then

$$y(t) = T[x(t)] \Rightarrow x(t) = T_i[y(t)]. \tag{2.64}$$

Hence, $T_i[\,\cdot\,]$ denotes the inverse transformation. If a system is invertible, we can find the unique $x(t)$ for each $y(t)$ in (2.64). We illustrate an invertible system in Figure 2.39. In this figure,

$$z(t) = T_2[y(t)] = T_{i1}(T_1[x(t)]) = x(t), \tag{2.65}$$

where $T_2(\,\cdot\,) = T_{i1}(\,\cdot\,)$, the inverse of system $T_1(\,\cdot\,)$.

A simple example of the inverse of a system is an ideal amplifier with gain 5. Note that we can obtain the inverse system by solving for $x(t)$ in terms of $y(t)$:

$$y(t) = T[x(t)] = 5x(t) \Rightarrow x(t) = T_i[y(t)] = 0.2y(t). \tag{2.66}$$

The inverse system is an ideal amplifier with gain 0.2.

A transducer is a physical device used in the measurement of physical variables. For example, a *thermistor* (a temperature-sensitive resistor) is one device used to measure temperature. To determine the temperature, we measure the resistance of a thermistor and use the known temperature-resistance characteristic of that thermistor to determine the temperature.

The discussion in the preceding paragraph applies to transducers in general. In measuring physical variables (signals), we measure the *effect* of the physical variable on the transducer. We must be able to determine the input to the transducer (the physical variable) by measuring the transducer's output (the effect of the physical

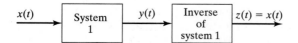

Figure 2.39   Inverse system.

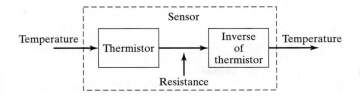

Figure 2.40 Temperature sensor.

variable). This cause-and-effect relationship must be invertible. A *sensor* is a transducer followed by its inverse system and is illustrated in Figure 2.40.

> A glass-bulb thermometer is a second example of a transducer. The glass bulb is the system, and the scale attached to the bulb is the inverse system. A change in temperature causes a change in the density of the liquid in the bulb. As a result, the level in the column of liquid changes. The calibrated scale then converts the liquid level to units of temperature.

The output signal of the inverse system seldom has the same units as the system input signal; however, the amplitudes of the two signals are equal.

Causality

A system is causal if the output at any time t_0 is dependent on the input only for $t \leq t_0$.

A causal system is also called a *nonanticipatory system. All physical systems are causal*.

> A *filter* is a physical device (system) for removing certain unwanted components from a signal. We can design better filters for a signal if both all past values and all future values of the signal are available. In real time (as the signal occurs in the physical system), we never know the future values of a signal. However, if we record a signal and then filter it, the "future" values of the signal are available. Thus, we can design better filters if the filters are to operate only on recorded signals; of course, the filtering is not performed in real time.

A system described by

$$y(t) = x(t - 2), \tag{2.67}$$

with t in seconds, is causal, since the present output is equal to the input of 2 s ago. For example, we can realize this system by recording the signal $x(t)$ on magnetic tape. The playback head is then placed 2 s downstream on the tape from the recording head. A system described by (2.67) is called an *ideal time delay*. The form of the signal is not altered; the signal is simply delayed.

A system described by

$$y(t) = x(t + 2) \tag{2.68}$$

is not causal, since, for example, the output at $t = 0$ is equal to the input at $t = 2$ s. This system is an *ideal time advance*, which is not physically realizable.

Another example of a causal system is illustrated in Figure 2.41. In this system, a time delay of 30 s is followed by a time advance of 25 s. Hence, the total system is

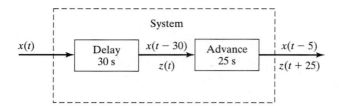

Figure 2.41 Causal system with a noncausal component.

causal and can be realized physically. However, the time-advance part of the system is not causal, but it can be realized if preceded by a time delay of at least 25 s. An example of this type of system is the non–real-time filtering described earlier.

Stability

We now define *stability*. Many definitions exist for the stability of a system; we give the *bounded-input bounded-output* (BIBO) definition.

> **BIBO Stability**
> A system is stable if the output remains bounded for any bounded input.

By definition, a signal $x(t)$ is bounded if there exists a *number M* such that

$$|x(t)| \leqq M \text{ for all } t. \tag{2.69}$$

Hence, a system is bounded-input bounded-output stable if, for a *number R*,

$$|y(t)| \leqq R \text{ for all } t \tag{2.70}$$

for *all* $x(t)$ such that (2.69) is satisfied. Bounded $x(t)$ and $y(t)$ are illustrated in Figure 2.42. To determine BIBO stability for a given system, given any value M in (2.69), a value R (in general a function of M) must be found such that (2.70) is satisfied.

There are a number of common misconceptions in the determination of BIBO stability. First, we point out that it is only the *amplitude* of the input and output signals that must be finite. The time index t runs from $-\infty$ to ∞ because both signals are defined for all time. Second, it is important to recognize that if the input $x(t)$ is unbounded, the output of even a BIBO stable system can be expected to become unbounded. In a sense, BIBO stability means that if a system is used responsibly (i.e., the input is bounded), then the system will behave predictably (i.e., not blow up). We will see a simple test for BIBO stability in Chapter 3 (3.45).

An ideal amplifier as shown in Figure 2.31 where $y(t) = 10x(t)$ is stable because in (2.69) and (2.70)

$$|y(t)| \leq R = 10M.$$

In this system, if $x(t)$ is bounded, then $y(t)$ can never be more than 10 times the value of x(t). Thus, this system is BIBO stable.

A squaring circuit

$$y(t) = x^2(t) \tag{2.71}$$

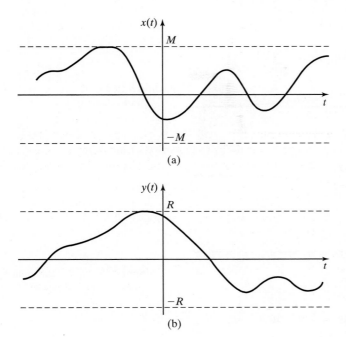

$x(t)$

M

$-M$

(a)

$y(t)$

R

$-R$

(b)

Figure 2.42   Bounded functions.

is stable because $y(t)$ is bounded for any bounded input. In (2.69) and (2.70),

$$|y(t)| \le R = M^2.$$

Consider the system modeled by the circuit shown in Figure 2.33 with $R = 0$. If we consider the ideal voltage source $v(t)$ as the input and the circuit current $i(t)$ to be the output, then the output is described by the equation

$$i(t) = \frac{1}{L} \int_{-\infty}^{t} v(\tau)d\tau. \tag{2.72}$$

If the input is $v(t) = u(t)$, a unit step, the output is $i(t) = \frac{1}{L}tu(t)$, a ramp function, which is unbounded. In this example, the input is bounded as required by (2.69), but the output is unbounded and thus does not satisfy the condition for BIBO stability given in (2.70):

$$\lim_{t \to \infty} y(t) = \lim_{t \to \infty} \frac{1}{L}t = \infty \,.$$

Therefore, the circuit is not BIBO stable. However, if the input to the circuit is the decaying exponential

$$v(t) = e^{-at}u(t), \quad a > 0$$

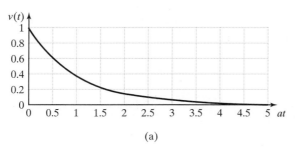

(a)

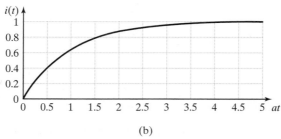

(b)

Figure 2.43 (a) System input signal;
(b) bounded output signal.

shown in Figure 2.43(a), then

$$i(t) = \frac{1}{aL}(1 - e^{-at})u(t),$$

which is shown in Figure 2.43(b) (for the particular case that $aL = 1$) and is bounded for all time. Therefore, the system response is bounded for some inputs, but unbounded for others. A system with this characteristic is *not* BIBO stable, but is sometimes said to be *marginally stable*.

Stability is a basic property required of almost all physical systems. Generally, a system that is not stable cannot be controlled and is of no value. An example of an unstable system is a public address system that has broken into oscillation; the output of this system is unrelated to its input. A second example of an unstable system has been seen several times in television news segments: the first stage of a space booster or a missile that went out of control (unstable) and had to be destroyed.

Time Invariance
A system is said to be time invariant if a time shift in the input signal results *only* in the same time shift in the output signal.

For a time-invariant system for which the input $x(t)$ produces the output $y(t)[x(t) \rightarrow y(t)]$, then $x(t - t_0)$ produces $y(t - t_0)$. That is,

$$x(t - t_0) \rightarrow y(t - t_0)$$

for all t_0. In other words, a time-invariant system does not change with time; if it is used today, it will behave the same way as if were used next week or next year. A time-invariant system is also called a *fixed system*.

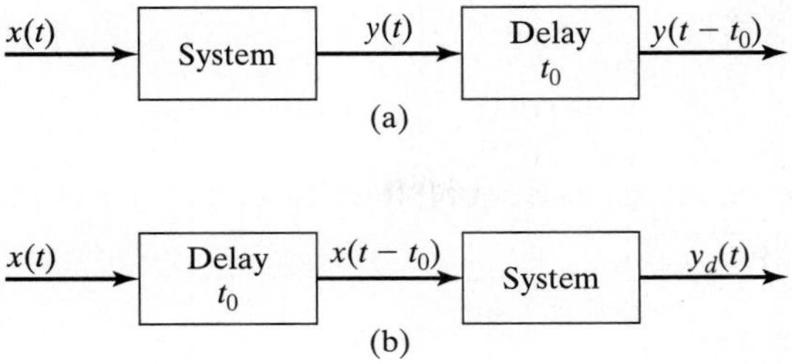

Figure 2.44   Test for time invariance.

A test for time invariance is given by

$$y(t)\bigg|_{t-t_0} = y(t)\bigg|_{x(t-t_0)}, \tag{2.73}$$

provided that $y(t)$ is expressed as an explicit function of $x(t)$. This test is illustrated in Figure 2.44. The signal $y(t - t_0)$ is obtained by delaying $y(t)$ by t_0 seconds. Define $y_d(t)$ as the system output for the delayed input $x(t - t_0)$, such that

$$x(t - t_0) \rightarrow y_d(t).$$

The system in Figure 2.44 is time invariant, provided that

$$y(t - t_0) = y_d(t). \tag{2.74}$$

A system that is not time invariant is *time varying*.

As an example of time invariance, consider the system

$$y(t) = e^{x(t)}.$$

From (2.73) and (2.74),

$$y_d(t) = y(t)\bigg|_{x(t-t_0)} = e^{x(t-t_0)} = y(t)\bigg|_{t-t_0},$$

and the system is time invariant.

Consider next the system

$$y(t) = e^{-t}x(t).$$

In (2.73) and (2.74),

$$y_d(t) = y(t)\bigg|_{x(t-t_0)} = e^{-t}x(t - t_0)$$

and

$$y(t)\Big|_{t-t_0} = e^{-(t-t_0)}x(t - t_0).$$

The last two expressions are not equal; therefore, (2.74) is not satisfied, and the system is time varying.

EXAMPLE 2.18   **Test for Time invariance**

Figure 2.45(a) illustrates the test for time invariance (2.73) for a system that performs a time reversal on the input signal. The input signal chosen for the test is a unit step function, $x(t) = u(t)$. In the top branch of Figure 2.45(a), we first reverse $u(t)$ to obtain $y(t) = u(-t)$

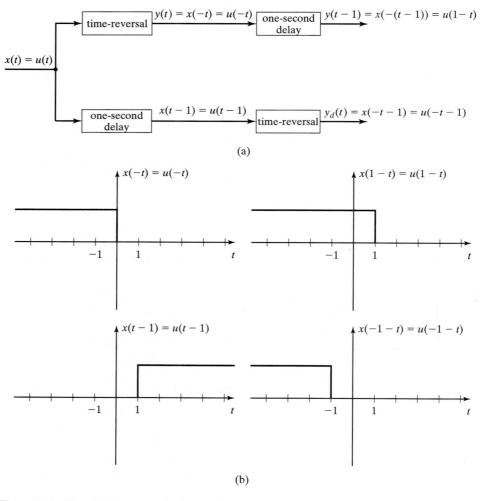

(a)

(b)

Figure 2.45   Time invariance test for Example 2.18.

and then delay it by 1 s to form $y(t - 1) = u(-(t - 1)) = u(1 - t)$. In the bottom branch of the diagram, we first delay the input by 1 s to form $u(t - 1)$ and then reverse in time to form $y_d(t) = u(-t - 1)$. The signals for this system are shown in Figure 2.45(b). Because $y_d(t) \neq y(t - 1)$, the time-reversal operation is not time invariant. Intuitively, this makes sense because a time shift to the right before a time reversal will result in a time shift to the left after the time reversal.   ■

A time-varying system has characteristics that vary with time. The manner in which this type of system responds to a particular input depends on the time that the input is applied.

> An example of a time-varying physical system is the booster stage of the NASA shuttle. Newton's second law applied to the shuttle yields
>
> $$f_z(t) = M(t)\frac{d^2z(t)}{dt^2},$$
>
> where the force $f_z(t)$ is the engine thrust in the z-axis direction developed by burning fuel, $M(t)$ is the mass of the shuttle, and $z(t)$ is the position along the z-axis. The mass decreases as the fuel burns and, hence, is a function of time.

> In designing controllers for automatic control systems, we sometimes intentionally use time-varying gains to improve the system characteristics. Of course, time-varying gains result in a time-varying system. For example, the control system that automatically lands aircraft on U.S. Navy aircraft carriers uses time-varying gains [6]. The time-varying gains are based on time-to-touchdown, and the gains increase as the time-to-touchdown decreases. This increase in gain results in a decreased time-of-response in correcting errors in the plane's flight path.

Linearity

The property of *linearity* is one of the most important properties that we consider. Once again, we define the system input signal to be $x(t)$ and the output signal to be $y(t)$.

> **Linear System**
> A system is linear if it meets the following two criteria:
>
> **1.** Additivity: If $x_1(t) \rightarrow y_1(t)$ and $x_2(t) \rightarrow y_2(t)$, then
>
> $$x_1(t) + x_2(t) \rightarrow y_1(t) + y_2(t) \tag{2.75}$$
>
> **2.** Homogeneity: If $x_1(t) \rightarrow y_1(t)$, then
>
> $$ax_1(t) \rightarrow ay_1(t) \tag{2.76}$$
>
> where a is a constant. The criteria must apply for all $x_1(t)$ and $x_2(t)$ and for all a.

These two criteria can be combined to yield the *principle of superposition*. A system satisfies the principle of superposition if, with the inputs and outputs as just defined,

$$a_1x_1(t) + a_2x_2(t) \rightarrow a_1y_1(t) + a_2y_2(t), \tag{2.77}$$

where a_1 and a_2 are constants. A system is linear if it satisfies the principle of super-position.

No physical system is linear under all operating conditions. However, a physical system can be tested using (2.77) to determine ranges of operation for which the system is approximately linear.

An example of a linear system is an ideal amplifier, described by $y(t) = Kx(t)$. An example of a nonlinear system is the squaring circuit mentioned earlier:

$$y(t) = x^2(t).$$

For inputs of $x_1(t)$ and $x_2(t)$, the outputs are

$$x_1(t) \rightarrow x_1^2(t) = y_1(t)$$

and

$$x_2(t) \rightarrow x_2^2(t) = y_2(t). \tag{2.78}$$

However, the input $[x_1(t) + x_2(t)]$ produces the output

$$x_1(t) + x_2(t) \rightarrow [x_1(t) + x_2(t)]^2 = x_1^2(t) + 2x_1(t)x_2(t)$$
$$+ x_2^2(t) = y_1(t) + y_2(t) + 2x_1(t)x_2(t), \tag{2.79}$$

and $[x_1(t) + x_2(t)]$ does not produce $[y_1(t) + y_2(t)]$. Hence, the squaring circuit is nonlinear.

A linear time-invariant (LTI) system is a linear system that is also time invari-ant. The LTI system, for both continuous-time and discrete-time systems, is the type that is emphasized in this book.

An important class of continuous-time LTI systems is that which is modeled by linear differential equations with constant coefficients. An example of this type of system is the *RL* circuit of Figure 2.33. which is modeled by

[eq(2.53)] $$L\frac{di(t)}{dt} + Ri(t) = v(t).$$

This type of system is discussed in detail in Chapter 3.

EXAMPLE 2.19 **Determining the properties of a particular system**

The characteristics for the system

$$y(t) = \sin 2t\, x(t)$$

are now investigated. Note that this system can be considered to be an amplifier with a time-varying gain that varies between -1 and 1, that is, with the gain $K(t) = \sin 2t$ and $y(t) = K(t)x(t)$. The characteristics are as follows:

1. This system is *memoryless* because the output is a function of the input at only the present time.
2. The system is *not invertible* because, for example, $y(\pi) = 0$, regardless of the value of the input. Hence, the system has no inverse.
3. The system is *causal* because the output does not depend on the input at a future time.
4. The system is *stable*, the output is bounded for all bounded inputs because the multiplier sin $(2t)$ has a maximum value of 1. If $|x(t)| \leq M, |y(t)| \leq M$ also
5. The system is *time varying*. From (2.73) and (2.74),

$$y_d(t) = y(t)\Big|_{x(t-t_0)} = \sin 2t x(t - t_0)$$

and

$$y(t)\Big|_{t-t_0} = \sin 2(t - t_0)x(t - t_0).$$

6. The system is *linear*, since

$$a_1 x_1(t) + a_2 x_2(t) \rightarrow \sin 2t[a_1 x_1(t) + a_2 x_2(t)] = a_1 \sin 2t x_1(t) + a_2 \sin 2t x_2(t)$$
$$= a_1 y_1(t) + a_2 y_2(t). \quad \blacksquare$$

EXAMPLE 2.20   **Testing for linearity using superposition**

As a final example, consider the system described by the equation $y(t) = 3x(t)$, a linear amplifier. This system is easily shown to be linear using superposition. However, the system $y(t) = [3x(t) + 1.5]$, an amplifier that adds a dc component, is nonlinear. By superposition,

$$y(t) = 3[a_1 x_1(t) + a_2 x_2(t)] + 1.5 \neq a_1 y_1(t) + a_2 y_2(t).$$

This system is not linear because a part of the output signal is independent of the input signal. $\quad \blacksquare$

Analysis similar to that of Example 2.20 shows that systems with non-zero initial conditions are not linear. They can be analyzed as linear systems only if the non-zero initial conditions are treated as inputs to the system.

In this section, several important properties of continuous-time systems have been defined; these properties allow us to classify systems. For example, probably the most important general system properties are linearity and time invariance, since the analysis and design procedures for LTI systems are simplest and easiest to apply. We continually refer back to these general system properties for the remainder of this book.

SUMMARY

In this chapter, we introduce continuous-time signals and systems, with emphasis placed on the modeling of signals and the properties of systems.

First, three transformations of the independent variable time are defined: reversal, scaling, and shifting. Next, the same three transformations are defined with respect to the amplitude of signals. A general procedure is developed for handling all six transformations. These transformations are important with respect to time signals, and as we will see in Chapter 5, they are equally important as transformations in frequency.

The signal characteristics of evenness, oddness, and periodicity are defined next. These three characteristics appear often in the study of signals and systems.

TABLE 2.4 Key Equations of Chapter Two

Equation Title	Equation Number	Equation
Independent-variable transformation	(2.6)	$y(t) = x(at + b)$
Signal-amplitude transformation	(2.8)	$y(t) = Ax(t) + B$
Even part of a signal	(2.13)	$x_e(t) = \frac{1}{2}[x(t) + x(-t)]$
Odd part of a signal	(2.14)	$x_o(t) = \frac{1}{2}[x(t) - x(-t)]$
Definition of periodicity	(2.15)	$x(t) = x(t + T), \quad T > 0$
Fundamental frequency in hertz and radians/second	(2.16)	$f_0 = \frac{1}{T_0}$ Hz, $\quad \omega_0 = 2\pi f_0 = \frac{2\pi}{T_0}$ rad/s
Exponential function	(2.18)	$x(t) = C e^{at}$
Euler's relation	(2.19)	$e^{j\theta} = \cos\theta + j\sin\theta$
Cosine equation	(2.21)	$\cos\theta = \dfrac{e^{j\theta} + e^{-j\theta}}{2}$
Sine equation	(2.22)	$\sin\theta = \dfrac{e^{j\theta} - e^{-j\theta}}{2j}$
Complex exponential in polar form	(2.23) and (2.24)	$e^{j\theta} = 1 \angle \theta$ and $\arg e^{j\theta} = \tan^{-1}\left[\dfrac{\sin\theta}{\cos\theta}\right] = \theta$
Unit step function	(2.32)	$u(\tau) = \begin{cases} 1, & \tau > 0 \\ 0, & \tau < 0 \end{cases}$
Unit impulse function	(2.40)	$\delta(t - t_0) = 0, \quad t \neq t_0;$ $\int_{-\infty}^{\infty} \delta(t - t_0)dt = 1$
Sifting property of unit impulse function	(2.41)	$\int_{-\infty}^{\infty} f(t)\delta(t - t_0)dt = f(t_0)$
Multiplication property of unit impulse function	(2.42)	$f(t)\delta(t - t_0) = f(t_0)\delta(t - t_0)$
Test for time invariance	(2.73)	$y(t)\Big|_{t-t_0} = y(t)\Big|_{x(t-t_0)}$
Test for linearity	(2.77)	$a_1 x_1(t) + a_2 x_2(t) \rightarrow a_1 y_1(t) + a_2 y_2(t)$

Models of common signals that appear in physical systems are then defined. The signals considered are exponential signals and sinusoids whose amplitudes vary exponentially. Singularity functions are defined; the unit step and the unit impulse functions are emphasized. These two functions will prove to be very useful not only in the time domain, but also in the frequency domain.

Procedures are developed for expressing certain types of signals as mathematical functions. Nonsinusoidal periodic functions are introduced, and a technique for writing equations for these periodic signals is presented.

A general technique is then given for expressing the output of a continuous-time system that is an interconnection of systems. An example is given of a feedback-control system. Several important properties of systems are defined, and procedures are given to determine if a system possesses these properties.

This chapter is devoted to continuous-time signals and systems. In Chapter 9, the same topics as related to discrete-time signals and systems are developed. Many of the topics are identical; however, in some cases there are significant differences. (See Table 2.4.)

REFERENCES

1. G. Carlson, *Signal and Linear System Analysis*, 2d ed. New York, John Wiley & Sons, 1998.
2. G. Doetsch, *Guide to the Applications of the Laplace and z-Transforms*. London: Van Nostrand Reinhold, 1971.
3. W. Kaplan, *Operational Methods for Linear Systems*. Reading, MA: Addison–Wesley, 1962.
4. R. V. Churchill, *Operational Mathematics*, 3d ed. New York: McGraw–Hill, 1977.
5. G. Doetsch, *Guide to the Applications of Laplace Transforms*. London: Van Nostrand Reinhold, 1961.
6. G. Doetsch, *Introduction to the Theory and Application of the Laplace Transform*. New York: Springer–Verlag, 1974.
7. R. F. Wigginton, *Evaluation of OPS-II Operational Program for the Automatic Carrier Landing System*. Saint Inigoes, MD: Naval Electronic Systems Test and Evaluation Facility, 1971.

PROBLEMS

2.1. The signals in Figure P2.1 are zero except as shown.

    (a) For the signal $x(t)$ of Figure P2.1(a), plot

        (i) $x(-t/3)$ **(ii)** $x(-t)$

        (iii) $x(3 + t)$ **(iv)** $x(2 - t)$

    Verify your results by checking at least two points.

    (b) Repeat (a) for the signal $x(t)$ of Figure P2.1(b).

    (c) Repeat (a) for the signal $x(t)$ of Figure P2.1(c).

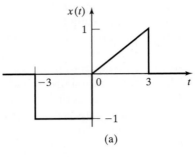

(a)

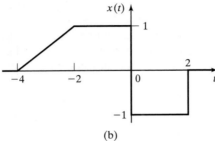

(b)

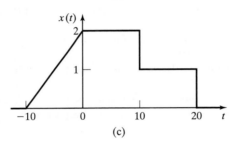

(c)

Figure P2.1

2.2. The signals in Figure P2.1 are zero except as shown.

    (a) For the signal $x(t)$ of Figure P2.1(a), plot

        (i) $4x(t) - 2$ **(ii)** $2x(t) + 2$

        (iii) $2x(2t) + 2$ **(iv)** $-4x(t) + 2$

    Verify your results by checking at least two points.

    (b) Repeat part (a) for the signal $x(t)$ of Figure P2.1(b).

    (c) Repeat part (a) for the signal $x(t)$ of Figure P2.1(c).

2.3. Given the two signals in Figure P2.3,

    (a) Express $x_2(t)$ as a function of $x_1(t)$.

    (b) Verify your result by checking at least three points in time.

2.4. Given the signals $x_1(t)$ and $x_2(t)$ in Figure P2.4,

    (a) Express $x_2(t)$ as a function of $x_1(t)$.

    (b) Verify your results by checking at least three points in time.

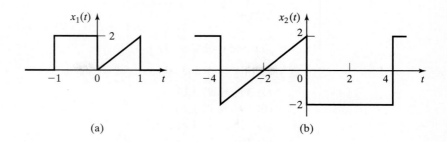

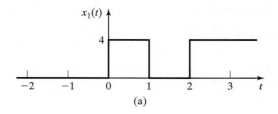

Figure P2.3

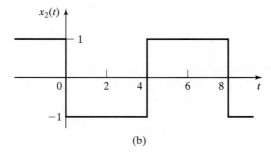

Figure P2.4

(c) Express $x_1(t)$ as a function of $x_2(t)$.

(d) Verify the results of part (c) by checking at least three points in time.

2.5. Given the signal $x(t)$ in Figure P2.5, find and sketch

$$y(t) = x(2t - 1).$$

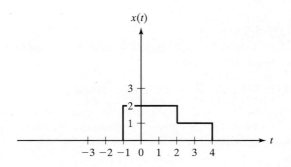

Figure P2.5

2.6. Given

$$x(t) = 3u(t + 3) - u(t) + 3u(t - 3) - 5u(t - 6),$$

find and sketch $x(-3t - 6)$.

2.7. Plot the even and odd parts of the signal of

  (a) Figure P2.1(a)
  (b) Figure P2.1(b)
  (c) Figure P2.1(c)
  (d) Figure P2.4(a)
  (e) Verify your results using (2.11).

2.8. For each of the signals given, determine mathematically if the signal is even, odd, or neither. Sketch the waveforms to verify your results.

  (a) $x(t) = -4t$
  (b) $x(t) = e^{-|t|}$
  (c) $x(t) = 5\cos 3t$
  (d) $x(t) = \sin(3t - \frac{\pi}{2})$
  (e) $x(t) = u(t)$

2.9. The average value A_x of a signal $x(t)$ is given by

$$A_x = \lim_{T \to \infty} \frac{1}{2T} \int_{-T}^{T} x(t)dt.$$

Let $x_e(t)$ be the even part and $x_o(t)$ be the odd part of $x(t)$.

  (a) Show that

$$\lim_{T \to \infty} \frac{1}{2T} \int_{-T}^{T} x_o(t)dt = 0.$$

  (b) Show that

$$\lim_{T \to \infty} \frac{1}{2T} \int_{-T}^{T} x_e(t)dt = A_x.$$

  (c) Show that $x_o(0) = 0$ and $x_e(0) = x(0)$.

2.10. Give proofs of the following statements:

  (a) The sum of two even functions is even.
  (b) The sum of two odd functions is odd.
  (c) The sum of an even function and an odd function is neither even nor odd.
  (d) The product of two even functions is even.
  (e) The product of two odd functions is even.
  (f) The product of an even function and an odd function is odd.

2.11. Given in Figure P2.11 are the parts of a signal $x(t)$ and its even part $x_e(t)$, for $t \geq 0$ only; that is, $x(t)$ and $x_e(t)$ for $t < 0$ are not given. Complete the plots of $x(t)$ and $x_e(t)$, and give a plot of the odd part, $x_o(t)$, of $x(t)$. Give the equations used for plotting each part of the signals.

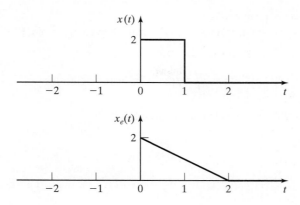

Figure P2.11

2.12. Prove mathematically that the signals given are periodic. For each signal, find the fundamental period T_0 and the fundamental frequency ω_0.

(a) $x(t) = 7\sin 3t$

(b) $x(t) = \sin(8t + 30°)$

(c) $x(t) = e^{j2t}$

(d) $x(t) = \cos t + \sin 2t$

(e) $x(t) = e^{j(5t+\pi)}$

(f) $x(t) = e^{-j10t} + e^{j15t}$

2.13. (a) Consider the signal

$$x(t) = 3\cos(15t + 30°) + \sin 20t.$$

If this signal is periodic, find its fundamental period T_0 and its fundamental frequency ω_0. Otherwise, prove that the signal is not periodic.

(b) Repeat Part (a) for the signal

$$x(t) = \cos 5t + \sin\pi t.$$

(c) Repeat Part (a) for the signal

$$x(t) = \cos 5t + 3e^{-10t}.$$

2.14. Suppose that $x_1(t)$ is periodic with period T_1 and that $x_2(t)$ is periodic with period T_2.

(a) Show that the sum

$$x(t) = x_1(t) + x_2(t)$$

is periodic only if the ratio T_1/T_2 is equal to a ratio of two integers k_2/k_1.

(b) Find the fundamental period T_0 of $x(t)$, for $T_1/T_2 = k_2/k_1$.

2.15. For each signal, if it is periodic, find the fundamental period T_0 and the fundamental frequency ω_0. Otherwise, prove that the signal is not periodic.

(a) $x(t) = \cos 3t + \sin 5t$.

(b) $x(t) = \cos 6t + \sin 8t + e^{j2t}$.

(c) $x(t) = \cos t + \sin \pi t$.

(d) $x(t) = x_1(t) + x_2(3t)$ where $x_1(t) = \sin(\frac{\pi t}{6})$ and $x_2(t) = \sin(\frac{\pi t}{9})$.

2.16. Find

$$\int_{-\infty}^{\infty} \delta(at - b)\sin^2(t - 4)dt,$$

where $a > 0$. *Hint*: Use a change of variables.

2.17. Given a signal

$$x(t) = u(t + 8),$$

let

$$y(t) = x(-5t + 7).$$

Find values for a (where a is NOT EQUAL to -5) and b (where b is NOT EQUAL to 7) in the time transformation

$$z(t) = x(at + b),$$

so that

$$x(t) = y(t).$$

Note: There are an infinite number of solutions to this problem.

2.18. Express the following in terms of $x(t)$:

$$y(t) = \frac{1}{2}\int_{-\infty}^{\infty} x(\tau)[\delta(\tau-2) + \delta(\tau + 2)]d\tau.$$

2.19. Consider the triangular pulse of Figure P2.19(a)

(a) Write a mathematical function for this waveform.

(b) Verify the results of Part (a) using the procedure of Example 2.12.

(c) Write a mathematical function for the triawngular wave of Figure P2.19(b), using the results of Part (a).

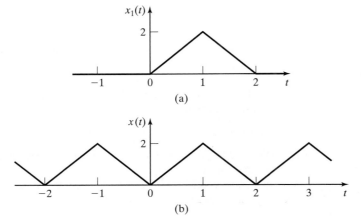

(a)

(b)   **Figure P2.19**

2.20. **(a)** Prove the time-scaling relation in Table 2.3:

$$\int_{-\infty}^{\infty} \delta(at)dt = \frac{1}{|a|}\int_{-\infty}^{\infty} \delta(t)dt.$$

(*Hint*: Use a change of variable.)

(b) Prove the following relation from Table 2.3:

$$u(t - t_0) = \int_{-\infty}^{t} \delta(\tau - t_0)d\tau.$$

(c) Evaluate the following integrals:

(i) $\displaystyle\int_{-\infty}^{\infty} \sin(-3t)\delta(t)dt$

(ii) $\displaystyle\int_{-\infty}^{\infty} \sin(-3t)\delta(t - 4)dt$

(iii) $\displaystyle\int_{-\infty}^{\infty} \sin[-3(t - 4)]\delta(t - 4)dt$

(iv) $\displaystyle\int_{-\infty}^{\infty} \sin[3(t - 1)]\delta(t + 2)dt$

(v) $\displaystyle\int_{-\infty}^{\infty} \sin[3(t - 1)]\delta(2t + 4)dt$

2.21. Express the following functions in the general form of the unit step function $u(\pm t - t_0)$:

(a) $u(\frac{t}{4} - 4)$

(b) $u(\frac{t}{4} + 4)$

(c) $u(-3t + 6)$

(d) $u(3t + 6)$

In each case, sketch the function derived.

2.22. Express the following signals in terms of $u(t - t_0)$. Sketch each expression to verify the results.

(a) $u(-t)$

(b) $u(3 - t)$

(c) $tu(-t)$

(d) $(t - 3)u(3 - t)$

2.23. **(a)** Express the output $y(t)$ as a function of the input and the system transformations, in the form of (2.56), for the system of Figure P2.23(a).

(b) Repeat Part (a) for the system of Figure P2.23(b).

(c) Repeat Part (a) for the case that the summing junction with inputs $y_3(t)$ and $y_5(t)$ is replaced with a multiplication junction, such that its output is the product of these two signals.

(d) Repeat Part (b) for the case that the summing junction with inputs $y_3(t)$, $y_4(t)$, and $y_5(t)$ is replaced with a multiplication junction, such that its output is the product of these three signals.

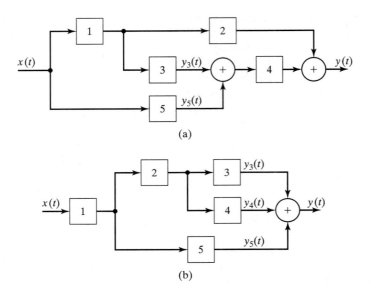

(a)

(b)

Figure P2.23

2.24. Consider the feedback system of Figure P2.24. Express the output signal as a function of the transformation of the input signal, in the form of (2.58).

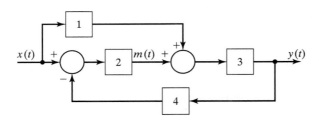

Figure P2.24

2.25. Consider the feedback system of Figure P2.25. Express the output signal as a function of the transformation of the input signal, in the form of (2.58). The minus sign at the summing junction indicates that the signal is subtracted.

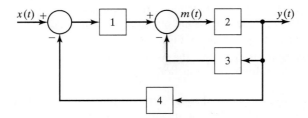

Figure P2.25

2.26. (a) Determine whether the system described by

$$y(t) = \int_{t}^{t+1} x(\tau - \alpha)d\tau,$$

(where α is a constant) is

(i) memoryless,   **(ii)** invertible,

(iii) stable,   **(iv)** time invariant, and

(v) linear.

(b) For what values of the constant α is the system causal?

2.27. Determine whether the system described by

$$y(t) = 3x(3t + 3)$$

is

(i) memoryless,   **(ii)** invertible,

(iii) stable,   **(iv)** time invariant, and

(v) linear.

2.28. You are given an LTI system. The response of the system to an input

$$x_1(t) = u(t) - u(t - 1)$$

is a function $y_1(t)$. What is the response of the system to the input $x_2(t)$ in Figure P2.28 in terms of $y_1(t)$?

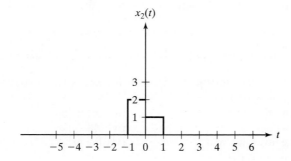

Figure P2.28

2.29. (a) Determine whether the system described by

$$y(t) = \cos{[x(t - 1)]}$$

is

(i) memoryless,   **(ii)** invertible,

(iii) causal,   **(iv)** stable,

(v) time invariant, and   **(vi)** linear.

(b) Repeat Part (a) for

$$y(t) = \ln{[x(t)]}.$$

(c) Repeat Part (a) for

$$y(t) = e^{tx(t)}.$$

(d) Repeat Part (a) for

$$y(t) = 7x(t) + 6.$$

(e) Repeat Part (a) for

$$y(t) = \int_{-\infty}^{t} x(5\tau)d\tau.$$

(f) Repeat Part (a) for

$$y(t) = e^{-j\omega t} \int_{-\infty}^{\infty} x(\tau)e^{-j\omega\tau}d\tau.$$

(g) Repeat Part (a) for

$$y(t) = \int_{t-1}^{t} x(\tau)d\tau.$$

2.30. (a) Determine whether *the ideal time delay*

$$y(t) = x(t - t_0)$$

is

(i) memoryless,   **(ii)** invertible,
(iii) causal,   **(iv)** stable,
(v) time invariant, and   **(vi)** linear.

2.31. Let $h(t)$ denote the response of a system for which the input signal is the unit impulse function $\delta(t)$. Suppose that $h(t)$ for a *causal* system has the given even part $h_e(t)$ for $t > 0$:

$$h_e(t) = t[u(t) - u(t - 1)] + u(t - 1), t > 0.$$

Find $h(t)$ for all time, with your answer expressed as a mathematical function.

2.32. (a) Sketch the characteristic y versus x for the system $y(t) = |x(t)|$. Determine whether this system is

(i) memoryless,   **(ii)** invertible,
(iii) causal,   **(iv)** stable,
(v) time invariant, and   **(vi)** linear.

(b) Repeat Part (a) for

$$y(t) = \begin{cases} x(t), & x \geq 0 \\ 0, & x < 0 \end{cases}.$$

(c) Repeat Part (a) for

$$y(t) = \begin{cases} -10, & x < -1 \\ 10x(t), & |x| \leq 1. \\ 10, & x > 1 \end{cases}$$

(d) Repeat Part (a) for

$$y(t) = \begin{cases} 2, & 2 < x \\ 1, & 1 < x \leq 2 \\ 0, & 0 < x \leq 1 \\ -1, & -1 < x \leq 0 \\ -2, & x \leq -1 \end{cases}.$$

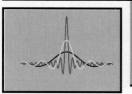

3

CONTINUOUS-TIME LINEAR TIME-INVARIANT SYSTEMS

In Chapter 2, several characteristics of continuous-time systems were defined. We now restate two of these characteristics; systems with these two characteristics are investigated extensively in this chapter.

Consider a system described by

$$x(t) \rightarrow y(t). \tag{3.1}$$

This system is *time invariant* if a time shift of the input results in the same time shift of the output—that is, if

$$x(t - t_0) \rightarrow y(t - t_0), \tag{3.2}$$

where t_0 is an arbitrary constant.

For the system of (3.1), let

$$x_1(t) \rightarrow y_1(t), \quad x_2(t) \rightarrow y_2(t). \tag{3.3}$$

This system is *linear*, provided that the principle of superposition applies:

$$a_1 x_1(t) + a_2 x_2(t) \rightarrow a_1 y_1(t) + a_2 y_2(t). \tag{3.4}$$

This property applies for all constants a_1 and a_2 and for all signals $x_1(t)$ and $x_2(t)$.

In this chapter, we consider continuous-time systems that are both linear and time invariant. We will refer to these systems as continuous-time LTI (linear time-invariant) systems. We have several reasons for emphasizing these systems:

1. Many physical systems can be modeled accurately as LTI systems. For example, the basic electric-circuit models of the resistance, inductance, and capacitance are LTI models.

2. We can solve mathematically the equations that model LTI systems for both continuous-time and discrete-time systems. No general procedures exist for the mathematical solution of the describing equations of non-LTI systems.

3. Much information is available for both the analysis and design of LTI systems. This is especially true for system design. In fact, in preliminary design stages for non-LTI physical systems, we often fit an LTI model to the physical system so as to have a starting point for the design. The LTI model may not be very accurate, but the use of an LTI model allows us to initiate the design process with standard design procedures.

    4. We can sometimes model a general signal $x(t)$ as a sum of functions

$$x(t) = x_1(t) + x_2(t) + \cdots.$$

The functions $x_1(t)$, $x_2(t)$, ... are standard functions for which an LTI system response is much easier to find than is the response to $x(t)$. The system response is then the sum of the responses to the standard functions,

$$y(t) = y_1(t) + y_2(t) + \cdots,$$

where $y_i(t)$ is the response to $x_i(t)$, $i = 1, 2, \ldots$.

3.1   IMPULSE REPRESENTATION OF CONTINUOUS-TIME SIGNALS

In this section, a relationship is developed that expresses a general signal $x(t)$ as a function of an impulse function. This relationship is useful in deriving general properties of continuous-time linear time-invariant (LTI) systems.

    Recall that two definitions of the impulse function are given in Section 2.4. The first definition is, from (2.40),

$$\delta(t - t_0) = 0, \quad t \neq t_0$$

$$\int_{-\infty}^{\infty} \delta(t - t_0)\, dt = 1, \tag{3.5}$$

and the second one is, from (2.41),

$$\int_{-\infty}^{\infty} x(t)\delta(t - t_0)\, dt = x(t_0). \tag{3.6}$$

The second definition requires that $x(t)$ be continuous at $t = t_0$. Based on (3.6), the sifting property of impulse functions is, from (2.42),

$$x(t)\delta(t - t_0) = x(t_0)\delta(t - t_0). \tag{3.7}$$

We now derive the desired relationship. From (3.7), with $t_0 = \tau$,

$$x(t)\delta(t - \tau) = x(\tau)\delta(t - \tau).$$

From (3.5), we use the preceding result to express $x(t)$ as an integral involving an impulse function:

$$\int_{-\infty}^{\infty} x(\tau)\delta(t - \tau)\,d\tau = \int_{-\infty}^{\infty} x(t)\delta(t - \tau)\,d\tau$$

$$= x(t) \int_{-\infty}^{\infty} \delta(t - \tau)\,d\tau = x(t)$$

We rewrite this equation as

$$x(t) = \int_{-\infty}^{\infty} x(\tau)\delta(t - \tau)\,d\tau. \tag{3.8}$$

This equation is the desired result, in which a general signal $x(t)$ is expressed as a function of an impulse function. We use this expression for $x(t)$ in the next section.

We now derive (3.8) by a much longer procedure; this procedure is useful in later results. From Section 2.4, we can consider the unit impulse function $\delta(t - t_0)$ to be the limit as Δ approaches zero of the pulse $p_\Delta(t - t_0)$ in Figure 3.1; that is,

$$\delta(t - t_0) = \lim_{\Delta \to 0} p_\Delta(t - t_0), \tag{3.9}$$

where

$$p_\Delta(t - t_0) = \frac{1}{\Delta}[u(t - t_0) - u(t - t_0 - \Delta)].$$

This pulse is now used to derive (3.8).

Consider the signal $x(t)$ in Figure 3.2. For simplicity, the range for which the signal is nonzero is limited. As an approximation, $x(t)$ can be represented as $x_p(t)$, the sum of the three rectangular pulses shown:

$$x_p(t) = x(-\Delta)p_\Delta(t + \Delta)\Delta + x(0)p_\Delta(t)\Delta + x(\Delta)p_\Delta(t - \Delta)\Delta$$

$$= \sum_{k=-1}^{1} x(k\Delta)p_\Delta(t - k\Delta)\Delta \approx x(t). \tag{3.10}$$

Note that $p_\Delta(t)$ is multiplied by its width, Δ, such that $p_\Delta(t)\Delta$ has an amplitude of unity.

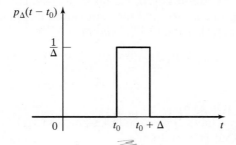

Figure 3.1 Rectangular pulse.

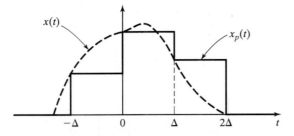

Figure 3.2   Signal represented as pulses.

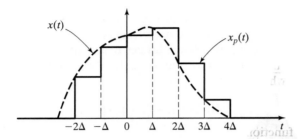

Figure 3.3   Effects of reducing Δ.

Suppose that the value of Δ is reduced by a factor of 2 to yield a more accurate approximation of $x(t)$, as shown in Figure 3.3. Six pulses are now used in $x_p(t)$ to approximate $x(t)$:

$$x_p(t) = \sum_{k=-2}^{3} x(k\Delta)p_\Delta(t - k\Delta)\Delta \approx x(t).$$

We extend this procedure to a general function $x(t)$ such that $x_p(t)$ becomes

$$x_p(t) = \sum_{k=-\infty}^{\infty} x(k\Delta)p_\Delta(t - k\Delta)\Delta \approx x(t). \tag{3.11}$$

To improve the approximation, we let Δ approach zero. The rectangular pulse becomes the impulse function, as stated in (3.9):

$$\lim_{\Delta \to 0} p_\Delta(t - k\Delta) = \delta(t - k\Delta). \tag{3.12}$$

In (3.11), as Δ approaches zero, $k\Delta$ approaches a continuous variable that we denote as τ, Δ approaches $d\tau$, and the summation becomes an integral, by the definition of an integral [1]. Then (3.11) becomes

$$x(t) = \int_{-\infty}^{\infty} x(\tau)\delta(t - \tau)\, d\tau, \tag{3.13}$$

which is identical to (3.8).

In (3.11), we have approximated a function $x(t)$ that is difficult to specify by a function $x_p(t)$ that is easy to specify. Note the relationship of the approximation

(3.11) to the Euler rule for numerical integration given in Section 1.3. In Euler integration, we use the approximation of (3.11) to replace a general function that is difficult to integrate with a function that we can easily integrate.

The approximation of a complex signal (or system) with a simpler signal (or system) that is easier to manipulate is basic to engineering. No model of a signal or system is exact; we can *always* improve the accuracy of a model by making it more complex. A common problem in engineering is to find a simple model of *adequate accuracy* for the problem at hand. We must avoid models that are too simple (too inaccurate) or too complex (more difficult to use).

3.2   CONVOLUTION FOR CONTINUOUS-TIME LTI SYSTEMS

An equation relating the output of a continuous-time LTI system to its input is developed in this section. We begin the development by considering the system shown in Figure 3.4, for which

$$x(t) \rightarrow y(t).$$

A unit impulse function $\delta(t)$ is applied to the system input. Recall the description (3.5) of this input signal; the input signal is zero at all values of time other than $t = 0$, at which time the signal is unbounded.

With the input an impulse function, we denote the LTI system response in Figure 3.4 as $h(t)$, that is,

$$\delta(t) \rightarrow h(t). \tag{3.14}$$

Because the system is time invariant, the response to a time-shifted impulse function, $\delta(t - t_0)$, is given by

$$\delta(t - t_0) \rightarrow h(t - t_0).$$

The notation $h(\cdot)$ will *always* denote the *unit impulse response.*

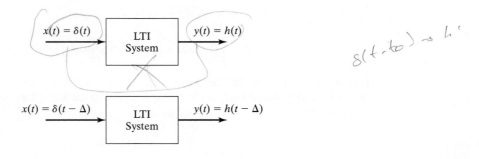

Figure 3.4   Impulse response of an LTI system.

We now derive an expression for the output of an LTI system in terms of its unit impulse response $h(t)$ of (3.14).

We show that a system's response to any input signal, $x(t)$, is expressed as an integral involving only the input function and the system's impulse response function, $h(t)$. This interaction between the input signal and the impulse response function is called *convolution*.

Convolution can be visualized as an extended application of superposition. For an LTI system, the system's response to an impulse input is always the same form, regardless of the time that the impulse is applied. As shown in Figure 3.4 the impulse response is shifted in time by Δ seconds to correspond to the time that the impulse input is applied.

According to the principle of superposition, an LTI system's total response to a sum of inputs is the sum of the responses to each individual input. It follows that if the input is a sum of weighted, time-shifted impulses

$$x(t) = \sum_{k=0}^{\infty} \Delta \delta(t - k\Delta), \tag{3.15}$$

then the output signal is a sum of weighted, time-shifted impulse responses

$$y(t) = \sum_{k=0}^{\infty} \Delta h(t - k\Delta). \tag{3.16}$$

EXAMPLE 3.1   **Sum of impulse responses**

Consider the system with the impulse response $h(t) = e^{-t}u(t)$ as shown in Figure 3.5(a). This system's response to an input of $x(t) = \delta(t - 1)$ would be $y(t) = h(t - 1) = e^{-(t-1)}u(t - 1)$ as shown in Figure 3.5(b). If the input signal is a sum of weighted, time-shifted impulses as described by (3.15), separated in time by $\Delta = 0.1$ (s), so that

$$x(t) = \sum_{k=0}^{\infty} 0.1\delta(t - 0.1k)$$

as shown in Figure 3.5(c), then, according to (3.16), the output is

$$y(t) = \sum_{k=0}^{\infty} 0.1 h(t - 0.1k) = 0.1 \sum_{k=0}^{\infty} e^{-(t-0.1k)}u(t - 0.1k).$$

This output signal is plotted in Figure 3.5(d).

Now consider that the weights of the impulses in the input function vary as a function of time so that

$$x(t) = \sum_{k=-\infty}^{\infty} v(k\Delta)\delta(t - k\Delta)\Delta.$$

Applying the linearity property of scalar multiplication, we see that the output signal in response to this input is

$$y(t) = \sum_{k=-\infty}^{\infty} v(k\Delta)h(t - k\Delta)\Delta.$$

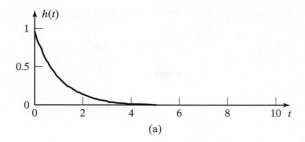

(a)

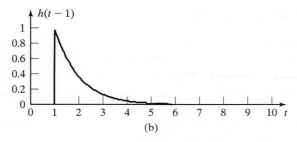

(b)

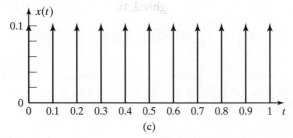

(c)

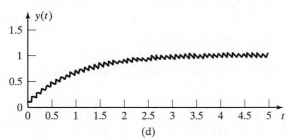

(d)

Figure 3.5 Impulse responses for the system of Example 3.1.

If the time shift between impulses approaches zero, the discrete time variable $k\Delta$ becomes a continuous-time variable that we can denote by the symbol τ, the time increment Δ becomes the time-differential $d\tau$, and the summation becomes an integral so that

$$y(t) = \lim_{\Delta \to 0} \sum_{k=-\infty}^{\infty} v(k\Delta)h(t - k\Delta)\Delta = \int_{-\infty}^{\infty} v(\tau)h(t - \tau)d\tau.$$

In addition, the input signal becomes

$$x(t) = \lim_{\Delta \to 0} \sum_{k=-\infty}^{\infty} v(k\Delta)\delta(t - k\Delta)\Delta = \int_{-\infty}^{\infty} v(\tau)\delta(t - \tau)d\tau. \tag{3.17}$$

We can also use the sifting property to write this input signal as

$$x(t) = \int_{-\infty}^{\infty} v(\tau)\delta(t - \tau)d\tau = v(t).$$

So we can now write

$$y(t) = \int_{-\infty}^{\infty} x(\tau)h(t - \tau)d\tau, \tag{3.18}$$

which is the result that we are seeking.   ■

This result is fundamental to the study of LTI systems, and its importance cannot be overemphasized. The system response to any input $x(t)$ is expressed as an integral involving only the input function and the system response to an impulse function $h(t)$. From this result, we see the importance of impulse functions to the investigation of LTI systems.

The result in (3.18) is called the *convolution integral*. We denote this integral with an asterisk, as in the following notation:

$$y(t) = \int_{-\infty}^{\infty} x(\tau)h(t - \tau)\,d\tau = x(t)*h(t). \tag{3.19}$$

Next we derive an important property of the convolution integral by making a change of variables in (3.18); let $s = (t - \tau)$. Then $\tau = (t - s)$ and $d\tau = -ds$. Equation (3.18) becomes

$$y(t) = \int_{-\infty}^{\infty} x(\tau)h(t - \tau)\,d\tau = \int_{\infty}^{-\infty} x(t - s)h(s)[-ds]$$

$$= \int_{-\infty}^{\infty} x(t - s)h(s)\,ds.$$

Next we replace s with τ in the last integral, and thus the convolution can also be expressed as

$$y(t) = \int_{-\infty}^{\infty} x(\tau)h(t - \tau)\,d\tau = \int_{-\infty}^{\infty} x(t - \tau)h(\tau)\,d\tau. \tag{3.20}$$

The convolution integral is symmetrical with respect to the input signal $x(t)$ and the impulse response $h(t)$, and we have the property

$$y(t) = x(t)*h(t) = h(t)*x(t). \tag{3.21}$$

An additional property of the convolution integral is derived by considering the convolution integral for a unit impulse input; that is, for $x(t) = \delta(t)$,

$$y(t) = \delta(t)*h(t).$$

By definition, this output is equal to $h(t)$, the impulse response:

$$y(t) = \delta(t)*h(t) = h(t). \tag{3.22}$$

This property is independent of the functional form of $h(t)$. Hence, the convolution of any function $g(t)$ with the unit impulse function yields that function $g(t)$. Because of the time-invariance property, the general form of (3.22) is given by

$$y(t - t_0) = \delta(t - t_0)*h(t) = h(t - t_0).$$

This general property may be stated in terms of a function $g(t)$:

$$\delta(t)*g(t) = g(t)$$

and   (3.23)

$$\delta(t - t_0)*g(t) = g(t - t_0)*\delta(t) = g(t - t_0).$$

The second relationship is based on (3.21).

Do not confuse convolution with multiplication. From Table 2.3, the multiplication property (sifting property) of the impulse function is given by

$$\delta(t - t_0)g(t) = g(t_0)\delta(t - t_0)$$

and

$$g(t - t_0)\delta(t) = g(-t_0)\delta(t).$$

The convolution integral signifies that the impulse response of an LTI discrete system, $h(t)$, contains a *complete input–output description* of the system. If this impulse response is known, the system response to any input can be found, using (3.21).

The results thus far are now summarized:

1. A general signal $x(t)$ can be expressed as a function of an impulse function:

    [eq(3.13)] $$x(t) = \int_{-\infty}^{\infty} x(\tau)\delta(t - \tau)\, d\tau.$$

2. By definition, for a continuous-time LTI system,

    $$\delta(t) \rightarrow h(t). \tag{3.24}$$

The system response $y(t)$ for a general input signal $x(t)$ can be expressed as

    [eq(3.20)] $$y(t) = \int_{-\infty}^{\infty} x(\tau)h(t - \tau)\, d\tau$$

    $$= \int_{-\infty}^{\infty} x(t - \tau)h(\tau)\, d\tau.$$

Examples are now given that illustrate the convolution integral for certain systems.

EXAMPLE 3.2 **Impulse response of an integrator**

Consider the system of Figure 3.6. The system is an integrator, in which the output is the integral of the input:

$$y(t) = \int_{-\infty}^{t} x(\tau)\, d\tau. \tag{3.25}$$

This equation is the mathematical model of the system. We use the integral symbol in a block to denote the integrator. The system is practical and can be realized as an electronic circuit with an operational amplifier, a resistor, and a capacitor, as described in Section 1.2. Integrating amplifiers of this type are used extensively in analog signal processing and in closed-loop control systems.

We see that the impulse response of this system is the integral of the unit impulse function, which is the unit step function:

$$h(t) = \int_{-\infty}^{t} \delta(\tau)\, d\tau = u(t) = \begin{cases} 0, & t < 0 \\ 1, & t > 0 \end{cases}.$$

We will now use the convolution integral to find the system response for the unit ramp input, $x(t) = tu(t)$. From (3.20),

$$y(t) = x(t)*h(t) = tu(t)*u(t) = \int_{-\infty}^{\infty} \tau u(\tau) u(t - \tau)\, d\tau.$$

In this integral, t is considered to be constant. The unit step $u(\tau)$ is zero for $\tau < 0$; hence, the lower limit on the integral can be increased to zero with $u(\tau)$ removed from the integrand:

$$y(t) = \int_{0}^{\infty} \tau u(t - \tau)\, d\tau.$$

In addition, the unit step $u(t - \tau)$ is defined as

$$u(t - \tau) = \begin{cases} 0, & \tau > t \\ 1, & \tau < t \end{cases}.$$

The upper limit on the integral can then be reduced to t, and thus,

$$y(t) = \int_{0}^{t} \tau\, d\tau = \frac{\tau^2}{2}\Big|_{0}^{t} = \frac{t^2}{2} u(t).$$

Integrator

$$x(t) \longrightarrow \boxed{\int} \longrightarrow y(t)$$

$$y(t) = \int_{-\infty}^{t} x(\tau)\, d\tau$$

Figure 3.6 System for Example 3.2.

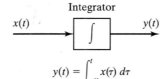

This result is easily verified from the system equation (3.25):

$$y(t) = \int_{-\infty}^{t} x(\tau)\, d\tau = \int_{-\infty}^{t} \tau u(\tau)\, d\tau = \int_{0}^{t} \tau\, d\tau = \frac{t^2}{2} u(t). \qquad \blacksquare$$

Example 3.2 illustrates that the convolution integral can be used to find the output of an LTI system given its input. The example also illustrates that as a practical matter, other methods are used; the convolution integral is seldom the most efficient procedure for finding the response of a system. Instead, the convolution integral is normally used in developing the properties of LTI systems and in developments involving the use of LTI systems. In practice, computer solutions of the system equations, called *system simulations*, are used to find system responses. Four additional examples in convolution will now be given.

EXAMPLE 3.3   **Convolution for the system of Example 3.1**

Consider the system of Example 3.1 for the case that the time increment between impulses approaches zero. If we let $\Delta = 0.1$, the input signal in Example 3.1 can be written as

$$x(t) = \sum_{k=0}^{\infty} 0.1\delta(t - 0.1k) = \sum_{k=-\infty}^{\infty} u(k\Delta)\delta(t - k\Delta)\,(\Delta),$$

because $u(k\Delta) = 0$ for $k < 0$ and $u(k\Delta) = 1$, for $k \geq 1$. From (3.17), as $\Delta \to 0$, the input signal becomes

$$x(t) = \int_{-\infty}^{\infty} u(\tau)\delta(t - \tau)d\tau = u(t).$$

From (3.18), the output signal is calculated from

$$y(t) = \int_{-\infty}^{\infty} x(\tau)h(t - \tau)d\tau = \int_{-\infty}^{\infty} u(\tau)e^{-(t-\tau)}u(t - \tau)d\tau.$$

Because $u(\tau) = 0$ for $\tau < 0$, and $u(t - \tau) = 0$ for $\tau > t$, this convolution integral can be rewritten as

$$y(t) = \int_{0}^{t} e^{-(t-\tau)}d\tau = e^{-t}\int_{0}^{t} e^{\tau}d\tau = (1 - e^{-t})u(t).$$

The output signal is plotted in Figure 3.7. Compare this result with the summation result shown in Figure 3.5(d).

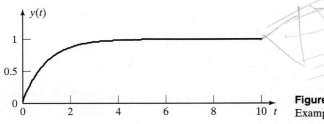

Figure 3.7   System output signal for Example 3.3. $\blacksquare$

EXAMPLE 3.4   **Graphical evaluation for the response of an integrator**

This example is a continuation of Example 3.2. The output for that system (an integrator) will be found graphically. The graphical solution will indicate some of the important properties of convolution. Recall from Example 3.2 that the impulse response of the system is the unit step function, that is, $h(t) = u(t)$. To find the system output, we evaluate the convolution integral

$$y(t) = \int_{-\infty}^{\infty} x(\tau)h(t - \tau)\, d\tau.$$

Note that the integration is with respect to τ; hence, t is considered to be constant. Note also that the impulse response is time reversed to yield $h(-\tau)$, and then time shifted to yield $h(t - \tau)$. These signal manipulations are illustrated in Figure 3.8.

The function $h(t - \tau)$ can also be found directly by the variable-transformation procedures of Section 2.1. To obtain $h(t - \tau)$, first change the variable of the impulse response to be s, yielding $h(s)$ as shown in Figure 3.8(a). We use the variable s, since the variables t and τ are already used. Next we solve for τ:

$$t - \tau = s \Rightarrow \tau = t - s.$$

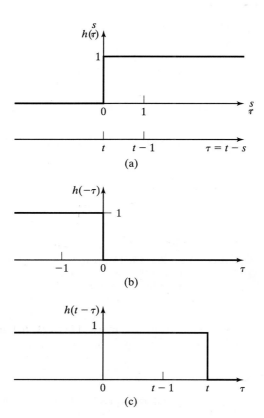

(a)

(b)

(c)

Figure 3.8 Impulse response factor for convolution.

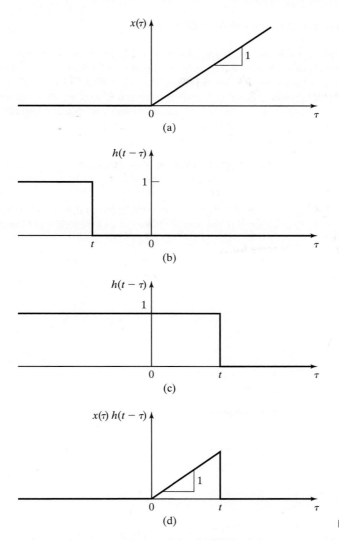

Figure 3.9 Convolution for Example 3.4.

The τ-axis is plotted in Figure 3.8(a) and is the same as the axis of $h(t - \tau)$ in Figure 3.8(c).

Shown in Figure 3.9(a) is the first term of the convolution integral, $x(\tau) = \tau u(\tau)$. Figure 3.9(b) shows the second term of the integral, $h(t - \tau)$, for $t < 0$. The product of these two functions is zero; hence, the value of the integral [and of $y(t)$] is zero for $t < 0$. Figure 3.9(c) shows the second term of the convolution integral for $t > 0$, and Figure 3.9(d) shows the product of the functions, $x(\tau)h(t - \tau)$, of Figure 3.9(a) and (c). Therefore, from the convolution integral, $y(t)$ is the area under the function in Figure 3.9(d). Because the product function is triangular, the area is equal to one-half the base times the height:

$$y(t) = \frac{1}{2}(t)(t) = \frac{t^2}{2}, \quad t > 0.$$

This value is the same as that found in Example 3.2.   ■

EXAMPLE 3.5   **A system with a rectangular impulse response**

As a third example, let the impulse response for an LTI system be rectangular, as shown in Figure 3.10. We will later use this system in the study of certain types of sampling of continuous-time signals. Furthermore, this system is used in the modeling of digital-to-analog converters. Note that one realization of this system is an integrator, an ideal time delay, and a summing junction as shown in Figure 3.11. The reader can show that the impulse response of the system is the rectangular pulse of Figure 3.10.

The input to this system is specified as

$$x(t) = \delta(t + 3) + 3e^{-0.5t}u(t)$$
$$= x_1(t) + x_2(t)$$

and is also plotted in Figure 3.10. We have expressed the input as the sum of two functions; by the linearity property, the response is the sum of the responses to each input function. Hence,

$$\left.\begin{array}{c} x_1(t) \rightarrow y_1(t) \\ x_2(t) \rightarrow y_2(t) \end{array}\right\}, \quad x(t) \rightarrow y_1(t) + y_2(t).$$

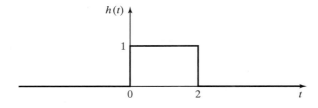

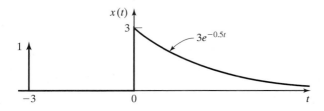

Figure 3.10   Input signal and impulse response for Example 3.5.

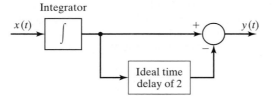

Figure 3.11   System for Example 3.5.

The system response to the impulse function is obtained from (3.23):

$$y_1(t) = h(t)*\delta(t + 3) = h(t + 3) = u(t + 3) - u(t + 1).$$

To determine the response to $x_2(t)$, we plot $h(t - \tau)$ as shown in Figure 3.12(a). This plot is obtained by time reversal and time shifting. Three different integrations must be performed to evaluate the convolution integral:

1. The first integration applies for $t \le 0$ as shown in Figure 3.12(a) and is given by

$$y_2(t) = \int_{-\infty}^{\infty} x_2(\tau)h(t - \tau)\,d\tau = \int_{-\infty}^{0} (0)h(t - \tau)\,d\tau$$

$$+ \int_{0}^{\infty} x_2(\tau)\,(0)\,d\tau = 0, \quad t \le 0.$$

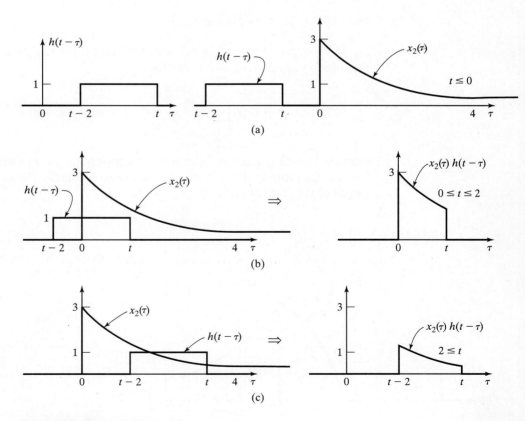

Figure 3.12 Signals for Example 3.5.

2. The second integration applies for $0 \leq t \leq 2$ and is illustrated in Figure 3.12(b):

$$y_2(t) = \int_{-\infty}^{\infty} x_2(\tau)h(t - \tau)\,d\tau = \int_{-\infty}^{0}(0)h(t - \tau)\,d\tau$$

$$+ \int_{0}^{t} 3e^{-0.5\tau}\,d\tau + \int_{t}^{\infty} x_2(\tau)(0)\,d\tau$$

$$= \frac{3e^{-0.5\tau}}{-0.5}\Big|_{0}^{t} = 6(1 - e^{-0.5t}), \quad 0 \leq t \leq 2.$$

3. Figure 3.12(c) applies for $2 \leq t \leq \infty$:

$$y_2(t) = \int_{t-2}^{t} 3e^{-0.5\tau}\,d\tau = \frac{3e^{-0.5\tau}}{-0.5}\Big|_{t-2}^{t} = 6(e^{-0.5(t-2)} - e^{-0.5t})$$

$$= 6e^{-0.5t}(e^1 - 1) = 10.31e^{-0.5t}, \quad 2 \leq t < \infty.$$

The output $y(t)$ is plotted in Figure 3.13.

     In Example 3.5 the time axis is divided into three ranges. Over each range, the convolution integral reduces to the form

$$y(t) = \int_{t_a}^{t_b} x(\tau)h(t - \tau)\,d\tau, \quad t_i \leq t \leq t_j, \tag{3.26}$$

where the limits t_a and t_b are either constants or functions of time t. The integral applies for the range $t_i \leq t \leq t_j$ where t_i and t_j are constants. Hence, three different integrals of the form of (3.26) are evaluated.

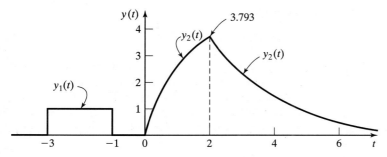

Figure 3.13   Output signal for Example 3.5.

EXAMPLE 3.6   **A system with a time-delayed exponential impulse response**

Consider a system with an impulse response of $h(t) = e^{-t}u(t - 1)$ and an input signal $x(t) = e^{t}u(-1 - t)$. The system's impulse response and the input signal are shown in Figure 3.14(a) and (b), respectively. The system's output is $y(t) = x(t)^{*}h(t)$ from (3.21).

To calculate the system output we perform a quasi-graphical convolution technique. First sketch the time-reversed, time-shifted impulse response function, $h(t - \tau)$, as shown in Figure 3.14(c). From this sketch, it is seen that the signal $h(t - \tau)$ will be nonzero over the interval $-\infty < \tau \le (t - 1)$ and for other values of τ, $h(t - \tau)$ will be zero. The input signal, $x(\tau)$, is zero for $\tau > -1$, so both functions in the product $x(\tau)h(t - \tau)$ will be nonzero only for $t \le 0$. Next, evaluate the convolution integral for the time interval $-\infty < \tau \le (t - 1)$, $t \le 0$. During this interval, both exponential functions are nonzero and the output is given by

$$y(t) = \int_{-\infty}^{t-1} e^{\tau}e^{-(t-\tau)}d\tau = \int_{-\infty}^{t-1} e^{-t}e^{2\tau}d\tau = \frac{e^{-2}e^{t}}{2}, -\infty < t \le 0.$$

Because $x(\tau)$ is zero for $\tau > -1$, for $t > 0$, the output signal is given by

$$y(t) = \int_{-\infty}^{-1} e^{\tau}e^{-(t-\tau)}d\tau = e^{-t}\int_{-\infty}^{-1} e^{2\tau}d\tau = \frac{e^{-2}e^{-t}}{2}, t > 0.$$

The output signal, $y(t)$, is plotted in Figure 3.14(d). It is an even function of time because in this example, $x(t) = h(-t)$.   ■

In this section, the convolution integral for continuous-time LTI systems is developed. This integral is fundamental to the analysis and design of LTI systems and is also used to develop general properties of these systems. Convolution is illustrated first by evaluating the integral using a strictly mathematical approach and then by evaluating the integral using a quasi-graphical approach.

3.3   PROPERTIES OF CONVOLUTION

The convolution integral of (3.18) has three important properties; these properties will now be presented:

1. *Commutative property*. As stated in (3.20), the convolution integral is symmetrical with respect to $x(t)$ and $h(t)$:

$$x(t)^{*}h(t) = h(t)^{*}x(t). \tag{3.27}$$

An illustration of this property is given in Figure 3.15. This figure uses a common representation of an LTI system as a block containing the impulse response. The outputs for the two systems in Figure 3.15 are equal, from (3.27).

2. *Associative property*. The result of the convolution of three or more functions is independent of the order in which the convolution is performed. For example,

$$[x(t)^{*}h_1(t)]^{*}h_2(t) = x(t)^{*}[h_1(t)^{*}h_2(t)] = x(t)^{*}[h_2(t)^{*}h_1(t)]. \tag{3.28}$$

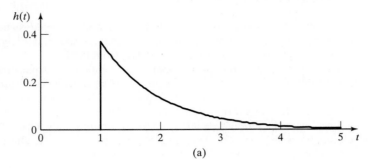

(a)

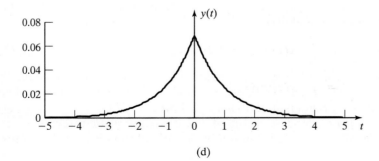

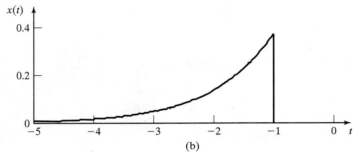

(b)

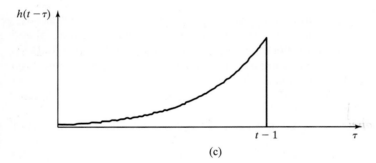

(c)

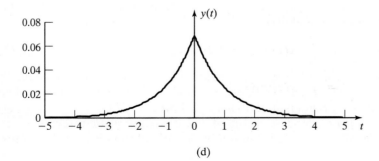

(d)

Figure 3.14 Impulse response for Example 3.6.

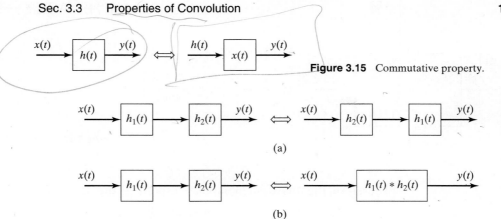

Figure 3.15   Commutative property.

(a)

(b)

Figure 3.16   Associative property.

The proof of this property involves forming integrals and a change of variables and is not given here. (See Problem 3.8.) This property is illustrated with the two cascaded systems in Figure 3.16(a). For cascaded LTI systems, the order of the connection may be changed with no effect on the system impulse response (the input–output characteristics). Note that the two cascaded systems in Figure 3.16(b) may be replaced with a single system with the impulse response $h(t)$, given by, from (3.27) and (3.28),

$$h(t) = h_1(t)*h_2(t) = h_2(t)*h_1(t). \tag{3.29}$$

It follows that for m cascaded systems, the impulse response of the total system is given by

$$h(t) = h_1(t)*h_2(t)* \cdots *h_m(t). \tag{3.30}$$

3.   *Distributive property.* The convolution integral satisfies the following relationship:

$$x(t)*h_1(t) + x(t)*h_2(t) = x(t)*[h_1(t) + h_2(t)]. \tag{3.31}$$

This property is developed directly from the convolution integral, (3.20),

$$
\begin{aligned}
x(t)*h_1(t) + x(t)*h_2(t) &= \int_{-\infty}^{\infty} x(\tau)h_1(t - \tau)\, d\tau + \int_{-\infty}^{\infty} x(\tau)h_2(t - \tau)\, d\tau \\
&= \int_{-\infty}^{\infty} x(\tau)[h_1(t - \tau) + h_2(t - \tau)]\, d\tau \\
&= x(t)*[h_1(t) + h_2(t)]. \tag{3.32}
\end{aligned}
$$

The two systems in parallel in Figure 3.17 illustrate this property, with the output given by

$$y(t) = x(t)*h_1(t) + x(t)*h_2(t) = x(t)*[h_1(t) + h_2(t)]. \tag{3.33}$$

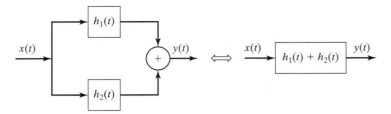

Figure 3.17 Distributive property.

Therefore, the total system impulse response is the sum of the impulse responses:

$$h(t) = h_1(t) + h_2(t). \tag{3.34}$$

In summary,

1.   the impulse response completely describes the input–output characteristics of a continuous-time LTI system, and
2.   the commutative, associative, and distributive properties give the rules for determining the impulse response of an interconnection of LTI systems.

Note that all results in this section were proved for LTI systems only. An example of the use of these properties will now be given.

Impulse response for an interconnection of systems

We wish to determine the impulse response of the system of Figure 3.18(a) in terms of the impulse responses of the four subsystems. First, from (3.34), the impulse response of the parallel systems 1 and 2 is given by

$$h_a(t) = h_1(t) + h_2(t)$$

as shown in Figure 3.18(b). From (3.29), the effect of the cascaded connection of system a and system 3 is given by

$$h_b(t) = h_a(t) * h_3(t) = [h_1(t) + h_2(t)] * h_3(t)$$

as shown in Figure 3.18(c). We add the effect of the parallel system 4 to give the total-system impulse response, as shown in Figure 3.18(d):

$$h(t) = h_b(t) + h_4(t) = [h_1(t) + h_2(t)] * h_3(t) + h_4(t). \qquad\blacksquare$$

This section gives three properties of convolution. Based on these properties, a procedure is developed for calculating the impulse response of an LTI system composed of subsystems, where the impulse responses of the subsystems are known. This procedure applies only for linear time-invariant systems. An equivalent and simpler procedure for finding the impulse response of the total system in terms of its subsystem impulse responses is the transfer-function approach. The

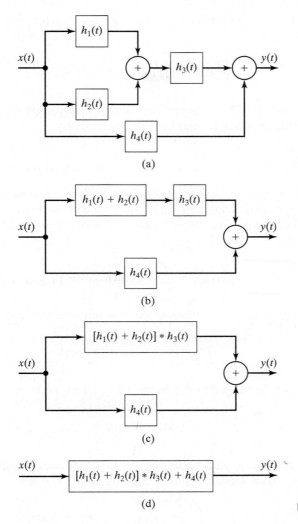

(a)

(b)

(c)

(d)

Figure 3.18 System for Example 3.7.

transfer-function approach is introduced later in this chapter and is generalized using the Fourier transform of Chapters 5 and 6 and the Laplace transform of Chapter 7.

3.4 PROPERTIES OF CONTINUOUS-TIME LTI SYSTEMS

In Section 2.7, several properties of continuous-time systems are defined. In this section, we investigate these properties as related to linear time-invariant systems.

The input–output characteristics of a continuous-time LTI system are completely described by its impulse response $h(t)$. Hence, all properties of a system can be determined from $h(t)$. From (3.20), for the input signal $x(t)$, the output signal $y(t)$ is given by

$$y(t) = \int_{-\infty}^{\infty} x(\tau)h(t - \tau)\, d\tau. \tag{3.35}$$

This characteristic of LTI systems will be used to derive certain properties for these systems. We begin by considering the memory property.

Memoryless Systems

Recall that a memoryless (static) system is one whose current value of output depends only on the current value of input; that is, the current value of the output does not depend on either past values or future values of the input. Let the present time be t_1. From (3.35),

$$y(t_1) = \int_{-\infty}^{\infty} x(\tau)h(t_1 - \tau)\, d\tau. \tag{3.36}$$

We compare this with the definition of the unit impulse function, (3.6):

[eq(3.6)] $$\int_{-\infty}^{\infty} x(t)\delta(t - t_0)\, dt = x(t_0).$$

This equation is not changed by replacing $\delta(t - t_0)$ with $\delta(t_0 - t)$ (Table 2.3):

$$\int_{-\infty}^{\infty} x(t)\delta(t_0 - t)\, dt = x(t_0).$$

Comparing this equation with (3.36), we see that $h(t)$ must be equal to the impulse function $K\delta(t)$, where K is a constant, resulting in

$$y(t_1) = \int_{-\infty}^{\infty} x(\tau)K\delta(t_1 - \tau)\, d\tau = Kx(t_1).$$

Hence, an LTI system is memoryless if and only if $h(t) = K\delta(t)$, that is, if $y(t) = Kx(t)$. A memoryless LTI system can be considered to be an ideal amplifier, with $y(t) = Kx(t)$. If the gain is unity ($K = 1$), the identity system results.

Invertibility

A continuous-time LTI system with the impulse response $h(t)$ is invertible if its input can be determined from its output. An invertible LTI system is depicted in Figure 3.19. For this system,

$$x(t)*h(t)*h_i(t) = x(t), \tag{3.37}$$

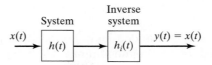

Figure 3.19   LTI invertible system.

where $h_i(t)$ is the impulse response of the inverse system. From (3.22),

$$x(t)*\delta(t) = x(t). \tag{3.38}$$

From (3.37) and (3.38),

$$h(t)*h_i(t) = \delta(t). \tag{3.39}$$

Thus, an LTI system with the impulse response $h(t)$ is invertible only if the function $h_i(t)$ can be found that satisfies (3.39). Then the inverse system has the impulse response $h_i(t)$.

We do not present a procedure for determining $h_i(t)$, given $h(t)$. This manipulation is most efficiently performed using the Fourier transform of Chapter 5 or the Laplace transform of Chapter 7.

Causality

A continuous-time LTI system is causal if the current value of the output depends on only the current and past values of the input. Because the unit impulse function $\delta(t)$ occurs at $t = 0$, the impulse response $h(t)$ of a causal system *must be zero* for $t < 0$. In addition, a signal that is zero for $t < 0$ is called a *causal signal*. The convolution integral for a causal LTI system can then be expressed as

$$y(t) = \int_{-\infty}^{\infty} x(t - \tau)h(\tau)\,d\tau = \int_{0}^{\infty} x(t - \tau)h(\tau)\,d\tau. \tag{3.40}$$

If the impulse response is expressed as $h(t - \tau)$, this function is zero for $(t - \tau) < 0$, or for $\tau > t$. The second form of the convolution integral can then be expressed as

$$y(t) = \int_{-\infty}^{\infty} x(\tau)h(t - \tau)\,d\tau = \int_{-\infty}^{t} x(\tau)h(t - \tau)\,d\tau. \tag{3.41}$$

Notice that (3.41) makes it clear that for a causal system, the output, $y(t)$, depends on values of the input only up to the present time, t, and not on future inputs.

In summary, for a causal continuous-time LTI system, the convolution integral can be expressed in the two forms

$$y(t) = \int_{0}^{\infty} x(t - \tau)h(\tau)\,d\tau = \int_{-\infty}^{t} x(\tau)h(t - \tau)\,d\tau. \tag{3.42}$$

Stability

Recall that a system is bounded-input bounded-output (BIBO) stable if the output remains bounded for any bounded input. The boundedness of the input can be expressed as

$$|x(t)| < M \quad \text{for all } t,$$

where M is a real constant. Then from (3.20), we can write

$$
\begin{aligned}
|y(t)| &= \left| \int_{-\infty}^{\infty} x(t - \tau)h(\tau)\, d\tau \right| \leq \int_{-\infty}^{\infty} |x(t - \tau)h(\tau)|\, d\tau \\[2mm]
&= \int_{-\infty}^{\infty} |x(t - \tau)||h(\tau)|\, d\tau \\[2mm]
&\leq \int_{-\infty}^{\infty} M|h(\tau)|\, d\tau = M \int_{-\infty}^{\infty} |h(\tau)|\, d\tau,
\end{aligned}
\tag{3.43}
$$

since

$$\left| \int_{-\infty}^{\infty} x_1(t)x_2(t)\, dt \right| \leq \int_{-\infty}^{\infty} |x_1(t)x_2(t)|\, dt.$$

Thus, because M is finite, $y(t)$ is bounded if

$$\int_{-\infty}^{\infty} |h(t)|\, dt < \infty. \tag{3.44}$$

If $h(t)$ satisfies this condition, it is said to be *absolutely integrable*. It can be shown that this requirement is also sufficient. (See Problem 3.15.) Thus, for an LTI system to be BIBO stable, the impulse response $h(t)$ must be absolutely integrable, as in (3.44). For an LTI *causal* system, this criterion reduces to

$$\int_{0}^{\infty} |h(t)|\, dt < \infty. \tag{3.45}$$

EXAMPLE 3.8　　**Stability for an LTI system derived**

We will determine the stability of the causal LTI system that has the impulse response given by

$$h(t) = e^{-3t}u(t).$$

In (3.45),

$$\int_{-\infty}^{\infty} |h(t)|\, dt = \int_{0}^{\infty} e^{-3t}\, dt = \left. \frac{e^{-3t}}{-3} \right|_{0}^{\infty} = \frac{1}{3} < \infty,$$

and this system is stable. For $h(t) = e^{3t}u(t)$,

$$\int_{-\infty}^{\infty} |h(t)|\, dt = \int_{0}^{\infty} e^{3t}\, dt = \left.\frac{e^{3t}}{3}\right|_{0}^{\infty},$$

which is unbounded at the upper limit. Hence, this system is not stable.   ■

EXAMPLE 3.9   **Stability for an integrator derived**

As a second example, consider an LTI system such that $h(t) = u(t)$. From Example 3.2 this system is an integrator, with the output equal to the integral of the input:

$$y(t) = \int_{-\infty}^{t} x(\tau)\, d\tau.$$

We determine the stability of this system from (3.45), since the system is causal:

$$\int_{0}^{\infty} |h(t)|\, dt = \int_{0}^{\infty} dt = \left. t \right|_{0}^{\infty}.$$

This function is unbounded, and thus the system is not BIBO stable.   ■

Unit Step Response

As has been stated several times, the impulse response of a system, $h(t)$, completely specifies the input–output characteristics of that system; the convolution integral,

[eq(3.35)] $$y(t) = \int_{-\infty}^{\infty} x(\tau)h(t - \tau)\, d\tau$$

allows the calculation of the output signal $y(t)$ for any input signal $x(t)$.

   Suppose that the system input is the unit step function $u(t)$. From (3.35), with $s(t)$ denoting the unit step response, we obtain

$$s(t) = \int_{-\infty}^{\infty} u(\tau)h(t - \tau)\, d\tau = \int_{0}^{\infty} h(t - \tau)\, d\tau \qquad (3.46)$$

because $u(\tau)$ is zero for $\tau < 0$. If the system is causal, $h(t - \tau)$ is zero for $(t - \tau) < 0$, or for $\tau > t$, and

$$s(t) = \int_{0}^{t} h(\tau)\, d\tau. \qquad (3.47)$$

We see, then, that the unit step response can be calculated directly from the unit impulse response, using either (3.46) or (3.47).

If (3.46) or (3.47) is differentiated (see Leibnitz's rule, Appendix B), we obtain

$$h(t) = \frac{ds(t)}{dt}. \tag{3.48}$$

Thus, the unit impulse response can be calculated directly from the unit step response, and we see that the unit step response also completely describes the input–output characteristics of an LTI system.

| EXAMPLE 3.10 | **Step response from the impulse response** |

Consider again the system of Example 3.8, which has the impulse response given by

$$h(t) = e^{-3t}u(t).$$

Note that this system is causal. From (3.47), the unit step response is then

$$s(t) = \int_0^t h(\tau)\,d\tau = \int_0^t e^{-3\tau}\,d\tau = \frac{e^{-3t}}{-3}\bigg|_0^t = \frac{1}{3}(1 - e^{-3t})u(t).$$

This result can be verified by differentiating $s(t)$ to obtain the impulse response. From (3.48), we get

$$h(t) = \frac{ds(t)}{dt} = \frac{1}{3}(1 - e^{-3t})\delta(t) + \frac{1}{3}(-e^{-3t})(-3)u(t)$$

$$= e^{-3t}u(t).$$

Why does $(1 - e^{-3t})\delta(t) = 0$? [See (2.42).] ∎

In this section, the properties of memory, invertibility, causality, and stability are considered with respect to LTI systems. Of course, by definition these systems are linear and time invariant. An important result is that the BIBO stability can always be determined from the impulse response of a system, using (3.44). It is then shown that the impulse response of an LTI system can be determined from its unit step response.

3.5 DIFFERENTIAL-EQUATION MODELS

Some properties of LTI continuous-time systems were developed in earlier sections of this chapter, with little reference to the actual equations that are used to model these systems. We now consider the most common model for LTI systems. Continuous-time LTI systems are usually modeled by *ordinary linear differential equations with constant coefficients*. We emphasize that we are considering the *models* of physical systems, not the physical systems themselves.

In Section 2.3, we considered the system model given by

$$\frac{dy(t)}{dt} = ay(t), \tag{3.49}$$

where a is constant. The system input $x(t)$ usually enters this model in the form

$$\frac{dy(t)}{dt} - ay(t) = bx(t), \tag{3.50}$$

where a and b are constants and $y(t)$ is the system output signal. The *order* of the system is the order of the differential equation that models the system. Hence, (3.50) is a *first-order system.*

Equation (3.50) is an ordinary linear differential equation with constant coefficients. The equation is ordinary, since no partial derivatives are involved. The equation is linear, since the equation contains the dependent variable and its derivative to the first degree only. One of the coefficients in the equation is equal to unity, one is $-a$, and one is b; hence, the equation has constant coefficients.

We now test the linearity of (3.50) using superposition. Let $y_i(t)$ denote the solution of (3.50) for the excitation $x_i(t)$, for $i = 1, 2$. By this, we mean that

$$\frac{dy_i(t)}{dt} - ay_i(t) = bx_i(t), \quad i = 1, 2. \tag{3.51}$$

We now show that the solution $[a_1y_1(t) + a_2y_2(t)]$ satisfies (3.50) for the excitation $[a_1x_1(t) + a_2x_2(t)]$, by direct substitution into (3.50):

$$\frac{d}{dt}[a_1y_1(t) + a_2y_2(t)] - a[a_1y_1(t) + a_2y_2(t)] = b[a_1x_1(t) + a_2x_2(t)].$$

This equation is rearranged to yield

$$a_1\left[\frac{dy_1(t)}{dt} - ay_1(t) - bx_1(t)\right] + a_2\left[\frac{dy_2(t)}{dt} - ay_2(t) - bx_2(t)\right] = 0. \tag{3.52}$$

Because, from (3.51), each term is equal to zero, the differential equation satisfies the principle of superposition and hence is linear.

Next we now test the model for time invariance. In (3.50), replacing t with $(t - t_0)$ results in the equation

$$\frac{dy(t - t_0)}{dt} - ay(t - t_0) = bx(t - t_0). \tag{3.53}$$

Delaying the input by t_0 delays the solution by the same amount; this system is then time invariant.

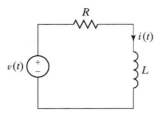

Figure 3.20 RL circuit.

A simple example of an ordinary linear differential equation with constant coefficients is the first-order differential equation

$$\frac{dy(t)}{dt} + 2y(t) = x(t). \tag{3.54}$$

This equation could model the circuit of Figure 3.20, namely,

$$L\frac{di(t)}{dt} + Ri(t) = v(t),$$

with $L = 1H$, $R = 2\Omega$, $y(t) = i(t)$, and $x(t) = v(t)$.

The general form of an nth-order linear differential equation with constant coefficients is

$$a_n\frac{d^n y(t)}{dt^n} + a_{n-1}\frac{d^{n-1}y(t)}{dt^{n-1}} + \cdots + a_1\frac{dy(t)}{dt} + a_0 y(t)$$
$$= b_m\frac{d^m x(t)}{dt^m} + b_{m-1}\frac{d^{m-1}x(t)}{dt^{m-1}} + \cdots + b_1\frac{dx(t)}{dt} + b_0 x(t),$$

where $a_0, \ldots, a_n$ and $b_0, \ldots, b_m$ are constants and $a_n \neq 0$. We limit these constants to having real values. This equation can be expressed in the more compact form

$$\sum_{k=0}^{n} a_k\frac{d^k y(t)}{dt^k} = \sum_{k=0}^{m} b_k\frac{d^k x(t)}{dt^k}. \tag{3.55}$$

It is easily shown by the preceding procedure that this equation is both linear and time invariant. Many methods of solution exist for (3.55); in this section, we review briefly one of the classical methods. In subsequent chapters, the solution by transform methods is developed.

Solution of Differential Equations

The method of solution of (3.55) presented here is called the *method of undetermined coefficients* [2] and requires that the general solution $y(t)$ be expressed as the sum of two functions:

$$y(t) = y_c(t) + y_p(t). \tag{3.56}$$

In this equation, $y_c(t)$ is called the *complementary function* and $y_p(t)$ is a *particular solution*. For the case that the differential equation models a system, the complementary function is usually called the *natural response*, and the particular solution, the *forced response*. We will use this notation. We only outline the method of solution; this method is presented in greater detail in Appendix E for readers requiring more review. The solution procedure is given as three steps:

1. *Natural response.* Assume $y_c(t) = Ce^{st}$ to be the solution of the homogeneous equation. Substitute this solution into the homogeneous equation [(3.55) with the right side set to zero] to determine the required values of s.

2. *Forced response.* Assume that $y_p(t)$ is a weighted sum of the mathematical form of $x(t)$ and its derivatives that are different in form from $x(t)$. Three examples are given:

$$x(t) = 5 \Rightarrow y_p(t) = P;$$
$$x(t) = 5e^{-7t} \Rightarrow y_p(t) = Pe^{-7t};$$
$$x(t) = 2 \cos 3t \Rightarrow y_p(t) = P_1 \cos 3t + P_2 \sin 3t.$$

This solution procedure can be applied only if $y_p(t)$ contains a finite number of terms.

3. *Coefficient evaluation.* Solve for the unknown coefficients P_i of the forced response by substituting $y_p(t)$ into the differential equation (3.55). Then use the general solution (3.56) and the initial conditions to solve for the unknown coefficients C_i of the natural response.

An example is now given.

EXAMPLE 3.11   **System response for a first-order LTI system**

As an example we consider the differential equation given earlier in the section, but with $x(t)$ constant; that is,

$$\frac{dy(t)}{dt} + 2y(t) = 2$$

for $t \geq 0$, with $y(0) = 4$. In Step 1 we assume the natural response $y_c(t) = Ce^{st}$. Then we substitute $y_c(t)$ into the homogeneous equation:

$$\frac{dy_c(t)}{dt} + 2y_c(t) = 0 \Rightarrow (s + 2)Ce^{st} = 0 \Rightarrow s = -2.$$

The natural response is then $y_c(t) = Ce^{-2t}$, where C is yet to be determined.

Because the forcing function is constant, and since the derivative of a constant is zero, the forced response in Step 2 is assumed to be

$$y_p(t) = P,$$

where P is an unknown constant. Substitution of the forced response $y_p(t)$ into the differential equation yields

$$\frac{dP}{dt} + 2P = 0 + 2P = 2$$

or $y_p(t) = P = 1$. From (3.56), the general solution is

$$y(t) = y_c(t) + y_p(t) = Ce^{-2t} + 1.$$

We now evaluate the coefficient C. The initial condition is given as $y(0) = 4$. The general solution $y(t)$ evaluated at $t = 0$ yields

$$y(0) = y_c(0) + y_p(0) = [Ce^{-2t} + 1]\Big|_{t=0}$$

$$= C + 1 = 4 \Rightarrow C = 3.$$

The total solution is then

$$y(t) = 1 + 3e^{-2t}.$$

This solution is verified with the MATLAB program

```
dsolve('Dy+2*y=2, y(0)=4')
ezplot(y)
```

If this example is not clear, the reader should study Appendix E.   ■

EXAMPLE 3.12   **Verification of the response of Example 3.11**

We now verify the solution in Example 3.11 by substitution into the differential equation. Thus,

$$\frac{dy(t)}{dt} + 2y(t)\Big|_{y=1+3e^{-2t}} = -6e^{-2t} + 2(1 + 3e^{-2t}) = 2,$$

and the solution checks. In addition,

$$y(0) = (1 + 3e^{-2t})\Big|_{t=0} = 1 + 3 = 4,$$

and the initial condition checks. Hence, the solution is verified.   ■

General Case

We now consider the natural response for the nth-order system:

[eq(3.55)] $$\sum_{k=0}^{n} a_k \frac{d^k y(t)}{dt^k} = \sum_{k=0}^{m} b_k \frac{d^k x(t)}{dt^k}.$$

The homogeneous equation is formed from (3.55) with the right side set to zero. That is,

$$a_n \frac{d^n y(t)}{dt^n} + a_{n-1} \frac{d^{n-1} y(t)}{dt^{n-1}} + \cdots + a_1 \frac{dy(t)}{dt} + a_0 y(t) = 0, \qquad (3.57)$$

with $a_n \neq 0$. The natural response $y_c(t)$ must satisfy this equation.

We assume that the solution of the homogeneous equation is of the form $y_c(t) = Ce^{st}$. Note that, in (3.57),

$$y_c(t) \quad = Ce^{st};$$
$$\frac{dy_c(t)}{dt} = Cse^{st};$$

$$\frac{d^2 y_c(t)}{dt^2} = Cs^2 e^{st};$$

$$\vdots \ ;$$

$$\frac{d^n y_c(t)}{dt^n} \quad = Cs^n e^{st}. \qquad (3.58)$$

Substitution of these terms into (3.57) yields

$$(a_n s^n + a_{n-1} s^{n-1} + \cdots + a_1 s + a_0) Ce^{st} = 0. \qquad (3.59)$$

If we assume that our solution $y_c(t) = Ce^{st}$ is nontrivial ($C \neq 0$), then, from (3.59),

$$a_n s^n + a_{n-1} s^{n-1} + \cdots + a_1 s + a_0 = 0. \qquad (3.60)$$

This equation is called the *characteristic equation*, or the *auxiliary equation*, for the differential equation (3.55). The polynomial may be factored as

$$a_n s^n + a_{n-1} s^{n-1} + \cdots + a_1 s + a_0$$

$$= a_n(s - s_1)(s - s_2) \cdots (s - s_n) = 0. \qquad (3.61)$$

Hence, n values of s, denoted as $s_i, 1 \leq i \leq n$, satisfy this equation; that is, $y_{ci}(t) = C_i e^{s_i t}$ for the n values of s_i in (3.61) satisfies the homogeneous equation (3.57). Since the differential equation is linear, the sum of these n solutions is also a solution. For the case of no repeated roots, the solution of the homogeneous equation (3.57) may be expressed as

$$y_c(t) = C_1 e^{s_1 t} + C_2 e^{s_2 t} + \cdots + C_n e^{s_n t}. \qquad (3.62)$$

See Appendix E for the case that the characteristic equation has repeated roots.

Relation to Physical Systems

We wish now to relate the general solution of differential-equation models to the response of physical systems. We, of course, assume that the differential-equation model of a physical system is reasonably accurate.

From these developments, we see that the natural-response part of the general solution of the linear differential equation with constant coefficients is independent of the forcing function $x(t)$; the natural response is dependent only on the structure of the system [the left side of (3.55)], and hence the term *natural response*. It is also called the *unforced response*, or the *zero-input response*. This component of the response is always present, independent of the manner in which the system is excited. However, the amplitudes C_i of the terms depend on both the initial conditions and the excitation. The factors $e^{s_i t}$ in the natural response are called the *modes* of the system.

The forced response of (3.56), $y_p(t)$, is also called the *zero-state response* of the system. In this application, the term *zero state* means *zero initial conditions*. The forced response is a function of both the system structure *and* the excitation, but is independent of the initial conditions.

For almost all LTI models of physical systems, the natural response approaches zero with increasing time; then only the forced part of the response remains. (The requirement for this to occur is that the system be BIBO stable.) For this reason, we sometimes refer to the natural response as the *transient response* and the forced response as the *steady-state response*. When we refer to the steady-state response of a stable LTI system, we are speaking of the forced response of the system differential equation. For a stable system, the steady-state response is the system response for the case that the input signal has been applied for a very long time.

EXAMPLE 3.13   **Time constant for the system of Example 3.11**

For the first-order system of Example 3.11, the solution is given by

$$y(t) = y_c(t) + y_p(t) = 3e^{-2t} + 1.$$

The natural response is the term $3e^{-2t}$, and the system has one mode, e^{-2t}. The steady-state response is the term 1. The system response is plotted in Figure 3.21. The time constant in this response is $\tau = \frac{1}{2} = 0.5$. (See Section 2.3.) Therefore, the natural response can be ignored after approximately 2.0 units of time (4τ), leaving only the steady-state response. (See Section 2.3.) ∎

In this section, we consider systems modeled by linear differential equations with constant coefficients. A classical-solution procedure for these equations is reviewed. The components of the solution are then related to attributes of the response of a physical system.

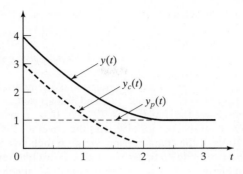

Figure 3.21   Response of a first-order system.

TERMS IN THE NATURAL RESPONSE

We now relate the terms of the natural response (complementary function) of a continuous-time LTI system to the signals that were studied in Section 2.3. The mathematical forms of the terms in the natural response are determined by the roots of the characteristic equation:

[eq(3.61)]
$$a_n s^n + a_{n-1} s^{n-1} + \cdots + a_1 s + a_0$$
$$= a_n (s - s_1)(s - s_2) \cdots (s - s_n) = 0.$$

With the roots distinct, the natural response is given by

[eq(3.62)]
$$y_c(t) = C_1 e^{s_1 t} + C_2 e^{s_2 t} + \cdots + C_n e^{s_n t}.$$

Hence, the general term is given by $C_i e^{s_i t}$, where $e^{s_i t}$ is called a *system mode*. The root s_i of the characteristic equation may be real or complex. However, since the coefficients of the characteristic equation are real, complex roots must occur in complex conjugate pairs. We now consider some of the different forms of the modes that can appear in the natural response.

s_i **Real**
If s_i is real, the resulting term in the natural response is exponential in form.

s_i **Complex**
If s_i is complex, we let

$$s_i = \sigma_i + j\omega_i$$

and the mode is given by

$$C_i e^{s_i t} = C_i e^{(\sigma_i + j\omega_i)t} = C_i e^{\sigma_i t} e^{j\omega_i t}. \tag{3.63}$$

Because the natural response $y_c(t)$ must be real, two of the terms of $y_c(t)$ can be expressed as, with $C_i = |C_i| e^{j\theta_i}$,

$$C_i e^{s_i t} + (C_i e^{s_i t})^* = |C_i| e^{j\theta_i} e^{\sigma_i t} e^{j\omega_i t} + |C_i| e^{-j\theta_i} e^{\sigma_i t} e^{-j\omega_i t}$$

$$= |C_i| e^{\sigma_i t} e^{j(\omega_i t + \theta_i)} + |C_i| e^{\sigma_i t} e^{-j(\omega_i t + \theta_i)}$$

$$= 2|C_i| e^{\sigma_i t} \cos(\omega_i t + \theta_i) \qquad (3.64)$$

by Euler's relation. If σ_i is zero, this function is an undamped sinusoid. If σ_i is negative, the function in (3.64) is a damped sinusoid and approaches zero as t approaches infinity; the envelope of the function is $2|C_i| e^{\sigma_i t}$. If σ_i is positive, the function becomes unbounded as t approaches infinity, with the envelope $2|C_i| e^{\sigma_i t}$.

Real roots of the characteristic equation then give real exponential terms in the natural response, while complex roots give sinusoidal terms. These relationships are illustrated in Figure 3.22, in which the symbols "×" denote characteristic-equation root locations.

We see, then, that the terms that were discussed in Section 2.3 appear in the natural response of an LTI system. These terms are independent of the type of excitation applied to the system.

Stability

We now consider the stability of a *causal* continuous-time LTI system. As stated earlier, the general term in the natural response is of the form $C_i e^{s_i t}$, where s_i is a root of the system characteristic equation. The magnitude of this term is given by $|C_i| |e^{\sigma_i t}|$, from (3.64). If σ_i is negative, the magnitude of the term approaches zero as t approaches infinity. However, if σ_i is positive, the magnitude of the term becomes unbounded as t approaches infinity. Hence, $\sigma_i > 0$ denotes instability.

Recall that the total solution of a constant-coefficient linear differential equation is given by

[eq(3.56)] $$\qquad\qquad y(t) = y_c(t) + y_p(t).$$

Recall also that, for stable systems, the forced response $y_p(t)$ is of the same mathematical form as the input $x(t)$. Hence, if $x(t)$ is bounded, $y_p(t)$ is also bounded. If the real parts of all roots of the characteristic equation satisfy the relation $\sigma_i < 0$, each term of the natural response is also bounded. Consequently, the *necessary and*

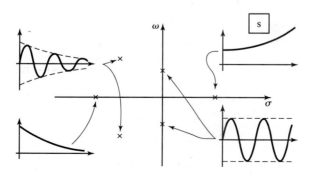

Figure 3.22 Characteristic equation root locations.

sufficient condition for a causal continuous-time LTI system to be BIBO stable is that the roots of the system characteristic equation all lie in the left half of the *s*-plane.

EXAMPLE 3.14   **Stability of an R-L circuit**

Consider the resistance and inductance (R-L) circuit shown in Figure 3.20. A differential equation model for this circuit is

$$L\frac{di(t)}{dt} + Ri(t) = v(t).$$

The characteristic equation of this system model is $s + R/L = 0$. The characteristic root is $s = -R/L$, and the system mode is $e^{-\frac{R}{L}t}$. From the system mode (or the characteristic root), we can see that the system is stable for $R > 0$. If $R = 0$, then $\sigma = 0$ and the system mode does not approach zero with increasing time and the system is not stable. When the real part of a characteristic root of a system is zero, the system is said to be *marginally stable*. The response of a marginally stable system may become unbounded for some inputs. For this marginally stable system, the model becomes

$$L\frac{di(t)}{dt} = v(t),$$

which has the solution

$$i(t) = \frac{1}{L}\int_{-\infty}^{t} v(\tau)d\tau.$$

From this result, we can see that if the input to this marginally stable system is an impulse, then the response is a constant current. If the input voltage is sinusoidal, then the response is also sinusoidal. However, if the input is a unit step function, the response is a current that becomes unbounded as time increases. We further illustrate the determination of stability with an example. ■

EXAMPLE 3.15   **Modes of a second-order LTI system**

Suppose that a causal system is described by the differential equation

$$\frac{d^2y(t)}{dt^2} + 1.25\frac{dy(t)}{dt} + 0.375y(t) = x(t).$$

From (3.57) and (3.60), the system characteristic equation is

$$s^2 + 1.25s + 0.375 = (s + 0.75)(s + 0.5) = 0.$$

This system is stable because the two roots, -0.75 and -0.5, are in the left half of the *s*-plane. The natural response is given by

$$y_c(t) = C_1e^{-0.75t} + C_2e^{-0.5t},$$

and this function approaches zero as *t* approaches infinity.

Consider a second causal system described by the differential equation

$$\frac{d^2y(t)}{dt^2} + 0.25\frac{dy(t)}{dt} - 0.375y(t) = x(t).$$

The system characteristic equation is given by

$$s^2 + 0.25s - 0.375 = (s + 0.75)(s - 0.5) = 0.$$

This system is unstable since one of the roots, at $s = 0.5$, is in the right half-plane. The natural response is given by

$$y_c(t) = C_1 e^{-0.75t} + C_2 e^{0.5t}.$$

The instability is evident in the mode $e^{0.5t}$. ∎

Note the information contained in the system characteristic equation. The first system of Example 3.15 has two time constants, with $\tau_1 = 1/0.75 = 1.33$ and $\tau_2 = 1/0.5 = 2$ and no oscillations in the transient response (no complex roots). The larger time constant is τ_2; the transient response of the system will die out in approximately $4\tau_2 = 8$ units of time. The characteristic equation for the system of the second example denotes instability.

For high-order systems, a computer must be used to find the roots of the system characteristic equation. For example, suppose that a system characteristic equation is given by

$$s^3 + 7s^2 + 14s + 8 = 0.$$

A MATLAB program that calculates the equation roots is given by

```
n = [1 7 14 8];
r = roots (n)
result: r = -4 -2 -1
```

Hence, the system is stable with three time constants: 0.25 s, 0.5 s, and 1 s.
A second example is the characteristic equation

$$s^3 + s^2 + 2s + 8 = 0.$$

A MATLAB program for this system is then

```
n = [1 1 2 8];
r = roots (n)
result: r = -2 0.5 + 1.9365i 0.5 - 1.9365i
```

This system is unstable, because the two roots $s = 0.5 \pm j1.9365$ are in the right half of the s-plane.

In this section, the terms of the natural response of an LTI system are shown to be of the form of the signals studied in Section 2.3. Then relationships of both the transient response and BIBO stability to characteristic-equation zero locations are developed.

3.7   SYSTEM RESPONSE FOR COMPLEX-EXPONENTIAL INPUTS

First in this section, we consider further the linearity property for systems. Then the response of continuous-time LTI systems to a certain class of input signals is derived.

Linearity

Consider the continuous-time LTI system depicted in Figure 3.23. This system is denoted by

$$x(t) \rightarrow y(t). \tag{3.65}$$

For an LTI system, (3.65) can be expressed as the convolution integral

$$y(t) = \int_{-\infty}^{\infty} x(t - \tau)h(\tau)d\tau. \tag{3.66}$$

The functions $x(t)$, $y(t)$, and $h(t)$ are all real for models of physical systems.
    Suppose now that we consider two real inputs $x_i(t)$, $i = 1, 2$. Then, in (3.65),

$$x_i(t) \rightarrow y_i(t), \quad i = 1, 2. \tag{3.67}$$

Thus, $y_i(t)$, $i = 1, 2$, are real, from (3.66). Because the system of (3.65) is linear, the principle of superposition applies, and it follows that

$$a_1 x_1(t) + a_2 x_2(t) \rightarrow a_1 y_1(t) + a_2 y_2(t). \tag{3.68}$$

No restrictions exist on the constants a_1 and a_2 in (3.68); hence, these constants may be chosen to be complex. For this development, we choose the constants to be

$$a_1 = 1, \quad a_2 = j = \sqrt{-1}.$$

With this choice, the superposition property of (3.68) becomes

$$x_1(t) + jx_2(t) \rightarrow y_1(t) + jy_2(t). \tag{3.69}$$

This result may be stated as follows: For a linear system model with a complex input signal, the real part of the input produces the real part of the output, and the imaginary part of the input produces the imaginary part of the output.

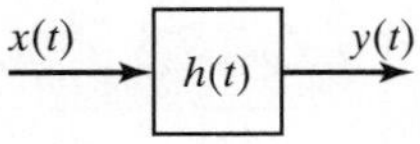

$$x(t) \longrightarrow \boxed{h(t)} \xrightarrow{\;y(t)\;}$$

Figure 3.23   LTI system.

Complex Inputs for LTI Systems

The response of an LTI system to the complex-exponential input

$$x(t) = Xe^{st} \tag{3.70}$$

is now investigated. For the general case, both X and s are complex. We investigate the important case that the system of (3.66) and Figure 3.23 is stable and is modeled by an nth-order linear differential equation with constant coefficients. The exponential input of (3.70) is assumed to exist for all time; hence, the *steady-state system response* will be found. In other words, we will find the forced response for a differential equation with constant coefficients for a complex-exponential input signal.

The differential-equation model for an nth-order LTI system is

[eq(3.55)]
$$\sum_{k=0}^{n} a_k \frac{d^k y(t)}{dt^k} = \sum_{k=0}^{m} b_k \frac{d^k x(t)}{dt^k},$$

where all a_i and b_i are real constants and $a_n \neq 0$. For the complex-exponential excitation of (3.70), recall from Section 3.5 that the forced response (steady-state response) of (3.55) is of the same mathematical form; hence,

$$y_{ss}(t) = Ye^{st}, \tag{3.71}$$

where $y_{ss}(t)$ is the steady-state response and Y is a complex constant to be determined [s is known from (3.70)]. We denote the forced response as $y_{ss}(t)$ rather than $y_p(t)$, for clarity. From (3.70) and (3.71), the terms of (3.55) become

$$
\begin{aligned}
a_0 y_{ss}(t) &= a_0 Ye^{st} & b_0 x(t) &= b_0 Xe^{st} \\
a_1 \frac{d y_{ss}(t)}{dt} &= a_1 s Ye^{st} & b_1 \frac{dx(t)}{dt} &= b_1 s Xe^{st} \\
a_2 \frac{d^2 y_{ss}(t)}{dt^2} &= a_2 s^2 Ye^{st} & b_2 \frac{d^2 x(t)}{dt^2} &= b_2 s^2 Xe^{st} \\
&\ \ \vdots & &\ \ \vdots \\
a_n \frac{d^n y_{ss}(t)}{dt^n} &= a_n s^n Ye^{st} & b_m \frac{d^m x(t)}{dt^m} &= b_m s^m Xe^{st}
\end{aligned}
\tag{3.72}
$$

These terms are substituted into (3.55), resulting in the equation

$$(a_n s^n + a_{n-1} s^{n-1} + \cdots + a_1 s + a_0)Ye^{st}$$

$$= (b_m s^m + b_{m-1} s^{m-1} + \cdots + b_1 s + b_0)Xe^{st}. \tag{3.73}$$

The only unknown in the steady-state response $y_{ss}(t)$ of (3.71) is Y. In (3.73), the factor e^{st} cancels, and Y is given by

$$Y = \left[\frac{b_m s^m + b_{m-1} s^{m-1} + \cdots + b_1 s + b_0}{a_n s^n + a_{n-1} s^{n-1} + \cdots + a_1 s + a_0} \right] X = H(s)X. \tag{3.74}$$

It is standard practice to denote the ratio of polynomials as

$$H(s) = \frac{b_m s^m + b_{m-1} s^{m-1} + \cdots + b_1 s + b_0}{a_n s^n + a_{n-1} s^{n-1} + \cdots + a_1 s + a_0}. \tag{3.75}$$

We show below that this function is related to the impulse response $h(t)$. The function $H(s)$ is called a *transfer function* and is said to be nth order. The order of a transfer function is the same as that of the differential equation upon which the transfer function is based.

We will now summarize this development. Consider an LTI system with the transfer function $H(s)$ as given in (3.74) and (3.75). If the system excitation is the complex exponential $Xe^{s_1 t}$, the steady-state response is given by, from (3.71) and (3.74),

$$x(t) = Xe^{s_1 t} \rightarrow y_{ss}(t) = XH(s_1)e^{s_1 t}. \tag{3.76}$$

The complex-exponential solution in (3.76) also applies for the special case of sinusoidal inputs. Suppose that, in (3.76), $X = |X|e^{j\phi}$ and $s_1 = j\omega_1$, where ϕ and ω_1 are real. Then

$$x(t) = Xe^{s_1 t} = |X|e^{j\phi}e^{j\omega_1 t} = |X|e^{j(\omega_1 t + \phi)}$$
$$= |X|\cos(\omega_1 t + \phi) + j|X|\sin(\omega_1 t + \phi). \tag{3.77}$$

Since, in general, $H(j\omega_1)$ is also complex, we let $H(j\omega_1) = |H(j\omega_1)|e^{j\theta_H}$. The right side of (3.76) can be expressed as

$$y_{ss}(t) = XH(j\omega_1)e^{j\omega_1 t} = |X||H(j\omega_1)|e^{j(\omega_1 t + \phi + \theta_H)}$$
$$= |X||H(j\omega_1)|[\cos[\omega_1 t + \phi + \angle H(j\omega_1)] + j\sin[\omega_1 t + \phi + \angle H(j\omega_1)]],$$

with $\theta_H = \angle H(j\omega_1)$. From (3.69), since the real part of the input signal produces the real part of the output signal,

$$|X|\cos(\omega_1 t + \phi) \rightarrow |X||H(j\omega_1)|\cos[\omega_1 t + \phi + \angle H(j\omega_1)]. \tag{3.78}$$

This result is general for an LTI system and is fundamental to the analysis of LTI systems with periodic inputs; its importance cannot be overemphasized.

Suppose that a system is specified by its transfer function $H(s)$. To obtain the system differential equation, we reverse the steps in (3.72) through (3.75). In fact, in $H(s)$ the numerator coefficients b_i are the coefficients of $d^i x(t)/dt^i$, and the denominator coefficients a_i are the coefficients of $d^i y(t)/dt^i$; we can consider the transfer function to be a shorthand notation for a differential equation. Therefore, the system differential equation can be written directly from the transfer function $H(s)$; consequently, $H(s)$ is a complete description of the input–output characteristics of a system, regardless of the input function. For this reason, an LTI system can be represented by the block diagram in Figure 3.24 with the system transfer function given inside the block. It is common engineering practice to specify an LTI system in this manner.

$$x(t) \rightarrow \boxed{H(s)} \rightarrow y(t)$$

Figure 3.24 LTI system.

The form of the transfer function in (3.75), which is a ratio of polynomials, is called a *rational function*. The transfer function of a continuous-time LTI system described by a linear differential equation with constant coefficients, as in (3.55), will *always* be a rational function.

We now consider two examples to illustrate the preceding developments.

EXAMPLE 3.16   **Transfer function of a servomotor**

In this example, we illustrate the transfer function using a physical device. The device is a servomotor, which is a dc motor used in position control systems. An example of a physical position-control system is the system that controls the position of the read/write heads on a computer hard disk. In addition, the audio compact-disk (CD) player has three position-control systems. (See Section 1.3.)

The input signal to a servomotor is the armature voltage $e(t)$, and the output signal is the motor-shaft angle $\theta(t)$. The commonly used transfer function of a servomotor is second order and is given by [3]

$$H(s) = \frac{K}{s^2 + as},$$

where K and a are motor parameters and are determined by the design of the motor. This motor can be represented by the block diagram of Figure 3.25, and the motor differential equation is

$$\frac{d^2\theta(t)}{dt^2} + a\frac{d\theta(t)}{dt} = Ke(t).$$

This common model of a servomotor is second order and is of adequate accuracy in most applications. However, if a more accurate model is required, the model order is usually increased to three [3]. The second-order model ignores the inductance in the armature circuit, while the third-order model includes this inductance.

Servomotor

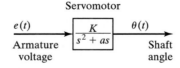

$$e(t) \rightarrow \boxed{\frac{K}{s^2 + as}} \rightarrow \theta(t)$$

Armature   Shaft
voltage   angle   **Figure 3.25** System for Example 3.16. ∎

EXAMPLE 3.17   **Sinusoidal response of an LTI system**

In this example, we calculate the system response of an LTI system with a sinusoidal excitation. Consider a system described by the second-order differential equation

$$\frac{d^2y(t)}{dt^2} + 3\frac{dy(t)}{dt} + 2y(t) = 10x(t).$$

Hence, from (3.75), the transfer function is given by

$$H(s) = \frac{10}{s^2 + 3s + 2}.$$

Suppose that the system is excited by the sinusoidal signal $x(t) = 5\cos(2t + 40°)$. In (3.78),

$$H(s)\bigg|_{s=j2} X = \frac{10}{s^2 + 3s + 2}\bigg|_{s=j2} (5\underline{/40°}) = \frac{50\underline{/40°}}{-4 + j6 + 2}$$

$$= \frac{50\underline{/40°}}{-2 + j6} = \frac{50\underline{/40°}}{6.325\underline{/108.4°}} = 7.905e^{-j68.4°}.$$

Thus, from (3.78), the system response is given by

$$y_{ss}(t) = 7.905\cos(2t - 68.4°).$$

Note the calculation required:

$$H(j2) = \frac{10}{(j2)^2 + 3(j2) + 2} = 1.581\underline{/-108.4°}.$$

From (3.78), the steady-state response can be written directly from this numerical value for the transfer function:

$$y_{ss}(t) = (1.581)(5)\cos(2t + 40° - 108.4°)$$
$$= 7.905\cos(2t - 68.4°).$$ ∎

A MATLAB program for the calculations in this example is given by

```
n = [0 0 10];d = [1 3 2];
h = polyval (n,2*j) / polyval (d,2*j);
ymag = 5*abs (h)
yphase = 40 + angle (h)*180/pi
results: ymag = 7.9057 yphase = -68.4349
```

Consider now the case in which the input function is a sum of complex exponentials:

$$x(t) = \sum_{k=1}^{N} X_k e^{s_k t}. \tag{3.79}$$

By superposition, from (3.76), the response of an LTI system with the transfer function $H(s)$ is given by

$$y_{ss}(t) = \sum_{k=1}^{N} X_k H(s_k) e^{s_k t}. \tag{3.80}$$

We illustrate this result with an example.

EXAMPLE 3.18   **Transfer function used to calculate LTI system response**

Suppose that the input to the stable LTI system in Figure 3.24, with the transfer function $H(s)$, is given by

$$x(t) = 8 - 5e^{-6t} + 3\cos(4t + 30°).$$

In terms of a complex-exponential input Xe^{st}, the first term in the sum is constant ($s = 0$), the second term is a real exponential ($s = -6$), and the third term is the real part of a complex exponential with $s = j4$ from (3.77). From (3.80), the steady-state response is given by

$$y_{ss}(t) = 8H(0) - 5H(-6)e^{-6t} + 3\text{Re}[H(j4)e^{j(4t+30°)}].$$

The sinusoidal-response term can also be simplified somewhat from (3.78), with the resulting output given by

$$y_{ss}(t) = 8H(0) - 5H(-6)e^{-6t} + 3|H(j4)|\cos[4t + 30° + \angle H(j4)].$$

This expression is the forced response for the differential equation (system) whose transfer function is $H(s)$. Note that the order of the differential equation does not significantly increase the complexity of the procedure; the only effect is that the numerical evaluation of $H(s)$ for a given value of s is somewhat more difficult for higher-order systems. ■

Impulse Response

Recall that when the impulse response of an LTI system was introduced, the notation $h(\cdot)$ was reserved for the impulse response. In (3.76), the notation $H(\cdot)$ is used to describe the transfer function of an LTI system. It will now be shown that the transfer function $H(s)$ is directly related to the impulse response $h(t)$, and $H(s)$ can be calculated directly from $h(t)$.

For the excitation $x(t) = e^{st}$, the convolution integral (3.20) yields the system response:

$$y(t) = \int_{-\infty}^{\infty} h(\tau)x(t - \tau)\, d\tau = \int_{-\infty}^{\infty} h(\tau)e^{s(t-\tau)}\, d\tau$$

$$= e^{st}\int_{-\infty}^{\infty} h(\tau)e^{-s\tau}\, d\tau. \tag{3.81}$$

In (3.76), the value of s_1 is not constrained and can be considered to be the variable s. From (3.76) and (3.81),

$$y(t) = e^{st}\int_{-\infty}^{\infty} h(\tau)e^{-s\tau}\, d\tau = H(s)e^{st},$$

and we see that the impulse response and the transfer function of a continuous-time LTI system are related by

$$H(s) = \int_{-\infty}^{\infty} h(t)e^{-st}\, dt$$

Figure 3.26 LTI system.

TABLE 3.1 Input–Output Functions for an LTI System

$$H(s) = \int_{-\infty}^{\infty} h(t)e^{-st}\, dt$$

$$Xe^{s_1 t} \rightarrow XH(s_1)e^{s_1 t}; \qquad X = |X|e^{j\phi}$$

$$|X|\cos(\omega_1 t + \phi) \rightarrow |X||H(j\omega_1)|\cos[\omega_1 t + \phi + \underline{/H(j\omega_1)}]$$

$$H(s) = \int_{-\infty}^{\infty} h(t)e^{-st}\, dt. \tag{3.82}$$

This equation is the desired result. Table 3.1 summarizes the results developed in this section.

We can express these developments in system notation:

$$e^{st} \rightarrow H(s)e^{st}. \tag{3.83}$$

We see that a complex exponential input signal produces a complex exponential output signal.

It is more common in practice to describe an LTI system by the transfer function $H(s)$ rather than the impulse response $h(t)$. However, we can represent LTI systems with either of the block diagrams given in Figure 3.26, with $H(s)$ and $h(t)$ related by (3.82).

Those readers familiar with the bilateral Laplace transform will recognize $H(s)$ in (3.82) as the Laplace transform of $h(t)$. Furthermore, with $s = j\omega$, $H(j\omega)$ in (3.82) is the Fourier transform of $h(t)$. We see then that both the Laplace transform (covered in Chapter 7) and the Fourier transform (covered in Chapter 5) appear naturally in the study of LTI systems.

We considered the response of LTI systems to complex-exponential inputs in this section, which led us to the concept of transfer functions. Using the transfer function approach, we can easily find the system response to inputs that are constant, real exponential, and sinusoidal. As a final point, the relationship between the transfer function of a system and its impulse response is derived.

3.8   BLOCK DIAGRAMS

Figure 3.26 gives two block-diagram representations of a system. In this section, we consider a third block-diagram representation. The purpose of this block diagram is to give an *internal structure* to systems, in addition to the usual input–output description

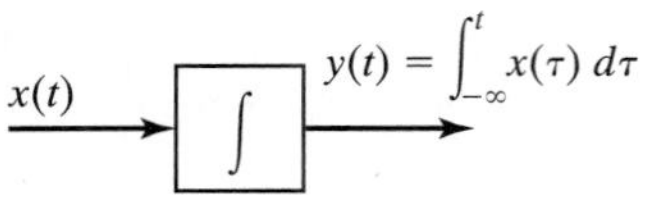

Figure 3.27   Representation of integration.

of Figure 3.26. This block diagram is used in cases in which we are interested not only in the output signal, but also in the internal operation of a system. The block diagram is also useful in physical implementations of analog filters.

The block-diagram representation of the differential equations that model a system requires that we have a block to denote integration. As was done in Example 3.2, we use a block containing an integral sign for this purpose, as shown in Figure 3.27. The input–output description of this block is given by

$$x(t) \rightarrow y(t) = \int_{-\infty}^{t} x(\tau)\, d\tau. \tag{3.84}$$

The procedure developed here involves finding a block diagram, constructed of certain defined elements, including integrators, that satisfies a given differential equation. An example will now be given to illustrate this procedure. Then a general procedure will be developed.

EXAMPLE 3.19   **Simulation diagram for a first-order LTI system**

Earlier in this chapter, we considered a continuous-time system described by the differential equation

$$\frac{dy(t)}{dt} + 2y(t) = x(t).$$

Hence, the system transfer function is given by

$$H(s) = \frac{1}{s + 2}.$$

A block diagram that satisfies this differential equation will now be constructed. First, we write the differential equation as

$$\frac{dy(t)}{dt} = -2y(t) + x(t). \tag{3.85}$$

If the *output* of an integrator is $y(t)$, its *input* must be $dy(t)/dt$. We then draw a block diagram containing an integrator such that the integrator input is the right side of (3.85); then the output of the integrator is $y(t)$. The result is given in Figure 3.28(a). Hence, this block diagram satisfies (3.85). The block diagram of Figure 3.28(a) gives both the input–output model and an internal model of the system, while that of Figure 3.28(b) gives only the input–output model of the same system. Both representations are used in practice.

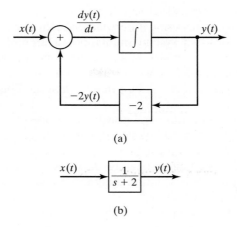

Figure 3.28   System for Example 3.19.

In Example 3.19 a block diagram for a differential equation, based on an integrator, was constructed for a first-order differential equation. Figure 3.28(a) shows the three components used in constructing block diagrams of this type:

1. integrators,
2. summing devices (the circle),
3. gains (the block with a gain of −2).

Figure 3.28(a) is also useful in realizing analog filters. If the result of an analog-filter design is the transfer function $H(s) = 1/(s + 2)$, the filter can be implemented as a physical device using an integrator and amplifiers (including summing amplifiers), as shown in Figure 3.28(a).

An *analog computer* is an electronic device that is used to solve differential equations by the interconnection of electronics circuits that (1) integrate signals, (2) sum signals, and (3) amplify (or attenuate) signals. The system of Figure 3.28(a) can be programmed directly on an analog computer, and the result will be the solution of the system differential equation for an applied voltage input function $x(t)$. A *simulation* is a machine solution of the equations that model a system. The analog computer is used for *analog simulations* of continuous-time systems.

If the integrator of Figure 3.28(a) is replaced with a numerical integrator, the resulting equations can be programmed on a digital computer, yielding a numerical solution of the differential equation. In this case, we have a machine solution, called a *digital simulation*, of the differential-equation model. For these reasons, block diagrams of the type given in Figure 3.28(a) are sometimes called *simulation diagrams*. One procedure for constructing either an analog simulation or a digital simulation of a system is first to draw a simulation diagram that is based on integrators.

In general, we construct simulation diagrams using integrators; we can also construct these diagrams using *differentiators*. However, we generally do not use differentiators to construct system simulations. Differentiators amplify any high-frequency noise present in a simulation, while integrators amplify any low-frequency noise. Generally, high-frequency noise is more of a problem than low-frequency noise. These characteristics will be evident when the Fourier transform is covered in Chapter 5.

A particular procedure will now be developed for the construction of simulation diagrams. First, it should be stated that given a differential equation, there is no unique simulation diagram for that equation. It can be shown that an unbounded number of simulation diagrams can be drawn for a given differential equation. Only two standard forms for simulation diagrams are given here. This topic is investigated in greater detail in Chapter 8.

As stated in Section 3.7, an nth-order linear differential equation with constant coefficients can be expressed as

[eq(3.55)]
$$\sum_{k=0}^{n} a_k \frac{d^k y(t)}{dt^k} = \sum_{k=0}^{m} b_k \frac{d^k x(t)}{dt^k},$$

where $x(t)$ is the excitation, $y(t)$ is the response, and $a_0, \ldots, a_n$ and $b_0, \ldots, b_m$ are constants, with $a_n \neq 0$.

As an example of the two standard forms for simulation diagrams to be covered, first a second-order differential equation will be considered. Then we develop the two forms for the nth-order equation of (3.55).

We consider a second-order differential equation of the form

$$a_2 \frac{d^2 y(t)}{dt^2} + a_1 \frac{dy(t)}{dt} + a_0 y(t) = b_2 \frac{d^2 x(t)}{dt^2} + b_1 \frac{dx(t)}{dt} + b_0 x(t). \qquad (3.86)$$

To develop the simulation diagrams, we note that the double integral of the second derivative of $y(t)$ with respect to time is $y(t)$. We write this double integral as

$$y_{(2-2)}(t) = \int_{-\infty}^{t} \left[\int_{-\infty}^{\tau} \frac{d^2 y(\sigma)}{d\sigma^2} d\sigma \right] d\tau = y(t),$$

where the notation $y_{(k-n)}(t)$ indicates the nth integral of the kth derivative of $y(t)$. With this notation, the double integral of the three terms on the left side of (3.86) can be written as

$$y_{(2-2)}(t) = y_{(0)}(t) = \int_{-\infty}^{t} \int_{-\infty}^{\tau} \frac{d^2 y(\sigma)}{d\sigma^2} \, d\sigma \, d\tau = y(t);$$

$$y_{(1-2)}(t) = y_{(-1)}(t) = \int_{-\infty}^{t} \int_{-\infty}^{\tau} \frac{dy(\sigma)}{d\sigma} d\sigma \, d\tau = \int_{-\infty}^{t} y(\tau) \, d\tau;$$

$$y_{(0-2)}(t) = y_{(-2)}(t) = \int_{-\infty}^{t} \int_{-\infty}^{\tau} y(\sigma) d\sigma \, d\tau.$$

Note then that $y_{(-i)}(t)$ is the ith integral of $y(t)$.

Direct Form I

The double integral of the differential equation of (3.86) yields

$$a_2 y(t) + a_1 y_{(-1)}(t) + a_0 y_{(-2)}(t) = b_2 x(t) + b_1 x_{(-1)}(t) + b_0 x_{(-2)}(t). \tag{3.87}$$

We denote the right side of this equation as $w(t)$:

$$w(t) = b_2 x(t) + b_1 x_{(-1)}(t) + b_0 x_{(-2)}(t).$$

and realize $w(t)$ by the block diagram shown in Figure 3.29(a). Then (3.87) becomes

$$a_2 y(t) + a_1 y_{(-1)}(t) + a_0 y_{(-2)}(t) = w(t).$$

Solving this equation for $y(t)$ yields

$$y(t) = \frac{1}{a_2} [w(t) - a_1 y_{(-1)}(t) - a_0 y_{(-2)}(t)]. \tag{3.88}$$

This equation is realized by the system of Figure 3.29(b). The total realization is the series (cascade) connection of the systems of Figure 3.29(a) and (b), as shown in Figure 3.29(c). This simulation diagram realizes (3.86) and is called either the *direct form I* realization or the direct form I simulation diagram.

Direct Form II

A second standard form for realizing a differential equation with integrators will be derived by manipulating the direct form I shown in Figure 3.29(c). This system is seen to be two cascaded systems, where one system realizes $w(t)$ and the other realizes $y(t)$ as a function of $w(t)$. Because the systems are linear, the order of the two systems can be reversed without affecting the input–output characteristics (see Figure 3.13); the result of this reversal is shown in Figure 3.30(a). In this figure, the same signal is integrated by the two sets of cascaded integrators; hence, the outputs of the integrators labeled 1 are equal, as are the outputs of the integrators labeled 2. Thus, one set of the cascaded integrators can be eliminated. The final system is given in Figure 3.30(b), and only two integrators are required. This form for the simulation diagram is called the *direct form II* realization.

*n*th-Order Realizations

Consider again the *n*th-order differential equation:

[(eq(3.55)] $$\sum_{k=0}^{n} a_k \frac{d^k y(t)}{dt^k} = \sum_{k=0}^{m} b_k \frac{d^k x(t)}{dt^k}.$$

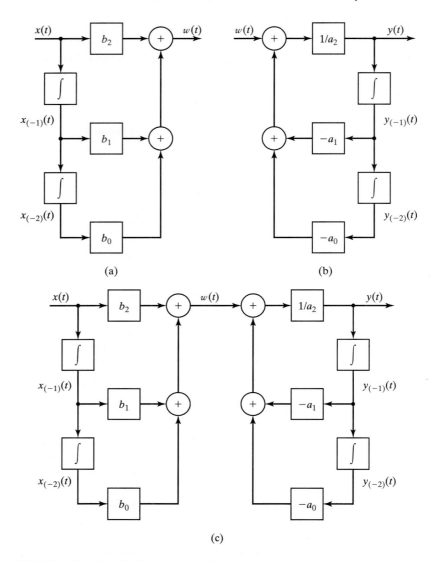

Figure 3.29   Direct form I realization of a second-order system.

To derive an integrator form of the simulation diagram, each side is integrated n times, resulting in the equation

$$\sum_{k=0}^{n} a_k y_{(k-n)}(t) = \sum_{k=0}^{m} b_k x_{(k-n)}(t).$$

Solving this equation for $y(t)$ yields

$$y(t) = \frac{1}{a_n}\left[\sum_{k=0}^{m} b_x x_{(k-n)}(t) - \sum_{k=0}^{n-1} a_k y_{(k-n)}(t)\right]. \tag{3.89}$$

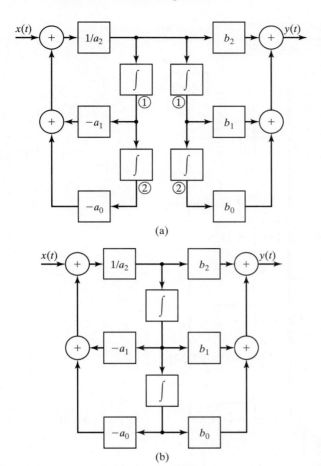

(a)

(b)

Figure 3.30 Direct form II realization of a second-order system.

Using the procedure illustrated with this second-order system, we draw the simulation diagrams for the direct form I and the direct form II in Figures 3.31 and 3.32, respectively, for $m = n$.

It can be shown that at least n integrators are required to realize an nth-order differential equation. If a realization has n integrators, the realization is *minimal*. Otherwise, the realization is *nonminimal*. The direct form I is nonminimal, and the direct form II is minimal.

Practical Considerations

We now discuss some practical considerations. Suppose that, for example, the form I simulation diagram of Figure 3.29(c) is constructed for a second-order mechanical system and that $y(t)$ is the position variable. Then velocity is $dy(t)/dt$ and acceleration is $d^2y(t)/dt^2$. Generally, we must solve for these variables in a simulation, and we obtain this information by requiring the outputs of two integrators to be these

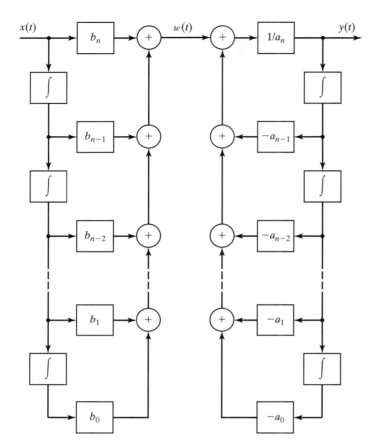

Figure 3.31 Direct form I for an nth-order system.

variables. However, neither form I nor form II meet these requirements. Forms I and II are useful conceptually for a general system; however, in practice these two forms are seldom used in writing simulations of systems. Instead, simulations are constructed such that the output of each integrator is a physical variable, as much as is possible. This topic is discussed further in Chapter 8.

A second important practical consideration was mentioned earlier in this section. We can construct simulations using either differentiators or integrators, or combinations of the two. However, differentiators amplify high-frequency noise, while integrators attenuate high-frequency noise. Hence, integrators are almost always used, either as electronic circuits or as numerical algorithms. Also, because of these noise problems, we try to avoid the use of differentiators *in any applications* in physical systems. Sometimes we cannot avoid using differentiators; these systems usually have noise problems.

Two procedures for representing the internal model of a system, given its input–output description, are presented in this section. These models are useful in realizing analog filters, and in developing analog and digital simulations of a system. However, in simulating a system, we prefer an internal model such that the outputs of the integrators represent physical variables, as much as possible.

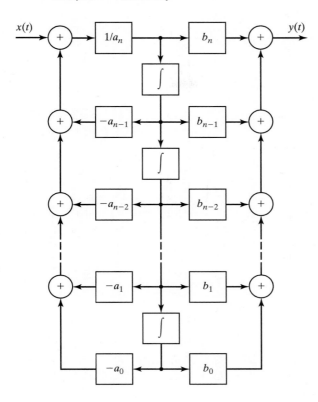

Figure 3.32 Direct form II for an *n*th-order system.

SUMMARY

In this chapter, we considered continuous-time linear time-invariant (LTI) systems. First, it was shown that a continuous-time signal can be expressed as a function of an impulse function. This representation was in the form of an integral and allowed us to describe the input–output characteristics of an LTI system in terms of its impulse response.

Describing a system by its impulse response is basic to the analysis and design of LTI systems; the impulse response gives a complete input–output description of an LTI system. It is shown that the input $x(t)$, the impulse response $h(t)$, and the output $y(t)$ are related by the convolution integral:

$$y(t) = \int_{-\infty}^{\infty} x(\tau)h(t - \tau)d\tau = \int_{-\infty}^{\infty} x(t - \tau)h(\tau)dt.$$

The importance of the impulse response of an LTI system cannot be overemphasized. It is also shown that the impulse response of an LTI system can be derived from its step response. Hence, the input–output description of a system is also contained in its step response.

TABLE 3.2 Key Equations of Chapter 3.

Equation Title	Equation Number	Equation
Unit impulse response	(3.14)	$\delta(t) \rightarrow h(t)$
Convolution integral	(3.20)	$y(t) = \int_{-\infty}^{\infty} x(\tau)h(t-\tau)\,d\tau = \int_{-\infty}^{\infty} x(t-\tau)h(\tau)\,d\tau$
Convolution with a unit impulse	(3.23)	$\delta(t)*g(t) = g(t)$
Convolution integral of an inverse system	(3.37)	$x(t)*h(t)*h_i(t) = x(t)$
Convolution integral of a causal system	(3.42)	$y(t) = \int_{-\infty}^{\infty} x(\tau)h(t-\tau)\,d\tau = \int_{-\infty}^{t} x(\tau)h(t-\tau)\,d\tau$
Condition on impulse response for BIBO stability	(3.44)	$\int_{-\infty}^{\infty} |h(t)|\,dt \leq \infty$
Derivation of step response from impulse response	(3.46)	$s(t) = \int_{-\infty}^{\infty} u(\tau)h(t-\tau)\,d\tau = \int_{0}^{\infty} h(t-\tau)\,d\tau$
Derivation of impulse response from step response	(3.48)	$h(t) = \dfrac{ds(t)}{dt}$
Linear differential equation with constant coefficients	(3.55)	$\sum_{k=0}^{n} a_k \dfrac{d^k y(t)}{dt^k} = \sum_{k=0}^{m} b_k \dfrac{d^k x(t)}{dt^k}$
Characteristic equation	(3.60)	$a_n s^n + a_{n-1}s^{n-1} + \cdots + a_1 s + a_0 = 0$
Solution of homogeneous equation	(3.62)	$y_c(t) = C_1 e^{s_1 t} + C_2 e^{s_2 t} + \cdots + C_n e^{s_n t}$
Transfer function	(3.75)	$H(s) = \dfrac{b_m s^m + b_{m-1}s^{m-1} + \cdots + b_1 s + b_0}{a_n s^n + a_{n-1}s^{n-1} + \cdots + a_1 s + a_0}$
Steady-state response to a complex exponential input	(3.76)	$x(t) = Xe^{s_1 t} \rightarrow y_{ss}(t) = XH(s_1)e^{s_1 t}$
Input expressed as sum of complex exponentials	(3.79)	$x(t) = \sum_{k=1}^{N} X_k e^{s_k t}$
Output expressed as sum of complex exponentials	(3.80)	$y_{ss}(t) = \sum_{k=1}^{N} X_k H(s_k)e^{s_k t}$
Transfer function expressed as integral of impulse response	(3.82)	$H(s) = \int_{-\infty}^{\infty} h(t)e^{-st}\,dt$

Next, the general system properties of an LTI system are investigated. These include memory, invertibility, causality, and stability.

A general procedure for solving linear differential equations with constant coefficients is reviewed. This procedure leads to a test that determines the BIBO stability for a causal LTI system.

The most common method of modeling LTI systems is by ordinary linear differential equations with constant coefficients; many physical systems can be modeled accurately by these equations. The concept of representing system models by simulation diagrams is developed. Two simulation diagrams, direct forms I and II, are given. However, an unbounded number of simulation diagrams exist for a given LTI system. This topic is considered further in Chapter 8.

A procedure for finding the response of differential-equation models of LTI systems is given for the case that the input signal is a complex-exponential function. Although this signal cannot appear in a physical system, the procedure has wide application in models of physical systems.

REFERENCES

1. F. B. Hildebrand, *Advanced Calculus and Applications*, 2d ed. Upper Saddle River, NJ: Prentice–Hall, 1976.
2. G. Birkhoff and G.-C. Rota, *Ordinary Differential Equations*, 4th ed. New York: Wiley, 1994.
3. C. L. Phillips and R. D. Harbor, *Feedback Control Systems*, 4th ed. Upper Saddle River, NJ: Prentice Hall, 1999.
4. G. Doetsch, *Guide to the Applications of the Laplace and z-Transforms*. London: Van Nostrand Reinhold, 1971.

PROBLEMS

3.1. Consider the integrator in Figure P3.1. This system is described in Example 3.1 and has the impulse response $h(t) = u(t)$.

    (a) Using the convolution integral, find the system response to the input $x(t)$, where

        (i) $u(t - 2)$ **(ii)** $e^{5t}u(t)$

        (iii) u(t)   **(iv)** $(t + 1)u(t + 1)$

    (b) Verify the results of Part (a) using the system equation

$$y(t) = \int_{-\infty}^{t} x(\tau)d\tau.$$

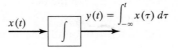

Figure P3.1

3.2. Suppose that the system of Figure P3.2(a) has the input $x(t)$ given in Figure P3.2(b). The impulse response is the unit step function $h(t) = u(t)$. Find and sketch the system output $y(t)$.

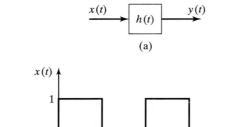

(a)

(b)

Figure P3.2

3.3. For the system of Figure P3.2(a), let $x(t) = u(t - t_0)$ and $h(t) = u(t - t_1)$, with $t_1 > t_0$. Find and plot the system output $y(t)$.

3.4. For the system of Figure P3.2(a), the input signal is $x(t)$, the output signal is $y(t)$, and the impulse response is $h(t)$. For each of the cases that follow, find and plot the output $y(t)$. The referenced signals are given in Figure P3.4.

    (a) $x(t)$ in (a), $h(t)$ in (b)
    (b) $x(t)$ in (a), $h(t)$ in (c)
    (c) $x(t)$ in (a), $h(t)$ in (d)
    (d) $x(t)$ in (a), $h(t)$ in (a).
    (e) $x(t)$ in (α), $h(t)$ in (β), where α and β are assigned by your instructor.

3.5. For the system of Figure P3.2(a), suppose that $x(t)$ and $h(t)$ are identical and as shown in Figure P3.4(b).

    (a) Find the output $y(t)$ only at the times $t = 0, 1, 2$, and 2.667. Solve this problem by inspection.
    (b) To verify the results in Part (a), solve for and sketch $y(t)$ for all time.

3.6. A continuous-time LTI system has the input $x(t)$ and the impulse response $h(t)$, as shown in Figure P3.6. Note that $h(t)$ is a delayed function.

    (a) Find the system output $y(t)$ for only $4 \leq t \leq 5$.
    (b) Find the maximum value of the output.
    (c) Find the ranges of time for which the output is zero.
    (d) Solve for and sketch $y(t)$ for all time, to verify all results.

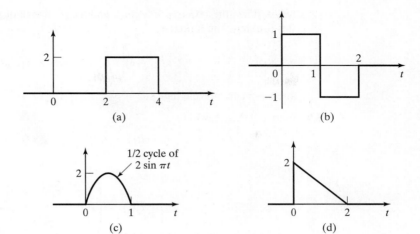

Figure P3.4

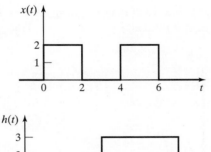

Figure P3.6

3.7. For the system of Figure P3.2(a), the input signal is $x(t)$, the output signal is $y(t)$, and the impulse response is $h(t)$. For each of the following cases find $y(t)$:

    (a) $x(t) = e^t u(-t)$ and $h(t) = 2u(t) - u(t - 1) - u(t - 2)$.

    (b) $x(t) = e^{-t} u(t)$ and $h(t) = u(t - 1) - u(t - 3)$.

    (c) $x(t) = u(1 - t)$ and $h(t) = e^{-t} u(t - 1)$.

    (d) $x(t) = e^{-at}[u(t) - u(t - 2)]$ and $h(t) = u(t)$.

    (e) $x(t) = u(-t)$ and $h(t) = e^{-t}[u(t) - u(t - 400)]$.

    (f) $x(t) = e^t u(-t)$ and $h(t) = 2u(1 - t)$.

3.8. Show that the convolution of three signals can be performed in any order by showing that

$$[f(t)*g(t)]*h(t) = f(t)*[g(t)*h(t)].$$

(*Hint:* Form the required integrals, and use a change of variables. In one approach to this problem, the function

$$\int_{-\infty}^{\infty} g(\tau) \left[\int_{-\infty}^{\infty} h(t - \tau - \sigma) f(\sigma) d\sigma \right] dt$$

appears in an intermediate step.)

3.9. Find $x_1(t)*x_2(t)$, where

$$x_1(t) = 2u(t + 2) - 2u(t - 2)$$

and

$$x_2(t) = \begin{cases} 0, & t < -4 \\ e^{-|t|}, & -4 \le t \le 4 \\ 0, & t > 4. \end{cases}$$

3.10. Find and sketch

$$u(t)*u(t - 5).$$

3.11. For the system of Figure P3.2(a), the input signal is $x(t)$ in Figure P3.11 (note that the signal is not symmetric) and $h(t) = e^{-|t|}u(-t)$. Find the system output $y(t)$.

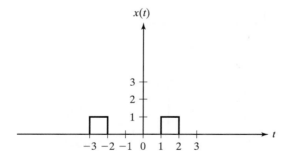

Figure P3.11

3.12. **(a)** Consider the two-LTI system cascaded in Figure P3.12. The impulse responses of the two systems are identical, with $h_1(t) = h_2(t) = e^{-t}u(t)$. Find the impulse response of the total system.
  (b) Repeat Part (a) for the case that $h_1(t) = h_2(t) = \delta(t)$.
  (c) Repeat Part (a) for the case that $h_1(t) = h_2(t) = \delta(t - 1)$.
  (d) Repeat Part (a) for the case that $h_1(t) = h_2(t) = u(t - 2) - u(t - 4)$.

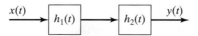

Figure 3.12

3.13. We define a new signal $z(t)$ of two signals $x(t)$ and $h(t)$ as

$$z(t) = \int_{-\infty}^{\infty} x(-\tau + a)h(t + \tau)d\tau.$$

Express $z(t)$ in terms of $y(t) = x(t)*h(t)$, the convolution of $x(t)$ and $h(t)$.

3.14. Suppose that the system of Figure P3.2(a) is described by each of the following system equations. For each case, find the impulse response of the system by letting $x(t) = \delta(t)$ to obtain $y(t) = h(t)$;

(a) $y(t) = x(t - 7)$

(b) $y(t) = \int_{-\infty}^{t} x(\tau-7)d\tau$

(c) $y(t) = \int_{-\infty}^{t} \left[\int_{-\infty}^{\sigma} x(\tau-7)d\tau \right] d\sigma$

3.15. It is shown in Section 3.4 that the necessary condition for a continuous-time LTI system to be bounded-input bounded-output stable is that the impulse response $h(t)$ must be absolutely integrable; that is,

$$\int_{-\infty}^{\infty} |h(t)|dt < \infty .$$

Show that any system that does not satisfy this condition is not BIBO stable; that is, show that this condition is also sufficient. [*Hint*: Assume a bounded input.]

$$x(t - \tau) = \begin{cases} 1, & h(\tau) > 0 \\ -1, & h(\tau) < 0. \end{cases}$$

3.16. Consider the LTI system of Figure P3.16.

(a) Express the system impulse response as a function of the impulse responses of the subsystems.

(b) Let

$$h_1(t) = h_4(t) = u(t)$$

and

$$h_2(t) = h_3(t) = 5\delta(t), \quad h_5(t) = e^{-2t}u(t).$$

Find the impulse response of the system.

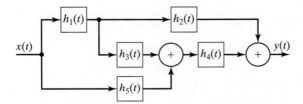

Figure P3.16

3.17. Consider the LTI system of Figure P3.17.

  (a) Express the system impulse response as a function of the impulse responses of the subsystems.

  (b) Let

$$h_3(t) = h_4(t) = h_5(t) = u(t)$$

  and

$$h_1(t) = h_2(t) = 5\delta(t).$$

  Find the impulse response of the system.

  (c) Give the characteristics of each block in Figure P3.17. For example, block 1 is an amplifier with a gain of 5.

  (d) Let $x(t) = \delta(t)$. Give the time function at the output of each block.

  (e) Use the result in Part (b) to verify the result in (d).

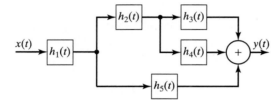

Figure P3.17

3.18. Consider the LTI system of Figure P3.18. Let

$$h_1(t) = u(t), h_2(t) = \delta(t).$$

Hence, system 1 is an integrator and system 2 is the identity system. Use a convolution approach to find the differential-equation model of this system.

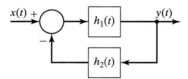

Figure P3.18

3.19. An LTI system has the impulse response

$$h(t) = e^t u(-t).$$

  (a) Determine whether this system is causal.

  (b) Determine whether this system is stable.

  (c) Find and sketch the system response to the unit step input $x(t) = u(t)$.

  (d) Repeat Parts (a), (b), and (c) for $h(t) = e^t u(t)$.

3.20. Consider a system described by the equation

$$y(t) = \sin(t)x(t).$$

  (a) Is this system linear?

  (b) Is this system time invariant?

(c) Determine the response to the input $\delta(t)$.

(d) Determine the response to the input $\delta(t - 1)$. From examining this result, it is evident that this system is not time invariant.

3.21. Determine the stability and the causality for the LTI systems with the following impulse responses.

(a) $h(t) = e^{-t}u(t - 1)$

(b) $h(t) = e^{t}u(-t + 1)$

(c) $h(t) = e^{-t}u(-t - 1)$

(d) $h(t) = \sin(5t)u(-t)$

(e) $h(t) = e^{t}u(-t)$

(f) $h(t) = e^{t}\sin(5t)u(-t)$

3.22. Consider an LTI system with the input and output related by

$$y(t) = \int_{0}^{\infty} e^{-\tau}x(t - \tau)d\tau.$$

(a) Find the system impulse response $h(t)$ by letting $x(t) = \delta(t)$.

(b) Is this system causal? Why?

(c) Determine the system response $y(t)$ for the input shown in Figure P3.22(a).

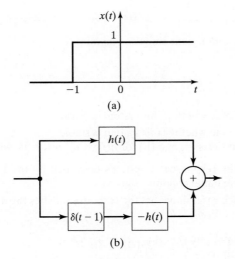

(a)

(b)   **Figure 3.22**

(d) Consider the interconnections of the LTI systems given in Figure P3.22(b), where $h(t)$ is the function found in Part (a). Find the impulse response of the total system.

(e) Solve for the response of the system of Part (d) to the input of Part (c) by doing the following:

(i) Using the results of Part (c). This output can be written by inspection.

(ii) Using the results of Part (d) and the convolution integral.

3.23. **(a)** Given an LTI system with the output

$$y(t) = \int_{-\infty}^{t} e^{-2(t-\tau)}x(\tau-1)d\tau,$$

  (i) Find the impulse response of this system by letting $x(t) = \delta(t)$.
  (ii) Is this system causal?
  (iii) Is this system stable?

b. Given an LTI system with the output

$$y(t) = \int_{-\infty}^{\infty} e^{-2(t-\tau)}x(\tau-1)d\tau,$$

  (i) Find the impulse response of this system by letting $x(t) = \delta(t)$.
  (ii) Is this system causal?
  (iii) Is this system stable?

3.24. An LTI system has the impulse response

$$h(t) = u(t + 1) - u(t - 3)$$

  (a) Determine whether this system is causal.
  (b) Determine whether this system is stable.
  (c) Find and sketch the system response to the input

$$x(t) = \delta(t - 1) - 2\delta(t + 1).$$

3.25. An LTI system has the impulse response

$$h(t) = e^{-at}u(t - 1),$$

where $a > 0$.

  (a) Determine whether this system is causal.
  (b) Determine whether this system is stable.
  (c) Repeat Parts (a) and (b) for $h(t) = e^{-at}u(t + 1)$, where $a < 0$.

3.26. **(a)** Find the responses of systems described by the following differential equations with the initial conditions given.
  (b) For each case, show that the response satisfies the differential equation and the initial conditions:

  (i) $\dfrac{dy(t)}{dt} + 2y(t) = 3u(t), \quad y(0) = -1$

  (ii) $\dfrac{dy(t)}{dt} + 2y(t) = 3e^{-t}u(t), \quad y(0) = 1$

  (iii) $\dfrac{dy(t)}{dt} + 2y(t) = 3\sin t\, u(t), \quad y(0) = 2$

  (iv) $-.7\dfrac{d}{dt}y(t) + y(t) = 6e^{3t}, \quad y(0) = 0$

  (v) $\dfrac{d}{dt}y(t) + 2y(t) = 4e^{2t}, \quad y(0) = 0$

3.27. Indicate whether the following transfer functions for LTI systems are stable:

(a) $H(s) = \dfrac{7}{s + 10}$

(b) $H(s) = \dfrac{10(s + 3)}{(s + 1)(s + 2)(s + 4)} = \dfrac{10s + 30}{s^3 + 7s^2 + 14s + 8}.$

(c) $H(s) = \dfrac{1}{s^2 - 2.5s + 1}$

3.28. Suppose that the following differential equation is a model of a physical system:

$$\frac{d^2}{dt^2}y(t) - 2.5\frac{d}{dt}y(t) + y(t) = x(t).$$

Find its modes. Is this system stable?

3.29. Suppose that the differential equations in Problem 3.26 are models of physical systems.

(a) For each system, give the system modes.

(b) For each system, give the time constants of the system modes.

(c) A unit step $u(t)$ is applied to each system. After how long in time will the system output become approximately constant? How did you arrive at your answer?

(d) Repeat Parts (a), (b), and (c) for the transfer function of Problem 3.27(b).

3.30. A system has the transfer function

$$H(s) = \frac{1}{0.01s^2 + 1}.$$

(a) Find the system modes. These modes are not real, even though the system is a model of a physical system.

(b) Express the natural response as the sum of the modes of Part (a) and as a real function.

(c) The input $e^{-t}u(t)$ is applied to the system, which is initially at rest (zero initial conditions). Find an expression for the system output.

(d) Show that the result in Part (c) satisfies the system differential equation and the initial conditions.

3.31. (a) Consider the system of Figure P3.31. The input signal $x(t) = 4$ is applied at $t = 0$. Find the value of $y(t)$ at a very long time after the input is applied.

(i) $H(s) = \dfrac{5}{s + 4}$ (ii) $H(s) = \dfrac{2s + 10}{s^2 + 2s + 10}$

(b) Repeat Part (a) for the input signal $x(t) = 4e^{3t}$.

(c) Repeat Part (a) for the input signal $x(t) = 4\cos 3t$. Use MATLAB to check your calculations.

(d) Repeat Part (a) for the input signal $x(t) = 4e^{j3t}$.

(e) Repeat Part (a) for the input signal $x(t) = 4\sin 3t$. Use MATLAB to check your calculations.

(f) How are the responses of Parts (c) and (e) related?

(g) (i) Find the time constants of the two systems in Part (a).

(ii) In this problem, quantify the expression "a very long time."

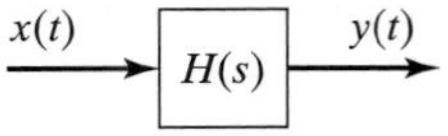

$$x(t) \xrightarrow{\hspace{1cm}} \boxed{H(s)} \xrightarrow{\hspace{1cm}} y(t)$$

Figure P3.31

3.32. For the system of Figure 3.31, the transfer function is known to be of the form

$$H(s) = \frac{K}{s + a}.$$

With the notation $x(t) \rightarrow y(t)$, the following steady-state response is measured:

$$2 \cos 4t \rightarrow 5 \cos (4t - 45°).$$

(a) Find the transfer-function parameters K and a.
(b) Verify the results in Part (a) using MATLAB.

3.33. Draw the direct form I and the direct form II block diagrams for each of the following system equations:

(a) $2\dfrac{dy(t)}{dt} + 5y(t) = 3x(t)$

(b) $\dfrac{dy(t)}{dt} = x(t) + 2\dfrac{dx(t)}{dt} + 3\displaystyle\int_{-\infty}^{t} x(\tau)d\tau$

(c) $\dfrac{d^2y(t)}{dt^2} + 0.5\dfrac{dy(t)}{dt} + 0.01y(t) = 2\dfrac{d^2x(t)}{dt^2} + x(t)$

(d) $\dfrac{d^3y(t)}{dt^3} + 2\dfrac{d^2y(t)}{dt^2} + 3.5\dfrac{dy(t)}{dt} + 4.25y(t) = 2\dfrac{d^3x(t)}{dt^3} + 5\dfrac{d^2x(t)}{dt^2} + 6\dfrac{dx(t)}{dt} + 8x(t)$

3.34. Consider the system simulation diagram of Figure P3.34. This figure shows a simula-tion-diagram form used in the area of automatic control.

(a) Find the differential equation of the system.
(b) Is this one of the two forms given in Section 3.8? If so, which one?

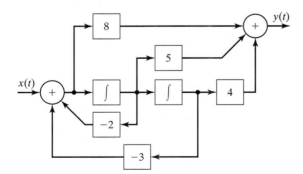

Figure P3.34

3.35. **(a)** For the LTI system of Figure P3.35(a), show that the system transfer function $H(s)$ is given by

$$H(s) = H_1(s)H_2(s),$$

where the transfer functions are as defined in (3.82).

(b) For the LTI system of Figure P3.35(b), show that the system transfer function $H(s)$ is given by

$$H(s) = H_1(s) + H_2(s),$$

where the transfer functions are as defined in (3.82).

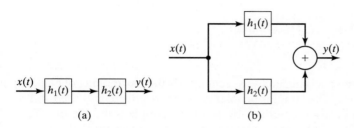

(a)   (b)   **Figure P3.35**

3.36. Using the results of Problem 3.35, do the following:

(a) Find the transfer function for the system of Figure P3.16.
(b) Find the transfer function for the system of Figure P3.17.
(c) Find the transfer function for the system of Figure P3.18.

3.37. Assume that the systems involved are LTI, with the ith system having the impulse response $h_i(t)$. Using the results of Problem 3.35,

(a) Find the transfer function for the system of Figure P2.24.
(b) Find the transfer function for the system of Figure P2.25.

4 FOURIER SERIES

A common engineering analysis technique is the partitioning of complex problems into simpler ones. The simpler problems are then solved, and the total solution becomes the sum of the simpler solutions. One example is the use of a Taylor's series expansion, in which a function $f(t)$ is expressed as a constant, plus a ramp function, plus a parabolic function, and so on:

$$f(t) = f(0) + f'(0)t + f''(0)\frac{t^2}{2!} + \cdots. \tag{4.1}$$

In this equation,

$$f'(0) = \left.\frac{df(t)}{dt}\right|_{t=0}; \quad f''(0) = \left.\frac{d^2f(t)}{dt^2}\right|_{t=0}.$$

The problem involving $f(t)$ is solved considering only the constant, then considering only the ramp function, and so on. The final solution is the sum of these solutions.

Three requirements must be satisfied for a solution as described earlier to be both valid and useful. First, we must be able to express the problem as a number of simpler problems. Next, the problem must be *linear*, such that the solution for the sum of functions is equal to the sum of the solutions considering only one function at a time.

The third requirement is that the contributions of the simpler solutions to the total solution must become negligible after considering only a few terms; otherwise, the advantage may be lost if a very large number of simple solutions is required.

This method of partitioning a complex problem into simpler problems sometimes has an additional advantage. We may have insight into the simpler problems and can thus gain insight into the complex problem; in some cases, we are interested only in this insight and may not actually generate the simpler solutions.

As an example, consider again the Taylor's series. Suppose that the input signal to a physical system that has a linear model is a complicated function, for which the response is difficult to calculate. However, it may be that we can, with *adequate accuracy*, consider the input function to be the constant term plus the ramp function of the Taylor's series and that we can calculate rather easily the system response for a constant input and for a ramp input. Hence, we have the approximate system response for the complex function. For example, the procedure just described is used in investigating steady-state errors in feedback-control systems [1].

In the foregoing paragraph, the term *adequate accuracy* is used. As engineers, we must never lose sight of the fact that the mathematics that we employ is a means to an end and is not the end itself. Engineers apply mathematical procedures to the analysis and design of physical systems. A physical system cannot be modeled exactly. Hence, even though the mathematical equations that model physical systems can sometimes be solved exactly, these exact results will apply only approximately to a given physical system. If care is used in employing approximations in the mathematics, the results are both useful and accurate in the application to physical systems.

In this chapter, we consider one of the most important procedures in signal and linear time-invariant (LTI) system analysis; this procedure is used to express a complicated periodic signal as a sum of simpler signals. The simpler signals are sinusoids, and the resulting sum is called the *Fourier series*, or the *Fourier expansion*. As we shall see, the requirement of signal periodicity is relaxed in Chapter 5, where the Fourier series is modified to yield the Fourier transform.

4.1   APPROXIMATING PERIODIC FUNCTIONS

In the study of Fourier series, we consider the independent variable of the functions involved to be time; for example, we consider a function $x(t)$, where t represents time. However, all the procedures developed in this chapter apply if the independent variable is other than time.

In this section, we consider two topics. First, periodic functions are investigated in some detail. Next, the approximation of a nonsinusoidal periodic function with a sinusoid is investigated.

Periodic Functions

We define a function $x(t)$ to be *periodic*, with the period T, if the relationship $x(t) = x(t + T)$ is satisfied for all t. For example, the function $\cos \omega t$ is periodic $(\omega = 2\pi f = 2\pi/T)$, because

$$\cos \omega(t + T) = \cos (\omega t + \omega T) = \cos (\omega t + 2\pi) = \cos \omega t.$$

Another example is shown in Figure 4.1, where the function is constructed of connected straight lines. Periodic functions have the following properties (see Section 2.2):

1. Periodic functions are assumed to exist for all time; in the equation $x(t) = x(t + T)$, we do not limit the range of t.

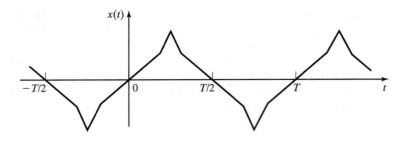

Figure 4.1 Periodic function.

2. A function that is periodic with period T is also periodic with period nT, where n is any integer. Hence for a periodic function,

$$x(t) = x(t + T) = x(t + nT), \qquad (4.2)$$

with n any integer.

3. We define *the fundamental period T_0* as the minimum value of the period $T > 0$ that satisfies $x(t) = x(t + T)$. The *fundamental frequency* is defined as $\omega_0 = 2\pi f_0 = 2\pi/T_0$. For the units of T_0 in seconds, the units of ω_0 are radians per second (rad/s) and of f_0 are hertz (Hz).

The second property is seen for the function of Figure 4.1 and also for $\cos \omega t$, since

$$\cos \omega(t + nT) = \cos(\omega t + n\omega T) = \cos (\omega t + n2\pi) = \cos \omega t.$$

We usually choose the period T to be the fundamental period T_0; however, any value nT_0, with n an integer, satisfies the definition of periodicity.

Approximating Periodic Functions

Consider again the periodic signal of Figure 4.1. Suppose that this function is the input to a stable LTI system and we wish to find the steady-state response. It is quite difficult to find the exact steady-state response for this input signal, since the mathematical description of the signal is a sum of ramp functions. However, we may be able, with *adequate accuracy*, to approximate this signal by a sinusoid, as indicated in Figure 4.2. Then the approximate response will be the sinusoidal steady-state response of the linear system, which is relatively easy to calculate. [See (3.78).]

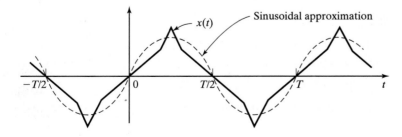

Figure 4.2 Sinusoidal approximation for a periodic function.

Generally, in approximations of the type just discussed, we choose the sinusoidal function that is "best" in some sense. The word *best* is in quotation marks because we are free to define the criterion that determines best. A common procedure for choosing a best approximation is to minimize a function of the difference between the original signal and its approximation, that is, to minimize the error in the approximation. In the approximation of $x(t)$ in Figure 4.2 by the signal $B_1 \sin \omega_0 t$, we define the error of the approximation as

$$e(t) = x(t) - B_1 \sin \omega_0 t \qquad (4.3)$$

and choose B_1 such that some function of this error is minimized.

We denote the function of error to be minimized as $J[e(t)]$. A very common function used in minimization is the *mean-square error*. By definition, minimization of the mean-square error of (4.3) is the minimization of the function

$$J[e(t)] = \frac{1}{T_0} \int_0^{T_0} e^2(t)\, dt = \frac{1}{T_0} \int_0^{T_0} [x(t) - B_1 \sin \omega_0 t]^2\, dt. \qquad (4.4)$$

The minimization process involves choosing the value of B_1 such that $J[e(t)]$ is minimized. The function $J[e(t)]$ is called the *cost function*, or simply the *cost*. The result of its minimization is a value for B_1 such that *no other value* will result in a smaller error function (4.4).

For minimization, we differentiate (4.4), using Leibnitz's rule from Appendix B and set the result to zero:

$$\frac{\partial J[e(t)]}{\partial B_1} = 0 = \frac{1}{T_0} \int_0^{T_0} 2[x(t) - B_1 \sin \omega_0 t](-\sin \omega_0 t)\, dt. \qquad (4.5)$$

This equation is rearranged to yield

$$\int_0^{T_0} x(t) \sin \omega_0 t\, dt = B_1 \int_0^{T_0} \sin^2 \omega_0 t\, dt$$

$$= B_1 \int_0^{T_0} \tfrac{1}{2}(1 - \cos 2\omega_0 t)\, dt = \frac{B_1}{2} \int_0^{T_0} dt = \frac{B_1 T_0}{2}, \qquad (4.6)$$

since the integral of a sinusoid over an integer number of periods is zero. In (4.6), we used the trigonometric identity from Appendix A:

$$\sin^2 \alpha = \tfrac{1}{2}(1 - \cos 2\alpha).$$

From (4.5), the second derivative of the cost function is given by

$$\frac{\partial^2 J[e(t)]}{\partial B_1^2} = \frac{2}{T_0} \int_0^{T_0} \sin^2 \omega_0 t\, dt > 0. \qquad (4.7)$$

The second derivative is positive; thus if (4.6) is satisfied, the mean-square error is minimized (not maximized).

Solving (4.6) for B_1 yields

$$B_1 = \frac{2}{T_0} \int_0^{T_0} x(t) \sin \omega_0 t \, dt. \tag{4.8}$$

This value minimizes the mean-square error. Note that this result is general and is not limited by the functional form of $x(t)$. Given a periodic function $x(t)$ with the period T_0, the best approximation, in a mean-square-error sense, of any periodic function $x(t)$ by the sinusoid $B_1 \sin \omega_0 t$ is to choose B_1 to satisfy (4.8), where $\omega_0 = 2\pi/T_0$. We next consider an example.

EXAMPLE 4.1   **Mean-square minimization**

We now find the best approximation, in a mean-square sense, of the square wave of Figure 4.3(a) by a sine wave. From (4.8),

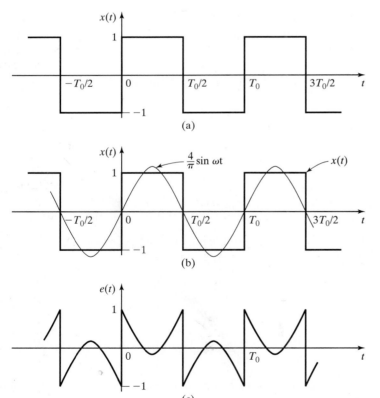

(a)

(b)

(c)

Figure 4.3   Functions for Example 4.1.

$$B_1 = \frac{2}{T_0} \int_0^{T_0} x(t) \sin \omega_0 t \, dt$$

$$= \frac{2}{T_0} \int_0^{T_0/2} (1) \sin \omega_0 t \, dt + \frac{2}{T_0} \int_{T_0/2}^{T_0} (-1) \sin \omega_0 t \, dt,$$

since

$$x(t) = \begin{cases} 1, & 0 < t < T_0/2 \\ -1, & T_0/2 < t < T_0. \end{cases}$$

Therefore,

$$B_1 = \frac{2}{T_0} \left[\frac{-\cos \omega_0 t}{\omega_0} \bigg|_0^{T_0/2} + \frac{\cos \omega_0 t}{\omega_0} \bigg|_{T_0/2}^{T_0} \right]$$

$$= \frac{1}{\pi} (-\cos \pi + \cos 0 + \cos 2\pi - \cos \pi) = \frac{4}{\pi},$$

since

$$\omega_0 t \bigg|_{t=T_0/2} = \left(\frac{2\pi}{T_0} \right) \left(\frac{T_0}{2} \right) = \pi, \quad \omega_0 T_0 = 2\pi.$$

Hence, the best approximation of a square wave with unity amplitude by the function $B_1 \sin \omega_0 t$ is to choose $B_1 = 4/\pi$; we used the minimum mean-square error as the criterion for best. Shown in Figure 4.3(b) is the square wave and the approximating sine wave, and Figure 4.3(c) shows the error in the approximation. This error is given by

$$e(t) = x(t) - \frac{4}{\pi} \sin \omega_0 t = \begin{cases} 1 - \dfrac{4}{\pi} \sin \omega_0 t, & 0 < t < T_0/2 \\ -1 - \dfrac{4}{\pi} \sin \omega_0 t, & T_0/2 < t < T_0 \end{cases} \quad \blacksquare$$

Figure 4.3(c) illustrates very well why we minimize the average squared error rather than the average error. In this figure, we see that the average error is zero, while the approximation is not an especially good one. Furthermore, *any* value of B_1 will give an *average error of zero*, with large values of negative error canceling large values of positive error. However, the squared error is a nonnegative function, and no cancellation can occur.

Other error functions can give reasonable results in the minimization procedure; one example is the average value of the magnitude of $e(t)$:

$$J_m[e(t)] = \frac{1}{T_0} \int_0^{T_0} |e(t)| \, dt.$$

However, we choose to use the mean-square error, which is mathematically tractable and leads us to the Fourier series; the Fourier series is defined in the next section.

4.2 FOURIER SERIES

To introduce the Fourier series, we consider the sum

$$x(t) = 10 + 3\cos \omega_0 t + 5\cos(2\omega_0 t + 30°) + 4 \sin 3\omega_0 t. \tag{4.9}$$

This signal is easily shown to be periodic with period $T_0 = 2\pi/\omega_0$. We now manipulate this signal into a different mathematical form, using Euler's relation from Appendix D:

$$x(t) = 10 + \frac{3}{2}[e^{j\omega_0 t} + e^{-j\omega_0 t}]$$

$$+ \frac{5}{2}[e^{j(2\omega_0 t + 30°)} + e^{-j(2\omega_0 t + 30°)}] + \frac{4}{2j}[e^{j3\omega_0 t} - e^{-j3\omega_0 t}],$$

or

$$x(t) = (2e^{j\pi/2})e^{-j3\omega_0 t} + (2.5e^{-j\pi/6})e^{-j2\omega_0 t} + 1.5e^{-j\omega_0 t}$$

$$+ 10 + 1.5e^{j\omega_0 t} + (2.5e^{j\pi/6})e^{j2\omega_0 t} + (2e^{-j\pi/2})e^{j3\omega_0 t}. \tag{4.10}$$

This equation can be expressed in the compact form

$$x(t) = C_{-3}e^{-j3\omega_0 t} + C_{-2}e^{-j2\omega_0 t} + C_{-1}e^{-j\omega_0 t} + C_0 + C_1 e^{j\omega_0 t} + C_2 e^{j2\omega_0 t} + C_3 e^{j3\omega_0 t}$$

$$= \sum_{k=-3}^{3} C_k e^{jk\omega_0 t}.$$

The coefficients C_k for this series of complex exponential functions are listed in Table 4.1. Note that $C_k = C_{-k}^*$, where the asterisk indicates the complex conjugate.

We see then that a sum of sinusoidal functions can be converted to a sum of complex exponential functions. Note that even though some of the terms are complex, the sum is real. As shown next, (4.10) is one form of the Fourier series.

TABLE 4.1 Coefficients for Example 4.2

k	C_k	C_{-k}
0	10	—
1	1.5	1.5
2	2.5∕30°	2.5∕−30°
3	2∕−90°	2∕90°

Fourier Series

Given a real periodic signal $x(t)$, a *harmonic series* for this signal is defined as

$$x(t) = \sum_{k=-\infty}^{\infty} C_k e^{jk\omega_0 t}; \quad C_k = C_{-k}^*. \tag{4.11}$$

The frequency ω_0 is called the *fundamental frequency* or the *first harmonic*, and the frequency $k\omega_0$ is called the *k*th *harmonic*. If the coefficients C_k and the signal $x(t)$ in (4.11) are related by an equation to be developed later, this harmonic series is a *Fourier series*. For this case, the summation (4.11) is called the *complex exponential form*, or simply the exponential form, of the Fourier series; the coefficients C_k are called the *Fourier coefficients*. Equation (4.10) is an example of a Fourier series in the exponential form. We next derive a second form of the Fourier series.

The general coefficient C_k in (4.11) is complex, as indicated in Table 4.1, with C_{-k} equal to the conjugate of C_k. The coefficient C_k can be expressed as

$$C_k = |C_k| e^{j\theta_k},$$

with $-\infty < k < \infty$. Since $C_{-k} = C_k^*$, it follows that $\theta_{-k} = -\theta_k$. For a given value of k, the sum of the two terms of the same frequency $k\omega_0$ in (4.11) yields

$$C_{-k}e^{-jk\omega_0 t} + C_k e^{jk\omega_0 t} = |C_k|e^{-j\theta_k}e^{-jk\omega_0 t} + |C_k|e^{j\theta_k}e^{jk\omega_0 t}$$

$$= |C_k|[e^{-j(k\omega_0 t+\theta_k)} + e^{j(k\omega_0 t+\theta_k)}]$$

$$= 2|C_k|\cos(k\omega_0 t + \theta_k). \tag{4.12}$$

Hence, given the Fourier coefficients C_k, we can easily find the *combined trigonometric form* of the Fourier series:

$$x(t) = C_0 + \sum_{k=1}^{\infty} 2|C_k|\cos(k\omega_0 t + \theta_k). \tag{4.13}$$

A third form of the Fourier series can be derived from (4.13). From Appendix A, we have the trigonometric identity

$$\cos(a + b) = \cos a \cos b - \sin a \sin b. \tag{4.14}$$

The use of this identity with (4.13) yields

$$x(t) = C_0 + \sum_{k=1}^{\infty} 2|C_k|\cos(k\omega_0 t + \theta_k)$$

$$= C_0 + \sum_{k=1}^{\infty}[2|C_k|\cos\theta_k \cos k\omega_0 t - 2|C_k|\sin\theta_k \sin k\omega_0 t]. \tag{4.15}$$

From Euler's relationship, we define the coefficients A_k and B_k implicity via the formula

$$2C_k = 2|C_k|e^{j\theta_k}$$

$$= 2|C_k|\cos\theta_k + j2|C_k|\sin\theta_k = A_k - jB_k, \qquad (4.16)$$

where A_k and B_k are real. Substituting (4.16) into (4.15) yields the *trigonometric form* of the Fourier series

$$x(t) = A_0 + \sum_{k=1}^{\infty}[A_k \cos k\omega_0 t + B_k \sin k\omega_0 t], \qquad (4.17)$$

with $A_0 = C_0$. The original work of Joseph Fourier (1768–1830) involved the series in this form.

The three forms of the Fourier series [(4.11), (4.13), (4.17)] are listed in Table 4.2. Also given is the equation for calculating the coefficients; this equation is developed later. From (4.16), the coefficients of the three forms are related by

$$2C_k = A_k - jB_k; \quad C_k = |C_k|e^{j\theta_k}; \quad C_0 = A_0. \qquad (4.18)$$

Recall that A_k and B_k are real and, in general, C_k is complex.

Fourier Coefficients

Next, the calculation of the Fourier coefficients C_k, given $x(t)$, is considered. Several approaches may be taken to deriving the equation for C_k. We take the approach of assuming that the complex-exponential form of the Fourier series is valid, that is, that the coefficients C_k can be found that satisfy the equation

TABLE 4.2 Forms of the Fourier Series

Name	Equation
Exponential	$\displaystyle\sum_{k=-\infty}^{\infty} C_k e^{jk\omega_0 t}; \quad C_k = |C_k|e^{j\theta_k}, C_{-k} = C_k^*$
Combined trigonometric	$\displaystyle C_0 + \sum_{k=1}^{\infty} 2|C_k|\cos(k\omega_0 t + \theta_k)$
Trigonometric	$\displaystyle A_0 + \sum_{k=1}^{\infty}(A_k\cos k\omega_0 t + B_k\sin k\omega_0 t)$
	$2C_k = A_k - jB_k, C_0 = A_0$
Coefficients	$\displaystyle C_k = \frac{1}{T_0}\int_{T_0} x(t)e^{-jk\omega_0 t}dt$

[eq(4.11)]
$$x(t) = \sum_{k=-\infty}^{\infty} C_k e^{jk\omega_0 t},$$

where $x(t)$ is a periodic function with the fundamental frequency ω_0. We consider the convergence of the right side of (4.11) in the next section. First each side of equation (4.11) is multiplied by $e^{-jn\omega_0 t}$, with n an integer, and then integrated from $t = 0$ to $t = T_0$:

$$\int_0^{T_0} x(t)e^{-jn\omega_0 t}\,dt = \int_0^{T_0}\left[\sum_{k=-\infty}^{\infty} C_k e^{jk\omega_0 t}\right]e^{-jn\omega_0 t}\,dt.$$

Interchanging the order of summation and integration on the right side yields the equation

$$\int_0^{T_0} x(t)e^{-jn\omega_0 t}\,dt = \sum_{k=-\infty}^{\infty} C_k\left[\int_0^{T_0} e^{j(k-n)\omega_0 t}\,dt\right]. \tag{4.19}$$

The general term in the summation on the right side can be expressed as, using Euler's relation,

$$C_k\int_0^{T_0} e^{j(k-n)\omega_0 t}\,dt = C_k\int_0^{T_0}\cos(k-n)\omega_0 t\,dt + jC_k\int_0^{T_0}\sin(k-n)\omega_0 t\,dt. \tag{4.20}$$

The second term on the right side of this equation is zero, since the sine function is integrated over an integer number of periods. The same is true for the first term on the right side, except for $k = n$. For this case,

$$C_k\int_0^{T_0}\cos(k-n)\omega_0 t\,dt\bigg|_{k=n} = C_n\int_0^{T_0} dt = C_n T_0. \tag{4.21}$$

Hence, the right side of (4.19) is equal to $C_n T_0$, and (4.19) can be expressed as

$$\int_0^{T_0} x(t)e^{-jn\omega_0 t}\,dt = C_n T_0.$$

This equation is solved for C_n:

$$C_n = \frac{1}{T_0}\int_0^{T_0} x(t)e^{-jn\omega_0 t}\,dt. \tag{4.22}$$

This is the desired relation between $x(t)$ and the Fourier coefficients C_n. It can be shown that this equation minimizes the mean-square error defined in Section 4.1 [2].

The equation for the Fourier coefficients is simple, because, in (4.19),

$$C_k \int_0^{T_0} e^{jk\omega_0 t} e^{-jn\omega_0 t}\, dt = 0, \quad k \neq n.$$

These two complex-exponential functions are said to be *orthogonal* over the interval $(0, T_0)$. In general, two functions $f(t)$ and $g(t)$ are said to be *orthogonal* over the interval (a, b) if

$$\int_a^b f(t)g(t)\, dt = 0.$$

Orthogonal functions other than complex exponentials exist, and these functions may be used in a manner similar to that in (4.11) and (4.22) to express a general function as a series [2]. We do not consider this general topic further.

Equation (4.22) is the desired result and gives the coefficients of the exponential form of the Fourier series as a function of the periodic signal $x(t)$. The coefficients of the two trigonometric forms of the Fourier series are given in Table 4.2. Because the integrand in (4.22) is periodic with the fundamental period T_0, the limits on the integral can be generalized to t_1 and $t_1 + T_0$, where t_1 is arbitrary. We express this by the notation of writing T_0 at the lower limit position of the integral and leaving the upper limit position blank:

$$C_k = \frac{1}{T_0} \int_{T_0} x(t) e^{-jk\omega_0 t}\, dt. \tag{4.23}$$

We now consider the coefficient C_0. From (4.23),

$$C_0 = \frac{1}{T_0} \int_{T_0} x(t)\, dt.$$

Hence, C_0 is the *average value* of the signal $x(t)$. This average value is also called the *dc value*, a term that originated in circuit analysis. For some waveforms, the dc value can be found by inspection.

The exponential form and the combined trigonometric form of the Fourier series are probably the most useful forms. The coefficients of the exponential form are the most convenient to calculate, while the amplitudes of the harmonics are directly available in the combined trigonometric form. We will usually calculate C_k from (4.23); if the amplitudes of the harmonics are required, these amplitudes are equal to $2|C_k|$, except that the dc amplitude is C_0.

In this section, we defined the Fourier series for a periodic function, and derived the equation for calculating the Fourier coefficients. Table 4.2 gives the three forms for the Fourier series. Examples of calculating the Fourier coefficients are given in the next section.

4.3   FOURIER SERIES AND FREQUENCY SPECTRA

In this section, we present three examples of Fourier series. These examples lead us to the important concept of the *frequency spectra* of periodic signals. Then a table of Fourier series for some common signals is given.

EXAMPLE 4.2   **Fourier series of a square wave**

Consider the square wave of Figure 4.4. This signal is common in physical systems. For example, this signal appears in many electronic oscillators, as an intermediate step in the generation of a sinusoid.

We now calculate the Fourier coefficients of the square wave. Because

$$x(t) = \begin{cases} V, & 0 < t < T_0/2 \\ -V, & T_0/2 < t < T_0 \end{cases},$$

from (4.23), it follows that

$$C_k = \frac{1}{T_0} \int_{T_0} x(t) e^{-jk\omega_0 t} \, dt$$

$$= \frac{V}{T_0} \int_0^{T_0/2} e^{-jk\omega_0 t} \, dt - \frac{V}{T_0} \int_{T_0/2}^{T_0} e^{-jk\omega_0 t} \, dt$$

$$= \frac{V}{T_0(-jk\omega_0)} \left[e^{-jk\omega_0 t} \Big|_0^{T_0/2} - e^{-jk\omega_0 t} \Big|_{T_0/2}^{T_0} \right].$$

The values at the limits are evaluated as

$$\omega_0 t \Big|_{t=T_0/2} = \frac{2\pi}{T_0} \frac{T_0}{2} = \pi; \; \omega_0 T_0 = 2\pi.$$

Therefore,

$$C_k = \frac{jV}{2\pi k} (e^{-jk\pi} - e^{-j0} - e^{-jk2\pi} + e^{-jk\pi})$$

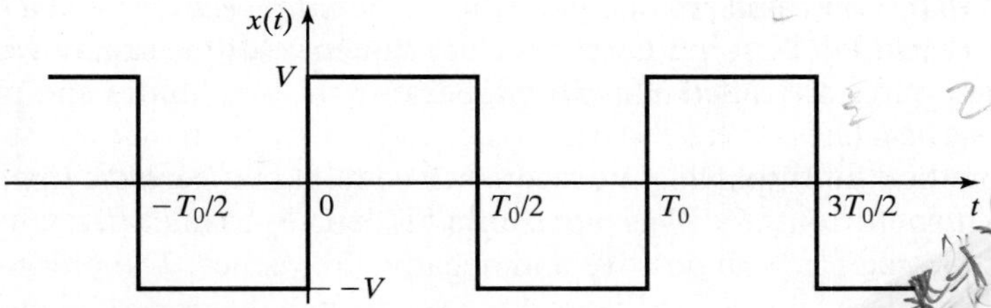

Figure 4.4 Square wave with amplitude V.

$$= \begin{cases} -\dfrac{2jV}{k\pi} = \dfrac{2V}{k\pi}\angle{-90°}, & k \text{ odd} \\ 0, & k \text{ even} \end{cases}, \tag{4.24}$$

with $C_0 = 0$. The value of C_0 is seen by inspection, since the square wave has an average value of zero. Also, C_0 can be calculated from (4.24) using L'Hôpital's rule, Appendix B.

The exponential form of the Fourier series of the square wave is then

$$x(t) = \sum_{\substack{k=-\infty \\ k \text{ odd}}}^{\infty} \frac{2V}{k\pi} e^{-j\pi/2} e^{jk\omega_0 t}. \tag{4.25}$$

The combined trigonometric form is given by

$$x(t) = \sum_{\substack{k=1 \\ k \text{ odd}}}^{\infty} \frac{4V}{k\pi} \cos(k\omega_0 t - 90°) \tag{4.26}$$

from (4.13). Hence, the first harmonic has an amplitude of $4V/\pi$, the third harmonic $4V/3\pi$, the fifth harmonic $4V/5\pi$, and so on. The calculation of C_1 is verified with the MATLAB program

```
syms C1 ker t
w0=2*pi; k=1;
ker=exp (-j*k*w0*t);
C1=int (ker, 0, 0.5) +int (-ker, 0.5, 1)
double (C1)
```

Equation (4.26) is easily converted to the trigonometric form, because $\cos(a - 90°) = \sin a$. Hence,

$$x(t) = \sum_{\substack{k=1 \\ k \text{ odd}}}^{\infty} \frac{4V}{k\pi} \sin k\omega_0 t. \qquad\blacksquare$$

Frequency Spectra

For the square wave, the amplitude of the harmonics decrease by the factor $1/k$, where k is the number of the harmonic. For a graphical display of the harmonic content of a periodic signal, we plot a *frequency spectrum* of the signal. A frequency spectrum is generally a graph that shows, in some manner, the amplitudes $(2|C_k|)$ and the phases $(\arg C_k)$ of the harmonic terms of a periodic signal. Given in Figure 4.5 is a frequency spectrum that shows a magnitude plot (the *magnitude spectrum*) and a phase plot (the *phase spectrum*) of $2C_k = 2|C_k|e^{j\theta_k}$ versus frequency for the square wave of Example 4.2. These plots are called *line spectra*, because the amplitudes and phases are indicated by vertical lines.

A second method for displaying the frequency content of a periodic signal is a plot of the Fourier coefficients C_k. This plot shows $|C_k|$ and θ_k as line spectra versus frequency and is plotted for both positive and negative frequency. The plot for the square wave of Example 4.2 is given in Figure 4.6. Even though the plots of Figures 4.5

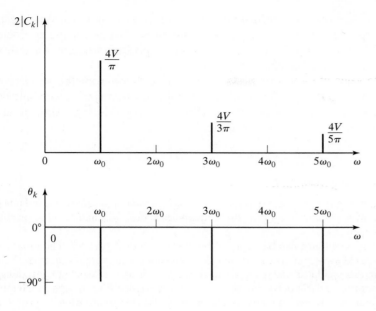

Figure 4.5   Frequency spectrum for a square wave.

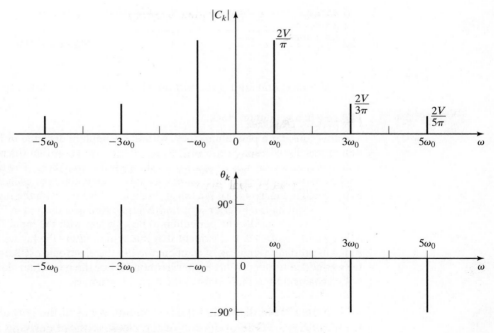

Figure 4.6   Frequency spectrum for a square wave.

and 4.6 are different, the same information is given. If harmonics for both positive frequency and negative frequency are shown, the plots must be for the coefficients C_k of the complex exponentials. If harmonics for only positive frequency are shown, the plots are of the coefficients $2C_k$ of the sinusoids.

The usefulness of the frequency spectrum is evident from the square-wave spectrum of Figure 4.5. It was mentioned in Example 4.2 that some electronic oscillators generate a square wave as an intermediate step to producing a sinusoidal signal. We now discuss this case as an example.

EXAMPLE 4.3   **Filtering for an electronic oscillator**

From Figure 4.5, we see that a sinusoid of frequency ω_0 is present in a square wave. Hence, in the oscillator we must remove (filter out) the frequencies at $3\omega_0$, $5\omega_0$, and so on to produce a sinusoid of frequency of ω_0.

The electronic oscillator can be depicted by the system of Figure 4.7. The input signal to the filter is a square wave, and its output signal is a sinusoid of the same frequency. The engineers designing the oscillator of Figure 4.7 have the two design tasks of (1) designing the square-wave generator and (2) designing the filter. We introduce filter design in Chapter 6. The system of Figure 4.7 is used in many oscillators, because the square wave is easy to generate and the higher frequencies are not difficult to filter out. A filter of the type required is analyzed mathematically in Section 4.5.

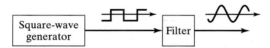

Figure 4.7   Electronic oscillator.   ■

We next consider a second physical system that illustrates frequency spectra.

EXAMPLE 4.4   **Filtering in a pendulum clock**

We now consider a pendulum clock, with the pendulum depicted in Figure 4.8(a). The pendulum was discussed in Section 1.2. From physics, we know that the motion of a simple pendulum $\theta(t)$ approximates a sinusoid, as shown in Figure 4.8(b). The mainspring of the clock applies periodic pulses of force at one of the extreme points of each swing of the pendulum, as indicated in Figure 4.8(a). We can approximate this force with the signal $f(t)$ of Figure 4.8(c). This force will have a Fourier series, with a spectrum as indicated in Figure 4.8(d).

We now consider the pendulum to be a system with the input $f(t)$ and the output $\theta(t)$, as shown in Figure 4.9. It is evident that this system filters the higher harmonics of the force signal to produce a sinusoid of frequency ω_0. In the electronic oscillator of Example 4.3, a filter was added to remove the higher harmonics. For the pendulum clock, the pendulum itself is a mechanical filter that removes the higher harmonics.   ■

Note that in discussing the pendulum, we used the *concept* of frequency spectra to *understand* the operation of the clock, without deriving a model or assigning numbers to the system. We can see the importance of concepts in engineering. Of

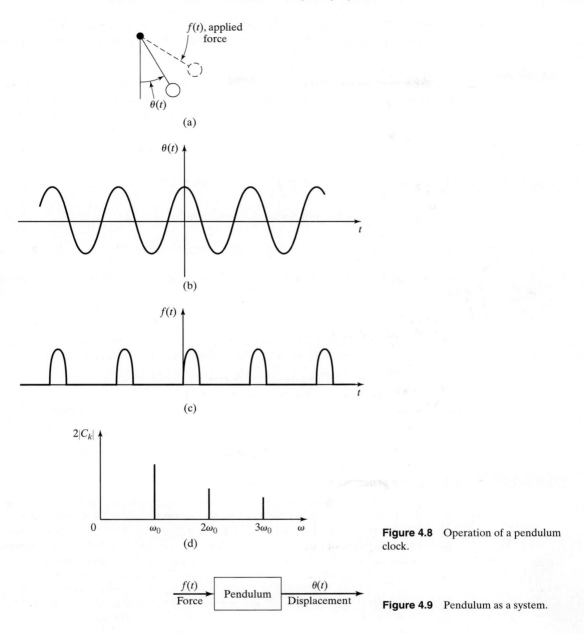

Figure 4.8 Operation of a pendulum clock.

Figure 4.9 Pendulum as a system.

course, for design we *must* have mathematical models and numerical values for parameters.

In the analysis of a system with a periodic input of the type shown in Figure 4.8(c), we may be able to approximate the pulses as impulse functions, with the weight of each impulse function equal to the area under a pulse. As a second example of calculating Fourier coefficients, we consider a train of impulse functions.

EXAMPLE 4.5   **Fourier series for an impulse train**

The Fourier series for the impulse train shown in Figure 4.10 will be calculated. From (4.23),

$$C_k = \frac{1}{T_0} \int_{T_0} x(t) e^{-jk\omega_0 t}\, dt$$

$$= \frac{1}{T_0} \int_{-T_0/2}^{T_0/2} \delta(t) e^{-jk\omega_0 t}\, dt = \frac{1}{T_0} e^{-jk\omega_0 t} \Big|_{t=0} = \frac{1}{T_0}.$$

$\delta(t) = k\omega_0$

This result is based on the property of the impulse function

[eq(2.41)] $\int_{-\infty}^{\infty} f(t)\delta(t - t_0)\, dt = f(t_0),$

provided that $f(t)$ is continuous at $t = t_0$. The exponential form of the Fourier series is given by

$$x(t) = \sum_{k=-\infty}^{\infty} \frac{1}{T_0} e^{jk\omega_0 t}. \tag{4.27}$$

A line spectrum for this function is given in Figure 4.11. Because the Fourier coefficients are real, no phase plot is given. From (4.13), the combined trigonometric form for the train of impulse functions is given by

$$x(t) = \frac{1}{T_0} + \sum_{k=1}^{\infty} \frac{2}{T_0} \cos k\omega_0 t.$$

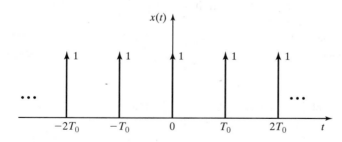

Figure 4.10   Impulse train.

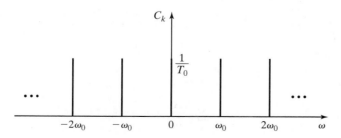

Figure 4.11   Frequency spectrum for an impulse train.

Note that this is also the trigonometric form.

A comparison of the frequency spectrum of the square wave (Figure 4.6) with that of the train of impulse functions (Figure 4.11) illustrates an important property of impulse functions. For the square wave, the amplitude of the harmonics decrease by the factor $1/k$, where k is the harmonic number. Hence, we expect that the higher harmonics can be ignored in most situations and that a finite sum of the harmonics is usually adequate to represent a square wave. This statement cannot be applied to the impulse train. The amplitudes of the harmonics remain constant for all harmonic frequencies. Hence, usually all harmonics must be considered for a train of impulse functions. This point is considered further in the next section.

Given in Table 4.3 are the Fourier coefficients of seven periodic signals that are important in engineering applications. Since the coefficient C_0 is the average, or dc, value of the signal, this value is not unique for a particular form of a periodic signal. For example, if we add a constant value to a sawtooth signal, the result is still a sawtooth signal, with only the average value C_0 changed. This point is covered in greater detail in Section 4.6.

A MATLAB program that verifies the first three coefficients of the triangular wave in Table 4.3 is

```
syms Ck ker t
for k=1 : 3
    w0=2*pi;
    ker=exp (-j*k*w0*t);
    Ck=int (2*t*ker, 0, 0.5) + int (2* (1 - t) * ker, 0.5, 1);
    simplify (Ck)
end
```

This program can also be written in the general variable k, but the results must be simplified. The coefficients of the remaining signals in Table 4.3 can be derived by altering this program in an appropriate manner.

As a final example in this section, we consider the important case of a train of rectangular pulses.

EXAMPLE 4.6   **Frequency spectrum of a rectangular pulse train**

For this example, the frequency spectrum of the rectangular pulse train of Figure 4.12 will be plotted. This waveform is common in engineering. The clock signal in a digital computer is a rectangular pulse train of this form. Also, in communications, one method of modulation is to vary the amplitudes of the rectangular pulses in a pulse train according to the information to be transmitted. This method of modulation, called pulse-amplitude modulation, is described in Sections 1.3 and 6.6.

From Table 4.3, the Fourier series for this signal is given by

$$x(t) = \sum_{k=-\infty}^{\infty} \frac{TX_0}{T_0} \text{sinc} \frac{Tk\omega_0}{2} e^{jk\omega_0 t}, \tag{4.28}$$

TABLE 4.3 Fourier Series for Common Signals

Name	Waveform	C_0	$C_k, k \neq 0$	Comments
1. Square wave		0	$-j\dfrac{2X_0}{\pi k}$	$C_k = 0,$ k even
2. Sawtooth		$\dfrac{X_0}{2}$	$j\dfrac{X_0}{2\pi k}$	
3. Triangular wave		$\dfrac{X_0}{2}$	$\dfrac{-2X_0}{(\pi k)^2}$	$C_k = 0,$ k even
4. Full-wave rectified		$\dfrac{2X_0}{\pi}$	$\dfrac{-2X_0}{\pi(4k^2 - 1)}$	
5. Half-wave rectified		$\dfrac{X_0}{\pi}$	$\dfrac{-X_0}{\pi(k^2 - 1)}$	$C_k = 0,$ k odd, except $C_1 = -j\dfrac{X_0}{4}$ and $C_{-1} = j\dfrac{X_0}{4}$
6. Rectangular wave		$\dfrac{TX_0}{T_0}$	$\dfrac{TX_0}{T_0}$ sinc $\dfrac{Tk\omega_0}{2}$	$\dfrac{Tk\omega_0}{2} = \dfrac{\pi Tk}{T_0}$
7. Impulse train		$\dfrac{X_0}{T_0}$	$\dfrac{X_0}{T_0}$	

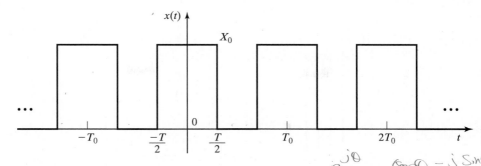

Figure 4.12   Rectangular pulse train.

where T is the width of the rectangular pulses and sinc x is defined as

$$\text{sinc } x = \frac{\sin x}{x}.$$

(4.29)

The coefficients of the combined trigonometric form of the Fourier series are given by

$$2|C_k| = \frac{2T X_0}{T_0}\left|\text{sinc }\frac{Tk\omega_0}{2}\right|, \quad \theta_k = \begin{cases} 0, & \text{sinc } Tk\omega_0/2 > 0 \\ 180°, & \text{sinc } Tk\omega_0/2 < 0 \end{cases}$$

We obtain the *envelope* of the magnitude characteristic by replacing $k\omega_0$ with ω:

$$\text{envelope} = \frac{2T X_0}{T_0}\left|\text{sinc }\frac{T\omega}{2}\right|.$$

Note that, by L'Hôpital's rule, $\lim\limits_{x\to o}\text{sinc } x = 1$. The first zero of the envelope occurs at

$$\sin T\omega/2 = \sin \pi$$

or at $\omega = 2\pi/T$. This value of ω is not necessarily a harmonic frequency. The frequency spectrum for this pulse train is plotted in Figure 4.13. ∎

The function sinc x appears often in signal and system analysis; we now investigate this function further. The definition

$$\text{sinc } x = \frac{\sin x}{x}$$

(4.30)

is not standard; however, mathematicians normally use this definition. [The definition sinc $x = \sin(\pi x)/\pi x$ is also used in engineering.] As stated, the sinc x function has a value of unity at $x = 0$. For $x \neq 0$, sinc x has zeros at the points at which $\sin x$ is zero, that is, at $x = \pm\pi, \pm 2\pi, \dots$. The magnitude and the angle characteristics for sinc x are plotted in Figure 4.14.

In Figure 4.14, the angle of C_k is zero for sinc x positive and 180° (or −180°) for sinc x negative. Table 4.4 gives the maximum values of the magnitude characteristic in each half-cycle, for $|x| < 5\pi$. For x large, since the maximum magnitude of

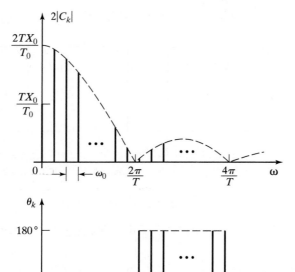

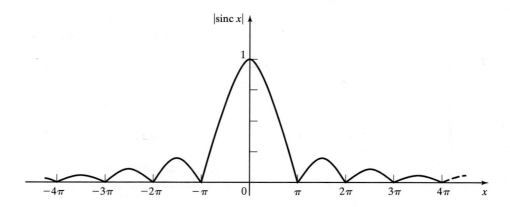

Figure 4.13 Spectrum for a rectangular pulse train.

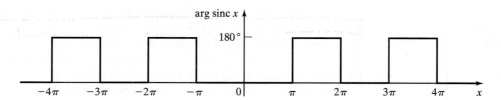

Figure 4.14 Plot of sinc x.

TABLE 4.4 Peak Values of Sinc x

x	$\lvert\text{sinc x}\rvert$	$1/x$
0.0	1.0	—
4.493	0.2172	0.2226
7.725	0.1284	0.1294
10.90	0.0913	0.0917
14.07	0.0709	0.0711

sin x is unity, the maximum magnitude of sinc x is approximately $1/x$. For example, in the vicinity of $x = 25$, the maximum magnitude of sinc x is approximately $1/25 = 0.04$. Values of $1/x$ are also given in Table 4.4.

In this section, we introduce one of the basic concepts of signals and systems engineering, that of frequency spectra. Many analysis and design procedures for signals and systems are based on this concept. We expand on this concept in the remainder of this chapter and in Chapter 5 extend this concept to aperiodic signals.

4.4 PROPERTIES OF FOURIER SERIES

In this section, some properties of the Fourier series are stated. These properties are then discussed, using examples for illustration.

Any single-valued periodic function $x(t)$ that satisfies the *Dirichlet* conditions can be expanded into a Fourier series. The Dirichlet conditions are [2]

1. $x(t)$ has at most a finite number of discontinuities in one period;
2. $x(t)$ has at most a finite number of maxima and minima in one period;
3. $x(t)$ is bounded.

The third condition has been expanded to include singularity functions and may be stated as [3]

3a. $\displaystyle\int_{T_0} \lvert x(t)\rvert \, dt < \infty.$

Any function of time that appears in physical systems will satisfy these conditions.

Several properties of the Fourier series will now be given. The readers interested in the proofs of these properties should see Refs. 2, 4, and 5. For $x(t)$ satisfying the Dirichlet conditions 1, 2, and 3, the following are true:

1. The Fourier series converges to the value of $x(t)$ at every point of continuity where $x(t)$ has a right-hand and a left-hand derivative, whether these derivatives are the same or different. The right-hand derivative of $x(t)$ at $t = t_a$ is defined as the derivative as t approaches t_a from the right. The left-hand derivative is the derivative as t approaches t_a from the left.

2. If $x(t)$ has a discontinuity at a point, the Fourier series converges to the mean of the limits approached by $x(t)$ from the right and from the left; that is, at every point t_a,

$$\sum_{k=-\infty}^{\infty} C_k e^{jk\omega_0 t_a} = \frac{x(t_a^-) + x(t_a^+)}{2}, \tag{4.31}$$

where $x(t_a^-)$ is the limiting value of $x(t)$ as t approaches t_a from the left and $x(t_a^+)$ is the limiting value from the right. Note that (4.31) is also satisfied for t_a a point of continuity.

3. Almost any *continuous function* $x(t)$ of period T_0 can be uniformly approximated by a truncated Fourier series with any preassigned degree of accuracy, where the series is given by

$$x_N(t) = \sum_{k=-N}^{N} C_k e^{jk\omega_0 t} = C_0 + \sum_{k=1}^{N} 2|C_k|\cos(k\omega_0 t + \theta_k). \tag{4.32}$$

This property applies to any continuous periodic function that we might encounter in the practice of engineering. We define the error of the approximation by the truncated series of (4.32) as

$$e(t) = x(t) - x_N(t). \tag{4.33}$$

This property states that the magnitude of the error, $|e(t)|$, may be bounded by any nonzero value by choosing N sufficiently large.

4. Consider further the error of approximation in (4.33), with the coefficients given by (4.23). We minimize the mean-square error, defined as

$$\text{mean-square error} = \frac{1}{T_0}\int_{T_0} e^2(t)\, dt. \tag{4.34}$$

That is, *no other choice* of coefficients in the harmonic series (4.32) will produce a smaller mean-square error in (4.34). This property was discussed in Section 4.2.

5. A sum of trigonometric functions of $\omega_0 t$ that is periodic is its own Fourier series.

6. The Fourier coefficient of the kth harmonic for $x(t)$ always decreases in magnitude at least as fast as $1/k$, for sufficiently large k. If $x(t)$ has one or more discontinuities in a period, the coefficients can decrease no faster than this. If the nth derivative of $x(t)$ is the first derivative that contains a discontinuity and if all derivatives through the nth satisfy the Dirichlet conditions, the Fourier coefficients approach zero as $1/k^{n+1}$, for sufficiently large k.

7. The Fourier series of a sum of periodic functions is equal to the sum of the Fourier series for the functions. Of course, the sum of the periodic functions must be periodic; if not, the sum does not have a Fourier series.

The first property is illustrated by the triangular wave of Table 4.3. The Fourier series converges to $x(t)$ for every value of t, although at two points each cycle, the left-hand and right-hand derivatives are not equal (at $t = 0$ and at $t = T_0/2$, for example).

The second property is illustrated by the square wave of Example 4.2 and Table 4.3. In Example 4.2, the Fourier series was calculated to be

$$x(t) = \sum_{\substack{k=1 \\ k \text{ odd}}}^{\infty} \frac{4V}{k\pi} \cos(k\omega_0 t - 90°) = \sum_{\substack{k=1 \\ k \text{ odd}}}^{\infty} \frac{4V}{k\pi} \sin k\omega_0 t. \tag{4.35}$$

At $t = nT_0/2$ with n an integer, $\sin k\omega_0 t = \sin kn\pi = 0$ for all k. Thus, the Fourier series is equal to zero at $t = nT_0/2$, which is the average value of the discontinuities at these points.

The third property is important, since it states that almost any *continuous periodic function* can be approximated by a truncated Fourier series to any degree of accuracy.

Figure 4.15 illustrates the errors for the square wave of (4.35). Shown is a positive half-cycle of the square wave and (a) the first harmonic, (b) the sum of the first and third harmonics, and (c) the sum through the ninth harmonic. The reduction in error by adding higher harmonics is evident.

Figure 4.15 also illustrates the *Gibbs phenomenon*. The ripples in the waveform of the series become narrower as the number of terms used becomes larger. However, the amplitudes of the ripple nearest each discontinuity do not approach zero, but instead, approach approximately 9% of the height of the discontinuity [2].

We now illustrate the fifth property with an example. The function

$$x(t) = \cos 2t + 3\cos 4t$$

is periodic and hence has a Fourier series. Property 5 states that this sum of harmonic sinusoids is its own Fourier series. If we calculate the Fourier coefficients for $x(t)$ using (4.23), we find that $C_1 = C_{-1} = 0.5$, $C_2 = C_{-2} = 1.5$, and $C_k = 0$ for all other k. (See Problem 4.1.)

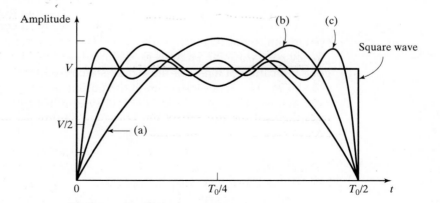

Figure 4.15   Truncated sums for a half-cycle of a square wave.

Property 6 is illustrated by the square wave of Table 4.3 and Example 4.2, where the Fourier coefficients were calculated to be

$$
C_k = \begin{cases} \dfrac{2V}{j\pi k}, & k \text{ odd} \\ 0, & k \text{ even} \end{cases}.
$$

The square wave has discontinuities, and the harmonics approach zero as $1/k$, as k approaches infinity.

However, the Fourier coefficients for the train of impulse functions of Example 4.5 and Table 4.3 do not satisfy Property 6. These coefficients were calculated in Example 4.5 to be $C_k = 1/T_0$ for all k and do not approach zero as k approaches infinity. This is not surprising, since an impulse function is not bounded. Hence, the Dirichlet Condition 3 is violated, and Property 6 does not apply for this periodic function.

In this section, seven properties of Fourier series are given, without proof. These properties are useful in the application of Fourier series in both analysis and design.

4.5 SYSTEM ANALYSIS

In this section, we consider the analysis of stable LTI systems with periodic inputs. It is assumed that the system inputs are periodic for all time and can be represented by Fourier series. Since the systems are stable, the natural responses can be ignored; only the *steady-state responses* are determined.

The system linearity allows the use of superposition. Since a periodic input signal can be represented as a sum of complex exponential functions, the system response can be represented as the sum of steady-state responses to these complex exponential functions. We can also represent the periodic input signal as a sum of sinusoidal functions; the system response is then a sum of steady-state sinusoidal responses. In either case, it will be shown that we must consider the variation of the system sinusoidal response with frequency. This variation is called the *system frequency response*. The analysis procedure developed in this section does not give a good indication of a plot of the steady-state response as a function of time; instead, it yields the frequency spectrum of the output signal.

We begin by considering the LTI system of Figure 4.16 and use the standard notation for systems:

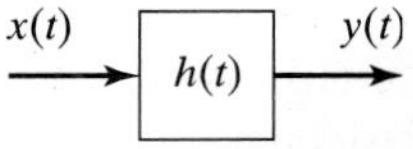

Figure 4.16 LTI system.

TABLE 4.5 Input–Output Functions for an LTI System

$$H(s) = \int_{-\infty}^{\infty} h(t)e^{-st}\, dt$$

$$Xe^{s_1 t} \rightarrow XH(s_1)e^{s_1 t}; \quad X = |X|e^{j\phi}$$

$$|X|\cos(\omega_1 t + \phi) \rightarrow |X||H(j\omega_1)|\cos[\omega_1 t + \phi + \underline{/H(j\omega_1)}]$$

$$x(t) \rightarrow y(t). \tag{4.36}$$

Table 3.1, Section 3.7, gives the steady-state input–output functions for a complex-exponential input or a sinusoidal input to an LTI system with the transfer function $H(s)$; we repeat these functions in Table 4.5.

From Table 4.5, for a complex-exponential input,

$$Xe^{s_1 t} \rightarrow XH(s_1)e^{s_1 t}, \tag{4.37}$$

where both s_1 and X may be complex. For a periodic input $x(t)$, we can represent $x(t)$ by its Fourier series in the exponential form. From (4.37), with $s_1 = jk\omega_0$, the system representation of (4.36) becomes

$$x(t) = \sum_{k=-\infty}^{\infty} C_{kx}e^{jk\omega_0 t} \rightarrow y_{ss}(t) = \sum_{k=-\infty}^{\infty} H(jk\omega_0)C_{kx}e^{jk\omega_0 t}, \tag{4.38}$$

where $y_{ss}(t)$ is the steady-state output signal. In general, both C_{kx} and $H(jk\omega_0)$ are complex. Thus, the periodic output can be represented as a Fourier series with the coefficients

$$y_{ss}(t) = \sum_{k=-\infty}^{\infty} C_{ky}e^{jk\omega_0 t}, \quad C_{ky} = H(jk\omega_0)C_{kx}. \tag{4.39}$$

This equation gives the Fourier coefficients of the output signal $y(t)$ as a function of the Fourier coefficients of the input signal $x(t)$ and the system transfer function $H(j\omega)$. $H(j\omega)$ is also called the *system frequency response*.

From (4.13), the combined trigonometric form of the Fourier series for the input signal is

$$x(t) = C_{0x} + \sum_{k=1}^{\infty} 2|C_{kx}|\cos(k\omega_0 t + \theta_{kx}). \tag{4.40}$$

From Table 4.5, the steady-state sinusoidal response for an LTI system is expressed as

$$|X|\cos(\omega_1 t + \phi) \rightarrow |X||H(j\omega_1)|\cos[\omega_1 t + \phi + \underline{/H(j\omega_1)}]. \tag{4.41}$$

Hence, by superposition, the steady-state system output for the periodic input signal (4.40) is given by

$$y_{ss}(t) = C_{0y} + \sum_{k=1}^{\infty} 2|C_{ky}|\cos(k\omega_0 t + \theta_{ky}), \tag{4.42}$$

where, from (4.40) and (4.41),

$$C_{ky} = |C_{ky}|\underline{/\theta_{ky}} = H(jk\omega_0)C_{kx}. \tag{4.43}$$

This equation is identical to the one given in (4.39), since both equations give C_{ky}. An example is now given to illustrate these relationships.

EXAMPLE 4.7   **LTI system response for a square-wave input**

Suppose that for the LTI system of Figure 4.16, the impulse response and the transfer function are given by

$$h(t) = e^{-t}u(t) \Leftrightarrow H(s) = \frac{1}{s+1}.$$

For example, the interested reader can show that the circuit of Figure 4.17 has this transfer function, with $x(t) = v_i(t)$ and $y(t) = v_o(t)$. Suppose that the input signal $x(t)$ is the square wave of Figure 4.18, where t is in seconds. Since the fundamental period is $T_0 = 2\pi$, the fundamental frequency is $\omega_0 = 2\pi/T_0 = 1$ rad/s and $k\omega_0 = k$. From Table 4.3, the Fourier series of $x(t)$ is given by

$$x(t) = C_{0x} + \sum_{\substack{k=-\infty \\ k\neq 0}}^{\infty} C_{kx}e^{jk\omega_0 t} = 2 + \sum_{\substack{k=-\infty \\ k \text{ odd}}}^{\infty} \frac{4}{\pi k} e^{-j\pi/2}e^{jkt}.$$

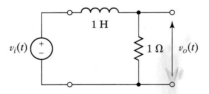

1 H

$v_i(t)$ 1 Ω $v_o(t)$

Figure 4.17   *RL* circuit.

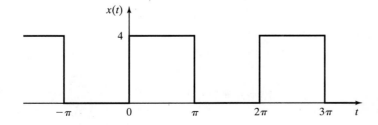

Figure 4.18   Input signal for Example 4.7.

TABLE 4.6 Fourier Coefficients for Example 4.7

k	$H(jk\omega_0)$	C_{kx}	C_{ky}	$|C_{kx}|$	$|C_{ky}|$
0	1	2	2	2	2
1	$\dfrac{1}{\sqrt{2}}\angle-45°$	$\dfrac{4}{\pi}\angle-90°$	$\dfrac{4}{\pi\sqrt{2}}\angle-135°$	1.273	0.900
3	$\dfrac{1}{\sqrt{10}}\angle-71.6°$	$\dfrac{4}{3\pi}\angle-90°$	$\dfrac{4}{3\pi\sqrt{10}}\angle-161.6°$	0.424	0.134
5	$\dfrac{1}{\sqrt{26}}\angle-78.7°$	$\dfrac{4}{5\pi}\angle-90°$	$\dfrac{4}{5\pi\sqrt{26}}\angle-168.7°$	0.255	0.050

Now,

$$H(jk\omega_0)|_{\omega_0=1} = \frac{1}{1+jk} = \frac{1}{\sqrt{1+k^2}}\angle\tan^{-1}(-k).$$

For k odd, from (4.39),

$$C_{ky} = H(jk\omega_0)C_{kx} = \frac{1}{\sqrt{1+k^2}}\left[\frac{4}{\pi k}\right]\angle-\pi/2 - \tan^{-1}(k)$$

for $k \neq 0$, and

$$C_{0y} = H(j0)C_{0x} = (1)(2) = 2.$$

Table 4.6 gives the first four nonzero Fourier coefficients of the output $y(t)$. The Fourier coefficients for the output signal decrease in magnitude as $1/k^2$ for large k, since for this case both $|C_{kx}|$ and $|H(jk\omega_0)|$ decrease as $1/k$. Hence, the system attenuates the high harmonics relative to the low harmonics. A system with this characteristic is called a *low-pass system*.

A MATLAB program that implements the complex calculations of Table 4.6 is

```
n = [0 1];
d = [1 1];
w = 1:2:5;
h = freqs (n, d, w);
ckx = 4 ./ (pi*w) .* exp(-j*pi/2);
cky = h .* ckx;
ckymag = abs(cky);
ckyphase = angle(cky)*180/pi;
results: ckymag = 0.9003 0.1342 0.0499
         ckyphase = -135.0000 -161.5651 -168.6901
```

The period followed by a mathematical operator indicates the operation on the two arrays, element by element. These symbols must be bracketed by spaces. ∎

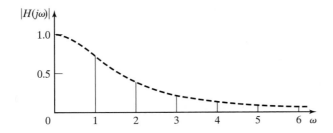

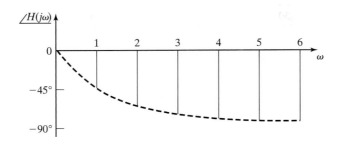

Figure 4.19   Frequency response for Example 4.7.

The system in Example 4.7 has the transfer function $H(s) = 1/(s + 1)$ and is low pass. At the harmonic frequencies $k\omega_0 = k$, from Example 4.7,

$$H(jk\omega_0)|_{\omega_0=1} = \frac{1}{\sqrt{1 + k^2}} \angle \tan^{-1}(-k). \tag{4.44}$$

The plot of this function in Figure 4.19 shows graphically the system frequency response. This plot can be verified with the following MATLAB program:

```
n = [0 1];
d = [1 1];
w = 0:0.25:5;
h = freqs(n, d, w);
hmag = abs(h);
hphase = angle(h)*180/pi;
plot(w, hmag)
plot(w, hphase)
[w', hmag', hphase']
```

The last statement gives a table of frequency, and the magnitude and phase of the frequency response.

The low-pass nature of the system is evident from Figure 4.19, because $|H(jk\omega_0)|$ (called the system *gain* at the frequency $k\omega_0$) approaches zero as $k\omega_0$ approaches infinity. As shown in Figure 4.19, only isolated points have meaning for periodic inputs. However, as shown in Chapters 5 and 6, the total frequency response has meaning with respect to aperiodic inputs. Table 4.7 gives the frequency response for the system of Example 4.7, as calculated by the MATLAB program.

TABLE 4.7 Frequency Response for Example 4.7

ω	$|H(j\omega)|$	$\angle H(j\omega)$
0	1.0000	0
0.2500	0.9701	−14.0362
0.5000	0.8944	−26.5651
0.7500	0.8000	−36.8699
1.0000	0.7071	−45.0000
1.2500	0.6247	−51.3402
1.5000	0.5547	−56.3099
1.7500	0.4961	−60.2551
2.0000	0.4472	−63.4349
2.2500	0.4061	−66.0375
2.5000	0.3714	−68.1986
2.7500	0.3417	−70.0169
3.0000	0.3162	−71.5651
3.2500	0.2941	−72.8973
3.5000	0.2747	−74.0546
3.7500	0.2577	−75.0686
4.0000	0.2425	−75.9638
4.2500	0.2290	−76.7595
4.5000	0.2169	−77.4712
4.7500	0.2060	−78.1113
5.0000	0.1961	−78.6901

EXAMPLE 4.8   **Plot of the response for Example 4.7**

In this example, we plot the system time response of Example 4.7. From this example, from Table 4.6, the response $y(t)$ for the square-wave input was found to be

$$y(t) = 2 + 1.800\cos(t - 135°) + 0.268\cos(3t - 161.6°) + 0.100\cos(5t - 168.7°)$$

for terms through the fifth harmonic. This function was plotted using the MATLAB program

```
t = 0:.1:7;
y = 2 + 1.8*cos(t - 2.356) + .268*cos(3*t - 2.82) ...
                          + .1*cos(5*t - 2.944);
plot(t, y)
```

(MATLAB requires that the cosine argument be in radians.) The result is plotted in Figure 4.20. The dashed curve is explained later. Note that the form of this function is not evident from the Fourier series.

The analysis in Examples 4.7 and 4.8 gives the approximate system steady-state response. The time constant τ of the system (see Section 2.3) is determined from

$$h(t) = e^{-t}u(t) = e^{-t/\tau}u(t). \tag{4.45}$$

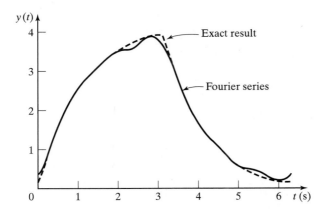

Figure 4.20 Steady-state response for the system of Example 4.7.

Hence, $\tau = 1$ s, and the system reaches approximate steady state in $4\tau = 4$ s. Because the period of the square wave in Figure 4.18 is 2π seconds, the system reaches approximate steady state in less than one period after the square wave is applied. ∎

We now consider the exact response for the system of Example 4.7. The characteristic equation of the system is the denominator of the transfer function, set to zero:

$$s + 1 = 0 \Rightarrow s = -1.$$

Hence, from (3.61), the natural response for the system is given by

$$y_c(t) = Ce^{st}\Big|_{s=-1} = Ce^{-t}. \tag{4.46}$$

With the square wave of Figure 4.18 applied, the input signal over a half-cycle is either constant (equal to 4) or zero. Hence, from Section 3.5, the total response *over any half-cycle* is of the form

$$y(t) = y_c(t) + y_p(t) = Ce^{-t} + P, \tag{4.47}$$

where the initial value of $y(t)$ is that value of $y(t)$ at the end of the preceding half-cycle, and P is either 4 or 0. The response is then of the form shown in Figure 4.21, where Y_{max} is the steady-state maximum value and Y_{min} is the steady-state minimum value. It is assumed that the square wave is applied at $t = 0$, and hence the response includes the transient response. This system was simulated, and the simulation results for the steady state are shown in Figure 4.20, along with the steady-state output determined in Example 4.7 via the Fourier series. We see then the effects of ignoring the higher harmonics in the Fourier series for this example.

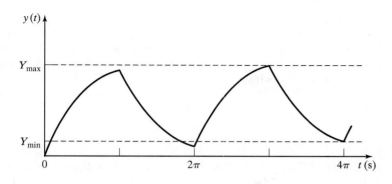

Figure 4.21 Total response for the system of Example 4.7.

Three points are now given concerning Example 4.8:

1. The Fourier-series approach gives the amplitudes and phases of the sinusoidal components (the frequency spectrum) of the output signal. However, the variation of the output signal with time is not evident, as illustrated in the examples.

2. Even with this simple system, numerical integration (a system simulation) is the simplest method for finding the system output as a function of time. For complex systems, simulations are almost always used to determine a system's time response.

3. For a physical signal, a *spectrum analyzer* can be used to determine the signal spectrum. A spectrum analyzer is an electronic instrument designed to determine signal spectra.

This section develops a steady-state analysis procedure for LTI systems with periodic inputs. The procedure does not give a plot of the time response; instead, the frequency spectra of the output is calculated.

This procedure introduces us to the frequency response of LTI systems, which is one of the most important concepts of LTI system analysis and design. The frequency-response concept is extended to aperiodic signals in Chapters 5 and 6.

4.6   FOURIER SERIES TRANSFORMATIONS

Table 4.3 gives the Fourier coefficients for seven common signals. We now give two procedures that extend the usefulness of this table. In developing these procedures, we will use the notation of (4.38),

$$x(t) = \sum_{k=-\infty}^{\infty} C_{kx}e^{jk\omega_0 t}.$$

Amplitude Transformations

As stated earlier, a constant offset in the amplitude of a periodic function will affect only the C_0 coefficient of the Fourier series. For example, the Fourier series for the triangular wave of Table 4.3 is given by

$$x(t) = \frac{X_0}{2} + \sum_{\substack{k=-\infty \\ k \text{ odd}}}^{\infty} \frac{-2X_0}{(\pi k)^2} e^{jk\omega_0 t}. \tag{4.48}$$

The triangular waveform, shifted down in amplitude by $X_0/2$, is shown in Figure 4.22. We denote this signal as $y(t)$, and its Fourier series is then

$$y(t) = x(t) - \frac{X_0}{2} = \sum_{\substack{k=-\infty \\ k \text{ odd}}}^{\infty} \frac{-2X_0}{(\pi k)^2} e^{jk\omega_0 t}. \tag{4.49}$$

The average value of this signal is zero; this is verified by inspection of Figure 4.22.

If the amplitude of a signal in Table 4.3 is offset by a constant amount, the Fourier coefficients for the offset signal are those given in Table 4.3 with C_0 adjusted by the value of the offset.

Recall from Section 2.1 that, from (2.8), the general amplitude transformation of a signal $x(t)$ is given by

$$y(t) = Ax(t) + B, \tag{4.50}$$

with A and B constants and $y(t)$ the transformed signal. As indicated earlier, for a Fourier series, B affects only the average value C_0. The constant A affects all coefficients, since the substitution of the Fourier series for $x(t)$ into (4.50) yields

$$y(t) = C_{0y} + \sum_{\substack{k=-\infty \\ k \neq 0}}^{\infty} C_{ky} e^{jk\omega_0 t}$$

$$= A\left[C_{0x} + \sum_{\substack{k=-\infty \\ k \neq 0}}^{\infty} C_{kx} e^{jk\omega_0 t}\right] + B = (AC_{0x} + B) + \sum_{\substack{k=-\infty \\ k \neq 0}}^{\infty} AC_{kx} e^{jk\omega_0 t}, \tag{4.51}$$

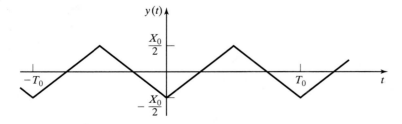

Figure 4.22 Triangular wave.

where C_{kx} denotes the Fourier coefficients for $x(t)$ and C_{ky} denotes those for $y(t)$. Therefore,

$$C_{0y} = AC_{0x} + B$$

and

$$C_{ky} = AC_{kx}, \quad k \neq 0, \tag{4.52}$$

and the effects of the amplitude transformation of (4.50) are given by (4.52). An example illustrating an amplitude transformation will now be given.

EXAMPLE 4.9   **Amplitude transformation for a Fourier series**

Consider the sawtooth signal $x(t)$ of Figure 4.23(a). From Table 4.3, the Fourier series is given by

$$x(t) = \frac{X_0}{2} + \sum_{\substack{k=-\infty \\ k \neq 0}}^{\infty} \frac{X_0}{2\pi k} e^{j\pi/2} e^{jk\omega_0 t}.$$

We wish to find the Fourier series for the sawtooth signal $y(t)$ of Figure 4.23(b). First, note that the total amplitude variation of $x(t)$ (the maximum value minus the minimum value) is X_0, while the total variation of $y(t)$ is 4. Also note that we invert $x(t)$ to get $y(t)$, yielding $A = -4/X_0$ in (4.50). (The division by X_0 normalizes the amplitude variation to unity.) In addition, if $x(t)$ is multiplied by $-4/X_0$, this signal must be shifted up in amplitude by one unit to form $y(t)$. Thus, in (4.50),

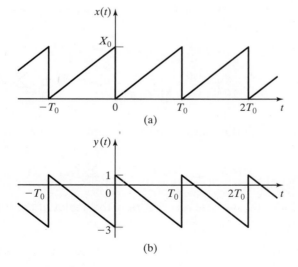

(a)

(b)

Figure 4.23   Sawtooth wave.

$$y(t) = Ax(t) + B = -\frac{4}{X_0}x(t) + 1.$$

This relation can be checked by testing it for several values of $x(t)$. Then, from (4.52),

$$C_{0y} = AC_{0x} + B = \left(-\frac{4}{X_0}\right)\frac{X_0}{2} + 1 = -1$$

and

$$C_{ky} = AC_{kx} = \left(-\frac{4}{X_0}\right)\frac{X_0}{2\pi k}e^{j\pi/2} = \frac{2}{\pi k}e^{-j\pi/2}, \quad k \neq 0.$$

Note that the value C_{0y} can be checked by inspection, from the symmetry of the signal.   ■

Time Transformations

Next we investigate the effects on the Fourier coefficients of certain time transformations. In this section, we will use the symbol τ to represent time for the original signal such as those shown in Table 4.3. We will reserve t to represent time for the transformed signal. Hence, the original signal will be denoted as $x(\tau)$ and the transformed signal as $y(t)$. From (2.6), the general time transformation is

$$\tau = at + b, \tag{4.53}$$

where a and b are constants. Then

$$y(t) = x(at + b). \tag{4.54}$$

This general transformation is considered in Problem 4.28; here we consider only the two cases of $a = -1, b = 0$ and $a = 1, b = -t_0$.

For $a = -1$ and $b = 0$, we have $y(t) = x(-t)$, or time reversal. Then

$$y(t) = \sum_{k=-\infty}^{\infty} C_{ky}e^{jk\omega_0 t} = x(-t)$$

$$= \sum_{k=-\infty}^{\infty} C_{kx}e^{-jk\omega_0 t}. \tag{4.55}$$

To express this series in the standard form, we replace k with $-k$ to obtain

$$y(t) = \sum_{k=-\infty}^{\infty} C_{-kx}e^{jk\omega_0 t} = \sum_{k=-\infty}^{\infty} C_{kx}^{*}e^{jk\omega_0 t}, \tag{4.56}$$

since $C_{-k} = C_{k}^{*}$. Hence, for time reversal, only the angles of the Fourier coefficients are affected, with

$$C_{ky} = C_{kx}^{*}. \tag{4.57}$$

The case of $a = 1$ and $b = -t_0$ in (4.54) results in a time shift. A time delay results for $t_0 > 0$ and a time advance results for $t_0 < 0$. Now

$$y(t) = x(t - t_0) = x(\tau)\Big|_{\tau=t-t_0} = \sum_{k=-\infty}^{\infty} C_{kx} e^{jk\omega_0 \tau}\Big|_{\tau=t-t_0}$$

$$= \sum_{k=-\infty}^{\infty} C_{kx} e^{jk\omega_0(t-t_0)} = \sum_{k=-\infty}^{\infty} (C_{kx} e^{-jk\omega_0 t_0}) e^{jk\omega_0 t}. \tag{4.58}$$

Hence, the effect of a time delay of t_0 on the Fourier coefficients is given by

$$C_{ky} = C_{kx} e^{-jk\omega_0 t_0}. \tag{4.59}$$

The effect of a time shift is to change the angles of the Fourier coefficients. This result is expected, since time shifting a sinusoid affects only its phase.

The effects on Fourier coefficients of amplitude and time transformations are listed in Table 4.8. In this table, the Fourier coefficients C_{kx} are known and C_{ky} are the coefficients of the transformed signal. An example is given now.

EXAMPLE 4.10   **Amplitude and time transformations for a Fourier series**

We again consider the sawtooth waveform of Example 4.9 and Figure 4.23. Hence,

$$x(t) = \frac{X_0}{2} + \sum_{\substack{k=-\infty \\ k \neq 0}}^{\infty} \frac{X_0}{2\pi k} e^{j\pi/2} e^{jk\omega_0 t}.$$

We first ignore the amplitude transformation and consider the signal $y(t)$ in Figure 4.23(b) to be time reversed to yield $y_1(t) = x(-t)$; thus, from Table 4.8,

$$C_{ky1} = C_{kx}^* = \frac{X_0}{2\pi k} e^{-j\pi/2}, \quad k \neq 0.$$

For $k = 0, C_{0y1} = C_{0x} = X_0/2$. The amplitude scaling of $y_1(t)$ is expressed as

$$y(t) = \frac{4}{X_0} y_1(t) - 3 = A y_1(t) + B.$$

TABLE 4.8   Amplitude and Time Transformations

Amplitude	$y(t) = Ax(t) + B$
	$C_{0y} = AC_{0x} + B$
	$C_{ky} = AC_{kx}, k \neq 0$
Time	$\tau = -t \Rightarrow C_{ky} = C_{kx}^*$
	$\tau = t - t_0 \Rightarrow C_{ky} = C_{kx} e^{-jk\omega_0 t_0}$

This result can be verified by testing the signal at several different values of time. Then, from Table 4.8,

$$C_{0y} = AC_{0y1} + B = \left(\frac{4}{X_0}\right)\frac{X_0}{2} - 3 = -1$$

and

$$C_{ky} = AC_{ky1} = \frac{4}{X_0}\frac{X_0}{2\pi k}e^{-j\pi/2} = \frac{2}{\pi k}e^{-j\pi/2}, \quad k \neq 0.$$

These results check those of Example 4.9, where these coefficients were calculated using only an amplitude-reversal approach.   ■

In this section, we consider amplitude and time transformations of periodic signals, so as to extend the usefulness of Table 4.3. A second procedure for extending the usefulness of the table was stated in the Section 4.4; the Fourier series of a sum of periodic signals is equal to the sum of the Fourier series of the signals, provided that the sum is periodic. If the sum is not periodic, it does not have a Fourier series.

SUMMARY

In this chapter, we introduced the Fourier series, which is a representation of a periodic signal by an infinite sum of harmonically related sinusoids. The Fourier series can be written in three forms: the exponential form, the combined trigonometric form, and the trigonometric form.

Several properties of the Fourier series were discussed. The Fourier series was introduced using the property that the Fourier series minimizes the mean-square error between a periodic function and its series.

The Fourier series of seven periodic functions that occur in engineering practice were given in Table 4.3. Procedures were then used to expand the usefulness of this table. These procedures included the independent variable transformations and the amplitude transformations discussed in Chapter 2.

Frequency spectra give graphical representations of Fourier series. The spectra are generated by plotting the coefficients of either the exponential form or the combined trigonometric form of the Fourier series versus frequency. The plots for the combined trigonometric form give the amplitudes of the sinusoidal harmonic components directly.

As the final topic, the analysis of linear time-invariant systems with periodic inputs was presented. The basis of this analysis is the sinusoidal steady-state response of a system. The system response is the sum of the responses for each harmonic, by superposition. This analysis procedure yields the frequency spectrum of the output signal; however, it does not give a good indication of a plot of this signal as a function of time. See Table 4.9.

TABLE 4.9 Key Equations of Chapter 4

Equation Title	Equation Number	Equation
Exponential form of Fourier series	(4.11)	$x(t) = \sum_{k=-\infty}^{\infty} C_k e^{jk\omega_0 t}; \quad C_k = C_{-k}^*$
Combined trigonometric form of Fourier series	(4.13)	$x(t) = C_0 + \sum_{k=1}^{\infty} 2|C_k|\cos(k\omega_0 t + \theta_k)$
Trigonometric form of Fourier series	(4.17)	$x(t) = A_0 + \sum_{k=1}^{\infty}[A_k\cos k\omega_0 t + B_k \sin k\omega_0 t],$ $A_k + jB_k = 2C_k, A_0 = C_0$
Relation of different forms of Fourier coefficients	(4.18)	$2C_k = A_k + jB_k; \quad C_k = |C_k|e^{j\theta_k}; \quad C_0 = A_0$
Fourier series coefficients formula	(4.23)	$C_k = \frac{1}{T_0}\int_{T_0} x(t)e^{-jk\omega_0 t}\, dt$
Sinc function	(4.29)	$\text{sinc } x = \frac{\sin x}{x}$
Steady-state output expressed as Fourier series	(4.38)	$x(t) = \sum_{k=-\infty}^{\infty} C_{kx} e^{jk\omega_0 t} \rightarrow y_{ss}(t) = \sum_{k=-\infty}^{\infty} H(jk\omega_0)C_{kx}e^{jk\omega_0 t}$
Fourier coefficients of output signal	(4.39)	$y_{ss}(t) = \sum_{k=-\infty}^{\infty} C_{ky}e^{jk\omega_0 t}, \quad C_{ky} = H(jk\omega_0)C_{kx}$

REFERENCES

1. C. L. Phillips and R. D. Harbor, *Feedback Control Systems*, 4th ed. Upper Saddle River, NJ: Prentice Hall, 1999.
2. D. Jackson, *Fourier Series and Orthogonal Polynomials*. Menosha, WI: Collegiate Press, 1981.
3. A. V. Oppenheim and A. S. Willsky, *Signals and Systems*. Upper Saddle River, NJ: Prentice-Hall, 1996.
4. W. Kaplan, *Operational Methods for Linear Systems*. Reading, MA: Addison–Wesley, 1962.
5. R. V. Churchill, *Operational Mathematics*, 2d ed. New York: McGraw–Hill, 1972.

PROBLEMS

4.1. For the harmonic series

$$f(t) = \cos 2t + 3\cos 4t,$$

show that the Fourier coefficients of the exponential form of $f(t)$ are given by the following:

(a) $C_0 = 0$
(b) $C_1 = 0.5$

(c) $C_2 = 1.5$

(d) $C_k = 0, k \geqslant 3$

4.2. Consider the Fourier series for the periodic functions given.

   (i) $x(t) = \sin 4t + \cos 8t + 7$
   (ii) $x(t) = \cos^2 t$
   (iii) $x(t) = \cos t + \sin 2t$
   (iv) $x(t) = \sin^2 2t + 2\cos t$
   (v) $x(t) = \cos 2t$
   (vi) $x(t) = (\cos t)(\sin 2t)$

   (a) Find the Fourier coefficients of the exponential form for each signal.
   (b) Find the Fourier coefficients of the combined trigonometric form for each signal.

4.3. **(a)** Determine whether the functions given can be represented by a Fourier series.

   (i) $x(t) = \cos(3t) + \sin(5t)$
   (ii) $x(t) = \cos(6t) + \sin(8t) + e^{j2t}$
   (iii) $x(t) = \cos(t) + \sin(\pi t)$
   (iv) $x_3(t) = x_1(t) + x_2(3t)$ where $x_1(t) = \sin\left(\frac{\pi t}{6}\right)$ and $x_2(t) = \sin\left(\frac{\pi t}{9}\right)$.

   (b) For those signals in Part (a) that can be represented by a Fourier series, find only the coefficient of the first harmonic, expressed in the exponential form.

4.4. A periodic signal $x(t)$ is expressed as an exponential Fourier Series:

$$x(t) = \sum_{k=-\infty}^{\infty} C_k e^{jk\omega_o t}.$$

Show that the Fourier Series for $\hat{x}(t) = x(t - t_o)$ is given by

$$\hat{x}(t) = \sum_{k=-\infty}^{\infty} \hat{C}_k e^{jk\omega_o t},$$

in which

$$|\hat{C}_k| = |C_k| \quad \text{and} \quad \angle \hat{C}_k = \angle C_k - k\omega_o t_o.$$

4.5. For a real periodic signal $x(t)$, the *trigonometric form* of its Fourier series is given by

$$x(t) = A_0 + \sum_{k=1}^{\infty} [A_k \cos k\omega_o t + B_k \sin k\omega_o t].$$

Express the exponential form Fourier coefficients C_k in terms of A_k and B_k.

4.6. This problem will help illustrate the orthogonality of exponentials. Calculate the following integrals:

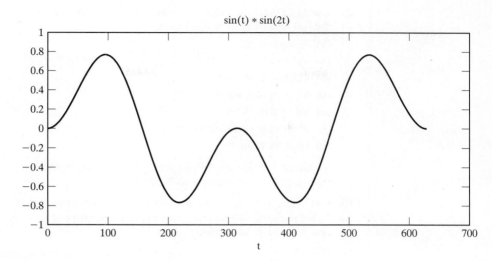

Figure P4.6

(a) $\displaystyle\int_0^{2\pi} sin^2(t)dt$

(b) $\displaystyle\int_0^{2\pi} sin^2(2t)dt$

(c) $\displaystyle\int_0^{2\pi} sin(t)sin(2t)dt$

   (Hint: In Figure P4.6, one period of sin(t) sin(2t) is plotted.)

4.7. (a) Find all integer values of m and n such that $\cos m\omega_0 t$ and $\cos n\omega_0 t$ are orthogonal over the range T_0, where $\omega_0 = 2\pi/T_0$.

   (b) Find all integer values of m and n such that $\cos m\omega_0 t$ and $\sin n\omega_0 t$ are orthogonal over the range T_0.

   (c) Find all integer values of m and n such that $\sin m\omega_0 t$ and $\sin n\omega_0 t$ are orthogonal over the range T_0.

4.8. Find the combined trigonometric form of the Fourier series for the following signals in Table 4.3:

   (a) Square wave
   (b) Sawtooth wave
   (c) Triangular wave
   (d) Rectangular wave
   (e) Full-wave rectified wave
   (f) Half-wave rectified wave
   (g) Impulse train

4.9. Use (4.23) and the integral tables in Appendix A to verify the Fourier coefficients for the following signals in Table 4.3:

(a) Square wave

(b) Sawtooth wave

(c) Triangular wave

(d) Rectangular wave

(e) Full-wave rectified wave

(f) Half-wave rectified wave

(g) Impulse train

Verify each preceding result, using the symbolic mathematics of MATLAB. Simplify each expression to agree with Table 4.3.

4.10. Use (4.23) to calculate the Fourier coefficients for the signals in Figure P4.10. Evaluate C_0 for each waveform, and verify these values directly from the waveform; L'Hôpital's rule is useful in some cases.

4.11. Using Table 4.3, find the Fourier coefficients for the exponential form for the signals of Figure P4.11. Evaluate all coefficients.

4.12. Consider the signals of Figure P4.11(a) and (d).

(a) Change the period of $x_a(t)$ to $T_0 = 2\pi$. Use Table 4.3 to find the Fourier coefficients of the exponential form for this signal.

(b) Use Table 4.3 to find the Fourier coefficients of the exponential form for $x_d(t)$.

(c) Consider the signal

$$x(t) = a_1 x_a(t) + b_1 x_d(t),$$

where $x_a(t)$ is defined in part (a). By inspection of Figure P4.11(a) and (d), find a_1 and b_1 such that $x(t)$ is constant for all time; that is, $x(t) = A$, where A is a constant. In addition, evaluate A.

(d) Use the results of Parts (a) and (b) to show that all the Fourier coefficients of $x(t)$ in part (c) are zero except for $C_0 = A$.

4.13. Let $x_a(t)$ be the half-wave rectified signal in Table 4.3. Let $x_b(t)$ be the same signal delayed by $T_0/2$.

(a) Find the coefficients in the exponential form for $x_b(t)$. *Hint:* Consider time delay.

(b) Show that the Fourier coefficients of the sum $[x_a(t) + x_b(t)]$ are those of the full-wave rectified signal in Table 4.3.

4.14. (a) Use Table 4.3 to find the exponential form of the Fourier series of the impulse train in Figure P4.14. The magnitude of the weight of each impulse function is unity, with the signs of the weights alternating.

(b) Verify the results of Part (a) by calculating the Fourier coefficients using (4.23).

4.15. Consider the waveforms $x_c(t)$ and $x_d(t)$ in Figure P4.10. Let the sum of these signals be $x_s(t)$. Then $x_s(t)$ is the same waveform of Example 4.2, except for the average value. Show that the Fourier coefficients of $x_s(t)$ are the equal to those of $x(t)$ in Example 4.2, except for the average value.

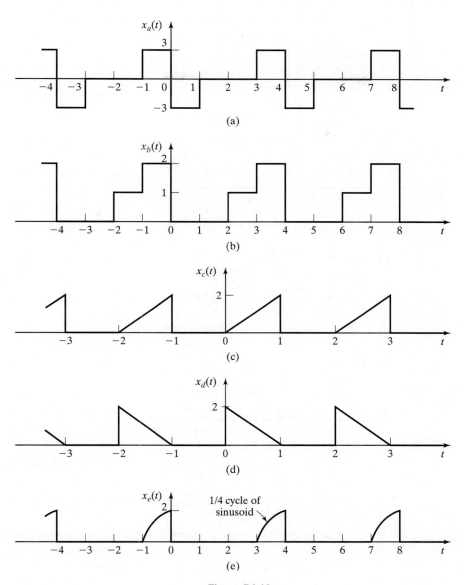

Figure P4.10

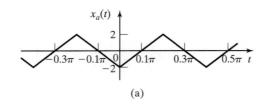

(a)

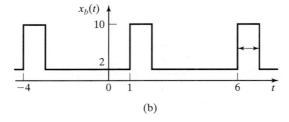

(b)

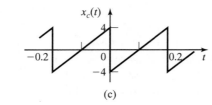

(c)

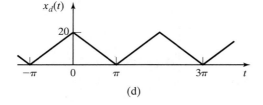

(d)

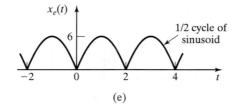

(e)

Figure P4.11

4.16. A signal has half-wave symmetry if $x(t - T_0/2) = -x(t)$. For example, $\sin \omega_0 t$ has half-wave symmetry, as does the triangular wave of Figure P4.11(a). Show that a signal with half-wave symmetry has no even harmonics; that is, $C_k = 0, k = 0, 2, 4, 6, \ldots$.

4.17. Consider the signals in Figure P4.11. For k sufficiently large, the Fourier coefficient of the kth harmonic decreases in magnitude at the rate of $1/k^m$. Use the properties in Section 4.4 to find m for the signals shown in the following figures:

   (a) Figure P4.11(a)
   (b) Figure P4.11(b)

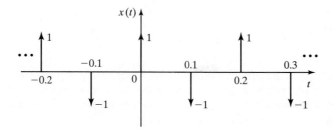

Figure P4.14

(c) Figure P4.11(c)
(d) Figure P4.11(d)
(e) Figure P4.11(e)

Check your results using Table 4.3.

4.18. (a) Sketch the frequency spectrum for the square wave of Table 4.3, for $X_0 = 10$.
  (b) Repeat Part (a) for the sawtooth wave.
  (c) Repeat Part (a) for the triangular wave.
  (d) Repeat Part (a) for the full-wave rectified signal.
  (e) Repeat Part (a) for the half-wave rectified signal.
  (f) Repeat Part (a) for the rectangular wave.
  (g) Repeat Part (a) for the impulse train.

4.19. (a) Sketch the frequency spectrum for the signal of Figure P4.10(a), showing the dc component and the first four harmonics.
  (b) Repeat Part (a) for the signal of Figure P4.10(b).
  (c) Repeat Part (a) for the signal of Figure P4.10(c).
  (d) Repeat Part (a) for the signal of Figure P4.10(d).
  (e) Repeat Part (a) for the signal of Figure P4.10(e).

4.20. (a) Sketch the frequency spectrum for the signal of Figure P4.11(a), showing the dc component and the first four harmonics.
  (b) Repeat Part (a) for the signal of Figure P4.11(b).
  (c) Repeat Part (a) for the signal of Figure P4.11(c).

4.21. Given $\omega_0 = \pi$, $C_0 = 2$, $C_1 = 1$, $C_3 = \frac{1}{2}e^{\frac{j\pi}{4}}$, and $C_{-3} = \frac{1}{2}e^{\frac{-j\pi}{4}}$, find the signal $x(t)$ with these Fourier coefficients. This is an example of signal *synthesis*.

4.22. Find the Fourier coefficients of $x(t) = \sum_{k=-\infty, \, k \, even}^{\infty}[u(t - k) - u(t - 1 - k)]$.

4.23. Given the periodic signal $x(t) = \sum_{n=-\infty, \, n \, even}^{\infty}[u(t - 3n) - u(t - 3n - 1)]$,
  (a) find its fundamental frequency;
  (b) find the Fourier coefficients of the exponential form.

4.24. Consider the system of Figure P4.24, with

$$H(s) = \frac{10}{s + 5}.$$

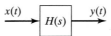

$x(t)$ → $H(s)$ → $y(t)$

Figure P4.24

(a) The input $x(t)$ is the square wave of Table 4.3, with $X_0 = 20$ and $T_0 = 3$. For the output $y(t)$, find the numerical values for the dc component and the first, second, and third harmonics of the combined trigonometric form of the Fourier series.

(b) Verify the results in Part (a) using MATLAB.

(c) Repeat Parts (a) and (b) for the sawtooth wave of Table 4.3.

(d) Repeat Parts (a) and (b) for the triangular wave of Table 4.3.

(e) Repeat Parts (a) and (b) for the full-wave rectified signal of Table 4.3.

(f) Repeat Parts (a) and (b) for the half-wave rectified signal of Table 4.3.

(g) Repeat Parts (a) and (b) for the rectangular wave of Table 4.3. Let the width of the pulse be unity, that is, $w = 1$.

(h) Repeat Parts (a) and (b) for the impulse train of Table 4.3.

4.25. Consider the system of Figure P4.24, with

$$H(s) = \frac{20}{s + 2}.$$

(a) The square wave of Table 4.3 is applied to this system, with $T_0 = 1$ s. Find the ratio of the amplitude of the first harmonic in the output signal to the amplitude of the first harmonic in the input without solving for the amplitude of the first harmonic in the output.

(b) The square wave of Table 4.3 is applied to this system, with $T_0 = 1$ s. Without solving for the first and third harmonics in the output signal, find the ratio of the amplitudes of these harmonics.

(c) Verify the results using MATLAB.

(d) Repeat Parts (a) and (c) with $T_0 = 0.1$ s.

(e) Repeat Parts (b) and (c) with $T_0 = 0.1$ s.

(f) Repeat Parts (a) and (c) with $T_0 = 10$ s.

(g) Repeat Parts (b) and (c) with $T_0 = 10$ s.

(h) Comment on the differences in the results of Parts (a), (d), and (f).

(i) Comment on the differences in the results of Parts (b), (e), and (g).

4.26. Consider the RC circuit of Figure P4.26:

(a) The square wave of Table 4.3 is applied to the input of this circuit, with $T_0 = 2\pi$ s and $X_0 = 10$ V. Solve for the frequency spectrum of the output signal. Give numerical values for the amplitudes and phases of the first three nonzero sinusoidal harmonics.

(b) Verify the results in Part (a) using MATLAB.

(c) Let the input of the circuit be as in Part (a), but with a dc value of 20 V added to the square wave. Solve for the frequency spectrum of the output signal. Give numerical values for the dc component and first three nonzero sinusoidal harmonics.

(d) Is the circuit low pass? Why?

(e) The period of the square wave is changed to $T_0 = \pi$. State the effects of this change on the answers to Parts (a) and (c), without solving these parts again. Give the reasons for your answers.

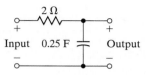

Input   0.25 F   Output

Figure P4.26

4.27. Consider the RL circuit of Figure P4.27.

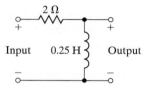

Input   0.25 H   Output

Figure P4.27

(a) The square wave of Table 4.3 is applied to the input of this circuit, with $T_0 = 0.5\pi s$ and $X_0 = 10$ V. Solve for the frequency spectrum of the output signal. Give numerical values for the amplitudes and phases of the first three nonzero sinusoidal harmonics.

(b) Verify the results in Part (a) using MATLAB.

(c) Let the input of the circuit be as in Part (a), but with a dc value of 20 V added to the square wave. Solve for the frequency spectrum of the output signal. Give numerical values for the dc component and the first three nonzero sinusoidal harmonics.

(d) Is the circuit low pass? Why?

(e) The period of the square wave is changed to $T_0 = 2\pi$. State the effects of this change on the answers to Parts (a) and (c), without solving these parts again. Give the reasons for your answers.

4.28. Consider that for the general time transformation,

$$y(t) = x(at + b).$$

Show that the Fourier coefficients for the signal $y(t)$ are given by

$$C_{ky} = \begin{cases} C_{kx}e^{jkw_0b}, & a > 0 \\ [C_{kx}e^{jkw_0b}]^*, & a < 0 \end{cases},$$

where C_{kx} are the Fourier coefficients for $x(t)$.

4.29. **(a)** For the full-wave rectified signal in Table 4.3, prove that for $y(t) = x(-t)$, the Fourier coefficients are unaffected.

(b) For the full-wave rectified signal in Table 4.3, find the Fourier coefficients for $y(t) = x(t - T_0/2)$.

4.30. Consider the system of P4.30, with $h(t) = e^{-\alpha t}u(t)$.

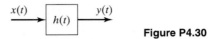

Figure P4.30

(a) For what values of α will the system be BIBO stable?

(b) Assume that the system is BIBO stable. The input signal is $x(t) = \sin t + \cos 3t$. Find $y(t)$.

4.31. Consider the system of P4.30, with $h(t) = \alpha e^{-\alpha t}u(t)$, $\alpha > 0$. This is a *low-pass filter*.

(a) The input signal is $x(t) = \sin^2 2t$. Find $y(t)$. Notice that the higher frequency components are *attenuated* more than the dc component.

(b) Repeat Part (a) with $x(t) = 1 + \cos t + \cos 8t$. Again, notice how the higher frequency component ($\cos 8t$) is attenuated more than the lower frequency components ($\cos t$ and the dc term).

4.32. Consider the system of P4.30, with $h(t) = e^{-\alpha t}u(t)$, where $\alpha > 0$. The input signal is $x(t) = \sum_{k=1}^{\infty} \cos(kt)$. Find $y(t)$.

5 THE FOURIER TRANSFORM

The Fourier transform is a method of representing mathematical models of signals and systems in the frequency domain. We begin to get a hint of this process as we represent periodic time-domain signals in terms of their harmonic frequency components using the Fourier series. The Fourier transform is an extension of this concept.

Engineers use the Fourier transform to simplify the mathematical analysis of signals and systems and for explaining physical phenomena mathematically. It is widely used in the field of electrical engineering, especially in the study of electronic communication signals and systems. For this reason, every student of electrical engineering should become familiar with the Fourier transform and its applications.

In this chapter, the Fourier transform is introduced in a way that will give each student an understanding of its mathematical basis and a glimpse at its utility in the analysis and design of linear signals and systems. The relationship between the Fourier transform and the Fourier series is presented with intent to give the reader an intuitive feeling for the Fourier transform. Mathematical properties of the Fourier transform are presented with emphasis on application of the properties rather than formal, mathematical proof.

5.1 DEFINITION OF THE FOURIER TRANSFORM

We approach the definition of the Fourier transform by first considering the Fourier series, which was described in Chapter 4, where the Fourier series was defined, in the *exponential form*, as

[eq(4.11)]
$$f(t) = \sum_{k=-\infty}^{\infty} C_k e^{jk\omega_0 t},$$

where

[eq(4.23)]
$$C_k = \frac{1}{T_0} \int_{T_0} f(t) e^{-jk\omega_0 t} dt.$$

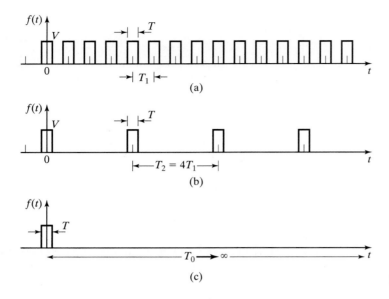

Figure 5.1 Rectangular pulse trains.

In previous chapters, we have denoted a general, continuous-time signal as $x(t)$. It is important for engineers to be able to work with a variety of notations. Therefore, for a variation, in this chapter we denote the general signal as $f(t)$.

Previously, we have considered how a periodic signal such as the one shown in Figure 5.1(a) can be represented by its harmonic components in a Fourier series. Now we consider the consequences of lengthening the period of the periodic signal as shown in Figure 5.1(b) and (c). Note that as indicated in Figure 5.1, the process we are considering is simply making the period longer and longer until finally the period becomes infinite and the waveform in Figure 5.1(c) is never repeated.

Consider the exponent of the exponential function contained in the integrand of (4.23). The quantity $k\omega_0$ changes by an amount ω_0 as k is incremented. Let us define this incremental change in frequency as

$$\Delta\omega = (k + 1)\omega_0 - k\omega_0 = \omega_0.$$

Because $\omega_0 = 2\pi/T_0$, the incremental change in frequency becomes smaller as T_0 the period of the waveform, grows longer. In the limit as T_0 approaches infinity, the frequency difference $\Delta\omega$ becomes the frequency differential $d\omega$:

$$\lim_{T_0 \to \infty} \frac{2\pi}{T_0} = d\omega.$$

Also, the quantity $k\omega_0 = 2\pi k/T_0$ approaches $kd\omega$ as T_0 becomes infinite. Since k is infinitely variable over integer values, the product $kd\omega$ becomes the continuous frequency variable ω. Now we can rewrite (4.23) as

$$C_{k\infty} = \lim_{T_0 \to \infty} \frac{1}{2\pi} \frac{2\pi}{T_0} \int_{-T_0/2}^{T_0/2} f(t)e^{-jk2\pi t/T_0} dt$$

$$= \frac{1}{2\pi} \left[\int_{-\infty}^{\infty} f(t)e^{-j\omega t}\, dt \right] d\omega.$$

The function in brackets in the preceding equation is defined as the *Fourier transform* and is written as

$$\mathcal{F}\{f(t)\} = F(\omega) = \int_{-\infty}^{\infty} f(t)e^{-j\omega t}\, dt. \tag{5.1}$$

We can write $C_{k\infty} = (1/2\pi)F(\omega)d\omega$ and, therefore, from (4.11),

$$f(t) = \sum_{k=-\infty}^{\infty} \frac{1}{2\pi} F(\omega)d\omega e^{jk\omega_0 t} = \frac{1}{2\pi} \sum_{k=-\infty}^{\infty} F(\omega)e^{jk\omega_0 t} d\omega.$$

Under these conditions the summation becomes an integral, and the equation for $f(t)$ can be rewritten as

$$f(t) = \frac{1}{2\pi} \int_{-\infty}^{\infty} F(\omega)e^{j\omega t}\, d\omega = \mathcal{F}^{-1}\{F(\omega)\}, \tag{5.2}$$

where we use the relationship $\lim_{T_0 \to \infty} k\omega_0 = \omega$ as before.

Equations (5.1) and (5.2) define the *Fourier transform* and the *inverse Fourier transform*, respectively:

[eq(5.1)] $$\mathcal{F}\{f(t)\} = F(\omega) = \int_{-\infty}^{\infty} f(t)e^{-j\omega t}\, dt$$

and

[eq(5.2)] $$\mathcal{F}^{-1}\{F(\omega)\} = f(t) = \frac{1}{2\pi} \int_{-\infty}^{\infty} F(\omega)e^{j\omega t}\, d\omega.$$

Together, these equations are called a *transform pair*, and their relationship is often represented in mathematical notation as

$$f(t) \overset{\mathcal{F}}{\longleftrightarrow} F(\omega).$$

EXAMPLE 5.1   **Physical significance of the Fourier transform**

To appreciate the physical significance of the derivation of the Fourier transform pair (5.1) and (5.2), we consider the rectangular pulse train of Figure 5.1. In Example 4.6, we considered the exponential Fourier series representation of the periodic signal

$$f(t) = \sum_{k=-\infty}^{\infty} C_k e^{jk\omega_0 t} = \sum_{k=-\infty}^{\infty} \frac{T}{T_0} V \operatorname{sinc}\left(\frac{Tk\omega_0}{2}\right) e^{jk\omega_0 t}, \tag{5.3}$$

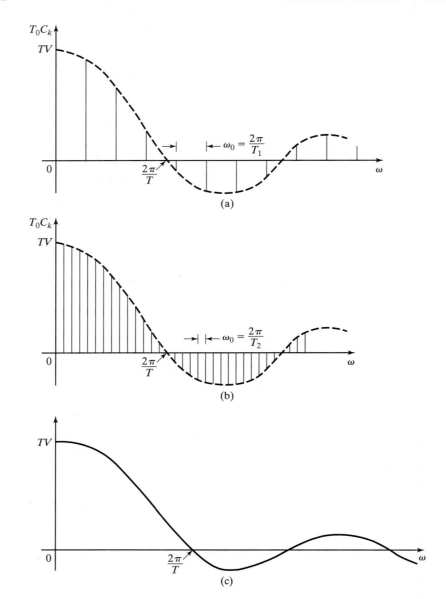

Figure 5.2 Frequency spectra of rectangular pulse trains.

where $\text{sinc}(x) = \sin(x)/x$. The magnitude of each harmonic is seen to vary according to the sinc function as shown in Figure 5.2(a), in which $T_0 C_k$ is plotted versus the frequency ω. The reason for plotting $T_0 C_k$ will become evident later.

Now consider the effect of increasing the period of the signal from $T_0 = T_1$ to $T_0 = T_2 = 4T_1$. Because $\omega_0 = 2\pi/T_0$, the components of the spectrum become closer together along the frequency axis as the period increases and therefore $\Delta\omega = \omega_0$ decreases. This is illustrated in Figure 5.2(b).

As $T_0 \to \infty$ the separation between the components becomes infinitesimally small; in other words, $\Delta\omega$ becomes $d\omega$. Therefore, the summation shown in (5.3) becomes an integration, and the frequency spectrum becomes a continuous curve.

The complex coefficient of each harmonic in the frequency spectrum of each signal is

$$C_k = \frac{T}{T_0} V \operatorname{sinc}\left(\frac{Tk\omega_0}{2}\right).$$

The magnitude of each coefficient is inversely proportional to the period of the signal. As the period approaches infinity, the magnitude of each harmonic will approach zero; however, for each of the signals shown, the relative magnitude of the signal's harmonics is determined by the sinc function. To illustrate this more clearly, Figure 5.2 actually shows plots of

$$T_0 C_k = TV \operatorname{sinc}\left(\frac{Tk\omega_0}{2}\right)$$

versus ω (for $\omega \geq 0$) rather than the true frequency spectra of the signals.

Note in Figure 5.2 that the envelopes of all the plots are the same. The plots are changed only by the frequency components of the signals coming closer together as the periods increase. It should be noted that $\operatorname{sinc}(kT\omega/2)$ goes to zero only when the argument of the sinc function is an integer multiple of π. Therefore, the zero crossings of the envelopes occur at frequencies of $\omega = 2\pi n/T, n = 1, 2, 3, \ldots$, regardless of the period T_0.   ■

The continuous frequency spectrum shown in Figure 5.3 is a graphical representation of the Fourier transform of a single rectangular pulse of amplitude V and duration T (which can also be considered to be a periodic pulse of infinite period). The analytical expression for the Fourier transform is found by using (5.1). The rectangular pulse can be described mathematically as the sum of two step functions:

$$f(t) = Vu(t + T/2) - Vu(t - T/2).$$

To simplify the integration in (5.1), we can recognize that $f(t)$ has a value of V during the period $-T/2 < t < + T/2$ and is zero for all other times. Then

$$F(\omega) = \int_{-T/2}^{+T/2} Ve^{-j\omega t}\, dt = V\left[\frac{e^{-j\omega t}}{-j\omega}\bigg|_{-T/2}^{+T/2}\right]$$

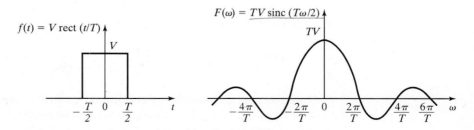

Figure 5.3   A rectangular pulse and its Fourier transform.

$$= V\left[\frac{e^{-jT\omega/2} - e^{+jT\omega/2}}{-j\omega}\right] = \frac{TV}{\omega T/2}\left[\frac{e^{jT\omega/2} - e^{-jT\omega/2}}{j2}\right]$$

$$= TV\left[\frac{\sin(T\omega/2)}{T\omega/2}\right] = TV \operatorname{sinc}(T\omega/2),$$

and we have derived our first Fourier transform:

$$\mathcal{F}\{V[u(t + T/2) - u(t - T/2)]\} = TV \operatorname{sinc}(T\omega/2).$$

Note that the Fourier transform of the nonperiodic rectangular pulse has the same form as the envelope of the Fourier series representation of the periodic rectangular pulse train derived in Example 4.6 and Table 4.3.

The waveforms of Example 5.1, the rectangular pulse and the sinc function, play important roles in signal representation and analysis. Many important waveforms, such as a digital "1" or a radar pulse, can be approximated by a rectangular pulse similar to the one used in the example. Because of the frequent use of the rectangular pulse in the study of communication signals it is often defined with a special function name such as

$$\operatorname{rect}(t/T) = [u(t + T/2) - u(t - T/2)].$$

Therefore, in our table of transform pairs we will list

$$\operatorname{rect}(t/T) \xleftrightarrow{\mathcal{F}} T \operatorname{sinc}(T\omega/2) \tag{5.4}$$

as representing the transform pair shown in Figure 5.3.

The transform pair (5.4) is valid even though we have not yet taken into consideration the fact that some waveforms do not have Fourier transforms.

Sufficient conditions for the existence of the Fourier transform are similar to those given earlier for the Fourier series. They are the *Dirichlet conditions*:

1. On any finite interval,
   a. $f(t)$ is bounded;
   b. $f(t)$ has a finite number of maxima and minima; and
   c. $f(t)$ has a finite number of discontinuities.

2. $f(t)$ is absolutely integrable; that is,

$$\int_{-\infty}^{\infty} |f(t)|\, dt < \infty.$$

Note that these are *sufficient* conditions and not *necessary* conditions. Use of the Fourier transform for the analysis of many useful signals would be impossible if these were necessary conditions.

Any useful signal $f(t)$ that meets the condition

$$E = \int_{-\infty}^{\infty} |f(t)|^2 \, dt < \infty \qquad (5.5)$$

is absolutely integrable. In (5.5), E is the energy associated with the signal, which can be seen if we consider the signal $f(t)$ to be the voltage across a 1-Ω resistor. The power delivered by $f(t)$ is then

$$p(t) = |f(t)|^2/R = |f(t)|^2,$$

and the integral of power over time is energy.

A signal that meets the condition of containing finite energy is known as an *energy signal*. Energy signals generally include nonperiodic signals that have a finite time duration (such as the rectangular function, which is considered in several examples) and signals that approach zero asymptotically so that $f(t)$ approaches zero as t approaches infinity.

An example of a mathematical function that does not have a Fourier transform, because it does not meet the Dirichlet condition of absolute integrability is $f(t) = e^{-t}$. However, the frequently encountered signal, $f(t) = e^{-t}u(t)$, does meet the Dirichlet conditions and does have a Fourier transform.

We have mentioned the use of Fourier transforms of useful signals that do not meet the Dirichlet conditions. Many signals of interest to electrical engineers are not energy signals and are, therefore, not absolutely integrable. These include the unit step function, the signum function, and all periodic functions. It can be shown that signals that have infinite energy, but contain a finite amount of power, and meet the other Dirichlet conditions do have valid Fourier transforms [1–3].

A signal that meets the condition

$$P = \lim_{T \to \infty} \frac{1}{T} \int_{-T/2}^{T/2} |f(t)|^2 \, dt < \infty \qquad (5.6)$$

is called a *power signal*.

The power computed using Equation (5.6) is called *normalized average power*. In electrical signal analysis, normalized power is defined as the power that a signal delivers to a 1Ω load. Using the normalized power definition, the signal $f(t)$ in (5.6) can represent either voltage or current as an electrical signal because

$$P = V_{rms}^2 = I_{rms}^2 \text{ when } R = 1\Omega.$$

The concept of normalized average power is often used to describe the strength of communication signals.

The step function, the signum function, and periodic functions that meet the Dirichlet conditions except for absolute integrability are power signals. We will see that the Fourier transforms that we derive for power signals contain impulse functions in the frequency domain. This is a general characteristic of power signals and

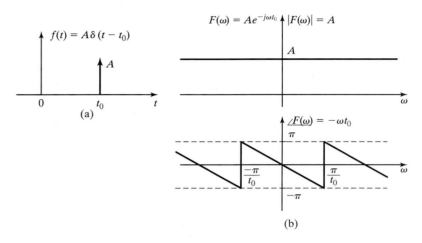

Figure 5.4 An impulse function and its frequency spectra.

can serve to distinguish the frequency spectrum of a power signal from that of an energy signal.

For practical purposes, for the signals or functions we may wish to analyze as engineers, we can use the rule that if we can draw a picture of the waveform $f(t)$, it has a Fourier transform. In fact, some waveforms that we cannot accurately draw pictures of (such as the impulse function) have Fourier transforms. *All physically realizable signals have Fourier transforms.*

The impulse function, in fact, provides a building block for several of the more important transform pairs. Consider the waveform

$$f(t) = A\delta(t - t_0)$$

that represents an impulse function of weight A that is nonzero only at time $t = t_0$, as illustrated in Figure 5.4(a). (See Section 2.4.) The Fourier transform of this waveform is

$$F(\omega) = \mathcal{F}\{A\delta(t - t_0)\} = \int_{-\infty}^{\infty} A\delta(t - t_0)e^{-j\omega t}\, dt.$$

Recall the sifting property of the impulse function described in (2.41), namely, that

$$\int_{-\infty}^{\infty} f(t)\delta(t - t_0)\, dt = f(t_0),$$

for $f(t)$ continuous at $t = t_0$. Using this property of the impulse function to evaluate the Fourier transform integral, we find that

$$\mathcal{F}\{A\delta(t - t_0)\} = Ae^{-j\omega t_0}. \tag{5.7}$$

This transform pair is shown in Figure 5.4, where it can be seen that $F(\omega)$ has a constant magnitude, A, at all frequencies and a phase angle that is a linear function of frequency with a slope of $-t_0$. The sawtooth look of the phase plot is a result of the phase angle being plotted modulo 2π (i.e., $-\omega t_0 = -\omega t_0 + 2\pi n$ for any integer value of n).

A special case of the impulse function considered previously is the unit impulse function occurring at $t = 0$:

$$f(t) = \delta(t).$$

From (5.7), with $A = 1$ and $t_0 = 0$, it is seen that

$$\delta(t) \xleftrightarrow{\mathscr{F}} 1. \tag{5.8}$$

While we are dealing with the impulse function, let's consider the case of an impulse function in the frequency domain. We have

$$F(\omega) = \delta(\omega - \omega_0),$$

which is defined in the same way as (2.40) $\delta(t)|_{t=\omega}$. Therefore, $\delta(\omega)$ has the same properties as described for $\delta(t)$ in Section 2.4:

$$\delta(\omega - \omega_0) = \begin{cases} \text{undefined}, & \omega = \omega_0 \\ 0, & \omega \neq \omega_0 \end{cases},$$

$$\int_{-\infty}^{\infty} \delta(\omega - \omega_0)d\omega = 1,$$

$$F(\omega)\delta(\omega - \omega_0) = F(\omega_0)\delta(\omega - \omega_0),$$

$$\int_{-\infty}^{\infty} F(\omega)\delta(\omega - \omega_0)d\omega = F(\omega_0), \text{ etc.}$$

The inverse Fourier transform of this impulse function is found from Equation (5.2):

$$f(t) = \mathscr{F}^{-1}\{F(\omega)\} = \mathscr{F}^{-1}\{\delta(\omega - \omega_0)\} = \frac{1}{2\pi}\int_{-\infty}^{\infty} \delta(\omega - \omega_0)e^{j\omega t}d\omega.$$

After applying the sifting property of the impulse function, we have

$$f(t) = \mathscr{F}^{-1}\{\delta(\omega - \omega_0)\} = \frac{1}{2\pi}e^{j\omega_0 t},$$

which is recognized to be a complex phasor of constant magnitude that rotates in phase at a frequency of ω_0 rad/s.

The Fourier transform pair

$$e^{j\omega_0 t} \xleftrightarrow{\mathscr{F}} 2\pi\delta(\omega - \omega_0) \tag{5.9}$$

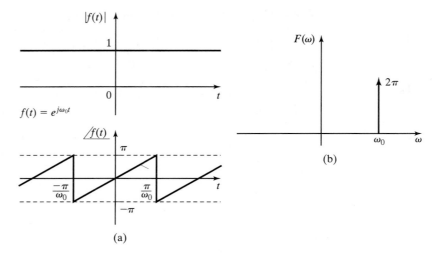

Figure 5.5   Time-domain plots and frequency spectra of $e^{j\omega_0 t}$.

is illustrated in Figure 5.5, where the rotating phasor in the time domain is represented in Figure 5.5(a) by separate plots of magnitude and phase angle.

Equations (5.7), (5.8), and (5.9) provide three more transform pairs that can be placed in a Fourier transform table, along with the transform pair from (5.4), for future use. Additional transform pairs can often be developed more easily by using those already known. Before deriving any additional transform pairs for our table, however, we will consider some special properties of the Fourier transform.

5.2   PROPERTIES OF THE FOURIER TRANSFORM

The Fourier transform has several properties that can greatly simplify its use in signal and system analysis. Table 5.1 gives a listing of properties of the Fourier transform that are commonly used by engineers. Selected properties are explained in this section.

The Fourier transform properties are usually stated here with examples of their use, but without further proof. Proofs of selected properties are given as problems at the end of this chapter. References 3 through 9 are recommended for those who wish to study the formal mathematical proofs of any or all properties.

Linearity

Because the Fourier transform (5.1) is an integral of $f(t)$ and its inverse (5.2) is an integral of $F(\omega)$, and because integration is a linear operation, it can be reasoned that the Fourier transform is a linear operation. The linearity property of the Fourier transform states that if we are given the transform pairs

$$f_1(t) \xleftrightarrow{\mathcal{F}} F_1(\omega) \quad \text{and} \quad f_2(t) \xleftrightarrow{\mathcal{F}} F_2(\omega),$$

TABLE 5.1   Fourier Transform Properties

Operation	Time Function	Fourier Transform
Linearity	$af_1(t) + bf_2(t)$	$aF_1(\omega) + bF_2(\omega)$
Time shift	$f(t - t_0)$	$F(\omega)e^{-j\omega t_0}$
Time scaling	$f(at)$	$\dfrac{1}{|a|}F\left(\dfrac{\omega}{a}\right)$
Time transformation	$f(at - t_0)$	$\dfrac{1}{|a|}F\left(\dfrac{\omega}{a}\right)e^{-j\omega t_0/a}$
Duality	$F(t)$	$2\pi f(-\omega)$
Frequency shift	$f(t)e^{j\omega_0 t}$	$F(\omega-\omega_0)$
Convolution	$f_1(t)*f_2(t)$	$F_1(\omega)F_2(\omega)$
	$f_1(t)f_2(t)$	$\dfrac{1}{2\pi}F_1(\omega)*F_2(\omega)$
Differentiation	$\dfrac{d^n[f(t)]}{dt^n}$	$(j\omega)^n F(\omega)$
	$(-jt)^n f(t)$	$\dfrac{d^n[F(\omega)]}{d\omega^n}$
Integration	$\displaystyle\int_{-\infty}^{t} f(\tau)d\tau$	$\dfrac{1}{j\omega}F(\omega) + \pi F(0)\delta(\omega)$

then

$$[af_1(t) + bf_2(t)] \xleftrightarrow{\mathscr{F}} [aF_1(\omega) + bF_2(\omega)], \qquad (5.10)$$

where a and b are constants. In words, the principle of superposition applies to the Fourier transform.

EXAMPLE 5.2

The linearity property of the Fourier transform

We can make use of the property of linearity to find the Fourier transforms of some types of waveforms. For example, consider

$$f(t) = B \cos \omega_0 t.$$

Using Euler's relation,

$$\cos \alpha = \frac{e^{j\alpha} + e^{-j\alpha}}{2},$$

we can rewrite the expression for $f(t)$ as

$$f(t) = \frac{B}{2}[e^{j\omega_0 t} + e^{-j\omega_0 t}] = \frac{B}{2}e^{j\omega_0 t} + \frac{B}{2}e^{-j\omega_0 t}.$$

Now we recognize that $f(t)$ is a linear combination of two rotating phasors. Equation (5.9) provides the Fourier transform pair for a rotating phasor; therefore, we use (5.9) and the property of linearity to find that

$$\mathcal{F}\{B \cos \omega_0 t\} = \frac{B}{2}\mathcal{F}\{e^{j\omega_0 t}\} + \frac{B}{2}\mathcal{F}\{e^{-j\omega_0 t}\}$$

$$= \pi B \delta(\omega - \omega_0) + \pi B \delta(\omega + \omega_0)$$

or

$$B \cos \omega_0 t \xleftrightarrow{\mathcal{F}} \pi B[\delta(\omega - \omega_0) + \delta(\omega + \omega_0)], \qquad (5.11)$$

which adds another entry to our list of Fourier transform pairs.   ∎

Time Scaling

The time-scaling property provides that if

$$f(t) \xleftrightarrow{\mathcal{F}} F(\omega),$$

then, for a constant scaling factor a,

$$f(at) \xleftrightarrow{\mathcal{F}} \frac{1}{|a|}F\left(\frac{\omega}{a}\right). \qquad (5.12)$$

This is proved by considering the defining equation (5.1) for the Fourier transform with the appropriate substitution of variables:

$$\mathcal{F}\{f(at)\} = \int_{-\infty}^{\infty} f(at)e^{-j\omega t}\, dt.$$

If the substitution $\tau = at$ is made, with $a > 0$, then $d\tau = a\, dt$ and the equation can be rewritten as

$$\mathcal{F}\{f(\tau)\} = \frac{1}{a}\int_{-\infty}^{\infty} f(\tau)e^{-j(\omega/a)\tau}\, d\tau.$$

Comparison of this result with equation (5.1) yields

$$\mathcal{F}\{f(at)\} = \frac{1}{a}F(\omega/a).$$

The absolute value sign on the scaling factor, a, allows (5.12) to be applicable when a has either positive or negative values. (See Problem 5.13.)

EXAMPLE 5.3   **The time-scaling property of the Fourier transform**

We now find the Fourier transform of the rectangular waveform

$$g(t) = \text{rect}(2t/T_1).$$

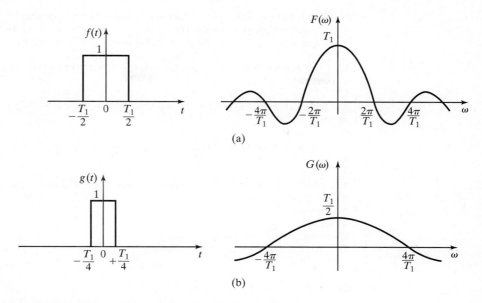

Figure 5.6   Rectangular pulses and their frequency spectra.

From the result of Example 5.1,

[eq(5.4)] $V \operatorname{rect}(t/T) \xleftrightarrow{\mathscr{F}} TV \operatorname{sinc}(T \omega/2),$

it is seen that $g(t)$ is simply the particular case where $T = T_1$, the time-scaling factor $a = 2$, and $V = 1$, as shown in Figure 5.6(a). Applying the time-scaling property to the transform pair obtained in (5.4) results in

$$g(t) = f(2t),$$

where

$$f(t) = \operatorname{rect}(t/T_1).$$

Therefore, from (5.12),

$$G(\omega) = \frac{1}{2} F(\omega/2) = \frac{T_1}{2} \operatorname{sinc}\left(\frac{\omega T_1}{4}\right).$$

This result is illustrated in Figure 5.6(b).
The following MATLAB program illustrates the duration–bandwidth relationship:

```
%   This program displays the relationship of the magnitude frequency spectra of two
%   rectangular pulses.
%
```

```
%   The width of each pulse width is selected by the user.
d1 = input ('Type in the desired pulse width for the first pulse in seconds.')
d2 = input ('Type in the desired pulse width for the second pulse in seconds.')
%
%   The time vector is limited to twice the duration of the widest pulse.
tlim=max([d1 d2]);
%
%   The time step is 1/1000 the maximum time span.
dt=2*tlim/1000;
%
%   Create the time vector.
t=[-tlim:dt:tlim];
%
%   Create the vectors which represent the two rectangular pulses.
p1=zeros(size(t));
p2=zeros(size(t));
w1=d1/dt;
w2=d2/dt;
for  n=501-round(w1/2)/:501+round(w1/2);
    p1(n)=1;
end
for  n=501-round(w2/2)/:501+round(w2/2);
    p2(n)=1;
end
%
%   Plot the two pulses.
subplot(2,2,1), plot(t,p1), xlabel('time (s)'), ylabel ('first pulse')
subplot(2,2,2), plot(t,p2,'g'), xlabel('time (s)'), ylabel ('second pulse')
    ylabel ('second pulse'),
%
%   Approximate the Fourier transform of each pulse using the fft function.
F1=fft(p1);
F2=fft(p2);
%
%   Create a vector of frequency values.
wlim=round(3*1001*dt/min([d1 d2]));
w=(2*pi/(1001*dt))*[-wlim:1:wlim];
%
%   Prepare to plot the magnitudes of the Fourier transform approximations.
P1=zeros(size(w));
P2=zeros(size(w));
mid=fix(lenght(w)/2);
for n=1:mid;
    P1(n+mid)=F1(n);
    P2(n+mid)=F2(n);
    P1(mid+1-n)=F1(n);
    P2(mid+1-n)=F2(n);
end
```

```
P1p=dt*abs(P1);
P2p=dt*abs(P2);
%
%  Plot the magnitude frequency spectra of the two pulses.
subplot(2,2,3), plot(w, P1p), xlabel('Frequency (rad/s)')
subplot(2,2,4), plot(w, P2p, 'g'), xlabel('Frequency (rad/s)')
```
■

Example 5.3 and the waveforms shown in Figure 5.6 give insight into an important physical relationship between the time domain and the frequency domain, which is implied by the time-scaling property. Notice how the frequency spectrum of the signal spreads as the time-domain waveform is compressed. This implies that a pulse with a short time duration contains frequency components with significant magnitudes over a wider range of frequencies than a pulse with longer time duration does. In the study of communication systems, this reciprocal relationship between time-domain waveforms and their frequency spectra is an important consideration. This is known as the *duration-bandwidth* relationship and is discussed in greater detail in Chapter 6.

Time Shifting

The property of time shifting previously appeared in the Fourier transform of the *impulse function* (5.7) derived in Section 5.1, although it was not recognized at that time. This property is stated mathematically as

$$f(t - t_0) \xleftrightarrow{\mathscr{F}} F(\omega)e^{-j\omega t_0}, \qquad (5.13)$$

where the symbol t_0 represents the amount of shift in time.

EXAMPLE 5.4   **The time-shifting property of the Fourier transform**

We now find the Fourier transform of the impulse function, which occurs at time zero. From (5.8),

$$\mathscr{F}\{\delta(t)\} = \int_{-\infty}^{\infty} \delta(t)e^{-j\omega t}dt = e^{-j\omega t}\Big|_{t=0} = 1.$$

If the impulse function is shifted in time so that it occurs at time t_0 instead of at $t = 0$, we see from the time-shifting property (5.13) that

$$\mathscr{F}\{\delta(t - t_0)\} = (1)e^{-j\omega t_0} = e^{-j\omega t_0},$$

which is recognized as the same result obtained in (5.7).
■

EXAMPLE 5.5   **Fourier transform of a time-delayed sinusoidal signal**

Consider the time-shifted cosine wave of frequency $\omega = 200\pi$ and a delay of 1.25 ms in its propagation:

$$x(t) = 10 \cos [200\pi(t - 1.25 \times 10^{-3})].$$

This signal can be viewed as a phase-shifted cosine wave where the amount of phase shift is $\pi/4$ radians:

$$x(t) = 10 \cos (200\pi t - \pi/4).$$

Using the linearity and time-shifting property, we find the Fourier transform of this delayed cosine wave:

$$\begin{aligned}
\mathcal{F}\{x(t)\} = X(\omega) &= 10\mathcal{F}\{\cos (200\pi t)\}e^{-j.00125\omega} \\
&= 10\pi[\delta(\omega-200\pi) + \delta(\omega + 200\pi)]e^{-j.00125\omega} \\
&= 10\pi[\delta(\omega-200\pi)e^{-j\pi/4} + \delta(\omega + 200\pi)e^{j\pi/4}].
\end{aligned}$$

The rotating phasor, $e^{-j.00125\omega}$, is reduced to the two fixed phasors shown in the final equation because the frequency spectrum has zero magnitude except at $\omega = 200\pi$ and $\omega = -200\pi$. Recall, from Table 2.3, that

$$F(\omega)\delta(\omega-\omega_0) = F(\omega_0)\delta(\omega-\omega_0).$$

Notice that the phase shift of $-\pi/4$ radians, which is the result of the 1.25-ms delay in the propagation of the cosine wave, is shown explicitly in the frequency spectrum.   ■

Time Transformation

The properties of time scaling and time shifting can be combined into a more general property of time transformation. The concept of the time-transformation property was introduced in Section 2.2 and for the Fourier series in Section 4.6.

Let

$$\tau = at - t_0,$$

where a is a scaling factor and t_0 is a time shift. Application of the time-scaling property (5.12) gives

$$f(at) \xleftrightarrow{\mathcal{F}} \frac{1}{|a|}F\left(\frac{\omega}{a}\right).$$

Application of the time-shift property (5.13) to this time-scaled function gives us the time-transformation property:

$$f(at - t_0) \xleftrightarrow{\mathcal{F}} \frac{1}{|a|}F\left(\frac{\omega}{a}\right)e^{-jt_0(\omega/a)}. \tag{5.14}$$

EXAMPLE 5.6	**The time-transformation property of the Fourier transform**

Consider the rectangular pulse shown in Figure 5.7(a). We will find the Fourier transform of this function using a known Fourier transform and the time-transformation property. Given the rectangular pulse of Figure 5.7(b), we easily determine the Fourier transform from (5.4) to be

$$F(\omega) = \text{sinc}(\omega/2).$$

The magnitude and phase $F(\omega)$ are plotted separately in Figure 5.7(c). From Figure 5.7(a) and (b), we write

$$g(t) = 3 \, \text{rect}\,[(t - 4)/2] = 3f(0.5t - 2).$$

Then, using the time-transformation property (5.14) with $a = 0.5$ and $t_0 = 2$ and the linearity property to account for the magnitude scaling, we can write

$$G(\omega) = 6 \, \text{sinc}(\omega)e^{-j4\omega}.$$

The magnitude and phase plots of $G(\omega)$ are shown in Figure 5.7(d) for comparison with the plots of $F(\omega)$. Note the effect of the time shift on the phase of $G(\omega)$. The step changes of π radians in the phase occur because of the changes in the algebraic sign of the sinc function. ∎

For signals such as $G(\omega)$ in Example 5.6 that have a phase angle that changes continuously with frequency, it is usually desirable to plot the magnitude and phase of the Fourier transform separately. These plots simplify the sketch and display the information in a way that is easier to interpret. These separate plots are called the *magnitude spectrum* and *phase spectrum*, respectively, of the signal.

Duality

The symmetry of the Fourier transformation and its inverse in the variables t and ω can be seen by comparing equations (5.1) and (5.2):

[eq.(5.1)] $$\mathcal{F}\,\{f(t)\} = F(\omega) = \int_{-\infty}^{\infty} f(t)e^{-j\omega t}\,dt;$$

[eq.(5.2)] $$\mathcal{F}^{-1}\,\{F(\omega)\} = f(t) = \frac{1}{2\pi}\int_{-\infty}^{\infty} F(\omega)e^{j\omega t}d\omega.$$

The duality property, which is sometimes known as the *symmetry* property, is stated as

$$F(t) \xleftrightarrow{\ \mathcal{F}\ } 2\pi f(-\omega) \quad \text{when} \quad f(t) \xleftrightarrow{\ \mathcal{F}\ } F(\omega). \tag{5.15}$$

This property states that if the mathematical function $f(t)$ has the Fourier transform $F(\omega)$ and a function of time exists such that

$$F(t) = F(\omega)\Big|_{\omega=t} ,$$

then $\mathcal{F}\,\{F(t)\} = 2\pi f(-\omega)$, where $f(-\omega) = f(t)\Big|_{t=-\omega} .$

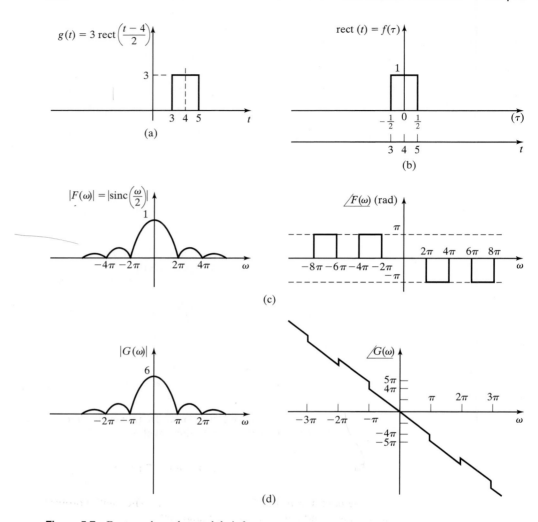

Figure 5.7   Rectangular pulses and their frequency spectra.

EXAMPLE 5.7   **The duality property of the Fourier transform**

We now find the inverse Fourier transform of the waveform shown in Figure 5.8(a), using the transform pair derived in Example 5.1. Figure 5.8(a) shows a rectangular waveform, but in the frequency domain rather than in the time domain.

Our task now is to find the time-domain waveform that has such a frequency spectrum. The waveform of Figure 5.8(a) can be described as

$$F(\omega) = A[u(\omega + \beta) - u(\omega - \beta)],$$

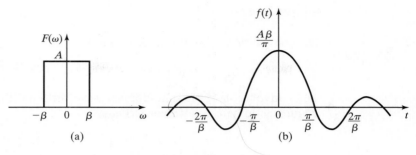

Figure 5.8 A rectangular pulse in the frequency domain.

or

$$F(\omega) = A \, \text{rect}(\omega/2\beta).$$

Compare this waveform description with the transform pair (5.4):

$$V \, \text{rect}(t/T) \xleftrightarrow{\mathscr{F}} TV \, \text{sinc}(\omega T/2).$$

According to the duality property, we write

$$TV \, \text{sinc}(Tt/2) \xleftrightarrow{\mathscr{F}} 2\pi V \, \text{rect}(-\omega/T).$$

Because the waveform is an even function of frequency, in other words, $F(-\omega) = F(\omega)$, we can rewrite the equation that describes the waveform as

$$2\pi f(-\omega) = 2\pi A \, \text{rect}(\omega/2\beta),$$

where we have substituted $T = 2\beta$ and $V = A$. The duality property can be used with these values substituted into (5.4) to determine that

$$F(t) = 2\beta \, A \, \text{sinc}(\beta t).$$

The transform pair

$$\frac{A\beta}{\pi} \, \text{sinc}(\beta t) \xleftrightarrow{\mathscr{F}} A \, \text{rect}(\omega/2\beta)$$

is shown in Figures 5.8(b) and (a), respectively.   ∎

The duality property can be quite useful for the derivation of new transform pairs based on the knowledge of established transform pairs, as shown in Example 5.7.

Convolution

The convolution property states that if

$$f_1(t) \xleftrightarrow{\mathscr{F}} F_1(\omega) \quad \text{and} \quad f_2 \xleftrightarrow{\mathscr{F}} F_2(\omega),$$

then convolution of the time-domain waveforms has the effect of multiplying their frequency-domain counterparts. Thus,

$$f_1(t)*f_2(t) \xleftrightarrow{\mathscr{F}} F_1(\omega)F_2(\omega), \tag{5.16}$$

where

$$f_1(t)*f_2(t) = \int_{-\infty}^{\infty} f_1(\tau)f_2(t-\tau)d\tau = \int_{-\infty}^{\infty} f_1(t-\tau)f_2(\tau)d\tau.$$

Also, by applying the duality property to (5.16), it is shown that multiplication of time-domain waveforms has the effect of convolving their frequency-domain representations. This is sometimes called the *multiplication property*,

$$f_1(t)f_2(t) \xleftrightarrow{\mathscr{F}} \frac{1}{2\pi}F_1(\omega)*F_2(\omega), \tag{5.17}$$

where

$$F_1(\omega)*F_2(\omega) = \int_{-\infty}^{\infty} F_1(\lambda)F_2(\omega-\lambda)d\lambda = \int_{-\infty}^{\infty} F_1(\omega-\lambda)F_2(\lambda)d\lambda.$$

Engineers make frequent use of the convolution property in analyzing the interaction of signals and systems.

EXAMPLE 5.8

The time-convolution property of the Fourier transform

Chapter 3 discusses the response of linear time-invariant systems to input signals. A block diagram of a linear system is shown in Figure 5.9(a). If the output of the system in response to an impulse function at the input is described as $h(t)$, then $h(t)$ is called the *impulse response* of the system. The output of the system in response to any input signal can then be determined by convolution of the impulse response, $h(t)$, and the input signal, $x(t)$:

$$y(t) = x(t)*h(t) = \int_{-\infty}^{\infty} x(\tau)h(t-\tau)d\tau.$$

Using the convolution property of the Fourier transform, we can find the frequency spectrum of the output signal from

$$Y(\omega) = X(\omega)H(\omega),$$

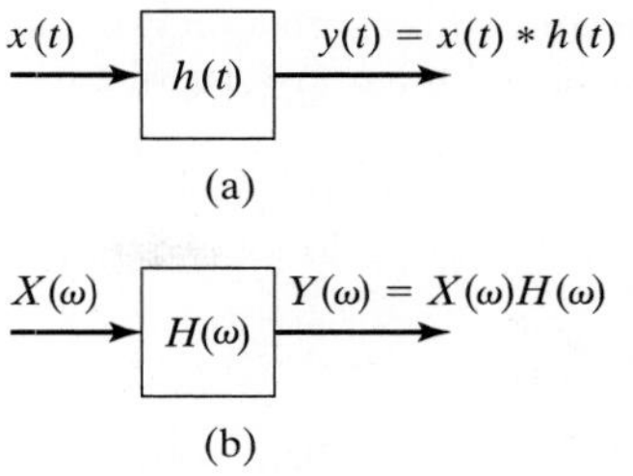

(a)

(b)

Figure 5.9   A linear time-invariant system.

where

$$h(t) \xleftrightarrow{\mathscr{F}} H(\omega), x(t) \xleftrightarrow{\mathscr{F}} X(\omega), \quad \text{and} \quad y(t) \xleftrightarrow{\mathscr{F}} Y(\omega).$$

The function $H(\omega)$ is the system transfer function discussed in Section 4.5. A block diagram of the signal/system relationship in the frequency domain is shown in Figure 5.9(b).   ■

   The application described in Example 5.8 and other applications of the convolution property will be explored more fully in Chapter 6.

Frequency Shifting

The frequency shifting property is stated mathematically as

$$x(t)e^{j\omega_0 t} \xleftrightarrow{\mathscr{F}} X(\omega - \omega_0). \tag{5.18}$$

This property was demonstrated in the derivation of Equation (5.9) without recognizing it.

EXAMPLE 5.9   **The frequency-shift property of the Fourier transform**   *Time Shifting*

In the generation of communication signals, often two signals such as   *p. 215*

$$g_1(t) = 2\cos(200\pi t) \quad \text{and} \quad g_2(t) = 5\cos(1000\pi t)$$

are multiplied together to give

$$g_3(t) = g_1(t)g_2(t) = 10\cos(200\pi t)\cos(1000\pi t).$$

We can use the frequency-shifting property to find the frequency spectrum of $g_3(t)$. The product waveform. $g_3(t)$, is rewritten using Euler's identity on the second cosine factor:

$$g_3(t) = 10\cos(200\pi t)\frac{e^{j1000\pi t} + e^{-j1000\pi t}}{2}$$

$$= 5\cos(200\pi t)e^{j1000\pi t} + 5\cos(200\pi t)e^{-j1000\pi t}.$$

The Fourier transform of this expression is found using the properties of linearity (5.10), frequency shifting (5.18), and the transform of $\cos(\omega_0 t)$ from (5.11):

$$G_3(\omega) = 5\pi[\delta(\omega - 200\pi - 1000\pi) + \delta(\omega + 200\pi - 1000\pi)]$$
$$+ 5\pi[\delta(\omega - 200\pi + 1000\pi) + \delta(\omega + 200\pi + 1000\pi)].$$

In final form, we write

$$G_3(\omega) = 5\pi[\delta(\omega - 1200\pi) + \delta(\omega - 800\pi) + \delta(\omega + 800\pi) + \delta(\omega + 1200\pi)].$$

The frequency spectra of $g_1(t)$, $g_2(t)$, and $g_3(t)$ are shown in Figure 5.10.

It is of interest to engineers to note that the inverse Fourier transform of $G_3(\omega)$ is

$$g_3(t) = \mathcal{F}^{-1}\{5\pi[\delta(\omega - 1200\pi) + \delta(\omega + 1200\pi)]\}$$
$$+ \mathcal{F}^{-1}\{5\pi[\delta(\omega - 800\pi) + \delta(\omega + 800\pi)]\}$$
$$= 5\cos 1200\pi t + 5\cos 800\pi t.$$

The product of two sinusoidal signals has produced a sum of two sinusoidal signals. (This is also seen from trigonometric identities.) One has the frequency that is the sum of the frequencies of the two original signals, whereas the other has the frequency that is the difference of the two original frequencies. This characteristic is often used in the process of generating signals for use in communication systems and in applications such as radar and sonar. These results can be confirmed using the following MATLAB program.

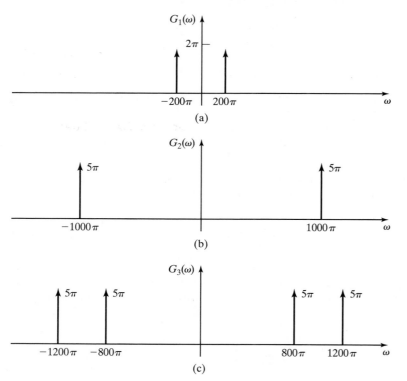

(a)

(b)

(c)

Figure 5.10 The frequency spectrum of $10\cos(200\pi t)\cos(1000\pi t)$.

```
% This MATLAB program finds the Fourier transform of the product
% of two sinusoidal signals using the symbolic math function
% "fourier".
%
syms t
g1=2*cos(200*pi*t)
g2=5*cos(1000*pi*t)
'Multipy the two sinusoidal signals.','g3 = g1*g2='
g3=g1*g2
'Compute the Fourier transform.','G3=fourier(g3)'
G3=fourier(g3)
```

∎

Time Differentiation

If

$$f(t) \xleftrightarrow{\mathscr{F}} F(\omega),$$

then

$$\frac{d[f(t)]}{dt} \xleftrightarrow{\mathscr{F}} j\omega F(\omega). \tag{5.19}$$

The differentiation property can be stated more generally for the nth derivative as

$$\frac{d^n[f(t)]}{dt^n} \xleftrightarrow{\mathscr{F}} (j\omega)^n F(\omega). \tag{5.20}$$

This property can be proven easily by differentiating both sides of Equation (5.2) with respect to time. This proof is assigned as Problem 5.4(e).

EXAMPLE 5.10   **Fourier transform of the signum function**

We now find the Fourier transform of the signum function shown in Figure 5.11(a):

$$f(t) = \text{sgn}(t).$$

The derivation of $F(\omega)$ is simplified by using the differentiation property. The time derivative of sgn(t) is shown in Figure 5.11(b) and is given by

$$\frac{d[f(t)]}{dt} = 2\delta(t).$$

Because $\delta(t) \xleftrightarrow{\mathscr{F}} 1$ (5.8)

$$j\omega F(\omega) = 2.$$

From this result, it is determined that

$$F(\omega) = \frac{2}{j\omega} + k\delta(\omega),$$

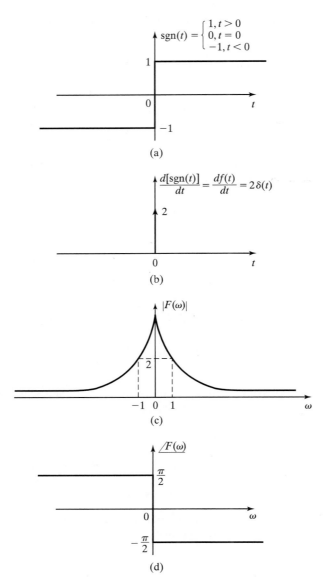

Figure 5.11 Finding the frequency spectrum of the signum function.

where the term $k\delta(\omega)$ is nonzero only at $\omega = 0$ and accounts for the time-averaged value of $f(t)$. In the general case, this term must be included because the time-derivative operation implied by the expression $j\omega F(\omega)$ would cause a loss of this information about the time-averaged value of $f(t)$. In this particular case, the time-averaged value of $sgn(t)$ is zero. Therefore, $k = 0$ in our expression for the Fourier transform of $sgn(t)$. This gives another pair for our Fourier transform table:

$$sgn(t) \xleftrightarrow{\mathcal{F}} \frac{2}{j\omega}. \qquad\qquad (5.21)$$

Figures 5.11(c) and (d) are sketches of the magnitude and phase frequency spectrum, respectively, of the signum function.   ∎

EXAMPLE 5.11   **The time-differentiation property of the Fourier transform**

The frequency spectrum of the signal shown in Figure 5.12(a) will be found. The figure shows a waveform $w(t)$ for which we have not previously determined a Fourier transform. The differentiation property of the Fourier transform can be used to simplify the process. Figure 5.12(b) shows $x(t)$, the first derivative of $w(t)$ with respect to time. This waveform can be described as a set of three rectangular pulses; however, the problem can be simplified even further by taking a second derivative with respect to time to get $y(t)$, the result shown in Figure 5.12(c). An equation for this waveform can easily be written

$$y(t) = \frac{A}{b-a}\delta(t+b) - \frac{bA}{a(b-a)}\delta(t+a)$$

$$+ \frac{bA}{a(b-a)}\delta(t-a) - \frac{A}{(b-a)}\delta(t-b),$$

and the Fourier transform found from Table 5.2:

$$Y(\omega) = \frac{A}{b-a}e^{j\omega b} - \frac{A}{b-a}e^{-j\omega b}$$

$$- \frac{bA}{a(b-a)}e^{j\omega a} + \frac{bA}{a(b-a)}e^{-j\omega a}.$$

Using Euler's identity, this can be rewritten as

$$Y(\omega) = \frac{j2A}{b-a}\sin(b\omega) - \frac{j2bA}{a(b-a)}\sin(a\omega)$$

$$= \frac{j\omega 2Ab}{b-a}\frac{\sin(b\omega)}{b\omega} - \frac{j\omega 2bA}{b-a}\frac{\sin(a\omega)}{a\omega}.$$

Because $y(t) = d[x(t)]/dt$ the time differentiation property yields

$$Y(\omega) = j\omega X(\omega)$$

or

$$X(\omega) = \frac{1}{j\omega}Y(\omega) + k\delta(\omega), \tag{5.22}$$

where, because the time-averaged value of $x(t)$ is seen by inspection to be zero, $k = 0$. Thus,

$$X(\omega) = \frac{2Ab}{b-a}[\,\text{sinc}(b\omega) - \text{sinc}(a\omega)].$$

Similarly,

$$W(\omega) = \frac{1}{j\omega}X(\omega) + k\delta(\omega), \tag{5.23}$$

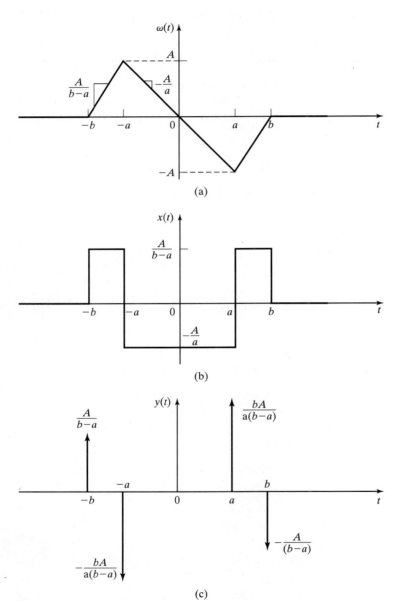

(a)

(b)

(c)

Figure 5.12 A waveform, $w(t)$, and its derivatives.

function

TABLE 5.2   Fourier Transform Pairs

Time Domain Signal	Fourier Transform
$f(t)$	$\displaystyle\int_{-\infty}^{\infty} f(t)e^{-j\omega t}\,dt$
$\displaystyle\frac{1}{2\pi}\int_{-\infty}^{\infty} F(\omega)e^{j\omega t}\,d\omega$	$F(\omega)$
$\delta(t)$	1
$A\delta(t + t_0)$	$Ae^{-j\omega t_0}$
$u(t)$	$\pi\delta(\omega) + \dfrac{1}{j\omega}$
1	$2\pi\delta(\omega)$
K	$2\pi K\delta(\omega)$
$\text{sgn}(t)$	$\dfrac{2}{j\omega}$
$e^{j\omega_0 t}$	$2\pi\delta(\omega - \omega_0)$ $e^{-j\omega_0 t} = 2\pi\delta(\omega + \omega_0)$
$\cos\omega_0 t$	$\pi[\delta(\omega - \omega_0) + \delta(\omega + \omega_0)]$
$\sin\omega_0 t$	$\dfrac{\pi}{j}[\delta(\omega - \omega_0) - \delta(\omega + \omega_0)]$
$\text{rect}(t/T)$	$T\,\text{sinc}(\omega T/2)$
$\cos(\omega_0 t)u(t)$	$\dfrac{\pi}{2}[\delta(\omega - \omega_0) + \delta(\omega + \omega_0)] + \dfrac{j\omega}{\omega_0^2 - \omega^2}$
$\sin(\omega_0 t)u(t)$	$\dfrac{\pi}{2j}[\delta(\omega - \omega_0) - \delta(\omega + \omega_0)] + \dfrac{\omega_0}{\omega_0^2 - \omega^2}$
$\text{rect}(t/T)\cos(\omega_0 t)$	$\dfrac{T}{2}\left[\text{sinc}\left(\dfrac{(\omega - \omega_0)T}{2}\right) + \text{sinc}\left(\dfrac{(\omega + \omega_0)T}{2}\right)\right]$
$\dfrac{\beta}{\pi}\,\text{sinc}(\beta t)$ $\,\,g(t\cdot t_0)$	$\text{rect}(\omega/2\beta)$
$\text{tri}(t/T)$	$T\,\text{sinc}^2(T\omega/2)$
$\text{sinc}^2(Tt/2)$	$\dfrac{2\pi}{T}\,\text{tri}(\omega/T)$
$e^{-at}u(t),\,\text{Re}\{a\} > 0$	$\dfrac{1}{a + j\omega}$
$te^{-at}u(t),\,\text{Re}\{a\} > 0$	$\left(\dfrac{1}{a + j\omega}\right)^2$
$t^{n-1}e^{-at}u(t),\,\text{Re}\{a\} > 0$	$\dfrac{(n - 1)!}{(a + j\omega)^n}$
$e^{-a|t|},\,\text{Re}\{a\} > 0$	$\dfrac{2a}{a^2 + \omega^2}$
$\displaystyle\sum_{n=-\infty}^{\infty} g(t - nT_0)$	$\displaystyle\sum_{n=-\infty}^{\infty} \omega_0 G(n\omega_0)\delta(\omega - n\omega_0),\,\omega_0 = \dfrac{2\pi}{T_0}$
$\displaystyle\sum_{n=-\infty}^{\infty} g(t - nT_0) = \sum_{k=-\infty}^{\infty} C_k e^{jk\omega_0 t}$	$\displaystyle 2\pi\sum_{k=-\infty}^{\infty} C_k\delta(\omega - k\omega_0),\,\omega_0 = \dfrac{2\pi}{T_0},\,C_k = \dfrac{1}{T_0}\int_{T_0} g(t)e^{-jk\omega_0 t}\,dt$
$\delta_T(t)$	$\displaystyle\sum_{k=-\infty}^{\infty} \omega_0\delta(\omega - k\omega_0)$

where again for this signal the time-average value, $k = 0$, and therefore,

$$W(\omega) = \frac{2Ab}{j\omega(a - b)}[\,\text{sinc}(a\omega) - \text{sinc}(b\omega)]$$

$$= \frac{2Ab}{\omega(a - b)}[\,\text{sinc}(a\omega) - \text{sinc}(b\omega)]e^{-j\pi/2}. \qquad \blacksquare$$

One must be careful (as we were) in using the time-differentiation property as in Example 5.11. Suppose that the waveform $w(t)$ were imposed on a nonzero dc level (time-averaged value). The time derivative of that waveform is exactly the same, as shown in Figure 5.12(b). Because no information about the time-averaged value of $w(t)$ remains once the derivative is taken, the method used earlier would give erroneous results unless one is careful to account for the time-averaged value of the original function. This problem will be investigated further later.

The approach used in Example 5.11 suggests a method by which engineers can write an equation for an existing physical waveform by approximating it with straight-line segments. Information about the frequency spectrum of the waveform can then be determined by use of the Fourier transform and its properties.

Time Integration

If

$$f(t) \xleftrightarrow{\mathscr{F}} F(\omega),$$

then

$$g(t) = \int_{-\infty}^{t} f(\tau)d\tau \xleftrightarrow{\mathscr{F}} \frac{F(\omega)}{j\omega} + \pi F(0)\delta(\omega) = G(\omega), \qquad (5.24)$$

where

$$F(0) = F(\omega)\Big|_{\omega=0} = \int_{-\infty}^{\infty} f(t)dt,$$

from (5.1). If $f(t)$ has a nonzero time-averaged value (dc value), then $F(0) \neq 0$. In this case, using the relationship

$$K \xleftrightarrow{\mathscr{F}} 2\pi K\delta(\omega),$$

we see from Table 5.2 that $g(t)$ has a dc value of $F(0)/2$. We also see that $g(t)$ is not an energy signal (5.5), but is a power signal (5.6); this explains the presence of the impulse function in $G(\omega)$.

The integration property was suggested by the method used in Examples 5.10 and 5.11, where we in effect integrated when we divided by $j\omega$ in (5.21), (5.22), and (5.23). However, in those examples we had no dc component to consider. Therefore, the value of $F(0)$ was zero and the second term on the right-hand side of (5.24) was not needed in those particular cases.

The time-integration property of the Fourier transform will now be proved. Consider the convolution of a generic waveform $f(t)$ with a unit step function:

$$f(t) * u(t) = \int_{-\infty}^{\infty} f(\tau)u(t - \tau)d\tau.$$

The unit step function $u(t - \tau)$ has a value of zero for $t < \tau$ and a value of 1 for $t > \tau$. This can be restated as

$$u(t - \tau) = \begin{cases} 1, & \tau < t \\ 0, & \tau > t \end{cases},$$

and, therefore,

$$f(t) * u(t) = \int_{-\infty}^{t} f(\tau)d\tau. \qquad (5.25)$$

The convolution property yields

$$f(t) * u(t) \xleftrightarrow{\mathscr{F}} F(\omega)\left[\pi\delta(\omega) + \frac{1}{j\omega}\right], \qquad (5.26)$$

where $\mathscr{F}\{u(t)\} = \pi\delta(\omega) + 1/j\omega$ from Table 5.2. (The derivation of this transform pair is provided in Section 5.3.)

By combining Equations (5.25) and (5.26), we write

$$\int_{-\infty}^{t} f(\tau)d\tau \xleftrightarrow{\mathscr{F}} \frac{F(\omega)}{j\omega} + \pi F(0)\delta(\omega).$$

The factor $F(0)$ in the second term on the right follows from the sifting property (2.42) of the impulse function.

EXAMPLE 5.12   **The time-integration property of the Fourier transform**

Figure 5.13(a) shows a linear system that consists of an integrator. As discussed in Section 1.2, this can be physically realized electronically by a combination of an operational amplifier, resistors, and capacitors. The input signal is a pair of rectangular pulses as shown in Figure 5.13(b). Using time-domain integration we can see that the output signal would be a triangular waveform, as shown in Figure 5.13(c). We wish to know the frequency spectrum of the output signal. We have not derived the Fourier transform of a triangular wave; however, we do know the Fourier transform of a rectangular pulse such as is present at the input of the system. Using the properties of linearity and time shifting, we can write the input signal as

$$x(t) = A \text{ rect}\left[\frac{t + t_1/2}{t_1}\right] - A \text{ rect}\left[\frac{t - t_1/2}{t_1}\right]$$

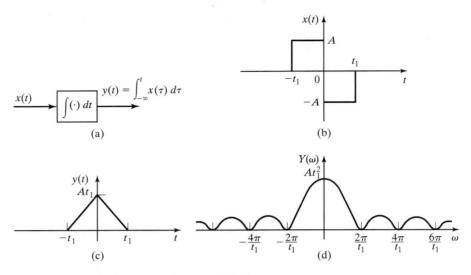

Figure 5.13   System and waveforms for Example 5.12.

and its Fourier transform as

$$
\begin{aligned}
X(\omega) &= At_1 \, \text{sinc}(t_1\omega/2)[e^{j\omega t_1/2} - e^{-j\omega t_1/2}] \\
&= 2jAt_1 \, \text{sinc}(t_1\omega/2)\sin(t_1\omega/2) \\
&= j\omega At_1^2 \, \text{sinc}(t_1\omega/2)\left[\frac{\sin(t_1\omega/2)}{t_1\omega/2}\right] \\
&= j\omega At_1^2 \, \text{sinc}^2(t_1\omega/2).
\end{aligned}
$$

Next, we use the time-integration property to find

$$
Y(\omega) = \frac{1}{j\omega}X(\omega) + \pi X(0)\delta(\omega),
$$

where $X(0) = 0$, as can be determined from the previous equation or by finding the time-average value of the signal shown in Figure 5.13(b). Therefore, the frequency spectrum of the output signal is given by

$$
Y(\omega) = At_1^2 \, \text{sinc}^2(t_1\omega/2).
$$

This frequency spectrum is sketched in Figure 5.13(d).   ∎

Triangular waveforms, such as the one shown in Figure 5.13(c), are sometimes generalized and named as functions. There is no universally accepted nomenclature for these triangular waveforms. We will define the triangular pulse as

$$
\text{tri}(t/T) = \begin{cases} 1 - \dfrac{|t|}{T}, & |t| < T \\ 0, & |t| \geq T \end{cases}.
$$

Therefore, the triangular pulse shown in Figure 5.13(c) is written as At_1 tri(t/t_1).

Notice that the triangular pulse tri(t/T) has a time duration of $2T$ seconds, in contrast to the rectangular pulse rect(t/T) which has a time duration of T seconds. From the result of Example 5.12 we write another Fourier transform pair:

$$\text{tri}(t/T) \xleftrightarrow{\mathscr{F}} T \text{ sinc}^2(T\,\omega/2). \tag{5.27}$$

Frequency Differentiation

The time-differentiation property given by (5.20) has a dual for the case of differentiation in the frequency domain. If

$$f(t) \xleftrightarrow{\mathscr{F}} F(\omega),$$

then

$$(-jt)^n f(t) \xleftrightarrow{\mathscr{F}} \frac{d^n F(\omega)}{d\omega^n}. \tag{5.28}$$

This is easily shown by differentiating both sides of the equation that defines the Fourier transform (5.1) with respect to ω:

$$\frac{dF(\omega)}{d\omega} = \int_{-\infty}^{\infty} [(-jt)f(t)]e^{-j\omega t}\, dt.$$

This equation then defines the Fourier transform pair

$$(-jt)f(t) \xleftrightarrow{\mathscr{F}} \frac{d[F(\omega)]}{d\omega}.$$

This result is easily extended to yield (5.28).

Summary

Several useful, key properties of the Fourier transform have been described and used in examples. Additional examples of the application of these properties will be given in subsequent sections of this chapter, as we use them to derive Fourier transforms for time-domain signals.

A concise listing of Fourier transform properties is given by Table 5.1. Proofs of these and other properties of the Fourier transform can be found in Refs. 3 to 9.

5.3   FOURIER TRANSFORMS OF TIME FUNCTIONS

In Sections 5.1 and 5.2, we have defined the Fourier transform and its inverse. We have listed and applied several important properties of the Fourier transform, and in the process we have derived the Fourier transforms of several time-domain signals. In this section, we derive additional Fourier transform pairs for future reference.

DC Level

Equation (5.9) gives the transform pair

$$e^{j\omega_0 t} \xleftrightarrow{\mathscr{F}} 2\pi\delta(\omega-\omega_0).$$

If we allow $\omega_0 = 0$, we have

$$1 \xleftrightarrow{\mathscr{F}} 2\pi\delta(\omega), \tag{5.29}$$

which, along with the linearity property, allows us to write the Fourier transform of a dc signal of any magnitude:

$$K \xleftrightarrow{\mathscr{F}} 2\pi K\delta(\omega). \tag{5.30}$$

By comparing this transform pair with that of an impulse function in the time domain,

[eq(5.8)]
$$\delta(t) \xleftrightarrow{\mathscr{F}} 1,$$

we see another illustration of the duality property (5.15).

Unit Step Function

The Fourier transform of the unit step function can be derived easily by considering the Fourier transform of the signum function developed in (5.21):

[eq(5.21)]
$$\mathrm{sgn}(t) \xleftrightarrow{\mathscr{F}} \frac{2}{j\omega}.$$

As illustrated in Figure 5.14, the unit step function can be written in terms of the signum function:

$$u(t) = \tfrac{1}{2}[1 + \mathrm{sgn}(t)].$$

Combining the linearity property with (5.21) and (5.30) yields

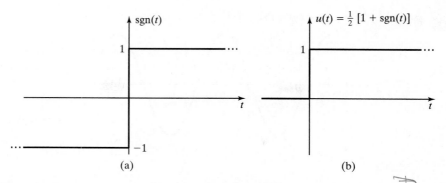

Figure 5.14   The signum function and the unit step function.

$$u(t) \xleftrightarrow{\mathscr{F}} \pi\delta(\omega) + \frac{1}{j\omega}. \tag{5.31}$$

Switched Cosine

The switched cosine as shown in Figure 5.15 is simply a cosine wave that is "turned on" at $t = 0$. Mathematically, this waveform can be described as the product of a cosine (which exists for all time, $-\infty < t < \infty$) and a unit step function (which is nonzero only for $t > 0$):

$$f(t) = \cos(\omega_0 t)u(t).$$

By applying Euler's identity, we can rewrite this function as

$$f(t) = \frac{e^{j\omega_0 t} + e^{-j\omega_0 t}}{2}u(t) = \tfrac{1}{2}e^{j\omega_0 t}u(t) + \tfrac{1}{2}e^{-j\omega_0 t}u(t).$$

We now apply the linearity property (5.10) and the frequency, shifting property (5.18) to the Fourier transform of the unit step function (5.31) to yield

$$\cos(\omega_0 t)u(t) \xleftrightarrow{\mathscr{F}} \frac{\pi}{2}[\delta(\omega - \omega_0) + \delta(\omega + \omega_0)] + \frac{j\omega}{\omega_0^2 - \omega^2}. \tag{5.32}$$

Pulsed Cosine

The pulsed cosine is shown in Figure 5.16. This waveform is encountered in various electronic communication systems and in detection systems such as radar and sonar. It can be expressed as the product of a rectangular function and a cosine wave,

$$f(t) = \text{rect}(t/T)\cos(\omega_0 t),$$

where $\text{rect}(t/T)$ is indicated by the dashed lines in Figure 5.16(a).

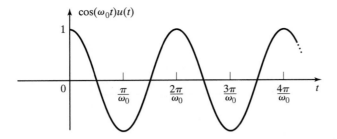

Figure 5.15   A switched cosine waveform.

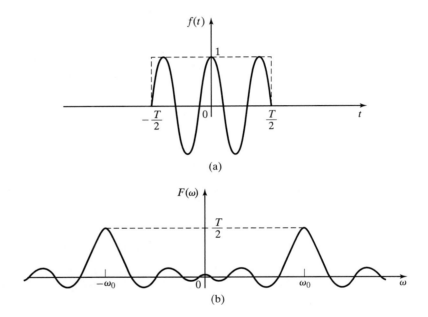

Figure 5.16   A pulsed cosine waveform and its frequency spectrum.

The derivation of this transform pair could be conducted much like that of the switched cosine, using the frequency-shift property. However, for the sake of variety we will use a different method for this one. Because the time-domain signal is described as the product of two signals for which we already know the Fourier transforms, the convolution property can be used to find the transform we are seeking.

From (5.4), we have

$$\operatorname{rect}(t/T) \xleftrightarrow{\;\mathcal{F}\;} T \operatorname{sinc}(\omega T/2),$$

and from (5.11),

$$\cos(\omega_0 t) \xleftrightarrow{\mathscr{F}} \pi[\delta(\omega - \omega_0) + \delta(\omega + \omega_0)].$$

Then, applying the convolution equation (5.17), we have

$$F(\omega) = \frac{T}{2} \int_{-\infty}^{\infty} [\delta(\omega - \lambda - \omega_0) + \delta(\omega - \lambda + \omega_0)]\text{sinc}(\lambda T/2)d\lambda.$$

Because of the sifting property of the impulse function, (2.41), the integrand has a nonzero value only when $\lambda = \omega - \omega_0$ and $\lambda = \omega + \omega_0$. Therefore, the convolution integral is easily evaluated, and the transform pair is found to be

$$\text{rect}(t/T)\cos(\omega_0 t) \xleftrightarrow{\mathscr{F}} \frac{T}{2}\left[\text{sinc}\frac{(\omega - \omega_0)T}{2} + \text{sinc}\frac{(\omega + \omega_0)T}{2} \right]. \qquad (5.33)$$

The frequency spectrum of the signal is shown in Figure 5.16(b). The sinc waveforms generated by the rectangular pulse are shifted in frequency so that one sinc pulse is centered at ω_0 and another at $-\omega_0$. Each of the sinc pulses has one-half the magnitude of the single sinc function, which represents the Fourier transform of rect(t/T). Since each of the sinc waveforms has nonzero frequency components over an infinite range of frequencies, there will be some overlap of frequency components from the two sinc waveforms in $F(\omega)$. However, if $\omega_0 \gg 2\pi/T$, the effect of the overlap is usually negligible in practical applications.

Exponential Pulse

The signal $f(t) = e^{-at}u(t)$, $a > 0$, is shown in Figure 5.17(a). The Fourier transform of this signal will be derived directly from the defining Equation (5.1):

$$F(\omega) = \int_{-\infty}^{\infty} e^{-at}u(t)e^{-j\omega t}dt = \int_{0}^{\infty} e^{-(a+j\omega)t}dt = \frac{1}{a + j\omega}.$$

The frequency spectra of this signal are shown in Figures 5.17(b) and (c).

It can be shown that this derivation applies also for a complex, with $\text{Re}\{a\} > 0$. Therefore, the transform pair can be written as

$$e^{-at}u(t), \quad \text{Re}\{a\} > 0 \xleftrightarrow{\mathscr{F}} \frac{1}{a + j\omega}. \qquad (5.34)$$

Fourier Transforms of Periodic Functions

In Chapter 4, we determined that a periodic function of time could be represented by its Fourier series,

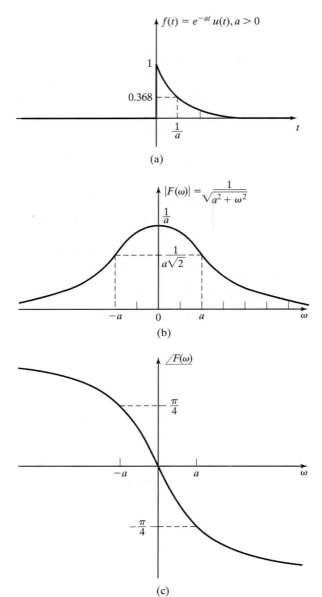

Figure 5.17 An exponential waveform and its frequency spectrum.

[eq(4.11)]

$$f(t) = \sum_{k=-\infty}^{\infty} C_k e^{jk\omega_0 t},$$

where

[eq(4.23)]

$$C_k = \frac{1}{T_0} \int_{T_0} f(t) e^{-jk\omega_0 t} \, dt,$$

We now will derive a method of determining the Fourier transform of periodic signals.

Using (5.1), the Fourier transform of (4.11) yields

$$F(\omega) = \int_{-\infty}^{\infty} \left[\sum_{k=-\infty}^{\infty} C_k e^{jk\omega_0 t} \right] e^{-j\omega t} \, dt = \sum_{k=-\infty}^{\infty} C_k \int_{-\infty}^{\infty} (e^{jk\omega_0 t}) e^{-j\omega t} \, dt.$$

From (5.9) and (5.10),

$$\sum_{k=-\infty}^{\infty} C_k e^{jk\omega_0 t} \overset{\mathscr{F}}{\longleftrightarrow} 2\pi \sum_{k=-\infty}^{\infty} C_k \delta(\omega - k\omega_0). \tag{5.35}$$

This gives us the important result that the frequency spectrum of a periodic signal is a series of impulse functions in the frequency domain located at integer multiples (harmonics) of the fundamental frequency of the periodic wave. The weight of each impulse function is the complex coefficient of that harmonic in the Fourier series multiplied by 2π.

We now express (5.35) in a different form. We find that we can use the Fourier transform to determine the complex coefficient, C_k, for a periodic function $f(t)$.

First, we define another function, which we will call the *generating function*, $g(t)$, such that

$$g(t) = \begin{cases} f(t), & -T_0/2 \leq t \leq T_0/2 \\ 0 & \text{elsewhere} \end{cases}, \tag{5.36}$$

where $T_0 = 2\pi/\omega_0$ is the fundamental period of the waveform. In other words, $g(t)$ is equal to $f(t)$ for one period of the wave, centered about $t = 0$, but is not repeated. This allows us to express the periodic function, $f(t)$, as an infinite summation of the time-shifted *generating function*, $g(t)$:

$$f(t) = \sum_{n=-\infty}^{\infty} g(t - nT_0). \tag{5.37}$$

Because from (3.23),

$$g(t) * \delta(t - t_0) = g(t - t_0),$$

(5.37) can be expressed as

$$f(t) = \sum_{n=-\infty}^{\infty} g(t)*\delta(t - nT_0) = g(t)* \sum_{n=-\infty}^{\infty} \delta(t - nT_0).$$

The train of impulse functions is expressed by its Fourier series

$$\sum_{n=-\infty}^{\infty} \delta(t - nT_0) = \sum_{n=-\infty}^{\infty} C_n e^{jn\omega_0 t},$$

where

$$C_n = \frac{1}{T_0}\int_{-T_0/2}^{T_0/2}\left[\sum_{m=-\infty}^{\infty} \delta(t - mT_0)\right]e^{-jn\omega_0 t}dt.$$

Within the limits of integration, the impulse function will be nonzero only for $m = 0$. Therefore,

$$C_n = \frac{1}{T_0}\int_{-T_0/2}^{T_0/2}\delta(t)dt = \frac{1}{T_0}.$$

Hence, according to the convolution property of the Fourier transform (5.16),

$$F(\omega) = G(\omega)\mathscr{F}\left\{\frac{1}{T_0}\sum_{n=-\infty}^{\infty} e^{jn\omega_0 t}\right\} = \frac{2\pi}{T_0}G(\omega)\sum_{n=-\infty}^{\infty} \delta(\omega - n\omega_0),$$

where $G(\omega)$ is the transform of the generating signal described by (5.36). Because of the sifting property of the impulse function, $F(\omega)$ will have nonzero values only when ω is an integer multiple of the fundamental frequency of the periodic signal

$$\omega = n\omega_0, \quad n = 0, \pm 1, \pm 2, \ldots.$$

Therefore, we can rewrite the equation for the Fourier transform of a periodic signal,

$$f(t) = \sum_{n=-\infty}^{\infty} g(t - nT_0),$$

as

$$F(\omega) = \sum_{n=-\infty}^{\infty} \omega_0 G(n\omega_0)\delta(\omega - n\omega_0). \tag{5.38}$$

We see then that frequency spectra of periodic signals are made up of discrete frequency components in the form of impulses occurring at integer multiples (harmonics) of the fundamental frequency of the signal. The weight of each impulse is found by multiplying the Fourier transform of the generating function, evaluated at that harmonic frequency, by the fundamental frequency of the periodic signal. Hence, the Fourier transform of a periodic signal is given by both (5.35) and (5.38).

EXAMPLE 5.13   **The Fourier transform of a periodic signal**

We now find the Fourier transform of the periodic train of rectangular pulses shown in Figure 5.18(a). The generating function in this case can be recognized as a familiar function from previous examples:

$$g(t) = A \operatorname{rect}(t/T).$$

From (5.4),

$$G(\omega) = AT \operatorname{sinc}(T\omega/2).$$

Substituting this into (5.38) yields the Fourier transform of the periodic train of rectangular pulses:

$$F(\omega) = \sum_{k=-\infty}^{\infty} AT\omega_0 \operatorname{sinc}(k\omega_0 T/2)\delta(\omega - k\omega_0).$$

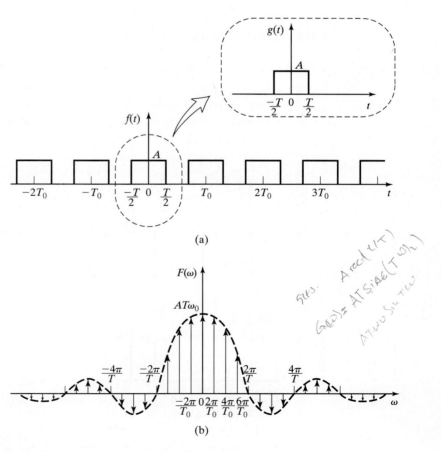

(a)

(b)

Figure 5.18   A periodic signal and its frequency spectrum.

This frequency spectrum is sketched in Figure 5.18(b). The dashed curve indicates the *weights* of the impulse functions. Note that in the distribution of impulses in frequency, Figure 5.18(b) shows the particular case that $T_0 = 4T$. ∎

EXAMPLE 5.14   **The frequency spectrum of a periodic impulse signal**

The frequency spectrum of the train of impulses shown in Figure 5.19(a) will be found. This signal is described mathematically as

$$f(t) = \sum_{n=-\infty}^{\infty} \delta(t - nT_0).$$

The generating function is

$$g(t) = \delta(t),$$

and therefore,

$$G(\omega) = 1.$$

Inserting this into (5.38), we get

$$F(\omega) = \sum_{n=-\infty}^{\infty} \omega_0 \delta\left(\omega - n\omega_0\right) = \sum_{n=-\infty}^{\infty} \frac{2\pi}{T_0} \delta\left(\omega - \frac{2\pi n}{T_0}\right),$$

which is a train of impulses in the frequency domain as shown in Figure 5.19(b).

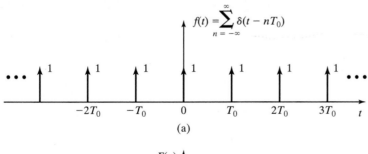

(a)

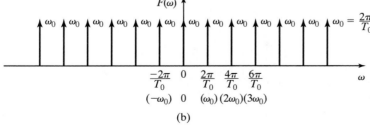

(b)

Figure 5.19   A train of impulses and its frequency spectrum. ∎

Summary

In this section we have calculated the Fourier transform of several time functions. The results of these derivations and the Fourier transforms of other time functions which are encountered in engineering practice are summarized in Table 5.2.

5.4 SAMPLING CONTINUOUS-TIME SIGNALS

The sampling of continuous-time signals is an important topic. The analysis and design of digital communication systems, digital controllers, etc. requires an understanding of the sampling process and its results.

Consider a continuous-time signal $f(t)$. We define sampling as the generation of an ordered number sequence by taking values of $f(t)$ at specified instants of time. Hence, sampling generates a number sequence $f(t_1), f(t_2), f(t_3), \ldots$, where the t_m are the instants at which sampling occurs. Note that we are considering $f(t)$ only at a set of fixed points in time.

A good physical example of the sampling operation under consideration is that implemented in hardware by an analog-to-digital converter (A/D or ADC). An analog-to-digital converter is an electronic device used to sample physical voltage signals. A common application of an analog-to-digital converter is the sampling of signals for processing by a digital computer.

An A/D is illustrated in Figure 5.20. The computer initiates the sampling operation by sending the A/D a control signal in the form of a pulse. The input signal $f(t)$ is sampled at the instant, t_m, that the pulse arrives, and the sample is converted into a binary number. Hence, the continuous-time signal $f(t)$ is now represented by a discrete-time binary code. This code is then transmitted to the computer for processing.

In most cases, continuous-time signals are sampled at equal increments of time. The sample increment, called the *sample period*, is usually denoted as T_s. Hence, the sampled signal values available in the computer are $f(nT_s)$, where n is an integer.

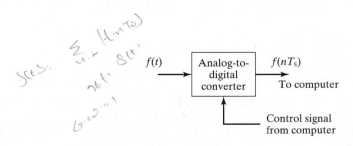

Figure 5.20 An analog-to-digital converter.

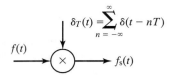

$$\delta_T(t) = \sum_{n=-\infty}^{\infty} \delta(t - nT)$$

$f(t)$

$f_s(t)$

Figure 5.21   Impulse sampling.

Impulse Sampling

The *ideal impulse sampling* operation is modeled by Figure 5.21 and is seen to be a modulation process (modulation will be discussed in Chapter 6), in which the carrier signal $\delta_T(t)$ is defined as the train of impulse functions:

$$\delta_T(t) = \sum_{n=-\infty}^{\infty} \delta(t - nT_S). \tag{5.39}$$

An illustration of $\delta_T(t)$ appears as Figure 5.19(a) if T_0 in the figure is replaced by T_s. (We justify this model later.) The output of the modulator, denoted by $f_s(t)$ is given by

$$f_s(t) = f(t)\delta_T(t) = f(t)\sum_{n=-\infty}^{\infty} \delta(t - nT_S) = \sum_{n=-\infty}^{\infty} f(nT_S)\delta(t - nT_S). \tag{5.40}$$

Ideal sampling is illustrated in Figure 5.22. Figures 5.22(a) and (b) show a continuous-time signal, $f(t)$, and the ideal sampling function, $\delta_T(t)$, respectively. The sampled signal $f_s(t)$ is illustrated in Figure 5.22(c) where the heights of the impulses are varied to imply graphically their variation in weight. Actually, all impulses have unbounded height, but each impulse in the sampled signal has its weight determined by the value of $f(t)$ at the instant that the impulse occurs.

We make two observations relative to $f_s(t)$. First, because impulse functions appear in this signal, it is not the exact model of a physical signal. The second observation is that the mathematical sampling operation does correctly result in the desired sampled sequence $f(nT_s)$ as weights of a train of impulses. It is shown in Chapter 6 that the modeling of the sampling operation using impulse functions is mathematically valid, even though $f_s(t)$ cannot appear in a physical system.

To investigate the characteristics of the sampling operation in Figure 5.21 and (5.40), we begin by taking the Fourier transform of $f_s(t)$. From Example 5.14,

$$\delta_T(t) = \sum_{n=-\infty}^{\infty} \delta(t - nT_S) \xleftrightarrow{\mathcal{F}} \omega_S \sum_{k=-\infty}^{\infty} \delta(\omega - k\omega_S), \tag{5.41}$$

where $\omega_s = 2\pi/T_s$ is the sampling frequency in radians/second. The sampling frequency in hertz is given by $f_s = 1/T_s$; therefore, $\omega_s = 2\pi f_s$.

Recall that the Fourier transform of an impulse function in time is not an impulse function in frequency; however, the Fourier transform of a train of periodic

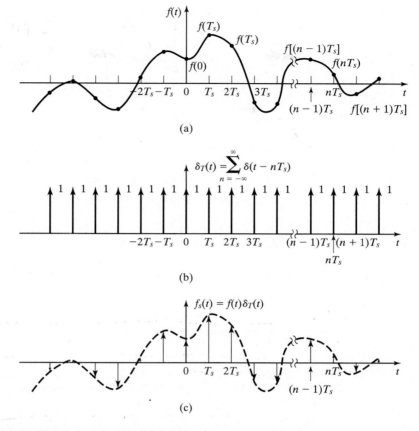

Figure 5.22   Generation of a sampled-data signal.

impulse functions in time is a train of impulse functions in frequency, just as the Fourier transform of any periodic signal is a sequence of impulses in frequency.

From Table 5.1, multiplication in the time domain results in convolution in the frequency domain. Then, from (5.40) and (5.41),

$$F_s(\omega) = \frac{1}{2\pi}F(\omega)*\left[\omega_s \sum_{k=-\infty}^{\infty} \delta(\omega - k\omega_S)\right] = \frac{1}{T_s} \sum_{k=-\infty}^{\infty} F(\omega)*\delta(\omega - k\omega_S). \quad (5.42)$$

Recall that because of the convolution property of the impulse function [see (3.23)],

$$F(\omega)*\delta(\omega - k\omega_S) = F(\omega - k\omega_S).$$

Thus, the Fourier transform of the impulse-modulated signal (5.40) is given by

$$F_s(\omega) = \frac{1}{T_s} \sum_{k=-\infty}^{\infty} F(\omega - k\omega_S). \quad (5.43)$$

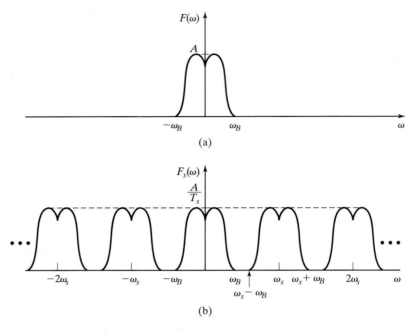

Figure 5.23 The frequency spectrum of a sampled-data signal.

Frequency domain characteristics of the sampling operation are now derived from this result.

We first let the frequency spectrum of the signal $f(t)$ be limited such that $F(\omega) = 0$ for $|\omega| > \omega_B$. [This is illustrated in Figure 5.23(a).] We assume that the highest frequency in $F(\omega)$ is less than one-half the sampling frequency; that is, $\omega_B < \omega_S/2$. From (5.43), we see that the effect of sampling $f(t)$ is to replicate the frequency spectrum of $F(\omega)$ about the frequencies $k\omega_S, k = \pm 1, \pm 2, \pm 3, \ldots$. This result is illustrated in Figure 5.23(b) for the signal of Figure 5.23(a). For this case, we can, theoretically, recover the signal $f(t)$ exactly from its samples using an ideal low-pass filter. We call the recovery of a signal from its samples *data reconstruction*. Data reconstruction will be discussed, as an application of the Fourier transform, in Chapter 6.

The frequency $\omega_S/2$ is called the *Nyquist frequency*. One of the requirements for sampling is that the sampling frequency must be chosen such that $\omega_S > 2\omega_M$, where ω_M is the highest frequency in the frequency spectrum of the signal to be sampled. This is stated in Shannon's sampling theorem [1]:

Shannon's sampling theorem

A function of time $f(t)$, which contains no frequency components greater than f_M hertz is determined uniquely by the values of $f(t)$ at any set of points spaced

$T_M/2$ ($T_M = 1/f_M$) seconds apart. Hence, according to Shannon's sampling theorem, we must take at least two samples per cycle of the highest frequency component in $f(t)$.

Figure 5.24 illustrates the requirement that the sampling frequency must be properly chosen. In Figure 5.24 the continuous sine wave represents a signal $f(t) = \sin(0.9\pi t)$. The black dots on stems represent the sampled values obtained by sampling the signal with a sampling period, $T_S = 2.5$ seconds; therefore, $\omega_S = 2\pi/2.5 = 0.8\pi$ (rad/s). Note that $\omega_S < 2\omega_M$. The dashed sine wave represents the signal that would be recovered from the sample data. It is seen that because the signal was sampled with a sampling frequency less than the Nyquist frequency ($\omega_S < 2\omega_M$), the signal recovered from the sample data is not the original signal. Instead, a signal $g(t) = \sin(0.3\pi t)$ is recovered from the sample data even though that signal was not part of the original signal, $f(t)$.

This phenomenon, wherein an erroneous signal is recovered from sample data because the sampling frequency was too low, is called *aliasing*. A familiar example of aliasing is seen in a movie or television program when the wheels of a car that is traveling at high speed appear to be rotating at a slower speed. This is caused by the relatively low sampling rate of the camera.

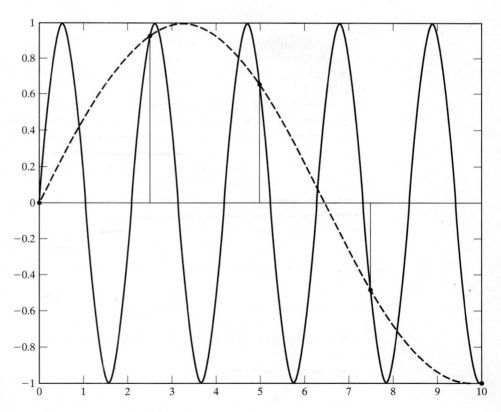

Figure 5.24 An illustration of aliasing.

Practical sampling

Shannon's theorem is not directly applicable to practical situations, because it requires samples of the signal for all time, both past and future. In the practical case, generally the sample period T_S is chosen to be much less than $T_M/2$ ($\omega_S \gg 2\omega_M$).

A signal limited in frequency cannot be limited in time; hence, a *physical* signal cannot be limited in frequency as shown in Figure 5.23(a), since it must then exist for all time. However, physical signals do exist such that the amplitude of the frequency spectrum above a certain frequency is so small as to be negligible. The practical requirement for sampling a signal is that the magnitude of the frequency spectrum of the signal be insignificant at frequencies greater than $\omega_S/2$.

Note the importance of the Fourier transform in determining the characteristics of sampling. The frequency domain clearly shows the effects of sampling a signal and the requirement for selecting the sampling frequency so that the signal can be recovered from the sample data.

5.5   APPLICATION OF THE FOURIER TRANSFORM

Frequency Response of Linear Systems

Fourier transforms can be used to simplify the calculation of the response of linear systems to input signals. For example, Fourier transforms allow the use of algebraic equations to analyze systems that are described by linear, time-invariant differential equations.

Consider the simple circuit shown in Figure 5.25(a), where $v_1(t)$ is the input signal and $v_2(t)$ is the output signal. This circuit can be described by the differential equations

$$v_1(t) = Ri(t) + L\frac{di(t)}{dt} \quad \text{and} \quad v_2(t) = L\frac{di(t)}{dt}.$$

If we take the Fourier transform of each equation, using the properties of linearity and time derivative, we get

$$V_1(\omega) = RI(\omega) + j\omega LI(\omega) \quad \text{and} \quad V_2(\omega) = j\omega LI(\omega).$$

From the first equation, we solve algebraically for $I(\omega)$:

$$I(\omega) = \frac{1}{R + j\omega L}V_1(\omega).$$

Substituting this result into the second equation yields

$$V_2(\omega) = \frac{j\omega L}{R + j\omega L}V_1(\omega),$$

which relates the output voltage of the system to the input voltage.

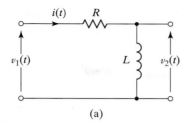

(a)

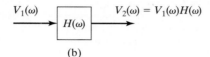

(b)

Figure 5.25   An electrical network and its block diagram.

We define a function

$$H(\omega) = \frac{j\omega L}{R + j\omega L} \tag{5.44}$$

and write the input–output relationship for the system as

$$V_2(\omega) = H(\omega)V_1(\omega), \tag{5.45}$$

or

$$H(\omega) = \frac{V_2(\omega)}{V_1(\omega)}. \tag{5.46}$$

Because the quantity $H(\omega)$ determines the output of the circuit for any given input signal, it is commonly called the *transfer function* of the system. The relationship of (5.45) is illustrated in Figure 5.25(b).

The function $H(\omega)$ in (5.44) gives mathematically the variation of the input–output relationship of the circuit with frequency. Therefore, $H(\omega)$ is also called the *frequency response function* of the system.

The frequency response function $H(\omega)$ is the same as the transfer function of (3.75) when $j\omega$ is substituted for s.

The frequency response can be determined experimentally, and somewhat laboriously, by applying a sinusoidal signal to the input of the circuit and measuring the magnitude and phase of the input and output signals. This process is repeated for different frequencies so that a large set of measurements is acquired over a wide range of frequencies. Since (5.46) can be expressed in polar form as

$$H(\omega) = |H(\omega)|\angle\phi(\omega) = \frac{|V_2(\omega)|\angle V_2}{|V_1(\omega)|\angle V_1} = \frac{|V_2(\omega)|}{|V_1(\omega)|}\angle V_2 - \angle V_1,$$

a plot of the ratio of the magnitudes of the input and output signals $|V_2(\omega)|/|V_1(\omega)|$ versus frequency yields a plot of $|H(\omega)|$. A plot of the difference between the recorded phase angles $(\underline{/V_2} - \underline{/V_1})$ versus frequency yields a plot of $\phi(\omega)$.

EXAMPLE 5.15   **The frequency response of a system**

An engineering professor required a student to determine the frequency response of the circuit shown in Figure 5.26(a). The student decided to spend some time in the laboratory collecting data using the experimental system shown in Figure 5.26(b). The student proceeded by measuring the input and output signals with a dual-trace oscilloscope as shown in Figure 5.26(c). A series of measurements were taken at a progression of frequency settings on the function generator that produced the input signal. For each frequency setting, the student set the amplitude of $v_1(t)$ to 1 V. Therefore, the input waveform at each frequency setting, ω_x, can be written as

$$v_1(t) = \cos \omega_x t.$$

The student then measured the amplitude of $v_2(t)$ using the oscilloscope and recorded it. Because a sinusoidal input to a linear system forces a sinusoidal output of the same frequency, but generally differing in amplitude and phase, the student knew that the output signal would be of the form

$$v_2(t) = A_2 \cos (\omega_x t + \phi_2).$$

Thus, the amplitude ratio is

$$|H(\omega_x)| = \frac{|V_2(\omega_x)|}{|V_1(\omega_x)|} = A_2.$$

By measuring and recording the time lag between zero-crossing points on the output sinusoid relative to the input, the student was able to calculate the phase difference using the equation

$$\phi_2 = \frac{t_1 - t_2}{T_x} \times 360°.$$

As shown in Figure 5.26(c), t_1 and t_2 are the time of the zero crossings of the input and output waveforms, respectively, and T_x is the period of the waveforms at the frequency currently being used.

To confirm that the data were correct, the student reviewed his circuit analysis notes and then derived the transfer function of the network using *sinusoidal-steady-state* circuit analysis techniques:

$$H(\omega) = \frac{R}{R + j\omega L + 1/j\omega C} = \frac{j(R/L)\omega}{-\omega^2 + 1/LC + j(R/L)\omega}$$

$$= \frac{j10^4\omega}{-\omega^2 + 10^8 + j10^4\omega}$$

$$= \frac{1}{1 + j(\omega^2 - 10^8)/10^4\omega}.$$

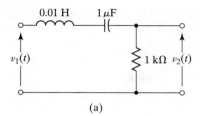

(a)

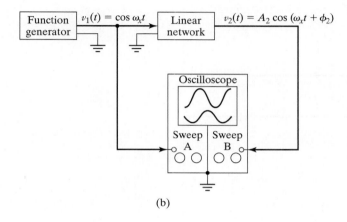

(b)

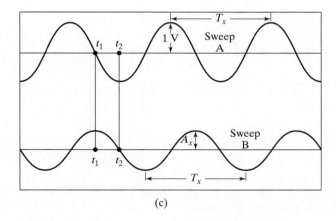

(c)

Figure 5.26 Illustrations for Example 5.15.

The transfer function can be written in polar form as

$$H(\omega) = |H(\omega)|e^{j\phi(\omega)},$$

where

$$|H(\omega)| = \frac{1}{[1 + [(\omega^2 - 10^8)/10^4\omega]^2]^{1/2}}$$

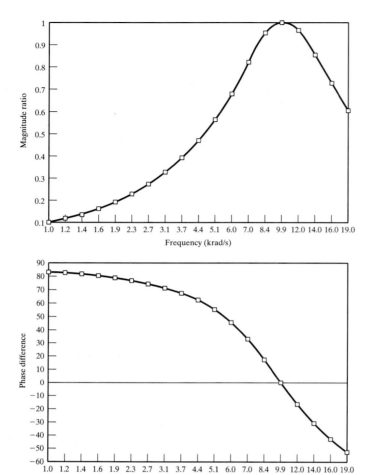

Figure 5.27   Magnitude ratio and phase difference plots.

and

$$\phi(\omega) = -\tan^{-1}[(\omega^2 - 10^8)/10^4\omega].$$

The student then plotted the magnitude and phase versus frequency as shown in Figure 5.27.

The experimental data from the physical system compared reasonably well with the calculated data, but did not coincide exactly with the plots of Figure 5.27. This is to be expected since the mathematical model used to compute the data can never describe the physical system perfectly.   ■

A widely used technique for displaying the frequency response of systems is the Bode plot. Bode plots most commonly consist of one semi-logarithmic plot of the magnitude frequency response (in decibels) and a separate semi-logarithmic plot of the phase frequency response (in degrees), both plotted vs. frequency on $\log_{10}$ scale.

EXAMPLE 5.16   **Bode plots of a system's frequency response**

The frequency response function used in Example 5.15 can be written as

$$H(\omega) = \frac{10^4(j\omega)}{(j\omega)^2 + 10^4(j\omega) + 10^8}.$$

With $H(\omega)$ written in this form, the numerator and denominator coefficient vectors for use in computing the frequency response using MATLAB are

$$num = [10^4 \quad 0] \quad \text{and} \quad den = [1 \quad 10^4 \quad 10^8].$$

Using the results of Example 5.15, we choose to calculate and plot the frequency response over the frequency range $10^3 \le \omega \le 10^5 (rad/s)$. Since the plot will range over two decades of frequency, we choose to plot 200 points of data. For use in MATLAB, we write the vector

$$flim = [3, 5, 200],$$

where the first two numbers establish the range of frequencies and the third entry is the number of points of data to be calculated.
The MATLAB command

$$\textbf{bodec}(\texttt{num,den,flim})$$

causes the MATLAB m-file listed to execute and produce the Bode plots shown in Figure 5.28.

```
function bodec (num,den,flim)
% This function plots the magnitude and phase Bode plots of the frequency
% response for an LTIC system. The system is modeled by a transfer function
% with a numerator polynomial represented by the vector of coefficients "num"
% and a denominator polynomial represented by the vector of coefficients "den".
%
% The frequency range, is entered as an array of the form: flim= [d1,d2,N] where
% d1 and d2 are the lower and upper limits of the frequency range to be
% used expressed as integer exponents of 10. e.g. For a lower limit of
% 0.01 rad/s, enter -2 for d1. For an upper limit of 100 rad/s, enter 2
% for d2. Enter 200 for N to generate 200 points of data. For this example
% enter >>flim= [-2,2,200].
%
 w=logspace (flim(1,1), flim(1,2), flim(1,3));
for k=1:length (w);
nx (k) =polyval (num, j*w (k));
dx (k) =polyval (den, j*w (k));
x (k) =nx (k)/dx (k);
end
  mdb=20*log (abs (x));
  px=180*angle (x)/pi;
  figure (1), semilogx (w,mdb), grid, title ('Magnitude Bode
Plot'), xlabel ('Frequency (rad/s)'), ylabel ('Magnitude (dB)')
  figure (2), semilogx (w,px), grid, title ('Phase Bode
Plot'), xlabel ('Frequency (rad/s)'), ylabel ('Phase Angle (deg)')
```

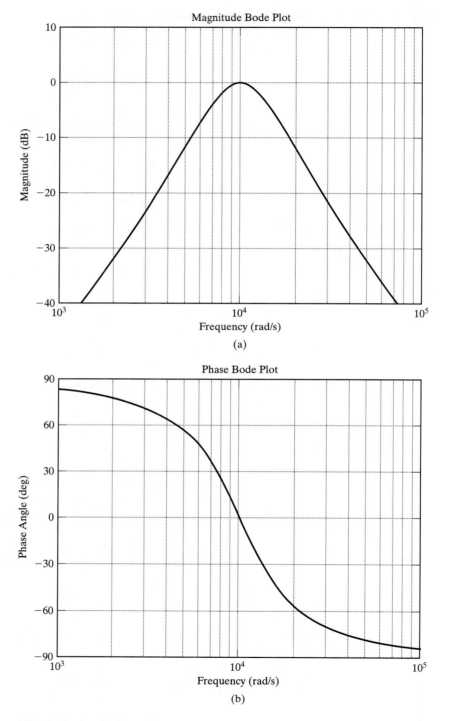

Figure 5.28 Bode plots for Example 5.16.

We have established that the output of a linear system can be expressed in the frequency domain as the product of the system's transfer function and the Fourier transform of the input signal

[eq(5.45)] $$V_2(\omega) = H(\omega)V_1(\omega).$$

Consider now the result when the input is an impulse function $v_1(t) = \delta(t)$. In this case, $V_1(\omega) = 1$ and, therefore, $V_2(\omega) = H(\omega)$. In other words, the transform of the output is exactly equal to the transfer function when the input is an impulse. Considering this, we call the inverse Fourier transform of the transfer function the *impulse response*, denoted by $h(t)$. Thus,

$$h(t) \xleftrightarrow{\mathcal{F}} H(\omega).$$

Using the convolution property to find the inverse Fourier transform of (5.45) yields

$$v_2(t) = v_1(t)*h(t) = \int_{-\infty}^{\infty} v_1(\tau)h(t - \tau)d\tau. \tag{5.47}$$

Of course, this agrees with the convolution integral in (3.18).

If the impulse response or its Fourier transform, the transfer function of a linear system is known, the output for any given input can be found by evaluating either (5.45) or (5.47).

EXAMPLE 5.17   **Using the Fourier transform to find the response of a system to an input signal**

Consider the system shown in Figure 5.29(a). The electrical network in the diagram responds to an impulse of voltage at the input, $x(t) = \delta(t)$, with an output of $h(t) = (1/RC)e^{-t/RC}u(t)$ as shown in Figure 5.29(b). Our task is to determine the frequency spectrum of the output of this system for a step function input of voltage. Thus, $x(t) = Vu(t)$, as shown in Figure 5.29(c). Because the impulse response of this linear network is known, the output response to any input can be determined by evaluating the convolution integral:

$$y(t) = x(t)*h(t) = \int_{-\infty}^{\infty} x(\tau)h(t - \tau)d\tau.$$

This looks like an onerous task; however, we are saved by the convolution property of the Fourier transform. From (5.14),

$$Y(\omega) = \mathcal{F}\{x(t)*h(t)\} = X(\omega)H(\omega).$$

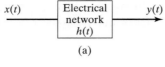

(a)

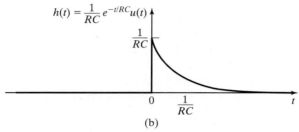

(b)

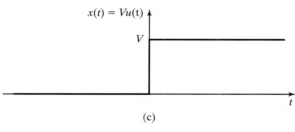

(c)

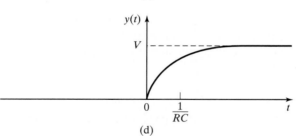

(d)

Figure 5.29   Illustrations for Example 5.17.

From Table 5.2, the Fourier transforms of $h(t)$ and $x(t)$ are found to be

$$H(\omega) = \mathcal{F}\{(1/RC)e^{-t/RC}u(t)\} = \frac{-1}{1 + j\omega RC}$$

and

$$X(\omega) = \mathcal{F}\{Vu(t)\} = V\left[\frac{1}{j\omega} + \pi\delta(\omega)\right].$$

Therefore,

$$Y(\omega) = V\left[\frac{1}{1 + j\omega RC}\right]\left[\frac{1}{j\omega} + \pi\delta(\omega)\right] = V\left[\frac{1}{j\omega(1 + j\omega RC)} + \frac{\pi\delta(\omega)}{1 + j\omega RC}\right].$$

Partial fraction expansion of the first term in brackets (see Appendix F) gives

$$Y(\omega) = V\left[\frac{-RC}{1 + j\omega RC} + \frac{1}{j\omega} + \frac{\pi\delta(\omega)}{1 + j\omega RC}\right].$$

The impulse function in the third term in brackets has a nonzero value only at $\omega = 0$; at $\omega = 0$ the denominator has a value of 1, and hence the equation is further simplified to

$$Y(\omega) = V\left[\frac{-RC}{1 + j\omega RC} + \frac{1}{j\omega} + \pi\delta(\omega)\right].$$

This equation describes the frequency spectrum of the output signal of the network when the input is a step function of magnitude V.

The time-domain representation of the output can now be found:

$$y(t) = \mathscr{F}^{-1}\{Y(\omega)\} = V\left[\mathscr{F}^{-1}\left\{\frac{1}{j\omega} + \pi\delta(\omega)\right\} - \mathscr{F}^{-1}\left\{\frac{1}{(1/RC) + j\omega}\right\}\right].$$

Using the transform pairs listed in Table 5.2, we find that

$$u(t) \xleftrightarrow{\mathscr{F}} \frac{1}{j\omega} + \pi\delta(\omega) \quad \text{and} \quad e^{-t/RC}u(t) \xleftrightarrow{\mathscr{F}} \frac{1}{(1/RC) + j\omega}.$$

Therefore, the time-domain expression for the output of the network is

$$y(t) = V(1 - e^{-t/RC})u(t).$$

The output waveform is shown in Figure 5.29(d).

This example, which yields what is probably a familiar result for those who have studied electrical circuit analysis, shows the utility of the convolution property in system analysis. ∎

Frequency Spectra of Signals

The Fourier transform can be used to analyze the frequency spectrum of any physical signal that can be described mathematically. The following example illustrates the procedure for a signal that is often present in electronic systems.

EXAMPLE 5.18   **The frequency spectrum of a rectified sinusoidal signal**

The half-wave rectifier described in Section 1.2 is shown in Figure 5.30. The rectifier circuit is a nonlinear system and therefore cannot be described by a frequency response function. However, the signal at the output of the rectifier, shown in Figure 5.31(c), can be described by a Fourier transform.

The half-wave rectified signal, $v_1(t)$, shown in Figure 5.31(c) is described by the product of the full cosine wave, $v_s(t)$, shown in Figure 5.31(a) and the train of rectangular pulses, $f(t)$, shown in Figure 5.31(b).

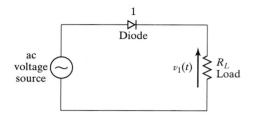

Figure 5.30   Half-wave rectifier.

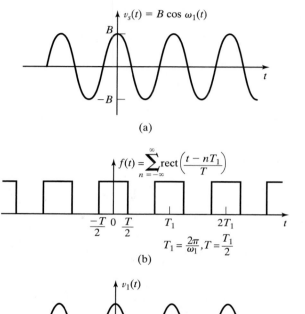

Figure 5.31   Waveforms for Example 5.18.

After some relatively simple algebraic manipulation, we find $V_1(\omega)$ by using the frequency shift and linearity properties of the Fourier transform:

$$v_1(t) = f(t)B \cos (\omega_1 t) = f(t)\frac{B}{2}[e^{j\omega_1 t} + e^{-j\omega_1 t}]$$

$$= \frac{B}{2}f(t)e^{j\omega_1 t} + \frac{B}{2}f(t)e^{-j\omega_1 t}.$$

Using Table 5.1, we see that

$$V_1(\omega) = \frac{B}{2}F(\omega - \omega_1) + \frac{B}{2}F(\omega + \omega_1).$$

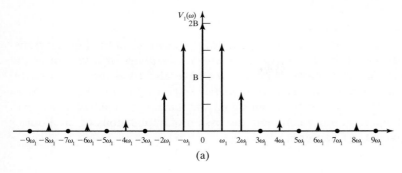

(a)

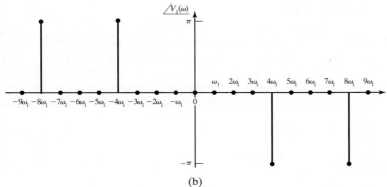

(b)

Figure 5.32 Frequency spectra for
Example 5.18.

In Example 5.13, we found the Fourier transform of a train of rectangular pulses such as $f(t)$
to be

$$F(\omega) = \sum_{n=-\infty}^{\infty} AT\,\omega_0 \operatorname{sinc}(nT\omega_0/2)\delta(\omega - n\omega_0).$$

For the particular signal we are now considering, $A = 1$, $T = T_1/2$, and $\omega_0 = \omega_1 = 2\pi/T_1$;
therefore,

$$F(\omega) = \sum_{n=-\infty}^{\infty} \pi \operatorname{sinc}(n\pi/2)\delta(\omega - n\omega_1).$$

Thus,

$$V_1(\omega) = \frac{B\pi}{2} \sum_{n=-\infty}^{\infty} \operatorname{sinc}(n\pi/2)[\delta(\omega - \omega_1 - n\omega_1) + \delta(\omega + \omega_1 - n\omega_1)].$$

The frequency spectrum of the half-wave rectified signal is shown in Figure 5.32.   ■

It should be noted that the availability of an ideal rectifier is assumed in the
previous example. Measured voltages from actual circuits cannot be expected to
match exactly the results of Example 5.18.

Summary

We have considered two of the many engineering applications of the Fourier transform. In this section, we have defined the *frequency response* of a system as the
Fourier transform of the *impulse response*. We have also used the term *transfer
function* as a pseudonym for the system *frequency response function* because it describes the relationship between the input signal and the output signal of a system in
the frequency domain. A more complete discussion of applications of the Fourier
transform is provided in Chapter 6.

5.6 ENERGY AND POWER DENSITY SPECTRA

In this section, we define and show application for the *energy spectral density function* and the *power spectral density function*. These two functions are used to determine the energy distribution of an energy signal or the power distribution of a
power signal in the frequency spectrum. Knowledge of the energy or power distribution of a signal can be quite valuable in the analysis and design of communication
systems, for example.

Energy Density Spectrum

An energy signal is defined in Section 5.1 as a waveform, $f(t)$, for which

[eq(5.5)]
$$E = \int_{-\infty}^{\infty} |f(t)|^2 dt < \infty,$$

where E is the energy associated with the signal. It was noted that energy signals
generally include aperiodic signals that have a finite time duration and signals that
approach zero asymptotically as t approaches infinity.

If the signal is written in terms of its Fourier transform,

$$f(t) = \frac{1}{2\pi} \int_{-\infty}^{\infty} F(\omega)e^{j\omega t} d\omega,$$

its energy equation can be rewritten as

$$E = \int_{-\infty}^{\infty} f(t)\left[\frac{1}{2\pi} \int_{-\infty}^{\infty} F(\omega)e^{j\omega t} d\omega \right] dt.$$

The order of integration can be rearranged so that

$$E = \frac{1}{2\pi} \int_{-\infty}^{\infty} F(\omega)\left[\int_{-\infty}^{\infty} f(t)e^{j\omega t} dt \right] d\omega.$$

The term in brackets is similar to the defining equation for the Fourier transform (5.1); the difference is the sign of the exponent. Substitution of $-\omega$ for ω in the Fourier transform equation yields

$$F(-\omega) = \int_{-\infty}^{\infty} f(t)e^{j\omega t}dt.$$

Substituting this result into the energy equation (5.5) yields

$$E = \frac{1}{2\pi}\int_{-\infty}^{\infty} F(\omega)F(-\omega)d\omega.$$

For signals $f(t)$ that are real valued (this includes all voltage and current waveforms that can be produced by a physical circuit),

$$F(-\omega) = F^*(\omega),$$

where $F^*(\omega)$ is the complex conjugate of the function $F(\omega)$. Hence,

$$E = \frac{1}{2\pi}\int_{-\infty}^{\infty} F(\omega)F^*(\omega)d\omega = \frac{1}{2\pi}\int_{-\infty}^{\infty} |F(\omega)|^2 d\omega.$$

The final, important result that we want to recognize is that

$$E = \int_{-\infty}^{\infty} |f(t)|^2 dt = \frac{1}{2\pi}\int_{-\infty}^{\infty} |F(\omega)|^2 d\omega. \tag{5.48}$$

The relationship described by equation (5.48) is known as *Parseval's theorem*. It can be shown that (5.48) is valid for both real- and complex-valued signals.

Because the function $|F(\omega)|^2$ is a real and even function of frequency, we can rewrite the energy equation in the frequency spectrum as

$$E = \frac{1}{2\pi}\int_{-\infty}^{\infty} |F(\omega)|^2 d\omega = \frac{1}{\pi}\int_{0}^{\infty} |F(\omega)|^2 d\omega.$$

The *energy spectral density* function of the signal $f(t)$ is defined as

$$\mathscr{E}_f(\omega) \equiv \frac{1}{\pi}|F(\omega)|^2 = \frac{1}{\pi}F(\omega)F(\omega)^* \tag{5.49}$$

and describes the distribution of signal energy over the frequency spectrum. With the energy density function thus defined, the energy equation (5.48) can be rewritten

$$E = \int_{0}^{\infty} \mathscr{E}_f(\omega)d\omega. \tag{5.50}$$

EXAMPLE 5.19   **Energy spectral density of a rectangular pulse**

For the rectangular waveform shown in Figure 5.33(a), we have previously found the frequency spectrum to be described by the sinc function shown in Figure 5.33(b). We now find the energy spectral density. The magnitude of this curve is squared and divided by 2π to form the frequency spectrum of the energy from (5.48); the result is shown in Figure 5.33(c). Next we fold the energy frequency spectrum about the $\omega = 0$ axis and add the frequency components as they overlap. This result is shown in Figure 5.33(d), which is a plot of the energy spectral density, $\mathscr{E}_f(\omega)$, of the rectangular waveform.

The energy contained in some band of frequencies of particular interest is found by finding the area under the energy spectral density curve over that band of frequencies. For example, in Figure 5.33(d), the amount of energy contained in the band of frequencies between ω_1 and ω_2 is the shaded area under the curve. This energy content can be found

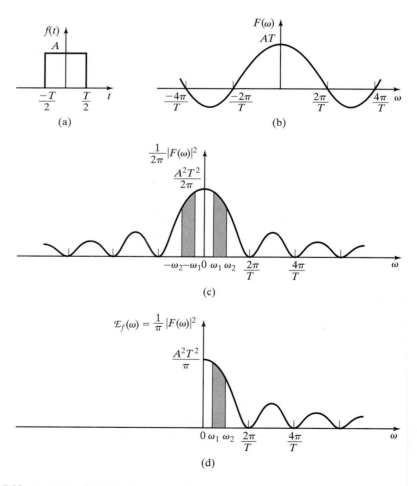

Figure 5.33   A rectangular voltage pulse and its energy spectrum.

mathematically by evaluating

$$E_B = \int_{\omega_1}^{\omega_2} \mathcal{E}_f(\omega)d\omega.$$ ∎

Power Density Spectrum

Next we consider signals that have infinite energy but contain a finite amount of power. For these signals, the normalized average signal power is finite:

$$P = \lim_{T \to \infty} \frac{1}{T} \int_{-\frac{T}{2}}^{\frac{T}{2}} |f(t)|^2 dt < \infty. \tag{5.51}$$

Such signals are called *power signals*.

The step function, the signum function, and all periodic functions are examples of power signals. As the reader might already have reasoned, power signals are often employed in real-life applications because many physical systems make use of periodic waveforms.

A problem with working in the frequency domain in the case of power signals arises from the fact that power signals have infinite energy and, therefore, may not be Fourier transformable. To overcome this problem, a version of the signal that is truncated in time is employed. The signal $f_T(t)$ shown in Figure 5.34(c) is a truncated

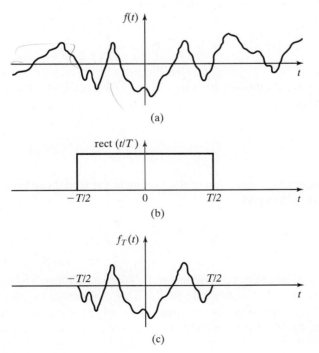

Figure 5.34   The time truncation of a power signal.

version of the signal $f(t)$. The truncation can be achieved by multiplying the signal $f(t)$ as shown in Figure 5.34(a) by a rectangular pulse having unity amplitude and duration T as shown in Figure 5.34(b). The truncated signal

$$f_T(t) = f(t) \, \text{rect}(t/T)$$

has finite energy. This signal meets the other Dirichlet conditions and, therefore, has a Fourier transform:

$$f_T(t) \xleftrightarrow{\mathscr{F}} F_T(\omega).$$

In working with power signals, it is often desirable to know how the total power of the signal is distributed in the frequency spectrum. This can be determined by developing a *power spectral density* function similar to the energy spectral density function considered earlier. We begin by writing the power equation in terms of the truncated signal:

$$P = \lim_{T \to \infty} \frac{1}{T} \int_{-\infty}^{\infty} |f_T(t)|^2 dt.$$

Note that the limits of integration have been changed from (5.51). This is justified because $f_T(t)$ has zero magnitude for $|t| > T/2$.

Because $f_T(t)$ has finite energy, the integral term can be recognized as the total energy contained in the truncated signal:

$$E = \int_{-\infty}^{\infty} |f_T(t)|^2 dt.$$

The energy can be expressed in terms of $f_T(t)$ by applying Parseval's theorem (5.48), to get

$$E = \int_{-\infty}^{\infty} |f_T(t)|^2 dt = \frac{1}{2\pi} \int_{-\infty}^{\infty} |F_T(\omega)|^2 d\omega.$$

The frequency-domain expression of the energy in the signal can be substituted into the power equation to yield

$$P = \lim_{T \to \infty} \frac{1}{2\pi T} \int_{-\infty}^{\infty} |F_T(\omega)|^2 d\omega. \tag{5.52}$$

As the duration of the rectangular pulse increases, it can be seen that the energy of the signal will also increase. In the limit, as T approaches infinity, the energy will become infinite also. For the average power of the signal to remain finite, the energy of the signal must increase at the same rate as T, the duration of the signal.

Under this condition, it is permissible to interchange the order of the limiting action on T and the integration over ω so that

$$P = \frac{1}{2\pi} \int_{-\infty}^{\infty} \lim_{T \to \infty} \frac{1}{T} |F_T(\omega)|^2 d\omega.$$

In this form of the average power equation, the integrand is called the *power spectral density* and is denoted by the symbol

$$\mathcal{P}_f(\omega) \equiv \lim_{T \to \infty} \frac{1}{T} |F_T(\omega)|^2. \tag{5.53}$$

In terms of the power spectral density function, the equation for normalized average signal power is

$$P = \frac{1}{2\pi} \int_{-\infty}^{\infty} \mathcal{P}_f(\omega) d\omega = \frac{1}{\pi} \int_{0}^{\infty} \mathcal{P}_f(\omega) d\omega, \tag{5.54}$$

because $\mathcal{P}_f(\omega)$ is an even function.

For periodic signals, the normalized average power can be determined from the Fourier series as

$$P = \sum_{k=-\infty}^{\infty} |C_k|^2 = C_0^2 + 2\sum_{k=1}^{\infty} |C_k|^2. \tag{5.55}$$

Using the relationship shown in (5.35),

$$|F(k\omega_0)| = 2\pi |C_k|,$$

the normalized average power of a signal $f(t)$ is written in terms of the Fourier transform as

$$P = \left(\frac{1}{2\pi}\right)^2 \sum_{k=-\infty}^{\infty} |F(k\omega_0)|^2 = \frac{1}{4\pi^2} |F(0)|^2 + \frac{1}{2\pi^2} \sum_{k=1}^{\infty} |F(k\omega_0)|^2. \tag{5.56}$$

It is seen that for a periodic signal, the power distribution over any band of frequencies can be determined from the Fourier transform of the signal.

EXAMPLE 5.20   **Power spectral density of a periodic signal**

The magnitude frequency spectrum of a periodic signal is shown in Figure 5.35(a). According to (5.56), the power spectral density can be displayed by squaring the magnitude of each discrete frequency component and dividing by $4\pi^2$. This result is shown in Figure 5.35(b). It

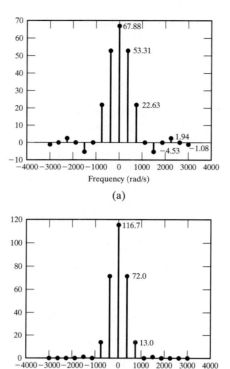

Figure 5.35 Power spectral density of a periodic signal.

should be noted that the values of the spectral components at frequencies above 1000 rad/s are small in magnitude but not zero as may be implied by Figure 5.35(b).

The normalized average power in the frequency band $|\omega| \leq 1000$ rad/s is found by summing the power of the discrete frequency components in that range:

$$116.7 + 2(72.0 + 13.0) = 286.7 \text{ W.}\qquad\blacksquare$$

Power and Energy Transmission

The input–output relationship of a system

$$G(\omega) = H(\omega)F(\omega) \qquad (5.57)$$

can also be expressed in terms of the energy or power spectral densities of the input and output signals. First, we conjugate both sides of (5.57):

$$G^{*}(\omega) = [H(\omega)F(\omega)]^{*} = H^{*}(\omega)F^{*}(\omega).$$

Multiplying both sides of the equation by (5.57) yields

$$G(\omega)G^*(\omega) = H(\omega)H^*(\omega)F(\omega)F^*(\omega)$$

or

$$|G(\omega)|^2 = |H(\omega)|^2|F(\omega)|^2. \tag{5.58}$$

If both sides on this expression are divided by π and the equivalents from (5.49) are substituted, we have an expression that describes the transmission of energy through a linear system:

$$\mathcal{E}_g(\omega) = |H(\omega)|^2\mathcal{E}_f(\omega). \tag{5.59}$$

For the case that the input to a system is a power signal, the time-averaging operation can be applied to both sides of (5.58):

$$\lim_{T\to\infty}\frac{1}{T}|G_T(\omega)|^2 = |H(\omega)|^2\lim_{T\to\infty}\frac{1}{T}|F_T(\omega)|^2,$$

Then, from (5.53),

$$\mathcal{P}_g(\omega) = |H(\omega)|^2\mathcal{P}_f(\omega). \tag{5.60}$$

Usually, the exact content of an information signal in a communications system cannot be predicted; however, its power spectral density can be determined statistically. Thus, (5.58) is often used in the analysis and design of these systems.

EXAMPLE 5.21   **Power spectral density of a system's output signal**

A signal $x(t)$ with power spectral density shown in Figure 5.36(a) is the input to a linear system with the frequency response plotted in Figure 5.36(b). The power spectral density of the output signal, $y(t)$, is determined by application of Equation (5.60).

Because the power spectral density of the input signal is a discrete-frequency function,

$$\mathcal{P}_y(\omega) = \mathcal{P}_x(\omega)|H(\omega)|^2$$

can be determined by evaluating the equation at only those frequencies where the power spectral density of $x(t)$ is nonzero. For example, to determine the power density in the output signal at $\omega = 60$ (rad/s), from the frequency response of the linear system, we find that $|H(60)| = 0.7071$; therefore, $|H(60)|^2 = 0.5$. From Figure 5.36(a), $\mathcal{P}_x(60) = 7.84$. We calculate $\mathcal{P}_y(\omega) = 3.92$. These calculations are repeated for all frequencies of interest. The power spectral density of the output signal is plotted in Figure 5.36(c).

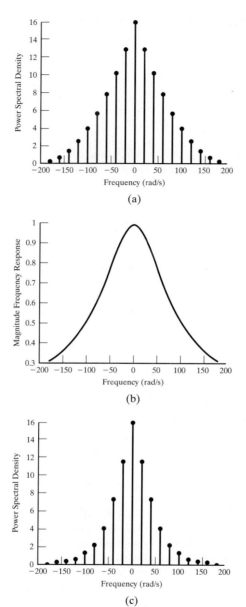

Figure 5.36   Plots for Example 5.21.   ■

Summary

In this section, we have defined the *energy spectral density function*

[eq(5.49)] $$\mathcal{E}_f(\omega) \equiv \frac{1}{\pi}|F(\omega)|^2,$$

which describes the distribution of the energy in an *energy signal* in the frequency spectrum, and the *power spectral density function*

[eq(5.53)]
$$\mathscr{P}_f(\omega) \equiv \lim_{T \to \infty} \frac{1}{T} |F_T(\omega)|^2,$$

which describes the distribution of the power in a *power signal* in the frequency spectrum.

The input–output relationship for an *energy signal* transmitted through a linear system was determined to be

[eq(5.59)]
$$\mathscr{E}_g(\omega) = |H(\omega)|^2 \mathscr{E}_f(\omega),$$

where $\mathscr{E}_f(\omega)$ is the energy spectral density of the input signal and $\mathscr{E}_g(\omega)$ is the energy spectral density of the output signal.

The input–output relationship for a power signal transmitted through a linear system was shown to be

[eq(5.60)]
$$\mathscr{P}_g(\omega) = |H(\omega)|^2 \mathscr{P}_f(\omega),$$

where $\mathscr{P}_f(\omega)$ is the power spectral density of the input signal and $\mathscr{P}_g(\omega)$ is the power spectral density of the output signal.

SUMMARY

In this chapter, we defined the Fourier transform (5.1) and the inverse Fourier transform (5.2).

The sufficient conditions for the existence of the integral (5.1) are called the Dirichlet conditions. In general, the Fourier transform of $f(t)$ exists if it is reasonably well behaved (if we could draw a picture of it) and if it is absolutely integrable. These conditions are sufficient, but not necessary. It is shown that many practical signals that do not fit these conditions do, in fact, have Fourier transforms.

The Fourier transform of a time-domain signal is called the frequency spectrum of the signal. The frequency spectrum is often plotted in two parts. $|F(\omega)|$ is plotted as the *magnitude spectrum* and $\arg[F(\omega)]$ is plotted as the *phase spectrum*. A third representation of the frequency spectrum is a plot of $|F(\omega)|^2$, which is called the *energy spectrum*.

Several useful properties of the Fourier transform were introduced and are listed in Table 5.1. Fourier transforms of several time-domain functions were derived and listed in Table 5.2.

The Fourier transform of the *impulse response* of a linear system was shown to be the system's *frequency response*, which is also the *transfer function* of the system in the frequency domain. This led to the important result that if a system with transfer function $H(\omega)$ has an input signal $X(\omega)$, the Fourier transform of the output signal $Y(\omega)$ is given by the product of the transfer function and the Fourier transform of the input function.

Energy and power spectral densities were defined by (5.49) and (5.53), respectively.

The usefulness of the energy and power spectral density functions in the analysis of systems and signals was discussed. An important application is the study of power-signal transmission through a linear system. In this case, the equation for the power spectral density of the output signal was given in terms of the system transfer function and the power spectral density of the input signal.

Several additional applications of the Fourier transform are discussed in Chapter 6. See Table 5.3.

TABLE 5.3 Key Equations of Chapter 5

Equation Title	Equation Number	Equation
Fourier transform	(5.1)	$\mathcal{F}\{f(t)\} = F(\omega) = \int_{-\infty}^{\infty} f(t)e^{-j\omega t}dt$
Inverse Fourier transform	(5.2)	$f(t) = \dfrac{1}{2\pi}\int_{-\infty}^{\infty} F(\omega)e^{j\omega t}d\omega = \mathcal{F}^{-1}\{F(\omega)\}$
Fourier transform of *rect* function	(5.4)	$\mathrm{rect}(t/T) \xleftrightarrow{\mathcal{F}} T\,\mathrm{sinc}(T\,\omega/2)$
Time-transformation property	(5.14)	$f(at - t_0) \xleftrightarrow{\mathcal{F}} \dfrac{1}{|a|}F\left(\dfrac{\omega}{a}\right)e^{-jt_0(\omega/a)}$
Duality property	(5.15)	$F(t) \xleftrightarrow{\mathcal{F}} 2\pi f(-\omega)$ when $f(t) \xleftrightarrow{\mathcal{F}} F(\omega)$
Convolution property	(5.16)	$f_1(t)*f_2(t) \xleftrightarrow{\mathcal{F}} F_1(\omega)F_2(\omega)$
Multiplication property	(5.17)	$f_1(t)f_2(t) \xleftrightarrow{\mathcal{F}} \dfrac{1}{2\pi}F_1(\omega)*F_2(\omega)$
Frequency-shifting property	(5.18)	$x(t)e^{j\omega_0 t} \xleftrightarrow{\mathcal{F}} X(\omega-\omega_0)$
Fourier transform of periodic signal	(5.35)	$\displaystyle\sum_{k=-\infty}^{\infty} C_k e^{jk\omega_0 t} \xleftrightarrow{\mathcal{F}} 2\pi \sum_{k=-\infty}^{\infty} C_k\delta(\omega-k\omega_0)$
Sampled signal	(5.40)	$f_s(t) = f(t)\delta_T(t) = f(t)\displaystyle\sum_{n=-\infty}^{\infty} \delta(t - nT_s) = \sum_{n=-\infty}^{\infty} f(nT_s)\delta(t - nT_s)$
Ideal sampling function	(5.41)	$\delta_T(t) = \displaystyle\sum_{n=-\infty}^{\infty} \delta(t - nT_s) \xleftrightarrow{\mathcal{F}} \omega_s \sum_{k=-\infty}^{\infty} \delta(\omega-k\omega_s)$
Fourier transform of sampled signal	(5.42)	$F_s(\omega) = \dfrac{1}{2\pi}F(\omega)*[\omega_s \displaystyle\sum_{k=-\infty}^{\infty} \delta(\omega-k\omega_s)] = \dfrac{1}{T_s}\sum_{k=-\infty}^{\infty} F(\omega)*\delta(\omega-k\omega_s)$
Frequency response		$Y(\omega) = X(\omega)H(\omega),\; h(t) \leftrightarrow H(\omega)$

REFERENCES

1. R. Bracewell, *The Fourier Transform and Its Applications*, 2d ed. New York: McGraw–Hill, 1986.
2. M. J. Lighthill, *Fourier Analysis and Generalised Functions*. Cambridge: Cambridge University Press, 1958.
3. A. Papoulis, *The Fourier Integral and Its Applications*. New York: McGraw–Hill, 1962.
4. R. A. Gabel and R. A. Roberts, *Signals and Linear Systems*. New York: Wiley, 1987.
5. H. P. Hsu, *Fourier Analysis*. New York: Simon & Schuster, 1970.
6. B. P. Lathi, *Signals, Systems and Communication*. New York: Wiley, 1965.
7. N. K. Sinha, *Linear Systems*. New York: Wiley, 1991.
8. S. S. Soliman and M. D. Srinath, *Continuous and Discrete Signals and Systems*, 2d ed. Upper Saddle River, NJ: Prentice Hall, 1997.
9. A. V. Oppenheim and A. S. Willsky, *Signals and Systems*, 2d ed. Upper Saddle River, NJ: Prentice Hall, 1996.

PROBLEMS

5.1. Find the Fourier transform for each of the following signals, using the Fourier integral:

 (a) $x(t) = 2[u(t) - u(t - 4)]$

 (b) $x(t) = e^{-3t}[u(t) - u(t - 4)]$

 (c) $x(t) = 2t[u(t) - u(t - 4)]$

 (d) $x(t) = \cos(4\pi t)[u(t + 2) - u(t - 2)]$

5.2. Use the definition of the Fourier transform (5.1) to find the transform of the following time signals:

 (a) $f(t) = (1 - e^{-bt})\, u(t)$

 (b) $f(t) = A \cos(\omega_0 t + \phi)$

 (c) $f(t) = e^{at}u(-t),\, a > 0$

 (d) $f(t) = C\delta(t + t_0)$

5.3. Use the table of Fourier transforms (Table 5.2) and the table of properties (Table 5.1) to find the Fourier transform of each of the following signals. *Do not* use the Fourier integral (5.1). (Note that the signals are the same as in Problem 5.1.)

 (a) $x(t) = 2[u(t) - u(t - 4)]$

 (b) $x(t) = e^{-3t}[u(t) - u(t - 4)]$

 (c) $x(t) = 2t[u(t) - u(t - 4)]$

 (d) $x(t) = \cos(4\pi t)[u(t + 2) - u(t - 2)]$

5.4. Prove mathematically that the following properties of the Fourier transform described in Table 5.1 are valid:

 (a) Linearity

 (b) Time shifting

(c) Duality

(d) Frequency shifting

(e) Time differentiation

(f) Time convolution

(g) Time-scale property

5.5. The Fourier transform of $\sin(\omega_0 t)$ is given in Table 5.2. Derive the Fourier transform of $\cos(\omega_0 t)$ using the following:

(a) the differentiation property

(b) the time-shifting property

5.6. Find and sketch the Fourier transform of the following time-domain signals.

(a) $Ae^{-\beta t}\cos(\omega_0 t)u(t)$, $\mathrm{Re}\{\beta\} > 0$

(b) $A\sin(\omega_1 t) + B\cos(\omega_2 t)$

(c) $6\,\mathrm{sinc}(0.5t)$

(d) $6\,\mathrm{rect}[(t - 4)/3]$

5.7. Find the frequency spectra of the signals shown in Figure P5.7.

5.8. Given

$$e^{-|t|} \xleftrightarrow{\ \mathcal{F}\ } \frac{2}{\omega^2 + 1},$$

find the Fourier transform of the following:

(a) $\dfrac{d}{dt}e^{-|t|}$

(b) $\dfrac{1}{2\pi(t^2 + 1)}$

(c) $\dfrac{4\cos(2t)}{t^2 + 1}$

5.9. Find and compare the frequency spectra of the trapezoidal waveforms shown in Figure P5.9.

5.10. Use the time-derivative property to find the Fourier transform of the triangular waveform shown in Figure P5.10.

5.11. Consider a linear, time-invariant system with impulse response

$$h(t) = \frac{0.5\sin(2t)}{t}.$$

Find the system output $y(t)$ if the input is $x(t) = \cos(t) + \sin(3t)$.

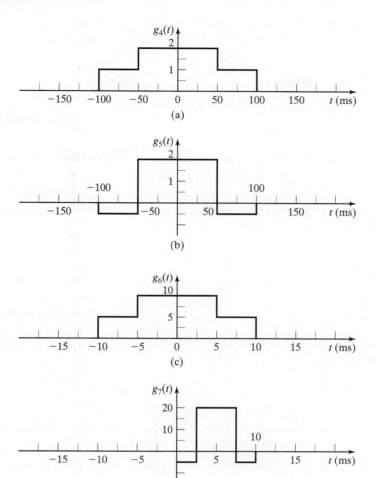

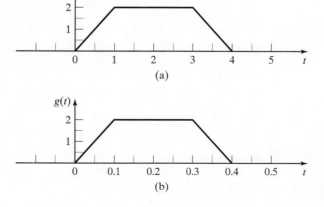

Figure P5.7

Figure P5.9

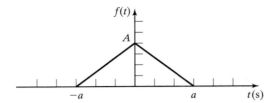

Figure P5.10

5.12. For the electrical network shown in Figure P5.12, complete the following:

(a) Determine the frequency response function.

(b) Sketch the magnitude and phase frequency response.

(c) Find the impulse response function for this network.

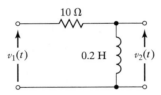

Figure P5.12

5.13. Show that the time-scaling property of the Fourier transform, with a constant, is valid. That is, show that

$$f(at) \xleftrightarrow{\mathscr{F}} \frac{F(\omega/a)}{|a|}.$$

5.14. Determine the Fourier transforms of the signals shown in Figure P5.14. (Use the property tables to minimize the effort.)

5.15. Determine the time-domain functions that have the frequency spectra shown in Figure P5.15.

5.16. The signal $g(t)$ has the Fourier transform

$$G(\omega) = \frac{j\omega}{-\omega^2 + 7j\omega + 6}.$$

Find the Fourier transform of the following functions:

(a) $g(4t)$

(b) $g(6t - 12)$

(c) $\dfrac{dg(t)}{dt}$

(d) $g(-t)$

(e) $e^{-j200t}g(t)$

(f) $\displaystyle\int_{-\infty}^{t} g(\tau)d\tau$

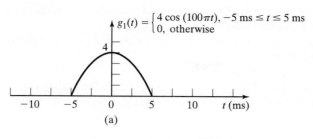

$$g_1(t) = \begin{cases} 4\cos(100\pi t), & -5\text{ ms} \le t \le 5\text{ ms} \\ 0, & \text{otherwise} \end{cases}$$

(a)

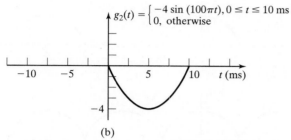

$$g_2(t) = \begin{cases} -4\sin(100\pi t), & 0 \le t \le 10\text{ ms} \\ 0, & \text{otherwise} \end{cases}$$

(b)

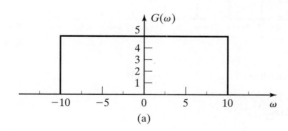

$$g_3(t) = \begin{cases} 4\cos(1000\pi t - \pi/2), & -1\text{ ms} \le t \le 0 \\ 0, & \text{otherwise} \end{cases}$$

(c)

Figure P5.14

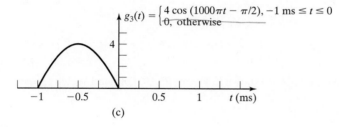

(a)

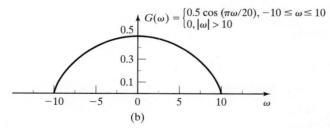

$$G(\omega) = \begin{cases} 0.5\cos(\pi\omega/20), & -10 \le \omega \le 10 \\ 0, & |\omega| > 10 \end{cases}$$

(b)

Figure P5.15

5.17. (a) Find and sketch the frequency spectrum of the half-wave rectified cosine wave-form shown in Figure P5.17(a).

(b) Find and sketch the frequency spectrum of the full-wave rectified cosine wave form shown in Figure P5.17(b).

(c) Compare the results of Parts (a) and (b).

(d) How would the frequency spectra be changed if the period of each waveform in Parts (a) and (b) was halved?

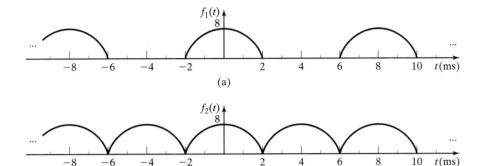

(a)

(b)

Figure P5.17

5.18. The periodic signal $g_p(t)$ is shown in Figure P5.18. Find and sketch $G_p(\omega)$.

Figure P5.18

5.19. Find the Fourier transform of the following signals:

(a) $x(t) = \dfrac{1}{2\pi}\dfrac{1}{(a - jt)^2}$ where $a > 0$

(b) $x(t) = e^{-a|t|}$ where $a > 0$

(c) $f(t) = \beta sinc\left(\dfrac{\beta t}{2}\right)$

(d) $f(t) = sinc^2 t$

5.20. Find the following convolutions:

(a) $sinc(t)*sinc(t)$

(b) $sinc^2(t)*sinc(t)$

(c) $sinc(t)*e^{j2t}sinc(t)$

5.21. Given $x_1(t) = sinc^2(t)$ and $x_2(t) = e^{jAt}x_1(t)$, specify the range of values of A, where A is a real number and $A \in (-\infty, \infty)$, such that $x_1(t)*x_2(t)$ is nonzero.

5.22. The signal $v_1(t) = \sin(50t)$ volts is applied to the input terminals of the electrical network shown in Figure P5.12. Find the frequency spectrum of the signal $v_2(t)$ that is produced at the output terminals of the circuit.

5.23. The pulsed cosine signal shown in Figure P5.23 is "on" for two cycles and then "off" for a period of time equivalent to 18 cycles of the cosine wave. The signal is periodic and the frequency of the cosine wave is 2000π rad/s. Sketch the frequency spectrum for this signal.

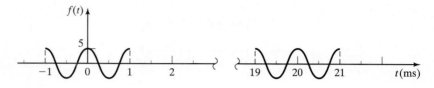

Figure P5.23

5.24. Show that the inverse Fourier Transform of $X(\omega) = \displaystyle\sum_{k=-\infty}^{\infty} 2\pi C_k \delta(\omega - k\omega_0)$ is

$$x(t) = \sum_{k=-\infty}^{\infty} C_k e^{jk\omega_0 t}.$$

5.25. The signals with the frequency spectra shown in Figure P5.25(a) and (b) are sampled using an ideal sampler with $\omega_s = 200$ rad/s.

   (a) Sketch the frequency spectra of the sampled signals.
   (b) Compare and discuss the results of Part (a).

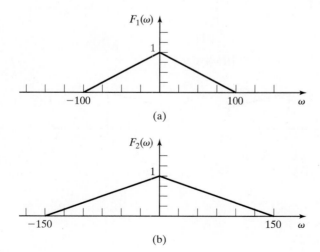

Figure P5.25

5.26. The signal $v_1(t) = 10 \, \text{rect}(t)$ is applied to the input of the network with the frequency response $H(\omega) = \text{rect}(\omega/4\pi)$ as shown in Figure P5.26. Determine and sketch the frequency spectrum of the output signal $V_2(\omega)$. (This filter is not physically realizable. See Section 6.1.)

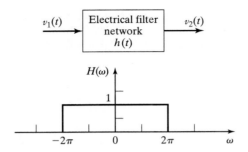

Figure P5.26

5.27. What percentage of the total energy in the energy signal $f(t) = e^{-t}u(t)$ is contained in the frequency band $-7 \, \text{rad/s} \leq \omega \leq 7 \, \text{rad/s}$?

5.28. A power signal with the power spectral density shown in Figure P5.28(a) is the input to a linear system with the frequency response shown in Figure P5.28(b). Calculate and sketch the power spectral density of the system's output signal.

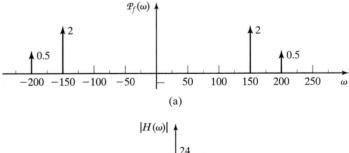

Figure P5.28

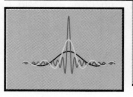

6

APPLICATIONS OF THE FOURIER TRANSFORM

In this chapter, several engineering applications of the Fourier transform are considered. The mathematical basis and several properties of the Fourier transform are presented in Chapter 5. We present additional examples of how the Fourier transform, and the frequency domain in general, can be used to facilitate the analysis and design of signals and systems.

6.1 IDEAL FILTERS

The concept of the transfer function, which is one of the ways the Fourier transform is applied to the analysis of systems, was discussed in Chapter 5,

[eq(5.46)]
$$H(\omega) = \frac{V_2(\omega)}{V_1(\omega)},$$

where $V_1(\omega)$ is the Fourier transform of the input signal to a system and $V_2(\omega)$ is the Fourier transform of the output signal. Consideration of this concept leads us to the idea of developing transfer functions for special purposes. Filtering is one of those special purposes that is often applied in electronic signal processing. Figure 6.1 shows the frequency-response characteristics of the four basic types of filters: *the ideal low-pass filter*, the *ideal high-pass filter*, the *ideal bandpass filter*, and the *ideal bandstop filter*.

These *ideal* filters have transfer functions such that the frequency components of the input signal that fall within the *passband* are passed to the output without modification, whereas the frequency components of the input signal that fall into the *stopband* are completely eliminated from the output signal.

Consider the frequency response shown in Figure 6.1(a). This is the magnitude frequency spectrum of an *ideal low-pass filter*. As can be seen, this filter has a unity magnitude frequency response for frequency components such that $|\omega| \leqq \omega_c$ and zero frequency response for $|\omega| > \omega_c$. The range of frequencies $|\omega| \leqq \omega_c$ is called the *passband* of the filter and the range of frequencies $|\omega| > \omega_c$ is called the

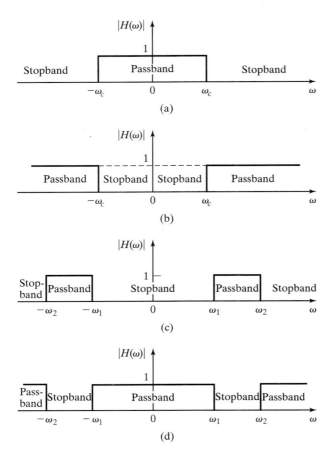

Figure 6.1 Frequency responses of four types of ideal filters.

stopband. The output of this filter will consist only of those frequency components of the input signal that are within the passband. Figure 6.2 illustrates the effect of the ideal low-pass filter on an input signal.

Filters are used to eliminate unwanted components of signals. For example, the high-frequency noise shown to be present in $V_1(\omega)$ in Figure 6.2(b) is outside the passband (in the *stopband*). Therefore, this noise is not passed through the filter to $V_2(\omega)$, and the desired portion of the signal is passed unaltered by the filter. This filtering process is illustrated in Figure 6.2(c) and (d).

It should be noted that the filters described previously are called *ideal* filters. As with most things we call ideal, they are not physically attainable. However, the concept of the ideal filter is very helpful in the analysis of linear system operation because it greatly simplifies the mathematics necessary to describe the process.

That ideal filters are not possible to construct physically can be demonstrated by reconsideration of the frequency response of the ideal low-pass filter. The transfer function of this filter can be written as

$$H(\omega) = \text{rect}(\omega/2\omega_c).$$

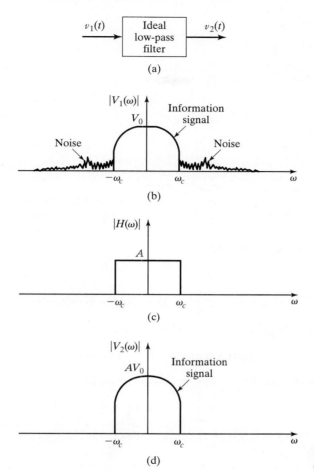

Figure 6.2 is described below.

Figure 6.2   An ideal low-pass filter used to eliminate noise.

Therefore, from Table 5.2, its impulse response is

$$h(t) = \mathcal{F}^{-1}\{H(\omega)\} = (\omega_c/\pi)\,\text{sinc}(\omega_c t),$$

as sketched in Figure 6.3. It is seen that the impulse response for this ideal filter begins long before the impulse occurs at $t = 0$ (theoretically at $t = -\infty$). Systems such as this, which respond to an input before the input is applied, are called *noncausal systems*, as discussed in Chapters 2 and 3. Of course, the physical existence of noncausal systems is impossible. However, the concept of noncausal systems, such as ideal filters, can be useful during the initial stages of a design or analysis effort. The following examples illustrate some applications of the *ideal filter* concept.

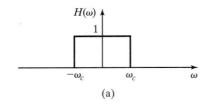

(a)

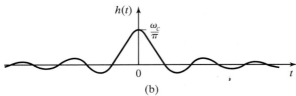

(b)

Figure 6.3   The impulse response of an ideal low-pass filter.

EXAMPLE 6.1   **Application of an ideal high-pass filter**

Two signals,

$$g_1(t) = 2\cos(200\pi t) \quad \text{and} \quad g_2(t) = 5\cos(1000\pi t),$$

have been multiplied together as described in Example 5.9. The product is the signal

$$g_3(t) = 5\cos(1200\pi t) + 5\cos(800\pi t).$$

For this example, assume that a certain application requires

$$g_4(t) = 3\cos(1200\pi t).$$

This can be obtained from $g_3(t)$ by using a high-pass filter. The Fourier transform of $g_4(t)$ is found, using Table 5.2, to be

$$G_4(\omega) = 3\pi[\delta(\omega - 1200\pi) + \delta(\omega + 1200\pi)].$$

Similarly, the Fourier transform of $g_3(t)$ is found using Table 5.2 and the linearity property:

$$G_3(\omega) = 5\pi[\delta(\omega - 800\pi) + \delta(\omega + 800\pi)]$$
$$+ 5\pi[\delta(\omega - 1200\pi) + \delta(\omega + 1200\pi)].$$

The frequency spectra of $g_4(t)$ and $g_3(t)$ are shown in Figure 6.4(a) and (b), respectively. It can be seen that if the frequency components of $G_3(\omega)$ at $\omega = \pm1200\pi$ are multiplied by 0.6, and if the frequency components at $\omega = \pm800\pi$ are multiplied by zero, the result will be the desired signal, $G_4(\omega)$. An ideal high-pass filter that will accomplish this is shown in Figure 6.4(c). The filtering process can be written mathematically as

$$G_4(\omega) = G_3(\omega)H_1(\omega),$$

where

$$H_1(\omega) = 0.6[1 - \text{rect}(\omega/2\omega_c)], \quad 800\pi < \omega_c < 1200\pi.$$

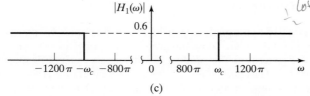

(a)

(b)

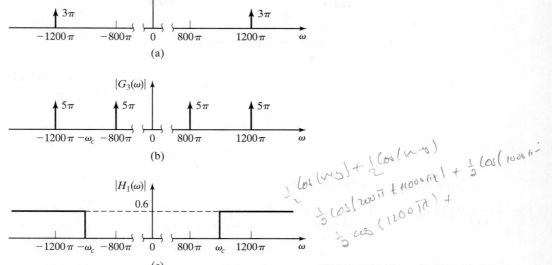

(c)

Figure 6.4   Figure for Example 6.1.   ∎

EXAMPLE 6.2   **Application of an ideal low-pass filter**

We work with the signals from Example 6.1. Assume this time that an application requires a signal

$$g_5(t) = 4\cos(800\pi t)$$

This can be obtained from $g_3(t)$ by using a low-pass filter. The Fourier Transform of $g_5(t)$ is

$$G_5(\omega) = 4\pi[\delta(\omega - 800\pi) + \delta(\omega + 800\pi)]$$

The frequency spectrum of $g_5(t)$ is shown in Figure 6.5. To pass the frequency components of $G_3(\omega)$ at $\omega = +- 800\pi$ with an output amplitude of 4π requires a gain of 0.8 as shown in the ideal low-pass filter in Figure 6.6. Again, the filtering process is written as

$$G_5(\omega) = G_3(\omega)H_2(\omega)$$

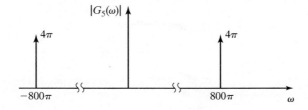

Figure 6.5   Frequency spectrum for Example 6.2.

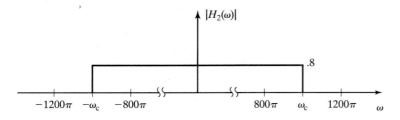

Figure 6.6 Frequency response for filter of Example 6.2.

where

$$H_2(\omega) = .8\,\text{rect}(\omega/2\omega_c),\ \omega_c > 800\pi \qquad\blacksquare$$

EXAMPLE 6.3 **Multiplication of two signals**

The multiplication of two signals as considered in Examples 6.1 and 6.2 is a simple mathematical concept and there are some integrated circuit devices that will accomplish this function. One way of generating the product of two signals using basic electronic components is shown in Figure 6.7. The *square-law device* shown in Figure 6.7 is an approximation of the effect of passing the signal through a nonlinear device such as an amplifier biased near the saturation level [1]. The output of the square-law device is the square of the input signal. In the system shown, the input to the device is

$$w(t) = C_1\cos(\omega_1 t) + C_2\cos(\omega_2 t).$$

Therefore, the output signal is

$$x(t) = w^2(t) = C_1^2\cos^2(\omega_1 t) + 2C_1 C_2\cos(\omega_1 t)\cos(\omega_2 t) + C_2^2\cos^2(\omega_2 t).$$

Using the trigonometric identity of Appendix A, namely,

$$\cos^2\phi = \tfrac{1}{2}[1 + \cos 2\phi],$$

we can rewrite the output of the square-law device as

$$\begin{aligned}x(t) = \tfrac{1}{2}C_1^2\,[1 + \cos(2\omega_1 t)] &+ 2C_1 C_2\cos(\omega_1 t)\cos(\omega_2 t)\\ &+ \tfrac{1}{2}C_2^2[1 + \cos(2\omega_2 t)].\end{aligned}$$

As shown in Example 5.9 [or by use of another trigonometric identity: $2\cos\alpha\cos\beta = \cos(\alpha-\beta) + \cos(\alpha + \beta)$], the second term of this expression can be rewritten as

$$2C_1 C_2\cos(\omega_1 t)\cos(\omega_2 t) = C_1 C_2\cos[(\omega_1 + \omega_2)t] + C_1 C_2\cos[(\omega_1 - \omega_2)t].$$

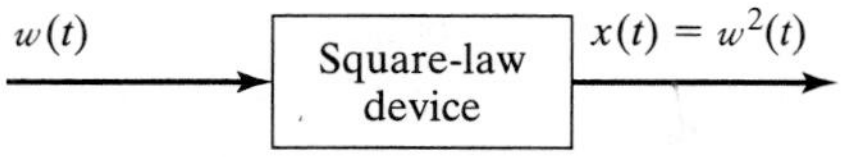

$w(t)$ ⟶ | Square-law device | ⟶ $x(t) = w^2(t)$

Figure 6.7 Block diagram of a square-law device.

Now the output of the square-law device is given by

$$x(t) = \tfrac{1}{2}[C_1^2 + C_2^2] + \tfrac{1}{2}C_1^2\cos(2\omega_1 t) + \tfrac{1}{2}C_2^2\cos(2\omega_2 t)$$
$$+ C_1 C_2\cos[(\omega_1 + \omega_2)t] + C_1 C_2\cos[(\omega_1 - \omega_2)t].$$

The Fourier transform of this signal is found from Table 5.2 and the linearity property to be

$$X(\omega) = \pi[C_1^2 + C_2^2]\delta(\omega) + \frac{\pi}{2}C_1^2[\delta(\omega-2\omega_1) + \delta(\omega + 2\omega_1)]$$

$$+ \frac{\pi}{2}C_2^2[\delta(\omega - 2\omega_2) + \delta(\omega + 2\omega_2)]$$
$$+ C_1 C_2\,\pi[\delta(\omega - \omega_1 + \omega_2) + \delta(\omega + \omega_1 - \omega_2)]$$
$$+ C_1 C_2\,\pi[\delta(\omega - \omega_1 - \omega_2) + \delta(\omega + \omega_1 + \omega_2)].$$

If the signals considered in Example 6.1 are added together to form the input to the system in Figure 6.7, then

$$w(t) = g_1(t) + g_2(t) = 2\cos(200\pi t) + 5\cos(1000\pi t)$$

and

$$x(t) = 4\cos^2(200\pi t) + 20\cos(200\pi t)\cos(1000\pi t) + 25\cos^2(1000\pi t).$$

The Fourier transform of the signal $x(t)$ is

$$X(\omega) = 29\pi(\omega) + 2\pi[\delta(\omega - 400\pi) + \delta(\omega + 400\pi)]$$
$$+ 12.5\pi[\delta(\omega - 2000\pi) + \delta(\omega + 2000\pi)]$$
$$+ 10\pi[\delta(\omega - 800\pi) + \delta(\omega + 800\pi)]$$
$$+ 10\pi[\delta(\omega - 1200\pi) + \delta(\omega + 1200\pi)].$$

The frequency spectrum of $x(t)$ is shown graphically in Figure 6.8(a).
If we desire the output of the system to be

$$y(t) = 3\cos(800\pi t),$$

as in Example 6.1, we require that

$$Y(\omega) = 3\pi[\delta(\omega - 800\pi) + \delta(\omega + 800\pi)].$$

This result can be achieved by multiplying $X(\omega)$ by a transfer function $H(\omega)$ such that the frequency components in the ranges $400\pi < \omega < 1200\pi$ and $-1200\pi < \omega < -400\pi$ are multiplied by 0.3 and the frequency components outside those ranges are multiplied by zero. The frequency response of an ideal bandpass filter that accomplishes this multiplication is shown in Figure 6.8(b). Figure 6.8(c) shows the frequency spectrum of the system's output signal.

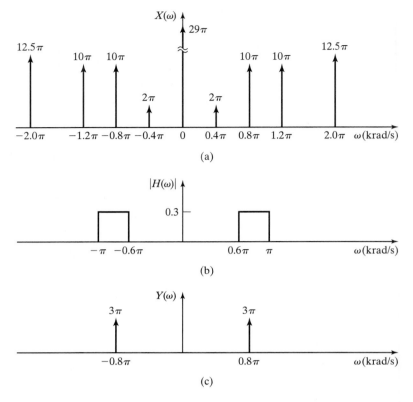

Figure 6.8 Figure for Example 6.3.   ∎

Examples 6.1 and 6.2 illustrate the concept of modifying signals by the application of filters. The filters considered in this section are *ideal filters*. It was shown that ideal filters are not physically realizable. Therefore, the results shown in these examples are not achievable with physical systems. However, the results can be approximated by physical systems, and the concept of the ideal filter is useful for simplifying the analysis and design processes.

6.2 REAL FILTERS

The ideal filters described in Section 6.1 are not physically realizable. This was shown by examining the inverse Fourier transform of the frequency-domain functions that describe the ideal low-pass filter frequency response. The impulse response of the ideal low-pass filter implies a noncausal system. Similar analyses could be used to show that the ideal bandpass, ideal high-pass, and ideal bandstop filters are also physically unrealizable.

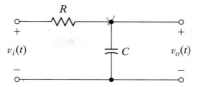

Figure 6.9   An RC low-pass filter.

RC Low-Pass Filter

Figure 6.9 shows the schematic diagram of an *RC* low-pass filter. We now find the frequency response function of this electrical network and show that it is an approximation of the ideal low-pass filter.

   To find the frequency-response function, we begin by writing the differential equations that describe the voltages and current in the circuit:

$$v_i(t) = Ri(t) + \frac{1}{C}\int_{-\infty}^{t} i(\tau)d\tau; \quad v_o(t) = \frac{1}{C}\int_{-\infty}^{t} i(\tau)d\tau.$$

After finding the Fourier transform of each equation, term by term, we have

$$V_i(\omega) = RI(\omega) + \frac{1}{j\omega C}I(\omega), V_o(\omega) = \frac{1}{j\omega C}I(\omega).$$

Therefore, the frequency-response function that describes the relationship between the input voltage $v_i(t)$ and the output voltage $v_o(t)$ in the frequency domain is

$$H(\omega) = \frac{V_o(\omega)}{V_i(\omega)} = \frac{1}{1 + j\omega RC}.$$

If we define the *cutoff frequency* of this simple filter as

$$\omega_c = \frac{1}{RC},$$

the frequency response function can be rewritten as

$$H(\omega) = \frac{1}{1 + j\omega/\omega_c} = |H(\omega)|e^{j\Phi(\omega)}. \tag{6.1}$$

The magnitude and phase frequency spectra of the filter are described by the equations

$$|H(\omega)| = \frac{1}{\sqrt{1 + (\omega/\omega_c)^2}} \quad \text{and} \quad \Phi(\omega) = -\arctan(\omega/\omega_c),$$

respectively. The magnitude frequency spectrum of the filter is shown in Figure 6.10.

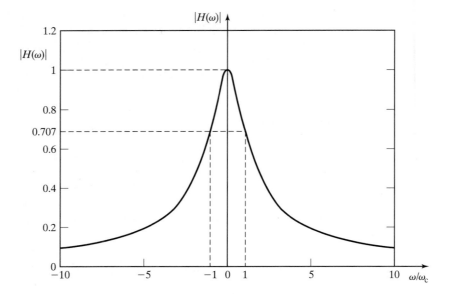

Figure 6.10   Frequency spectrum of an RC low-pass filter.

It should be noted that at the frequency $\omega = \omega_c$, the magnitude ratio between the input and output signals is

$$|H(\omega_c)| = \frac{|V_o(\omega_c)|}{|V_i(\omega_c)|} = \frac{1}{\sqrt{2}}.$$

The ratio of the normalized power (normalized average power is defined in Section 5.1) of the input and output signals is given by

$$|H(\omega_c)|^2 = \frac{|V_o(\omega_c)|^2}{|V_i(\omega_c)|^2} = \frac{1}{2}.$$

Because of this relationship the *cutoff frequency* of this type of filter is often called the *half-power frequency*.

Let's compare the polar form of $H(\omega)$ as given in (6.1) with the form of the Fourier transform of a time-shifted function

[eq(5.13)] $f(t - t_0) \xleftrightarrow{\mathscr{F}} F(\omega)e^{-j\omega t_0}.$

We might suspect that the phase angle, $\Phi(\omega)$, is somehow related to a time shift caused by this circuit as a signal is processed through it. This is indeed true. The time shift involved is called the *phase delay* and it is a function of the signal frequency. This time delay (or phase shift if we consider the frequency-domain manifestation) is a characteristic of all physically realizable filters. Generally, the more closely a physical filter approximates an ideal filter, the more time delay (negative phase shift) is present in the output signal.

It is apparent from a comparison of the magnitude frequency spectrum of the *RC* low-pass filter, shown in Figure 6.10, with those of the ideal low-pass filter, shown in Figure 6.11, that the *RC* low-pass filter is a relatively crude approximation of the ideal low-pass filter. The magnitude ratio of the *RC* low-pass filter decreases gradually as the frequency increases instead of remaining flat for $0 \leq \omega \leq \omega_c$. Also, the magnitude ratio is finite instead of zero for $\omega > \omega_c$. Engineers who design analog electronic filters try to achieve a "good enough" approximation to the ideal filter by designing for a "flat enough" frequency response in the passband and a "steep enough" roll-off at the cutoff frequency. One of the filter designs that is commonly used to satisfy these criteria is the *Butterworth filter*.

Butterworth Filter

The general form of the magnitude frequency-response function for the Butterworth filter is

$$|H(\omega)| = \frac{1}{\sqrt{1 + (\omega/\omega_c)^{2N}}}, \tag{6.2}$$

where N is called the "order" of the filter. In other words, N is the order of the differential equation needed to describe the dynamic behavior of the filter in the time domain. By comparing (6.1) and (6.2), we can see that the *RC* low-pass filter is a first-order Butterworth filter.

Figure 6.12(a) and (b) show *RLC* realizations of second- and third-order Butterworth filters, respectively. The values of the electrical components are determined by the desired cutoff frequency, ω_c, as indicated in the figure. It is left as an

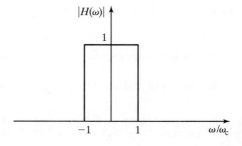

Figure 6.11   Frequency spectrum of an ideal low-pass filter.

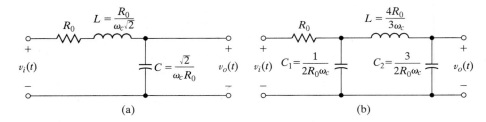

(a)   (b)

Figure 6.12   Butterworth filters.

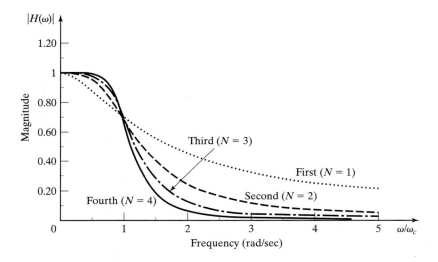

Figure 6.13   Frequency spectra of Butterworth filters.

exercise for the student to confirm that the circuits shown in Figure 6.12 have magnitude spectra described by (6.2).

Magnitude spectra for low-pass Butterworth filters of various orders are shown in Figure 6.13. It can be seen from the figure that for each value of N, the magnitude spectrum has the same cutoff frequency, ω_c.

EXAMPLE 6.4   **MATLAB program to show frequency response of Butterworth filters**

The frequency response of normalized Butterworth filters of various orders can be generated by the following MATLAB program:

```
% This MATLAB program generates a Butterworth filter of
% specified order and displays the Bode plot of the
% magnitude frequency response.
```

```
N=input...
         ('Specify the order of the filter:')
z_p_k='The zeros, poles and multiplying constant.'
[z,p,k]=buttap(N), pause
num_den='The numerator and denominator coefficients.'
[num,den]=zp2tf(z,p,k),pause
[mag,phase,w]=bode(num,den);
plot(w,mag)
title([Magnitude Bode plot ',num2str(N),...
    'th order Butterworth filter'])
xlabel('omega')
ylabel('Magnitude')
```

■

EXAMPLE 6.5   **Design of a second-order Butterworth filter**

The input signal, $v_1(t)$, to the filter network shown in Figure 6.14 is a rectified cosine voltage signal with a peak amplitude of 33.94 V and a frequency of 377 rad/s. Using the results of Example 5.18 the half-wave rectified cosine signal is found to have the frequency spectrum

$$V_1(\omega) = 53.31 \sum_{n=-\infty}^{\infty} \text{sinc}(n\pi/2)[\delta(\omega - (n + 1)377) + \delta(\omega - (n - 1)377)].$$

The frequency spectrum of the rectifier output signal is plotted in Figure 6.15(a). We now design a physically realizable filter to minimize all frequency components except the dc component at $\omega = 0$. The filter and load circuit shown in Figure 6.14 will be designed as a second-order Butterworth filter with a cutoff frequency of 100 rad/s. Because we are dealing with a filter that is implemented with a physically realizable electrical circuit, the impedance of the filter will distort the rectified cosine signal if the filter is connected directly to the rectifier circuit. In order to simplify the following discussion we will assume that the rectified cosine signal at the input to the filter circuit is the output of an isolation amplifier, as discussed in Section 2.6 and shown in Figure 2.36.

A second-order Butterworth filter has a magnitude frequency response described by

$$|H(\omega)| = \frac{1}{\sqrt{1 + (\omega/\omega_c)^4}} = \frac{\omega_c^2}{\sqrt{\omega^4 + \omega_c^4}}. \tag{6.3}$$

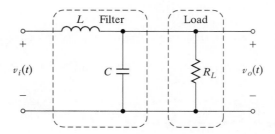

Figure 6.14   A practical filter.

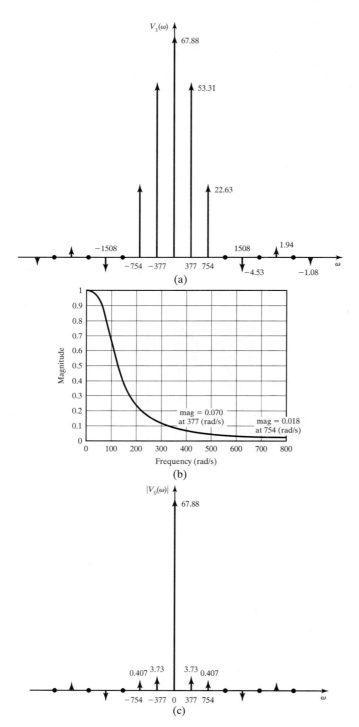

Figure 6.15　Figure for Example 6.5.

The frequency response function of the filter circuit is

$$H(\omega) = \frac{1}{1 - \omega^2 LC + j\omega L/R},$$

which has the magnitude frequency response

$$|H(\omega)| = \frac{1/LC}{\sqrt{\omega^4 + (1/LC)^2 + \omega^2 \left[(1/RC)^2 - \dfrac{2}{LC}\right]}}. \tag{6.4}$$

By comparing (6.3) and (6.4), we can see that the RLC filter will match the Butterworth form if we choose

$$\omega_c = \sqrt{\frac{1}{LC}} \quad \text{and} \quad L = 2R^2C.$$

If we assume a load resistance

$$R = 1\,\text{k}\Omega$$

and calculate the inductor and capacitor values to give a cutoff frequency of 100 rad/s, we find that

$$L = 14.14\,\text{H} \quad \text{and} \quad C = 7.07\,\mu\text{F}.$$

The frequency response of this filter is plotted in Figure 6.15(b). Figure 6.15(c) shows the magnitude frequency response of the filter's output signal,

$$|V_o(\omega)| = |H(\omega)||V_1(\omega)|.$$

It can be seen that non-dc components have been reduced in magnitude by the Butterworth filter, but not completely eliminated as they would be if an ideal filter were available. (See the results in Example 6.1.) ∎

The following example of an application of a Butterworth filter shows the effect of filtering in both the time and frequency domains.

EXAMPLE 6.6   **Butterworth Filter Simulation**

Figure 6.16(a) shows a SIMULINK simulation of a simple system. The Signal Generator block is set generate a square wave with magnitude of 1 V and fundamental frequency of 100 rad/s. The Analog Butterworth Filter block is set to simulate a fourth-order low-pass Butterworth filter with a cutoff frequency of 150 rad/s. The filter transfer function is derived using the MATLAB command, **butter(N,Wn,'s')** with N = 4 and Wn = 150.

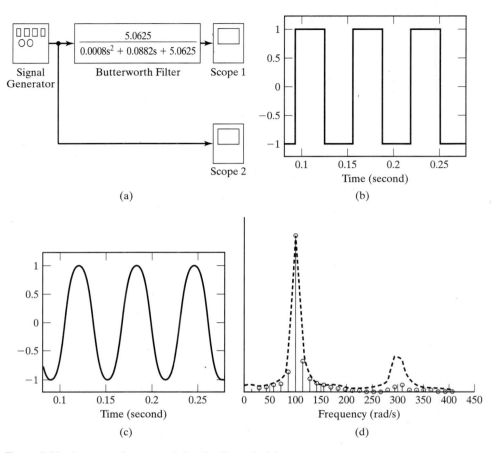

Figure 6.16 SIMULINK diagram and plots for Example 6.6.

Scope 2 displays the filter's input signal. Scope 1 displays the filter's output signal. The filter input and output signals are shown in Figure 6.16(b) and (c), respectively.

Figure 6.16(d) shows an approximation of the magnitude frequency spectra of the filter input and output. The stem plot shows the magnitude frequency spectrum of the output signal. The dashed curve indicates the magnitude frequency spectrum of the input signal. From the plots in Figure 6.16(d), it is seen that the high-frequency components ($\omega > 200$ rad/s) of the output signal are reduced in magnitude by the filter. ∎

It can be confirmed that each of the low-pass Butterworth filters described can be converted to a high-pass Butterworth filter with the same cutoff frequency by replacing each capacitor with an inductor so that

$$L_i = \frac{1}{C_i \omega_c^2}, \qquad\qquad (6.5)$$

and each inductor with a capacitor so that

$$C_j = \frac{1}{L_j \omega_c^2}. \tag{6.6}$$

EXAMPLE 6.7   **Design of a third-order high-pass Butterworth filter**

A high-pass filter with load resistance 1 kΩ and cutoff frequency 2 kHz is to be designed. A third-order Butterworth filter will be used. We can convert the circuit of Figure 6.12(b) to a third-order Butterworth high-pass filter by making the substitutions described in (6.5) and (6.6). Substituting the values specified for the load resistance and cutoff frequency into the circuit element expressions (6.5) and (6.6), we find that

$$L_1 = 2R_0/\omega_c = 0.159\,\text{H}, \quad L_2 = 2R_0/3\omega_c = 0.053\,\text{H}, \quad \text{and} \quad C_1 = 3/4R_0\omega_c = 60\,\text{nF}.$$

After making the substitutions described, we obtain the desired high-pass filter circuit as shown in Figure 6.17.

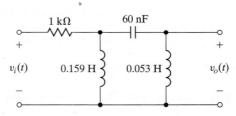

Figure 6.17   Filter for Example 6.7.   ■

Chebyschev and Elliptic Filters

Other standard methods of approximating an ideal filter are the *Chebyschev* and *elliptic filters*. For some applications, these filters provide a better approximation of the ideal filter than the Butterworth does, because for filters of equal order, they exhibit a steeper roll-off at the cutoff frequency. However, Chebyschev and elliptic filters are usually more complex to realize than Butterworth filters of equal order. Next, we describe some characteristics of the Chebyschev filter.

Chebyschev filters have magnitude frequency spectra described by

$$|H(\nu)| = \frac{1}{\sqrt{1 + \epsilon^2 C_n^2(\nu)}}, \tag{6.7}$$

where ϵ is a constant less than 1 and ν is the normalized frequency variable:

$$\nu = \frac{\omega}{\omega_c}.$$

C_n is called the Chebyschev polynomial of order n:

$$C_n = \cos(n \cos^{-1}(v)).$$

(6.8)

From (6.8), it can easily be determined that

$$C_0 = 1$$

and

$$C_1 = v.$$

(6.9)

To develop an algorithm for finding higher-order Chebyschev polynomials, we define a new variable,

$$\phi = \cos^{-1}(v),$$

so that we can now write

$$C_n = \cos(n\phi).$$

(6.10)

Now from Appendix A, we can use the trigonometric identities

$$\cos(\alpha \pm \beta) = \cos(\alpha)\cos(\beta) \mp \sin(\alpha)\sin(\beta)$$

to determine that

$$\cos[(n+1)\phi] + \cos[(n-1)\phi] = 2\cos(n\phi)\cos(\phi).$$

Using (6.10), we rewrite this as

$$C_{n+1}(v) = 2\cos(n\phi)\cos(\phi) - C_{n-1}(v).$$

Making use of (6.9), we write this result in a form that is more useful for determining Chebyschev polynomials:

$$C_{n+1}(v) = 2vC_n(v) - C_{n-1}(v).$$

(6.11)

Using (6.11) and (6.9), we can generate a list of Chebyschev polynomials as shown in Table 6.1.

Figure 6.18(a) shows the magnitude frequency responses of four orders of normalized Chebyschev filters. It is seen that the magnitude response decreases more rapidly near the cutoff frequency as the order is increased. Figure 6.18(b) provides a comparison of the frequency responses of fourth-order Butterworth and Chebyschev filters.

EXAMPLE 6.8   **MATLAB program to show the frequency response of Chebyschev filters**

The following MATLAB program can be used to generate the Chebyschev filters of various orders:

TABLE 6.1 Chebyschev
Polynomials

$C_0 = 1$
$C_1 = 32\nu$
$C_2 = 2\nu^2 - 1$
$C_3 = 4\nu^3 - 3\nu$
$C_4 = 8\nu^4 - 8\nu^2 + 1$
$C_5 = 16\nu^5 - 20\nu^3 + 5\nu$
$C_6 = 32\nu^6 - 48\nu^4 + 18\nu^2 - 1$

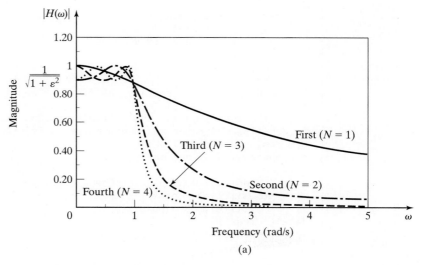

(a)

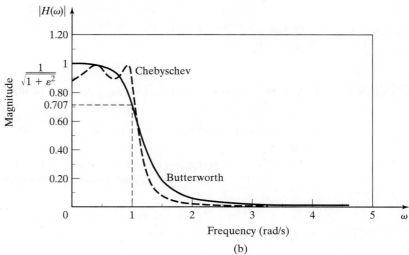

(b)

Figure 6.18 Frequency spectra of Chebyschev filters.

```
% This MATLAB program generates a Chebyschev filter
% of specified type and order. The magnitude frequency
% response is displayed.
typ=input...
        ('Specify the type of Chebyschev filter (1 or 2) ')
N=input...
        ('Specify the order of the filter:')
'('For a type 1 filter the passband ripple should be a positive value'
'('For a type 2 filter the stopband ripple is entered as a positive value
that represents a negative dB level, e.g., 40 dB. ')'
R=input...
        ('Specify the passband/stopband ripple (dB). ')
if typ==1,
    [z,p,k]=cheb1ap(N,R),pause
        elseif typ==2,
        [z,p,k]=cheb2ap(N,R),pause
        R=-R
else
        error...
            ('You must specify the filter type (1 or 2).'),
end
z_p_k='The zeros, poles, and multiplying constant.'
num_den='The numerator and denominator coefficients.'
[num,den]=zp2tf(z,p,k),pause
[mag,phase,w]=bode(num,den);
plot(w,mag)
title(['Magnitude Bode plot of Chebyschev type ',...
        num2str(typ),' order ',num2str(N),' with ',...
        num2str(R),'dB ripple.'])
xlabel('omega')
ylabel('Magnitude')                                                    ■
```

The elliptic filter has characteristics similar to the Chebyschev filter, but its magnitude frequency response contains zeros and it has a steeper roll-off at the cut-off frequency.

EXAMPLE 6.9   **MATLAB program to show the frequency response of elliptic filters**

The following MATLAB program can be used to generate elliptic filters of various order for comparison:

```
% This MATLAB program generates an elliptic filter
% of specified order. The magnitude frequency response
% is displayed.
N=input...
        ('Specify the order of the filter:')
Rp=input...
```

```
          ('Specify passband ripple magnitude in dB. ')
Rs=input...
          ('Specify stopband ripple in dB below passband gain.')
z_p_k='The zeros, poles, and multiplying constant.'
[z,p,k]=ellipap(N,Rp,Rs),pause
num_den='The numerator end denominator coefficients.'
[num,den]=zp2tf(z,p,k),pause
[mag,phase,w]=bode(num,den);
plot(w,mag)
title(['Elliptic filter frequency response. Order',...
              num2str(N), ' Rp = ',num2str(Rp),...
              'dB Rs = -',num2str(Rs),'dB'])
xlabel('omega')
ylabel('Magnitude')
```
■

Handbooks on filter design and textbooks on filter theory provide detailed instructions on the design and implementation of Butterworth, Chebyschev, and elliptic filters [2–5].

Bandpass Filters

The filters we have examined are described primarily as low-pass filters. By using a nonlinear frequency transformation, low-pass filter designs can be transformed into bandpass filters [2]. The frequency-response function of the bandpass filter can be found from

$$H_B(\omega) = H_L(\omega_L)\Big|_{\omega_L=\omega_c(\omega^2-\omega_u\omega_l)/\omega(\omega_u-\omega_l)}, \tag{6.12}$$

where $H_L(\omega_L)$ is the frequency response function of the low-pass filter to be transformed. The frequency variable of the low-pass filter has been designated as ω_L; ω_c is the cutoff frequency of the low-pass filter. The upper and lower cutoff frequencies of the bandpass filter are denoted by ω_u and ω_l, respectively.

EXAMPLE 6.10   **Transformation of a low-pass Butterworth filter into a bandpass filter**

We will now apply the transformation equation (6.12) to the design of a bandpass filter with an upper cutoff frequency of 4 krad/s and a lower cutoff frequency of 100 rad/s. We will transform the first-order Butterworth (*RC* low-pass) filter. From (6.1),

$$H_L(\omega) = \frac{1}{1 + j\omega_L/\omega_c}.$$

Therefore, from (6.12),

$$H_B(\omega) = \frac{1}{1 + j(\omega^2 - \omega_u\omega_l)/\omega(\omega_u - \omega_l)} = \frac{1}{1 + j(\omega^2 - 4 \times 10^5)/\omega(3.9 \times 10^3)}.$$

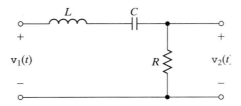

Figure 6.19   Figure for Example 6.10.

With a little algebraic manipulation, the transfer function can be written as

$$H_B(\omega) = \frac{1}{1 + j(2.56 \times 10^{-4})\omega + 1/j(9.75 \times 10^{-3})\omega}.$$

The *RLC* circuit shown in Figure 6.19 has a frequency response function of the form $H_B(\omega)$. For that circuit,

$$H(\omega) = \frac{V_2(\omega)}{V_1(\omega)} = \frac{R}{R + j\omega L + 1/j\omega C} = \frac{1}{1 + j\omega L/R + 1/j\omega RC}.$$

We now see that the bandpass filter can be realized by choosing appropriate values for *R, L,* and *C* so that the terms of the denominator of $H(\omega)$ match the corresponding terms in the denominator of $H_B(\omega)$. For example, if we choose $L = 1$ H, we can calculate the other component values as

$$R = 3.9 \text{ k}\Omega \quad \text{and} \quad C = 2.5 \,\mu\text{F.} \qquad\blacksquare$$

Summary

The *ideal* filters considered in Section 6.1 are not physically realizable. However, the concept of the ideal filter is useful in the initial stages of system analysis and design.

The filter design process can be viewed as an attempt to approximate the frequency response of an ideal filter with a physical system. Two standard methods of achieving this approximation are the Butterworth and Chebyschev filter designs.

Physically realizable systems must be *causal*; that is, their impulse response cannot begin before the impulse occurs. To approximate the amplitude frequency response of ideal (*noncausal*) filters, the impulse response of the physical filter must be similar to the impulse response of the ideal filter, but delayed in time. This time delay results in a negative phase shift in the frequency response of the physical filter.

6.3 BANDWIDTH RELATIONSHIPS

One of the concerns of an engineer in designing an electronic system is the frequency bandwidth requirement. The Fourier transform provides a means of determining the bandwidth of signals and systems. You will recall that we have sometimes called

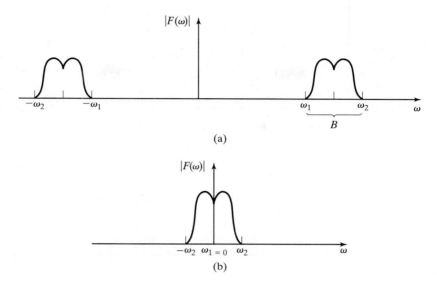

(a)

(b)

Figure 6.20   Absolute bandwidth of signals.

the Fourier transform of a signal its frequency spectrum. By examination of the frequency spectrum of a signal, the bandwidth can be determined. However, there are several definitions of bandwidth from which to choose. We will describe three definitions that may be useful in our study. The definitions explained later are equally valid, and any one of them may be the best to use for a particular application.

   Absolute bandwidth is $B = \omega_2 - \omega_1$, as illustrated by Figure 6.20. The frequency spectrum is nonzero only within the band of frequencies, $\omega_1 \leq \omega \leq \omega_2$. Note that ω_1 and ω_2 are both taken as positive frequencies for the *bandpass* signal in Figure 6.20(a). For the *baseband* signal shown in Figure 6.20(b), $\omega_1 = 0$. Therefore, for both bandpass and baseband signals, the bandwidth is defined by the range of positive frequencies for which the frequency spectrum of the signal is nonzero. As we have seen from our earlier derivations of frequency spectra, absolute bandwidth is not applicable for some signals because they are nonzero over an infinite range of frequencies. For example, the Fourier transform of the rectangular pulse has nonzero values over the entire frequency spectrum.

   Three-dB bandwidth or *half-power bandwidth* is illustrated by Figure 6.21. It is defined as the range of frequencies for which the magnitude of the frequency spectrum is no less than $1/\sqrt{2}$ times the maximum value within the range. The term 3-dB bandwidth comes from the relationship

$$20\log_{10}\left(\frac{1}{\sqrt{2}}\right) = -3 \text{ dB},$$

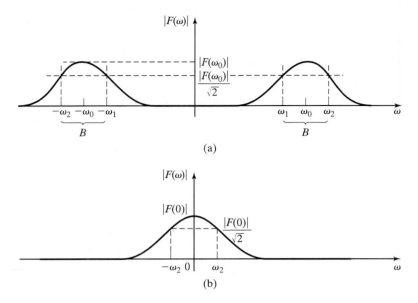

Figure 6.21   3-dB or half-power bandwidth.

where dB is the abbreviation for decibel. The term *half-power* refers to the fact that if the magnitude of voltage or current is divided by $\sqrt{2}$, the power delivered to a load by that signal is halved, because

$$P = \frac{V_{\text{rms}}^2}{R} = I_{\text{rms}}^2\, R.$$

This is a widely used definition of bandwidth and one that most electrical engineering students are familiar with from their circuit analysis courses.

Null-to-null bandwidth or zero-crossing bandwidth is shown in Figure 6.22. It is defined as the range of frequencies $B = \omega_2 - \omega_1$. The frequency of the first null (zero magnitude) in the frequency spectrum above ω_m is labeled ω_2, and for bandpass signals, ω_1, is the frequency of the first null below ω_m, where ω_m is the frequency at which the spectrum has its maximum magnitude. For baseband signals such as the one shown in Figure 6.22(b), $\omega_1 = 0$. For baseband signals, this definition of bandwidth is sometimes called the *first-null bandwidth*.

Null-to-null bandwidth is applicable only to cases where there is a definite zero value (null) in the magnitude frequency spectrum. However, it is a useful definition because some widely used waveforms, the rectangular pulse for example, have nulls in their frequency spectra.

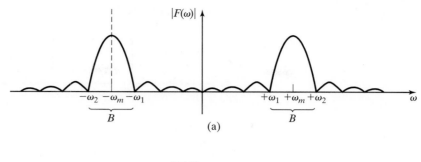

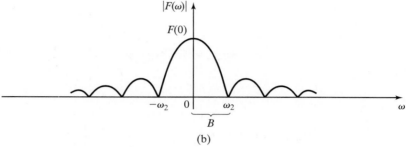

Figure 6.22   Null-to-null and first-null bandwidth.

| EXAMPLE 6.11 | **Bandwidth of a rectangular pulse** |

Determine the bandwidth of the rectangular pulse shown in Figure 6.23(a). The first step is to find the frequency spectrum by applying the Fourier transform. Examination of the frequency spectrum, shown in Figure 6.23(b), leads us quickly to the conclusion that the *absolute bandwidth* as described previously is not applicable for this waveform. However, we may choose to use either the *half-power bandwidth* or the *null-to-null bandwidth* to describe this signal as long as we are consistent and remember which definition we are using. Figure 6.23(c) shows both half-power and null-to-null bandwidths for this signal:

$$B_{\text{null}} = \omega_2 = 2\pi/T;$$
$$B_{3\text{dB}} = \omega_1.$$

∎

Note that whichever definition of bandwidth we use in Example 6.11, the bandwidth increases as the duration of the rectangular pulse decreases. In other words, the bandwidth of the signal is an inverse function of its time duration. This is generally true, and this is a key point for engineering students to remember. *The time duration of a signal and its frequency bandwidth are inversely related.* Also, any time that a signal makes a sudden change of magnitude in the time domain, it has a wide bandwidth in frequency domain. Conversely, if a signal must have a narrow bandwidth, its values must change gradually in time. Two signals that illustrate the limits of this principle are the impulse function, which has zero time duration and infinite bandwidth,

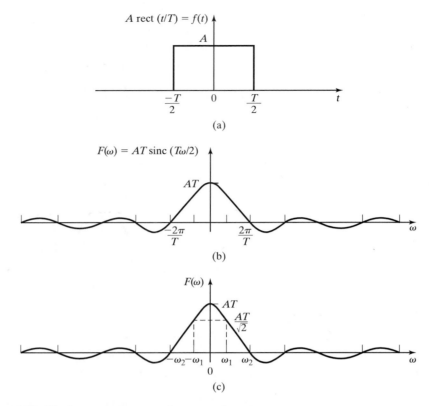

Figure 6.23   The frequency spectrum of a rectangular pulse.

and the sinusoid, which changes values very smoothly and gradually in time (all of its derivatives exist) and has a single frequency component (zero bandwidth).

6.4   RECONSTRUCTION OF SIGNALS FROM SAMPLE DATA

In many systems, continuous-time signals are sampled to create a sequence of discrete-time values for use in digital systems as discussed in Section 5.4. In many applications, there also exists the need to convert discrete-time sample data into continuous-time signals. For example, as discussed in Chapter 1, an audio compact disk (CD) player converts binary data stored on the CD into a continuous-time signal to drive the speakers. The process of converting discrete-time sample data into a continuous-time signal is called *signal reconstruction*.

The frequency-domain result of sampling an analog signal with an ideal sampling function was discussed in Section 5.4. If a continuous-time signal with the frequency spectrum $F(\omega)$ shown in Figure 6.24(a) is sampled by multiplying it by an

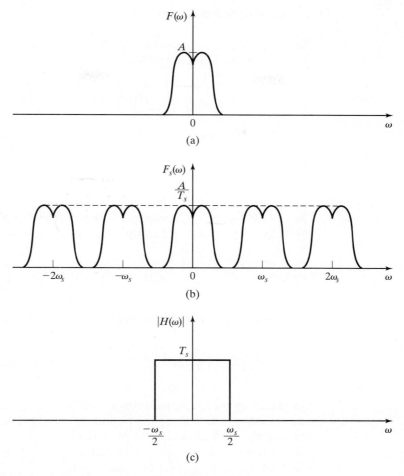

Figure 6.24   The frequency spectrum of a sampled-data signal.

ideal sampling function (5.39), the frequency spectrum of the sampled-data signal is shown in Figure 6.24(b). Since, as shown in Figure 6.24(b), the sampled signal frequency spectrum is made up of an infinite number of copies of the frequency spectrum of the original continuous signal, modified only in magnitude, the information contained in the original signal can, theoretically, be recovered from the sampled signal by filtering the sample data signal, $f_S(t)$, with an ideal low-pass filter. The frequency response of such a filter is shown in Figure 6.24(c). The effect of this filter is to multiply the frequency components of the input signal that fall into the filter's pass band by the factor T_S and to eliminate all frequency components that fall outside the passband. The output of the filter is then a baseband continuous signal containing all the information of the original signal.

Interpolating function

The frequency response function of the ideal low-pass filter with the frequency spectrum shown in Figure 6.24(c) is

$$H(\omega) = T_s \operatorname{rect}\left(\omega / \omega_s\right).$$

Using Table 5.3, we find that the impulse response of this filter is

$$h(t) = \operatorname{sinc}\left(\omega_s t / 2\right).$$

The function $\operatorname{sinc}(\omega_s t/2)$ is called the ideal *interpolating function*. The convolution of this ideal interpolating function with the impulse samples,

[eq(5.40)] $$x_S(t) = \sum_{n=-\infty}^{\infty} x(nT_s)\delta(t - nT_s),$$

results in perfect reconstruction of the original continuous-time signal:

$$x(t) = x_S(t)*\operatorname{sinc}\left(\omega_s t / 2\right) = \sum_{n=-\infty}^{\infty} x(nT_S)\operatorname{sinc}\left(\omega_S(t - nT_s) / 2\right).$$

Figure 6.25 shows the partial reconstruction of a signal, $x(t) = (1 - e^{-t/10})u(t)$, from samples taken with a sampling period, $T_S = 1(s)$. The first five interpolating functions are shown. The bold curve plots the sum of the first 20 interpolating functions. The bold curve would approach an exact plot of $x(t)$ as the number of interpolating functions summed together became larger.

As discussed in Section 6.1, the ideal low-pass filter is noncausal and its impulse response is of infinite duration. Therefore, it is physically unrealizable. It follows that a system with the ideal interpolating function as its impulse response is also physically unrealizable. However, like ideal filters, the ideal interpolating function can be approximated by physical systems.

 This concept of recovering the information from a sampled-data signal by use of an ideal low-pass filter (or ideal interpolating function) is valid only under the condition that there is no overlap in the sampled signal's frequency spectrum such as that shown in Figure 6.26(b). If there is overlap in frequency, then some spectral components will be added to the original signal and it will be distorted. A low-pass filter can be used to reconstruct a continuous-time signal, but it will not reproduce the original signal. The reconstructed signal will contain extraneous frequency components that were not present in the original signal. This erroneous information caused by the overlapping frequency components is called aliasing. Aliasing was discussed in Section 5.4.

Reconstruction of $x(t)$ from samples

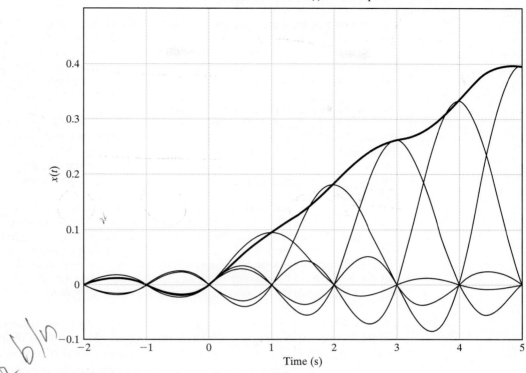

Figure 6.25   Signal reconstruction with interpolating functions.

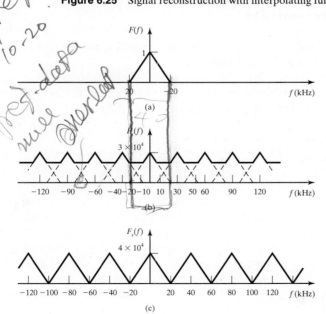

Figure 6.26   Frequency spectra for Example 6.12.

EXAMPLE 6.12   **Sampling frequency selection**

An electronic signal is to be sampled so that the discrete-time values can be recorded and re-covered for later analysis. It is known that the highest-frequency component that can ever be present in the signal is 20,000 Hz.

Initially, a sampling rate of 30 kHz was chosen and the sampled data were recorded. An attempt was made to recover the signal using an ideal low-pass filter with a cutoff fre-quency of 20 kHz. It was discovered that the recovered signal contained distortion caused by unexpectedly strong high-frequency components.

The problem can be explained using the frequency spectrum sketch shown in Figure 6.26(a). When this signal is sampled with a sampling frequency of 30 kHz, the sampled data signal has the Fourier transform sketched in Figure 6.26(b). The frequency components be-tween 10 and 20 kHz overlap and add together to increase the magnitude of the frequency spectrum in that frequency range. The same effect is seen in the frequency ranges 40 to 50 kHz, −10 to −20 kHz, and so on.

By increasing the sampling rate to 40 kHz, which is twice the highest-frequency com-ponent of the signal, the "overlap" in the frequency spectrum of the sampled-data signal is eliminated. The pattern seen in Figure 6.26(c) shows that when the sampling frequency is 40 kHz or greater, the form of the original continuous signal's spectrum is repeated about each integer multiple of the sampling frequency. Because there is now no overlap, the original sig-nal can be recovered accurately using an ideal low-pass filter. (As described in Section 1.3, the sampling frequency for music signals recorded on compact disks is 44.1 kHz.)   ■

According to Shannon's sampling theorem [7], to avoid aliasing and allow for complete reconstruction of the continuous signal, the sampling frequency must be greater than twice ω_M, the highest-frequency component of the signal to be sam-pled; that is, $\omega_S > 2\omega_M$. The frequency $2\omega_M$ is known as the Nyquist rate.

Digital-to-analog conversion

Consider the signal reconstruction illustrated in Figure 6.27(a) and (b). We can write the equation for the D/A output as

$$x_p(t) = \dots + x(0)[u(t) - u(t - T_S)] + x(T_S)[u(t - T_S) - u(t - 2T_S)]$$
$$+ x(2T_S)[u(t - 2T_S) - u(t - 3T_S)] + \dots . \tag{6.13}$$

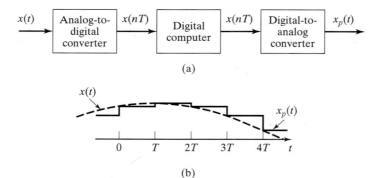

(a)

(b)

Figure 6.27   A sampled-data system and signals.

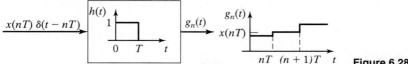

Figure 6.28   Signal reconstruction.

This signal can be expressed as the summation

$$x_p(t) = \sum_{n=-\infty}^{\infty} x(nT_S)[u(t - nT_S) - u(t - (n + 1)T_S)]. \tag{6.14}$$

Consider now the general term of $x_p(t)$, which we will denote as $g_n(t)$; that is,

$$g_n(t) = x(nT_S)[u(t - nT_S) - u(t - (n + 1)T_S)]. \tag{6.15}$$

This signal is realized by the system output in Figure 6.28, where the system has the impulse response

$$h(t) = u(t) - u(t - T_S). \tag{6.16}$$

For the system of Figure 6.28, the input signal is the impulse function $x(nT_S)\delta(t - nT_S)$ and the output is $g_n(t)$ in (6.15). If the input signal is the sampled data signal,

$$x_S(t) = \sum_{n=-\infty}^{\infty} x(nT_S)\delta(t - nT_S),$$

the system output is, by superposition, Equation (6.14). Thus, the system of Figure 6.29(a) can be modeled by an impulse modulator cascaded with the system with the impulse response of (6.16) as shown in Figure 6.29(b). The signal-reconstruction system in this figure is called a *zero-order hold*.

The system of Figure 6.29(b) is an accurate model of the system of Figure 6.29(a). However, the impulse modulator does not model the A/D, and the zero-order hold does not model the D/A. This point is discussed further below.

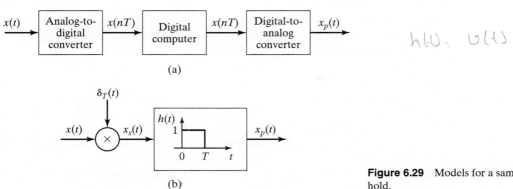

(a)

(b)

$h(t). \quad u(t)$

Figure 6.29   Models for a sampler and data hold.

As a final point, consider the output of the impulse modulator:

$$x_S(t) = \sum_{n=-\infty}^{\infty} x(nT_S)\delta(t - nT_S).$$

From Table 5.2, the Fourier transform for a time-shifted impulse function is

$$\delta(t - t_0) \xleftrightarrow{\mathcal{F}} e^{-j\omega t_0}.$$

Hence, using the linearity property of the Fourier transform, we find the Fourier transform of $x_S(t)$ to be

$$X_S(\omega) = \sum_{n=-\infty}^{\infty} x(nT_S)e^{-jnT_S\omega}. \tag{6.17}$$

Here we note that (6.17) is an alternative expression for the same Fourier transform given by (5.43). The relationship between the two forms of $X_S(\omega)$ is such that (6.17) is to (5.43) as the Fourier series in time is to a signal that is periodic in time.

The impulse response of the zero-order hold of Figure 6.29(b) is given by (6.16). Using the time-shift property of the Fourier transform, the frequency response function of the zero-order hold is found to be

$$H(\omega) = \frac{1 - e^{-jT_S\omega}}{j\omega}.$$

Using Euler's relation and algebraic manipulation, we can rewrite (6.18) as

$$H(\omega) = T_S \operatorname{sinc}\left(\pi\omega / \omega_S\right)e^{-j\pi\omega/\omega_S}. \tag{6.18}$$

The output of the sampler and data-hold is, then, from Figure 6.29(b),

$$X_p(\omega) = X_S(\omega)H(\omega) = \sum_{n=-\infty}^{\infty} x(nT_S)e^{-jnT_S\omega} \cdot T_S \operatorname{sinc}\left(\pi\omega / \omega_S\right)e^{-j\pi\omega/\omega_S}. \tag{6.19}$$

The magnitude frequency response of the zero-order hold is plotted in Figure 6.30. From (6.18) and Figure 6.30, it is seen that the zero-order hold is not a close approximation of an ideal low-pass filter. Frequency components of $X_S(\omega)$ above $\omega_S/2$ will be diminished by the zero-order hold, but not eliminated. Additional low-pass filtering can be used make $x_p(t)$ a closer approximation of $x(t)$.

The sampler and data-hold can be modeled as shown in Figure 6.31. The device shown as a switch is called an ideal sampler. The output of the ideal sampler is the train of impulse functions $x_S(t)$ and is not a physical signal. However, the input $x(t)$ and the output $x_p(t)$ of the overall system are physical signals, and the complete system accurately models the physical hardware.

In summary, the representation of the physical sampling and signal-reconstruction system of Figure 6.29(a) by the model of Figures 6.28(b) and 6.29 is mathematically valid. Hence, we may use the impulse model for the sampling of physical signals for the case that the sampling results in the number sequence

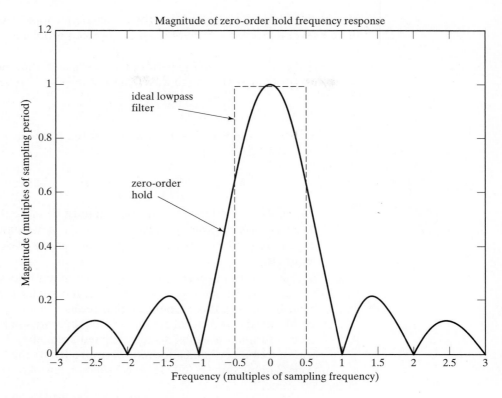

Figure 6.30   Magnitude frequency response of a zero-order hold.

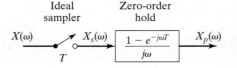

Figure 6.31   Transform models for a sampler and data hold.

$x(nT_S)$, $n = 0, \pm 1, \pm 2, \pm 3, \ldots$. The signal can be reconstructed as shown in Figure 6.27(b). The device that reconstructs sampled signals in this manner is called a zero-order hold. Other types of data holds can be constructed [2] and are equally valid for the modeling process. The only change in Figure 6.31 is in the transfer function for the data hold. Further discussion of these data holds is beyond the scope of this book.

6.5 SINUSOIDAL AMPLITUDE MODULATION

Modulation is the process used to shift the frequency of an information signal so that the resulting signal is in the desired frequency band. There are several reasons why it is important to do so. One reason is that the human voice is dominated by frequency components of less than 1 kHz. If we were to attempt to transmit a human voice signal by the propagation of electromagnetic (radio) waves, we would encounter several problems. Two of the more obvious problems are as follows:

1. *Antenna length requirement.* For efficient radiation, an antenna must be longer than $\lambda/10$. λ is the wavelength of the signal to be radiated, given by

$$\lambda = c/f_c,$$

where c is the speed of light and f_c is the frequency of the signal. For a signal of 1 kHz,

$$\lambda = 3 \times 10^8 (m/s)/1 \times 10^3 (1/s) = 300\,km.$$

Therefore, the antenna for this system must be more than 30 km in length! After considering the wavelength formula, it becomes apparent that by increasing the frequency of the signal we can decrease the antenna length required.

2. *Interference from other signals.* If two communicators wished to transmit messages at the same time in the same geographical area using the baseband frequency, there would be interference between the two signals. To avoid this, the two signal sources can be separated in frequency by shifting each information signal to an assigned frequency band. In the United States, frequency band allocation is controlled by the Federal Communications Commission (FCC). Table 6.2 shows a few of the FCC frequency band assignments.

It can be seen that a solution to both of these problems is to shift the frequency of the information signal to some higher, assigned frequency for radio transmission. The process for doing this is called *modulation*.

We consider some of the commonly used methods of modulation. Probably the simplest method in concept is double-sideband, suppressed-carrier, amplitude modulation (DSB/SC-AM). This modulation technique will be studied first.

DSB/SC-AM is accomplished by multiplying the information (message) signal, $m(t)$, by a sinusoidal signal called the carrier signal, $c(t)$, which is at the desired frequency for efficient radio transmission. This process is illustrated in Figure 6.32. Figure 6.32(d) shows the signal out of the multiplier circuit. It is simply a cosine

TABLE 6.2 FCC Frequency Band Assignments

Frequency Band	Designation	Typical Uses
3–30 kHz	Very low frequency (VLF)	Long-range navigation
30–300 kHz	Low frequency (LF)	Marine communications
300–3000 kHz	Medium frequency (MF)	AM radio broadcasts
3–30 MHz	High frequency (HF)	Amateur radio; telephone
30–300 MHz	Very high frequency (VHF)	VHF television; FM radio
0.3–3 GHz	Ultrahigh frequency (UHF)	UHF television; radar
3–30 GHz	Superhigh frequency (SHF)	Satellite communications

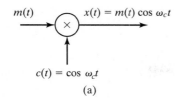

(a)

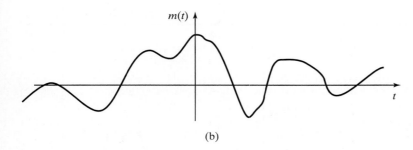

(b)

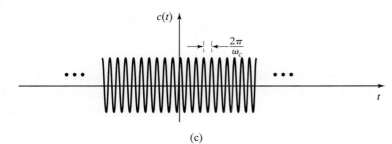

(c)

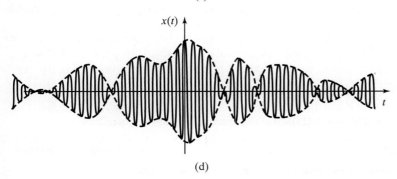

(d)

Figure 6.32   DSB/SC-AM modulation.

wave with the frequency of $c(t)$ and with the amplitude varying directly with the information signal $m(t)$:

$$x(t) = m(t)c(t) = m(t)\cos(\omega_c t).$$

Using Euler's identity for the cosine wave, this is written as

$$x(t) = m(t) \times \tfrac{1}{2}[e^{j\omega_c t} + e^{-j\omega_c t}]$$
$$= \tfrac{1}{2}m(t)e^{j\omega_c t} + \tfrac{1}{2}m(t)e^{-j\omega_c t}$$

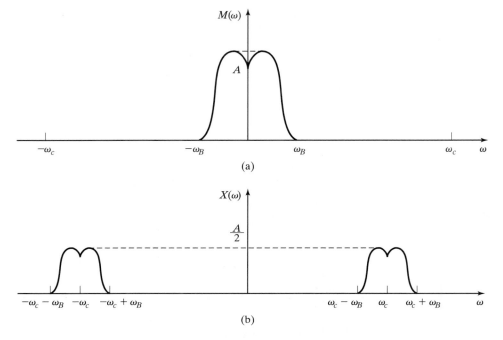

Figure 6.33 The frequency spectrum of the DSB/SC-AM signal.

Using the frequency-shifting theorem of the Fourier transform (5.18), we write the frequency spectrum of the modulated signal as

$$X(\omega) = \tfrac{1}{2}[M(\omega - \omega_c) + M(\omega + \omega_c)]. \tag{6.20}$$

The magnitude frequency spectra of $m(t)$ and $x(t)$ are shown in Figure 6.33. It can be seen that $H(\omega)$ contains the spectral distribution of $M(\omega)$, except that the magnitude is divided by 2 and centered about ω_c and $-\omega_c$ rather than all being centered about $\omega = 0$.

EXAMPLE 6.13   **MATLAB program to generate a DSB/SC-AM signal and frequency spectrum**

The following MATLAB program generates a DSB/SC-AM signal and its frequency spectrum:

```
% This MATLAB program generates a DSB/SC-AM signal and
% its Fourier transform using symbolic math.
%
% Define symbolic variables to be used.
syms t w
% generate the message signal m(t)=2cos(20ft)+sin(4ft)
'The message signal is:'
m= 2*cos(20*pi*t) + sin(4*pi*t)
% modulate a carrier signal, cos(200ft, with m(t).
```

```
'The DSB/SC modulated signal at carrier frequency 100pi (rad/s) is:'
c = m.*cos(100*pi*t)
% use the symbolic "fourier" function to calculate the Fourier
% transform of the modulated signal.
'The Fourier transform of the DSB/SC signal is:'
Cw = fourier(c)
```

The following MATLAB program generates and plots a DSB/SC-AM signal and its frequency spectrum numerically:

```
% This MATLAB program generates and plots a DSB/SC-AM signal and
% its frequency spectrum numerically.
% generate a time vector of 256 elements in .01s steps.
t = .01*(1:256);
% generate the message signal m(t)=2cos(20ft)+sin(4ft)
m= 2*cos(20*pi*t) + sin(4*pi*t);
%Display the message signal
subplot(3,1,1), plot(t,m), title('The message signal m(t)'),grid
% modulate a carrier signal, cos(200ft, with m(t).
c = m.*cos(100*pi*t);
%Display the message signal
subplot(3,1,2), plot(t,c), title('DSB/SC-AM Signal'), grid
% use the fft function to approximate the Fourier
% transform of the modulated signal.
CF = fft(c,256);
%generate a frequency vector of 256 elements.
f = 2*pi*(1:256)/(256*.01);
% plot the magnitude spectrum of the modulated signal
%Display the message signal
subplot(3,1,3), plot(f, abs(Cf)), title('DSB/SC-AM Frequency
Spectrum'), grid, grid
```

■

To make the message suitable for human ears on the receiving end of the communication link, the modulation process must be reversed. This process is called *demodulation*.

Demodulation can be accomplished in much the same way that the modulation was done. First the received signal, which we will assume to be the same as the transmitted signal, $x(t)$, is multiplied by a "local oscillator" signal. The local oscillator is tuned to produce a sinusoidal wave at the same frequency as the carrier wave in the transmitter. Because of this requirement to match the local-oscillator frequency to the carrier frequency, this demodulation technique is classified as *synchronous detection*. As shown in Figure 6.34(a),

$$y(t) = x(t)\cos(\omega_c t).$$

Again using Euler's identity and the frequency-shifting property of the Fourier transform, we find that

$$Y(\omega) = \tfrac{1}{2}[X(\omega - \omega_c) + X(\omega + \omega_c)].$$

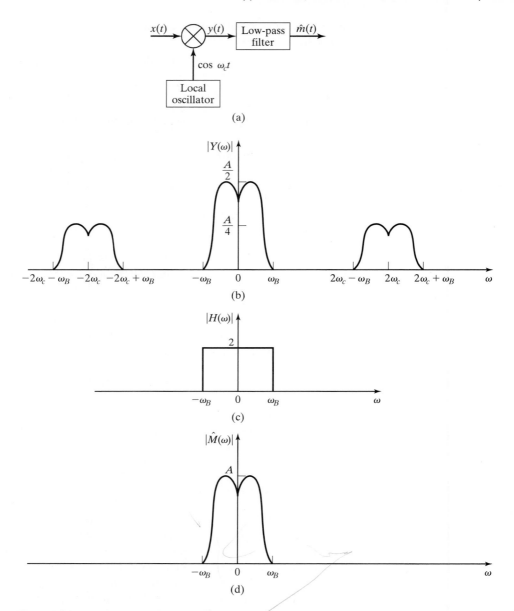

Figure 6.34   The synchronous demodulator.

Substituting in $X(\omega)$ in terms of $M(\omega)$ from (6.20), we have

$$Y(\omega) = \tfrac{1}{2}M(\omega) + \tfrac{1}{4}M(\omega - 2\omega_c) + \tfrac{1}{4}M(\omega + 2\omega_c),$$

which is illustrated in Figure 6.34(b). The remainder of the demodulation process is accomplished by passing $y(t)$ through a low-pass filter with an ideal frequency response as shown in Figure 6.34(c). The effect of the filter is to double the magnitude of the frequency components of the input signal that are within the passband of the filter, $-\omega_B \le \omega \le \omega_B$, and to eliminate all frequency components outside the passband. The output signal from the filter is, theoretically, an exact reproduction of the information signal, $m(t)$. This can be seen by comparing the frequency spectrum of the demodulated signal, $\hat{M}(\omega)$, shown in Figure 6.34(d) with $M(\omega)$ shown in Figure 6.33(a).

Another, more commonly encountered type of amplitude modulation is double-sideband modulation with a carrier component in the frequency spectrum of the modulated signal (DSB/WC). Commercial AM radio broadcasts use this method. Figure 6.35 illustrates a technique for DSB/WC-AM modulation. In this method, the modulated signal is described mathematically as

$$s(t) = [1 + k_a m(t)]c(t), \tag{6.21}$$

where $m(t)$ is the message signal and $c(t)$ is the carrier signal

$$c(t) = A_c \cos(\omega_c t).$$

The *amplitude sensitivity*, k_a, is chosen such that

$$1 + k_a m(t) > 0$$

at all times. The modulated signal can be rewritten as

$$s(t) = A_c \cos(\omega_c t) + k_a A_c m(t) \cos(\omega_c t),$$

and its frequency spectrum is given by

$$\begin{aligned} S(\omega) = {} & A_c \pi[\delta(\omega - \omega_c) + \delta(\omega + \omega_c)] \\ & + \frac{k_a A_c}{2}[M(\omega - \omega_c) + M(\omega + \omega_c)]. \end{aligned} \tag{6.22}$$

From Figure 6.36 it is seen that the modulated signal is a sinusoidal signal with an amplitude that varies in time according to the amplitude of the message signal $m(t)$.

From (6.22) and Figure 6.36(d), we see that the frequency spectrum of the modulated signal contains the carrier-signal frequency component in addition to the frequency-shifted message signal. Hence, the nomenclature of this modulation technique is *double-sideband, with-carrier, amplitude modulation* (DSB/WC-AM).

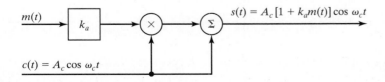

$m(t)$ k_a

$c(t) = A_c \cos \omega_c t$

$s(t) = A_c[1 + k_a m(t)] \cos \omega_c t$

Figure 6.35   A system for amplitude modulation.

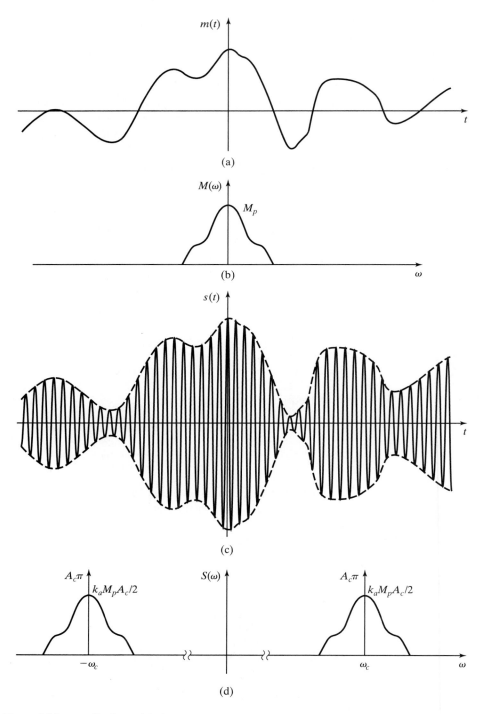

Figure 6.36   Amplitude modulation signals and spectra.

Because the use of this technique is so widespread, it is often called simply, but less precisely, AM.

The distinct frequency component at the carrier frequency in the AM signal contains no information and therefore can be considered a waste of power in the transmitted signal. However, the presence of this carrier frequency component makes it possible to demodulate and recover the message signal, $m(t)$, without requiring a "local oscillator" at the receiver to generate a carrier frequency signal as was described for DSB/SC-AM.

EXAMPLE 6.14   **MATLAB program to generate a DSB/WC-AM signal and frequency spectrum**

The following MATLAB program generates a DSB/WC-AM signal and its frequency spectrum:

```
This MATLAB program generates a DSB/WC-AM signal and displays it
in both the time and frequency domain.
% First, generate a time vector of N discrete times.
N=159;
dt=0.0001*2*pi;
t=dt*(1:N);
% Generate the message signal m(t)=3sin(500t)
m=3*sin(500*t);
%Display the message signal
subplot (2,2,1), plot(t,m), title ('The message signal m(t)'),grid
%Generate the carrier signal c(t)=1cos(1500t)
% c=cos(1500*t);
% Generate the amplitude modulated signal, x(t)=[1+km(t)]c(t),
% using a modulation index of ka=0.3
ka = 0.3;
x=(1+ka*m).*c;
%Plot the DSB/WC-AM signal
subplot(2,2,2), plot(t,x), tile('DSB/WC-AM Signal'),grid
% Use the fft function (explained in Chapter 12) to approximate
% the Fourier transform of the signals.
Mw=fft(m);
Cw=fft(c);
Xw=fft(x);
%Generate a frequency vector.
limw = 20;
w=2*pi*(0:2*limw)/(N*dt);
%
% Prepare the frequency spectra data for plotting (the magnitude
% is adjusted to approximate the weight of the impulses in the
% Fourier transform)
M=zeros(size(w));
X=zeros(size(w));
for n=1:length(w)
        M(n)=2/N*pi*abs(Mw(n));
        X(n)=2/N*pi*abs(Xw(n));
end
```

```
% Display the magnitude frequency spectrum of m(t).
subplot(2,2,3), stem(w,M), xlabel('Frequency Spectrum of
m(t)'),grid
```

```
% Display the magnitude frequency spectrum of x(t).
subplot(2,2,4), stem(w,X), xlabel('Frequency Spectrum of DSB/WC
signal)'),grid                                            ∎
```

Frequency-Division Multiplexing

As stated previously, commercial AM radio is an example of sinusoidal amplitude modulation. It is also an example of frequency-division multiplexing. Frequency-division multiplexing allows the transmission of multiple signals over a single medium by separating the signals in frequency.

Frequency-division multiplexing is illustrated in Figure 6.37. This figure shows a frequency band with three modulated signals representing three DSB/SC amplitude modulated signals.

It is assumed that each transmission $Y_i(\omega)$, $i = 1, 2, 3$, is assigned a carrier frequency that is sufficiently separated from the adjacent (in frequency) carrier frequencies, to prevent the signals from overlapping in the frequency domain.

Figure 6.37 shows three modulated signals at the carrier frequencies of ω_{c1}, ω_{c2}, and ω_{c3}. Each signal is represented with a different frequency spectrum to differentiate clearly among the three signals. The three modulation systems are as shown in Figure 6.38. The three modulating signals are denoted as $x_1(t)$, $x_2(t)$, and $x_3(t)$; the modulated signals are $y_1(t)$, $y_2(t)$, and $y_3(t)$. The frequency spectra of the three signals $Y_1(\omega)$, $Y_2(\omega)$, and $Y_3(\omega)$ are shown in Figure 6.37. *The three signals appear simultaneously in time, but are separated in frequency.*

The information signals are recovered at the receiving end by bandpass filters. The first signal $x_1(t)$ requires a bandpass filter with the center frequency of ω_{c1} and a bandwidth that is both *sufficiently wide* to recover $\hat{y}_1(t)$ and *sufficiently narrow* to reject the other signals. This filtering is illustrated in Figure 6.39 by $H_1(\omega)$, and $\hat{y}_1(t)$ is approximately equal to $y_1(t)$ for practical filtering. After the three signals are separated at the receiving end, the information signals can be recovered by synchronous demodulation as described earlier. Figure 6.39 depicts synchronous demodulation. In this figure, $\hat{x}_1(t)$ is approximately equal to $x_1(t)$. For ideal filtering, $\hat{y}_1(t) = y_1(t)$ and $\hat{x}_1(t) = x_1(t)$.

It is seen that a bandpass filter that can be tuned (the center frequency can be changed) is required for a receiver to allow the direct selection of one of several frequency-multiplexed signals. Tunable bandpass filters are difficult to implement. This problem is averted for commercial AM radio receivers by using *superheterodyning*. The principle of the superheterodyne receiver is shown in Figure 6.40. Rather than using a tunable bandpass filter, a tunable oscillator is used to shift the information from the carrier frequency to a chosen, constant frequency $\omega_i(t) = 2\pi f_i$. This frequency is called the *intermediate frequency*. For commercial AM radio, the standard intermediate frequency is $f_i = 455$ kHz. The receiver can then employ fixed bandpass filters with a center frequency of ω_i. For commercial AM radio receivers, two or more cascaded stages of bandpass filtering are usually

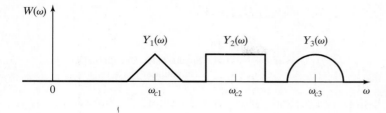

Figure 6.37 Frequency-division multiplexed signals.

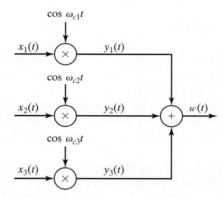

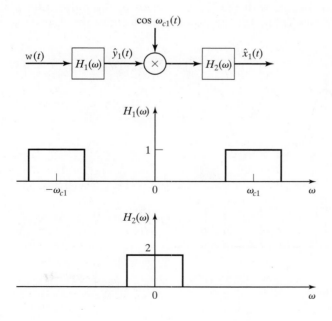

Figure 6.38 System for frequency-division multiplexing.

Figure 6.39 System for demodulation of FDM signals.

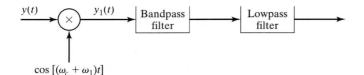

Figure 6.40 A superheterodyne
demodulator.

used, where each stage is a second-order bandpass filter. The information signal is demodulated from the intermediate-frequency signal, which has the same frequency band regardless of the carrier frequency of the selected incoming signal.

6.6 PULSE-AMPLITUDE MODULATION

In the previous section, we considered amplitude modulation with sinusoidal carrier signals. In this section, we present a modulation procedure, called *pulse-amplitude modulation* (PAM), based on a different type of carrier signal. The carrier signal is a train of rectangular pulses rather than a sinusoid. The sinusoidal carrier signal, $c(t)$, used in Section 6.5 is shown in Figure 6.41(a). Figure 6.41(b) shows a pulse carrier signal $p(t)$ as used in pulse-amplitude modulation, with the frequency of the carrier $f_c = 1/T_c$.

A system that implements pulse-amplitude modulation is given in Figure 6.42. Note that this system is identical to that presented in Section 6.5 for DSB/SC-AM, except that the pulse carrier signal $p(t)$ is used instead of the sinusoidal carrier signal, $c(t)$. A typical pulse-amplitude–modulated signal $y(t)$ is Figure 6.42(b). We analyze this system using both the Fourier series and the Fourier transform.

The analysis begins with the Fourier series of the pulse train $p(t)$ shown in Figure 6.41(b). From Table 4.3, the exponential form of the Fourier series for this

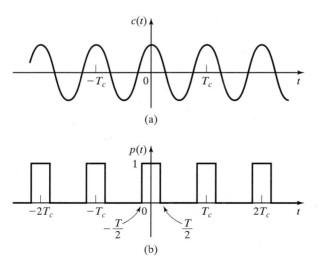

Figure 6.41 Carriers signals for amplitude modulation.

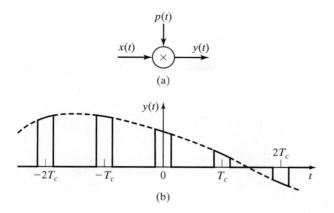

Figure 6.42   Pulse amplitude modulation.

signal is given by

$$p(t) = \sum_{k=-\infty}^{\infty} C_k e^{jk\omega_c t}, \quad C_k = \frac{T}{T_c} \text{ sinc } (k\omega_c T/2), \tag{6.23}$$

where $\omega_c = 2\pi/T_c$. The combined trigonometric form of this series is given by

$$p(t) = C_0 + \sum_{k=1}^{\infty} 2|C_k| \cos(k\omega_c t + \theta_k), \tag{6.24}$$

where $\theta_k = arg\, C_k$. It is seen from equation (6.24) that pulse-amplitude modulation can be considered to be a variation of sinusoidal amplitude modulation, with the carrier signal a sum of sinusoids rather than a single sinusoid. From (6.24), the modulator output in Figure 6.42 is

$$y(t) = x(t)p(t) = x(t)\left[C_0 + \sum_{k=1}^{\infty} 2|C_k| \cos(k\omega_c t + \theta_k)\right]$$

$$= C_0 x(t) + \sum_{k=1}^{\infty} 2|C_k| x(t) \cos(k\omega_c t + \theta_k). \tag{6.25}$$

Next, we assume that $x(t)$ is bandlimited, as shown in Figure 6.43(a) with $X(\omega) = 0$ for $\omega > \omega_M$. This assumption allows us to show in a clear manner the properties of pulse amplitude modulation. Just as seen previously for DSB/SC-AM, the effect of the multiplications by $\cos(k\omega_c t + \theta_k)$ in (6.25) is replication of the frequency spectrum $X(\omega)$ about the center frequencies $\pm k\omega_c$, $k = 1, 2, \ldots$. This effect is illustrated in Figure 6.43(b). The frequency spectrum of the pulse-amplitude–modulated signal $y(t)$ is multiplied by C_k, which is given in (6.23). Hence, the frequency spectrum of $Y(\omega)$ is that of Figure 6.43(a) multiplied by C_k. The final result is as given in Figure 6.43(c). Each replication is an undistorted version of $X(\omega)$, because C_k is constant for each value of k.

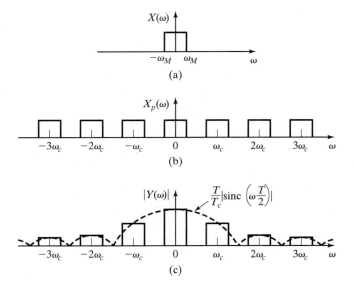

Figure 6.43 Frequency spectra for pulse amplitude modulation.

The mathematical derivation of the preceding results will now be given. The Fourier transform for $p(t)$ for both (6.23) and (6.24) is given by

$$P(\omega) = \sum_{k=-\infty}^{\infty} 2\pi C_k \delta(\omega - k\omega_c), \quad C_k = \frac{T}{T_c} \text{sinc}(k\omega_c T/2), \tag{6.26}$$

and the modulated signal is then a convolution in frequency:

$$Y(\omega) = \frac{1}{2\pi} X(\omega) * P(\omega) = \sum_{k=-\infty}^{\infty} C_k X(\omega - k\omega_c). \tag{6.27}$$

We see then that each term in Figure 6.43(b) is multiplied by C_k, resulting in Figure 6.43(c).

An important application of pulse-amplitude modulation is in the time-division multiplexing of signals, which is presented next.

Time-Division Multiplexing

In Section 6.5, we considered frequency-division multiplexing, in which multiple signals are transmitted simultaneously through the same channel. These signals are separated in frequency, but not in time. For this reason, the signals may be recovered with bandpass filters.

Here we consider a second method of multiplexing, which is called *time-division multiplexing*, or simply time multiplexing. For this procedure, multiple signals are transmitted in the same channel, with the signals separated in time, but not in frequency. Each signal is pulse-amplitude modulated as described earlier, with no overlap in time of the signals.

Figure 6.44(a) illustrates a simple system for the pulse-amplitude modulation and time-division multiplexing of three signals. The electronic switches are controlled by the signals $s_1(t)$, $s_2(t)$, and $s_3(t)$, which are depicted in Figure 6.44(b), so that the circuit is completed to the three modulation signals in order. At the receiving end of the transmission, another electronic switch, which is synchronized with the multiplexer switch, is used to demultiplex (separate) the time-division–multiplexed signal into three separate signals. The signal $y_i(t)$ is the pulse-amplitude–modulated signal of $x_i(t)$, $i = 1, 2, 3$. The demodulation of $y_i(t)$ is accomplished using a low-pass filter, as

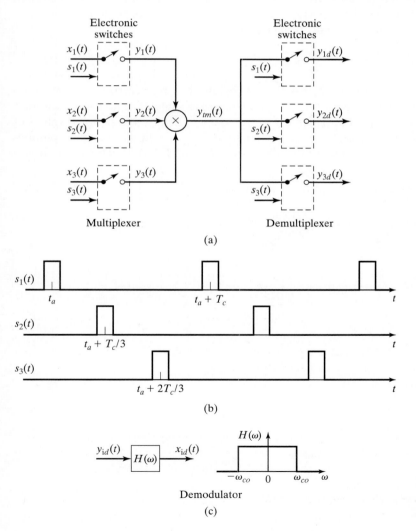

(a)

(b)

(c)

Figure 6.44   A system for time-division multiplexing.

shown in Figure 6.44(c). Note that no bandpass filters are required in the demodulation of pulse-amplitude–modulated signals. Note also that from Figure 6.43, each signal must be bandlimited such that $\omega_M < \omega_{c/2}$.

The telephone system in the United States uses time-division multiplexing. In Figure 6.45, the rate at which each conversation is sampled is 8 kHz. Audio signals (human conversations) have a spectrum that runs to approximately 20 kHz. To ensure proper separation of the pulse-modulated signal in the frequency domain, these conversations are passed through a bandpass filter with a lower cutoff frequency of 200 Hz and an upper cutoff frequency of 3.2 kHz. This filtering degrades the audio quality of telephone conversations. However, the frequency limitation allows the information content of the voice signal to be used to amplitude modulate a train of rectangular pulses with a relatively long period. The PAM signal can then be interleaved in time with PAM signals from several other conversations for transmission over a single communication circuit.

Flat-top PAM

Thus far, we have produced the PAM signal by multiplying a carrier signal made up of a train of rectangular pulses by an analog message signal. We now study another process for producing a PAM signal. In this case, we use discrete sampled values of the message signal to modulate the carrier signal. The result will be an amplitude-modulated train of rectangular pulses known as *flat-top* PAM. Practically, a *flat-top*

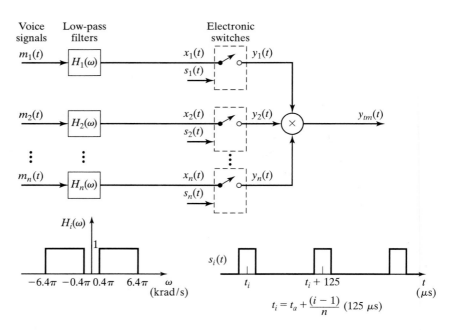

Figure 6.45   Pulse-amplitude modulation with TDM.

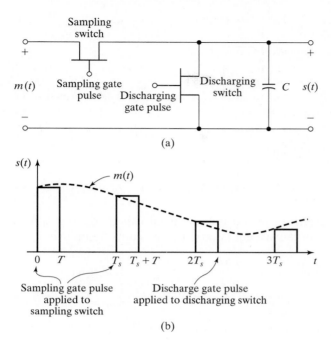

(a)

(b)

Figure 6.46   A flat-top pulse-amplitude modulation system.

PAM signal can be produced by an electronic sample-and-hold circuit such as that shown in Figure 6.46(a) [6].

The analog signal to be sampled is applied to the input terminals. A gating pulse causes the FET (field-effect transistor) *sampling switch* to conduct briefly, but long enough for the capacitor to charge up to the voltage level of the input signal. Once the gating pulse is terminated, the *sampling switch* is "closed," and the capacitor remains charged at a constant voltage level until a second gating pulse causes the *discharging switch* to conduct and provide a low-resistance discharge path. The output signal can be approximated as a train of rectangular pulses, as shown in Figure 6.46(b).

For mathematical analysis of *flat-top* PAM signal generation, we can model the *sample-and-hold* circuit as a linear system that has the impulse response shown in Figure 6.47:

$$h(t) = \text{rect}\left[\frac{t - T/2}{T}\right].$$

The input to the linear system is then modeled as the sampled-data signal (5.40):

$$m_S(t) = m(t)\delta_T(t) = \sum_{n=-\infty}^{\infty} m(nT_s)\delta(t - nT_s).$$

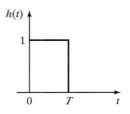

Figure 6.47   Impulse response of a data-hold circuit.

The output signal is given by

$$s(t) = h(t) * m_s(t) = h(t) * m(t) \sum_{n=-\infty}^{\infty} \delta(t - nT_s).$$

The frequency spectrum of the output signal can be determined by finding the Fourier transform of the last equation using the convolution property:

$$S(\omega) = H(\omega)\left[\frac{1}{2\pi} M(\omega) * \sum_{n=-\infty}^{\infty} \omega_s\delta(\omega - n\omega_s)\right]$$

$$= \frac{1}{T_S} H(\omega)\left[\sum_{n=-\infty}^{\infty} M(\omega - n\omega_s)\right]$$

$$= \frac{T}{T_s} \text{sinc}(\omega T/2)\left[\sum_{n=-\infty}^{\infty} M(\omega - n\omega_s)\right] e^{-j\omega T/2}.$$

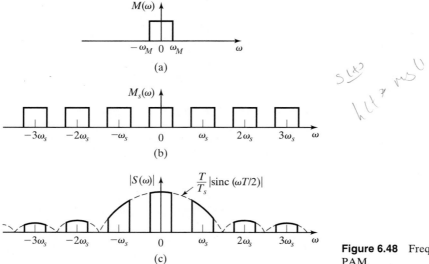

(a)

(b)

(c)

Figure 6.48   Frequency spectra for flat-top PAM.

TABLE 6.3 Key Equations of Chapter 6

Equation Title	Equation Number	Equation
Fourier transform of sampled signal	(6.17)	$X_S(\omega) = \sum\limits_{n=-\infty}^{\infty} x(nT_S)e^{-jnT_S\omega}$
Frequency spectrum of cosine-modulated signal	(6.20)	$X(\omega) = \frac{1}{2}[M(\omega - \omega_c) + M(\omega + \omega_c)]$

This frequency spectrum is illustrated in Figure 6.48. The reader is encouraged to compare the frequency spectrum of the *flat-top* PAM signal with that of the *natural-top* PAM signal shown in Figure 6.43(c).

Notice that the original continuous-time signal can be approximately recovered by filtering the sampled signal with a low-pass filter with cutoff frequency $\omega_M < \omega_c < \omega_s/2$.

SUMMARY

In this chapter, we have looked at several ways that the Fourier transform can be applied to the analysis and design of signals and systems. The applications considered here demonstrate use of the Fourier transform as an analysis tool.

We considered the *duration–bandwidth* relationship and found that the bandwidth of a signal is inversely proportional to its time duration. We saw that if a signal changes values rapidly in time, it has a wide bandwidth in frequency.

Four basic types of *ideal filters* were presented. Applications were shown for the concepts of the *ideal low-pass, ideal high-pass, ideal bandpass*, and *ideal band-stop* filters. Although these ideal filters are not physically realizable, it was shown that the concept of an ideal filter can simplify the early stages of a system analysis or design.

Butterworth and *Chebyschev filters* were presented as standard filter designs that provide physically realizable approximations of ideal filters. Examples were given to show how these filters can be realized by electrical circuits.

Signal reconstruction was presented as an application of filtering and as the process of convolving the sample-data signal with an interpolating function.

Two techniques of sinusoidal modulation (DSB/SC-AM, and DSB/WC-AM) and two types of pulse-amplitude modulation (natural and flat top) were presented to demonstrate applications of the Fourier transform to the study of communication systems and signals. See Table 6.3.

REFERENCES

1. L. W. Couch II, *Modern Communication Systems*, Upper Saddle River, NJ: Prentice-Hall, 1995.

2. G. E. Carlson, *Signal and Linear System Analysis*. Boston: Houghton Mifflin, 1992.

3. International Telephone and Telegraph Corporation, *Reference Data for Radio Engineers*, 5th ed. Indianapolis, IN: Howard W. Sams, 1973.

4. C. J. Savant, Jr., M. S. Roden, and G. L. Carpenter, *Electronic Design: Circuits and Systems*, 2d ed. Redwood City, CA: Benjamin/Cummings, 1991.

5. A. B. Williams, *Electronic Filter Design Handbook*. New York: McGraw–Hill, 1981.

6. S. Haykin, *An Introduction to Analog and Digital Communications*. New York: Wiley, 1989.

7. A. J. Jerri, "The Shannon Sampling Theorem—Its Various Extensions and Applications: A Tutorial Review," *Proceedings of IEEE*, vol. 65, pp. 1565–1596, 1977.

PROBLEMS

6.1. Show mathematically that the ideal high-pass filter is not physically realizable.

6.2. Show mathematically that the ideal bandpass filter is not physically realizable.

6.3. As illustrated in Figure P6.3, the periodic square wave is the input signal to an ideal low-pass filter with the frequency spectrum shown. Use a computer or programmable calculator to find the output signal if the input signal has a period of

    (a) 40 ms;

    (b) 20 ms;

    (c) 10 ms.

Sketch each of the output signals.

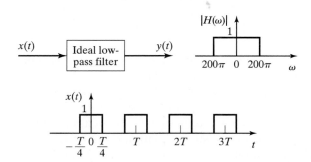

Figure P6.3

6.4. Use MATLAB and SIMULINK to show $y(t)$ if the ideal low-pass filter of Problem 6.3 is replaced by an *RC* low-pass filter with $\omega_c = 200\pi$.

6.5. Calculate the frequency response of the circuit shown in Figure P6.5 and determine what type of ideal filter is approximated by this circuit.

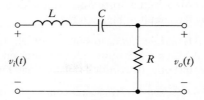

Figure P6.5

6.6. Calculate the frequency response of the circuit shown in Figure P6.6 and determine what type of ideal filter is approximated by this circuit.

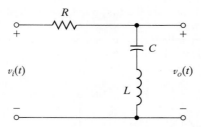

Figure P6.6

6.7. Show mathematically that the circuit shown in Figure 6.12(a) is a second-order Butterworth filter.

6.8. Find component values for the circuit shown in Figure 6.12(a) that make it a second-order Butterworth filter with a 3-dB bandwidth of 10 kHz.

6.9. **(a)** Show that the circuit shown in Figure P6.9 is a Butterworth filter.
   (b) Determine the order of the filter.
   (c) What is the 3-dB bandwidth of the filter?

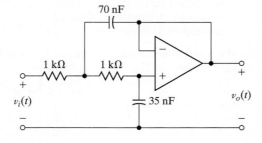

Figure P6.9

6.10. Design a high-pass Butterworth filter with a lower half-power frequency of 1000 rad/s by modifying the circuit of Figure 6.12(a).

6.11. Use MATLAB and SIMULINK to simulate an electrical power supply. The input signal can be generated as a full-wave rectified sinusoid. The power supply voltage is to be generated by filtering the rectified sinusoid. See Figure P6.11.

(a) Let $v_S(t) = 110\cos(377t)$ use a second-order Butterworth filter.
(b) Let $v_S(t) = 110\cos(377t)$ use a second-order Chebyschev (type 1) filter.

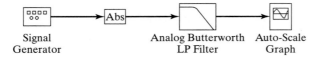

Signal   Analog Butterworth   Auto-Scale
Generator   LP Filter   Graph   **Figure P6.11**

6.12. **(a)** Find and compare the first-null bandwidth of the three triangular pulses shown in Figure P6.12.
(b) What general conclusions can be drawn from the time–bandwidth relationship of these signals?

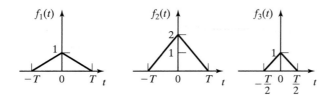

Figure P6.12

6.13. Find the minimum sampling frequency that can be used to obtain samples of each signal listed below. Assume ideal system components.

(a) $v(t) = \sin(200t)$
(b) $w(t) = \sin(100t) - 4\cos(100\pi t) + 30\cos(200t)$
(c) $x(t) = \operatorname{sinc}(200t)$
(d) $y(t) = 50\operatorname{sinc}^2(50\pi t)$

6.14. **(a)** Plot the ideally sampled signal and its frequency spectrum for the signal of Problem 6.13 (d) for sampling frequencies of 50, 100, and 200 Hz.
(b) Discuss the suitability of these sampling frequencies for the ideal system.

6.15. The signal with the amplitude frequency spectrum shown in Figure P6.15 is to be sampled using an ideal sampler.

(a) Sketch the spectrum of the resulting signal for $|\omega| \leq 120\pi$ rad/s when sampling periods of 10 and 50 ms are used.

(b) Which of the sampling frequencies is acceptable for use if the signal is to be reconstructed using an ideal low-pass filter?

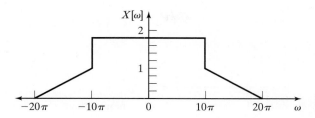

Figure P6.15

6.16. The signal $f(t) = \cos(\omega_c t)$ is sampled with an impulse train with period $T = \frac{4\pi}{3\omega_c}$. Find
and sketch the sampled spectrum.

6.17. The signal $x(t)$ with Fourier Transform $X(\omega) = \text{tri}(\frac{\omega}{\omega_c})$ is sampled with three different
impulse trains with periods $T_1 = \frac{\pi}{\omega_c}$, $T_2 = \frac{\pi}{2\omega_c}$, and $T_3 = \frac{2\pi}{\omega_c}$. Find and sketch the sampled spectrum for each case. Which case or cases experience aliasing?

6.18. The inverse Fourier Transform of the signal in the previous example is $x(t) = \frac{\omega_c}{2\pi}$
$\text{sinc}^2(\frac{\omega_c t}{2})$. Find and sketch the sampled signals using the sampling trains of the previ-
ous example $\left(T_1 = \frac{\pi}{\omega_c}, T_2 = \frac{\pi}{2\omega_c}, \text{and } T_3 = \frac{2\pi}{\omega_c}\right)$. Notice how aliasing appears in the time
domain.

6.19. A signal $x(t) = \cos\left(\frac{3\pi}{4}t\right)$ is sampled with an impulse train $p(t) = \frac{1}{\pi}\sum_{k=-\infty}^{\infty}\delta(t - 2k)$ to
form a signal $y(t) = x(t)p(t)$.

(a) Is the sampling theorem violated? Why or why not?

(b) The signal $y(t)$ is filtered with a filter $A(\omega)$ to form $\hat{Y}(\omega) = Y(\omega)A(\omega)$, where

$$A(\omega) = \begin{cases} 1, & |\omega| \leq \pi \\ 0, & |\omega| > \pi. \end{cases}$$

Find $\hat{y}(t)$, the inverse Fourier transform of $\hat{Y}(\omega)$.

6.20. The Fourier transform $X(\omega)$ of a signal $x(t)$ appears in Figure P6.20. The signal $x(t)$
is sampled with an impulse train $p(t)$ to form a new signal $\hat{x}(t) = x(t)p(t)$. The
Fourier transform of $p(t)$ is $P(\omega) = 4\sum_{k=-\infty}^{\infty}\delta(\omega - 4k)$. Sketch the Fourier transform of $\hat{x}(t)$.

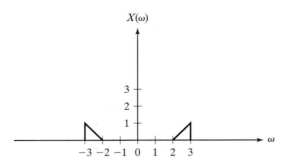

Figure P6.20

6.21. A signal $x(t) = \cos\frac{2\pi}{4}t$ is sampled with a periodic impulse train $p(t) = \frac{1}{\pi}\sum_{k=-\infty}^{\infty}\delta(t - kT)$ to form a signal $y(t) = x(t)p(t)$.

   (a) What constraint must be placed on T to avoid aliasing in the sampled signal $y(t)$?

   (b) The signal $x(t)$ is now sampled with a new periodic impulse train $p'(t) = \frac{1}{\pi}\sum_{k=-\infty}^{\infty}\delta(t - k)(t - k)$ to form a new signal $z(t) = x(t)p'(t)$. Sketch $Z(\omega)$, the Fourier transform of $z(t)$.

6.22. You are given an input signal $x(t)$, which is plotted in Figure P6.22. You are given two different filters: One is a low-pass filter and one is a high-pass filter. The input signal $x(t)$ is filtered with each of these two filters, and the two outputs are also plotted in Figure P6.22.

   (a) Is Filter A a low-pass or high-pass filter? Explain your answer.
   (b) Is Filter B a low-pass or high-pass filter? Explain your answer.

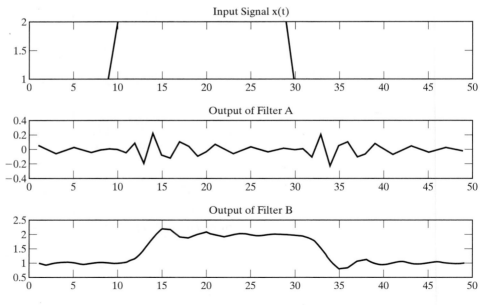

Figure P6.22

6.23. An input signal $x(t) = \cos \omega_c t$ is sampled using an ideal sampling function with $\omega_S = \frac{3}{2}\omega_c$. The sampled signal is then filtered with an ideal low-pass filter $H(\omega) = rect\left(\frac{\omega}{3/2\omega_c}\right)$ to form an output signal $y(t)$.

   (a) Sketch the Fourier Transform of the sampled signal.
   (b) Determine $y(t)$. Has aliasing occured?

6.24. An input signal $\sin \omega_c t$ is sampled with an ideal sampling function with ω_S, and a signal is reconstructed from its samples. Recall that $\omega_c = 2\pi f_c$ and $\omega_S = 2\pi f_s$. For each of the following cases, what frequency would the reconstructed signal have? Verify this using MATLAB.

   (a) $f_c = 40$ Hz and $f_s = 60$ Hz.
   (b) $f_c = 40$ Hz and $f_s = 120$ Hz.
   (c) $f_c = 149$ Hz and $f_s = 150$ Hz.

6.25. For the system of Figure P6.25, sketch $A(\omega)$, $B(\omega)$, $C(\omega)$, and $Y(\omega)$. Show all amplitudes and frequencies.

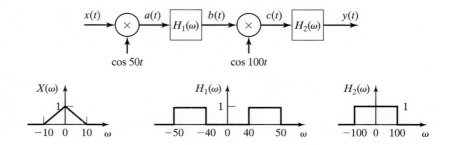

Figure P6.25

6.26. For the system of Figure P6.26, with $c_1(t) = c_2(t) = \cos(\omega_c t)$, sketch $Y(\omega)$ and $Z(\omega)$. Identify all amplitudes and frequencies of importance.

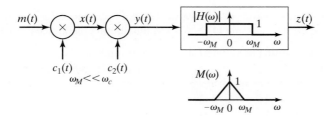

Figure P6.26

6.27. For the system of Figure P6.26, with $c_1(t) = \cos(\omega_c t)$ and $c_2(t) = \sin(\omega_c t)$, sketch $Y(\omega)$ and $Z(\omega)$. Identify all amplitudes and frequencies of importance.

6.28. In QAM [8], it is possible to send two signals on a single channel, which effectively doubles the bandwidth of the channel. QAM is used in the uplink (path from the house to the service provider) in today's 56,000 bits/second modems, in DSL modems, as well as in Motorola's Nextel cellular phones.

A block diagram of a QAM system is shown in Figure P6.28. Assume that $f_1(t)$ and $f_2(t)$ have bandwidth ω_0, where $\omega_0 \ll \omega_c$ and ω_c is the carrier frequency.

You will find the trigonometric identities in Appendix A useful for solving this problem.

We form the following signals as shown in Figure P6.28:

$$\phi(t) = f_1(t)\cos \omega_c t + f_2(t) \sin \omega_c t$$
$$g_1(t) = \phi(t) \cos \omega_c t$$
$$g_2(t) = \phi(t) \sin \omega_c t$$

(a) Determine the signal $g_1(t)$.

(b) Determine the signal $g_2(t)$.

(c) As shown in Figure P6.28, $g_1(t)$ and $g_2(t)$ are filtered by ideal low-pass filters with cutoff frequency of $2\omega_0$ and unit amplitude to form the output signals $e_1(t)$ and $e_2(t)$. Determine $e_1(t)$ and $e_2(t)$.

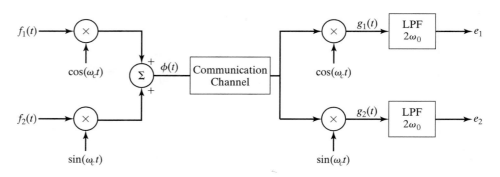

Figure P6.28

6.29. The triangular pulse waveform shown in Figure P6.29 modulates a sinusoidal carrier signal $c(t) = \cos(10^6 \pi t)$ using DSB/SC-AM modulation techniques.

(a) Sketch the resulting modulated signal, $s(t) = m(t)c(t)$.

(b) Derive the frequency spectrum of the modulated signal.

(c) Sketch the frequency spectrum, $S(\omega)$.

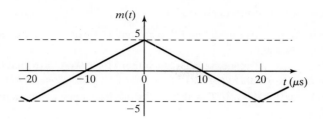

m(t)

Figure P6.29

6.30. The triangular pulse waveform shown in Figure P6.29 is used to modulate a carrier signal $\cos(10^6 \pi t)$ using DSB/WC-AM modulation techniques with $k_a = 0.05$.

   (a) Sketch the resulting modulated signal, $s(t) = m(t)c(t)$.
   (b) Derive the frequency spectrum of the modulated signal.
   (c) Sketch the frequency spectrum, $S(\omega)$.

6.31. The signal with the frequency spectrum shown in Figure P6.31(a) is used to pulse-amplitude modulate the signal $p(t)$ shown in Figure P6.31(b). Sketch the modulated signal's magnitude frequency spectrum.

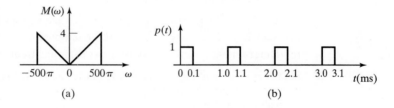

(a)   (b)

Figure P6.31

6.32. The system shown in Figure P6.32 is used, with a sampling rate of 2.4 MHz, to time-division multiplex a number of pulse-amplitude–modulated signals.

   (a) If the pulses are $8 \, \mu s$ in duration, how many PAM signals can be multiplexed?
   (b) If the pulses are $8 \, \mu s$ in duration, what is the first-null bandwidth of the TDM signal?

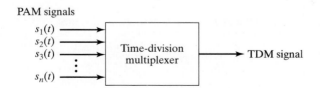

Figure P6.32

6.33. Consider the system shown in Figure P6.33.

    (a) Give the constraints on $x(t)$ and T such that $x(t)$ can be reconstructed (approximately) from $x_p(t)$.

    (b) Give the frequency response $H(\omega)$ such that $y(t) = x(t)$, provided that $x(t)$ and T satisfy the constraints in part (a).

    (c) Let $x(t) = \cos(200\pi t)$. If $T = 0.004$ s, list all frequency components of $x_p(t)$ less than 700 Hz.

    (d) Let $x(t) = \cos(2\pi f_x t)$. Find a value of $f_x \neq 100$Hz such that the same frequencies appear in $x_p(t)$ as in Part (c).

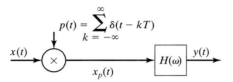

Figure P6.33

6.34. The continuous signal $m(t) = 2\cos(200\pi t) + 6\cos(800\pi t)$ is sampled using an ideal sampling function.

    (a) Sketch the frequency spectrum of the sampled-data signal if $\omega_S = 2000\pi$.

    (b) Sketch the frequency spectrum of the sampled-data signal if $\omega_S = 1000\pi$.

    (c) Determine the minimum satisfactory sampling frequency.

6.35. Sketch the frequency spectrum of the flat-topped PAM signal resulting from modulating the signal $p(t)$ shown in Figure P6.31(b) with a signal $m(t)$ with the frequency spectrum shown in Figure P6.31(a).

6.36. Find and sketch the frequency spectrum of the flat-topped PAM signal resulting from modulating the signal $p(t)$, shown in Figure P6.31(b), with a signal $m(t) = 4\,\text{sinc}(40\pi t)$.

6.37. In early modems (e.g. 300 bits/second data rates), binary data (bits) were sent over an analog channel using a modulation method known as Frequency Shift Key. To send a digital 0, the signal $s_0(t)$ is sent where:

$$s_0(t) = A\phi_0(t).$$

To send a digital 1, the signal $s_1(t)$ is sent where:

$$s_1(t) = A\phi_1(t).$$

The basis functions $\phi_0(t)$ and $\phi_1(t)$ are orthonormal, meaning that:

$$\int_0^T \phi_j(t)\phi_k(t)dt = \begin{cases} 1 & j = k \\ 0 & j \neq k \end{cases}$$

The receiver works by determining a constant r_i, $i = 0, 1$ as

$$r_i = \int_0^T r(t)\phi_i(t)dt$$

where $r(t)$ is the received signal (see Figure P6.37).

Assume that the received signal $r(t)$ is equal to the signal that is sent (i.e. you do not have to consider noise in the communication channel).

(a) If $s_0(t)$ is sent for a 0, determine r_0 and r_1.

(b) If $s_1(t)$ is sent for a 1, determine r_0 and r_1.

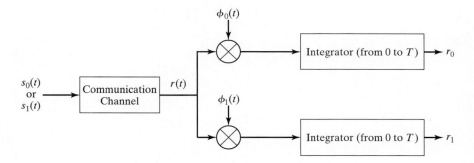

Figure P6.37

6.38. In Binary Phase Shift Key, binary data (bits) are sent over an analog channel using the following method.

A signal $s(t)$ is sent where

$$s(t) = (2a - 1)\phi(t)$$

where a is the binary bit being sent (0 or 1).

The receiver works by determining a constant r as

$$r = \int_0^T r(t)\phi(t)dt$$

where $r(t)$ is the received signal (see Figure P6.38) and

$$\int_0^T \phi(t)\phi(t)dt = 1.$$

Assume $r(t) = s(t)$ (i.e. you do not have to consider noise in the communication channel). The decision rule is if $r > 0$, a 1 was sent and if $r < 0$, a 0 was sent.

(a) What is r if a digital 0 is sent?
(b) What is r if a digital 1 is sent?

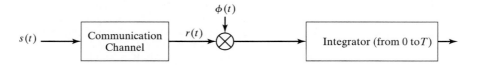

Figure P6.38

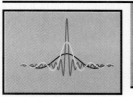

7 THE LAPLACE TRANSFORM

In this chapter, we study the *Laplace transform*, which is one of several important transforms used in linear-system analysis. One purpose of a transform is to convert operations of one type into operations of a different type. These different types of operations can offer certain advantages; for example, certain characteristics of the original operations may be more evident from the transformed operations. Another advantage may be that the transformed operations are simpler to perform.

The Laplace transform offers significant advantages. When possible, we model continuous-time physical systems with linear differential equations with constant coefficients (when the models are sufficiently accurate). The Laplace transform of these equations gives us a good description of the characteristics of the equations (the model) and, hence, of the physical system. In addition, the transformed differential equations are algebraic and, hence, are easier to manipulate; in particular, the transformed equations are easier to solve. When we use the Laplace transform to solve differential equations, the solutions are functions of the Laplace-transform variable s rather than of the time variable t. As a consequence, we must have a method for converting from functions of s back to functions of t; this procedure is called the *inverse Laplace transform*.

Several important properties of the Laplace transform are derived in this chapter. These derivations are not mathematically rigorous (see Ref. 1, p. 8); rigorous derivations are generally beyond the scope of this book. Hence, for some properties, certain constraints apply that are not evident from the derivations. However, these constraints will be stated; see Refs. 1 through 4 for rigorous mathematical derivations related to all aspects of the Laplace transform.

As a final point, we state once again that all mathematical procedures apply directly to the *models* of physical systems, not to the physical systems themselves. The relevancy of mathematical results to a particular physical system depends on the accuracy of the model. No equation models a physical system exactly; hence, we speak only of a model having sufficient accuracy. The term *sufficient* depends on the particular application.

7.1 DEFINITIONS OF LAPLACE TRANSFORMS

We begin by defining the direct Laplace transform and the inverse Laplace transform. We usually omit the term *direct* and call the direct Laplace transform simply the Laplace transform. By definition, the (*direct*) *Laplace transform F(s)* of a time function is $f(t)$ given by the integral

$$\mathcal{L}_b[f(t)] = F_b(s) = \int_{-\infty}^{\infty} f(t)e^{-st}dt, \tag{7.1}$$

where $\mathcal{L}_b[\cdot]$ indicates the Laplace transform. Definition (7.1) is called the *bilateral*, or *two-sided*, *Laplace transform*—hence, the subscript b. The *inverse Laplace transform* is given by

$$f(t) = \mathcal{L}^{-1}[F(s)] = \frac{1}{2\pi j}\int_{c-j\infty}^{c+j\infty} F(s)e^{st}\,ds,\, j = \sqrt{-1}, \tag{7.2}$$

where $\mathcal{L}^{-1}[\cdot]$ indicates the inverse Laplace transform. The reason for omitting the subscript on $F(s)$ in the inverse transform is given later. The parameter c in the limits of the integral in (7.2) is defined in Section 7.3. Equation (7.2) is called the *complex inversion integral*. Equations (7.1) and (7.2) are called the *bilateral Laplace-transform pair*. The bilateral Laplace transform is discussed more thoroughly in Section 7.8.

We now modify Definition (7.1) to obtain a form of the Laplace transform that is useful in many applications. First, we express (7.1) as

$$\mathcal{L}_b[f(t)] = F_b(s) = \int_{-\infty}^{0} f(t)e^{-st}dt + \int_{0}^{\infty} f(t)e^{-st}\,dt. \tag{7.3}$$

Next, we *define f(t)* to be zero for $t < 0$, such that the first integral in (7.3) is zero. The resulting transform, called the *unilateral*, or *single-sided Laplace transform*, is given by

$$\mathcal{L}[f(t)] = F(s) = \int_{0}^{\infty} f(t)e^{-st}\,dt, \tag{7.4}$$

where $\mathcal{L}[\cdot]$ denotes the unilateral Laplace transform. This transform is usually called simply the Laplace transform, and we follow this custom. We refer to the transform of (7.1) as the bilateral Laplace transform. We take the approach of making the unilateral transform a special case of the bilateral transform. This approach is not necessary; we could start with (7.4), with $f(t) = 0$ for $t < 0$, as a definition.

The equation for the inverse Laplace transform, (7.2), is the same for both the bilateral and unilateral Laplace transforms, and thus $F(s)$ is not subscripted. In addition, the inverse Laplace transform of the unilateral Laplace transform, (7.4), gives the function $f(t)$ for all time and, in particular, gives the value $f(t) = 0, t < 0$ [3]. Equations (7.2) and (7.4) form the *Laplace-transform pair*.

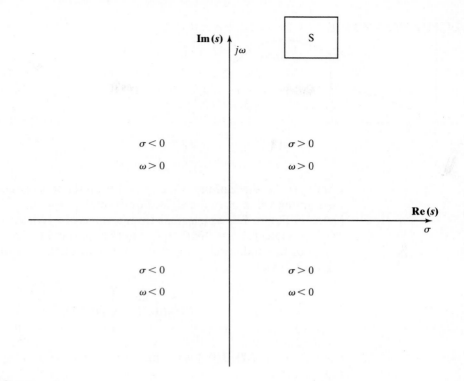

Figure 7.1   The s-plane.

The Laplace-transform variable s is complex, and we denote its real part as σ and its imaginary part as ω; that is,

$$s = \sigma + j\omega.$$

Figure 7.1 shows the complex plane commonly called the s-plane.

If $f(t)$ is Laplace transformable [if the integral in (7.4) exists], evaluation of (7.4) yields a function $F(s)$. Evaluation of the inverse transform with $F(s)$ using the complex inversion integral, (7.2), then yields $f(t)$. We denote this relationship with

$$f(t) \xleftrightarrow{\;\mathcal{L}\;} F(s). \tag{7.5}$$

As we see later, we seldom, if ever, use the complex inversion integral (7.2) to find the inverse transform, because of the difficulty in evaluating the integral. A simpler procedure is presented in Section 7.6.

If $f(t)$ has a discontinuity at $t = t_a$, the complex inversion integral gives the average of the discontinuity; that is,

$$f(t_a) = \frac{f(t_a^-) + f(t_a^+)}{2}, \tag{7.6}$$

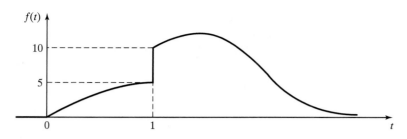

Figure 7.2 Function with a discontinuity.

where $f(t_a^-)$ is the limiting value of $f(t)$ from the left as t approaches t_a, and $f(t_a^+)$ is the limiting value from the right. For example, suppose that $f(t)$ is the function depicted in Figure 7.2, which steps from a value of 5 to a value of 10 at $t = 1$. Hence, $f(1^-) = 5$ and $f(1^+) = 10$; then $f(1)$ in (7.6) is equal to 7.5.

For the unilateral Laplace transform, evaluation of the complex inversion integral (7.2) yields

$$\mathcal{L}^{-1}[F(s)] = \begin{cases} f(t), & t > 0 \\ f(0^+)/2, & t = 0 \\ 0, & t < 0 \end{cases} \tag{7.7}$$

from (7.6). In (7.7), $f(0^+)$ is the limiting value of $f(t)$ as t approaches zero from the right.

Two important properties of the Laplace transform are now demonstrated. Consider the function $f(t) = f_1(t) + f_2(t)$. The Laplace transform of $f(t)$ is given by

$$\mathcal{L}[f(t)] = \mathcal{L}[f_1(t) + f_2(t)] = \int_0^\infty [f_1(t) + f_2(t)]e^{-st}dt$$
$$= \int_0^\infty f_1(t)e^{-st}dt + \int_0^\infty f_2(t)e^{-st} dt \tag{7.8}$$
$$= \mathcal{L}[f_1(t)] + \mathcal{L}[f_2(t)] = F_1(s) + F_2(s),$$

and we see that the Laplace transform of a sum of two functions is equal to the sum of the Laplace transforms of the two functions. This property is extended to the sum of any number of functions by replacing $f_2(t)$ in the previous derivation with the sum $f_3(t) + f_4(t)$, and so on.

A second property is derived by considering the Laplace transform of $f(t) = af_5(t)$, where a is any constant:

$$\mathcal{L}[af_5(t)] = \int_0^\infty af_5(t)e^{-st}dt = a \int_0^\infty f_5(t)e^{-st}dt$$

$$= a\mathcal{L}[f_5(t)] = aF_5(s). \tag{7.9}$$

Thus, the Laplace transform of a constant multiplied by a function is equal to the constant multiplied by the Laplace transform of the function. A transform with the properties (7.8) and (7.9) is said to be a *linear transform*; the Laplace transform is then a linear transform. These two properties are often stated as a single equation,

$$\mathcal{L}[a_1f_1(t) + a_2f_2(t)] = a_1F_1(s) + a_2F_2(s), \tag{7.10}$$

where a_1 and a_2 are any constants.

Conditions for the existence of the unilateral Laplace transform are now given, but not proved. For a given $f(t)$, if real constants M and α exist such that

$$|f(t)| < Me^{\alpha t}, \tag{7.11}$$

for t greater than some finite value t_0, $f(t)$ is called an exponential-order function; the unilateral Laplace transform exists if $f(t)$ is of exponential order [3].

In this section, the unilateral and bilateral Laplace transforms are defined. Care is taken in the definitions to reduce possible confusion in the use of the transforms. The definitions are mathematical; hence, we rely heavily on mathematicians for these definitions and for proper use of the transforms, as given in Refs. 1 through 4.

7.2   EXAMPLES

In this section, two examples of the derivation of Laplace transforms are presented followed by an example illustrating the use of the Laplace transform.

Before presenting the first example, we recall the unit step function, $u(t - t_0)$:

$$u(t - t_0) = \begin{cases} 1, & t > t_0 \\ 0, & t < t_0 \end{cases}. \tag{7.12}$$

This function is illustrated in Figure 7.3. As stated in Chapter 2, no standard exists for assigning the value of $u(t - t_0)$ at $t = t_0$; we use the definition of the unit step function in Ref. 2 and do not define the value at the instant that the step occurs. This reference states that this choice does not affect the Laplace integral (7.4).

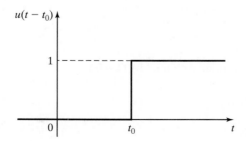

Figure 7.3   Unit step function.

EXAMPLE 7.1   **Laplace transform of a unit step function**

The Laplace transform of the unit step function is now derived for the step occurring at $t = 0$. From (7.4) and (7.12),

$$\mathcal{L}[u(t)] = \int_0^\infty u(t)e^{-st}dt = \int_0^\infty e^{-st}dt$$

$$= \frac{e^{-st}}{-s}\Big|_0^\infty = \frac{-1}{s}\left[\lim_{t\to\infty} e^{-st} - 1\right].$$

Hence, the Laplace transform of the unit step function exists *only* if the real part of s is greater than zero. We denote this by

$$\mathcal{L}[u(t)] = \frac{1}{s}, \quad \text{Re}(s) > 0,$$

where $\text{Re}(\cdot)$ denotes the real part of $(\cdot)$. We then have the transform pair

$$u(t) \xleftrightarrow{\mathcal{L}} \frac{1}{s}. \tag{7.13}$$

∎

EXAMPLE 7.2   **Laplace transform of an exponential function**

We next derive the Laplace transform of the exponential function $f(t) = e^{-at}$. From (7.4),

$$F(s) = \int_0^\infty e^{-at}e^{-st}dt = \int_0^\infty e^{-(s+a)t}dt$$

$$= \frac{e^{-(s+a)t}}{-(s+a)}\Big|_0^\infty = \frac{-1}{s+a}\left[\lim_{t\to\infty} e^{-(s+a)t} - 1\right].$$

This transform exists only if $\text{Re}(s + a)$ is positive. Hence,

$$\mathcal{L}[e^{-at}] = \frac{1}{s+a}, \quad \text{Re}(s + a) > 0,$$

and we have the Laplace transform pair

$$e^{-at} \xleftrightarrow{\mathcal{L}} \frac{1}{s+a}. \tag{7.14}$$

This transform is verified with the MATLAB program

```
syms f t a
f = exp (-a*t)
laplace (f)
```

∎

TABLE 7.1 Two Laplace Transforms

$f(t), t > 0$	$F(s)$
$u(t)$	$\dfrac{1}{s}$
e^{-at}	$\dfrac{1}{s+a}$

As seen from Examples 7.1 and 7.2, the Laplace transforms of the exponential function e^{-at} and the unit step function $u(t)$ have conditions for existence. The Laplace transform of any function $f(t)$, denoted as $F(s)$, generally has similar conditions for existence. The conditions for existence of a Laplace transform establish a region in the s-plane called the *region of convergence* (ROC). The parameter c in the inversion integral (7.2) must be chosen so that the path of integration for (7.2) lies in the ROC. Because we do not use (7.2), we generally omit stating the ROC. In addition, in the derivations that follow, conditions for the existence of integrals are usually not stated; these conditions are evident from the derivation. However, when we introduce the bilateral Laplace transform in Section 7.8, the region of convergence of the transforms must be considered.

A short table of Laplace transforms is constructed from Examples 7.1 and 7.2 and is given as Table 7.1. Note that the functions $f(t)$ are valid only for $t > 0$. From the complex-inversion integral (7.2), the inverse Laplace transform of $F(s)$ has a convergence value of zero for $t < 0$.

Generally, a table of Laplace transforms is used to find inverse Laplace transforms, rather than the inversion integral of (7.2). In any transform pair

$$f(t) \xleftrightarrow{\;\mathscr{L}\;} F(s),$$

given $f(t)$, the transform is $F(s)$; given $F(s)$, the inverse transform is $f(t)$. This operation requires that the transformation in either direction be unique. [See Ref. 3 for the rare (and insignificant) exceptions to uniqueness.] For example, in (7.14),

[eq(7.14)] $$e^{-at} \xleftrightarrow{\;\mathscr{L}\;} \frac{1}{s+a}.$$

The Laplace transform of e^{-at} is $1/(s + a)$; the inverse transform of $1/(s + a)$ is e^{-at} for $t > 0$. This procedure of using a table to find inverse Laplace transforms is illustrated in Section 7.3.

We now use the Laplace transform to solve a simple circuit. First, the Laplace transform of an operation, that of differentiation, must be derived. We begin with the Laplace transform of a function $f(t)$, (7.4):

[eq(7.4)] $$\mathscr{L}[f(t)] = \int_0^\infty f(t)e^{-st}\, dt.$$

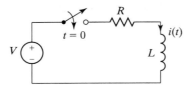

Figure 7.4 *RL* circuit.

We evaluate this integral by parts (see Appendix B), with

$$u = f(t), \quad dv = e^{-st} \, dt.$$

Then

$$du = \frac{df(t)}{dt} dt, \quad v = \frac{e^{-st}}{-s}.$$

Hence, in (7.4),

$$\mathcal{L}[f(t)] = F(s) = -\frac{1}{s} f(t)e^{-st} \bigg|_0^\infty + \int_0^\infty \frac{df(t)}{dt} \frac{e^{-st}}{s} \, dt$$

$$= \frac{1}{s}[-0 + f(0)] + \frac{1}{s} \int_0^\infty \frac{df(t)}{dt} e^{-st} dt.$$

The second term on the right side in this equation contains the Laplace transform of the derivative of $f(t)$ as a factor, and solving this equation for this factor yields

$$\mathcal{L}\left[\frac{df(t)}{dt}\right] = sF(s) - f(0).$$

Recall that, in general, $\mathcal{L}^{-1}[F(s)] = f(t)$ is discontinuous at $t = 0$. [See (7.7).] In addition, the value of $f(0)$ (value at a point) cannot affect the Laplace transform. For these reasons, this theorem must be stated in terms of $f(0^+)$ (see Ref. 3 for a discussion of this point):

$$\mathcal{L}\left[\frac{df(t)}{dt}\right] = sF(s) - f(0^+). \tag{7.15}$$

We now demonstrate the use of the Laplace transform in solving for the current in an electric circuit. Consider the *RL* circuit in Figure 7.4, where *V* is constant. The loop equation for this circuit is given by

$$L\frac{di(t)}{dt} + Ri(t) = Vu(t), \quad t > 0,$$

because the switch is closed at $t = 0$. The Laplace transform of this equation yields, from (7.15) and Table 7.1,

$$L[sI(s) - i(0^+)] + RI(s) = \frac{V}{s}, \tag{7.16}$$

since $\mathcal{L}[Vu(t)] = V\mathcal{L}[u(t)]$, with V constant. Note that the linearity property of the Laplace transform, given in (7.10), is used in deriving (7.16). We know that the initial current, $i(0^+)$, is zero; the current $i(t)$ is zero for negative time since the switch is open for $t < 0$, and the current in an inductance cannot change instantaneously.

Solving the loop equation of (7.16) for $I(s)$ yields

$$I(s) = \frac{V}{s(Ls + R)} = \frac{V/L}{s(s + R/L)}.$$

The inverse Laplace transform of $I(s)$ is the desired current $i(t), t > 0$. However, the transform for $I(s)$ is not given in Table 7.1. In cases such as this, we use *partial-fraction expansions* to express a Laplace transform as a sum of simpler terms that are in the table. We can express $I(s)$ as

$$I(s) = \frac{V/L}{s(s + R/L)} = \frac{a}{s} + \frac{b}{s + R/L}.$$

The expansion on the right side of this equation is called a partial-fraction expansion, where the values a and b that satisfy this equation are to be found. To find these values, we recombine terms on the right side to yield

$$I(s) = \frac{V/L}{s(s + R/L)} = \frac{as + aR/L + bs}{s(s + R/L)} = \frac{(a + b)s + aR/L}{s(s + R/L)}.$$

Equating numerator coefficients yields

$$a + b = 0 \qquad \therefore a = -b \,;$$

$$\frac{aR}{L} = \frac{V}{L} \qquad \therefore a = \frac{V}{R}.$$

Thus, the partial-fraction expansion for $I(s)$ is given by

$$I(s) = \frac{V/R}{s} - \frac{V/R}{s + R/L}.$$

This result is checked by recombining the terms. Table 7.1 gives $i(t)$ as

$$i(t) = \frac{V}{R}(1 - e^{-(R/L)t}), \quad t > 0.$$

The initial condition $i(0^+) = 0$ is satisfied by $i(t)$. Also, substitution of $i(t)$ into the differential equation satisfies that equation. Thus, we have solved a first-order differential equation via the Laplace transform.

Note the following points from the preceding example:

1. A differential equation with constant coefficients is transformed into an algebraic equation.
2. The algebraic equation is solved for $\mathcal{L}[i(t)] = I(s)$, which is a function of the Laplace transform variable s.
3. A table of transforms is used to find the inverse transform rather than using the inversion integral of (7.2).
4. In general, a partial-fraction expansion is required to expand complicated functions of s into the simpler functions that appear in tables of Laplace transforms.
5. The solution of a differential equation by the Laplace transform does not require separate solutions of the complementary functions and the particular integral (see Appendix E); the general solution is obtained directly.

We expand on these conclusions in the developments of the following sections.

7.3 LAPLACE TRANSFORMS OF FUNCTIONS

The unilateral Laplace transform is defined by

[eq(7.4)]
$$\mathcal{L}[f(t)] = F(s) = \int_0^\infty f(t)e^{-st}\, dt,$$

and the inverse Laplace transform by

[eq(7.2)]
$$f(t) = \frac{1}{2\pi j}\int_{c-j\infty}^{c+j\infty} F(s)e^{st}\, ds.$$

If $f_1(t)$ in (7.4) yields $F_1(s)$, then $F_1(s)$ in (7.2) yields the same $f_1(t)$. The value of c in the limits of the integral in (7.2) must be chosen to be real and in the region of convergence of the integral in (7.4). For example, Example 7.1 shows that for the Laplace transform of the unit step function $u(t)$, the region of convergence in the complex plane is $Re(s) > 0$; hence, in (7.2), c must be greater than zero. The minimum value of c for a particular transform is called its *abscissa of absolute convergence* [3].

As stated previously, we seldom use the integral in (7.2) to determine the inverse transform; hence, the region of convergence is of secondary importance to us. In fact, we seldom state the region of convergence when we give a Laplace transform. However, the reader should be aware that a particular Laplace transform does have a region of convergence.

We now derive several commonly used transforms. First, consider the impulse function, which was defined in Section 2.4. From (2.41), the rigorous definition of the unit impulse function $\delta(t)$ is

$$\int_{-\infty}^{\infty} f(t)\delta(t - t_0)\, dt = f(t_0), \tag{7.17}$$

with $f(t)$ continuous at $t = t_0$. From (2.40), a nonrigorous, but very useful, definition of the unit impulse function is

$$\int_{-\infty}^{\infty} \delta(t - t_0)dt = 1 \quad \text{with } \delta(t - t_0) = 0, \quad t \neq t_0. \tag{7.18}$$

From (7.17), for $t_0 \geq 0$ (see Ref. 3), the Laplace transform of the unit impulse function is given by

$$\mathcal{L}[\delta(t - t_0)] = \int_0^{\infty} \delta(t - t_0)e^{-st}dt = e^{-st}\Big|_{t=t_0} = e^{-t_0 s}.$$

Hence, we have the Laplace transform pair

$$\delta(t - t_0) \xleftrightarrow{\mathcal{L}} e^{-t_0 s}. \tag{7.19}$$

For the unit impulse function occurring at $t = 0$ ($t_0 = 0$),

$$\delta(t) \xleftrightarrow{\mathcal{L}} 1.$$

Next, we derive some other transform pairs. Recall the pair

[eq(7.14)] $e^{-at} \xleftrightarrow{\mathcal{L}} \dfrac{1}{s + a}.$

We now use this transform to find the transforms of certain sinusoidal functions. By Euler's relation,

$$\cos bt = \frac{e^{jbt} + e^{-jbt}}{2}.$$

Hence,

$$\mathcal{L}[\cos bt] = \tfrac{1}{2}[\mathcal{L}[e^{jbt}] + \mathcal{L}[e^{-jbt}]]$$

by the linearity property, (7.10). Then, from (7.14),

$$\mathcal{L}[\cos bt] = \frac{1}{2}\left[\frac{1}{s - jb} + \frac{1}{s + jb}\right] = \frac{s + jb + s - jb}{2(s - jb)(s + jb)} = \frac{s}{s^2 + b^2}.$$

By the same procedure, because $\sin bt = (e^{jbt} - e^{-jbt})/2j$,

$$\mathcal{L}[\sin bt] = \frac{1}{2j}[\mathcal{L}[e^{jbt}] - \mathcal{L}[e^{-jbt}]] = \frac{1}{2j}\left[\frac{1}{s - jb} - \frac{1}{s + jb}\right]$$

$$= \frac{s + jb - s + jb}{2j(s - jb)(s + jb)} = \frac{b}{s^2 + b^2}.$$

The foregoing procedure can also be used for sinusoids with exponentially varying amplitudes. Now,

$$e^{-at}\cos bt = e^{-at}\left[\frac{e^{jbt} + e^{-jbt}}{2}\right] = \frac{e^{-(a-jb)t} + e^{-(a+jb)t}}{2};$$

thus,

$$\mathscr{L}[e^{-at}\cos bt] = \frac{1}{2}\left[\frac{1}{s + a - jb} + \frac{1}{s + a + jb}\right]$$

$$= \frac{s + a + jb + s + a - jb}{2(s + a - jb)(s + a + jb)} = \frac{s + a}{(s + a)^2 + b^2}.$$

Note the two transform pairs:

$$\cos bt \xleftrightarrow{\mathscr{L}} \frac{s}{s^2 + b^2}$$

and

$$e^{-at}\cos bt \xleftrightarrow{\mathscr{L}} \frac{s + a}{(s + a)^2 + b^2}.$$

We see that for these two functions, the effect of multiplying a time function by the exponential function e^{-at} is to replace s with $(s + a)$ in the Laplace transform. We now show that this property is general; that is,

$$\mathscr{L}[e^{-at}f(t)] = \int_0^\infty e^{-at}f(t)e^{-st}dt = \int_0^\infty f(t)e^{-(s+a)t}dt$$

$$= F(s)\bigg|_{s \leftarrow s+a} = F(s + a), \tag{7.20}$$

where $F(s) = \mathscr{L}[f(t)]$ and the notation $s \leftarrow (s + a)$ indicates that s is replaced with $(s + a)$. Using the transform pair for $\sin bt$ and this theorem, we see that

$$\sin bt \xleftrightarrow{\mathscr{L}} \frac{b}{s^2 + b^2}.$$

Therefore,

$$e^{-at}\sin bt \xleftrightarrow{\mathscr{L}} \frac{b}{(s + a)^2 + b^2}.$$

The last transform is that of the product of two time functions. Note that

$$\mathscr{L}[e^{-at}\sin bt] \neq \mathscr{L}[e^{-at}]\mathscr{L}[\sin bt].$$

This result is general; that is,

$$\mathcal{L}[f_1(t)f_2(t)] \neq \mathcal{L}[f_1(t)]\mathcal{L}[f_2(t)].$$

The Laplace transform of the product of functions is not equal to the product of the transforms. We now derive an additional transform as an example.

EXAMPLE 7.3   **Laplace transform of a unit ramp function**

We now find the Laplace transform of the unit ramp function $f(t) = t$:

$$\mathcal{L}[t] = \int_0^\infty te^{-st}dt.$$

From the table of integrals, Appendix A,

$$\int ue^u du = e^u(u - 1) + C.$$

Then, letting $u = -st$, we get

$$\int_0^\infty te^{-st}dt = \frac{1}{(-s)^2}\int_0^\infty (-st)e^{(-st)} \, d(-st) = \frac{1}{s^2} e^{-st}(-st - 1)\Big|_0^\infty$$

$$= \frac{1}{s^2}[0 - (-1)] = \frac{1}{s^2}, \quad \text{Re}(s) > 0,$$

since, by L'Hôpital's rule, Appendix B, the function at the upper limit is zero:

$$\lim_{t \to \infty} \frac{t}{e^{st}} = \lim_{t \to \infty} \frac{1}{se^{st}} = 0.$$

Thus, in this example, we have developed the transform pair

$$t \xleftrightarrow{\mathcal{L}} \frac{1}{s^2}.$$

This transform is verified with the MATLAB program

```
syms f t
f=t;
laplace (f)
```
■

In this section, we have developed several Laplace transform pairs. These pairs, in addition to several others, are given in Table 7.2. The last column in this table gives the region of convergence (ROC) for each transform. In the next section,

TABLE 7.2 Laplace Transforms

f(t), t ≥ 0	F(s)	ROC
1. $\delta(t)$	1	All s
2. $u(t)$	$\dfrac{1}{s}$	$\text{Re}(s) > 0$
3. t	$\dfrac{1}{s^2}$	$\text{Re}(s) > 0$
4. t^n	$\dfrac{n!}{s^{n+1}}$	$\text{Re}(s) > 0$
5. e^{-at}	$\dfrac{1}{s+a}$	$\text{Re}(s) > -a$
6. te^{-at}	$\dfrac{1}{(s+a)^2}$	$\text{Re}(s) > -a$
7. $t^n e^{-at}$	$\dfrac{n!}{(s+a)^{n+1}}$	$\text{Re}(s) > -a$
8. $\sin bt$	$\dfrac{b}{s^2+b^2}$	$\text{Re}(s) > 0$
9. $\cos bt$	$\dfrac{s}{s^2+b^2}$	$\text{Re}(s) > 0$
10. $e^{-at}\sin bt$	$\dfrac{b}{(s+a)^2+b^2}$	$\text{Re}(s) > -a$
11. $e^{-at}\cos bt$	$\dfrac{s+a}{(s+a)^2+b^2}$	$\text{Re}(s) > -a$
12. $t\sin bt$	$\dfrac{2bs}{(s^2+b^2)^2}$	$\text{Re}(s) > 0$
13. $t\cos bt$	$\dfrac{s^2-b^2}{(s^2+b^2)^2}$	$\text{Re}(s) > 0$

we derive several properties for the Laplace transform. It is then shown that these properties allow additional transform pairs to be derived easily. Also, these properties aid us in solving linear differential equations with constant coefficients.

7.4 LAPLACE TRANSFORM PROPERTIES

In Sections 7.1 through 7.3, two properties were derived for the Laplace transform. These properties are

[eq(7.10)] $\mathcal{L}[a_1f_1(t) + a_2f_2(t)] = a_1F_1(s) + a_2F_2(s)$

and

[eq(7.20)] $\mathcal{L}[e^{-at}f(t)] = F(s)\Big|_{s\leftarrow s+a} = F(s+a).$

Equation (7.10) is the *linearity* property. Equation (7.20) is sometimes called the *complex shifting* property, since multiplication by e^{-at} in the time domain results in

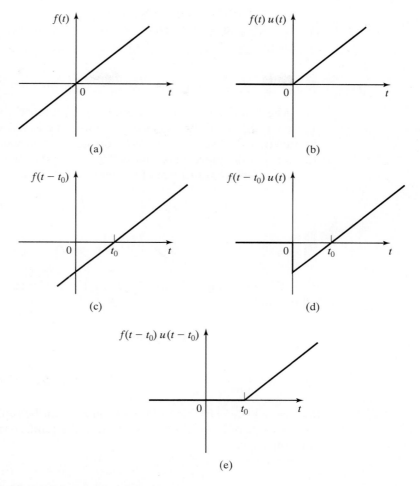

Figure 7.5 Examples of shifted functions.

a shift in the s-plane variable by the amount a. Of course, in general $s = \sigma + j\omega$ is complex, and a may also be complex.

Real Shifting

Next a property relating to shifting in the time domain is derived. Consider the time function $f(t)$ illustrated in Figure 7.5(a). We let $f(t)$ be a ramp function for simplicity, but the following results apply for any time function. We now consider various ways of shifting time functions.

Figure 7.5(b) is a plot of $f(t)u(t)$, where $u(t)$ is the unit step function. Hence,

$$f(t)u(t) = \begin{cases} f(t), & t > 0 \\ 0, & t < 0 \end{cases}.$$

Figure 7.5(c) shows a plot of the function $f(t - t_0)$, where t_0 is the amount of the shift in time, with $t_0 > 0$. The function $f(t - t_0)u(t)$ is shown in Figure 7.5(d), and the function $f(t - t_0)u(t - t_0)$ is given in Figure 7.5(e). For this last function,

$$f(t - t_0)u(t - t_0) = \begin{cases} f(t - t_0), & t > t_0 \\ 0, & t < t_0 \end{cases}.$$

The reader should note carefully the differences in the functions in Figure 7.5. As we have defined the Laplace transform, the Laplace transform of $f(t)$ requires the function in Figure 7.5(b). We now derive a property that relates the Laplace transform of the function of Figure 7.5(e) to that of the function of Figure 7.5(b).

The Laplace transform of the function of Figure 7.5(e) is given by

$$\mathscr{L}[f(t - t_0)u(t - t_0)] = \int_0^\infty f(t - t_0)u(t - t_0)e^{-st}\, dt$$

$$= \int_{t_0}^\infty f(t - t_0)e^{-st}\, dt.$$

We make the change of variable $(t - t_0) = \tau$. Hence, $t = (\tau + t_0)$, $dt = d\tau$, and it follows that

$$\mathscr{L}[f(t - t_0)u(t - t_0)] = \int_0^\infty f(\tau)e^{-s(\tau + t_0)}\, d\tau$$

$$= e^{-t_0 s}\int_0^\infty f(\tau)e^{-s\tau}\, d\tau. \tag{7.21}$$

Because τ is the variable of integration and can be replaced with t, the integral on the right side of the (7.21) is $F(s)$. Hence, the Laplace transform of the shifted time function is given by

$$\mathscr{L}[f(t - t_0)u(t - t_0)] = e^{-t_0 s}F(s), \tag{7.22}$$

where $t_0 \geq 0$ and $\mathscr{L}[f(t)] = F(s)$. This relationship, called the real-shifting, or real-translation, property, applies only for a function of the type shown in Figure 7.5(e); it is necessary that the shifted function be zero for time less than t_0, the amount of the shift. Three examples are now given to illustrate this property.

Laplace transform of a delayed exponential function

Consider the exponential function shown in Figure 7.6(a), which has the equation

$$f(t) = 5e^{-0.3t},$$

where t is in seconds. This function delayed by 2 s and multiplied by $u(t - 2)$ is shown in Figure 7.6(b); the equation for this delayed exponential function is given by

$$f_1(t) = 5e^{-0.3(t - 2)}u(t - 2).$$

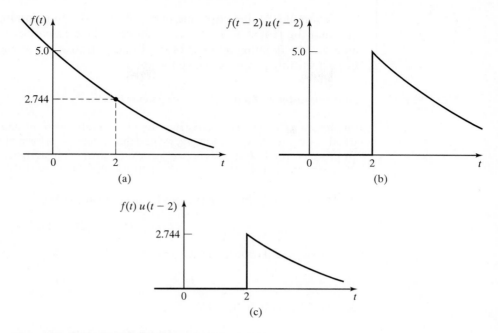

Figure 7.6 Shifted functions for Examples 7.4 and 7.5.

From Table 7.2 and (7.22),

$$\mathcal{L}[f_1(t)] = F_1(s) = e^{-2s}F(s) = \frac{5e^{-2s}}{s + 0.3}.\ \blacksquare$$

EXAMPLE 7.5   **Laplace transform of a more complex delayed function**

Consider now the function of Figure 7.6(c), which is the exponential of Figure 7.6(a) with no delay, but with a value of zero for $t < 2$ s. The equation for this function is

$$f_2(t) = 5e^{-0.3t}\, u(t - 2).$$

This equation is not of the form of (7.22), but can be manipulated into that form as follows:

$$\begin{aligned}
f_2(t) &= 5e^{-0.3t}\, u(t - 2)[e^{0.3(2)}e^{-0.3(2)}]\\
&= (5e^{-0.6})e^{-0.3(t-2)}u(t - 2) = 2.744e^{-0.3(t-2)}u(t - 2).
\end{aligned}$$

Hence, $f_2(t)$ is now of the form required in (7.22), and $F_2(s)$ is given by

$$F_2(s) = \mathcal{L}[f_2(t)] = \frac{2.744e^{-2s}}{s + 0.3}.\ \blacksquare$$

Example 7.5 illustrates the manipulation of a function into the form of (7.22) such that the real-shifting property applies. The alternative to this procedure is to integrate the defining integral of the Laplace transform. A third example of using the real-shifting property is given next.

Laplace transform of a straight-line-segments function

It is sometimes necessary to construct complex waveforms from simpler waveforms, as discussed in Chapter 2. As an example, we find the Laplace transform of the signal in Figure 7.7. The procedure of Section 2.5 is used to write the equation of this signal. We write this equation in four steps, in which $f_i(t)$, $i = 1, 2, 3$ are the results of the first three steps, respectively.

1. The slope of the function changes from 0 to 10 at $t = 1$:

$$f_1(t) = 10(t - 1)u(t - 1).$$

2. The slope of the function changes from 10 to 0 at $t = 2$:

$$f_2(t) = f_1(t) - 10(t - 2)u(t - 2).$$

3. The function steps by -3 at $t = 2$:

$$f_3(t) = f_2(t) - 3u(t - 2).$$

4. The function steps by -7 at $t = 3$:

$$\begin{aligned} f(t) &= f_3(t) - 7u(t - 3) \\ &= 10(t - 1)u(t - 1) - 10(t - 2)u(t - 2) - 3u(t - 2) - 7u(t - 3). \end{aligned} \quad (7.23)$$

We verify this function (as the sum of four terms) as follows:

$$\begin{aligned} t < 1, \qquad & f(t) = 0 - 0 - 0 - 0 = 0; \\ 1 < t < 2, \quad & f(t) = 10(t - 1) - 0 - 0 - 0 = 10(t - 1); \end{aligned}$$

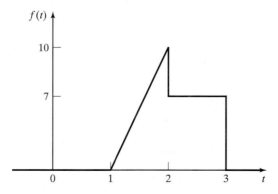

Figure 7.7 Complex waveform.

$$2 < t < 3, \quad f(t) = 10(t - 1) - 10(t - 2) - 3 - 0 = 7;$$
$$3 < t, \qquad f(t) = 7 - 7 = 0.$$

Hence, the equation agrees with the figure. Each term in $f(t)$, (7.23), is of the form required by the real-shifting property:

[eq(7.22)] $\mathcal{L}[f(t - t_0)u(t - t_0)] = e^{-t_0 s} F(s).$

The Laplace transform of $f(t)$ is, then, from (7.23) and Table 7.2,

$$F(s) = \frac{10e^{-s}}{s^2} - \frac{10e^{-2s}}{s^2} - \frac{3e^{-2s}}{s} - \frac{7e^{-3s}}{s}.$$

If desired, these terms can be combined to yield

$$F(s) = \frac{10e^{-s} - 10e^{-2s} - 3se^{-2s} - 7se^{-3s}}{s^2}. \qquad \blacksquare$$

We make two points relative to this example. First, complicated Laplace transforms can occur for complex waveforms. For signals of this type, the function $f(t)$ should be written as a sum of terms such that each term is of the form required in the real-shifting property, (7.22). Otherwise, the definition, (7.4), must be integrated to find the Laplace transform.

As the second point, note from Table 7.2 that all transforms listed are ratios of two polynomials in s. A ratio of polynomials is called a *rational function*. Any sum of the signals in Table 7.2 generally yields a rational function of higher order. The appearance of an exponential function of s in a Laplace transform generally results from delayed time functions.

Differentiation

We next consider two of the most useful properties of the Laplace transform, which are related to differentiation and integration. The differentiation property was derived in Section 7.2, and is, from (7.15),

$$\mathcal{L}\left[\frac{df(t)}{dt}\right] = sF(s) - f(0^+). \qquad (7.24)$$

Property (7.24) is now extended to higher-order derivatives. The Laplace transform of the second derivative of $f(t)$ can be expressed as

$$\mathcal{L}\left[\frac{d^2 f(t)}{dt^2}\right] = \mathcal{L}\left[\frac{df'(t)}{dt}\right], \quad f'(t) = \frac{df(t)}{dt}. \qquad (7.25)$$

Then, replacing $f(t)$ with $f'(t)$ in (7.24), we can express (7.25) as

$$\mathcal{L}\left[\frac{d^2 f(t)}{dt^2}\right] = s\mathcal{L}[f'(t)] - f'(0^+),$$

$$= S[\,S\,F(s) - f(0) - f'(0)\,]$$

$$= S^2 F(s) - S f(0) - f'(0)$$

where $f'(0^+)$ is the value of $df(t)/dt$ as $t \rightarrow 0^+$. Substituting (7.24) into the last equation yields the Laplace transform of the second derivative of a function:

$$\mathcal{L}\left[\frac{d^2 f(t)}{dt^2}\right] = s[sF(s) - f(0^+)] - f'(0^+) = s^2 F(s) - sf(0^+) - f'(0^+). \quad (7.26)$$

By the same procedure,

$$\mathcal{L}\left[\frac{d^3 f(t)}{dt^3}\right] = \mathcal{L}\left[\frac{df''(t)}{dt}\right] = s\mathcal{L}[f''(t)] - f''(0^+), \quad (7.27)$$

where $f''(0^+)$ is the value of $d^2 f(t)/dt^2$ as $t \rightarrow 0^+$. Then, from (7.26) and (7.27),

$$\mathcal{L}\left[\frac{d^3 f(t)}{dt^3}\right] = s^3 F(s) - s^2 f(0^+) - sf'(0^+) - f''(0^+). \quad (7.28)$$

It is seen that this procedure can be extended to the nth derivative of $f(t)$, with the result

$$\mathcal{L}\left[\frac{d^n f(t)}{dt^n}\right] = s^n F(s) - s^{n-1} f(0^+) - s^{n-2} f'(0^+)$$
$$- \cdots - sf^{(n-2)}(0^+) - f^{(n-1)}(0^+), \quad (7.29)$$

where $f^{(i)}(0^+)$ is the value of $d^i f(t)/dt^i$ as $t \rightarrow 0^+$.

A rigorous proof of (7.29) shows that $f'(0), \ldots, f^{(n-1)}(0)$ must exist and that $f^{(n)}(t)$ must also exist. [No discontinuities are allowed in $f^{(n-1)}(t)$.] In particular, this property, (7.29), does not apply to any derivatives of the unit step function [3].

Some problems can occur in (7.29) with the notation for initial conditions. The problems appear in systems in which initial conditions change instantaneously. For that case, the initial conditions in (7.29) are denoted as $f^{(i)}(0^-)$; this notation indicates the initial conditions *before* the instantaneous change, and $f^{(i)}(0^+)$ indicates the initial condition *after* the instantaneous change [3]. This topic is discussed further in Section 7.6, after material is covered that allows us to present an illustrative example.

EXAMPLE 7.7   **Illustration of the differentiation property**

As an illustration of the differentiation property, consider the Laplace transform of $\sin bt$, from Table 7.2:

$$\mathcal{L}[\sin bt] = \frac{b}{s^2 + b^2}.$$

Now, $\sin bt$ can also be expressed as

$$\sin bt = -\frac{1}{b}\frac{d}{dt}(\cos bt).$$

We now use this result to find $\mathscr{L}[\sin bt]$. From the differentiation property (7.24) and Table 7.2,

$$\mathscr{L}[\sin bt] = \mathscr{L}\left[-\frac{1}{b}\frac{d}{dt}(\cos bt) \right]$$

$$= -\frac{1}{b}\left[s\mathscr{L}[\cos bt] - \cos bt \Big|_{t\to 0^+} \right]$$

$$= -\frac{1}{b}\left[s\frac{s}{s^2 + b^2} - 1 \right] = \frac{b}{s^2 + b^2}.$$

Hence, we get the same transform for $\sin bt$ by both the direct transform and the differentiation property. ■

Example 7.7 illustrates the differentiation property; however, the principal use of this property is in the analysis and solution of differential equations, as is developed later.

Integration

The property for the integral of a function $f(t)$ is now derived. Let the function $g(t)$ be expressed by

$$g(t) = \int_0^t f(\tau)\, d\tau.$$

We wish to find the Laplace transform of $g(t)$ in terms of the Laplace transform of $f(t)$. Consider the Laplace transform of $g(t)$:

$$\mathscr{L}[g(t)] = \mathscr{L}\left[\int_0^t f(\tau)\, d\tau \right] = \int_0^\infty \left[\int_0^t f(\tau)\, d\tau \right] e^{-st}\, dt. \qquad (7.30)$$

We integrate this expression by parts (see Appendix B), with

$$u = \int_0^t f(\tau)\, d\tau, \quad dv = e^{-st}\, dt.$$

Using Leibnitz's rule of Appendix B to find du yields

$$du = f(t)\, dt, \quad v = \frac{e^{-st}}{-s}.$$

Thus, from (7.30),

$$\mathscr{L}\left[\int_0^t f(\tau)\, d\tau \right] = \frac{e^{-st}}{-s} \int_0^t f(\tau)\, d\tau \Big|_{t=0}^{\infty} + \frac{1}{s} \int_0^\infty f(t) e^{-st}\, dt$$

$$= -[0 - 0] + \frac{1}{s}F(s).$$

For the first term on the right side, the exponential function is zero at the upper limit, and the integral is zero at the lower limit. With $F(s) = \mathcal{L}[f(t)]$, the property for integration is then

$$\mathcal{L}\left[\int_0^t f(\tau)d\tau\right] = \frac{1}{s}F(s). \tag{7.31}$$

We illustrate this property with an example.

EXAMPLE 7.8   **Illustration of the integration property**

Consider the following relationship, for $t > 0$:

$$\int_0^t u(\tau)d\tau = \tau\Big|_0^t = t.$$

The Laplace transform of the unit step function is $1/s$, from Table 7.2. Hence, from (7.31),

$$\mathcal{L}[t] = \mathcal{L}\left[\int_0^t u(\tau)\,d\tau\right] = \frac{1}{s}\mathcal{L}[u(t)] = \frac{1}{s}\frac{1}{s} = \frac{1}{s^2},$$

which is the Laplace transform of $f(t) = t$. Note that this procedure can be extended to find the Laplace transform of t^n, for n any positive integer. ∎

Five properties of the Laplace transform have thus far been derived: linearity, complex shifting, real shifting, differentiation, and integration. Additional properties are derived in the next section.

7.5   ADDITIONAL PROPERTIES

Four additional properties of the Laplace transform are derived in this section; then a table of properties is given.

Multiplication by t

To derive the first property, consider

$$\mathcal{L}[tf(t)] = \int_0^\infty tf(t)e^{-st}\,dt. \tag{7.32}$$

With $F(s) = \mathcal{L}[f(t)]$, we can write, using Leibnitz's rule,

$$-\frac{dF(s)}{ds} = -\frac{d}{ds}\left[\int_0^\infty f(t)e^{-st}\,dt\right] = \int_0^\infty tf(t)e^{-st}\,dt. \tag{7.33}$$

From (7.32) and (7.33), we have the *multiplication-by-t property*:

$$\mathcal{L}[tf(t)] = -\frac{dF(s)}{ds}.$$ (7.34)

An example that illustrates this property is given next.

EXAMPLE 7.9   **Illustration of the multiplication-by-*t* property**

We now derive the transform of $f(t) = t\cos t$. From Table 7.2,

$$\mathcal{L}[\cos bt] = \frac{s}{s^2 + b^2}.$$

Then, from (7.34),

$$\mathcal{L}(t\cos bt) = -\frac{d}{ds}\left[\frac{s}{s^2 + b^2}\right]$$

$$= -\frac{(s^2 + b^2)(1) - s(2s)}{(s^2 + b^2)^2}$$

$$= \frac{s^2 - b^2}{(s^2 + b^2)^2},$$

which agrees with the transform given in Table 7.2. Two more examples of the value of this property are

$$\mathcal{L}[t^2] = -\frac{d}{ds}\,\mathcal{L}[t] = -\frac{d}{ds}\left[\frac{1}{s^2}\right] = \frac{2}{s^3} = \frac{2!}{s^3}$$

and

$$\mathcal{L}[t^3] = -\frac{d}{ds}\left[\frac{2}{s^3}\right] = \frac{2\cdot 3}{s^4} = \frac{3!}{s^4}.$$ ■

Initial Value

We define the initial value of $f(t)$, $f(0^+)$, as the limit of $f(t)$ as t approaches zero from the right. The initial-value property allows us to find $f(0^+)$ directly from $F(s)$, without first finding the inverse transform $f(t)$. To derive this property, consider the Laplace transform of the derivative of $f(t)$, from (7.24):

$$\mathcal{L}\left[\frac{df(t)}{dt}\right] = \int_0^\infty \frac{df(t)}{dt}\, e^{-st}dt = sF(s) - f(0^+).$$ (7.35)

We take the limit of this relation as s approaches infinity, for s real and positive:

$$\lim_{s\to\infty}\int_0^\infty \frac{df(t)}{dt}e^{-st}dt = \lim_{s\to\infty}[sF(s) - f(0^+)].$$

The limiting process for the left side can be taken inside the integral, and thus the integrand is zero because of the exponential function. As a result,

$$0 = \lim_{s\to\infty}[sF(s) - f(0^+)].$$

Because $f(0^+)$ is independent of s, we can write

$$f(0^+) = \lim_{s\to\infty} sF(s), \qquad (7.36)$$

with s real and positive [4]. This result is the *initial-value property* of the Laplace transform and is useful in linear system analysis. A rigorous-derivation of (7.36) shows that $f(t)$ must be continuous for $t \geq 0$, except for possibly a finite number of finite jumps over any finite interval [4].

Final Value

Consider again the Laplace transform of the derivative of $f(t)$, given in (7.35). We let s approach zero, with the result

$$\lim_{s\to 0}\mathcal{L}\left[\frac{df(t)}{dt}\right] = \lim_{s\to 0}\int_0^\infty \frac{df(t)}{dt}e^{-st}dt = \int_0^\infty \frac{df(t)}{dt}dt$$

$$= \lim_{t\to\infty}[f(t)] - f(0^+), \qquad (7.37)$$

where the limiting operation shown is taken inside the integral, with

$$\lim_{s\to 0} e^{-st} = 1.$$

Also, from (7.35),

$$\lim_{s\to 0}\mathcal{L}\left[\frac{df(t)}{dt}\right] = \lim_{s\to 0}[sF(s) - f(0^+)]. \qquad (7.38)$$

Equating the right sides of (7.37) and (7.38) yields

$$\lim_{t\to\infty} f(t) = \lim_{s\to 0} sF(s). \qquad (7.39)$$

This is the *final-value property*, and a rigorous proof requires that $f(t)$ have a final value and be continuous for $t \geq 0$ except for possibly a finite number of finite jumps over any finite interval [4]. If $f(t)$ does not have a final value, the right side of (7.39) may still give a finite value, which is incorrect. Hence, care must be used in applying (7.39).

EXAMPLE 7.10   **Illustrations of initial- and final-value properties**

Examples of the initial-value property and the final-value property are now given. Consider first the unit step function, which has both an initial value and a final value of unity. Because $\mathscr{L}[u(t)] = 1/s$, from (7.36), we have

$$f(0^+) = \lim_{s \to \infty} sF(s) = \lim_{s \to \infty} s\frac{1}{s} = \lim_{s \to \infty} 1 = 1$$

and from (7.39),

$$\lim_{t \to \infty} f(t) = \lim_{s \to 0} sF(s) = \lim_{s \to 0} 1 = 1.$$

Consider next the function $\sin bt$, where, from Table 7.2,

$$\mathscr{L}[\sin bt] = \frac{b}{s^2 + b^2}.$$

The initial value of $\sin bt$ is zero, and the final value is undefined. From (7.36), the initial value is

$$f(0^+) = \lim_{s \to \infty} sF(s) = \lim_{s \to \infty} \frac{bs}{s^2 + b^2} = \lim_{s \to \infty} \frac{bs}{s^2} = 0,$$

which is the correct value. Application of the final-value property (7.39) yields

$$\lim_{t \to \infty} f(t) = \lim_{s \to 0} sF(s) = \lim_{s \to 0} \frac{bs}{s^2 + b^2} = 0,$$

which is not correct. Recall that the final-value property is applicable only if $f(t)$ has a final value. This example illustrates that care must be exercised in applying the final-value property. ∎

Time Transformation

Time transformations were introduced in Section 2.1. We now consider the effect of these transformations on the Laplace transform of a function; the result is a combined property of *real shifting* and *time scaling*.

For a function $f(t)$, the general independent-variable transformation is given by $\tau = (at - b)$, yielding

$$f(at - b) = f(\tau)\Big|_{\tau = at - b} = f_t(t). \tag{7.40}$$

Since we are considering the single-sided Laplace transform, we require that $a > 0$ and $b \geqq 0$. As in real shifting, (7.22), we also require that $f(at - b)$ be multiplied by the shifted unit step function $u(at - b)$.

We wish to express $F_t(s)$ as a function of $F(s) = \mathscr{L}[f(t)]$. From (7.40),

$$F_t(s) = \mathscr{L}[f(at - b)u(at - b)]$$

$$= \int_0^\infty f(at - b)u(at - b)e^{-st}\, dt. \tag{7.41}$$

We make the change of variable

$$\tau = at - b \Rightarrow t = \frac{\tau + b}{a}; \quad dt = \frac{d\tau}{a}.$$

Then, from (7.41),

$$F_t(s) = \int_{-b}^{\infty} f(\tau)u(\tau)e^{-s(\tau+b)/a} \frac{d\tau}{a}$$

$$= \frac{e^{-sb/a}}{a} \int_{0}^{\infty} f(\tau)e^{-(s/a)\tau} d\tau = \frac{e^{-sb/a}}{a} F\left(\frac{s}{a}\right)$$

and the time-transformation property is shown by the transform pair

$$f(at - b)u(at - b) \longleftrightarrow \frac{e^{-\frac{bs}{a}}}{a} F(s/a) \tag{7.42}$$

This property is now illustrated with an example.

EXAMPLE 7.11   **Illustration of time shifting and time transformation**

Consider the function $\sin 3t$. From Table 7.2,

$$f(t) = \sin 3t \xleftrightarrow{\mathcal{L}} \frac{3}{s^2 + 9} = F(s).$$

We wish to find the Laplace transform of

$$f_t(t) = \sin\left[3\left(4t - \frac{\pi}{6}\right)\right]u\left(4t - \frac{\pi}{6}\right).$$

From (7.40), $a = 4$ and $b = \pi/6$. Then, from (7.42),

$$F_t(s) = \frac{e^{-s\pi/24}}{4} F\left(\frac{s}{4}\right) = \frac{e^{-s\pi/24}}{4} \frac{3}{(s/4)^2 + 9} = \frac{12e^{-s\pi/24}}{s^2 + 144}.$$

To check this result, consider

$$\sin\left[3\left(4t - \frac{\pi}{6}\right)\right]u\left(4t - \frac{\pi}{6}\right) = \sin\left[12\left(t - \frac{\pi}{24}\right)\right]u\left(t - \frac{\pi}{24}\right),$$

since $u(at - b) = u(t - b/a)$. From Table 7.2 and the real-shifting property, (7.22),

$$\mathcal{L}\left[\sin\left[3\left(4t - \frac{\pi}{6}\right)\right]u\left(4t - \frac{\pi}{6}\right)\right] = \mathcal{L}\left[\sin\left[12\left(t - \frac{\pi}{24}\right)\right]u\left(t - \frac{\pi}{24}\right)\right]$$

$$= e^{-s\pi/24} \mathcal{L}[\sin 12t] = \frac{12e^{-s\pi/24}}{s^2 + 144},$$

and the transform is verified.   ■

Several properties of the Laplace transform have been developed. These properties are useful in generating tables of Laplace transforms and in applying the Laplace transform to the solutions of linear differential equations with constant coefficients. Because we prefer to model continuous-time physical systems with linear differential equations with constant coefficients, these properties are useful in both the analysis and design of linear time-invariant physical systems. Table 7.3 gives the derived properties for the Laplace transform, plus some additional properties. The derivations of some of these additional properties are given as problems at the end of this chapter, or are derived later when the properties are used.

TABLE 7.3 Laplace Transform Properties

Name	Property
1. Linearity, (7.10)	$\mathcal{L}[a_1 f_1(t) + a_2 f_2(t)] = a_1 F_1(s) + a_2 F_2(s)$
2. Derivative, (7.15)	$\mathcal{L}\left[\dfrac{df(t)}{dt}\right] = sF(s) - f(0^+)$
3. nth-order derivative, (7.29)	$\mathcal{L}\left[\dfrac{d^n f(t)}{dt^n}\right] = s^n F(s) - s^{n-1} f(0^+)$ $-\cdots - s f^{(n-2)}(0^+) - f^{(n-1)}(0^+)$
4. Integral, (7.31)	$\mathcal{L}\left[\displaystyle\int_0^t f(\tau)d\tau\right] = \dfrac{F(s)}{s}$
5. Real shifting, (7.22)	$\mathcal{L}[f(t - t_0)u(t - t_0)] = e^{-t_0 s} F(s)$
6. Complex shifting, (7.20)	$\mathcal{L}[e^{-at} f(t)] = F(s + a)$
7. Initial value, (7.36)	$\displaystyle\lim_{t \to 0^+} f(t) = \lim_{s \to \infty} sF(s)$
8. Final value, (7.39)	$\displaystyle\lim_{t \to \infty} f(t) = \lim_{s \to 0} s\,F(s)$
9. Multiplication by t, (7.34)	$\mathcal{L}[t f(t)] = -\dfrac{dF(s)}{ds}$
10. Time transformation, (7.42) $(a > 0; b \geq 0)$	$\mathcal{L}[f(at - b)u(at - b)] = \dfrac{e^{-sb/a}}{a} F\!\left(\dfrac{s}{a}\right)$
11. Convolution	$\mathcal{L}^{-1}[F_1(s)F_2(s)] = \displaystyle\int_0^t f_1(t - \tau)f_2(\tau)\,d\tau$ $= \displaystyle\int_0^t f_1(\tau)f_2(t - \tau)d\tau$
12. Time periodicity	$\mathcal{L}[f(t)] = \dfrac{1}{1 - e^{-sT}} F_1(s)$, where
$[f(t) = f(t + T)], t \geq 0$	$F_1(s) = \displaystyle\int_0^T f(t)e^{-st}\,dt$

7.6 RESPONSE OF LTI SYSTEMS

In this section, we apply the Laplace transform to the calculation of time responses of LTI systems. This procedure is based on finding the inverse Laplace transform using partial fractions and transform tables. It is assumed that the reader is familiar with partial-fraction expansions; those who are unfamiliar with this topic are referred to Appendix F. We begin this section with an example of a problem with initial conditions that was mentioned in Section 7.4.

Initial Conditions

Consider the *RL* circuit of Figure 7.8(a) Let $R = 1\Omega$ and $L = 1H$. The loop equation for this circuit is given by

$$\frac{di(t)}{dt} + i(t) = v(t).$$

The Laplace transform of this equation yields

$$sI(s) - i(0^+) + I(s) = V(s), \tag{7.43}$$

where $i(0^+)$ is the initial current. Solving for the transformed current $I(s)$ yields

$$I(s) = \frac{V(s) + i(0^+)}{s + 1}. \tag{7.44}$$

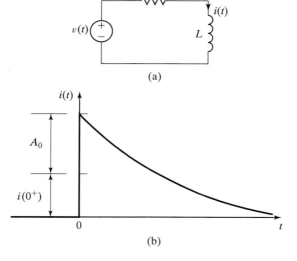

(a)

(b)

Figure 7.8   (a) *RL* circuit and (b) response.

We now let the voltage source in Figure 7.8(a) be an impulse function, with $v(t) = A_0\delta(t)$. Then $V(s) = A_0$, and for $t > 0$, the current $i(t)$ is given by

$$I(s) = \frac{A_0 + i(0^+)}{s + 1} \Rightarrow i(t) = [A_0 + i(0^+)]e^{-t}. \tag{7.45}$$

This current is plotted in Figure 7.8(b). From either the initial-value property and $I(s)$ or directly from $i(t)$, the current $i(0^+)$ is given by

$$i(0^+) = A_0 + i(0^+). \tag{7.46}$$

We see that

1. an inconsistency exists, since the two values denoted as $i(0^+)$ in (7.46) are not equal;
2. the effect of the impulse function is to change the current instantaneously.

We have a problem in defining what is meant by the term *initial conditions*. Doetsch in Ref. 3 suggests that the initial current of the differentiation property in (7.43) be denoted as $i(0^-)$ for the case that the initial condition changes instantaneously. The current $i(0^+)$ is then the value found by the initial-value property (7.36) and is the correct value. Note that no problem exists if the initial condition does not change instantaneously.

For clarity, we use the notation $t = 0^-$ to denote initial conditions *before* instantaneous changes occur. This notation results in (7.46) being expressed as

$$i(0^+) = A_0 + i(0^-).$$

If no instantaneous changes occur, we have no problem and all initial conditions are denoted as occurring at $t = 0^+$. Many authors state that the initial conditions in the differentiation property actually occur at $t = 0^-$ for all cases and change the lower limit of the unilateral-transform integral to $t = 0^-$. However, this approach leads to a different inconsistency. The inversion formula gives zero for all values for $t < 0$ [see (7.7)]. Hence, any variable evaluated using the Laplace-transform pair is zero at $t = 0^-$.

Next, we consider transfer functions.

Transfer Functions

As stated earlier, we prefer to model continuous-time systems with linear differential equations with constant coefficients. The models are then linear and time invariant. (See Section 3.5.) The general equation for the nth-order LTI model is given by

$$\sum_{k=0}^{n} a_k \frac{d^k y(t)}{dt^k} = \sum_{k=0}^{n} b_k \frac{d^k x(t)}{dt^k}, \tag{7.47}$$

where $x(t)$ is the input signal, $y(t)$ is the output signal, and the constants a_k, b_k, and n are parameters of the system.

We now derive the transfer-function model for (7.47). From (7.29), recall the differentiation property:

$$\mathcal{L}\left[\frac{d^k f(t)}{dt^k}\right] = s^k F(s) - s^{k-1}f(0^+) - \cdots - f^{(k-1)}(0^+).$$

Initial conditions must be ignored when deriving transfer functions because a system with non-zero initial conditions is not linear. The transfer function shows the relationship between the input signal and the output signal for a linear system. The differentiation property is then

$$\mathcal{L}\left[\frac{d^k f(t)}{dt^k}\right] = s^k F(s).$$

We use this property to take the transform of (7.47):

$$\sum_{k=0}^{n} a_k s^k Y(s) = \sum_{k=0}^{n} b_k s^k X(s).$$

Expanding this equation gives

$$[a_n s^n + a_{n-1}s^{n-1} + \cdots + a_1 s + a_0]Y(s)$$
$$= [b_n s^n + b_{n-1}s^{n-1} + \cdots + b_1 s + b_0]X(s). \tag{7.48}$$

The system transfer function $H(s)$ is defined as the ratio $Y(s)/X(s)$, from (7.48). Therefore, the transfer function for the model of (7.47) is given by

$$H(s) = \frac{Y(s)}{X(s)} = \frac{b_n s^n + b_{n-1}s^{n-1} + \cdots + b_1 s + b_0}{a_n s^n + a_{n-1}s^{n-1} + \cdots + a_1 s + a_0}. \tag{7.49}$$

For this case, the transfer function is a *rational function* (a ratio of polynomials). Note that this transfer function is identical to that derived in Chapter 3; however, the derivation in Chapter 3 applies only for a complex-exponential input signal. The transfer function (7.49) applies for any input that has a Laplace transform and, hence, is a generalization of that of Chapter 3. An example is now given.

EXAMPLE 7.12   **LTI system response using Laplace transforms**

Consider again the RL circuit of Figure 7.8, and let $R = 4\Omega$ and $L = 0.5H$. The loop equation for this circuit is given by

$$0.5\frac{di(t)}{dt} + 4i(t) = v(t).$$

The Laplace transform of the loop equation (ignoring initial conditions) is given by

$$(0.5s + 4)I(s) = V(s).$$

We define the circuit input to be the voltage $v(t)$ and the output to be the current $i(t)$; hence, the transfer function is

$$H(s) = \frac{I(s)}{V(s)} = \frac{1}{0.5s + 4}.$$

Note that we could have written the transfer function directly from the loop equation and Equations (7.47) and (7.49).

Now we let $v(t) = 12u(t)$. The transformed current is given by

$$I(s) = H(s)V(s) = \frac{1}{0.5s + 4}\frac{12}{s} = \frac{24}{s(s + 8)}.$$

The partial-fraction expansion of $I(s)$ is then

$$I(s) = \frac{24}{s(s + 8)} = \frac{k_1}{s} + \frac{k_2}{s + 8},$$

where (see Appendix F)

$$k_1 = s\left[\frac{24}{s(s + 8)}\right]_{s=0} = \frac{24}{s + 8}\bigg|_{s=0} = 3$$

and

$$k_2 = (s + 8)\left[\frac{24}{s(s + 8)}\right]_{s=-8} = \frac{24}{s}\bigg|_{s=-8} = -3.$$

Thus,

$$I(s) = \frac{24}{s(s + 8)} = \frac{3}{s} + \frac{-3}{s + 8},$$

and the inverse transform, from Table 7.2, yields

$$i(t) = 3[1 - e^{-8t}]$$

for $t > 0$.

This inverse transform is verified with the MATLAB program

```
syms F s
F=24 / ( s* (s+8) )
ilaplace (F)
```

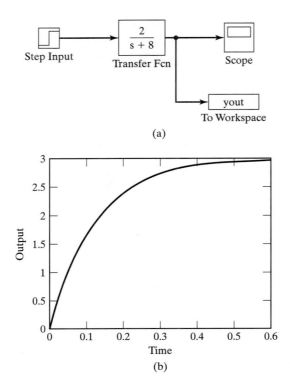

Figure 7.9 Simulink simulation for Example 7.12.

Note from the circuit that the initial current is $i(0) = 0$. The Laplace-transform solution gives $i(0^+) = 0$; this value is also found from the initial-value property:

$$i(0^+) = \lim_{s \to \infty} sI(s) = \lim_{s \to \infty} \frac{24}{s + 8} = 0.$$

Hence, the current does not change instantaneously. The solution can be verified by the substitution of $i(t)$ into the loop differential equation.

This system was simulated using SIMULINK. The block diagram from the simulation is given in Figure 7.9(a), and the response is given in Figure 7.9(b). We see that the system time constant is 0.125 s. Hence, the transient part of the response becomes negligible after approximately four times this time constant, or 0.5 s. Figure 7.9(b) shows this. In addition, the final value of $i(t)$ is 3, which is also evident in the figure. ∎

If the numerator and denominator polynomials in (7.49) are presented in product-of-sums form the transfer functions is shown as

$$H(s) = \frac{K(s - z_1)(s - z_2) \cdots (s - z_m)}{(s - p_1)(s - p_2) \cdots (s - p_n)}. \tag{7.50}$$

In (7.50), $K = b_m/a_n$, where b_m is the coefficient of the highest-order power of s in the numerator polynomial as shown in (7.49). In the transfer functions of many physical systems $m < n$ [i.e., in application of (7.49), b_n and often some of the other b_i

may have a value of zero]. If the function is evaluated with $s = z_i, 1 \le i \le m$, we find $H(z_i) = 0$. Therefore, the z_i are called *zeros* of the transfer function. If $H(s)$ is evaluated with $s = p_i, 1 \le i \le n$, we find that this causes a divide by zero and $H(p_i)$ is undefined. Because $H(p_i)$ becomes undefined, the p_i are called *poles* of the transfer function.

EXAMPLE 7.13   **Poles and zeros of a transfer function**

A transfer function is given in the form of (7.49) as

$$H(s) = \frac{4s + 8}{2s^2 + 8s + 6}.$$

The transfer function is rewritten in the form of (7.50) as

$$H(s) = \frac{2(s + 2)}{(s + 1)(s + 3)}.$$

We now see that this transfer function has one zero at $s = -2$ and two poles located at $s = -1$ and $s = -3$. The poles and the zero of the transfer function are plotted in the s-plane in Figure 7.10. It is standard practice to plot zeros with the symbol ● and poles with the symbol ×.

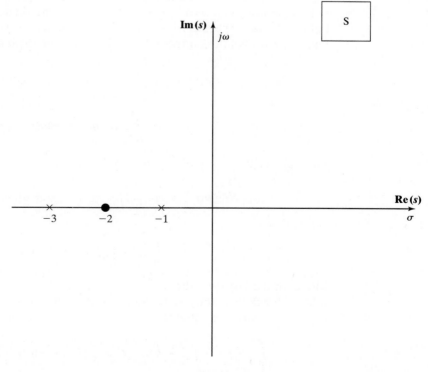

Figure 7.10

(a)

(b)   **Figure 7.11** LTI system.

Convolution

The transfer function of an LTI system described by an nth-order linear differential equation with constant coefficients was just derived. The result in (7.49) is a rational function. We next consider the more general case of the convolution integral, which was derived in Section 3.2. The convolution property of the single-sided Laplace transform is given in Table 7.3. We now derive this property.

From Section 3.2 the convolution integral is given by

$$y(t) = x(t)*h(t) = \int_{-\infty}^{\infty} x(\tau)h(t - \tau)\, d\tau, \tag{7.51}$$

where $x(t)$ is the system input, $y(t)$ is the system output, and $h(t)$ is the system impulse response, as shown in Figure 7.11(a). For the single-sided Laplace transform, both $x(t)$ and $h(t)$ are zero for t less than zero; hence, we consider only causal systems. The convolution integral for this case can be expressed as

$$x(t)*h(t) = \int_{-\infty}^{\infty} x(\tau)u(\tau)h(t - \tau)u(t - \tau)\, d\tau$$

$$= \int_{0}^{\infty} x(\tau)h(t - \tau)u(t - \tau)\, d\tau. \tag{7.52}$$

The Laplace transform of this integral is given by

$$\mathcal{L}[x(t)*h(t)] = \int_{0}^{\infty}\left[\int_{0}^{\infty} x(\tau)h(t - \tau)u(t - \tau)d\tau\right]e^{-st}dt$$

$$= \int_{0}^{\infty} x(\tau)\left[\int_{0}^{\infty} h(t - \tau)u(t - \tau)e^{-st}dt\right]d\tau, \tag{7.53}$$

where, in the last step, the order of integration has been reversed. The integral inside the brackets in the last expression is the Laplace transform of the delayed function $h(t - \tau)u(t - \tau)$; that is,

$$\int_{0}^{\infty} h(t - \tau)u(t - \tau)e^{-st}dt = \mathcal{L}[h(t - \tau)u(t - \tau)] = e^{-\tau s} H(s),$$

from Table 7.3. We can then write (7.53) as

$$\mathscr{L}[x(t)*h(t)] = \int_0^\infty x(\tau)H(s)e^{-s\tau}\,d\tau$$

$$= H(s)\int_0^\infty x(\tau)e^{-s\tau}\,d\tau = H(s)X(s), \tag{7.54}$$

and we see that convolution in the time-domain transforms into multiplication in the s-domain.

In the foregoing derivation, the convolution integral gives the response of an LTI system as depicted in Figure 7.11(a). For this system, from (7.54),

$$y(t) = h(t)*x(t) \Rightarrow Y(s) = H(s)X(s). \tag{7.55}$$

The block diagram for the transformed relationship is given in Figure 7.11(b); this block diagram is defined by (7.55).

It is seen from this development that the transfer function $H(s)$ is the Laplace transform of the system impulse response $h(t)$:

$$H(s) = \int_0^\infty h(t)e^{-st}\,dt. \tag{7.56}$$

Hence, the system impulse response $h(t)$ is the inverse Laplace transform of the transfer function $H(s)$. Consequently, we can specify an LTI continuous-time system by three mathematical relationships:

1. the system differential equation, as in (7.47);
2. the system transfer function $H(s)$;
3. the system impulse response $h(t)$.

Usually, in practice, the transfer function is specified. However, given any one of these three models, we can calculate the other two. We now give examples of calculating the time response of LTI systems using the transfer-function approach, in which the transfer function is not a rational function.

EXAMPLE 7.14   **Response of LTI system from the impulse response**

The unit step response is calculated for an LTI system with the impulse response $h(t)$ given in Figure 7.12(a). We express this function as

$$h(t) = u(t) - u(t - 1).$$

A practical application of a system with this transfer function is in digital-to-analog converters.

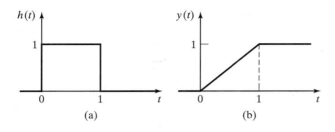

Figure 7.12 Signals for Example 7.14.

Using the real-shifting property, we find the Laplace transform of $h(t)$ to be

$$H(s) = \frac{1 - e^{-s}}{s}.$$

Note that this transfer function is not a rational function. From (7.55), the system output $Y(s)$ is then

$$Y(s) = H(s)X(s) = \frac{1 - e^{-s}}{s} \frac{1}{s} = \frac{1}{s^2}[1 - e^{-s}].$$

From the Laplace transform table and the real-shifting property, we find the system output to be

$$y(t) = tu(t) - [t - 1]u(t - 1).$$

This inverse transform is verified with the MATLAB program

```
syms F s
F = ( 1-exp (-s) ) / (s^2)
ilaplace (F)
```

The results of running this program contains the expression *Heaviside* $(t - 1)$, which is the MATLAB expression for the unit step function $u(t - 1)$.
     This response is shown in Figure 7.12(b).   ■

Transforms with Complex Poles

We next consider a transformed function that has a pair of complex poles. Suppose that $F(s)$ is nth-order with two complex poles. For convenience, we let the other $(n - 2)$ poles be real, so that

$$F(s) = \frac{N(s)}{(s - p_1)(s - p_2)(s - p_3)\cdots(s - p_n)},$$

where $N(s)$ is the numerator polynomial. Let $p_1 = a - jb$ and $p_2 = a + jb$; then, with the order of the numerator less than that of the denominator, the partial-fraction expansion for $F(s)$ can be written as

$$F(s) = \frac{k_1}{s - a + jb} + \frac{k_2}{s - a - jb} + \frac{k_3}{s - p_3} + \cdots + \frac{k_n}{s - p_n}. \tag{7.57}$$

The coefficients k_1 and k_2 can be evaluated by the usual partial-fraction expansion. These coefficients are complex valued, and k_2 is the conjugate of k_1. Thus, the inverse transform of (7.57) has two terms that are complex; however, the sum of these two terms must be real. This sum is not a convenient form. We now present a different procedure for finding the inverse transform, such that all terms are real.

In (7.57), we evaluate k_1 and k_2 from

$$k_1 = (s - a + jb)F(s)\Big|_{s=a-jb} = |k_1|e^{j\theta}$$

and

$$k_2 = (s - a - jb)F(s)\Big|_{s=a+jb} = |k_1|e^{-j\theta}. \qquad (7.58)$$

Let $f_1(t)$ be the sum of the inverse transforms of the first two terms of (7.57):

$$f_1(t) = |k_1|e^{j\theta}e^{(a-jb)t} + |k_1|e^{-j\theta}e^{(a+jb)t}.$$

Using Euler's relation, we can express this relation as

$$f_1(t) = 2|k_1|e^{at}\left[\frac{e^{-j(bt-\theta)} + e^{j(bt-\theta)}}{2}\right]$$

$$= 2|k_1|e^{at}\cos(bt - \theta). \qquad (7.59)$$

Note that, in (7.58), b should be chosen positive, such that b in (7.59) is positive. The sinusoidal expression in (7.59) is a more convenient form than the sum of complex exponential functions. An example illustrating this procedure is given next.

EXAMPLE 7.15   **Inverse transform involving complex poles**

We now find the response of a system with the transfer function

$$H(s) = \frac{3s + 1}{s^2 + 2s + 5}$$

to the input $x(t) = e^{-3t}$. We have

$$Y(s) = H(s)X(s) = \frac{3s + 1}{s^2 + 2s + 5}\frac{1}{s + 3}$$

$$= \frac{3s + 1}{[(s + 1)^2 + 2^2](s + 3)}$$

$$= \frac{k_1}{s + 1 + j2} + \frac{k_2}{s + 1 - j2} + \frac{k_3}{s + 3},$$

and $p_1 = a - jb = -1 - j2$, $p_2 = -1 + j2$, and $p_3 = -3$. For the pole at $s = -3$,

$$k_3 = \left.\frac{3s + 1}{s^2 + 2s + 5}\right|_{s=-3} = \frac{-9 + 1}{9 - 6 + 5} = -1.$$

For the complex poles, from (7.58), $p_1 = a - jb = -1 - j2$. Hence, $a = -1$, $b = 2$, and

$$k_1 = (s + 1 + j2)Y(s)\Big|_{s=-1-j2} = \left.\frac{3s + 1}{(s + 1 - j2)(s + 3)}\right|_{s=-1-j2}$$

$$= \frac{3(-1 - j2) + 1}{(-1 - j2 + 1 - j2)(-1 - j2 + 3)} = \frac{-2 - j6}{-j4(2 - j2)}.$$

Thus,

$$k_1 = \frac{6.325\angle{-108.4°}}{(4\angle{-90°}(2.828\angle{-45°})} = 0.559\angle{26.6°}.$$

Then, in (7.59), $2|k_1| = 1.118$, $\theta = 26.6°$, and for $t > 0$,

$$y(t) = 1.118e^{-t}\cos(2t - 26.6°) - e^{-3t}.$$

This result can be verified by finding its transform. The partial-fraction expansion can be verified by the MATLAB program

```
n = [0 0 3 1];
d1 = [1 2 5];
d2 = [1 3];
d = conv (d1,d2);
[r,p,k] = residue (n,d)
result:   r = -1 0.5-0.25j 0.5+0.25j
          p = -3 -1+2j -1-2j
          k = 0
k1mag = abs ( r(3) )
k1phase = angle ( r(3))*180/pi
result: k1mag = 0.5590 k1phase = 26.5651
```

The statement $d = $ conv(d1, d2) multiplies the two polynomials. Why was $r(3)$ chosen as k_1 rather than $r(2)$? This is an important point. ■

This MATLAB program finds $y(t)$ using the symbolic math toolbox:

```
syms Y s
Y=(3*s+1)/(((s+1)^2+2^2)*(s+3))
ilaplace(Y)
```

We wish to make two points relative to Example 7.15. First, the example illustrates one procedure for finding the inverse Laplace transform of functions that contain sinusoids. Other procedures are available; however, the foregoing procedure has the advantage that the amplitudes and phases of sinusoids are evident.

The second point is that complex poles in a transfer function result in sinusoidal terms (or complex-exponential terms) in the system's natural response. Of course, real poles in the transfer function result in real exponential terms in the system response.

Functions with Repeated Poles

We next illustrate the inverse transform of a function with a multiple pole.

EXAMPLE 7.16   **Inverse transform involving repeated poles**

The unit step response of a system with the third-order transfer function

$$\frac{Y(s)}{X(s)} = H(s) = \frac{4s^2 + 4s + 4}{s^3 + 3s^2 + 2s}$$

will be found. Hence, $x(t) = u(t)$ and $X(s) = 1/s$. The system output is then

$$Y(s) = H(s)X(s) = \frac{4s^2 + 4s + 4}{s^3 + 3s^2 + 2s}\left(\frac{1}{s}\right).$$

Because this function is not in Table 7.2, we must find its partial-fraction expansion:

$$Y(s) = \frac{4s^2 + 4s + 4}{s^2(s + 1)(s + 2)} = \frac{k_1}{s^2} + \frac{k_2}{s} + \frac{k_3}{s + 1} + \frac{k_4}{s + 2}.$$

We solve first for k_1, k_3, and k_4:

$$k_1 = \left.\frac{4s^2 + 4s + 4}{(s + 1)(s + 2)}\right|_{s=0} = \frac{4}{2} = 2;$$

$$k_3 = \left.\frac{4s^2 + 4s + 4}{s^2(s + 2)}\right|_{s=-1} = \frac{4 - 4 + 4}{(1)(1)} = 4;$$

$$k_4 = \left.\frac{4s^2 + 4s + 4}{s^2(s + 1)}\right|_{s=-2} = \frac{16 - 8 + 4}{(4)(-1)} = -3.$$

We calculate k_2 by using Equation (F.8) of Appendix F:

$$k_2 = \frac{d}{ds}[s^2 Y(s)]_{s=0} = \frac{d}{ds}\left[\frac{4s^2 + 4s + 4}{s^2 + 3s + 2}\right]_{s=0}$$

$$= \left.\frac{(s^2 + 3s + 2)(8s + 4) - (4s^2 + 4s + 4)(2s + 3)}{[s^2 + 3s + 2]^2}\right|_{s=0}$$

$$= \frac{(2)(4) - (4)(3)}{4} = -1.$$

The partial-fraction expansion is then

$$Y(s) = \frac{4s^2 + 4s + 4}{s^2(s^2 + 3s + 2)} = \frac{2}{s^2} + \frac{-1}{s} + \frac{4}{s + 1} + \frac{-3}{s + 2},$$

which yields the output signal

$$y(t) = 2t - 1 + 4e^{-t} - 3e^{-2t},$$

for $t > 0$. The following MATLAB program verifies the partial-fraction expansion:

```
n = [0 0 4 4 4];
d = [1 3 2 0 0];
[r,p,k] = residue (n,d)
result: r = −3 4 −1 2
        p = −2 −1 0 0
        k = 0
```

This MATLAB program verifies the results of Example 7.16 using the Symbolic Math Toolbox.

```
%
% The system transfer function:
H=(4*s^2+4*s+4)/(s^3+3*s^2+2*s)
% The Laplace transform of the input signal:
X=1/s
% Calculate the Laplace transform of the system output signal.
Y=H*X
% Determine the time-domain output signal by finding the inverse Laplace
transform.
y=ilaplace(Y)
```

In this section, the Laplace-transform solution of differential equations with constant coefficients is demonstrated. This procedure transforms these differential equations into algebraic equations. The algebraic equations are then solved, and partial-fraction expansions are used to transform the solutions back to the time domain. For the case that initial conditions are ignored, this method of solution leads us to the transfer-function representation of LTI systems. The transfer-function approach is a standard procedure for the analysis and design of LTI systems. An important use of transfer functions is in the determination of an LTI system's characteristics. For example, the transfer function gives us the modes of the system and, hence, the nature of the system's transient response, as we will show in the next section.

7.7   LTI SYSTEMS CHARACTERISTICS

In this section, we consider the properties of causality, stability, invertibility, and frequency response for LTI systems, relative to the Laplace transform.

Causality

The unilateral Laplace transform requires that any time function be zero for $t < 0$. Hence, the impulse response $h(t)$ must be zero for negative time. Because this is also the requirement for causality, the unilateral transform can be applied to causal systems only. The bilateral Laplace transform, introduced in Section 7.8, must be employed for noncausal systems.

Stability

We now relate bounded-input bounded-output (BIBO) stability to transfer functions. Recall the definition of BIBO stability:

> ### BIBO Stability
> A system is stable if the output remains bounded for all time for any bounded input.

We express the transfer function of an nth-order system as

$$[\text{eq(7.49)}] \qquad H(s) = \frac{Y(s)}{X(s)} = \frac{b_n s^n + b_{n-1} s^{n-1} + \cdots + b_1 s + b_0}{a_n s^n + a_{n-1} s^{n-1} + \cdots + a_1 s + a_0},$$

where $a_n \neq 0$. The denominator of this transfer function can be factored as

$$a_n s^n + a_{n-1} s^{n-1} + \cdots + a_1 s + a_0 = a_n(s - p_1)(s - p_2) \cdots (s - p_n). \quad (7.60)$$

The zeros of this polynomial are the *poles* of the transfer function, where, by definition, the poles of a function $H(s)$ are the values of s at which $H(s)$ is unbounded.

We can express the output $Y(s)$ in (7.49) as

$$Y(s) = \frac{b_n s^n + b_{n-1} s^{n-1} + \cdots + b_1 s + b_0}{a_n(s - p_1)(s - p_2) \cdots (s - p_n)} X(s)$$

$$= \frac{k_1}{s - p_1} + \frac{k_2}{s - p_2} + \cdots + \frac{k_n}{s - p_n} + Y_x(s), \qquad (7.61)$$

where $Y_x(s)$ is the sum of the terms in this expansion that originate in the poles of the input $X(s)$. Hence, $Y_x(s)$ is the forced response. We have assumed in (7.61) that $H(s)$ has no repeated poles. We have also assumed in the partial-fraction expansion of (7.61) that $b_n = 0$. If $b_n \neq 0$ a constant term, b_n/a_n will appear in the partial-fraction expansion. (See Appendix F.) As discussed in Section 7.6, in the mathematical models of many physical systems, $b_n = 0$.

The inverse transform of (7.61) yields

$$y(t) = k_1 e^{p_1 t} + k_2 e^{p_2 t} + \cdots + k_n e^{p_n t} + y_x(t)$$

$$= y_c(t) + y_x(t). \qquad (7.62)$$

The terms of $y_c(t)$ originate in the poles of the transfer function; $y_c(t)$ is called the system's natural response. (See Section 3.5.) The natural response is always present in the system output, independent of the form of the input signal $x(t)$. Each term of

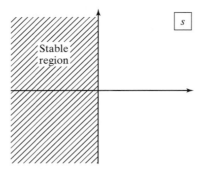

Figure 7.13 Stable region for poles of $H(s)$.

the natural response, $e^{p_i t}$, is called a *mode* of the system. In the classical solution of differential equations discussed in Section 3.5, $y_c(t)$ is called the complementary function and $y_x(t)$ is called the particular solution.

If the input $x(t)$ is bounded, the forced response $y_x(t)$ will remain bounded because $y_x(t)$ is of the functional form of $x(t)$, that is, $Y_x(s)$ has the same poles as $X(s)$. Thus, the output becomes unbounded only if at least one of the natural-response terms, $k_i e^{p_i t}$, becomes unbounded. This unboundedness can occur only if the real part of at least one pole p_i is non-negative.

We see from the preceding discussion that an LTI system is stable provided that all poles of the system transfer function are in the left half of the s-plane, that is, provided $\mathrm{Re}(p_i) < 0, i = 1, 2, \ldots, n$. Recall that we derived this result in Section 3.6 by taking a different approach. The stable region for the poles of $H(s)$ in the s-plane is illustrated in Figure 7.13.

The *system characteristic equation* is, by definition, the denominator polynomial of the transfer function set to zero; that is, the characteristic equation is (7.60), set to zero:

$$a_n s^n + a_{n-1} s^{n-1} + \cdots + a_1 s + a_0 = a_n (s - p_1)(s - p_2) \cdots (s - p_n) = 0. \quad (7.63)$$

Hence, an LTI system is stable provided that all roots of its characteristic equation (poles of its transfer function) are in the left half-plane. We now illustrate system stability with an example.

EXAMPLE 7.17   **Stability of an LTI system**

A much-simplified transfer function for the booster stage of the Saturn V rocket, used in trips to the moon, is given by

$$H(s) = \frac{0.9402}{s^2 - 0.0297} = \frac{0.9402}{(s + 0.172)(s - 0.172)},$$

where the system input was the engine thrust and the system output was the angle of the rocket relative to the vertical. The system modes are $e^{-0.172t}$ and $e^{0.172t}$; the latter mode is obviously unstable. A control system was added to the rocket, such that the overall system was stable and responded in an acceptable manner.   ■

Invertibility

We restate the definition of the inverse of a system from Section 2.7 in terms of transfer functions.

| **Inverse of a System**
| The inverse of an LTI system $H(s)$ is a second system $H_i(s)$ that, when cascaded with
| $H(s)$, yields the identity system.

Thus, $H_i(s)$ is defined by the equation

$$H(s)H_i(s) = 1 \Rightarrow H_i(s) = \frac{1}{H(s)}. \tag{7.64}$$

These systems are illustrated in Figure 7.14.

We now consider the characteristics of the inverse system, assuming that the transfer function $H(s)$ of a causal system can be expressed as in (7.49):

$$H(s) = \frac{b_n s^n + b_{n-1}s^{n-1} + \cdots + b_0}{a_n s^n + a_{n-1}s^{n-1} + \cdots + a_0}. \tag{7.65}$$

Hence, the inverse system has the transfer function

$$H_i(s) = \frac{a_n s^n + a_{n-1}s^{n-1} + \cdots + a_0}{b_n s^n + b_{n-1}s^{n-1} + \cdots + b_0}. \tag{7.66}$$

This inverse system is also causal because (7.66) is a unilateral transfer function. Note that the differential equation of the inverse system can easily be written from (7.66), since the coefficients of the transfer function are also the coefficients of the system differential equation. (See Section 7.6.)

Next we investigate the stability of the inverse system. For the system of (7.65) to be stable, the poles of the transfer function $H(s)$ must lie in the left half of the s-plane. For the inverse system of (7.66) to be stable, the poles of $H_i(s)$ [the zeros of

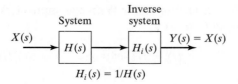

$$H_i(s) = 1/H(s)$$

Figure 7.14 System with its inverse.

$H(s)$] must also lie in the left half-plane. Thus, both a system and its inverse are stable provided that the poles and zeros of the system transfer function are in the left half-plane.

Frequency Response

Recall from (5.1) the definition of the Fourier transform:

$$F(\omega) = \mathscr{F}[f(t)] = \int_{-\infty}^{\infty} f(t)e^{-j\omega t}\, dt. \tag{7.67}$$

From Section 5.4, using the Fourier transform, we find that the transfer function for a causal system with the impulse response $h(t)$ is given by

$$H_f(\omega) = \mathscr{F}[h(t)] = \int_{0}^{\infty} h(t)e^{-j\omega t}\, dt. \tag{7.68}$$

Comparing this transfer function with that based on the Laplace transform, namely,

$$H_l(s) = \mathscr{L}[h(t)] = \int_{0}^{\infty} h(t)e^{-st}\, dt, \tag{7.69}$$

we see that the two transfer functions are related by

$$H_f(\omega) = H_l(s)\Big|_{s=j\omega} = H_l(j\omega). \tag{7.70}$$

Here we have subscripted the transfer functions for clarity, and we see a problem in notation. If we do not subscript the transfer functions, (7.70) is expressed as $H(\omega) = H(j\omega)$, which is inconsistent, to say the least. However, in using the Fourier transform, it is common to denote the frequency response as $H(\omega)$. When using the Laplace transform, it is common to denote the same frequency response as $H(j\omega)$. The reader should note this inconsistency; it is not likely to be changed. We will use the same confusing custom.

For the system of (7.49), the frequency response is given by

$$H(j\omega) = \frac{Y(j\omega)}{X(j\omega)} = \frac{b_n(j\omega)^n + b_{n-1}(j\omega)^{n-1} + \cdots + b_1(j\omega) + b_0}{a_n(j\omega)^n + a_{n-1}(j\omega)^{n-1} + \cdots + a_1(j\omega) + a_0}. \tag{7.71}$$

Recall from Section 3.7 that this frequency response can be measured experimentally on a stable physical system. With the input $x(t) = \cos \omega t$, from (3.78), the steady-state output is given by

$$y_{ss}(t) = |H(j\omega)| \cos (\omega t + \arg H(j\omega)), \tag{7.72}$$

where arg $H(j\omega)$ is the angle of the complex function $H(j\omega)$. Hence, by measuring the steady-state response for a sinusoidal input signal, we can obtain the frequency response of a system.

Generally, for a physical system, the frequency response approaches zero as the input frequency becomes very large. However, in (7.71),

$$\lim_{\omega \to \infty} H(j\omega) = \frac{b_n}{a_n}. \tag{7.73}$$

Hence, for the model of (7.71) to be accurate at higher frequencies, $b_n = 0$ and the order of the numerator of $H(s)$ must be less than the order of its denominator. For this reason, quite often we specify, for the general transfer function, that the numerator order is less than that of the denominator.

Because frequency response is covered in detail in Section 5.5, we do not repeat that coverage here. Readers not familiar with this material should study that section, while keeping (7.70) in mind.

EXAMPLE 7.18 **Plotting the frequency response from the transfer function**

The Bode plots of the frequency response of the system considered in Example 7.15

$$H(s) = \frac{3s + 1}{s^2 + 2s + 5}$$

can be calculated and plotted using the following MATLAB program.

```
% This MATLAB program computes the Bode Diagram of the frequency
% response of the system discussed in Examples 7.15 and 7.18.
% H(s)=(3s+1)/(s^2+2s+5)
num=[3 1];
den=[1 2 5];
flim=[-2,2,400];
bodec(num,den,flim)
% The resulting plots show the magnitude frequency response of the system
% in decibels (dB) and the phase frequency response in degrees
% both are plotted versus frequency in rad/sec on a logarithmic scale.
```

The MATLAB function bodec.m is presented in Section 5.5, Example 5.16.   ■

In this section, the characteristics of causality, stability, invertibility, and frequency response are investigated for LTI systems. It is shown that stability can always be determined for systems modeled by linear differential equations with constant coefficients. This type of system is stable if and only if the poles of its transfer function are all inside the left half-plane.

7.8 BILATERAL LAPLACE TRANSFORM

Recall the definition of the bilateral Laplace transform:

[eq(7.1)]
$$F_b(s) = \mathcal{L}_b[f(t)] = \int_{-\infty}^{\infty} f(t)e^{-st}\, dt$$

and

[eq(7.2)]
$$f(t) = \mathcal{L}_b^{-1}[F_b(s)] = \frac{1}{2\pi j}\int_{c-j\infty}^{c+j\infty} F_b(s)e^{st}\, ds.$$

The difference in the unilateral transform and the bilateral transform is that the bilateral transform includes negative time. For example, in system analysis, we choose $t = 0$ as that time that some significant event occurs, such as switching in an electrical circuit. Then we solve for the resulting system response, for $t \geq 0$. The unilateral Laplace transform is used for this case.

The bilateral Laplace transform is used when results are needed for negative time as well as for positive time. In this section, we consider the bilateral Laplace transform.

We introduce the bilateral Laplace transform by example. First, we find the transform for $f(t) = e^{-at}u(t)$, with a real. This signal is plotted in Figure 7.15(a), for $a > 0$. Of course, since $f(t) = 0$ for $t < 0$, the bilateral transform of this signal is identical to its unilateral transform:

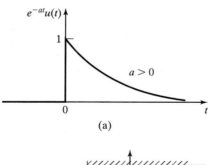

(a)

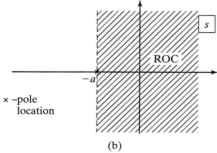

(b)

Figure 7.15   Signal and ROC for $e^{-at}u(t)$.

$$F_b(s) = \mathscr{L}_b[e^{-at}u(t)] = \int_{-\infty}^{\infty} e^{-at}u(t)e^{-st}\,dt$$

$$= \int_{-\infty}^{0} 0e^{-st}\,dt + \int_{0}^{\infty} e^{-at}e^{-st}\,dt = \frac{e^{-(s+a)t}}{-(s+a)}\bigg|_{0}^{\infty}$$

$$= \frac{1}{s+a}, \quad \mathrm{Re}(s+a) > 0 \quad \text{or} \quad \mathrm{Re}(s) > -a. \tag{7.74}$$

This transform exists for s in the half-plane defined by $\mathrm{Re}(s) > -a$. [If a is complex, this inequality is $\mathrm{Re}(s) > \mathrm{Re}(-a)$.] As we will see, we must state the region of convergence (ROC) for each bilateral Laplace transform. The ROC for (7.74) is plotted in Figure 7.15(b), with the function's pole also shown. Neither the boundary of the ROC ($\mathrm{Re}(s) = -a$) nor the pole is in the ROC.

Consider next the exponential function $f(t) = -e^{-at}u(-t)$, with a real. This function is plotted in Figure 7.16(a) for $a > 0$. For this case,

$$F_b(s) = \mathscr{L}_b[-e^{-at}u(-t)] = \int_{-\infty}^{\infty} -e^{-at}u(-t)e^{-st}\,dt$$

$$= \int_{-\infty}^{0} -e^{-at}e^{-st}\,dt + \int_{0}^{\infty} 0e^{-st}\,dt = \frac{e^{-(s+a)t}}{s+a}\bigg|_{-\infty}^{0}$$

$$= \frac{1}{s+a}, \quad \mathrm{Re}(s+a) < 0 \quad \text{or} \quad \mathrm{Re}(s) < -a. \tag{7.75}$$

This transform exists in the half-plane $\mathrm{Re}(s) < -a$. [$\mathrm{Re}(s) < \mathrm{Re}(-a)$ for a complex.] This ROC is plotted in Figure 7.16(b), along with the pole of $F(s)$. As described earlier, neither the boundary nor the pole is in the ROC.

Note the differences in the ROCs of the last two transforms, from Figures 7.15(b) and 7.16(b). The two transforms are equal, but the signals are not equal. Hence, a time function is defined by *both* the transform and the ROC. The ROC *must* be given for a bilateral transform.

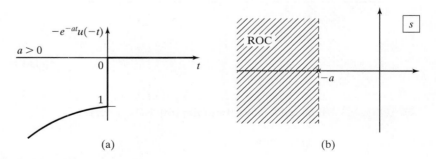

(a)   (b)

Figure 7.16 Signal and ROC for $-e^{-at}u(t)$.

Region of Convergence

As noted, the region of convergence of a bilateral transform is of primary importance. We first further illustrate ROCs with examples. Then some properties of ROCs are given.

EXAMPLE 7.19   **Bilateral Laplace transform of a signal**

Consider the function

$$f(t) = f_1(t) + f_2(t) = 2e^{-5t}u(t) + e^{-4t}u(-t).$$

From (7.74),

$$F_{b1}(s) = \mathcal{L}_b[2e^{-5t}u(t)] = \frac{2}{s + 5}, \quad \text{Re}(s) > -5;$$

and from (7.75),

$$F_{b2}(s) = \mathcal{L}_b[e^{-4t}u(-t)] = \frac{-1}{s + 4}, \quad \text{Re}(s) < -4.$$

For $F_b(s)$ to converge, both $F_{b1}(s)$ and $F_{b2}(s)$ must also converge; therefore, the ROC of $F_b(s)$ is the *intersection* of the ROC of $F_{b1}(s)$ and $F_{b2}(s)$. Thus,

$$F_b(s) = \frac{2}{s + 5} - \frac{1}{s + 4} = \frac{s + 3}{s^2 + 9s + 20}, \quad -5 < \text{Re}(s) < -4.$$

The ROC is shown in Figure 7.17, along with the poles of $F_b(s)$.

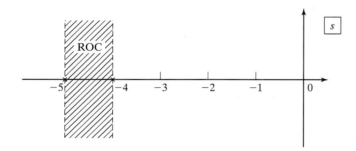

Figure 7.17   ROC for Example 7.19.   ■

EXAMPLE 7.20   **Bilateral Laplace transform with ROC different**

Consider now

$$f(t) = f_1(t) + f_2(t) = e^{-4t}u(t) + 2e^{-5t}u(-t).$$

Note that this function is similar to that of the last example, except now $2e^{-5t}$ is the function for negative time and e^{-4t} is the function for positive time. Thus, from (7.74) and (7.75), we have

$$F_{b1}(s) = \mathcal{L}_b[e^{-4t}u(t)] = \frac{1}{s+4}, \quad \text{Re}(s) > -4$$

and

$$F_{b2}(s) = \mathcal{L}_b[2e^{-5t}u(-t)] = \frac{-2}{s+5}, \quad \text{Re}(s) < -5.$$

Then $\mathcal{L}_b[f(t)]$ exists for $-4 < \text{Re}(s) < -5$; no values of s satisfy this inequality. Hence, the bilateral Laplace transform for $f(t)$ *does not exist*.

EXAMPLE 7.21   **Bilateral Laplace transform with ROC different**

Consider next

$$f(t) = f_1(t) + f_2(t) = 2e^{-5t}u(t) + e^{-4t}u(t).$$

For this function, the bilateral transform and the unilateral transform are equal. Therefore, from Table 7.2,

$$F_{b1}(s) = \mathcal{L}[2e^{-5t}u(t)] = \frac{2}{s+5}, \quad \text{Re}(s) > -5;$$

$$F_{b2}(s) = \mathcal{L}[e^{-4t}u(t)] = \frac{1}{s+4}, \quad \text{Re}(s) > -4;$$

and

$$F_b(s) = \frac{2}{s+5} + \frac{1}{s+4} = \frac{3s+13}{s^2+9s+20}, \quad -4 < \text{Re}(s).$$

The ROC is given in Figure 7.18, along with the poles of the function.

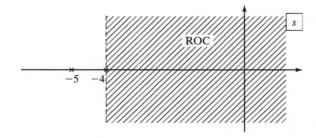

Figure 7.18 ROC for Example 7.21. ∎

The unilateral Laplace transform of a sum of functions exists if the transform of each term of the sum exists. Example 7.20 shows that the bilateral transform of a sum of functions may not exist, even though the transform of each term of the sum does exists.

Note in Figures 7.15 through 7.18 that the poles of the Laplace transforms are not in the ROCs. This is a general property; by definition, the Laplace integral does not converge at a pole of the transform. Hence, the poles of a transform always occur either on the boundary or outside of the ROC.

Bilateral Transform from Unilateral Tables

We next consider a procedure for finding bilateral Laplace transforms from a unilateral Laplace-transform table, such as Table 7.2. Any table of unilateral transforms can be used, provided that ROCs are included in the table.

We first give four definitions relative to functions:

1. A function $f(t)$ is *right sided* if $f(t) = 0$ for $t < t_0$, where t_0 can be positive or negative. For example, $u(t + 10)$ is right sided because

$$u(t + 10) = \begin{cases} 0, & t < -10 \\ 1, & t > -10 \end{cases}.$$

A second example is shown in Figure 7.19(a).

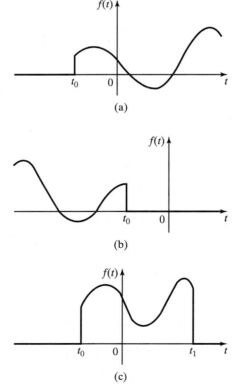

(a)

(b)

(c)

Figure 7.19 Signal types: (a) right sided; (b) left sided; (c) finite duration.

2. A function $f(t)$ is *left sided* if $f(t) = 0$ for $t > t_0$, where t_0 can be positive or negative. For example, $u(-t + 10)$ is left sided because

$$u(-t + 10) = \begin{cases} 1, & t < 10 \\ 0, & t > 10 \end{cases}.$$

A second example is shown in Figure 7.19(b).

3. A function $f(t)$ is *two sided* if it is neither right sided nor left sided. For example, $\cos t$ is two sided.

4. A function is of *finite duration* if it is both right sided and left sided. For example, $[u(t) - u(t - 10)]$ is of finite duration. A second example is illustrated in Figure 7.19(c).

For a function $f(t)$, we now define the two functions $f^+(t)$ and $f^-(t)$:

$$f^+(t) = f(t)u(t); \quad f^-(t) = f(t)u(-t).$$

Hence, $f^+(t)$ is right sided and $f^-(t)$ is left sided. Thus, $f(t)$ can be expressed as

$$f(t) = f^+(t) + f^-(t) \Rightarrow F_b(s) = F_b^+(s) + F_b^-(s), \tag{7.76}$$

provided that the transform of the sum exists. For the right-sided function $f^+(t)$, the bilateral transform $F_b^+(s)$ is that given in the unilateral tables, with the ROC as given.

For the left-sided function $f^-(t)$, we make the substitutions

$$\mathcal{L}_b[f^-(t)] = \int_{-\infty}^{0} f^-(t)e^{-st}\, dt \bigg|_{\substack{t=-t_c \\ s=-s_c}} = \int_{0}^{\infty} f^-(-t_c)e^{-s_c t_c}\, dt_c. \tag{7.77}$$

Thus, the bilateral transform of $f^-(t)$ is the unilateral transform of $f^-(-t)$ with s replaced with $-s$. If the ROC for the unilateral transform of $f^-(-t)$ is $\mathrm{Re}(s) > \alpha$, the ROC of the bilateral transform of $f^-(t)$ is $\mathrm{Re}(-s) > \alpha$, or $\mathrm{Re}(s) < -\alpha$. In summary, to find the bilateral Laplace transform of the left-sided function $f^-(t)$,

1. Find the unilateral Laplace transform of $F_{b1}^-(s) = \mathcal{L}[f^-(-t)]$, with ROC $\mathrm{Re}(s) > \alpha$.

2. The bilateral transform is then

$$F_b^-(s) = \mathcal{L}_b[f^-(t)u(-t)] = F_{b1}^-(-s), \quad \mathrm{Re}(s) < -\alpha.$$

The bilateral transform of $f(t)$ is the sum given in (7.76). We now consider an example of this procedure.

EXAMPLE 7.22   **Bilateral Laplace transform using general approach**

Consider again the function of Example 7.19.

$$f(t) = 2e^{-5t}u(t) + e^{-4t}u(-t).$$

From (7.76), $f^+(t) = 2e^{-5t}u(t)$, and from Table 7.2,

$$F_b^+(s) = \mathcal{L}[2e^{-5t}u(t)] = \frac{2}{s+5}, \quad \text{Re}(s) > -5.$$

From Step 1 of the foregoing procedure, $f^-(-t) = e^{4t}u(t)$, and from Table 7.2,

$$F_{b1}^-(-s) = \mathcal{L}[e^{4t}u(t)] = \frac{1}{s-4}, \quad \text{Re}(s) > 4.$$

From Step 2

$$F_b^-(s) = \mathcal{L}_b[e^{-4t}u(-t)] = \frac{1}{-s-4}, \quad \text{Re}(s) < -4.$$

The bilateral transform is then $F_b^+(s) + F_b^-(s)$:

$$F_b(s) = \frac{2}{s+5} - \frac{1}{s+4} = \frac{s+3}{s^2+9s+20}, \quad -5 < \text{Re}(s) < -4,$$

as derived in Example 7.19 by integration. ∎

Inverse Bilateral Laplace Transform

We have seen in this section that specifying a bilateral transform $F_b(s)$ is not sufficient; the ROC of the transform must also be given. The complex inversion integral for the inverse bilateral transform is given by

[eq(7.2)] $$f(t) = \mathcal{L}_b^{-1}[F_b(s)] = \frac{1}{2\pi j}\int_{c-j\infty}^{c+j\infty} F_b(s)e^{st}\, ds.$$

The value of c in the limits of the integral is chosen as a real value in the ROC. However, this integral is seldom used, except in derivations. As in the case for the unilateral Laplace transform, we use tables to evaluate the inverse bilateral Laplace transform.

We develop this procedure by considering again the functions of Example 7.22. In that example, the bilateral Laplace transform of the right-sided function $2e^{-5t}u(t)$ was found to be

$$F_{b1}(s) = \mathcal{L}_b[2e^{-5t}u(t)] = \frac{2}{s+5}, \quad \text{Re}(s) > -5.$$

$F_{b1}(s)$ has a pole at $s = -5$, which is to the left of the ROC. An examination of the unilateral Laplace-transform table, Table 7.2, shows that the poles of each s-plane function in this table are to the left of the ROCs. This is a general property:

1. The poles of the transform for a right-sided function are always to the left of the ROC of the transform. Figure 7.20 illustrates the poles and the ROC of a right-sided function.

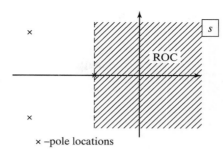

Figure 7.20 Pole locations for a right-sided function.

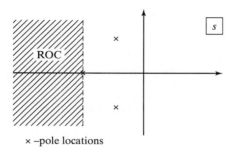

Figure 7.21 Pole locations for a left-sided function.

Next we consider the left-sided function $e^{-4t}u(-t)$. From Example 7.22,

$$F_{b2}(s) = \mathcal{L}_b[e^{-4t}u(-t)] = \frac{-1}{s+4}, \quad \mathrm{Re}(s) < -4.$$

$F_{b2}(s)$ has a pole at $s = -4$, which is to the right of the ROC. The procedure for finding the bilateral transform of a left-sided function, given before Example 7.22, shifts the ROC to the left of the poles of the transform. This is a general property:

2. The poles of the transform for a left-sided function are always to the right of the ROC of the transform. Figure 7.21 illustrates the poles and the ROC of a left-sided function.

In summary, knowing the poles and the ROC, a bilateral transform is expressed as a sum of functions using a partial-fraction expansion. The sum of those terms with poles to the left of the ROC form a right-sided function, and the sum of those terms with poles to the right of the ROC form a left-sided function. The inverse transform of each sum of functions is found from a unilateral-transform table. An example is given next.

EXAMPLE 7.23   **Inverse bilateral Laplace transform**

Consider the Laplace transform

$$F_b(s) = \frac{s+4}{s^2 + 3s + 2} = \frac{s+4}{(s+1)(s+2)}, \quad -2 < \mathrm{Re}(s) < -1.$$

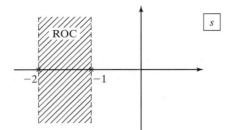

Figure 7.22 ROC for Example 7.23.

First, we must calculate the partial-fraction expansion of $F_b(s)$.

$$F_b(s) = \frac{s + 4}{(s + 1)(s + 2)} = \frac{3}{s + 1} + \frac{-2}{s + 2}, -2 < \text{Re}(s) < -1.$$

The poles and the ROC are plotted in Figure 7.22. The pole at $s = -1$ is to the right of the ROC and is the transform of the left-sided term $f^-(t)$. Thus,

$$f^-(t) = \mathcal{L}_b^{-1}\left[\frac{3}{s + 1}\right] = -3e^{-t}u(-t).$$

The pole at $s = -2$ is to the left of the ROC and is the transform of the right-sided term $f^+(t)$. Hence,

$$f^+(t) = \mathcal{L}_b^{-1}\left[\frac{-2}{s + 2}\right] = -2e^{-2t}u(t).$$

The inverse transform is then the sum of these two functions:

$$f(t) = -3e^{-t}u(-t) - 2e^{-2t}u(t). \qquad \blacksquare$$

In this section, we have defined the bilateral Laplace transform; then examples of the transform and the inverse transform are given. The ROC for a bilateral transform must always be given because the inverse transform is dependent on the ROC.

7.9   RELATIONSHIP OF THE LAPLACE TRANSFORM TO THE FOURIER TRANSFORM

In this section, we note a conflicting notation between the Laplace transform and the Fourier notation. The definition of the unilateral Laplace transform is given by

[eq(7.4)] $$\mathcal{L}[f(t)] = F(s) = \int_0^\infty f(t)e^{-st}\, dt.$$

We now let $f(t)$ equal zero for $t < 0$, that is, we can write $f(t) = f(t)u(t)$. Thus, we can express (7.4) as

$$\mathscr{L}[f(t)] = F(s) = \int_{-\infty}^{\infty} f(t)e^{-st}dt. \qquad (7.78)$$

As an aside, we note that this is the equation of the bilateral Laplace transform.

The definition of the Fourier transform is

[eq(5.4)] $\qquad\qquad \mathscr{F}[f(t)] = F(\omega) = \int_{-\infty}^{\infty} f(t)e^{-j\omega t}dt.$

Evaluating (7.78) for $s = j\omega$ yields

$$F(s)\Big|_{j\omega} = F(j\omega) = \int_{-\infty}^{\infty} f(t)e^{-j\omega t}\, dt. \qquad (7.79)$$

The integrals in (5.4) and (7.79) are equal, and we see an inconsistent notation in these two equations:

$$F(\omega) \overset{?}{=} F(j\omega). \qquad (7.80)$$

However, this conflicting notation is standard; the reader should be aware of it.

A second point is that, for $f(t)$ as defined earlier (7.78), we can write

$$\mathscr{F}[f(t)u(t)] = \mathscr{L}[f(t)u(t)]\Big|_{s=j\omega} \qquad (7.81)$$

provided that each transform exists.

SUMMARY

The unilateral and the bilateral Laplace transforms were introduced in this chapter. We took the approach of developing the unilateral transform as a special case of the bilateral transform. The unilateral transform was used in the analysis and design of linear time-invariant (LTI) continuous-time systems that are causal. This transform is especially useful in understanding the characteristics and in the design of these systems.

The unilateral transform was emphasized in this chapter. A table of transforms and a table of properties were developed for the unilateral transform. System analysis using the unilateral transform was then demonstrated. Next the unilateral Laplace transform of periodic signals was given.

The bilateral Laplace transform is useful in the steady-state analysis of LTI continuous-time systems, and in the analysis and design of noncausal systems. A procedure was developed for finding bilateral transforms from a unilateral transform table. Then some properties of the bilateral transform were derived. See Table 7.4.

TABLE 7.4 Key Equations of Chapter 7

Equation Title	Equation Number	Equation
Bilateral Laplace transform	(7.1)	$\mathcal{L}_b[f(t)] = F_b(s) = \displaystyle\int_{-\infty}^{\infty} f(t)e^{-st}\, dt$
Inverse Laplace transform	(7.2)	$f(t) = \mathcal{L}^{-1}[F(s)] = \dfrac{1}{2\pi j}\displaystyle\int_{c-j\infty}^{c+j\infty} F(s)e^{st}\, ds, \; j = \sqrt{-1}$
Unilateral Laplace transform	(7.4)	$\mathcal{L}[f(t)] = F(s) = \displaystyle\int_{0}^{\infty} f(t)e^{-st}\, dt$
Transfer function	(7.49)	$H(s) = \dfrac{Y(s)}{X(s)} = \dfrac{b_n s^n + b_{n-1}s^{n-1} + \cdots + b_1 s + b_0}{a_n s^n + a_{n-1}s^{n-1} + \cdots + a_1 s + a_0}$

REFERENCES

1. R. V. Churchill, *Operational Mathematics*, 3d ed. New York: McGraw–Hill, 1977.
2. G. Doetsch, *Introduction to the Theory and Application of the Laplace Transform*. New York: Springer–Verlag, 1970.
3. G. Doetsch, *Guide to the Applications of the Laplace and z-Transforms*. New York: Van Nostrand Reinhold, 1971.
4. W. Kaplan, *Operational Methods for Linear Systems*. Reading, MA: Addison–Wesley, 1962.

PROBLEMS

7.1. Use the definition of the Laplace transform, (7.4), and an integral table to verify the following Laplace transforms:

(a) $\mathcal{L}[t \sin bt] = \dfrac{2bs}{(s^2 + b^2)^2}$

(b) $\mathcal{L}[\cos bt] = \dfrac{b}{s^2 + b^2}$

(c) $\mathcal{L}[e^{at}] = \dfrac{1}{s - a}$

(d) $\mathcal{L}[te^{at}] = \dfrac{1}{(s - a)^2}$

(e) $\mathcal{L}[t] = \dfrac{1}{s^2}$

(f) $\mathcal{L}[te^{-at}] = \dfrac{1}{(s + a)^2}$

7.2. Sketch the time functions given. Then use the definition of the Laplace transform, (7.4), and the table of integrals in Appendix A to calculate the Laplace transforms of the time functions.

(a) $6u(t - 4)$

(b) $6[u(t - 1) - u(t - 3)]$

(c) $6u(t - 1)u(4 - t)$

(d) $6u(t - a)u(b - t)$, where $b > a > 0$

7.3. Consider the waveform $f(t)$ in Figure P7.3.

(a) Write a mathematical expression for $f(t)$.

(b) Find the Laplace transform for this waveform, using any method.

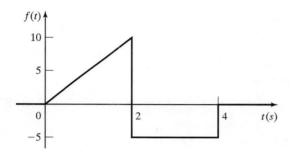

Figure P7.3

7.4. Consider the waveform $f(t)$ in Figure P7.4. This waveform is one cycle of a sinusoid for $0 \leq t \leq \pi$ s and is zero elsewhere.

(a) Write a mathematical expression for $f(t)$.

(b) Find the Laplace transform for this waveform, using (7.4), and the table of integrals in Appendix A.

(c) Use the real-shifting property to verify the results of Part (b).

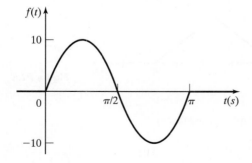

Figure P7.4

7.5. **(a)** Find the Laplace transform of $\cosh at = (e^{at} + e^{-at})/2$, using the Laplace transform tables.

(b) Use the Laplace transform of $\cos bt$ to verify the results in (a).

(c) Repeat Parts (a) and (b) for $\sinh at = (e^{at} - e^{-at})/2$.

7.6. (a) Sketch the time functions given.

    (i) $4e^{-4t}u(t-5)$ **(ii)** $-4e^{-4t}u(t-3)$

    (iii) $-5e^{-at}(t-b)$ **(iv)** $-5e^{-a(t-c)}u(t-b)$

(b) Use the definition of the Laplace transform, (7.4), and the table of integrals in Appendix A to calculate the Laplace transforms of these time functions.

(c) Use the real-shifting property and the transform table to find the Laplace transforms.

(d) Compare the results of Parts (b) and (c).

7.7. Use the real shifting property (7.22) and the transform table to find the Laplace transforms of the time functions given. Manipulate the time functions as required. Do not use the defining integral (7.4). Let $a > 0, b > 0, a < b$.

(a) $3u(t-3)u(6-t)$

(b) $(t-4)u(t-6)$

(c) $5u(t-5)u(7-t)$

(d) $t[u(t) - u(t-3)]$

(e) $3t[u(t-a) - u(t-b)]$

(f) $3e^{-bt}u(t-a)$

7.8. Consider the triangular voltage waveform $v(t)$ shown in Figure P7.8.

(a) Express $v(t)$ mathematically.

(b) Use the real-shifting property to find $\mathcal{L}[v(t)]$.

(c) Sketch the first derivative of $v(t)$.

(d) Find the Laplace transform of the first derivative of $v(t)$.

(e) Use the results of Part (d) and the integral property, (7.31), to verify the results of Parts (b) and (d).

(f) Use the derivative property, (7.15), and the result of (b) to verify the results of (b) and (d).

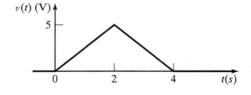

Figure P7.8

7.9. Given the Laplace transform

$$V(s) = \frac{s}{(s+2)(s+4)},$$

(a) Find the initial value of $v(t)$, $v(0^+)$, by

    (i)   the initial value property (Property 7 of Table 7.3);

    (ii) finding $v(t) = \mathcal{L}^{-1}[V(s)]$.

(b) Find the final value of $v(t)$ by
  (i) the final value property (Property 8 of Table 7.3);
  (ii) finding $v(t) = \mathcal{L}^{-1}[V(s)]$.
(c) Verify the partial-fraction expansions in Part (a) using MATLAB.

7.10. Given the Laplace transform

$$V(s) = \frac{2s + 1}{s^2 + 4},$$

(a) Find the initial value of $v(t)$, $v(0^+)$, by
  (i) the initial value property;
  (ii) finding $v(t) = \mathcal{L}^{-1}[V(s)]$.
(b) Find the final value of $v(t)$ by
  (i) the final value property;
  (ii) finding $v(t) = \mathcal{L}^{-1}[V(s)]$.

7.11. (a) Given $\mathcal{L}[u(t)] = 1/s$, use the multiplication-by-t property (Property 9 of Table 7.3) to find $\mathcal{L}[t]$.

(b) Repeat Part (a) for $\mathcal{L}[t \cos bt]$, given $\mathcal{L}[\cos bt] = s/(s^2 + b^2)$.
(c) Repeat Part (a) for $\mathcal{L}[t \times t^{n-1}]$, given $\mathcal{L}[t^{n-1}] = (n - 1)!/s^n$.

7.12. Use the derivative property (Property 2 of Table 7.3) to find $\mathcal{L}[\cos bt]$ from $\mathcal{L}[\sin bt] = b/(s^2 + b^2)$.

7.13. Find the inverse Laplace transforms of the functions given. Verify all partial-fraction expansions using MATLAB.

(a) $F(s) = \dfrac{5}{s(s + 2)}$

(b) $F(s) = \dfrac{s + 3}{s(s + 1)(s + 2)}$

(c) $F(s) = \dfrac{10(s + 3)}{s^2 + 25}$

(d) $F(s) = \dfrac{3}{s(s^2 + 2s + 5)}$

7.14. Find the inverse Laplace transforms of the functions given. Verify all partial-fraction expansions using MATLAB.

(a) $F(s) = \dfrac{10}{s(s + 1)^2}$

(b) $F(s) = \dfrac{s + 2}{s^2(s + 1)}$

(c) $F(s) = \dfrac{1}{s^2(s^2 + 4)}$

(d) $F(s) = \dfrac{30}{(s + 1)^2(s^2 + 6s + 25)}$

7.15. Find the inverse Laplace transforms of the functions given. Accurately sketch the time functions.

(a) $F(s) = \dfrac{3e^{-2s}}{s(s+3)}$

(b) $F(s) = \dfrac{3(1 - e^{-2s})}{s(s+3)}$

7.16. Sketch the periodic time function $f(t) = \mathcal{L}^{-1}[F(s)]$, if

(a) $F(s) = \dfrac{1 - e^{-s}}{s(s+1)}$

(b) $F(s) = \dfrac{e^{-2s} - e^{-3s}}{2}$

7.17. Consider the LTI systems described by the following differential equations.

(i) $\dfrac{d^2y(t)}{dt^2} + 6\dfrac{dy(t)}{dt} + 5y(t) = 4x(t)$

(ii) $\dfrac{d^2y(t)}{dt^2} + 6\dfrac{dy(t)}{dt} + 5y(t) = 4\dfrac{dx(t)}{dt} + 6x(t)$

(iii) $\dfrac{d^2y(t)}{dt^2} + 4y(t) = 2x(t)$

(iv) $\dfrac{d^3y(t)}{dt^3} - \dfrac{d^2y(t)}{dt^2} + 2y(t) = 2\dfrac{dx(t)}{dt} - 6x(t)$

(a) Find the unit impulse response $h(t)$ for each system.
(b) Find the unit step response $s(t)$ for each system.
(c) To verify your results, show that the functions in Parts (a) and (b) satisfy the equation

[eq(3.48)] $h(t) = \dfrac{ds(t)}{dt}.$

relating the impulse response $h(t)$ and the step response $s(t)$.
(d) Verify all partial-fraction expansions using MATLAB.

7.18. Consider a second-order system with the transfer function $H(s)$. For each part, include a pole-zero plot and the system modes. Give a transfer function $H(s)$ such that

(a) the system is stable;
(b) the system is unstable;
(c) the system's natural response contains a damped sinusoid;
(d) the system's natural response does not contain a damped sinusoid;
(e) the system's natural response contains an undamped sinusoid;
(f) the system's frequency response approaches zero at very high frequencies;
(g) the system's frequency response approaches a constant at very high frequencies.

7.19. For each of the systems of Problem 7.17, determine

(a) stability (use MATLAB as required);

(b) the inverse system's transfer function;

(c) the system modes.

7.20. You are given a linear, time-invariant (LTI) system that produces an output $y(t) = e^{-bt}u(t)$ to an input $x(t) = e^{-at}u(t)$ where $a > 0$ and $b > 0$. Find the impulse response $h(t)$ of the system.

7.21. Find the bilateral Laplace transforms of the following functions, giving the ROCs.

(a) $e^{-2t}\, u(t)$

(b) $e^{-2t}\, u(t - 4)$

(c) $-e^{2t}\, u(-t)$

(d) $-e^{2t}\, u(-t - 4)$

(e) $e^{-2t}u(t + 4)$

(f) $-e^{-2t}u(-t + 4)$

7.22. Sketch each waveform, and find its bilateral Laplace transform and its ROC for each of the signals given. If the transform does not exist, simply state that.

(a) $f(t) = e^{-3t}u(t) + e^{2t}u(-t)$

(b) $f(t) = e^{3t}u(t) + e^{-2t}u(-t)$

(c) $f(t) = e^{3t}u(t) + e^{2t}u(-t)$

(d) $f(t) = e^{-3t}u(t) + e^{-2t}u(-t)$

7.23. Consider the function

$$f(t) = \begin{cases} e^{-3t}, & -5 \le t \le 4 \\ 0, & \text{otherwise} \end{cases}.$$

(a) Calculate the bilateral Laplace transform of this function, using definition (7.1), and give its ROC.

(b) This function can be expressed as

$$f(t) = e^{-3t}[u(t + 5) - u(t - 4)].$$

Use tables to find its bilateral transform and its ROC.

(c) This function can be expressed as

$$f(t) = e^{-3t}[u(4 - t) - u(-5 - t)]$$

Use tables to find its bilateral transform and its ROC.

7.24. Find the inverse Laplace transform of the function

$$F(s) = \frac{s + 9}{s(s + 1)}$$

for the following regions of convergence:

(a) $\text{Re}(s) < -1$
(b) $\text{Re}(s) > 0$
(c) $-1 < \text{Re}(s) < 0$
(d) Give the final values of the functions of Parts (a), (b), and (c).

7.25. Given a Laplace Transform

$$X(s) = \frac{(s + 3)}{(s + 1)(s - 1)},$$

complete the following:

(a) Find all possible inverse bilateral Laplace Transforms.
(b) Sketch the region of convergence in each case.
(c) Label each time function as causal, noncausal, or two-sided.
(d) Label each time function as BIBO stable or not BIBO stable.
(e) Give the final values of the functions for each case.

7.26. You are given a transfer function

$$H(s) = \frac{1}{(s + a)(s + b)},$$

where $H(s)$ is the Laplace Transform of a time function $h(t)$.

(a) If $h(t)$ were causal, over what range of values of a and b would the system be BIBO stable? State the Region of Convergence.
(b) If $h(t)$ were two-sided, over what range of values of a and b would the system be BIBO stable? State the Region of Convergence.
(c) If $h(t)$ were noncausal, over what range of values of a and b would the system be BIBO stable? State the Region of Convergence.

7.27. Find the inverse Laplace Transform of

$$H(s) = \frac{s + 1}{s^2 + 6s + 8},$$

where $-4 < \text{Re}(s) < -2$.

7.28. You are given a Laplace Transform

$$H(s) = \frac{1}{(s + 10)(s + 5)(s - 3)},$$

where

$$-5 < Re(s) < 3.$$

Label each of the three poles as coming from a left-sided time function or right-sided time function.

7.29. In Chapter 3, direct convolution was used to solve for the output of LTI systems. For the following inputs and impulse responses, find the output using Laplace Transforms.

(a) Problem 3.1(ii) where $x(t) = e^{5t}u(t)$ and $h(t) = u(t)$.

(b) Problem 3.7(f) where $x(t) = e^t u(-t)$ and $h(t) = 2u(1 - t)$. You may find the Delay Property to be useful.

7.30. You are given a system with impulse response $h(t) = e^t u(t)$.

(a) Is the system bounded-input bounded-output stable?

(b) You now hook the system up into a *feedback* system as shown in Figure P7.30. Find the new system transfer function from the input $x(t)$ to the output $y(t)$.

(c) Finally, find the range of the parameter A such that the system is bounded-input bounded-output stable.

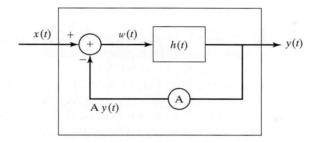

Figure P7.30

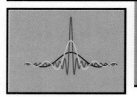

8 STATE VARIABLES FOR CONTINUOUS-TIME SYSTEMS

In Chapter 3, the modeling of continuous-time linear time-invariant (LTI) systems by linear differential equations with constant coefficients was presented. We considered the form I and the form II representations (simulation diagrams) for these systems. The form I representation requires $2n$ integrators for an nth-order system, while the form II representation requires n integrators.

In this chapter, we expand these representations. This leads us to a system model that is a set of n first-order coupled differential equations for an nth-order system. These models are called *state-variable models*, or more simply *state models*.

The state-variable model has several advantages:

1. An internal structure of the system is given, in addition to the input–output model. Thus, a state-variable model is more general than an input–output model, such as a transfer function.

2. Most numerical-integration algorithms are based on this type of model. Hence, a state model is usually required for the numerical solution of the system equations using a digital computer.

3. The modeling of nonlinear systems using state variables is a relatively simple extension of state-variable modeling for linear systems. Consequently, the digital-computer integration of nonlinear differential equations is rather easily performed.

4. Certain system analysis and design procedures are based on state-variable models. For example, many optimal system-design procedures require a state model.

Analysis and design via state-variable models require the use of matrix mathematics. The required mathematics is reviewed in Appendix G, and related terms used in this chapter are defined there.

8.1   STATE-VARIABLE MODELING

We introduce state-variable modeling with an example. Consider the RLC circuit of Figure 8.1. We consider the source voltage $v_i(t)$ to be the circuit input and the capacitor voltage $v_c(t)$ to be the circuit output. The circuit is described by the two equations

$$L\frac{di(t)}{dt} + Ri(t) + v_c(t) = v_i(t) \tag{8.1}$$

and

$$v_c(t) = \frac{1}{C}\int_{-\infty}^{t} i(\tau)d\tau . \tag{8.2}$$

The two unknown variables are $i(t)$ and $v_c(t)$.

We first convert (8.2) into a differential equation by differentiation, using Leibnitz's rule of Appendix B:

$$\frac{dv_c(t)}{dt} = \frac{1}{C}i(t). \tag{8.3}$$

Next we define two *state variables*, or more simply, *states*,

$$x_1(t) = i(t)$$

and

$$x_2(t) = v_c(t), \tag{8.4}$$

where $x_i(t)$ is the common notation for state variables. For this system, we choose as the two state variables the physical variables that represent the circuit's energy storage $[Li^2(t)/2$ and $Cv_c^2(t)/2]$. This is one procedure used for choosing the state variables for a system; other procedures are discussed later in this chapter.

Next, we substitute the state variables of (8.4) into the system differential equations (8.1) and (8.3), yielding

$$L\frac{dx_1(t)}{dt} + Rx_1(t) + x_2(t) = v_i(t)$$

$$\frac{dx_2(t)}{dt} = \frac{1}{C}x_1(t).$$

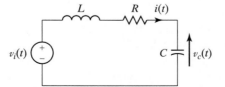

Figure 8.1   Example circuit.

These two equations are solved for the derivative terms:

$$\frac{dx_1(t)}{dt} = -\frac{R}{L}x_1(t) - \frac{1}{L}x_2(t) + \frac{1}{L}v_i(t)$$

and   (8.5)

$$\frac{dx_2(t)}{dt} = \frac{1}{C}x_1(t).$$

In addition to these coupled first-order differential equations, an equation that re-lates the system output to the state variables is required. Because the output signal is the capacitor voltage $v_c(t)$, the output equation is given by

$$y(t) = v_c(t) = x_2(t), \tag{8.6}$$

where we denote the system output with the common notation $y(t)$.

To simplify the notation, the overdot is used to indicate the first derivative; for example, $\dot{x}_1(t) = dx_1(t)/dt$. Then (8.5) is expressed as

$$\dot{x}_1(t) = -\frac{R}{L}x_1(t) - \frac{1}{L}x_2(t) + \frac{1}{L}v_i(t)$$

and   (8.7)

$$\dot{x}_2(t) = \frac{1}{C}x_1(t).$$

As a further simplification in notation, the state equations are written in a vector-matrix format. From (8.6) and (8.7),

$$\begin{bmatrix} \dot{x}_1(t) \\ \dot{x}_2(t) \end{bmatrix} = \begin{bmatrix} -\dfrac{R}{L} & -\dfrac{1}{L} \\ \dfrac{1}{C} & 0 \end{bmatrix} \begin{bmatrix} x_1(t) \\ x_2(t) \end{bmatrix} + \begin{bmatrix} \dfrac{1}{L} \\ 0 \end{bmatrix} v_i(t)$$

and   (8.8)

$$y(t) = \begin{bmatrix} 0 & 1 \end{bmatrix} \begin{bmatrix} x_1(t) \\ x_2(t) \end{bmatrix}.$$

These then are state equations for the circuit of Figure 8.1. As we discuss in Section 8.6, this set is not unique; that is, we can choose other variables to be the states of the system.

The standard form for the state equations of a continuous-time LTI system is given by

$$\dot{\mathbf{x}}(t) = \mathbf{A}\mathbf{x}(t) + \mathbf{B}u(t)$$

and   (8.9)

$$y(t) = \mathbf{C}\mathbf{x}(t) + \mathbf{D}u(t),$$

where boldface denotes vectors and matrices. The vector $\dot{\mathbf{x}}(t)$ is the time derivative of the vector $\mathbf{x}(t)$. In these equations,

$\mathbf{x}(t) = (n \times 1)$ state vector for an nth-order system;
$\mathbf{u}(t) = (r \times 1)$ input vector composed of the system input signals;
$\mathbf{y}(t) = (p \times 1)$ output vector composed of the defined output signals;
$\quad \mathbf{A} = (n \times n)$ system matrix;
$\quad \mathbf{B} = (n \times r)$ input matrix;
$\quad \mathbf{C} = (p \times n)$ output matrix;
$\quad \mathbf{D} = (p \times r)$ matrix that represents the direct coupling between the system inputs and the system outputs.

Expanding the vectors in (8.9) yields

$$\dot{\mathbf{x}}(t) = \begin{bmatrix} \dot{x}_1(t) \\ \dot{x}_2(t) \\ \vdots \\ \dot{x}_n(t) \end{bmatrix}; \quad \mathbf{x}(t) = \begin{bmatrix} x_1(t) \\ x_2(t) \\ \vdots \\ x_n(t) \end{bmatrix};$$

$$\dot{\mathbf{u}}(t) = \begin{bmatrix} u_1(t) \\ u_2(t) \\ \vdots \\ u_r(t) \end{bmatrix}; \quad \mathbf{y}(t) = \begin{bmatrix} y_1(t) \\ y_2(t) \\ \vdots \\ y_p(t) \end{bmatrix}. \tag{8.10}$$

It is standard notation to denote the input functions as $u_i(t)$. Unfortunately, this notation is also used for singularity functions, and confusion can result.

We illustrate the ith state equation in (8.9) as

$$\dot{x}_1(t) = a_{i1}x_1(t) + a_{i2}x_2(t) + \cdots + a_{in}x_n(t) \\ + b_{i1}u_1(t) + \cdots + b_{ir}u_r(t) \tag{8.11}$$

and the ith output equation in (8.9) as

$$y_i(t) = c_{i1}x_1(t) + c_{i2}x_2(t) + \cdots + c_{in}x_n(t) \\ + d_{i1}u_1(t) + \cdots + d_{ir}u_r(t). \tag{8.12}$$

We now define the *state* of a system:

The state of a system at any time t_0 is the information that, together with all inputs for $t \geq t_0$, determines the behavior of the system for $t \geq t_0$.

It will be shown that the state vector $\mathbf{x}(t)$ of the standard form (8.9) satisfies this definition. Note that for a differential equation, the initial conditions satisfy this definition.

We refer to the two matrix equations of (8.9) as the *state equations* of a system. The first equation, a differential equation, is called the *state equation*, and the second one, an algebraic equation, is called the *output equation*. The state equation is a first-order matrix differential equation, and the state vector $\mathbf{x}(t)$ is its solution. Given knowledge of $\mathbf{x}(t)$ and the input vector $\mathbf{u}(t)$, the output equation yields the output $\mathbf{y}(t)$. The output equation is a linear algebraic matrix equation.

In the state equations (8.9), only the first derivatives of the state variables may appear on the left side of the equation, and no derivatives of either the states or the inputs may appear on the right side. No derivatives may appear in the output equation. Valid first-order coupled equations that model an LTI system may be written without following these rules; however, those equations are not in the standard form.

The standard form of the state equations, (8.9), allows more than one input and more than one output. Systems with more than one input or more than one output are called *multivariable systems*. For a single-input system, the matrix $\mathbf{B}$ is an $(n \times 1)$ column vector and the input is the scalar $u(t)$. For a single-output system, the matrix $\mathbf{C}$ is a $(1 \times n)$ row vector and the output is the scalar $y(t)$. An example is now given to illustrate a multivariable system.

EXAMPLE 8.1   **State variables for a second-order system**

Consider the system described by the coupled differential equations

$$\dot{y}_1(t) + 2y_1(t) - 3y_2(t) = 4u_1(t) - u_2t$$

and

$$\dot{y}_2(t) + 2y_2(t) + y_1(t) = u_1(t) + 5u_2(t),$$

where $u_1(t)$ and $u_2(t)$ are system inputs and $y_1(t)$ and $y_2(t)$ are system outputs. We define the outputs as the states. Thus,

$$x_1(t) = y_1(t); \quad x_2(t) = y_2(t).$$

From the system differential equations, we write the state equations

$$\dot{x}_1(t) = -2x_1(t) + 3x_2(t) + 4u_1(t) - u_2(t)$$

and

$$\dot{x}_2(t) = -x_1(t) - 2x_2(t) + u_1(t) + 5u_2(t)$$

and the output equations

$$y_1(t) = x_1(t)$$

and

$$y_2(t) = x_2(t).$$

These equations may be written in vector-matrix form as

$$\dot{\mathbf{x}}(t) = \begin{bmatrix} -2 & 3 \\ -1 & -2 \end{bmatrix} \mathbf{x}(t) + \begin{bmatrix} 4 & -1 \\ 1 & 5 \end{bmatrix} \mathbf{u}(t)$$

and

$$\mathbf{y}(t) = \begin{bmatrix} 1 & 0 \\ 0 & 1 \end{bmatrix} \mathbf{x}(t).$$

Thus, we have derived a set of state equations for the system given.   ∎

In Example 8.1, suppose that the equations model a mechanical system. Suppose further that both $y_1(t)$ and $y_2(t)$ are displacements in the system. Then the state variables represent physical signals within the system. Generally, we prefer that the state variables be physically identifiable variables, but this is not necessary; this topic is developed further in Section 8.6.

When the equations of a physical system are written from physical laws, generally a set of coupled first- and second-order differential equations result. In this situation, we usually choose the state variables as illustrated in Example 8.1. Each first-order equation results in one state, and each second-order equation results in two states. For example, suppose that we have a five-loop circuit, in which three of the loop equations are second order and two are first order. This circuit is then modeled with eight state variables, with six from the three second-order equations and two from the two first-order equations.

We have introduced in this section the standard form of the state model of a continuous-time LTI system. We illustrated the state model with two systems described by differential equations. Next we consider a method for obtaining the state model directly from the system-transfer function $H(s)$. In general, this method does not result in the state variables being physical signals.

8.2   SIMULATION DIAGRAMS

In Section 8.1, we presented two methods of finding the state model of a system directly from the system differential equations. The procedure presented in those examples is very useful and is employed in many practical situations. If we write the system equations from the laws of physics, the result generally is a set of differential equations of the type displayed in Example 8. 1, that is, a set of coupled second-order differential equations. However, certain methods for obtaining models of physical systems, called *system-identification procedures* [1], result in a transfer function rather than differential equations. In this section, we present a procedure for obtaining a state model from a transfer function. Since a transfer function implies a single-input single-output system, the input is the scalar $u(t)$ and the output is the scalar $y(t)$.

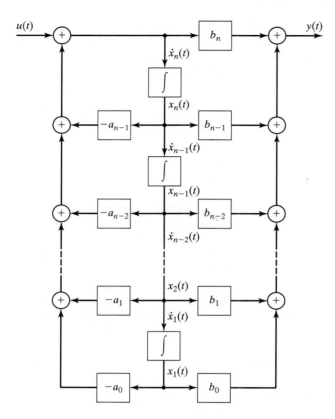

Figure 8.2 Direct form II for an nth-order system.

The procedure presented here is based on the form II representation, or simulation diagram, of a system; this form was developed in Section 3.8. This simulation diagram, repeated in Figure 8.2, represents an nth-order system with the transfer function

$$H(s) = \frac{Y(s)}{U(s)} = \frac{b_n s^n + b_{n-1} s^{n-1} + \cdots + b_1 s + b_0}{s^n + a_{n-1} s^{n-1} + \cdots + a_1 s + a_0}, \tag{8.13}$$

where $U(s)$ is the (transform of the) input and $Y(s)$ is the output. Note that the denominator coefficient a_n has been normalized to unity. If this coefficient is not unity for a given transfer function, we divide both numerator and denominator by a_n to obtain (8.13). Equation (8.13) can be expressed as

$$(s^n + a_{n-1} s^{n-1} + \cdots + a_1 s + a_0) Y(s)$$
$$= (b_n s^n + b_{n-1} s^{n-1} + \cdots + b_1 s + b_0) U(s). \tag{8.14}$$

Using either the procedures of Chapter 3 or the inverse Laplace transform, we write the system differential equation,

$$\sum_{k=0}^{n} a_k \frac{d^k y(t)}{dt^k} = \sum_{k=0}^{n} b_k \frac{d^k u(t)}{dt^k}, \tag{8.15}$$

with $a_n = 1$. Thus, the system modeled by the transfer function of (8.13) has the differential equation model of (8.15). Recall that initial conditions are ignored when using the Laplace transform to derive (8.14); hence, the initial conditions must be ignored when finding the inverse Laplace transform of (8.14).

We now give a procedure for writing the state model of the system of (8.13), (8.15), and Figure 8.2. First, we label each integrator output in Figure 8.2 as a state variable, as shown in the figure. If the output of an integrator is $x_i(t)$, its input must be $\dot{x}_i(t)$, also shown in Figure 8.2. We then write equations for the input signals to the integrators from Figure 8.2:

$$\dot{x}_1(t) = x_2(t);$$
$$\dot{x}_2(t) = x_3(t);$$
$$\vdots$$
$$\dot{x}_{n-1}(t) = x_n(t);$$
$$\dot{x}_n(t) = -a_0 x_1(t) - a_1 x_2(t) - \cdots - a_{n-2} x_{n-1}(t)$$
$$-a_{n-1} x_n(t) + u(t). \tag{8.16}$$

The equation for the output signal is, from Figure 8.2,

$$y(t) = (b_0 - a_0 b_n) x_1(t) + (b_1 - a_1 b_n) x_2(t)$$
$$+ \cdots + (b_{n-1} - a_{n-1} b_n) x_n(t) + b_n u(t). \tag{8.17}$$

We now write (8.16) and (8.17) as matrix equations:

$$\dot{x}(t) = \begin{bmatrix} 0 & 1 & 0 & \cdots & 0 & 0 \\ 0 & 0 & 1 & \cdots & 0 & 0 \\ \vdots & \vdots & \vdots & & \vdots & \vdots \\ 0 & 0 & 0 & \cdots & 0 & 1 \\ -a_0 & -a_1 & -a_2 & \cdots & -a_{n-2} & -a_{n-1} \end{bmatrix} x(t) + \begin{bmatrix} 0 \\ 0 \\ \vdots \\ 0 \\ 1 \end{bmatrix} u(t); \tag{8.18}$$

$$y(t) = [(b_0 - a_0 b_n)\ (b_1 - a_1 b_n) \cdots (b_{n-1} - a_{n-1} b_n)] x(t) + b_n u(t).$$

Note that the state equations can be written directly from the transfer function (8.13) or from the differential equation (8.15) because the coefficients a_i and b_i are given in these two equations. The intermediate step of drawing the simulation diagram is not necessary. An example using this procedure is now given.

EXAMPLE 8.2 | **State equations from a transfer function**

Suppose that we have a single-input single-output system with the transfer function

$$H(s) = \frac{Y(s)}{U(s)} = \frac{5s + 4}{s^2 + 3s + 2} = \frac{b_1 s + b_0}{s^2 + a_1 s + a_0}.$$

The state equations are written directly from (8.18):

$$\dot{\mathbf{x}}(t) = \begin{bmatrix} 0 & 1 \\ -2 & -3 \end{bmatrix} \mathbf{x}(t) + \begin{bmatrix} 0 \\ 1 \end{bmatrix} u(t);$$

$$y(t) = [4 \quad 5]\mathbf{x}(t). \tag{8.19}$$

The form II simulation diagram is given in Figure 8.3, with the numerical parameters and the state variables as shown. The state model (8.19) is verified directly from this diagram.   ■

The state model of (8.18) is required for certain design procedures for feedback-control systems. In that application, (8.18) is called the *control canonical form* [2, 3].

The procedure just used to write (8.18) from Figure 8.2 can be used to write the state equations of any system, given any form of a simulation diagram for that system; that is, the form II realization is not required. In this procedure, the integrator outputs are chosen as the states; hence, the integrator inputs are the derivatives of the states. The equations are then written for the integrator inputs $\dot{x}_i(t)$ and the system output $y(t)$ as functions of the system input $u(t)$, the integrator outputs $x_i(t)$, and the system parameters. We now discuss this procedure further.

In a simulation diagram, a signal is altered when transmitted through an integrator; hence, its designation is changed. We denote $\dot{x}_i(t)$ as the input to an integrator, and $x_i(t)$ as the integrator output. The state-equation procedure is simplified by omitting the integrators from the diagram; the equations are written directly from the remaining diagram. This omission is especially useful in complex simulation diagrams. The system of Figure 8.3, with the integrators omitted, is shown in Figure 8.4. Note that all signals are shown in this figure. In addition, the effects of the integrators are included, by the designations $x_i(t)$ and $\dot{x}_i(t)$.

The final step of the procedure is to write the equations for the integrator inputs and the system outputs on the diagram as functions of the system inputs, the integrator outputs, and the system parameters. The results for Figure 8.4 are given by

$$\dot{x}_1(t) = x_2(t),$$

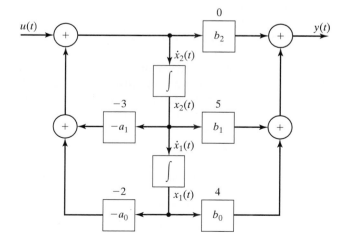

Figure 8.3   Second-order system.

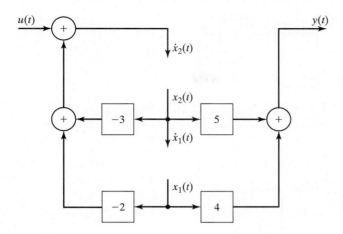

Figure 8.4   Second-order system.

$$\dot{x}_2(t) = -2x_1(t) - 3x_2(t) + u(t),$$

and

$$y(t) = 4x_1(t) + 5x_2(t).$$

These equations check those obtained in Example 8.2 using (8.18). A second example of this procedure will now be given.

EXAMPLE 8.3   **State equations from a simulation diagram**

We now write the state equations for the simulation diagram of Figure 8.5(a). Note that this diagram is not one of the two standard forms developed in Chapter 3. This form is sometimes used to realize analog filters, in which the integrators are implemented using operational amplifiers [2].

  The system has two integrators and is second order. First, we redraw the simulation diagram with the state variables shown and the integrators removed. The result is given in Figure 8.5(b). From this diagram, we write the state equations in matrix format:

$$\dot{\mathbf{x}}(t) = \begin{bmatrix} -2 & 3 \\ 0 & -4 \end{bmatrix}\mathbf{x}(t) + \begin{bmatrix} 1 \\ 2 \end{bmatrix}u(t); \;\; y(t) = [5 \;\; 6]\mathbf{x}(t). \qquad \blacksquare$$

  In Section 8.1, a procedure was given for writing state equations from differential equations. In this section, a procedure is developed for writing state equations directly from a transfer function. This procedure is then extended to writing state equations directly from a simulation diagram.

  A model of a physical system can be specified by

1. differential equations;
2. a transfer function;
3. a simulation diagram;
4. state equations.

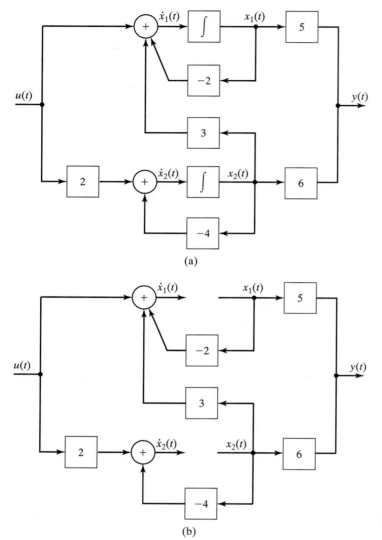

Figure 8.5 System for Example 8.3.

We have illustrated all four models in this section and Section 8.1. In Section 8.5, we give a procedure for obtaining the transfer function from the state equations. However, first we consider the solution of state equations.

8.3   SOLUTION OF STATE EQUATIONS

We have developed procedures for writing the state equations for an LTI system, given the system differential equation, the transfer function, or the simulation diagram. In this section, we present two methods for solving state equations.

Laplace-Transform Solution

The standard form for the state equation is given by

$$\dot{\mathbf{x}} = \mathbf{A}\mathbf{x}(t) + \mathbf{B}\mathbf{u}(t). \tag{8.20}$$

This equation will now be solved using the Laplace transform. Consider the first of the n scalar equations in (8.20), namely,

$$\dot{x}_1(t) = a_{11}x_1(t) + a_{12}x_2(t) + \cdots + a_{1n}x_n(t)$$
$$+ b_{11}u_1(t) + \cdots + b_{1r}u_r(t), \tag{8.21}$$

where a_{ij} and b_{ij} are the appropriate elements in the $\mathbf{A}$ and $\mathbf{B}$ matrices. Recall from Table 7.3 the Laplace transform of a derivative:

$$\mathscr{L}[\dot{x}_1(t)] = sX_1(s) - x_1(0).$$

Note that, for convenience, we have omitted the plus exponent on $x_1(0)$. (This is standard practice.) The Laplace transform of the first state equation, (8.21), is given by

$$sX_1(s) - x_1(0) = a_{11}X_1(s) + a_{12}X_2(s) + \cdots + a_{1n}X_n(s)$$
$$+ b_{11}U_1(s) + \cdots + b_{1r}U_r(s). \tag{8.22}$$

We will find the complete solution; hence, the initial condition $x_1(0)$ is included. The second equation in (8.20) is given by

$$\dot{x}_2(t) = a_{21}x_1(t) + a_{22}x_2(t) + \cdots + a_{2n}x_n(t) + b_{21}u_1(t) + \cdots + b_{2r}u_r(t),$$

which has the Laplace transform

$$sX_2(s) - x_2(0) = a_{21}X_1(s) + a_{22}X_2(s) + \cdots + a_{2n}X_n(s)$$
$$+ b_{21}U_1(s) + \cdots + b_{2r}U_r(s). \tag{8.23}$$

The Laplace transform of the remaining $(n - 2)$ equations in (8.20) yields the same form as (8.22) and (8.23). We see then that these transformed equations may be written in matrix form as

$$s\mathbf{X}(s) - \mathbf{x}(0) = \mathbf{A}\mathbf{X}(s) + \mathbf{B}\mathbf{U}(s).$$

We wish to solve this equation for $\mathbf{X}(s)$; to do this, we rearrange the last equation:

$$s\mathbf{X}(s) - \mathbf{A}\mathbf{X}(s) = \mathbf{x}(0) + \mathbf{B}\mathbf{U}(s). \tag{8.24}$$

It is necessary to factor $\mathbf{X}(s)$ in the left side to solve this equation. First, the term $s\mathbf{X}(s)$ must be written as $s\mathbf{I}\mathbf{X}(s)$, where $\mathbf{I}$ is the identity matrix. (See Appendix G.) Then

$$sIX(s) - AX(s) = (sI - A)X(s) = x(0) + BU(s). \tag{8.25}$$

This additional step is necessary since the subtraction of the matrix A from the scalar s is not defined; we cannot factor $X(s)$ directly in (8.24). Equation (8.25) may now be solved for $X(s)$:

$$X(s) = (sI - A)^{-1}x(0) + (sI - A)^{-1}BU(s). \tag{8.26}$$

The state vector $x(t)$ is the inverse Laplace transform of this equation.

To develop a general relationship for the solution, we define the *state transition matrix* $\Phi(t)$ as

$$\Phi(t) = \mathcal{L}^{-1}[\Phi(s)] = \mathcal{L}^{-1}[(sI - A)^{-1}]. \tag{8.27}$$

This matrix $\Phi(t)$ is also called *the fundamental matrix*. The matrix $\Phi(s) = (sI - A)^{-1}$ is called the *resolvant* of A [3]. Note that for an nth-order system, the state transition matrix is of order $(n \times n)$.

The inverse Laplace transform of a matrix, as in (8.27), is defined as the inverse transform of the elements of the matrix. Solving for $\Phi(t)$ in (8.27) in general is difficult, time consuming, and prone to error. A more practical procedure for calculating the state vector $x(t)$ is by computer simulation. Next an example is presented to illustrate the calculation in (8.27).

EXAMPLE 8.4   **State transition matrix for a second-order system**

We use the system of Example 8.2, described by the transfer function

$$H(s) = \frac{Y(s)}{U(s)} = \frac{5s + 4}{s^2 + 3s + 2}.$$

From Example 8.2, the state equations are given by

[eq(8.19)] $$\dot{x}(t) = \begin{bmatrix} 0 & 1 \\ -2 & -3 \end{bmatrix} x(t) + \begin{bmatrix} 0 \\ 1 \end{bmatrix} u(t);$$

$$y(t) = [4 \quad 5]x(t).$$

To find the state transition matrix, we first calculate the matrix $(sI - A)$:

$$sI - A = s\begin{bmatrix} 1 & 0 \\ 0 & 1 \end{bmatrix} - \begin{bmatrix} 0 & 1 \\ -2 & -3 \end{bmatrix} = \begin{bmatrix} s & -1 \\ 2 & s+3 \end{bmatrix}.$$

We next calculate the adjoint of this matrix (see Appendix G):

$$\text{Adj}(sI - A) = \begin{bmatrix} s+3 & 1 \\ -2 & s \end{bmatrix}.$$

The determinant of $(s\mathbf{I} - \mathbf{A})$ is given by

$$\det(s\mathbf{I} - \mathbf{A}) = s(s + 3) - (-1)(2)$$
$$= s^2 + 3s + 2 = (s + 1)(s + 2).$$

The inverse matrix is the adjoint matrix divided by the determinant:

$$\Phi(s) = (s\mathbf{I} - \mathbf{A})^{-1} = \begin{bmatrix} \dfrac{s + 3}{(s + 1)(s + 2)} & \dfrac{1}{(s + 1)(s + 2)} \\ \dfrac{-2}{(s + 1)(s + 2)} & \dfrac{s}{(s + 1)(s + 2)} \end{bmatrix}$$

$$= \begin{bmatrix} \dfrac{2}{s + 1} + \dfrac{-1}{s + 2} & \dfrac{1}{s + 1} + \dfrac{-1}{s + 2} \\ \dfrac{-2}{s + 1} + \dfrac{2}{s + 2} & \dfrac{-1}{s + 1} + \dfrac{2}{s + 2} \end{bmatrix}.$$

The state transition matrix is the inverse Laplace transform of this matrix:

$$\Phi(t) = \begin{bmatrix} 2e^{-t} - e^{-2t} & e^{-t} - e^{-2t} \\ -2e^{-t} + 2e^{-2t} & -e^{-t} + 2e^{-2t} \end{bmatrix}.$$

The following MATLAB program computes the resolvant and state-transition matrices for this example:

```
% This MATLAB program verifies the calculations of Example 8.4.
%
% Designate the symbolic variables to be used.
syms s t
% Enter the numerator and denominator coefficient arrays.
n=[5 4];
d=[1 3 2];
% Show the transfer function.
Hs=tf(n,d)
% Convert the system into a state-variable model.
[A,B,C,D]=tf2ss(n,d);
% Perform a similarity transformation to put the
% model in control-canonical form.
% (Similarity transformations are discussed in Section 8.6).
T=[0 1;1 0];
[A,B,C,D]=ss2ss(A,B,C,D,T)
% Form the Identity matrix.
I=[1 0;0 1];
% Compute Phi(s).
Phis=inv(s*I-A);
'The resolvant matrix, Phi(s)='
pretty(Phis)
% Compute phi(t), the state-transition matrix.
phit=ilaplace(Phis);
'The state-transition matrix, phi(t)='
pretty(phit).
```

■

Recall from Section 3.5 that the system characteristic equation is the denominator of the transfer function set to zero. For the system of Example 8.4, the characteristic equation is then

$$s^2 + 3s + 2 = (s + 1)(s + 2) = 0.$$

Hence, the modes of the system are e^{-t} and e^{-2t}. These modes are evident in the state transition matrix $\Phi(t)$. The modes of a system always appear in the state transition matrix; hence, the system's stability is evident from $\Phi(t)$.

Recall the complete solution of the state equation from (8.26):

$$\mathbf{X}(s) = (s\mathbf{I} - \mathbf{A})^{-1}\mathbf{x}(0) + (s\mathbf{I} - \mathbf{A})^{-1}\mathbf{B}U(s)$$
$$= \Phi(s)\mathbf{x}(0) + \Phi(s)\mathbf{B}U(s). \qquad (8.28)$$

Note that the resolvant matrix $\Phi(s)$ is basic to the solution of the state equations.

We illustrate the complete solution using an example; then a different form of the solution is presented.

EXAMPLE 8.5   **Total response for second-order state equations**

Consider again the system of Example 8.4. We have

$$\dot{\mathbf{x}}(t) = \begin{bmatrix} 0 & 1 \\ -2 & -3 \end{bmatrix} \mathbf{x}(t) + \begin{bmatrix} 0 \\ 1 \end{bmatrix} u(t)$$

and

$$y(t) = \begin{bmatrix} 4 & 5 \end{bmatrix}\mathbf{x}(t),$$

with the resolvant matrix

$$\Phi(s) = (s\mathbf{I} - \mathbf{A})^{-1} = \begin{bmatrix} \dfrac{s + 3}{(s + 1)(s + 2)} & \dfrac{1}{(s + 1)(s + 2)} \\ \dfrac{-2}{(s + 1)(s + 2)} & \dfrac{s}{(s + 1)(s + 2)} \end{bmatrix}.$$

Suppose that the input signal is a unit step function. Then $U(s) = 1/s$, and the second term in (8.28) becomes

$$(s\mathbf{I} - \mathbf{A})^{-1}\mathbf{B}U(s) = \begin{bmatrix} \dfrac{s + 3}{(s + 1)(s + 2)} & \dfrac{1}{(s + 1)(s + 2)} \\ \dfrac{-2}{(s + 1)(s + 2)} & \dfrac{s}{(s + 1)(s + 2)} \end{bmatrix} \begin{bmatrix} 0 \\ 1 \end{bmatrix} 1/s$$

$$= \begin{bmatrix} \dfrac{1}{s(s + 1)(s + 2)} \\ \dfrac{1}{(s + 1)(s + 2)} \end{bmatrix} = \begin{bmatrix} \dfrac{1/2}{s} + \dfrac{-1}{s + 1} + \dfrac{1/2}{s + 2} \\ \dfrac{1}{s + 1} + \dfrac{-1}{s + 2} \end{bmatrix}.$$

For $t > 0$, the inverse Laplace transform of this vector yields

$$\mathcal{L}^{-1}[(s\mathbf{I} - \mathbf{A})^{-1}\mathbf{B}U(s)] = \begin{bmatrix} \frac{1}{2} - e^{-t} + \frac{1}{2}e^{-2t} \\ e^{-t} - e^{-2t} \end{bmatrix}.$$

The state transition matrix was derived in Example 8.4. Hence, the complete solution of the state equations is, from (8.28) and Example 8.4,

$$\mathbf{x}(t) = \mathbf{\Phi}(t)\mathbf{x}(0) + \mathcal{L}^{-1}[(s\mathbf{I} - \mathbf{A})^{-1}\mathbf{B}U(s)]$$

$$= \begin{bmatrix} 2e^{-t} - e^{-2t} & e^{-t} - e^{-2t} \\ -2e^{-t} + 2e^{-2t} & -e^{-t} + 2e^{-2t} \end{bmatrix}\begin{bmatrix} x_1(0) \\ x_2(0) \end{bmatrix} + \begin{bmatrix} \frac{1}{2} - e^{-t} + \frac{1}{2}e^{-2t} \\ e^{-t} - e^{-2t} \end{bmatrix},$$

and the state variables are given by

$$x_1(t) = (2e^{-t} - e^{-2t})x_1(0) + (e^{-t} - e^{-2t})x_2(0) + \frac{1}{2}$$
$$- e^{-t} + \frac{1}{2}e^{-2t} = \frac{1}{2} + [2x_1(0) + x_2(0) - 1]e^{-t}$$
$$+ [-x_1(0) - x_2(0) + \frac{1}{2}]e^{-2t}.$$

In a like manner,

$$x_2(t) = (-2e^{-t} + 2e^{-2t})x_1(0) + (-e^{-t} + 2e^{-2t})x_2(0)$$
$$+ e^{-t} - e^{-2t} = [-2x_1(0) - x_2(0) + 1]e^{-t} + [2x_1(0) + 2x_2(0) - 1]e^{-2t}.$$

The output is given by

$$y(t) = 4x_1(t) + 5x_2(t) = 2 + [-2x_1(0) - x_2(0) + 1]e^{-t}$$
$$+ [6x_1(0) + 6x_2(0) - 3]e^{-2t}.$$

The first-row element of $\mathbf{\Phi}(t)\mathbf{x}(0)$ can be calculated directly with the MATLAB program

```
S=dsolve('Dx1 = x2,Dx2 = -2*x1-3*x2,x1(0) = x10,x2(0) = x20')
S.x1
S.x2
```

The second-row element can be calculated in the same manner.   ■

The solution of state equations is long and involved, even for a second-order system. The necessity for reliable machine solutions, such as digital-computer simulations, is evident. Almost all system analysis and design software have simulation capabilities; the simulations are usually based on state models. As an example, a SIMULINK simulation is now discussed.

EXAMPLE 8.6   **SIMULINK simulation**

The system of Example 8.5 was simulated using SIMULINK. The block diagram from the simulation in given in Figure 8.6(a), and the response $y(t)$ is given in Figure 8.6(b). From Example 8.5, we see that the two system time constants are 0.5 s and 1 s. Hence, the transient part of the response becomes negligible after approximately four times the larger time constant, or

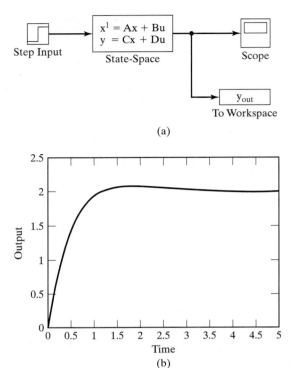

Figure 8.6 SIMULINK simulation for Example 8.5.

four seconds. Figure 8.6(b) shows this. In addition, the final value of $y(t)$ is 2, which is also evident in this figure. ∎

Convolution Solution

The complete solution of the state equations is expressed as

$$[eq(8.26)] \qquad \mathbf{X}(s) = (s\mathbf{I} - \mathbf{A})^{-1}\mathbf{x}(0) + (s\mathbf{I} - \mathbf{A})^{-1}\mathbf{B}U(s).$$

A second form for this solution is now developed. The second term in the right side of (8.26) is a product of two terms in the Laplace variable s. Thus, the inverse Laplace transform of this term can be expressed as a convolution integral. (See Table 7.3.) The inverse Laplace transform of (8.26) is then

$$\mathbf{x}(t) = \mathbf{\Phi}(t)\mathbf{x}(0) + \int_0^t \mathbf{\Phi}(t - \tau)\mathbf{Bu}(\tau)d\tau$$

$$= \mathbf{\Phi}(t)\mathbf{x}(0) + \int_0^t \mathbf{\Phi}(\tau)\mathbf{Bu}(t - \tau)d\tau. \qquad (8.29)$$

This solution has two terms. The first term is referred to as either the *zero-input term* or the *initial-condition term*, and the second term is called either the *zero-state term* or the *forced response*. Equation (8.29) is sometimes called the *convolution so-lution* of the state equation.

The solution in (8.29) is quite difficult to calculate except for the simplest of systems. The system of the preceding example is now used to illustrate this calculation.

EXAMPLE 8.7   **Convolution solution for second-order state equations**

In Example 8.5, the input was a unit step function. From (8.29), with $u(\tau) = 1$ for $\tau > 0$, the second term becomes

$$\int_0^t \Phi(t - \tau)\mathbf{B}\mathbf{u}(\tau)d\tau = \int_0^t \begin{bmatrix} 2e^{-(t-\tau)} - e^{-2(t-\tau)} & e^{-(t-\tau)} - e^{-2(t-\tau)} \\ -2e^{-(t-\tau)} + 2e^{-2(t-\tau)} & -e^{-(t-\tau)} + 2e^{-2(t-\tau)} \end{bmatrix} \begin{bmatrix} 0 \\ 1 \end{bmatrix} d\tau$$

$$= \begin{bmatrix} \int_0^t (e^{-(t-\tau)} - e^{-2(t-\tau)})d\tau \\ \int_0^t (-e^{-(t-\tau)} + 2e^{-2(t-\tau)})d\tau \end{bmatrix} = \begin{bmatrix} (e^{-t}e^\tau - \frac{1}{2}e^{-2t}e^{2\tau})_0^t \\ (-e^{-t}e^\tau + e^{-2t}e^{2\tau})_0^t \end{bmatrix}$$

$$= \begin{bmatrix} (1 - e^{-t}) - \frac{1}{2}(1 - e^{-2t}) \\ (-1 + e^{-t}) + (1 - e^{-2t}) \end{bmatrix} = \begin{bmatrix} \frac{1}{2} - e^{-t} + \frac{1}{2}e^{-2t} \\ e^{-t} - e^{-2t} \end{bmatrix}.$$

This result checks that of Example 8.5. Only the forced response is derived here. The initial-condition term of the solution is the function $\boldsymbol{\phi}(t)\mathbf{x}(0)$ in (8.29), which was evaluated in Example 8.5; it is not repeated here.   ∎

The complete solution to the state equations was derived in this section. This solution may be evaluated either by the Laplace transform or by a combination of the Laplace transform and the convolution integral. Either procedure is long, time consuming, and prone to errors.

Infinite Series Solution

As just shown, the state transition matrix can be evaluated using the Laplace transform. An alternative procedure for this evaluation is now developed.

One method of solution of homogeneous differential equations is to assume as the solution an infinite power series with unknown coefficients. The infinite series is then substituted into the differential equation to evaluate the unknown coefficients. This method is now used to find the state transition matrix as an infinite series.

We begin by considering all system inputs to be zero. Thus, from (8.9), the state equation may be written as

$$\dot{\mathbf{x}}(t) = \mathbf{A}\mathbf{x}(t), \tag{8.30}$$

with the solution

$$\mathbf{x}(t) = \boldsymbol{\phi}(t)\mathbf{x}(0), \tag{8.31}$$

from (8.29).

Because we are solving for the vector $\mathbf{x}(t)$, the state transmission matrix is assumed to be of the form

$$\boldsymbol{\Phi}(t) = \mathbf{K}_0 + \mathbf{K}_1 t + \mathbf{K}_2 t^2 + \mathbf{K}_3 t^3 + \cdots,$$

so that

$$\mathbf{x}(t) = (\mathbf{K}_0 + \mathbf{K}_1 t + \mathbf{K}_2 t^2 + \mathbf{K}_3 t^3 + \cdots)\mathbf{x}(0)$$

$$= \left[\sum_{i=0}^{\infty} \mathbf{K}_i t^i\right]\mathbf{x}(0) = \boldsymbol{\phi}(t)\mathbf{x}(0), \tag{8.32}$$

where the $n \times n$ matrices $\mathbf{K}_i$ are unknown and t is the scalar time. Differentiating this equation yields

$$\dot{\mathbf{x}}(t) = (\mathbf{K}_1 + 2\mathbf{K}_2 t + 3\mathbf{K}_3 t^2 + \cdots)\mathbf{x}(0). \tag{8.33}$$

Substituting (8.33) and (8.32) into (8.30) yields

$$\dot{\mathbf{x}}(t) = (\mathbf{K}_1 + 2\mathbf{K}_2 t + 3\mathbf{K}_3 t^2 + \cdots)\mathbf{x}(0)$$

$$= \mathbf{A}(\mathbf{K}_0 + \mathbf{K}_1 t + \mathbf{K}_2 t^2 + \mathbf{K}_3 t^3 + \cdots)\mathbf{x}(0). \tag{8.34}$$

Evaluating (8.32) at $t = 0$ yields $\mathbf{x}(0) = \mathbf{K}_0\mathbf{x}(0)$; hence, $\mathbf{K}_0 = \mathbf{I}$. We next equate the coefficients of t^i for $i = 0, 1, 2, \cdots$, in (8.34). The resulting equations are, with $\mathbf{K}_0 = \mathbf{I}$,

$$\mathbf{K}_1 = \mathbf{A}\mathbf{K}_0 \qquad \Rightarrow \mathbf{K}_1 = \mathbf{A};$$

$$2\mathbf{K}_2 = \mathbf{A}\mathbf{K}_1 = \mathbf{A}^2 \Rightarrow \mathbf{K}_2 = \frac{\mathbf{A}^2}{2!};$$

$$3\mathbf{K}_3 = \mathbf{A}\mathbf{K}_2 = \frac{\mathbf{A}^3}{2!} \Rightarrow \mathbf{K}_3 = \frac{\mathbf{A}^3}{3!};$$

$$\vdots \qquad\qquad\qquad \vdots \qquad . \tag{8.35}$$

Thus, from (8.32) and (8.35),

$$\boldsymbol{\Phi}(t) = \mathbf{I} + \mathbf{A}(t) + \mathbf{A}^2\frac{t^2}{2!} + \mathbf{A}^3\frac{t^3}{3!} + \cdots. \tag{8.36}$$

It can be shown that this series is convergent [4].

In summary, we can express the complete solution for state equations as

$$[\text{eq}(8.29)] \qquad \mathbf{x}(t) = \boldsymbol{\Phi}(t)\mathbf{x}(0) + \int_0^t \boldsymbol{\Phi}(t - \tau)\mathbf{B}\mathbf{u}(\tau)\, d\tau,$$

with the state transition matrix $\boldsymbol{\phi}(t)$ given either by (8.36) or by

[eq(8.27)] $$\boldsymbol{\Phi}(t) = \mathcal{L}^{-1}[\boldsymbol{\Phi}(s)] = \mathcal{L}^{-1}[(s\mathbf{I} - \mathbf{A})^{-1}].$$

Because of the similarity of (8.36) and the Taylor's series for the scalar exponential

$$e^{kt} = 1 + kt + k^2\frac{t^2}{2!} + k^3\frac{t^3}{3!} + \cdots, \tag{8.37}$$

the state transition matrix is often written, for notational purposes only, as the matrix exponential

$$\boldsymbol{\Phi}(t) = \exp \mathbf{A}t. \tag{8.38}$$

The *matrix exponential* is defined by (8.36) and (8.38). An example is given next that illustrates the calculation of the state transition matrix using the series in (8.36).

EXAMPLE 8.8   **Series solution for second-order state equations**

To give an example for which the series in (8.36) has a finite number of terms, we consider the movement of a rigid mass in a frictionless environment. The system model is given by

$$f(t) = M\frac{d^2x(t)}{dt^2},$$

where M is the mass, $x(t)$ the displacement, and $f(t)$ the applied force. For convenience, we let $M = 1$. We choose the state variables as the position and the velocity of the mass such that

$$x_1(t) = x(t),$$
$$x_2(t) = \dot{x}(t) = \dot{x}_1(t),$$

and

$$\dot{x}_2(t) = \ddot{x}(t) = f(t),$$

where the last equation is obtained from the system model. The state equations are then

$$\dot{\mathbf{x}}(t) = \begin{bmatrix} 0 & 1 \\ 0 & 0 \end{bmatrix}\mathbf{x}(t) + \begin{bmatrix} 0 \\ 1 \end{bmatrix}f(t).$$

Then, in (8.36),

$$\mathbf{A} = \begin{bmatrix} 0 & 1 \\ 0 & 0 \end{bmatrix}, \quad \mathbf{A}^2 = \begin{bmatrix} 0 & 1 \\ 0 & 0 \end{bmatrix}\begin{bmatrix} 0 & 1 \\ 0 & 0 \end{bmatrix} = \begin{bmatrix} 0 & 0 \\ 0 & 0 \end{bmatrix},$$

and

$$\mathbf{A}^3 = \mathbf{A}\mathbf{A}^2 = 0.$$

In a like manner,

$$\mathbf{A}^n = \mathbf{A}^2\mathbf{A}^{n-2} = \mathbf{0}; \quad n \geqq 3.$$

Thus, the state transition matrix is, from (8.36),

$$\boldsymbol{\Phi}(t) = \mathbf{I} + \mathbf{A}t = \begin{bmatrix} 1 & 0 \\ 0 & 1 \end{bmatrix} + \begin{bmatrix} 0 & 1 \\ 0 & 0 \end{bmatrix}t = \begin{bmatrix} 1 & t \\ 0 & 1 \end{bmatrix},$$

and the states are given by

$$\mathbf{x}(t) = \boldsymbol{\Phi}(t)\mathbf{x}(0) = \begin{bmatrix} 1 & t \\ 0 & 1 \end{bmatrix}\mathbf{x}(0) = \begin{bmatrix} x_1(0) + tx_2(0) \\ x_2(0) \end{bmatrix}.$$

This example was chosen to give a simple calculation. In general, evaluating the matrix exponential is quite involved. The calculation of $\boldsymbol{\Phi}(t)$ in this example is easily checked using Laplace transforms. (See Problem 8.18.) ∎

The series expansion of $\boldsymbol{\Phi}(t)$ is well suited to evaluation on a digital computer if $\boldsymbol{\Phi}(t)$ is to be evaluated at only a few instants of time. The series expansion is also useful in the analysis of digital control systems [5, 6]. However, as a practical matter, the time response of a system should be evaluated by simulation, such as is given in Example 8.6.

In this section, two expressions for the solution of state equations are derived. The first, (8.26), expresses the solution as a Laplace transform, while the second, (8.29), expresses the solution as a convolution. The state-transition matrix is found either by the Laplace transform, (8.27), or by the series (8.36).

8.4   PROPERTIES OF THE STATE-TRANSITION MATRIX

Three properties of the state-transition matrix will now be derived. First, for an unforced system, from (8.29),

$$\mathbf{x}(t) = \boldsymbol{\Phi}(t)\mathbf{x}(0) \Rightarrow \mathbf{x}(0) = \boldsymbol{\Phi}(0)\mathbf{x}(0); \tag{8.39}$$

hence, the first property is given by

$$\boldsymbol{\Phi}(0) = \mathbf{I}, \tag{8.40}$$

where $\mathbf{I}$ is the identity matrix. This property can be used in verifying the calculation of $\boldsymbol{\Phi}(t)$.

The second property is based on time invariance. From (8.39) with $t = t_1$,

$$\mathbf{x}(t_1) = \boldsymbol{\Phi}(t_1)\mathbf{x}(0). \tag{8.41}$$

Suppose that we consider $t = t_1$ to be the initial time and $\mathbf{x}(t_1)$ to be the initial conditions. Then, at t_2 seconds later than t_1, from (8.39), (8.41), and the time invariance of the system,

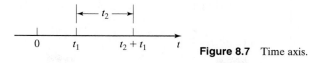

Figure 8.7 Time axis.

$$\mathbf{x}(t_2 + t_1) = \boldsymbol{\Phi}(t_2)\mathbf{x}(t_1) = \boldsymbol{\Phi}(t_2)\boldsymbol{\Phi}(t_1)\mathbf{x}(0).$$

Figure 8.7 shows the points on the time axis. Also, from (8.39), with $t = t_1 + t_2$,

$$\mathbf{x}(t_2 + t_1) = \boldsymbol{\Phi}(t_2 + t_1)\mathbf{x}(0).$$

From the last two equations, we see the second property:

$$\boldsymbol{\Phi}(t_1 + t_2) = \boldsymbol{\Phi}(t_1)\boldsymbol{\Phi}(t_2). \tag{8.42}$$

The third property can be derived from the second property, by letting $t_1 = t$ and $t_2 = -t$. Then, in (8.42),

$$\boldsymbol{\Phi}(t - t) = \boldsymbol{\Phi}(t)\boldsymbol{\Phi}(-t) = \boldsymbol{\Phi}(0) = \mathbf{I}$$

from (8.40). Thus,

$$\boldsymbol{\Phi}^{-1}(t) = \boldsymbol{\Phi}(-t). \tag{8.43}$$

It can be shown that $\boldsymbol{\Phi}^{-1}(t)$ always exists for t finite [7].

In summary, the three properties of the state transition matrix are given by

[eq(8.40)] $\boldsymbol{\Phi}(0) = \mathbf{I};$
[eq(8.42)] $\boldsymbol{\Phi}(t_1 + t_2) = \boldsymbol{\Phi}(t_1)\boldsymbol{\Phi}(t_2);$
[eq(8.43)] $\boldsymbol{\Phi}^{-1}(t) = \boldsymbol{\Phi}(-t).$

An example illustrating these properties will now be given.

EXAMPLE 8.9   **Illustrations of properties of the state transition matrix**

We use the state transition matrix from Example 8.4 to illustrate the three properties:

$$\boldsymbol{\Phi}(t) = \begin{bmatrix} 2e^{-t} - e^{-2t} & e^{-t} - e^{-2t} \\ -2e^{-t} + 2e^{-2t} & -e^{-t} + 2e^{-2t} \end{bmatrix}.$$

From (8.40),

$$\boldsymbol{\Phi}(0) = \begin{bmatrix} 2e^0 - e^0 & e^0 - e^0 \\ -2e^0 + 2e^0 & -e^0 + 2e^0 \end{bmatrix} = \begin{bmatrix} 1 & 0 \\ 0 & 1 \end{bmatrix} = \mathbf{I},$$

and the first property is satisfied. The second property, (8.41), yields

$$
\Phi(t_1)\Phi(t_2) = \begin{bmatrix} 2e^{-t_1} - e^{-2t_1} & e^{-t_1} - e^{-2t_1} \\ -2e^{-t_1} + 2e^{-2t_1} & -e^{-t_1} + 2e^{-2t_1} \end{bmatrix} \times \begin{bmatrix} 2e^{-t_2} - e^{-2t_2} & e^{-t_2} - e^{-2t_2} \\ -2e^{-t_2} + 2e^{-2t_2} & -e^{-t_2} + 2e^{-2t_2} \end{bmatrix}.
$$

The $(1, 1)$ element of the product matrix is given by

$$
\begin{aligned}
(1,1)\ \text{element} &= [2e^{-t_1} - e^{-2t_1}][2e^{-t_2} - e^{-2t_2}] \\
&\quad + [e^{-t_1} - e^{-2t_1}][-2e^{-t_2} + 2e^{-2t_2}] = [4e^{-(t_1+t_2)} - 2e^{-(2t_1+t_2)} \\
&\quad - 2e^{-(t_1+2t_2)} + e^{-2(t_1+t_2)}] + [-2e^{-(t_1+t_2)} + 2e^{-(2t_1+t_2)} \\
&\quad + 2e^{-(t_1+2t_2)} - 2e^{-2(t_1+t_2)}].
\end{aligned}
$$

Combining these terms yields

$$
(1,1)\ \text{element} = 2e^{-(t_1+t_2)} - e^{-2(t_1+t_2)}, \tag{8.44}
$$

which is the $(1, 1)$ element of $\Phi(t_1 + t_2)$. The other three elements of the product matrix can be verified in a like manner.

To illustrate the third property, (8.43), we assume that the property is true. Hence,

$$
\Phi(t)\Phi(-t) = \begin{bmatrix} 2e^{-t} - e^{-2t} & e^{-t} - e^{-2t} \\ -2e^{-t} + 2e^{-2t} & -e^{-t} + 2e^{-2t} \end{bmatrix} \times \begin{bmatrix} 2e^{t} - e^{2t} & e^{t} - e^{2t} \\ -2e^{t} + 2e^{2t} & -e^{t} + 2e^{2t} \end{bmatrix} = \mathbf{I}.
$$

As with the last property, we test only the $(1, 1)$ element of the product. This product is given in (8.44); in this equation, we let $t_1 = t$ and $t_2 = -t$, with the result

$$
(1,1)\ \text{element} = 2e^{-(t-t)} - e^{-2(t-t)} = 1
$$

as expected. In a like manner, the other three elements of the product matrix can be verified. ■

In this section, three properties of the state-transition matrix are developed. Property (8.40), $\Phi(0) = \mathbf{I}$, is easily applied as a check of the calculation of a state-transition matrix.

8.5 TRANSFER FUNCTIONS

A procedure was given in Section 8.2 for writing the state equations of a system from the system transfer function. In this section, we investigate the calculation of the transfer function from state equations.

The standard form of the state equations is given by

$$
\dot{\mathbf{x}}(t) = \mathbf{A}\mathbf{x}(t) + \mathbf{B}u(t)
$$

and

$$
y(t) = \mathbf{C}\mathbf{x}(t) + Du(t) \tag{8.45}
$$

for a single-input single-output system. The Laplace transform of the first equation in (8.45) yields [see (8.24)]

$$sX(s) = AX(s) + BU(s). \tag{8.46}$$

Because we are interested in the transfer function, the initial conditions are ignored. Collecting terms for $X(s)$ yields

$$(sI - A)X(s) = BU(s); \tag{8.47}$$

thus, $X(s)$ is given by

$$X(s) = (sI - A)^{-1}BU(s). \tag{8.48}$$

The Laplace transform of the output equation in (8.45) yields

$$Y(s) = CX(s) + DU(s). \tag{8.49}$$

From (8.48) and (8.49), the input–output relationship for the system is given by

$$Y(s) = [C(sI - A)^{-1}B + D]U(s). \tag{8.50}$$

Because the system transfer function is defined by the equation $Y(s) = H(s)U(s)$, from (8.50), we see that the transfer function is given by

$$H(s) = \frac{Y(s)}{U(s)} = C(sI - A)^{-1}B + D = C\Phi(s)B + D. \tag{8.51}$$

Because C is $(1 \times n), (sI - A)^{-1}$ is $(n \times n)$, and B is $(n \times 1)$, the product $C(sI - A)^{-1}B$ is (1×1), or a scalar, as required. An example is given to illustrate this result.

EXAMPLE 8.10   **Transfer function from state equations**

Consider the system of the earlier examples with the transfer function

$$H(s) = \frac{Y(s)}{U(s)} = \frac{5s + 4}{s^2 + 3s + 2}.$$

The state equations were found in Example 8.2 to be

[eq(8.19)]
$$\dot{x}(t) = \begin{bmatrix} 0 & 1 \\ -2 & -3 \end{bmatrix} x(t) + \begin{bmatrix} 0 \\ 1 \end{bmatrix} u(t);$$
$$y(t) = [4 \quad 5]x(t).$$

The resolvant matrix $(sI - A)^{-1}$ was calculated in Example 8.5. Then, from (8.51) and Example 8.5, with $D = 0$,

$$H(s) = \mathbf{C}(s\mathbf{I} - \mathbf{A})^{-1}\mathbf{B}$$

$$= \begin{bmatrix} 4 & 5 \end{bmatrix} \begin{bmatrix} \dfrac{s+3}{s^2+3s+2} & \dfrac{1}{s^2+3s+2} \\[2ex] \dfrac{-2}{s^2+3s+2} & \dfrac{s}{s^2+3s+2} \end{bmatrix} \begin{bmatrix} 0 \\ 1 \end{bmatrix}$$

$$= \begin{bmatrix} 4 & 5 \end{bmatrix} \begin{bmatrix} \dfrac{1}{s^2+3s+2} \\[2ex] \dfrac{s}{s^2+3s+2} \end{bmatrix} = \dfrac{5s+4}{s^2+3s+2}.$$

This transfer function checks the one given.   ∎

Although (8.51) does not appear to be useful in calculating the transfer function for higher order systems, relatively simple computer algorithms exist for evaluating the resolvant matrix $(s\mathbf{I} - \mathbf{A})^{-1}$ [3]. For many practical systems, the system differential equations are written from the laws of physics. State equations are then written from these differential equations. Then a digital-computer algorithm such as that mentioned is used to calculate the transfer function. Most system analysis and design software packages have programs for finding a state model from a transfer function and for finding the transfer function from a state model. Almost all analysis and design software use transfer functions or state models. The following MATLAB program solves Example 8.10:

```
A=[0 1;-2 -3];B[0;1];C=[4 5];D=0;
[n,d]=ss2tf(A,B,C,D)
Hs=tf(n,d)
```

Stability

We saw in Section 7.7 that bounded-input bounded-output (BIBO) stability can be determined from the transfer function of an LTI system. The transfer function of (8.51) can be expressed as a rational function:

$$H(s) = \mathbf{C}(s\mathbf{I} - \mathbf{A})^{-1}\mathbf{B} + D = \frac{b_n s^n + \cdots + b_1 s + b_0}{s^n + \cdots + a_1 s + a_0}. \qquad (8.52)$$

From Section 7.7, this system is BIBO stable provided that all poles of $H(s)$ are in the left half-plane. The poles of the transfer function are the zeros of the denominator polynomial in (8.52).

The transfer function $H(s)$ can be expressed as

$$\mathbf{C}(s\mathbf{I} - \mathbf{A})^{-1}\mathbf{B} + D = \mathbf{C}\left[\frac{\text{adj}(s\mathbf{I} - \mathbf{A})}{\det(s\mathbf{I} - \mathbf{A})}\right]\mathbf{B} + D. \qquad (8.53)$$

Hence, the denominator polynomial of $H(s)$ is the determinant of $(s\mathbf{I} - \mathbf{A})$; the poles of the transfer function are the roots of

$$\det(s\mathbf{I} - \mathbf{A}) = 0. \tag{8.54}$$

This equation is then the system characteristic equation. Note that the stability is a function only of the system matrix $\mathbf{A}$ and is not affected by $\mathbf{B}$, $\mathbf{C}$, or D. In fact, (8.54) is the characteristic equation of a multivariable system, which has more than one input and more than one output. We now consider an example illustrating stability.

EXAMPLE 8.11   **Stability from state equations**

We consider the second-order system of Figure 8.8, and we wish to find the range of the parameter a for which this system is stable. We write the state equations directly from Figure 8.8:

$$\dot{\mathbf{x}}(t) = \begin{bmatrix} -2 & -a \\ 1 & -4 \end{bmatrix}\mathbf{x}(t).$$

We have ignored the input and output terms because stability is independent of these terms. From (8.54), the characteristic equation is given by

$$\det(s\mathbf{I} - \mathbf{A}) = \det\begin{bmatrix} s+2 & a \\ -1 & s+4 \end{bmatrix} = s^2 + 6s + 8 + a = 0.$$

The zeros of this polynomial are given by

$$s = \frac{-6 \pm \sqrt{36 - 4(8+a)}}{2} = \frac{-6 \pm \sqrt{4 - 4a}}{2} = -3 \pm \sqrt{1-a}.$$

The system is stable, provided that the real parts of these two roots are negative—that is, provided that

$$\sqrt{1-a} < 3 \Rightarrow 1 - a < 9 \Rightarrow -8 < a,$$

or $a > -8$. We test this result by letting $a = -8$. The roots of the characteristic equation are then

$$s = -3 \pm \sqrt{1 - (-8)} = -3 \pm 3.$$

The roots are at 0 and -6. The root at $s = 0$ is on the stability boundary in the s-plane, as we would suspect. If we decrease the value of a from -8, this root moves into the right half-plane and the system is unstable. A MATLAB program that calculates the roots for $a = -8$ is given by

```
A=[-2 8;1 -4];
c=poly (A)
r=roots (c)
result: c=1 6 0 r=0 -6
```

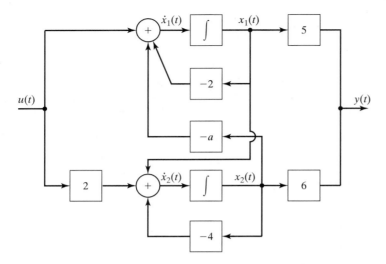

Figure 8.8 System for Example 8.11.

In this section, a procedure is developed for calculating the transfer function of a system from its state equations. The procedure can be implemented on a computer and is used extensively in the practice of engineering for calculating transfer functions of high-order systems. As a final point, it is shown that the characteristic polynomial of a system is equal to $\det(s\mathbf{I} - \mathbf{A})$; hence, this determinant is used to determine the modes of a system. The modes determine the stability of a system and the transient characteristics of a stable system.

8.6 SIMILARITY TRANSFORMATIONS

In this chapter, procedures have been presented for finding a state-variable model from system differential equations, from system transfer functions, and from system simulation diagrams. In this section, a procedure is given for finding a different state model from a given state model. It will be shown that a system has an unlimited number of state models. However, while the internal characteristics are different, each state model for a system will have the same input–output characteristics (same transfer function).

Transformations

The state model of an LTI single-input, single-output system is given by

$$\dot{\mathbf{x}}(t) = \mathbf{A}\mathbf{x}(t) + \mathbf{B}u(t)$$

and

$$y(t) = \mathbf{C}\mathbf{x}(t) + Du(t),$$

(8.55)

and the transfer function is given by, from (8.51),

$$H(s) = \frac{Y(s)}{U(s)} = \mathbf{C}(s\mathbf{I} - \mathbf{A})^{-1}\mathbf{B} + D. \tag{8.56}$$

As we show in this section, an unlimited number of combinations of the matrices **A**, **B**, and **C** and the scalar D will satisfy (8.56) for a given $H(s)$.

Suppose that we are given a state model for an nth-order system as in (8.55). We define an $n \times 1$ state vector $\mathbf{v}(t)$, such that the elements of $\mathbf{v}(t)$ are linear combinations of the elements of $\mathbf{x}(t)$, that is,

$$
\begin{aligned}
v_1(t) &= q_{11}x_1(t) + q_{12}x_2(t) + \cdots + q_{1n}x_n(t); \\
v_2(t) &= q_{21}x_1(t) + q_{22}x_2(t) + \cdots + q_{2n}x_n(t); \\
&\vdots \\
v_n(t) &= q_{n1}x_1(t) + q_{n2}x_2(t) + \cdots + q_{nn}x_n(t),
\end{aligned}
\tag{8.57}
$$

where the coefficients q_{jj} are constants. This equation can be written in matrix form as

$$\mathbf{v}(t) = \mathbf{Q}\mathbf{x}(t) = \mathbf{P}^{-1}\mathbf{x}(t), \tag{8.58}$$

where the matrix $\mathbf{Q} = [q_{ij}]$ has been defined as the inverse of a matrix **P**, to satisfy common notation. We require that **Q** have an inverse; the choice of the q_{ij} in (8.57) must result in the n equations being independent. From (8.58), the state vector $\mathbf{x}(t)$ can be expressed as

$$\mathbf{x}(t) = \mathbf{P}\mathbf{v}(t), \tag{8.59}$$

where the **P** matrix is called a transformation matrix, or simply a transformation. An example is given next.

EXAMPLE 8.12   **State-variable transformation for a second-order system**

Consider the system of Example 8.2, which has the transfer function

$$H(s) = \frac{Y(s)}{U(s)} = \frac{5s + 4}{s^2 + 3s + 2}.$$

From Example 8.2, the state equations are given by

[eq(8.19)]
$$\dot{\mathbf{x}}(t) = \begin{bmatrix} 0 & 1 \\ -2 & -3 \end{bmatrix}\mathbf{x}(t) + \begin{bmatrix} 0 \\ 1 \end{bmatrix}u(t);$$

$$y(t) = \begin{bmatrix} 4 & 5 \end{bmatrix}\mathbf{x}(t).$$

We arbitrarily define the elements of $\mathbf{v}(t)$ as

$$v_1(t) = x_1(t) + x_2(t)$$

and

$$v_2(t) = x_1(t) + 2x_2(t).$$

Thus, from (8.58),

$$\mathbf{v}(t) = \mathbf{Q}\mathbf{x}(t) = \begin{bmatrix} 1 & 1 \\ 1 & 2 \end{bmatrix}\mathbf{x}(t)$$

and

$$\mathbf{P}^{-1} = \mathbf{Q} = \begin{bmatrix} 1 & 1 \\ 1 & 2 \end{bmatrix} \Rightarrow \mathbf{P} = \begin{bmatrix} 2 & -1 \\ -1 & 1 \end{bmatrix}.$$

The components of $\mathbf{x}(t) = \mathbf{P}\mathbf{v}(t)$ are then

$$x_1(t) = 2v_1(t) - v_2(t);$$
$$x_2(t) = -v_1(t) + v_2(t).$$

It is seen from this example that given $\mathbf{v}(t)$ and the transformation $\mathbf{P} = \mathbf{Q}^{-1}$, we can solve for $\mathbf{x}(t)$. Or, given the vector $\mathbf{x}(t)$ and the transformation $\mathbf{Q}$, we can solve for the vector $\mathbf{v}(t)$. ■

Example 8.12 illustrates the transformation from one state vector to a different state vector. This transformation alters the internal model of the system (the state model) in such a manner as to leave the input–output model of the system (the transfer function) unchanged. This transformation is called a *similarity transformation*. The details of similarity transformations are now developed.

Assume that we are given the state equations (8.55) and a similarity transformation (8.59). Both equations are repeated:

[eq(8.55)]
$$\dot{\mathbf{x}}(t) = \mathbf{A}\mathbf{x}(t) + \mathbf{B}u(t);$$
$$y(t) = \mathbf{C}\mathbf{x}(t) + Du(t);$$

[eq(8.59)]
$$\mathbf{x}(t) = \mathbf{P}\mathbf{v}(t).$$

Substituting (8.59) into the state equation in (8.55) yields

$$\mathbf{P}\dot{\mathbf{v}}(t) = \mathbf{A}\mathbf{P}\mathbf{v}(t) + \mathbf{B}u(t). \tag{8.60}$$

Solving this equation for $\dot{\mathbf{v}}(t)$ results in the state model for the state vector $\mathbf{v}(t)$:

$$\dot{\mathbf{v}}(t) = \mathbf{P}^{-1}\mathbf{A}\mathbf{P}\mathbf{v}(t) + \mathbf{P}^{-1}\mathbf{B}u(t). \tag{8.61}$$

Using (8.59), we find that the output equation in (8.55) becomes

$$y(t) = \mathbf{C}\mathbf{P}\mathbf{v}(t) + Du(t). \tag{8.62}$$

We have the state equations expressed as a function of the state vector $\mathbf{x}(t)$ in (8.55) and as a function of the transformed state vector $\mathbf{v}(t)$ in (8.61) and (8.62).

The state equations as a function of $\mathbf{v}(t)$ can be expressed in the standard format as

$$\dot{\mathbf{v}}(t) = \mathbf{A}_v\mathbf{v}(t) + \mathbf{B}_v u(t)$$

and

$$y(t) = \mathbf{C}_v\mathbf{v}(t) + D_vu(t), \tag{8.63}$$

where the subscript indicates the transformed matrices. The matrices for the vector $\mathbf{x}(t)$ are not subscripted. From (8.61), (8.62), and (8.63), the transformed matrices are given by

$$\mathbf{A}_v = \mathbf{P}^{-1}\mathbf{AP}, \mathbf{B}_v = \mathbf{P}^{-1}\mathbf{B},$$
$$\mathbf{C}_v = \mathbf{CP}, \text{ and } D_v = D. \tag{8.64}$$

An example is given to illustrate the derivations.

EXAMPLE 8.13   **Similarity transformation for a second-order system**

Consider the system of Example 8.12. The state equations for the state vector $\mathbf{v}(t)$ will be derived. From Example 8.12,

$$\dot{\mathbf{x}}(t) = \mathbf{Ax}(t) + \mathbf{B}u(t) = \begin{bmatrix} 0 & 1 \\ -2 & -3 \end{bmatrix}\mathbf{x}(t) + \begin{bmatrix} 0 \\ 1 \end{bmatrix}u(t)$$

and

$$y(t) = \mathbf{Cx}(t) = [4 \quad 5]\mathbf{x}(t),$$

with the similarity transformation

$$\mathbf{P}^{-1} = \mathbf{Q} = \begin{bmatrix} 1 & 1 \\ 1 & 2 \end{bmatrix} \Rightarrow \mathbf{P} = \begin{bmatrix} 2 & -1 \\ -1 & 1 \end{bmatrix}.$$

From (8.64), the system matrices for $\mathbf{v}(t)$ are given by

$$\mathbf{A}_v = \mathbf{P}^{-1}\mathbf{AP} = \begin{bmatrix} 1 & 1 \\ 1 & 2 \end{bmatrix}\begin{bmatrix} 0 & 1 \\ -2 & -3 \end{bmatrix}\begin{bmatrix} 2 & -1 \\ -1 & 1 \end{bmatrix}$$

$$= \begin{bmatrix} -2 & -2 \\ -4 & -5 \end{bmatrix}\begin{bmatrix} 2 & -1 \\ -1 & 1 \end{bmatrix} = \begin{bmatrix} -2 & 0 \\ -3 & -1 \end{bmatrix};$$

$$\mathbf{B}_v = \mathbf{P}^{-1}\mathbf{B} = \begin{bmatrix} 1 & 1 \\ 1 & 2 \end{bmatrix}\begin{bmatrix} 0 \\ 1 \end{bmatrix} = \begin{bmatrix} 1 \\ 2 \end{bmatrix};$$

$$\mathbf{C}_v = \mathbf{CP} = [4 \quad 5]\begin{bmatrix} 2 & -1 \\ -1 & 1 \end{bmatrix} = [3 \quad 1].$$

The transformed state equations are then

$$\dot{\mathbf{v}}(t) = \mathbf{A}_v\mathbf{v}(t) + \mathbf{B}_vu(t) = \begin{bmatrix} -2 & 0 \\ -3 & -1 \end{bmatrix}\mathbf{v}(t) + \begin{bmatrix} 1 \\ 2 \end{bmatrix}u(t)$$

and   (8.65)

$$y(t) = \mathbf{C}_v\mathbf{v}(t) = [3 \quad 1]\mathbf{v}(t).$$

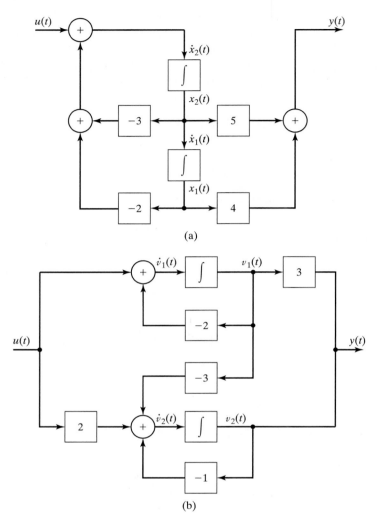

(a)

(b)

Figure 8.9 Simulation diagrams for Example 8.13.

A MATLAB Program that performs the matrix multiplications is given by

```
A=[0 1;-2 -3];B=[0;1];C=[4 5];D=0;T=[1 1;1 2];
sys=ss(A,B,C,D)
[n,d]=ss2tf(A,B,C,D);
Hs=tf(n,d)
[Av,Bv,Cv,Dv]=ss2ss(A,B,C,D,T)
```
∎

For the system in Example 8.13, the simulation diagram for the $\mathbf{x}(t)$-state vector is given in Figure 8.3 and is repeated in Figure 8.9(a). The simulation diagram for the $\mathbf{v}(t)$-state vector is given in Figure 8.9(b); this diagram satisfies the $\dot{\mathbf{v}}(t)$-state equations, (8.65). We show later that the transfer functions of the two simulation diagrams are equal; however, note the differences in the internal structures.

Example 8.13 gives two state models for the same system. If a different transformation matrix $\mathbf{P}$ (that has an inverse) had been chosen, a third model would result. In fact, for each different transformation $\mathbf{P}$ that has an inverse, a different state model results. Thus, an unlimited number of state models exists for a given system transfer function. The choice of the state model for a given system can be based on the natural state variables (position, velocity, current, voltage, etc.), on ease of analysis and design, and so on.

To check the state model developed in Example 8.13, we now derive the transfer function.

EXAMPLE 8.14   **Transfer function for the system of Example 8.13**

The transformed state equations of Example 8.13 are given by

$$\dot{\mathbf{v}}(t) = \mathbf{A}_v(t) + \mathbf{B}_v u(t) = \begin{bmatrix} -2 & 0 \\ -3 & -1 \end{bmatrix} \mathbf{v}(t) + \begin{bmatrix} 1 \\ 2 \end{bmatrix} u(t)$$

and

$$y(t) = \mathbf{C}_v \mathbf{v}(t) = \begin{bmatrix} 3 & 1 \end{bmatrix} \mathbf{v}(t).$$

From (8.56), the transfer function of this system is given by

$$H_v(s) = \mathbf{C}_v(s\mathbf{I} - \mathbf{A}_v)^{-1}\mathbf{B}_v.$$

First, we calculate $(s\mathbf{I} - \mathbf{A}_v)^{-1}$. Now,

$$s\mathbf{I} - \mathbf{A}_v = s\begin{bmatrix} 1 & 0 \\ 0 & 1 \end{bmatrix} - \begin{bmatrix} -2 & 0 \\ -3 & -1 \end{bmatrix} = \begin{bmatrix} s+2 & 0 \\ 3 & s+1 \end{bmatrix}.$$

Therefore,

$$\det(s\mathbf{I} - \mathbf{A}) = (s+2)(s+1) = s^2 + 3s + 2.$$

Then, letting $\det(s\mathbf{I} - \mathbf{A}) = \Delta(s)$ for convenience, we have

$$(s\mathbf{I} - \mathbf{A})^{-1} = \frac{\text{adj}(s\mathbf{I} - \mathbf{A})}{\det(s\mathbf{I} - \mathbf{A})} = \begin{bmatrix} \dfrac{s+1}{\Delta(s)} & 0 \\ \dfrac{-3}{\Delta(s)} & \dfrac{s+2}{\Delta(s)} \end{bmatrix},$$

and the transfer function is given by

$$H_v(s) = \mathbf{C}_v(s\mathbf{I} - \mathbf{A}_v)^{-1}\mathbf{B}_v$$

$$= \begin{bmatrix} 3 & 1 \end{bmatrix} \begin{bmatrix} \dfrac{s+1}{\Delta(s)} & 0 \\ \dfrac{-3}{\Delta(s)} & \dfrac{s+2}{\Delta(s)} \end{bmatrix} \begin{bmatrix} 1 \\ 2 \end{bmatrix}$$

$$= \begin{bmatrix} 3 & 1 \end{bmatrix} \begin{bmatrix} \dfrac{s+1}{\Delta(s)} \\[2ex] \dfrac{2s+1}{\Delta(s)} \end{bmatrix} = \dfrac{5s+4}{s^2+3s+2}.$$

This transfer function is the same as that used in Example 8.2 to derive the $\mathbf{x}(t)$-state model. The following MATLAB statements, when appended to the program in Example 8.13, verify the results of this example:

```
[n d]=ss2tf(Av,Bv,Cv,Dv)
Hs=tf(n,d)
```
∎

Properties

Similarity transformations have been demonstrated through an example. Certain important properties of these transformations are derived next. Consider first the determinant of $(s\mathbf{I} - \mathbf{A}_v)$. From (8.64),

$$\det(s\mathbf{I} - \mathbf{A}_v) = \det(s\mathbf{I} - \mathbf{P}^{-1}\mathbf{AP}) = \det(s\mathbf{P}^{-1}\mathbf{IP} - \mathbf{P}^{-1}\mathbf{AP})$$

$$= \det[\mathbf{P}^{-1}(s\mathbf{I} - \mathbf{A})\mathbf{P}]. \qquad (8.66)$$

For two square matrices,

$$\det \mathbf{R}_1\mathbf{R}_2 = \det \mathbf{R}_1 \det \mathbf{R}_2. \qquad (8.67)$$

Then (8.66) becomes

$$\det(s\mathbf{I} - \mathbf{A}_v) = \det \mathbf{P}^{-1}\det(s\mathbf{I} - \mathbf{A}) \det \mathbf{P}. \qquad (8.68)$$

For a matrix $\mathbf{R}$, $\mathbf{R}^{-1}\mathbf{R} = \mathbf{I}$. Then

$$\det \mathbf{R}^{-1}\mathbf{R} = \det \mathbf{R}^{-1}\det \mathbf{R} = \det \mathbf{I} = 1. \qquad (8.69)$$

Thus, (8.68) yields the first property:

$$\det(s\mathbf{I} - \mathbf{A}_v) = \det(s\mathbf{I} - \mathbf{A}) \det \mathbf{P}^{-1}\det \mathbf{P} = \det(s\mathbf{I} - \mathbf{A}). \qquad (8.70)$$

The roots of $\det(s\mathbf{I} - \mathbf{A})$ are the *characteristic values*, or the *eigenvalues*, of $\mathbf{A}$. (See Appendix G.) From (8.70), the eigenvalues of $\mathbf{A}_v$ are equal to those of $\mathbf{A}$. Because the transfer function is unchanged under a similarity transformation, and since the eigenvalues of $\mathbf{A}$ are the poles of the system transfer function, we are not surprised that they are unchanged.

A second property can be derived as follows: From (8.64),

$$\det \mathbf{A}_v = \det \mathbf{P}^{-1}\mathbf{AP} = \det \mathbf{P}^{-1} \det \mathbf{A} \det \mathbf{P} = \det \mathbf{A}. \qquad (8.71)$$

The determinant of $\mathbf{A}_v$ is equal to the determinant of $\mathbf{A}$. This property can also be seen from the fact that the determinant of a matrix is equal to the product of its eigenvalues. (See Appendix G.)

The third property of a similarity transformation can also be seen from the fact that the eigenvalues of $\mathbf{A}_v$, are equal to those of $\mathbf{A}$. Because the trace (sum of the diagonal elements) of a matrix is equal to the sum of the eigenvalues,

$$\text{tr } \mathbf{A}_v = \text{tr} \mathbf{A}. \tag{8.72}$$

A fourth property was demonstrated in Example 8.14. Because the transfer function is unchanged under a similarity transformation,

$$\mathbf{C}_v(s\mathbf{I} - \mathbf{A}_v)^{-1}\mathbf{B}_v + D_v = \mathbf{C}(s\mathbf{I} - \mathbf{A})^{-1}\mathbf{B} + D. \tag{8.73}$$

The proof of this property is left as an exercise. (See Problem 8.25.)

To summarize the properties of similarity transforms, we first let $\lambda_1, \lambda_2, \ldots, \lambda_n$ be the eigenvalues of the $n \times n$ matrix $\mathbf{A}$. Then, for the similarity transformation,

$$\mathbf{A}_v = \mathbf{P}^{-1}\mathbf{A}\mathbf{P}.$$

1.   The eigenvalues of $\mathbf{A}$ and $\mathbf{A}_v$ are equal:

$$\det(s\mathbf{I} - \mathbf{A}) = \det(s\mathbf{I} - \mathbf{A}_v)$$
$$= (s - \lambda_1)(s - \lambda_2)\cdots(s - \lambda_n). \tag{8.74}$$

2.   The determinants of $\mathbf{A}$ and $\mathbf{A}_v$ are equal:

$$\det\mathbf{A} = \det\mathbf{A}_v = \lambda_1\lambda_2\cdots\lambda_n. \tag{8.75}$$

3.   The traces of $\mathbf{A}$ and $\mathbf{A}_v$ are equal:

$$\text{tr}\mathbf{A} = \text{tr}\mathbf{A}_v = \lambda_1 + \lambda_2 + \cdots + \lambda_n. \tag{8.76}$$

4.   The following transfer functions are equal:

$$\mathbf{C}_v(s\mathbf{I} - \mathbf{A}_v)^{-1}\mathbf{B}_v + D_v = \mathbf{C}(s\mathbf{I} - \mathbf{A})^{-1}\mathbf{B} + D. \tag{8.77}$$

Property 4 was illustrated in Example 8.14. Properties 1 through 3 are illustrated in the following example.

EXAMPLE 8.15   **Illustrations of properties of similarity transformations**

The similarity transformation of Example 8.13 is used to illustrate the first three properties just developed. From Example 8.13, the matrices A and $\mathbf{A}_v$ are given by

$$\mathbf{A} = \begin{bmatrix} 0 & 1 \\ -2 & -3 \end{bmatrix}; \ \mathbf{A}_v = \begin{bmatrix} -2 & 0 \\ -3 & -1 \end{bmatrix}.$$

Then, for the first property, (8.74),

$$\det(s\mathbf{I} - \mathbf{A}) = \begin{vmatrix} s & -1 \\ 2 & s+3 \end{vmatrix} = s^2 + 3s + 2;$$

$$\det(s\mathbf{I} - \mathbf{A}_v) = \begin{vmatrix} s+2 & 0 \\ 3 & s+1 \end{vmatrix} = s^2 + 3s + 2.$$

Next, the eigenvalues are found from

$$\det(s\mathbf{I} - \mathbf{A}) = s^2 + 3s + 2 = (s+1)(s+2).$$

we obtain $\lambda_1 = -1, \lambda_2 = -2$.
For the second property, (8.75), the determinants of the two matrices are given by

$$|\mathbf{A}| = \begin{vmatrix} 0 & 1 \\ -2 & -3 \end{vmatrix} = 2; \quad |\mathbf{A}_v| = \begin{vmatrix} -2 & 0 \\ -3 & -1 \end{vmatrix} = 2,$$

and both determinants are equal to the product of the eigenvalues.
For the third property, (8.76), the traces of the two matrices are the sums of the diagonal elements

$$\text{tr}\mathbf{A} = 0 + (-3) = -3; \quad \text{tr}\mathbf{A}_v = -2 + (-1) = -3.$$

Thus, the traces are equal to the sum of the eigenvalues. ■

The eigenvalues of an $n \times n$ matrix can be found using the MATLAB program in Example 8.11. The following MATLAB program also calculates eigenvalues:

```
A=[0 1;-2 -3], Av=[-2 0;-3 -1]
'Compare the coeficients of the characteristic polynomials'
poly(A), poly(Av)
'Compare the eigenvalues of the two matrices'
eig(A), eig(Av)
'Compare the traces of the two matrices'
trace(A), trace(Av)
```

In this section, we develop similarity transformations for state equations. It is shown that any system has an unbounded number of state models. However, all state models for a given system have the same transfer function. As a final topic, four properties of similarity transformations for state equations are derived.

SUMMARY

In earlier chapters, we specified the model of a continuous-time LTI system by a differential equation or a transfer function. In both cases, the system input–output characteristics are given. In this chapter, a third model, the state-variable model, is developed. This model is a set of coupled first-order differential equations. The state model can be specified either by state equations or by a simulation diagram.

The state model gives an internal model of a system, in addition to the input–output characteristics. Methods are presented in this chapter to derive any one of the three models from any other one.

It is also demonstrated in this chapter that a state model for a given system is not unique. Similarity transformations may be used to develop any number of different state models from a given state model. The different state models have the same input–output characteristics, while the internal models are all different.

A state model of a continuous-time LTI system is required for applying certain types of analysis and design procedures [2]. State models are especially useful in computer-aided analysis and design.

See Table 8.1.

TABLE 8.1   Key Equations of Chapter 8

Equation Title	Equation Number	Equation
State equations of LTI system	(8.9)	$\dot{\mathbf{x}}(t) = \mathbf{A}\mathbf{x}(t) + \mathbf{B}\mathbf{u}(t)$
		$\mathbf{y}(t) = \mathbf{C}\mathbf{x}(t) + \mathbf{D}\mathbf{u}(t)$
State equations in matrix form	(8.18)	$\dot{x}(t) = \begin{bmatrix} 0 & 1 & 0 & \cdots & 0 & 0 \\ 0 & 0 & 1 & \cdots & 0 & 0 \\ \vdots & \vdots & \vdots & & \vdots & \vdots \\ 0 & 0 & 0 & \cdots & 0 & 1 \\ -a_0 & -a_1 & -a_2 & \cdots & -a_{n-2} & -a_{n-1} \end{bmatrix}\mathbf{x}(t) + \begin{bmatrix} 0 \\ 0 \\ \vdots \\ 0 \\ 1 \end{bmatrix}u(t)$
		$\mathbf{y}(t) = [(b_0 - a_0 b_n)(b_1 - a_1 b_n)\cdots(b_{n-1} - a_{n-1}b_n)]\mathbf{x}(t) + b_n u(t)$
State-transition matrix	(8.27)	$\boldsymbol{\phi}(t) = \mathcal{L}^{-1}[\boldsymbol{\phi}(s)] = \mathcal{L}^{-1}[(s\mathbf{I} - \mathbf{A})^{-1}]$
Solution of state equation	(8.28)	$\mathbf{X}(s) = (s\mathbf{I} - \mathbf{A})^{-1}\mathbf{x}(0) + (s\mathbf{I} - \mathbf{A})^{-1}\mathbf{B}U(s)$
		$= \boldsymbol{\Phi}(s)\mathbf{x}(0) + \boldsymbol{\Phi}(s)\mathbf{B}U(s)$
Convolution solution of state equation	(8.29)	$\mathbf{x}(t) = \boldsymbol{\Phi}(t)\mathbf{x}(0) + \displaystyle\int_0^t \boldsymbol{\Phi}(t-\tau)\mathbf{B}\mathbf{u}(\tau)d\tau$
		$= \boldsymbol{\Phi}(t)\mathbf{x}(0) + \displaystyle\int_0^t \boldsymbol{\Phi}(\tau)\mathbf{B}\mathbf{u}(t-\tau)d\tau$
Matrix exponential	(8.38)	$\boldsymbol{\Phi}(t) = \exp \mathbf{A}t$
Transfer function	(8.51)	$H(s) = \dfrac{Y(s)}{U(s)} = \mathbf{C}(s\mathbf{I} - \mathbf{A})^{-1}\mathbf{B} + D = \mathbf{C}\boldsymbol{\phi}(s)\mathbf{B} + D$
State vector transformation	(8.59)	$\mathbf{x}(t) = \mathbf{P}\mathbf{v}(t)$
State equations (similarity transformation)	(8.63)	$\dot{\mathbf{v}}(t) = \mathbf{A}_v\mathbf{v}(t) + \mathbf{B}_v u(t)$
		$y(t) = \mathbf{C}_v\mathbf{v}(t) + D_v u(t)$
Transformed matrices	(8.64)	$\mathbf{A}_v = \mathbf{P}^{-1}\mathbf{A}\mathbf{P},\ \mathbf{B}_v = \mathbf{P}^{-1}\mathbf{B}$
		$\mathbf{C}_v = \mathbf{C}\mathbf{P},\ \text{and}\ D_v = D$

REFERENCES

1. D. Graupe, *Identification of Systems*. Huntington, NY: Robert E. Kreiger, 1976.
2. C. L. Phillips and R. D. Harbor, *Feedback Control Systems*, 4th ed. Upper Saddle River, NJ: Prentice Hall, 1999.
3. B. Friedlander, *Control System Design*. New York: McGraw–Hill, 1986.
4. G. H. Golub and C. F. Van Loan, *Matrix Computations*, 2d ed. Baltimore, MD: Johns Hopkins University Press, 1996.
5. C. L. Phillips and H. T. Nagle, *Digital Control System Analysis and Design*, 3d ed. Upper Saddle River, NJ: Prentice Hall, 1996.
6. G. F. Franklin and J. D. Powell, *Digital Control of Dynamic Systems*. 3d ed. Reading, MA: Addison–Wesley, 1997.
7. W. L. Brogan, *Modern Control Theory*, 3d ed. Upper Saddle River, NJ: Prentice Hall, 1991.

PROBLEMS

8.1. Consider the *RL* circuit of Figure P8.1. The circuit input is the voltage $v_i(t)$.

    (a) Write the state equations for the circuit, with the state variable equal to the current $i(t)$ and the output equal to the resistance voltage $v_R(t)$.

    (b) Write the state equations for the circuit, with both the state variable and the output equal to $v_R(t)$.

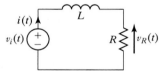

Figure P8.1

8.2. In Figure 8.1, replace the resistor with a capacitor of *C* farads and voltage $v_c(t)$. The input voltage is $v_i(t)$.

    (a) Write the state equations for the circuit, with the state variables equal to the current $i(t)$, the capacitor voltage $v_C(t)$, and the output equal to $v_C(t)$.

    (b) Write the state equations of the circuit, with the state variables equal to $i(t)$ and $v_C(t)$, and the output equal to $i(t)$.

8.3. Find a set of state equations for each of the systems described by the following differential equations:

    (a) $\dot{y}(t) + y(t) = u(t)$

    (b) $\ddot{y}(t) + 3\dot{y}(t) + 2y(t) = 4u(t)$

    (c) $5\ddot{y}(t) + 4\dot{y}(t) + y(t) = 3u(t)$

    (d) $\ddot{y}_1(t) + 5\dot{y}_1(t) + 6y_1(t) - y_2(t) = 4u_1(t) - u_2(t)$
        $\dot{y}_2(t) + y_2(t) + 8y_1(t) = u_1(t) - 3u_2(t)$

    (e) $\dot{y}_1(t) + y_1(t) + 2y_2(t) = u_1(t) - 5u_2(t)$
        $\ddot{y}_2(t) - 9\dot{y}_2 + 16y_2(t) - y_1(t) = 3u_1(t) + u_2(t)$

8.4. (a) Draw a simulation diagram for the system described by the transfer function

$$\frac{Y(s)}{U(s)} = H(s) = \frac{4}{s + 6}.$$

(b) Write the state equations for the simulation diagram of Part (a).
(c) Give the system differential equation for the simulation diagram of Part (a).
(d) Repeat Parts (a), (b), and (c) for the transfer function

$$\frac{Y(s)}{U(s)} = H(s) = \frac{50s}{s^2 + 3s + 1}.$$

(e) Repeat Parts (a) through (c) for the transfer function

$$\frac{Y(s)}{U(s)} = H(s) = \frac{100s + 800}{s^3 + 10s^2 + 60s + 120}.$$

8.5. (a) Draw a simulation diagram for the system described by the differential equation

$$2\dot{y}(t) + 4y(t) = 8u(t).$$

(b) Write the state equations for the simulation diagram of Part (a).
(c) Give the system transfer function for the simulation diagram of Part (a).
(d) Use MATLAB to verify the results in Part (c).
(e) Repeat Parts (a) through (d) for the differential equation

$$\ddot{y}(t) - 8\dot{y}(t) + 12y(t) = 40u(t).$$

(f) Repeat Parts (a) through (d) for the differential equation

$$\dddot{y}(t) + 20\ddot{y}(t) + 10\dot{y}(t) + 15y(t) = 50u(t).$$

8.6. (a) Write the state equations for the system modeled by the simulation diagram of Figure P8.6.
(b) Use the results of Part (a) to find the system-transfer function.
(c) Use MATLAB to verify the results in Part (b).
(d) Use the system transfer function to draw a simulation diagram that is different from that of Figure P8.6.

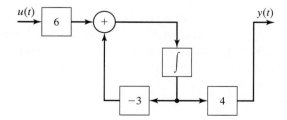

Figure P8.6

8.7. (a) Write the state equations for the system modeled by the simulation diagram of Figure P8.7.

(b) Use the results of Part (a) to find the system-transfer function.

(c) Use MATLAB to verify the results in Part (b).

(d) Use the system transfer function to draw a simulation diagram that is different from that of Figure P8.7.

(e) Write the state equations for the simulation diagram of Part (d).

(f) Use the results of Part (e) to calculate the system transfer function, which will verify the results of Part (e).

(g) Check the results of Part (f) using MATLAB.

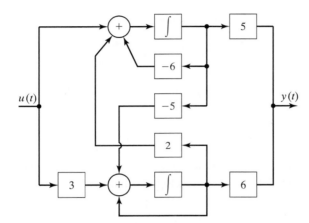

Figure P8.7

8.8. Figure P8.8 gives the simulation diagram of an automatic control system. The plant is the system that is controlled, and the compensator is an added system to give the closed-loop system certain specified characteristics.

(a) Write the state equations for the plant only, with input $m(t)$ and output $y(t)$.

(b) Give the transfer function $H_p(s)$ for the plant.

(c) Write the differential equation for the plant.

(d) Write the state equations for the compensator only, with input $e(t)$ and output $m(t)$.

(e) Give the transfer function $H_c(s)$ for the compensator.

(f) Write the differential equation for the compensator.

(g) Write the state equations for the closed-loop system, with input $u(t)$ and output $y(t)$. Choose as states those of Parts (a) and (d). These equations can be written directly from the simulation diagram.

(h) Give the transfer function $H(s)$ for the closed-loop system.

(i) Write the differential equation for the closed-loop system.

(j) Verify the results of Part (h) using MATLAB.

(k) It can be shown that the closed-loop transfer function is given by

$$\frac{Y(s)}{U(s)} = H(s) = \frac{H_c(s)H_p(s)}{1 + H_c(s)H_p}.$$

Verify your results in Part (h) by showing that this equation is satisfied by the derived transfer functions in Parts (b) and (e).

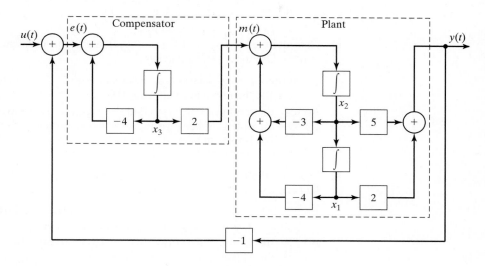

Figure P8.8

8.9. In Figure P8.8, let the compensator transfer function be $H_c(s) = 2$, a pure gain. Solve all parts of Problem 8.8 for this system.

8.10. Given a system described by the state equations

$$\dot{\mathbf{x}}(t) = \begin{bmatrix} -4 & 5 \\ 0 & 1 \end{bmatrix} \mathbf{x}(t) + \begin{bmatrix} 0 \\ 1 \end{bmatrix} u(t); \quad y(t) = \begin{bmatrix} 1 & 1 \end{bmatrix} \mathbf{x}(t) + 2u(t).$$

(a) Draw a simulation diagram of this system.
(b) Find the transfer function directly from the state equations.
(c) Use MATLAB to verify the results in Part (b).
(d) Draw a different simulation diagram of the system.
(e) Write the state equations of the simulation diagram of Part (d).
(f) Verify the simulation diagram of Part (d) by showing that its transfer function is that of Part (b)
(g) Verify the results of Part (f) using MATLAB.

(h) Repeat Parts (a) through (g) for the state equations

$$\dot{x}(t) = -4x(t) + 6u(t);$$
$$y(t) = x(t).$$

(i) Repeat Parts (a) through (g) for the state equations

$$\dot{\mathbf{x}}(t) = \begin{bmatrix} 0 & 1 & 0 \\ 0 & 0 & 1 \\ 1 & 0 & 1 \end{bmatrix} \mathbf{x}(t) + \begin{bmatrix} 2 \\ 0 \\ 0 \end{bmatrix} u(t); \quad y(t) = \begin{bmatrix} 0 & 0 & 1 \end{bmatrix} \mathbf{x}(t).$$

8.11. Consider the RL circuit of Figure P8.1. Parts of this problem are repeated from Problem 8.1. Use those results if available.

(a) Write the state equations of the circuit, with both the state variable and the output equal to the resistance voltage $v_R(t)$.
(b) Use the results of Part (a) to find the circuit transfer function.
(c) Use the s-plane impedance approach to verify the transfer function of Part (b). Recall that for the impedance approach, the impedance of the resistance is R and of the inductance is sL.

8.12. Consider the RL circuit of Figure P8.1. Parts of this problem are repeated from Problem 8.1. Use those results if available.

(a) Write the state equations of the circuit, with the state variable equal to the current $i(t)$ and the output equal to the resistance voltage $v_R(t)$.
(b) Use the results of Part (a) to find the circuit transfer function.
(c) Use the s-plane impedance approach to verify the transfer function of Part (b). Recall that for the impedance approach, the impedance of the resistance is R and of the inductance is sL.

8.13. Consider the LC circuit of Problem 8.2. Parts of this problem are repeated from Problem 8.2. Use those results if available.

(a) Write the state equations of the circuit, with the state variables equal to the current $i(t)$ and the capacitor voltage $v_C(t)$, and the output equal to the $v_C(t)$.
(b) Use the results of Part (a) to find the circuit transfer function.
(c) Use the s-plane impedance approach to verify the transfer function of Part (b). Recall that for the impedance approach, the impedance of the inductance is sL and of the capacitance is $1/(sC)$.

8.14. Consider the LC circuit of Problem 8.2. Parts of this problem are repeated from Problem 8.2. Use those results if available.

(a) Write the state equations of the circuit, with the state variables equal to the current $i(t)$ and the capacitor voltage $v_C(t)$, and the output equal to the $i(t)$.
(b) Use the results of Part (a) to find the circuit transfer function.
(c) Use the s-plane impedance approach to verify the transfer function of Part (b). Recall that for the impedance approach, the impedance of the inductance is sL and of the capacitance is $1/(sC)$.

8.15. Consider the system of Figure P8.6.

(a) Write the state equations, with the output of the integrator as the state.

(b) Find the state transition matrix.

(c) Find the system output for $u(t) = 0$ and the initial state given by $x(0) = 2$.

(d) Find the system unit step response, with $x(0) = 0$, using (8.28) and $y(t) = Cx(t) + Du(t)$.

(e) Verify the results of Part (d), using the Laplace transform and the system transfer function.

(f) Find the system response, with the initial conditions given in Part (c) and the input in Part (d).

(g) Use SIMULINK to verify the results of Part (f).

8.16. Consider the system of Figure P8.7.

(a) Write the state equations with the outputs of the integrators as the states.

(b) Find the state transition matrix.

(c) Find the system output for $u(t) = 0$ and the initial states given by $\mathbf{x}(0) = [1 \quad 0]^T$.

(d) Find the system unit step response, with $\mathbf{x}(0) = 0$, using (8.28) and $y(t) = \mathbf{Cx}(t) + Du(t)$.

(e) Verify the results of Part (d), using the Laplace transform and the system transfer function.

(f) Find the system response, with the initial conditions given in Part (c) and the input in Part (d).

(g) Use SIMULINK to verify the results of Part (f).

8.17. Consider the system of Problem 8.10, given by

$$\dot{\mathbf{x}}(t) = \begin{bmatrix} -4 & 5 \\ 0 & 1 \end{bmatrix} \mathbf{x}(t) + \begin{bmatrix} 0 \\ 1 \end{bmatrix} u(t); \quad y(t) = [1 \quad 1]\mathbf{x}(t) + 2u(t).$$

(a) Find the system output for $u(t) = 0$ and the initial states given by $\mathbf{x}(0) = [1 \quad 0]^T$.

(b) Find the unit step response of the system, with $\mathbf{x}(0) = 0$, using (8.28) and $y(t) = \mathbf{Cx}(t) + Du(t)$.

(c) Verify the results in Part (b), using the system transfer function and the Laplace transform.

(d) Use the transfer function to write the system differential equation.

(e) Verify the results of Part (b), by substituting the solution into the system differential equation.

(f) Find the system response with the initial conditions of Part (a) and the input of Part (b). Verify that the response has the correct initial condition.

(g) Use SIMULINK to verify the results of Part (f).

8.18. For the system of Example 8.8, the state equation is given by

$$\dot{\mathbf{x}}(t) = \begin{bmatrix} 0 & 1 \\ 0 & 0 \end{bmatrix} \mathbf{x}(t) + \begin{bmatrix} 0 \\ 1 \end{bmatrix} u(t).$$

Use the Laplace transform to show that the state transition matrix is given by

$$\Phi(t) \begin{bmatrix} 1 & t \\ 0 & 1 \end{bmatrix}.$$

8.19. Consider the system described by the state equations

$$\dot{x}(t) = \begin{bmatrix} 0 & 0 \\ 1 & 0 \end{bmatrix} x(t) + \begin{bmatrix} 1 \\ 1 \end{bmatrix} u(t); \quad y(t) = [0 \quad 1] x(t).$$

(a) Find the state transition matrix.
(b) Verify the results of Part (a), using the series of (8.36).
(c) Find the initial-condition response for $x(0) = [1 \quad 2]^T$.
(d) Verify the calculation of the state vector $x(t)$ in Part (c), by substitution in the equation $\dot{x}(t) = Ax(t)$.
(e) Calculate the system unit step response, with $x(0) = 0$, using (8.28) and $y(t) = Cx(t) + Du(t)$.
(f) Verify the results of Part (e), using the system-transfer function and the Laplace transform.
(g) Verify the results of Parts (c) and (e), using SIMULINK.

8.20. Consider the system described by the state equations

$$\dot{x}(t) = -2x(t) + 4u(t),$$

$$y(t) = 5x(t).$$

(a) Find the state transition matrix.
(b) Verify the results of Part (a), using the series of (8.36). Recall the series expansion of the exponential function in (8.37).
(c) Find the initial-condition response for $x(0) = 1$.
(d) Verify the calculation of the state $x(t)$ in Part (c), by substitution in the equation $\dot{x}(t) = Ax(t)$.
(e) Calculate the system unit step response, with $x(0) = 0$, using (8.28).
(f) Verify the results of Part (e), using the system transfer function and the Laplace transform.
(g) Verify the results of Parts (c) and (e), using SIMULINK.

8.21. Consider the system of Problem 8.10, given by

$$\dot{x}(t) = \begin{bmatrix} -4 & 5 \\ 0 & 1 \end{bmatrix} x(t) + \begin{bmatrix} 0 \\ 1 \end{bmatrix} u(t);$$

$$y(t) = [1 \quad 1] x(t) + 2u(t).$$

(a) Find the transfer function for this system.
(b) Use a similarity transformation to find a different state model for this system.
(c) Use MATLAB to check the results in Part (b).
(d) Calculate the transfer function of Part (b). This function should equal that of Part (a).
(e) Verify the results of Part (d), using MATLAB.
(f) You have just verified Property 4, (8.77), of similarity transformations. Verify the other three properties in (8.74), (8.75), and (8.76).

8.22. Consider the system of Problem 8.19, given by

$$\dot{\mathbf{x}}(t) = \begin{bmatrix} 0 & 0 \\ 1 & 0 \end{bmatrix}\mathbf{x}(t) + \begin{bmatrix} 1 \\ 1 \end{bmatrix}u(t); \quad y(t) = \begin{bmatrix} 0 & 1 \end{bmatrix}\mathbf{x}(t).$$

(a) Find the transfer function for this system.
(b) Use a similarity transformation to find a different state model for this system.
(c) Use MATLAB to check the results in Part (b).
(d) Calculate the transfer function of Part (b). This function should equal that of Part (a).
(e) Verify the results of Part (d), using MATLAB.
(f) You have just verified Property 4, (8.77), of similarity transformations. Verify the other three properties in (8.74), (8.75), and (8.76).

8.23. Consider the system of Problem 8.20, given by

$$\dot{x}(t) = -2x(t) + 4u(t);$$
$$y(t) = 5x(t).$$

(a) Find the transfer function for this system.
(b) Use a similarity transformation to find a different state model for this system.
(c) Use MATLAB to check the results in Part (b).
(d) Calculate the transfer function of Part (b). This function should equal that of Part (a).
(e) Verify the results of Part (d), using MATLAB.
(f) You have just verified Property 4, (8.77), of similarity transformations. Verify the other three properties in (8.74), (8.75), and (8.76).
(g) From the results of this problem, explain the general effects of a similarity transformation on a first-order system.

8.24. Consider the system of Problem 8.10(i), given by

$$\dot{\mathbf{x}}(t) = \begin{bmatrix} 0 & 1 & 0 \\ 0 & 0 & 1 \\ 1 & 0 & 1 \end{bmatrix}\mathbf{x}(t)\begin{bmatrix} 2 \\ 0 \\ 0 \end{bmatrix}u(t);$$

$$y(t) = \begin{bmatrix} 0 & 0 & 1 \end{bmatrix}\mathbf{x}(t).$$

(a) Find the transfer function for this system.

(b) Find a different state model for this system, using the transformation

$$\mathbf{P} = \begin{bmatrix} 1 & 1 & 0 \\ 0 & 0 & 1 \\ 1 & 0 & 0 \end{bmatrix}.$$

(c) Use MATLAB to check the results in Part (b).

(d) Calculate the transfer function of Part (b). This function should equal that of Part (a).

(e) You have just verified Property 4, (8.77), of similarity transformations. Verify the other three properties in (8.74), (8.75), and (8.76), without finding the eigenvalues.

(f) Verify the results of Part (d) using MATLAB.

8.25. Show that for the similarity transformation of (8.64),

$$\mathbf{C}_v(s\mathbf{I} - \mathbf{A}_v)^{-1}\mathbf{B}_v + D_v = \mathbf{C}(s\mathbf{I} - \mathbf{A})^{-1}\mathbf{B} + D.$$

8.26. Consider the system of Problem 8.21.

(a) Determine if this system is stable.
(b) Give the system modes.

8.27. Consider the system of Problem 8.23.

(a) Determine if this system is stable.
(b) Give the system modes.

8.28. Consider the system of Problem 8.24.

(a) Use MATLAB to determine if this system is stable.
(b) Give the system modes.

9

DISCRETE-TIME SIGNALS AND SYSTEMS

In this chapter, we first consider *discrete-time*, or more simply *discrete, signals*. A discrete-time signal is defined only at discrete instants of time. We denote a discrete-time signal as $x[n]$, where the independent variable n may assume only integer values. As a second topic in this chapter, we consider *discrete-time*, or simply *discrete, systems*. A discrete-time system is defined as one in which all signals are discrete time. This chapter follows closely the outline of Chapter 2.

As stated, a discrete signal is defined at only discrete instants of time. For example, suppose that a continuous-time signal $f(t)$ is to be processed by a digital computer. [This operation is called *digital signal processing* (DSP).] Because a computer can operate only on a number, the continuous-time signal must first be converted to a sequence of numbers. This conversion process is called *sampling*. If the signal is sampled at regular increments of time T, the number sequence $f(nT), n = \ldots, -2, -1, 0, 1, 2, \ldots$, results. The time increment T is called the *sampling period*. (Since there is little danger of confusion in this and following chapters, the symbol T is used to denote the sampling period instead of T_S as in Chapters 5 and 6.) The sampling process is illustrated in Figure 9.1(a), where each sample value is represented by a dot at the end of a vertical line.

The hardware normally used in sampling is represented in Figure 9.1(b). As described in Chapter 1, an analog-to-digital converter (A/D or ADC) is an electronic circuit that samples a voltage signal and converts each sample into a binary number; the binary numbers can then be transmitted to a digital computer for processing or for storage. Hence, an A/D is used to generate and transmit the number sequence $f(nT)$ to the computer. The instants that samples are taken are determined by timing pulses from the computer.

A word is in order concerning notation. The notation $f(t)$ indicates a continuous-time signal. The notation $f(nT)$ indicates the value of $f(t)$ at $t = nT$. The notation $f[n]$ denotes a *discrete-time signal* that is defined only for n an integer. Parentheses indicate continuous time; brackets indicate discrete time. However,

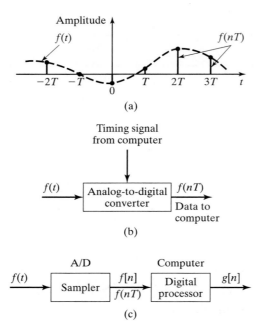

(a)

(b)

(c)

Figure 9.1 Hardware diagram for sampling and processing.

this notation is *not* universal; it is used here in an attempt to differentiate between $f(nT)$ and $f[n]$. If $f[n]$ is obtained from $f(t)$ by sampling every T seconds, then

$$f(nT) = f(t)|_{t=nT}$$

and

$$f[n] = f(t)|_{t=nT} \neq f(t)|_{t=n}. \tag{9.1}$$

Figure 9.1(c) illustrates a total system for digital signal processing. The sampler converts the continuous-time signal $f(t)$ into the discrete-time signal $f(nT) = f[n]$; the output of the processor is the signal $g[n]$. While $f(t)$ is defined for all time, $g[n]$ is defined only for n an integer; for example, $g[1.2]$ simply does not exist.

A discrete-time signal $x[n]$ can be a *continuous-amplitude signal*, for which the amplitude can assume any value $-\infty < x[n] < \infty$. A second class of discrete-time signals is a *discrete-amplitude signal*, for which $x[n]$ can assume only certain defined amplitudes. A discrete-amplitude discrete-time signal is also called a *digital signal*.

An example of a discrete-amplitude discrete-time signal is the output of an analog-to-digital converter. (See Figure 1.19.) For example, if the binary signal out of an analog-to-digital converter is represented by eight bits, the output-signal amplitude can assume only $2^8 = 256$ different values. A second example of a discrete-amplitude discrete-time signal is any signal internal to a digital computer.

In summary, a discrete-time signal is an ordered sequence of numbers. The sequence is usually expressed as $\{f[n]\}$, where this notation denotes the sequence $\ldots, f[-2], f[-1], f[0], f[1], f[2], \ldots$. We usually consider $f[n]$, for n a noninteger, to be undefined.

Some of the reasons that engineers are interested in discrete-time signals are as follows:

1. Sampling is required if we are to use digital signal processing, which is much more versatile than analog signal processing.

2. Many communication systems are based on the transmission of discrete-time signals, for a variety of reasons.

3. Sampling a signal allows us to store the signal in discrete memory.

4. The outputs of certain sensors that measure physical variables are discrete-time signals.

5. Complex strategies for automatically controlling physical systems require digital-computer implementation. The controlling signals from the computer are discrete time.

6. Many consumer products such as CDs, DVDs, digital cameras, and MP3 players use digital signals.

9.1 DISCRETE-TIME SIGNALS AND SYSTEMS

In this section, we introduce by example discrete-time signals. We use numerical integration as the example. Suppose that we wish to integrate a voltage signal, $x(t)$, using a digital computer. Integration by a digital computer requires that we use a numerical algorithm. In general, numerical algorithms are based on approximating a signal with an unknown integral with a signal that has a known integral. Hence, all integration algorithms are approximate in nature.

We use Euler's rule (discussed in Section 1.3), which is depicted in Figure 9.2. Euler's rule approximates the area under the curve $x(t)$ by the sum of the rectangular areas shown. In this figure, the step size H (the width of each rectangle) is called

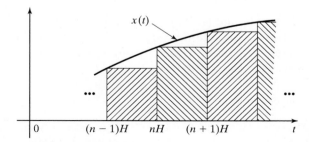

Figure 9.2   Euler integration.

the *numerical-integration increment*. The implementation of this algorithm requires that $x(t)$ be sampled every H seconds, resulting in the number sequence $x(nH)$, with n an integer.

Let $y(t)$ be the following integral of $x(t)$:

$$y(t) = \int_0^t x(\tau)d\tau. \tag{9.2}$$

The integral of $x(t)$ from $t = 0$ to $t = nH$ in Figure 9.2 can be expressed as the integral for $t = 0$ to $t = (n - 1)H$ plus the integral from $(n - 1)H$ to nH. Thus, in (9.2),

$$
\begin{aligned}
y(t)|_{t=nH} = y(nH) &= \int_0^{nH} x(\tau)d\tau \\
&= \int_0^{(n-1)H} x(\tau)d\tau + \int_{(n-1)H}^{nH} x(\tau)d\tau \\
&\approx y[(n-1)H] + Hx[(n-1)H].
\end{aligned} \tag{9.3}
$$

Ignoring the approximations involved, we expressed this equation as

$$y(nH) = y[(n-1)H] + Hx[(n-1)H]. \tag{9.4}$$

However, $y(nH)$ is only an approximation to the integral of $x(t)$ at $t = nH$.

In the notation for discrete-time signals discussed earlier, (9.4) is expressed as

$$y[n] = y[n-1] + Hx[n-1]. \tag{9.5}$$

An equation of this type is called a *difference equation*. A general Nth-order *linear difference equation* with *constant coefficients* is of the form

$$
\begin{aligned}
y[n] = b_1 y[n-1] + b_2 y[n-2] + \cdots + b_N y[n-N] \\
+ a_0 x[n] + a_1 x[n-1] + \cdots + a_N x[n-N],
\end{aligned} \tag{9.6}
$$

where the coefficients a_i and b_i, $i = 1, 2, \ldots, N$, are constants. Replacing n with $(n + N)$, we can also express this difference equation as

$$
\begin{aligned}
y[n+N] = b_1 y[n+N-1] + b_2 y[n+N-2] + \cdots + b_N y[n] \\
+ a_0 x[n+N] + a_1 x[n+N-1] + \cdots + a_N x[n].
\end{aligned} \tag{9.7}
$$

The formats of both (9.6) and (9.7) are used in specifying difference equations. In this chapter, we consider discrete-time signals of the type of $x[n]$ and $y[n]$ in (9.6) and (9.7) and discrete systems described by difference equations. However, we do not limit the difference equations to being linear.

EXAMPLE 9.1 **Difference-equation solution**

As an example of the solution of a difference equation, consider the numerical integration of a unit step function $u(t)$ using Euler's rule in (9.5). The continuous unit step function is defined as

$$u(t) = \begin{cases} 1, & t > 0 \\ 0, & t < 0. \end{cases}$$

We will assume that the initial condition $y(0)$ is zero, that is, $y(0) = 0$. Sampling a unit step function yields $x(nH) = 1$ for $n \geq 0$, and thus $x[n] = 1$ for $n \geq 0$. (We have assumed that $x[0] = 1$.) From (9.5), the difference equation to be solved is

$$y[n] = y[n-1] + Hx[n-1].$$

This equation is solved iteratively, beginning with $n = 1$:

$$y[1] = y[0] + Hx[0] = 0 + H = H;$$
$$y[2] = y[1] + Hx[1] = H + H = 2H;$$
$$y[3] = y[2] + Hx[2] = 2H + H = 3H;$$
$$\vdots$$
$$y[n] = y[n-1] + Hx[n-1] = (n-1)H + H = nH.$$

Thus, $y(nH) = nH$. The exact integral of the unit step function gives the result

$$y(t) = \int_0^t u(\tau)d\tau = \int_0^t d\tau = \tau \Big|_0^t = t, t > 0,$$

and $y(t)$ evaluated at $t = nH$ is equal to nH. Hence, Euler's rule gives the exact value for the integral of the unit step function. In general, the Euler rule is not exact. The reader may wish to consider why the results are exact, by constructing a figure of the form of Figure 9.2 for the unit step function. ∎

From Figure 9.2, we see that the integral of a general function $x(t)$ by Euler's rule yields the summation of $x[k]$ multiplied by the constant H:

$$y[n] = Hx[0] + Hx[1] + Hx[2] + \cdots + Hx[n-1] = H\sum_{k=0}^{n-1} x[k]. \quad (9.8)$$

Hence, we see that in this case, there is a relation between integration in continuous time and summation in discrete time. This relation carries over to many other situations.

Unit Step and Unit Impulse Functions

We begin the study of discrete-time signals by defining two signals. First, the *discrete-time unit step function* $u[n]$ is defined by

$$u[n] = \begin{cases} 1, & n \geq 0 \\ 0, & n < 0. \end{cases} \quad (9.9)$$

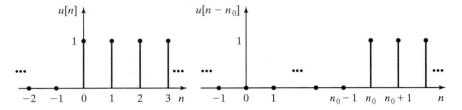

Figure 9.3 Discrete-time unit step functions.

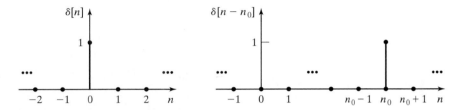

Figure 9.4 Discrete-time unit impulse functions.

Recall that this definition applies for n an integer only. The unit step function is illustrated in Figure 9.3. Dots at the end of a vertical line are used to denote the values of a discrete signal, as shown in Figure 9.3. The time-shifted unit step function is denoted as $u[n - n_0]$, where n_0 is an integer and

$$u[n - n_0] = \begin{cases} 1, & n \geq n_0 \\ 0, & n < n_0. \end{cases} \tag{9.10}$$

This function is also plotted in Figure 9.3 for n_0 positive.

The second signal to be defined is the *discrete-time unit impulse function* $\delta[n]$, also called the *unit sample function*. By definition, the discrete-time unit impulse function is given by

$$\delta[n] = \begin{cases} 1, & n = 0 \\ 0, & n \neq 0. \end{cases} \tag{9.11}$$

This function is plotted in Figure 9.4. Note that the discrete-time impulse function is well behaved mathematically and presents none of the problems of the continuous-time impulse function. In fact, the discrete-time unit impulse function can be expressed as the difference of two step functions:

$$\delta[n] = u[n] - u[n - 1]. \tag{9.12}$$

This result is seen by plotting $u[n]$ and $-u[n - 1]$. The shifted unit impulse function is defined by

$$\delta[n - n_0] = \begin{cases} 1, & n = n_0 \\ 0, & n \neq n_0 \end{cases} \tag{9.13}$$

and is also plotted in Figure 9.4, for $n_0 > 0$.

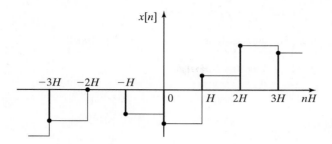

Figure 9.5 Summation yielding approximate integration.

Equivalent Operations

We now compare certain operations on discrete-time signals with equivalent operations on continuous-time signals. First, *integration* in continuous time is considered to be equivalent to *summation* in discrete time. This is illustrated in Figure 9.5, where the discrete-time signal is assumed to be generated by sampling a continuous-time signal. By Euler's rule, we see that

$$\text{i} \int_{-\infty}^{t} x(\tau)d\tau \Leftrightarrow H \sum_{k=-\infty}^{n} x[k], \qquad (9.14)$$

where $t = nH$.

In a like manner, we can approximate the slope of a continuous-time signal $x(t)$ with the samples $x[n]$, by the relation

$$\left. \frac{dx(t)}{dt} \right|_{t=kH} \approx \frac{x[k] - x[k-1]}{H}. \qquad (9.15)$$

This relation is illustrated in Figure 9.6. The numerator in the right side of (9.15) is called *the first difference*, which is considered to be the equivalent operation to the first derivative of a continuous-time signal:

$$\frac{dx(t)}{dt} \Leftrightarrow x[n] - x[n-1]. \qquad (9.16)$$

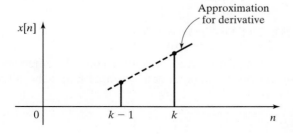

Figure 9.6 Approximate differentiation.

TABLE 9.1 Equivalent Operations

Continuous Time	Discrete Time
1. $\displaystyle\int_{-\infty}^{t} x(\tau)d\tau$	$\displaystyle\sum_{k=-\infty}^{n} x[k]$
2. $\dfrac{dx(t)}{dt}$	$x[n] - x[n-1]$
3. $x(t)\delta(t) = x(0)\delta(t)$	$x[n]\delta[n] = x[0]\delta[n]$
4. $\delta(t) = \dfrac{du(t)}{dt}$	$\delta[n] = u[n] - u[n-1]$
5. $u(t) = \displaystyle\int_{-\infty}^{t} \delta(\tau)d\tau$	$u[n] = \displaystyle\sum_{k=-\infty}^{n} \delta[k]$

Equivalent operations with impulse functions are given by

$$x(t)\delta(t) = x(0)\delta(t) \Leftrightarrow x[n]\delta[n] = x[0]\delta[n] \qquad (9.17)$$

[see (9.11)], where $\delta(t)$ is the continuous-time unit impulse function. Equivalent operations with impulse and step functions are given by

$$\delta(t) = \frac{du(t)}{dt} \Leftrightarrow \delta[n] = u[n] - u[n-1] \qquad (9.18)$$

[see (9.12)], and

$$u(t) = \int_{-\infty}^{t} \delta(\tau)d\tau \Leftrightarrow u[n] = \sum_{k=-\infty}^{n} \delta[k]. \qquad (9.19)$$

As an example of this summation, let $n = 3$:

$$u[3] = \sum_{k=-\infty}^{3} \delta[k] = \cdots + \delta[-1] + \delta[0] + \delta[1] + \delta[2] + \delta[3]$$
$$= \delta[0] = 1.$$

Recall that $\delta[0] = 1$ and $\delta[n] = 0$ for $n \neq 0$. These equivalent operations are summarized in Table 9.1.

In this section, we have introduced discrete-time signals and systems. In addition, a difference equation, which models an integrator, was solved. A general method for solving linear difference equations with constant coefficients is given in Chapter 10.

9.2   TRANSFORMATIONS OF DISCRETE-TIME SIGNALS

In this section, we consider six transformations on a discrete-time signal $x[n]$. Three transformations are on the independent variable n and the other three on the dependent variable $x[\,\cdot\,]$.

In naming the transformations for discrete signals, we continue to use the term *discrete time*, or simply *time*, for the discrete-increment variable *n*, because generally we are considering sampled signals. For sampled signals, we use *n* to denote the time $t = nT$, with T the sample period.

Time Transformations

First, we consider the three time transformations. In these transformations, for clarity we let *m* denote discrete time in the original signal and *n* denote discrete time in the transformed signal.

Time Reversal

To time-reverse a signal $x[m]$, we replace the independent variable *m* with $-n$. Hence, we are considering

$$y[n] = x[m]\bigg|_{m=-n} = x[-n], \qquad (9.20)$$

where $y[n]$ denotes the transformed signal. This operation has the effect of creating the mirror image of $x[m]$ about the vertical axis.

We will see in Chapter 10 that one application of time reversal is in calculating the responses for certain types of systems.

An example of time reversal is given in Figure 9.7, where the time variable *n* has been changed to *m* in the original signal. In this figure, from (9.20) $n = -m$, and the *n*-axis is shown directly below the *m*-axis. Hence, we plot the transformed signal by plotting $y[n]$ versus *n*. This plot is also given in Figure 9.7.

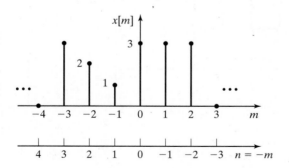

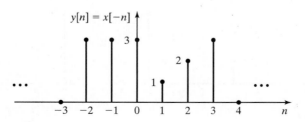

Figure 9.7 Signals illustrating time reversal.

Time Scaling

Given a signal $x[n]$, a time-scaled version of this signal is

$$y[n] = x[m]\Big|_{m=an} = x[an], \tag{9.21}$$

where we consider only the cases that $a = k$ or $a = 1/k$ for integer values of k. Note again that the time variable for the original signal has been changed to m, for clarity.

Figure 9.8(a) shows a signal $x[m]$, and we will plot the time-scaled signals $y_1[n] = x[2n]$ and $y_2[n] = x[\frac{n}{3}]$. For the first transformation,

$$m = 2n \Rightarrow n = \frac{m}{2}.$$

The $\frac{m}{2}$ axis is shown directly below the m-axis in Figure 9.8(a). Note that $m/2$ is not an integer for m odd; thus, the values of $x[m]$ for m odd do not appear in $y[n]$. In the time-scaled signal $x[an]$, signal information is lost for $a = k \geq 2$. This is illustrated in Figure 9.8(a) and (b), where the values of $x[m]$ for m odd do not appear in $x[2n]$.

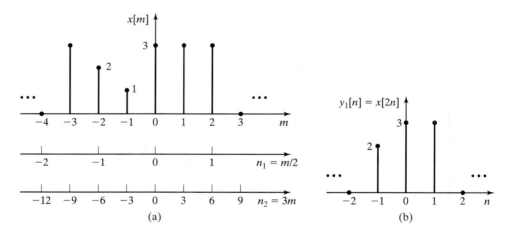

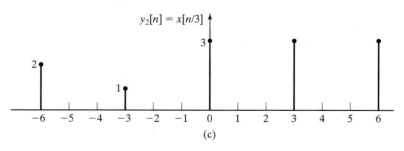

Figure 9.8 Signals illustrating time scaling.

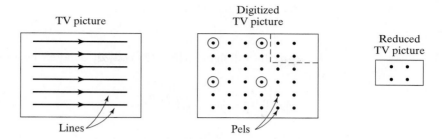

Figure 9.9 Television picture within a picture.

One practical application of this type of time scaling is the television *picture in a picture*, described in Section 1.3. We now review that description.

Consider Figure 9.9, where a much simplified TV picture is depicted as having six lines. Suppose that the picture is to be reduced in size by a factor of 3 and superimposed on another picture.

First, the lines of the picture, which are voltage signals, are sampled. In Figure 9.9, each line produces six samples, which are called *picture elements* (or pels). Both the number of lines and the number of samples per line must be reduced by three. Suppose that the samples retained for the reduced picture are the four circled in Figure 9.9.

Now let the digitized full picture represent a different picture in Figure 9.9; the four pels of the reduced picture then replace the four pels in the upper right-hand corner of the full picture. The inserted picture is outlined by the dashed lines. The generation of a picture in a picture is more complex than as described, but we can see the necessity to reduce the number of samples in a given line; the reduction shown in Figure 9.9 does illustrate a practical case of time scaling.

Next we plot the signal $y_2[n] = x[n/3]$ for the signal of Figure 9.8(a). Hence,

$$\frac{n}{3} = m \Rightarrow n = 3m.$$

The $3m$ axis is also shown below the m-axis in Figure 9.8(a). The signal $y_2[n]$ is plotted in Figure 9.8(c). It is seen then that for $a = 1/k$ in $x[an]$, the values of $x[n/k]$ are not defined at all discrete increments. For $x[n/3]$ in Figure 9.8(c), the values of $y_2[n]$ are undefined at $n = \pm1, \pm2, \pm4, \pm5$, and so on. If the signal $x[n/3]$ is to be used, all values must be defined; the missing values are usually assigned according to some logical rule, such as an *interpolation* scheme. However, if the signal $x[n/3]$ is used in real time, *extrapolation* must be used; at $n = 0$, for example, $y_2[3]$ is not known. The values $y_2[1]$ and $y_2[2]$ cannot be calculated by interpolation without knowing $y_2[3]$.

Time Shifting

Given a signal $x[m]$, a time-shifted version of this signal is $x[n - n_0]$, where n_0 is an integer constant. Hence,

$$y[n] = x[m]\Big|_{m=n-n_0} = x[n - n_0]. \tag{9.22}$$

As will be seen in Chapter 10, one application of time shifting is in the calculation of the responses of certain types of systems.

As an example of time shifting, consider

$$x[m] = a^m \cos (\pi m/4)u[m].$$

The time-shifted signal $x[n - 3]$ is $x[m]$ delayed by three sample periods and is given by

$$x[n - 3] = a^{n-3}\cos\left(\frac{\pi(n - 3)}{4}\right)u[n - 3] = a^{n-3}\cos\left(\frac{\pi n}{4} - \frac{3\pi}{4}\right)u[n - 3].$$

Shown in Figure 9.10 is another signal $x[n]$, along with $y_1[n] = x[n - 2]$ and $y_2[n] = x[n + 1]$. The n_1-axis and the n_2-axis are obtained from

$$m = n - 2 \Rightarrow n_1 = m + 2$$
$$m = n + 1 \Rightarrow n_2 = m - 1$$

Both the n_1-axis and the n_2-axis are plotted in Figure 9.10(a), and the transformed signals are given in Figure 9.10(b) and (c).

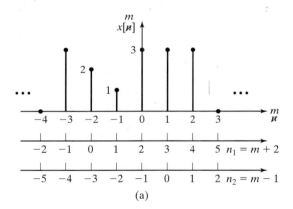

(a)

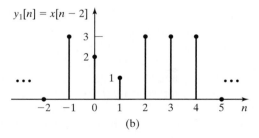

(b)

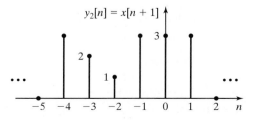

Figure 9.10 Time-shifted signals.

For the shifted signal $x[n - n_0]$, $x[0]$ occurs at $n = n_0$. Hence, if n_0 is positive, the signal is shifted to the right (delayed in time) on a plot of the signal; if n_0 is negative, the signal is shifted to the left (advanced in time).

Thus far, three transformations in time have been defined and are of the general form

$$y[n] = x[m]\Big|_{m=an+b} = x[an + b]. \tag{9.23}$$

In this equation, a is an integer or the reciprocal of an integer, and b is an integer. For example,

$$y[n] = x[an + b] = x[-3n + 2].$$

The value $a = -3$ yields time reversal (the minus sign) and time scaling ($|a| = 3$). The value $b = 2$ yields a time shift. An example of a time transformation will now be given.

EXAMPLE 9.2   **Time transformation of a discrete signal**

Consider the discrete signal $x[n]$ in Figure 9.11(a). We wish to plot the time-transformed signal for the transformation $m = 2 - n$, which has time reversal and time shifting. Then

$$y[n] = x[m]\Big|_{m=2-n} = x[2 - n].$$

The transformation is expressed as

$$m = 2 - n \Rightarrow n = 2 - m.$$

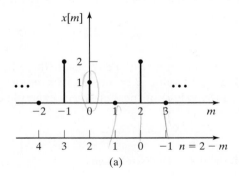

(a)

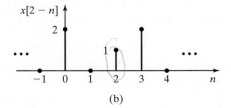

(b)

Figure 9.11 Signals for Example 9.2.

TABLE 9.2 Values for Example 9.2

m	$x[m]$	$n = 2 - m$	$y[n]$
−1	2	3	2
0	1	2	1
1	0	1	0
2	2	0	2

Table 9.2 gives values of n for significant values of m. Because only four points are involved, $y[n]$ is also included in this table. In Figure 9.11(a), the n-axis is shown directly below the m-axis. The n-axis values are given in the third column of Table 9.2.

Next we plot $y[n] = x[2 - n]$ versus n to show the transformed signal. This plot can be made directly from Figure 9.11(a) or from Table 9.2, and is shown in Figure 9.11(b). ∎

Example 9.2 illustrates two procedures for plotted time-transformed signals. We can draw the n-axis below the plot of the signal, as in Figure 9.11(a), or we can construct a table as in Table 9.2.

Amplitude Transformations

Next we consider the three transformations on the amplitude axis. Amplitude transformations follow the same rules as time transformations.

The three transformations in amplitude are of the general form

$$y[n] = Ax[n] + B, \tag{9.24}$$

where A and B are constants that are not necessarily integers; for example,

$$y[n] = -3.1x[n] - 5.75.$$

The value $A = -3.1$ yields *amplitude reversal* (the minus sign) and *amplitude scaling* ($|A| = 3.1$), and the value $B = -5.75$ gives *amplitude shifting* and changes the dc level (the average value) of the signal. An example of amplitude scaling is now given.

EXAMPLE 9.3 **Amplitude transformations of a discrete signal**

Consider again the signal of Example 9.2 and Figure 9.11(a). This signal is repeated in Figure 9.12(a). We will plot the transformed signal

$$y[n] = 3 - 2x[n].$$

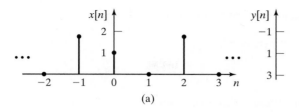

(a)

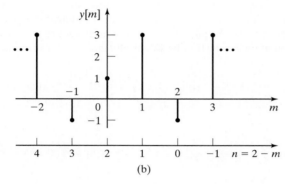

(b)

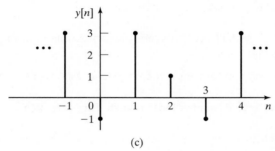

(c)

Figure 9.12   Signals for Examples 9.3 and 9.4.

TABLE 9.3   Values for Example 9.3

$x[n]$	$3-2x[n]$
0	3
1	1
2	−1

Hence, in Figure 9.12(a), the amplitude axis (the vertical axis) is replaced with the $(3 - 2x[n])$-axis, as shown in Figure 9.12(a). Table 9.3 gives the equivalent values of $y[n]$ for significant values of $x[n]$, and these values are shown on the vertical axis added to Figure 9.12(a). This information yields the desired plot of the transformed signal $y[m]$ in Figure 9.12(b).   ■

EXAMPLE 9.4   **Time and amplitude transformations of a discrete signal**

Next we consider the signal

$$y[n] = 3 - 2x[2 - n],$$

which has the time transformation of Example 9.2 and Figure 9.11 and the amplitude transformation of Example 9.3 and Figure 9.12. To plot this transformed signal, we can first transform the amplitude axis as shown in Figure 9.12(b) and then redraw the n-axis of Figure 9.11(a) in Figure 9.12(b) as shown. The signal is then plotted on the n-axis, as shown in Figure 9.12(c). This result can be verified by substituting values of n into the transformation. For example,

$$n = 0: \quad y[0] = 3 - 2x[2 - 0] = 3 - 2(2) = -1$$

and

$$n = 2: \quad y[2] = 3 - 2x[2 - 2] = 3 - 2(1) = 1,$$

which check two of the values in Figure 9.12(c).   ■

The following MATLAB program performs time and amplitude transformation for Example 9.4:

```
% This MATLAB program performs the time and amplitude transformation for
% Example 9.4.
% Establish vectors of sufficient length for both x(n) and n.
n=[-10:10];
x=zeros(1,length(n));
% Enter nonzero values for x(n).
x(10)=2; x(11)=1; x(13)=2;
% Plot x(n).
figure(1),stem(n(1,5:17),x(1,5:17),'fill'),grid,xlabel('n'),ylabel('x(n)')
for k = 3:21
    xt(k)=3-2*x(13-n(k));
end
figure(2),stem(n(1,5:17),xt(1,5:17),'fill'),grid,xlabel('n'),ylabel('xt(n)')
```

In summary, the six transformations defined are reversal, scaling, and shifting with respect to time, and reversal, scaling, and shifting with respect to amplitude. These operations are listed in Table 9.4.

TABLE 9.4 Transformations of Signals

Name	$y[n]$
Time reversal	$x[-n]$
Time scaling	$x[an]$
Time shifting	$x[n - n_0]$
Amplitude reversal	$-x[n]$
Amplitude scaling	$\lvert A \rvert x[n]$
Amplitude shifting	$x[n] + B$

9.3 CHARACTERISTICS OF DISCRETE-TIME SIGNALS

In Section 2.2, some useful characteristics of continuous-time signals were defined. We now consider the same characteristics for discrete-time signals.

Even and Odd Signals

In this section, we define even and odd signals (functions). A discrete-time signal $x_e[n]$ is *even* if

$$x_e[n] = x_e[-n], \tag{9.25}$$

and the signal $x_o[n]$ is *odd* if

$$x_o[n] = -x_o[-n]. \tag{9.26}$$

Any discrete-time signal $x[n]$ can be expressed as the sum of an even signal and an odd signal:

$$x[n] = x_e[n] + x_o[n]. \tag{9.27}$$

To show this, we replace n with $-n$ to yield

$$x[-n] = x_e[-n] + x_o[-n] = x_e[n] - x_o[n]. \tag{9.28}$$

The sum of (9.27) and (9.28) yields the even part of $x[n]$:

$$x_e[n] = \tfrac{1}{2}(x[n] + x[-n]). \tag{9.29}$$

The subtraction of (9.28) from (9.27) yields the odd part of $x[n]$:

$$x_o[n] = \tfrac{1}{2}(x[n] - x[-n]). \tag{9.30}$$

These two equations are used to find the even part and the odd part of a discrete-time signal. Note that the sum of (9.29) and (9.30) yields (9.27).

The *average* value, or *mean* value, of a discrete-time signal is given by

$$A_x = \lim_{N \to \infty} \frac{1}{2N+1} \sum_{k=-N}^{N} x[k]. \tag{9.31}$$

As is the case of continuous-time signals, the average value of a discrete-time signal is contained in its even part, and the average value of an odd signal is always zero. (See Problem 9.11.)

Even and odd signals have the following properties:

1. The sum of two even signals is even.
2. The sum of two odd signals is odd.
3. The sum of an even signal and an odd signal is neither even nor odd.
4. The product of two even signals is even.
5. The product of two odd signals is even.
6. The product of an even signal and an odd signal is odd.

These properties are easily proved. (See Problem 9.12.) An example of even and odd signals is now given.

EXAMPLE 9.5   **Even and odd functions**

The even and the odd parts of the discrete-time signal $x[n]$ of Figure 9.13(a) will be found. Since the signal has only six nonzero values, a strictly mathematical approach is used. Table 9.5 gives the solution, using (9.29) and (9.30). All values not given in this table are zero. The even and odd parts of $x[n]$ are plotted in Figure 9.13.

Three characteristics are evident from Table 9.5:

1. The sum of all values of $x_o[n]$ is zero, since, for any value of n, from (9.26), it follows that

$$x_o[n] + x_o[-n] = x_o[n] - x_o[n] = 0.$$

Note also that $x_o[0]$ is always zero.
2. Summed over all nonzero values,

$$\Sigma x[n] = \Sigma x_e[n] = 15.$$

(See Problem 9.11.)
3. The sum of $x_e[n]$ and $x_o[n]$ is equal to $x[n]$ for each value of n, from (9.27).

These three characteristics allow us to check the results of this example.

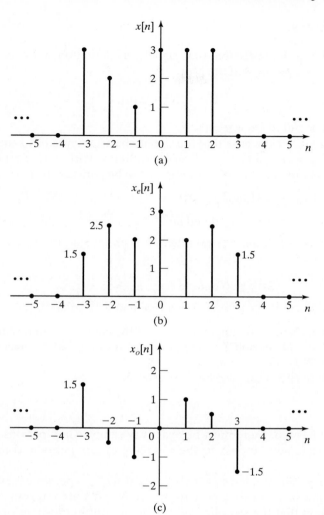

Figure 9.13 Signals for Example 9.5.

TABLE 9.5 Values for Example 9.5

n	$x[n]$	$x[-n]$	$x_e[n]$	$x_o[n]$
-3	3	0	1.5	1.5
-2	2	3	2.5	-0.5
-1	1	3	2	-1
0	3	3	3	0
1	3	1	2	1
2	3	2	2.5	0.5
3	0	3	1.5	-1.5

Signals Periodic in n

We now consider periodic discrete-time signals. By definition, a discrete-time signal $x[n]$ is *periodic* with period N if

$$x[n + N] = x[n]. \tag{9.32}$$

Of course, both n and N are integers.

We first consider the case that the signal $x[n]$ is obtained by sampling a sinusoidal signal $x(t) = \cos(\omega_0 t)$ every T seconds; that is, $x[n] = x(nT)$. [T is the sample period and not the period of $x(t)$.] For $x[n]$ to be periodic, from (9.32),

$$x[n] = \cos(n\omega_0 T) = x[n + N] = \cos[(n + N)\omega_0 T]$$
$$= \cos(n\omega_0 T + N\omega_0 T).$$

Hence, $N\omega_0 T$ must be equal to $2\pi k$, where k is an integer, because $\cos(\theta + 2\pi k) = \cos\theta$. Therefore,

$$2\pi k = N\omega_0 T = N\frac{2\pi}{T_0}T \Rightarrow \frac{k}{N} = \frac{T}{T_0}, \tag{9.33}$$

where $T_0 = 2\pi/\omega_0$ is the fundamental period of the continuous-time sinusoid. Thus, the ratio of the sample period T to the period of the sinusoid T_0 must be a ratio of integers; that is, T/T_0 must be *rational*.

The result in (9.33) can also be expressed as

$$NT = kT_0. \tag{9.34}$$

This relation states that there must be exactly N samples in k periods of the signal $\cos(\omega_0 t)$. This statement applies to the sampling of *any* periodic continuous-time signal $x(t)$.

In summary, the sampled signal $x[n] = \cos(n\omega_0 T)$ is periodic if exactly N samples are taken in exactly every k periods, where N and k are integers. Note the surprising conclusion that the sampling of a periodic continuous-time signal does not necessarily result in a periodic discrete-time signal. We now give an example.

EXAMPLE 9.6   **Sampling of a sinusoid**

In this example, we will consider sampling the periodic signal $x(t) = \sin\pi t$, which has the period $T_0 = 2\pi/\omega_0 = 2$ s. First we sample with the period $T = 0.5$ s. There are exactly four samples for each period of sinusoid; in (9.34), $kT_0 = (1)(2) = 2$ s and $NT = 4(0.5) = 2$ s. The signals are illustrated in Figure 9.14(a).

Next we sample with the period $T = \frac{3}{8}T_0 = 0.75$ s. In this case, we have exactly eight samples in every three periods ($8T = 3T_0$), or in every 6 s. These signals are illustrated in Figure 9.14(b).

As a final example, we sample a triangular wave that is periodic with a period of $T_0 = 2$ s. This signal is sampled with sample period $T = \frac{5}{4}T_0 = 2.5$ s, as shown in Figure 9.14(c). In this case, there is less than one sample per period of the triangular wave; however,

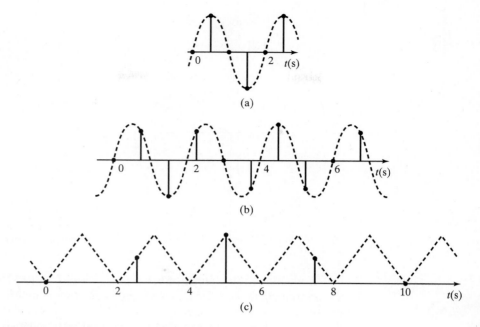

Figure 9.14   Periodic signals for Example 9.6.

the discrete-time signal is periodic, with four samples for every five periods of the triangular wave ($4T = 5T_0$). This last example illustrates that these results apply for any sampled periodic signal, and not just for sinusoidal signals. ∎

In the previous discussion, the sampled signals are real. However, the results apply directly for the sampled complex exponential signal

$$e^{j\omega_0 t}\Big|_{t=nT} = e^{j\omega_0 nT} = x[n].   \quad (9.35)$$

Then, for this signal to be periodic,

$$e^{jn\omega_0 T} = x[n] = x[n + N] = e^{j(n+N)\omega_0 T} = e^{jn\omega_0 T}e^{jN\omega_0 T}.   \quad (9.36)$$

Hence, for periodicity, $e^{jN\omega_0 T} = e^{j2\pi k}$ since $e^{j2\pi k}$ is equal to unity. The requirement that $N\omega_0 T = 2\pi k$ is the same as in (9.33).

We next consider the discrete-time complex exponential signal that is not necessarily obtained by sampling a continuous-time signal. We express the signal as

$$x[n] = e^{j\Omega_0 n} = 1\underline{/\Omega_0 n}.   \quad (9.37)$$

This signal can be represented in the complex plane as a vector of unity magnitude at the angles $\Omega_0 n$, as shown in Figure 9.15. The projection of this vector onto the real axis is $\cos(\Omega_0 n)$ and onto the imaginary axis is $\sin(\Omega_0 n)$, since

$$e^{j\Omega_0 n} = \cos(\Omega_0 n) + j\sin(\Omega_0 n).$$

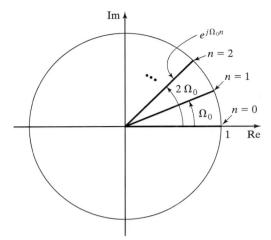

Figure 9.15 Representation of the complex exponential.

If we make the change of variable $\Omega_0 = \omega_0 T$ in (9.37), we have the complex exponential signal of (9.35), and all the preceding conclusions apply directly. The variable Ω has the units of radians; we refer to Ω as *normalized discrete frequency*, or simply *frequency*. For sampled signals, real frequency ω and discrete frequency Ω are related by $\omega T = \Omega$.

We now consider (9.37) in a different manner. The complex exponential signal of (9.37) is periodic, provided that

$$x[n] = e^{j\Omega_0 n} = x[n + N] = e^{j(\Omega_0 n + \Omega_0 N)} = e^{j(\Omega_0 n + 2\pi k)}, \tag{9.38}$$

where k is an integer. Thus, periodicity requires that

$$\Omega_0 N = 2\pi k \Rightarrow \Omega_0 = \frac{k}{N} 2\pi, \tag{9.39}$$

so that Ω_0 must be expressible as 2π multiplied by a rational number. For example, $x[n] = \cos(2n)$ is *not* periodic, since $\Omega_0 = 2$. The signal $x[n] = \cos(0.1\pi n)$ is periodic, since $\Omega_0 = 0.1\pi$. For this case, $k = 1$ and $N = 20$ satisfies (9.39).

As a final point, from (9.39), the complex exponential signal $e^{j\Omega_0 n}$ is periodic with N samples per period, provided that the integer N satisfies the equation

$$N = \frac{2\pi k}{\Omega_0}. \tag{9.40}$$

In this equation, k is the smallest positive integer that satisfies this equation, such that N is an integer greater than unity. For example, for the signal $x[n] = \cos(0.1\pi n)$, the number of samples per period is

$$N = \frac{2\pi k}{0.1\pi} = 20k = 20, \quad k = 1. \tag{9.41}$$

Plotting this signal as in Figure 9.15 yields 20 vectors ($N = 20$) per one revolution ($k = 1$).

For the signal $x[n] = \cos(5\pi n)$, the number of samples per period is

$$N = \frac{2\pi k}{5\pi} = 0.4k = 2, \quad k = 5.$$

For $k < 5$, N is not an integer. Plotting this signal as in Figure 9.15 yields two vectors ($N = 2$) per five revolutions ($k = 5$). Beginning at $n = 0$, the first two vectors are $1\underline{/0°}$ and $1\underline{/5\pi}$; the next vector is $1\underline{/10\pi} = 1\underline{/0°}$, which is the first vector repeated.

For a final example, consider the signal $x[n] = \cos(2\pi n)$. Then

$$N = \frac{2\pi k}{2\pi} = k, \tag{9.42}$$

and this equation is satisfied for $N = k = 1$. This signal can be expressed as

$$x[n] = \cos(2\pi n) = 1.$$

Hence, the discrete-time signal is constant.

Signals Periodic in Ω

The conditions for the complex exponential signal $e^{j\Omega_0 n}$ to be periodic in n were just developed. However, this signal is *always* periodic in the discrete-frequency variable Ω. Consider this signal with Ω_0 replaced with ($\Omega_0 + 2\pi$)—that is,

$$e^{j(\Omega_0+2\pi)n} = e^{j\Omega_0 n}e^{j2\pi n} = e^{j\Omega_0 n} \tag{9.43}$$

(since $e^{j2\pi n} = 1$). Hence, the signal $e^{j\Omega_0 n}$ is periodic in Ω with period 2π, independent of the value of Ω_0. Of course, the sinusoidal signal $\cos(\Omega_0 n + \theta)$ is also periodic in Ω with period 2π. This property has a great impact on the sampling of signals, as shown in Chapters 5 and 6.

Note that periodic continuous-time signals are not periodic in frequency. For example, for the complex exponential signal,

$$e^{j(\omega+a)t} = e^{j\omega t}e^{jat} \neq e^{j\omega t}, \quad a \neq 0.$$

To summarize, we have demonstrated two properties of discrete-time sinusoids that continuous-time sinusoids do not have. We now illustrate these properties. The continuous-time sinusoid $\cos(\omega t + 0)$ is always periodic, independent of ω. Also, $\cos(\omega_1 t)$ is equal to $\cos(\omega_2 t)$ only for $\omega_1 = \omega_2$. However, the discrete-time sinusoid $\cos(\Omega_0 n)$ has the following properties:

1. $\cos(\Omega_0 n)$ is periodic only if, from (9.39),

$$\frac{\Omega_0}{2\pi} = \frac{k}{N},$$

where k and N are integers.

2. $\cos(\Omega_0 n)$ is periodic in Ω with period 2π; that is, with k any integer,

$$\cos(\Omega_0 n) = \cos(\Omega_0 + 2\pi k)n.$$

Of course, the same properties apply to $\cos(\Omega_0 n + \theta)$ and $e^{j\Omega_0 n}$.

In this section, the properties of even and odd were defined with respect to discrete-time signals. These properties are useful in the applications of the discrete-time Fourier transform, as shown in Chapter 12. Next the properties of periodic discrete-time signal were investigated. The property that a discrete-time sinusoid is periodic in frequency has great implications with respect to the sampling of continuous-time signals, as shown in Chapters 5 and 6.

9.4   COMMON DISCRETE-TIME SIGNALS

In Section 2.3, we defined some common continuous-time signals that occur in the transient response of certain systems. In this section, equivalent discrete-time signals are introduced; these signals can appear in the transient response of certain discrete-time systems.

One such signal, the sinusoid, was mentioned in Section 9.3. For example, a digital computer can be programmed to output a discrete-time sinusoid to generate an audible tone of variable frequency. The discrete-time sinusoidal signal is transmitted from the computer to a digital-to-analog converter (D/A), which is an electronic circuit that converts binary numbers into a continuous-time voltage signal. (See Section 1.3.) This voltage is then applied through a power amplifier to a speaker. The operation is depicted in Figure 9.16. A timing chip in the computer is used to determine the sample period T and, hence, the frequency of the tone.

We now use an example of a system to introduce a common discrete-time signal. The block shown in Figure 9.17(a) represents a memory device that stores a number. Examples of this device are shift registers or memory locations in a digital computer. Every T seconds we shift out the number stored in the device. Then a different number is shifted into the device and stored. If we denote the number shifted into the device as $x[n]$, the number just shifted out must be $x[n - 1]$. A device used in this manner is called an *ideal time delay*. The term *ideal* indicates that the numbers are not altered in any way, but are only delayed.

Suppose that we connect the ideal time delay in the system shown in Figure 9.17(b). The number shifted out of the delay is multiplied by the constant a to form the next number to be stored, resulting in the system equation

$$x[n] = ax[n - 1]. \tag{9.44}$$

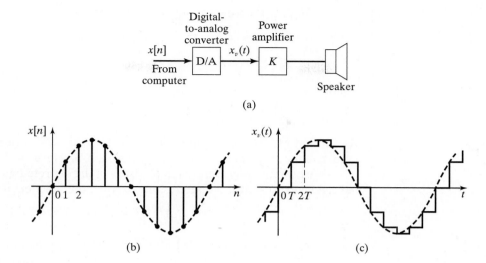

Figure 9.16   Computer generation of a tone.

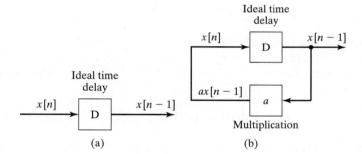

Figure 9.17   Discrete-time system.

Suppose that at the first instant (denoted as $n = 0$), the number unity is stored in the delay, that is, $x[0] = 1$. We now iteratively solve for $x[n], n > 0$, using (9.44) (recall that the ideal time delay outputs its number every T seconds):

$$x[1] = ax[0] = a;$$
$$x[2] = ax[1] = a^2;$$
$$x[3] = ax[2] = a^3;$$
$$\vdots;$$
$$x[n] = ax[n-1] = a^n.$$

Thus, this system generates the signal $x[n] = a^n$ for the initial condition $x[0] = 1$.

A MATLAB program that simulates this system for $a = 0.9$ is given by

```
a=0.9;
x(1)=1;
for m=2:5
    x(m)=a*x(m-1);
end
x
result:x=1 0.9 0.81 0.729 0.6561
```

Recall that MATLAB requires subscripts to be positive integers. Hence, this program evaluates the function $(0.9)^n$, for $n = 1, 2, 3, 4$ or $m = 2, 3, 4, 5$.

We now investigate the characteristics of the discrete signal $x[n] = a^n$. This signal can be expressed as a discrete-time exponential function, if we let $a = e^b$. Then

$$x[n] = a^n = (e^b)^n = e^{bn}. \tag{9.45}$$

For example, for the signal $x[n] = 0.9^n$, we solve the equation

$$0.9 = e^b \Rightarrow b = \ln 0.9 = -0.105,$$

and thus,

$$x[n] = 0.9^n = e^{-0.105n}.$$

We generally refer to the signal a^n as a *discrete-time exponential signal*.

Quite often discrete signals are generated by sampling continuous signals. Suppose that we sample an exponential signal $x(t) = e^{-\sigma t}$ every $t = T$ seconds, with $\sigma > 0$:

$$x[n] = e^{-\sigma nT} = (e^{-\sigma T})^n = (a)^n. \tag{9.46}$$

The continuous-time exponential signal has a time constant τ, where, from (2.25),

$$e^{-\sigma t} = e^{-t/\tau} \Rightarrow \tau = \frac{1}{\sigma}. \tag{9.47}$$

Hence, in (9.46),

$$x[n] = (e^{-T/\tau})^n = a^n. \tag{9.48}$$

The ratio of τ/T, the *number of samples per time constant*, is normally not an integer. From (9.48),

$$e^{-T/\tau} = a \Rightarrow \frac{\tau}{T} = \frac{-1}{\ln a}. \tag{9.49}$$

Thus, we can assign a time constant

$$\tau = \frac{-T}{\ln a}$$

to the discrete exponential signal a^n, provided that the discrete signal is based on a sampling process.

EXAMPLE 9.7   **Time constant of a discrete exponential signal**

For the discrete-time signal $x[n] = (0.8)^n$, from (9.49),

$$\frac{\tau}{T} = \frac{-1}{\ln 0.8} = 4.48 \Rightarrow \tau = 4.48T.$$

Hence, there are 4.48 samples per time constant. Assuming that an exponential decays to a negligible amplitude after four time constants (see Section 2.3), this signal can be ignored for

$$nT > 4\tau = 4(4.48T) \approx 18T$$

or for $n > 18$ samples.   ■

We now generalize the exponential signal

$$x[n] = Ca^n \qquad (9.50)$$

by considering the cases that both parameters, C and a, can be complex. Of course, complex signals cannot appear in nature. However, as is the case for differential equations, the solutions of many difference equations are simplified by assuming that complex signals can appear both as excitations and as solutions. Then, in translating the results back to the physical world, only the real parts or the imaginary parts of complex functions are used. We now consider three cases of the discrete complex exponential signal.

CASE 1

C and a Real

For the first case, consider the signal $x[n] = Ca^n$, with both C and a real. This signal is plotted in Figure 9.18 for both C and a positive. For $a > 1$, the signal increases exponentially with increasing n. For $0 < a < 1$, the signal decreases exponentially with increasing n. For $a = 1$, $x[n] = C(1)^n = C$ and the signal is constant.

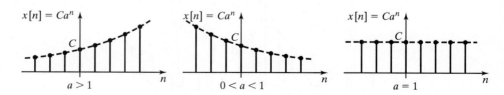

Figure 9.18   Discrete-time exponential signals.

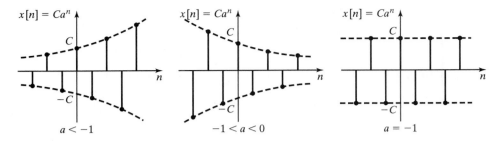

Figure 9.19 Discrete-time exponential signals.

Figure 9.19 gives the three cases for $a < 0$. Consider, for example, $x[n] = (-2)^n$. Beginning with $n = 0$, the number sequence for $x[n]$ is $1, -2, 4, -8, 16, -32, \ldots$. Hence, the number sequence is exponential with alternating sign. Letting $a = -\alpha$ with α positive, we obtain

$$x[n] = Ca^n = C(-\alpha)^n = C(-1)^n \alpha^n = (-1)^n x_a[n], \tag{9.51}$$

where $x_a[n]$ denotes the exponential signals plotted in Figure 9.18. Hence, the signals of Figure 9.19 have the same magnitudes as those of Figure 9.18, but with alternating signs.

CASE 2

C Complex, a Complex with Unity Magnitude

Next we consider the case that C and a are complex, with

$$C = Ae^{j\phi} = A\underline{/\phi}, \quad a = e^{j\Omega_0}, \tag{9.52}$$

where A, ϕ, and Ω_0 are real and constant. As defined in Section 9.3, Ω is the normalized discrete-frequency variable. The complex exponential signal $x[n]$ can be expressed as

$$x[n] = Ae^{j\phi}e^{j\Omega_0 n} = Ae^{j(\Omega_0 n + \phi)}$$

$$= A\cos(\Omega_0 n + \phi) + jA\sin(\Omega_0 n + \phi), \tag{9.53}$$

from Euler's relation in Appendix D. Recall from Section 9.3 that the sinusoids in (9.53) are periodic only for $\Omega_0 = 2\pi k/N$, with k and N integers. [See (9.39).] A plot of the real part of (9.53) is given in Figure 9.20, for $\phi = 0$.

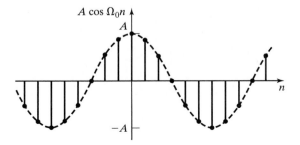

Figure 9.20 Undamped discrete-time sinusoidal signal.

CASE 3

Both *C* and *a* Complex

For this case, the complex exponential $x[n] = Ca^n$ has the parameters

$$C = Ae^{j\phi}, \quad a = e^{\Sigma_0 + j\Omega_0}. \tag{9.54}$$

The complex exponential signal $x[n]$ can then be expressed as

$$x[n] = Ae^{j\phi}e^{(\Sigma_0 + j\Omega_0)n} = Ae^{\Sigma_0 n}e^{j(\Omega_0 n + \phi)}$$

$$= Ae^{\Sigma_0 n}\cos(\Omega_0 n + \phi) + jAe^{\Sigma_0 n}\sin(\Omega_0 n + \phi). \tag{9.55}$$

Plots of the real part of (9.55) are given in Figure 9.21 for $\phi = 0$. Figure 9.21(a) shows the case that $\Sigma_0 > 0$ and Figure 9.21(b) shows the case that $\Sigma_0 < 0$. The *envelopes* of the sinusoids are given by $Ae^{\Sigma_0 n}$.

Consider again the exponential signal of (9.51). Since $\cos n\pi = (-1)^n$, (9.51) can be expressed as

$$x[n] = Ca^n = C\alpha^n(-1)^n = C\alpha^n\cos(n\pi) \tag{9.56}$$

with $\alpha = -a > 0$. Hence, the signal $x[n] = Ca^n$, with a real and negative, can be considered to be the result of sampling a sinusoidal signal exactly twice per cycle.

As a final point in this section, suppose that we sample the signal $x(t)$, with the result that

$$x[n] = e^{\sigma_0 t}\cos(\omega_0 t)\Big|_{t=nT} = e^{\sigma_0 nT}\cos(\omega_0 nT). \tag{9.57}$$

Comparing this result with the discrete signal in (9.55), we see that the continuous-time signal parameters are related to the discrete-time parameters by

$$\Sigma_0 = \sigma_0 T, \quad \Omega_0 = \omega_0 T. \tag{9.58}$$

We can consider the discrete complex exponential of (9.55) to be the result of sampling a continuous-time complex exponential signal of the form

$$x(t) = Ae^{(\sigma_0 + j\omega_0)t}$$

every T seconds, with the change of variables of (9.58).

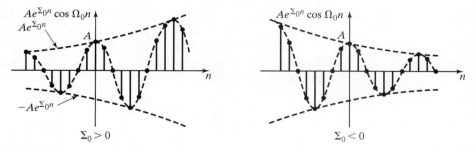

Figure 9.21   Real part of a complex exponential signal.

In this section, discrete-time signals were, for the most part, considered to be the result of sampling continuous-time signals. This is often the case. However, in many cases the discrete-time signals are generated directly in the discrete form. For example, the input to a digital filter is often obtained by sampling a continuous-time signal. However, the output of the filter is calculated from a difference equation and does not exist in a continuous form.

9.5 DISCRETE-TIME SYSTEMS

In this section, we define some general notation for discrete-time systems. Recall that we define a discrete-time system as one in which all signals are discrete time (number sequences). This section follows closely Section 2.6.

We begin by repeating the definition of a system given in Section 2.6:

System
A system is a process for which cause-and-effect relations exist.

For our purposes, the cause is the system input signal and the effect is the system output signal. Often we refer to the input signal and the output signal as simply the input and the output, respectively.

An example of a discrete-time system is the Euler integrator described in Section 9.1, with the difference equation

[eq(9.5)] $$y[n] = y[n-1] + Hx[n-1].$$

In this equation, $x[n]$ is the input signal to the numerical integrator and $y[n]$ is the output signal. A digital control system is a system controlled by a digital computer without the intervention of human beings. An example is an automatic landing system for commercial aircraft. Certain types of digital filters employed in digital control systems utilize an integrator as one of the basic components of the filter. The Euler integrator is used in some of these filters. A second popular integrator is based on the trapezoidal rule. (See Problem 9.22.)

For the integrator of (9.5), the input signal is $x[n]$ and the output signal is $y[n]$. We can represent this system by the *block diagram* of Figure 9.22. We can also represent this integrator as a transformation:

$$y[n] = T(x[n]). \tag{9.59}$$

This notation represents a transformation and not a function; that is, $T(x[n])$ is not a mathematical function into which we substitute $x[n]$ and directly calculate $y[n]$. The set of equations relating the input $x[n]$ and the output $y[n]$ is called a *mathematical model*, or simply a *model*, of the system. Given the input $x[n]$, this set of equations must be solved to obtain $y[n]$. For discrete-time systems, the model is usually a set of difference equations.

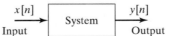

Figure 9.22 Block diagram for a discrete-time system.

As stated earlier, often we are careless in speaking of systems. Generally, when we use the word *system*, we are referring to the mathematical model of a physical system, not the physical system itself. This is common usage and is followed in this book. If we are referring to a physical system, we will call it a physical system if any confusion can occur. An example will now be given.

EXAMPLE 9.8   **A low-pass digital filter**

A low-pass digital filter is one that attenuates the higher frequencies relative to the lower frequencies in its input signal. In effect, the filter removes the higher frequencies in a signal, while passing the lower frequencies. This filter can be represented in general form by the transformation

$$y[n] = T(x[n]).$$

For example, the difference equation of a simple low-pass digital filter, called an α-filter [1], is given by

$$y[n] = (1 - \alpha)y[n - 1] + \alpha x[n], \tag{9.60}$$

where $0 < \alpha < 1$. Choices of the parameter α and the sample period T determine the range of frequencies that the filter will pass. One application of this filter is in the reduction of high-frequency noise in a radar signal. For example, this filter is employed in the automatic landing system for carrier-based aircraft [1]. One type of digital filter used in closed-loop control systems has the difference equation

$$y[n] = -a_1 y[n - 1] + b_0 x[n] + b_1 x[n - 1].$$

The filter parameters a_1, b_0, and b_1 are chosen by the system designer to give the control system certain desired characteristics. ∎

Interconnecting Systems

We define two basic connections for systems. The first, the *parallel* connection, is illustrated in Figure 9.23. The circle in this figure denotes the summation of signals.

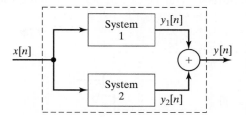

Figure 9.23 Parallel connection of systems.

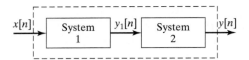

Figure 9.24   Series, or cascade, connection of systems.

Let the output of System 1 be $y_1[n]$ and that of System 2 be $y_2[n]$. The output signal of the total system, $y[n]$, is given by

$$y[n] = y_1[n] + y_2[n] = T_1(x[n]) + T_2(x[n]) = T(x[n]), \tag{9.61}$$

where $y[n] = T(x[n])$ is the notation for the total system.

The second basic connection for systems is illustrated in Figure 9.24. This connection is called the *series*, or *cascade*, connection. In this figure, the output signal of the first system is $y_1[n] = T_1(x[n])$, and the total system output signal is

$$y[n] = T_2(y_1[n]) = T_2(T_1(x[n])) = T(x[n]). \tag{9.62}$$

An example illustrating the interconnection of systems is now given.

EXAMPLE 9.9   **Interconnection of a discrete system**

Consider the system of Figure 9.25. Each block represents a system, with a number given to identify each system. We can write the following equations for the system:

$$y_3[n] = T_1(x[n]) + T_2(x[n])$$

and

$$y_4[n] = T_3(y_3[n]) = T_3(T_1(x[n]) + T_2(x[n])).$$

Thus,

$$\begin{aligned} y[n] &= y_2[n] + y_4[n] \\ &= T_2(x[n]) + T_3(T_1(x[n]) + T_2(x[n])) = T(x[n]). \end{aligned}$$

This equation denotes only the interconnection of the systems. The mathematical model of the total system will depend on the models of the individual subsystems.

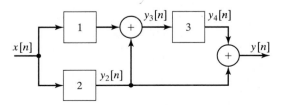

Figure 9.25   System for Example 9.9.   ■

9.6   PROPERTIES OF DISCRETE-TIME SYSTEMS

In Section 9.5, the Euler integrator and the α-filter were given as examples of discrete-time systems. In this section, we present some of the characteristics and properties of discrete-time systems.

In the following, $x[n]$ denotes the input of a system and $y[n]$ denotes the output. We show this relationship symbolically by the notation

$$x[n] \rightarrow y[n]. \tag{9.63}$$

As with continuous-time systems, we read this relation as $x[n]$ produces $y[n]$. Relationship (9.63) has the same meaning as

[eq(9.59)] $y[n] = T(x[n]).$

The definitions to be given are similar to those listed in Section 2.7 for continuous-time systems.

Systems with Memory

We first define a system that has memory:

> **Memory**
> A system has memory if its output at time n_0, $y[n_0]$, depends on input values other than $x[n_0]$. Otherwise, the system is memoryless.

For a discrete signal $x[n]$, time is represented by the discrete increment variable n. An example of a simple memoryless discrete-time system is the equation

$$y[n] = 5x[n].$$

A memoryless system is also called a *static system*.

A system with memory is also called a *dynamic system*. An example of a system with memory is the Euler integrator of (9.5):

$$y[n] = y[n - 1] + Hx[n - 1].$$

Recall from Section 9.1 and (9.8) that this equation can also be expressed as

$$y[n] = H \sum_{k=-\infty}^{n} x[k], \tag{9.64}$$

and we see that the output depends on all past values of the input.

A second example of a discrete system with memory is one whose output is the average of the last two values of the input. The difference equation describing this system is

$$y[n] = \tfrac{1}{2}[x[n] + x[n - 1]]. \tag{9.65}$$

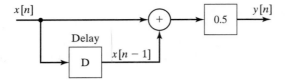

Figure 9.26   Averaging system.

This system can be represented as shown in Figure 9.26. Note the ideal delay in this figure. (See Figure 9.17.) This equation is an averaging filter; one application is in a system for generating a picture in a picture for television. (See references, Section 1.3.)

A third example of a discrete system with memory is one that calculates a 20-day running average of the Dow Jones industrial average for the U.S. stock market. The difference equation for the system is given by

$$y[n] = \frac{1}{20}\sum_{k=0}^{19} x[n - k]. \tag{9.66}$$

In this equation, $x[n]$ is the Dow Jones average for today, $x[n - 1]$ is the Dow Jones average for yesterday, and so on, and $y[n]$ is the average for the last 20 days. A block-diagram model of this system of the form of Figure 9.26 contains 19 delays. In a digital-computer implementation of this algorithm, the delays are realized by 19 memory locations.

Equation (9.66) can be considered to be a digital filter, with the output a filtered version of the daily average. All of the considerable theory of digital filtering can be employed to determine the *characteristics* of this system. For example, what are the effects of random fluctuations in the daily average on the 20-day average? If a significant change occurs in the daily average, what is the delay before this change becomes evident in the 20-day average? The definitions in this section allow us to classify systems so as to be better able to answer questions such as these. In addition, the discrete Fourier transform (covered in Chapter 12) allows us to determine the characteristics of systems such as this.

Invertibility

We now define *invertibility*:

> **Invertibility**
> A system is said to be invertible if distinct inputs result in distinct outputs.

A second definition of invertibility is that the input of an invertible system can be determined from its output. For example, the memoryless system described by

$$y[n] = |x[n]|$$

is not invertible. The inputs of $+2$ and -2 produce the same output of $+2$.

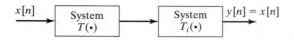

Figure 9.27 Identity system.

Inverse of a System

Invertibility is related to the *inverse* of a system.

> **Inverse of a System**
> The inverse of a system T is a second system T_i that, when cascaded with T, yields the identity system.

The identity system is defined by the equation $y[n] = x[n]$. Consider the two systems of Figure 9.27. System T_i is the inverse of system T if

$$y[n] = T_i[T(x[n])] = x[n]. \tag{9.67}$$

Causality

All physical systems are causal, whether continuous or discrete.

> **Causal Systems**
> A system is *causal* if the output at any time is dependent on the input only at the present time and in the past.

We have defined the unit delay as a system with an input of $x[n]$ and an output of $x[n-1]$, as shown in Figure 9.17. An example of a noncausal system is the unit advance, which has an input of $x[n]$ and an output of $x[n+1]$. Another example of a noncausal system is an averaging system given by

$$y[n] = \tfrac{1}{3}[x[n-1] + x[n] + x[n+1]]$$

and requires that we know a future value, $x[n+1]$, of the input signal in order to calculate the current value, $y[n]$, of the output signal.

We denote the unit advance with the symbol D^{-1}. A realizable system that contains a unit advance is the system of Figure 9.28 and is realized by first delaying a signal and then advancing it. However, we cannot advance a signal more than it has been delayed. Although this system may appear to have no application, this procedure is used in filtering signals "off line," or in nonreal time. If we store a signal in computer memory, we know "future" values of the signal relative to the value that

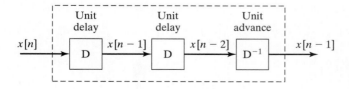

Figure 9.28 Realizable system with a unit advance.

we are considering at this instant. A second example is in modeling physical systems that contain more than one sampler, in which the samplers operate at the same frequency, but are not synchronized, or operate at different frequencies [1]. In cases such as these, we can speak of time advances. However, the signal out of a total system cannot be advanced relative to the signal into that system.

Stability

We give the same stability definition for discrete-time systems as for continuous-time systems.

> ### BIBO Stability
> A system is stable if the output remains bounded for any bounded input.

This is the bounded-input bounded-output (BIBO) definition of stability. By definition, a signal $x[n]$ is bounded if there exists a number M such that

$$|x[n]| \leq M \quad \text{for all } n. \tag{9.68}$$

Hence, a system is bounded-input bounded-output stable if, for a number R,

$$|y[n]| \leq R \quad \text{for all } n, \tag{9.69}$$

for all $x[n]$ such that (9.68) is satisfied. To determine BIBO stability, R [in general, a function of M in (9.68)] must be found such that (9.69) is satisfied.

Note that the Euler integrator of (9.5)

$$[eq(9.5)] \qquad\qquad y[n] = y[n-1] + Hx[n-1]$$

is not BIBO stable; if the signal to be integrated has a constant value of unity, the output increases without limit as n increases. [See (9.64).] It was shown in Chapter 2 that a continuous-time integrator is also not stable.

Time Invariance

The definition of time invariance is the same as that for continuous-time systems.

> ### Time-Invariant System
> A system is said to be time invariant if a time shift in the input results only in the same time shift in the output.

In this definition, the discrete increment n represents time. For a time-invariant system for which the input $x[n]$ produces the output $y[n]$, the input $x[n - n_0]$ produces $y[n - n_0]$, or

$$x[n] \rightarrow y[n]$$

and

$$x[n - n_0] \rightarrow y[n - n_0]. \tag{9.70}$$

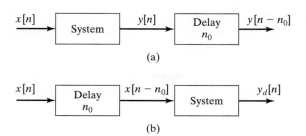

Figure 9.29   Test for time invariance.

A test for time invariance is given by

$$
y[n]\Big|_{n-n_0} = y[n]\Big|_{x[n-n_0]}, \tag{9.71}
$$

provided that $y[n]$ is expressed as an explicit function of $x[n]$. This test is illustrated in Figure 9.29. The signal $y[n - n_0]$ is obtained by delaying $y[n]$ by n_0. Define $y_d[n]$ as the system output for the delayed input $x[n - n_0]$, such that

$$
x[n - n_0] \rightarrow y_d[n].
$$

The system is time invariant, provided that

$$
y[n - n_0] = y_d[n]. \tag{9.72}
$$

A system that is not time-invariant is time varying. An example of a time-invariant system is $y_1[n] = e^{x[n]}$, whereas $y_2[n] = e^n x[n]$ is time varying.

Linearity

The property of *linearity* is one of the most important properties that we consider. Once again we define the system input signal to be $x[n]$ and the output signal to be $y[n]$.

> **Linear System**
> A system is linear if it meets the following two criteria:
>
> **1.** Additivity. If $x_1[n] \rightarrow y_1[n]$ and $x_2[n] \rightarrow y_2[n]$, then
>
> $$
> x_1[n] + x_2[n] \rightarrow y_1[n] + y_2[n]. \tag{9.73}
> $$
>
> **2.** Homogeneity. If $x[n] \rightarrow y[n]$, then, with a constant,
>
> $$
> ax[n] \rightarrow ay[n]. \tag{9.74}
> $$
>
> The criteria must be satisfied for all $x[n]$ and all a.

These two criteria can be combined to yield the *principle of superposition*. A system satisfies the principle of superposition if

$$
a_1 x_1[n] + a_2 x_2[n] \rightarrow a_1 y_1[n] + a_2 y_2[n], \tag{9.75}
$$

where a_1 and a_2 are arbitrary constants. A system is linear if it satisfies the principle of superposition. No physical system is linear under all operating conditions. However, a physical system can be tested using (9.75) to determine ranges of operation for which the system is approximately linear.

An example of a linear operation (system) is that of multiplication by a constant K, described by $y[n] = Kx[n]$. An example of a nonlinear system is the operation of squaring a signal,

$$y[n] = x^2[n].$$

For inputs of $x_1[n]$ and $x_2[n]$, the outputs of the squaring system are

$$x_1[n] \rightarrow y_1[n] = x_1^2[n]$$

and

$$x_2[n] \rightarrow y_2[n] = x_2^2[n]. \tag{9.76}$$

However, the input $(x_1[n] + x_2[n])$ produces the output

$$x_1[n] + x_2[n] \rightarrow (x_1[n] + x_2[n])^2 = x_1^2[n] + 2x_1[n]x_2[n]$$

$$+ x_2^2[n] = y_1[n] + y_2[n] + 2x_1[n]x_2[n]. \tag{9.77}$$

A linear time-invariant (LTI) system is a linear system that is also time invariant. LTI systems, for both continuous-time and discrete-time systems, are emphasized in this book.

An important class of LTI discrete-time systems are those that are modeled by linear difference equations with constant coefficients. An example of this type of system is the Euler integrator described earlier in this section:

$$y[n] - y[n - 1] = Hx[n - 1].$$

In this equation, $x[n]$ is the input, $y[n]$ is the output, and the numerical-integration increment H is constant.

The general forms of an nth-order linear difference equation with constant coefficients are given by (9.6) and (9.7). Equation (9.6) is repeated here:

$$y[n] = b_1 y[n - 1] + b_2 y[n - 2] + \cdots + b_N y[n - N]$$
$$+ a_0 x[n] + a_1 x[n - 1] + \cdots + a_N x[n - N].$$

This difference equation is said to be of order N. The second version of this general equation is obtained by replacing n with $(n + N)$. [See (9.7).]

EXAMPLE 9.10   **Illustrations of discrete-system properties**

The properties will be investigated for the system described by the equation

$$y[n] = \left[\frac{n + 2.5}{n + 1.5}\right]^2 x[n].$$

Note that the output $y[n]$ is equal to the input $x[n]$ multiplied by a value that varies with n (time). For $n = 0$, the multiplier is 2.778; as $|n|$ becomes large, the multiplier approaches unity.

1. This system is *memoryless*, since the output is a function of the input at the present time only.
2. The system is *invertible*, since we can solve for $x[n]$:

$$x[n] = \left[\frac{n + 1.5}{n + 2.5}\right]^2 y[n].$$

Given $y[n]$ for any n, we can find $x[n]$.
3. The system is *causal*, since the output does not depend on the input at a future time.
4. The system is *stable*, since the output is bounded for all bounded inputs. For $|x[n]| \leq M$, it is easily shown that (see Problem 9.26)

$$|y[n]| \leq R = 9M.$$

5. The system is *time varying*, since an input applied at $n = 0$ produces an output different from that of the same input applied at $n = 1$.
6. The system is *linear*, since, by superposition,

$$a_1 x_1[n] + a_2 x_2[n] \rightarrow \left[\frac{n + 2.5}{n + 1.5}\right]^2 (a_1 x_1[n] + a_2 x_2[n])$$

$$= a_1 \left[\frac{n + 2.5}{n + 1.5}\right]^2 x_1[n] \mid a_2 \left[\frac{n + 2.5}{n + 1.5}\right]^2 x_2[n]$$

$$= a_1 y_1[n] + a_2 y[n]. \qquad \blacksquare$$

This section defines certain properties for discrete-time systems. Probably the most important properties are linearity and time invariance. We can always determine the BIBO stability of LTI systems described by difference equations. No such statement can be made for other models. In addition, most digital-filter design procedures apply for LTI filters only.

SUMMARY

This chapter introduced discrete-time signals and systems. For a discrete-time signal $x[n]$, the discrete increment n represents time. First three transformations of the independent time variable n were defined: reversal, scaling, and shifting. Next the same three transformations were defined with respect to the amplitude of signals. A general procedure was developed for determining the effects of all six transformations. These transformations are important with respect to signals; they are equally important as transformations for functions of frequency. Frequency transformations will be covered when the discrete Fourier transformation is defined in Chapter 12.

The signal characteristics of evenness, oddness, and periodicity were defined next. These three characteristics appear often in the study of signals and systems.

Models of common signals that appear in certain types of physical systems were defined next. These signals included exponential signals and sinusoids whose amplitudes may vary exponentially. The impulse function was defined for discrete-time signals and was seen to be an ordinary function. It was shown that these discrete-time signals can be considered to be generated by sampling continuous-time signals. The study of periodic discrete signals was seen to be more complex than that of periodic continuous signals.

A general technique was given for expressing the output of a discrete-time system that is an interconnection of subsystems. As a final topic, some general properties of discrete-time systems were defined: memory, invertibility, causality, stability, time invariance, and linearity. For the remainder of this book, systems that are both linear and time invariant will be emphasized.

See Table 9.6.

TABLE 9.6 Key Equations of Chapter 9

Equation Title	Equation Number	Equation
DT unit step function	(9.9)	$u[n] = \begin{cases} 1, & n \geq 0 \\ 0, & n < 0 \end{cases}$
DT unit impulse function	(9.11)	$\delta[n] = \begin{cases} 1, & n = 0 \\ 0, & n \neq 0 \end{cases}$
DT independent-variable transformation	(9.23)	$y[n] = x[m]\Big|_{m=an+b} = x[an + b]$
DT signal-amplitude transformation	(9.24)	$y[n] = Ax[n] + B$
Even part of a DT signal	(9.29)	$x_e[n] = \frac{1}{2}(x[n] + x[-n])$
Odd part of a DT signal	(9.30)	$x_o[n] = \frac{1}{2}(x[n] - x[-n])$
Definition of DT periodicity	(9.32)	$x[n + N] = x[n]$
Requirement for DT periodicity	(9.39)	$\Omega_0 N = 2\pi k \Rightarrow \Omega_0 = \frac{k}{N}2\pi$
DT exponential function	(9.50)	$x[n] = Ca^n$
DT test for time invariance	(9.70)	$x[n] \rightarrow y[n]; \ x[n - n_0] \rightarrow y[n - n_0]$
DT test for linearity	(9.75)	$a_1 x_1[n] + a_2 x_2[n] \rightarrow a_1 y_1[n] + a_2 y_2[n]$

REFERENCES

C. L. Phillips and H. T. Nagle, *Digital Control System Analysis and Design*, 3d ed. Upper Saddle River, NJ: Prentice Hall, 1996.

PROBLEMS

9.1. Determine which of the following discrete-time functions is different:

(a) $x_1[n] = u[n + 1] - u[n - 2]$
(b) $x_2[n] = \sum_{k=-1}^{1} \delta[n - k]$
(c) $x_3[n] = 1$ for $n \in [-1, 1]$, and 0 otherwise
(d) $x_4[n] = \delta[n - 1] + \delta[n] + \delta[n + 1] + \delta[n + 2]$

9.2. The discrete-time signals in Figure P9.2 are zero except as shown.

(a) For the signal $x_a[n]$ of Figure P9.2(a), plot the following:

   (i) $x_a[2n]$ (ii) $x_a[-n/2]$
   (iii) $x_a[-n]$ (iv) $x_a[2 - n]$
   (v) $x_a[n - 2]$ (vi) $x_a[-2 - n]$

(b) Repeat Part (a) for the signal $x_b[n]$ of Figure P9.2(b).
(c) Repeat Part (a) for the signal $x_c[n]$ of Figure P9.2(c).
(d) Repeat Part (a) for the signal $x_d[n]$ of Figure P9.2(d).

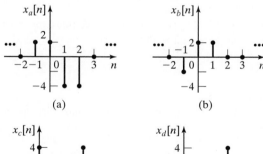

     (a)   (b)

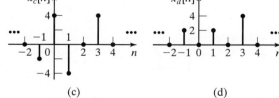

     (c)   (d)   **Figure P9.2**

9.3. The signals in Figure P9.2 are zero except as shown.

(a) For the signal $x_a[n]$ of Figure P9.2(a), plot the following:

   (i) $2 - 3x_a[n]$ (ii) $2x_a[-n]$
   (iii) $3x_a[n - 2]$ (iv) $3 - x_a[n]$
   (v) $1 + 2x_a[-2 + n]$ (vi) $2x_a[-n] - 4$

(b) Repeat Part (a) for the signal $x_b[n]$ of Figure P9.2(b).

(c) Repeat Part (a) for the signal $x_c[n]$ of Figure P9.2(c).

(d) Repeat Part (a) for the signal $x_d[n]$ of Figure P9.2(d).

9.4. The signals in Figure P9.2 are zero except as shown.

(a) For the signal $x_a[n]$ of Figure P9.2(a), plot the following:

   (i) $x_a[-n]u[n]$ **(ii)** $x_a[n]u[-n]$

   (iii) $x_a[n]u[n+2]$ **(iv)** $x_a[-n]u[-2-n]$

   (v) $x_a[n]\delta[n-2]$ **(vi)** $x_a[n](\delta[n+1]+\delta[n-1])$

(b) Repeat Part (a) for the signal $x_b[n]$ of Figure P9.2(b).

(c) Repeat Part (a) for the signal $x_c[n]$ of Figure P9.2(c).

(d) Repeat Part (a) for the signal $x_d[n]$ of Figure P9.2(d).

9.5. Given a signal $x[n] = \delta[n] + 2\delta[n-1] + 3\delta[n-2] \ +3\delta[n-3] + 2\delta[n-4] + \delta[n-5] \ + \delta[n-6] + 2\delta[n-7] + 3\delta[n-8]$, each of the following functions $y_i[n]$ can be written as a function of $x[n]$ with time scaling and time shifting; that is, $y_i[n] = x[a_i n + b_i]$. For each of the following parts, find the parameters a_i and b_i:

(a) $y_1[n] = \delta[n] + 3\delta[n-1] + \delta[n-2]$

(b) $y_2[n] = 2\delta[n] + 3\delta[n-1] + \delta[n-2] + 2\delta[n-3]$

(c) $y_3[n] = 3\delta[n-1] + \delta[n-2] + 3\delta[n-3]$

9.6. **(a)** For the general case of transformations of discrete signals, given $x[n]$, $x_t[n]$ can be expressed as

$$x_t[n] = Ax[an + n_0] + B,$$

where a is rational and n_0 is an integer. Solve this expression for $x[n]$.

(b) Suppose that for the signal of Figure P9.6,

$$x_1[n] = 0.5\,x_3[-n-1] + 2.$$

Sketch $x_3[n]$.

(c) Verify the results of Part (b) by checking at least three points in time.

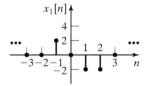 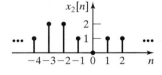

Figure P9.6

9.7. **(a)** Find the even and odd parts of $x_1[n]$ in Figure P9.6.

   (b) Find the even and odd parts of $x_2[n]$ in Figure P9.6.

9.8. **(a)** Plot the even and odd parts of the signal of Figure P9.2(a).

   (b) Repeat Part (a) for the signal of Figure P9.2(b).

(c) Repeat Part (a) for the signal of Figure P9.2(c).

(d) Repeat Part (a) for the signal of Figure P9.2(d).

9.9. (a) For each of the signals given, determine mathematically whether the signal is even, odd, or neither.

      (i) $x[n] = 3u[n - 2]$ (ii) $x[n] = n$

      (iii) $x[n] = (.7)^{|n|}$ (iv) $x[n] = 3 + .7^n + .7^{-n}$

      (v) $x[n] = \cos(n)$ (vi) $x[n] = \cos(n - \pi/6)$

(b) Sketch the signals and verify the results of Part (a).

(c) Find the even part and the odd part of each of the signals.

9.10. (a) Given in Figure P9.10 are the *parts* of a signal $x[n]$ and its even part $x_e[n]$, for *only* $n \geq 0$. Note that $x_e[n] = 2, n \geq 0$. Complete the plots of $x[n]$ and $x_e[n]$, and give a plot of the odd part, $x_o[n]$, of the signal. Give the equations used for plotting each part of the signals.

(b) In Figure P9.10, let $x[0] = 0$, with all other values unchanged. Give the changes in this case for the results of Part (a).

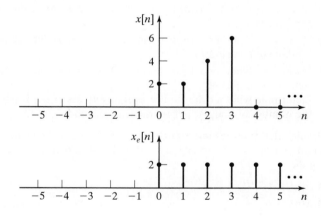

Figure P9.10

9.11. Let $x_e[n]$ and $x_o[n]$ be the even and odd parts, respectively, of $x[n]$.

(a) Show that $x_o[0] = 0$ and that $x_e[0] = x[0]$.

(b) Show that

$$\sum_{k=-\infty}^{\infty} x_o[n] = 0.$$

(c) Show that

$$\sum_{k=-\infty}^{\infty} x[n] = \sum_{k=-\infty}^{\infty} x_e[n].$$

(d) Do the results of Part (c) imply that

$$\sum_{k=n_1}^{n_2} x[n] = \sum_{k=n_1}^{n_2} x_e[n],$$

where n_1 and n_2 are any integers? Why?

9.12. Give proofs of the following statements:

(a) The sum of two even functions is even.
(b) The sum of two odd functions is odd.
(c) The sum of an even function and an odd function is neither even nor odd.
(d) The product of two even functions is even.
(e) The product of two odd functions is odd.
(f) The product of an even function and an odd function is odd.

9.13. Suppose that the signals $x_1[n]$, $x_2[n]$ and $x_3[n]$ are given by

$$x_1[n] = \cos(0.2\pi n), \quad x_2[n] = \cos(0.125\pi n), \text{ and } x_3[n] = \cos(.4\pi n).$$

(a) Determine whether $x_1[n]$ is periodic. If so, determine the number of samples per fundamental period.
(b) Determine whether $x_2[n]$ is periodic. If so, determine the number of samples per fundamental period.
(c) Determine whether $x_3[n]$ is periodic. If so, determine the number of samples per fundamental period.
(d) Determine whether the sum of $x_1[n]$, $x_2[n]$, and $x_3[n]$ is periodic. If so, determine the number of samples per fundamental period.

9.14. Consider the discrete-time signals that follow. For each signal, determine the fundamental period N_0 if the signal is periodic; otherwise, prove that the signal is not periodic.

(a) $x[n] = e^{j5\pi n/7}$
(b) $x[n] = e^{j5n}$
(c) $x[n] = e^{j2\pi n}$
(d) $x[n] = e^{j0.3n/\pi}$
(e) $x[n] = \cos(3\pi n/7)$
(f) $x[n] = e^{j0.3n}$
(g) $x[n] = e^{j5\pi n/7} + e^{j2\pi n}$
(h) $x[n] = e^{j5\pi n/7} + e^{j2\pi n} + \cos(3\pi n/7)$
(i) $x[n] = e^{j.3n} + e^{j2\pi n}$

9.15. (a) A continuous-time signal $x(t) = \cos 2\pi t$ is sampled every T seconds, resulting in the discrete-time signal $x[n] = x(nT)$. Determine whether the sampled signal is periodic for

(i) $T = 1\,s$ **(ii)** $T = 0.1\,s$
(iii) $T = 0.125\,s$ **(iv)** $T = 0.130\,s$
(v) $T = 5\,s$ **(vi)** $T = \frac{4}{3}\,s$

(b) For those sampled signals in Part (a) that are periodic, find the number of periods of $x(t)$ in one period of $x[n]$.

(c) For those sampled signals in Part (a) that are periodic, find the number of samples in one period of $x[n]$.

9.16. A continuous-time signal $x(t)$ is sampled at a 10-Hz rate, with the resulting discrete-time signals as given. Find the time constant τ for each signal, and the frequency ω of the sinusoidal signals.

(a) $x[n] = (.3)^n$

(b) $x[n] = (.3)^n \cos(n)$

(c) $x[n] = (-.3)^n$

(d) $x[n] = (.3)^n \sin(n + 1)$

9.17. (a) Determine which of the given signals are periodic:

(i) $x[n] = \cos(\pi n)$ **(ii)** $x[n] = -3\sin(0.01\pi n)$

(iii) $x[n] = \cos(3\pi n/2 + \pi)$ **(iv)** $x[n] = \sin(3.15n)$

(v) $x[n] = 1 + \cos(\pi n/2)$ **(vi)** $x[n] = \sin(3.15\pi n)$

(b) For those signals in Part (a) that are periodic, determine the number of samples per period.

9.18. (a) What is the difference between the unit step function $u[n + 2]$ and the time-scaled function $u[2n + 4]$?

(b) Repeat Part (a) for $u[n]$ and $u[n/3]$.

9.19. Consider the signals shown in Figure P9.2.

(a) Write an expression for $x_a[n]$. The expression will involve the sum of discrete impulse functions.

(b) Write an expression for $x_b[n]$.

(c) Write an expression for $x_c[n]$.

(d) Write an expression for $x_d[n]$.

9.20. (a) Draw a block diagram, as in Figure 9.25, for a system described by

$$y_a[n] = T(x[n]) = T_2(x[n] + T_1(x[n])) + T_3(T_1(x[n])).$$

(b) Repeat Part (a) for

$$y_b[n] = y_a[n] + x[n].$$

9.21. (a) Consider the feedback system of Figure P9.21. Express the output signal as a function of the transformation of the input signal, in the form of (9.61).

(b) Draw a block diagram, as in Figure P9.21, for a system described by

$$y[n] = T_2(T_1(x[n] - y[n]) - y[n]).$$

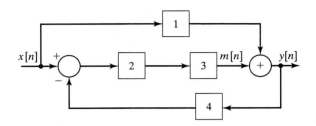

Figure P9.21

9.22. The trapezoidal rule for numerical integration is defined in Figure P9.22. The value of the integral at $t = kT$ is equal to its value at $t = (k - 1)T$ plus the trapezoidal area shown.

   (a) Write a difference equation relating $y[k]$, the numerical integral of $x(t)$, to $x[k]$ for this integrator.

   (b) Write a MATLAB program that integrates e^{-t}, $0 \le t \le 5$ s, with $T = 0.1$ s, using trapezoidal integration.

   (c) Run the program in Part (b), and verify the result.

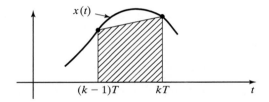

Figure P9.22

9.23. **(a)** Determine whether the system described by

$$y[n] = \cos(x[n + 2]) \text{ is}$$

   (i) memoryless;   **(ii)** invertible;

   (iii) causal;   **(iv)** stable;

   (v) time invariant;   **(vi)** linear.

   (b) Repeat Part (a) for $y[n] = x[-n]$

   (c) Repeat Part (a) for

$$y[n] = \frac{\sin x[n]}{x[n]}.$$

   Note that $\lim_{x \to 0}(\sin x)/x$ must be considered.

   (d) Repeat Part (a) for $y[n] = e^{x[n]}$.

   (e) Repeat Part (a) for $y[n] = e^{nx[n]}$.

   (f) Repeat Part (a) for $y[n] = 4x[n] - 2$

   (g) $y[n] = \sum_{k=-\infty}^{n-3} \sin(x[k])$

9.24. The system described by the linear difference equation

$$y[n] - 2y[n - 1] + y[n - 2] = x[n], \quad n \geq 0,$$

with constant coefficients can be shown to be invertible and unstable. Determine whether this system is

(a) memoryless;
(b) time invariant;
(c) linear.

9.25. (a) Determine whether the summation operation, defined by

$$y[n] = \sum_{k=-n}^{n} x[k + a], \text{ is}$$

where a is an integer.

    (i) memoryless;   **(ii)** invertible;
    (iii) causal;   **(iv)** stable;
    (v) time invariant;   **(vi)** linear.

    (b) Repeat Part (a) for the averaging filter

$$y[n] = \tfrac{1}{2}[x[n] + x[n - 1]]$$

    (c) Repeat part (a) for the running average filter (M > 0 is an integer)

$$y[n] = \frac{1}{M} \sum_{k=0}^{M-1} x[n - k].$$

9.26. For the system of Example 9.10, show that for $|x[n]| \leq M$, $|y[n]| \leq 9M$.

9.27. (a) Sketch the characteristic y versus x for the system $y[n] = -3|x[n]|$. Determine whether this system is

    (i) memoryless;   **(ii)** invertible;
    (iii) causal;   **(iv)** stable;
    (v) time invariant;   **(vi)** linear.

    (b) Repeat Part (a) for

$$y[n] = \begin{cases} 3x[n], & x < 0 \\ 0, & x \geq 0 \end{cases}.$$

    (c) Repeat Part (a) for

$$y[n] = \begin{cases} -10, & x < -1 \\ 10x[n], & |x| \leq 1. \\ 10, & x > 1 \end{cases}$$

(d) Repeat Part (a) for

$$y[n] = \begin{cases} 2, & 2 < x \\ 1, & 1 < x \le 2 \\ 0, & 0 < x \le 1. \\ -1, & -1 < x \le 0 \\ -2, & x \le -1 \end{cases}$$

9.28. Let $h[n]$ denote the response of a system for which the input signal is the unit impulse function $\delta[n]$. Suppose that $h[n]$ for a *causal* system has the given *even* part $h_e[n]$ for $n \ge 0$:

$$h_e[n] = \begin{cases} 0, & n = 0 \\ 2, & n = 1. \\ 4, & n \ge 2 \end{cases}$$

Find $h[n]$ for all time, with your answer expressed as a mathematical function.

10 DISCRETE-TIME LINEAR TIME-INVARIANT SYSTEMS

$\mathbf{I}$n Chapter 3, we developed important properties of continuous-time linear time-invariant (LTI) systems; those developments are applied to discrete-time LTI systems in this chapter. In one sense, discrete-time systems are easier to analyze and design, since difference equations are easier to solve than are differential equations. In a different sense, discrete-time systems are more difficult to analyze and design, since the system characteristics are periodic in frequency. (See Section 9.3.)

In Chapter 9, several properties of discrete-time systems were defined. We now restate two of these properties.

Consider first *time invariance*. We denote a discrete-time system with input $x[n]$ and output $y[n]$ by the notation

$$x[n] \rightarrow y[n]. \tag{10.1}$$

This system is time invariant if the only effect of a time shift of the input signal is the same time shift of the output signal; that is, in (10.1),

$$x[n - n_0] \rightarrow y[n - n_0], \tag{10.2}$$

where n_0 is an arbitrary integer.

Next *linearity* is reviewed. For the system of (10.1), suppose that

$$x_1[n] \rightarrow y_1[n], \quad x_2[n] \rightarrow y_2[n]. \tag{10.3}$$

This system is linear, provided that the principle of superposition applies:

$$a_1 x_1[n] + a_2 x_2[n] \rightarrow a_1 y_1[n] + a_2 y_2[n]. \tag{10.4}$$

This property applies for all constants a_1 and a_2, and for all signals $x_1[n]$ and $x_2[n]$.

In this chapter, we consider only discrete-time systems that are both linear and time invariant. We refer to these systems as discrete-time LTI systems. We have several reasons for emphasizing these systems:

1. Many physical systems can be modeled accurately as LTI systems. For example, most digital filters are designed to be both linear and time invariant.
2. We can solve the equations that model LTI systems, for both continuous-time and discrete-time systems. No general procedures exist for the solution of the describing equations of non-LTI systems.
3. Much information is available for both the analysis and the design of LTI systems. This is especially true for the design of LTI digital filters.

The developments in this chapter are analogous to those of Chapter 3 for continuous-time systems. Some of the developments are the same as those of Chapter 3, while others differ significantly. Because almost all signals and systems of this chapter are discrete time, we often omit this term in the descriptions.

10.1 IMPULSE REPRESENTATION OF DISCRETE-TIME SIGNALS

In this section, a relation is developed that expresses a general signal $x[n]$ as a function of impulse functions. This relation is useful in deriving properties of LTI discrete-time systems.

Recall the definition of the discrete-time impulse function (also called the unit sample function):

$$\delta[n - n_0] = \begin{cases} 1, & n = n_0 \\ 0, & n \neq n_0 \end{cases}. \tag{10.5}$$

An impulse function has a value of unity when its argument is zero; otherwise, its value is zero. From this definition, we see that

$$x[n]\delta[n - n_0] = x[n_0]\delta[n - n_0].$$

Consider the signal $x[n]$ in Figure 10.1(a). For simplicity, this signal has only three nonzero values. We define the following signal, using (10.5):

$$x_{-1}[n] = x[n]\delta[n + 1] = x[-1]\delta[n + 1] = \begin{cases} x[-1], & n = -1 \\ 0, & n \neq -1 \end{cases}.$$

In a like manner, we define the signals

$$x_0[n] = x[n]\delta[n] = x[0]\delta[n] = \begin{cases} x[0], & n = 0 \\ 0, & n \neq 0 \end{cases};$$

$$x_1[n] = x[n]\delta[n - 1] = x[1]\delta[n - 1] = \begin{cases} x[1], & n = 1 \\ 0, & n \neq 1 \end{cases}.$$

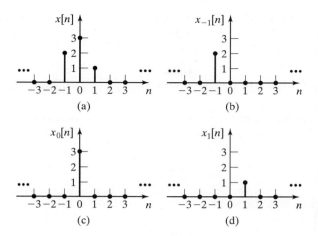

Figure 10.1   Representation of a signal with discrete-time impulse functions.

These three signals are also shown in Figure 10.1. The only nonzero values of $x[n]$ are contained in these three signals; hence, we can express the signal $x[n]$ as

$$
\begin{aligned}
x[n] &= x_{-1}[n] + x_0[n] + x_1[n] \\
&= x[-1]\delta[n + 1] + x[0]\delta[n] + x[1]\delta[n - 1] \\
&= \sum_{k=-1}^{1} x[k]\delta[n - k].
\end{aligned}
\tag{10.6}
$$

Next, we generalize this development, using the term

$$
x[k]\delta[n - k] = \begin{cases} x[k], & n = k \\ 0, & n \neq k \end{cases}.
\tag{10.7}
$$

The summation of terms for all k yields the general signal $x[n]$:

$$
x[n] = \sum_{k=-\infty}^{\infty} x[k]\delta[n - k].
\tag{10.8}
$$

This relation is useful in the developments in the following sections.

The function $\delta[n]$ is called either the unit sample function or the unit impulse function. We use the term *impulse function* to emphasize the symmetry of the relations between discrete-time impulse functions and systems and continuous-time impulse functions and systems.

10.2   CONVOLUTION FOR DISCRETE-TIME SYSTEMS

An equation relating the output of a discrete LTI system to its input will now be developed. Consider the system shown in Figure 10.2. A unit impulse function $\delta[n]$ is

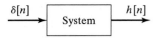

$\delta[n]$ System $h[n]$

Figure 10.2 Impulse response of a system.

applied to the system input. Recall that this input is unity for $n = 0$ and is zero at all other values of n.

With the input as described, the LTI system response in Figure 10.2 is denoted as $h[n]$; that is, in standard notation,

$$\delta[n] \rightarrow h[n]. \tag{10.9}$$

Since the system is time invariant,

$$\delta[n - k] \rightarrow h[n - k]. \tag{10.10}$$

The notation $h[\cdot]$ will *always* denote the unit impulse response. Because the system is linear, we can multiply each side of (10.10) by $x[k]$, resulting in the relation

$$x[k]\delta[n - k] \rightarrow x[k]h[n - k]. \tag{10.11}$$

Recall that the general input $x[n]$ can be expressed as a sum of impulse functions:

[eq(10.8)]
$$x[n] = \sum_{k=-\infty}^{\infty} x[k]\delta[n - k].$$

Because this input signal is a sum of impulse functions, the output signal is a sum of the impulse responses in (10.11) by the linearity property of (10.4), and it follows that

$$x[n] = \sum_{k=-\infty}^{\infty} x[k]\delta[n - k] \rightarrow y[n] = \sum_{k=-\infty}^{\infty} x[k]h[n - k]. \tag{10.12}$$

This result is called the *convolution sum* and is a basic result in the study of discrete-time LTI systems. We denote this sum with an asterisk:

$$y[n] = \sum_{k=-\infty}^{\infty} x[k]h[n - k] = x[n]*h[n]. \tag{10.13}$$

To illustrate a property of the convolution sum, we calculate the output at $n = 0$:

$$y[0] = \cdots + x[-2]h[2] + x[-1]h[1] + x[0]h[0]$$
$$+ x[1]h[-1] + x[2]h[-2] + \cdots. \tag{10.14}$$

Note that the discrete-time variable for $x[\cdot]$ increases as the discrete variable for $h[\cdot]$ decreases. The general output $y[n]$ is given by

$$
\begin{aligned}
y[n] = \cdots &+ x[-2]h[n+2] + x[-1]h[n+1] + x[0]h[n] \\
&+ x[1]h[n-1] + \cdots + x[n-1]h[1] + x[n]h[0] \\
&+ x[n+1]h[-1] + x[n+2]h[-2] + \cdots .
\end{aligned}
\tag{10.15}
$$

In this summation for $y[n]$, the sum of the arguments in each term is always n. Note the symmetry with respect to $x[\cdot]$ and $h[\cdot]$ in (10.15). We can then express the convolution sum as either of two relations:

$$
\begin{aligned}
y[n] &= \sum_{k=-\infty}^{\infty} x[k]h[n-k] = x[n]*h[n] \\
&= \sum_{k=-\infty}^{\infty} h[k]x[n-k] = h[n]*x[n].
\end{aligned}
\tag{10.16}
$$

This second form can also be derived from the first form by using a change of variables. (See Problem 10.1.)

We now note a property of the convolution sum. In the convolution sum (10.15) for $y[n]$, the general term is $x[k_1]h[k_2]$, with $(k_1 + k_2)$ *always* equal to n. For example, in calculating $y[5]$, $x[10]h[-5]$ is a term in the sum, but $x[10]h[-6]$ is not.

A second property of the convolution sum is derived by letting $x[n] = \delta[n]$. We have

$$
y[n] = \delta[n]*h[n] = h[n],
\tag{10.17}
$$

since, by definition, this output is the impulse response. Replacing n with $(n - n_0)$ in $h[n]$ yields

$$
\delta[n]*h[n - n_0] = h[n - n_0].
$$

Also, because of the time-invariance property, the general form of (10.17) is given by

$$
\delta[n - n_0]*h[n] = h[n - n_0].
$$

From the last two equations,

$$
\delta[n]*h[n - n_0] = h[n - n_0] = \delta[n - n_0]*h[n].
$$

No restrictions have been placed on $h[n]$; hence, $h[n]$ can be replaced with a general function $g[n]$, resulting in the second property of the convolution sum:

$$
\delta[n]*g[n - n_0] = \delta[n - n_0]*g[n] = g[n - n_0].
\tag{10.18}
$$

Do not confuse convolution with multiplication. The multiplication property of the impulse function is given by

$$\delta[n]g[n - n_0] = g[-n_0]\delta[n]$$

and

$$\delta[n - n_0]g[n] = g[n_0]\delta[n - n_0],$$

because $\delta[0]$ is the only nonzero value of $\delta[n]$, $-\infty < n < \infty$.

From the convolution sum (10.16), we see that if $h[n]$ is known, the system response for any input $x[n]$ can be calculated. Hence, the impulse response $h[n]$ of a discrete LTI system contains a complete *input–output description* of the system. We now give two examples that illustrate the use of the convolution sum.

EXAMPLE 10.1   **A finite impulse response system**

We consider the system depicted in Figure 10.3, in which the blocks labeled D are unit delays. We can write the system difference equation directly from the figure:

$$y[n] = (x[n] + x[n - 1] + x[n - 2])/3. \tag{10.19}$$

This system averages the last three inputs. It is a moving-average filter, which has many applications. We find the impulse response $h[n]$ for this system by applying the input $x[n] = \delta[n]$:

$$y[n] = h[n] = (\delta[n] + \delta[n - 1] + \delta[n - 2])/3. \tag{10.20}$$

Thus,

$$h[0] = (\delta[n] + \delta[n - 1] + \delta[n - 2])/3|_{n=0} = (1 + 0 + 0)/3 = 1/3;$$
$$h[1] = (\delta[n] + \delta[n - 1] + \delta[n - 2])/3|_{n=1} = (0 + 1 + 0)/3 = 1/3;$$
$$h[2] = (\delta[n] + \delta[n - 1] + \delta[n - 2])/3|_{n=2} = (0 + 0 + 1)/3 = 1/3;$$
$$h[n] = 0, \text{ all other } n.$$

This is a *finite impulse response* (FIR) system; that is, the impulse response contains a finite number of nonzero terms. As an exercise, the reader should trace the signal $x[n] = \delta[n]$ through the system in Figure 10.3 to verify $h[n]$. (Initially the numbers stored in the two delays must be zero. Otherwise the output also includes an initial-condition response, in addition to the impulse response, by superposition.)

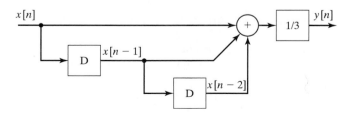

Figure 10.3   Discrete system.

EXAMPLE 10.2   **System response by convolution for an LTI system**

We continue Example 10.1 to illustrate the use of the convolution sum. Let the system input be given by

$$x[1] = 3; \quad x[2] = 4.5;$$
$$x[3] = 6; \quad x[n] = 0, \quad \text{all other } n.$$

The signal $x[n]$ is shown plotted in Figure 10.4(a). From (10.16), the response $y[n]$ is given by

$$y[n] = \sum_{k=-\infty}^{\infty} x[n-k]h[k].$$

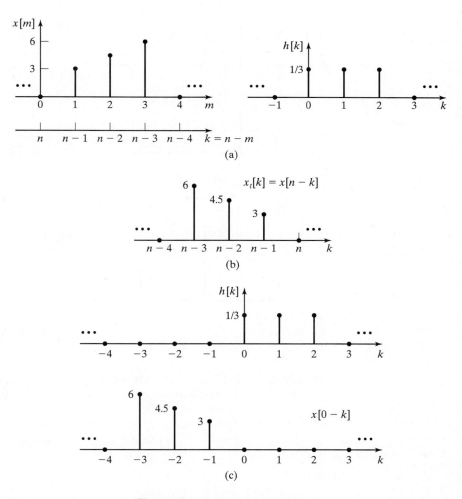

Figure 10.4   Signals for Example 10.2.

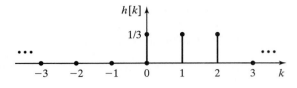

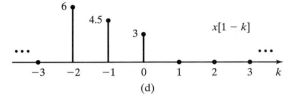

(d)

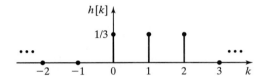

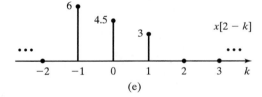

(e)

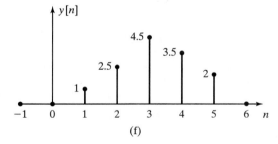

(f)

Figure 10.4 *(cont.)*

We now plot the factor $x[n - k]$. First, we change the time variable from n to m in Figure 10.4(a) to yield $x[m]$, to facilitate the plotting. Recall that in the convolution sum, n is considered constant. Now, from Section 9.2,

$$x_t[k] = x[m]\big|_{m=n-k} = x[n - k].$$

Thus,

$$m = n - k \Rightarrow k = n - m.$$

Next, the k-axis is plotted below the m-axis in Figure 10.4(a), resulting in the plot for $x[n - k]$ in Figure 10.4(b). The second factor in convolution summation, $h[k]$, was calculated in Example 10.1 and is also shown in Figure 10.4(a).

First, the two factors of the convolution summations, $x[n - k]$ and $h[k]$, are plotted in Figure 10.4(c) for $n = 0$. We see that for each nonzero value in one of the signals, the other signal has a value in zero. Hence, the product $x[n - k]h[k]$ is zero for $n = 0$ and all k, and it follows that $y[0] = 0$. Furthermore, we see that $y[n] = 0$ for all $n < 0$, since the plot of $x[n - k]$ is shifted to the left for $n < 0$.

Figure 10.4(d) gives $x[n - k]$ and $h[k]$ for $n = 1$. For this value of n,

$$x[n - k]h[k]|_{n=1,\, k=0} = x[1]h[0] = (1/3)(3) = 1,$$

and this product is zero for all other k. Thus, $y[1] = 1$.

Figure 10.4(e) shows $x[n - k]$ and $h[k]$ for $n = 2$. All product terms of the convolution sum are zero except for $k = 0$ and $k = 1$. Then

$$y[2] = x[2]h[0] + x[1]h[1] = 1.5 + 1 = 2.5.$$

Using the same procedure, we see that $y[3] = 4.5$, $y[4] = 3.5$, $y[5] = 2$, and $y[n] = 0$ for $n > 5$. The output is then

$$y[n] = \delta[n - 1] + 2.5\delta[n - 2] + 4.5\delta[n - 3]$$
$$+ 3.5\delta[n - 4] + 2\delta[n - 5].$$

The output signal $y[n]$ is plotted in Figure 10.4(f).

The output can also be calculated from the expansion of the convolution sum:

$$y[n] = \cdots + x[n - 3]h[3] + x[n - 2]h[2]$$
$$+ x[n - 1]h[1] + x[n]h[0] + x[n + 1]h[-1] + \cdots$$
$$= x[n - 2]h[2] + x[n - 1]h[1] + x[n]h[0].$$

This expansion has only three nonzero terms, because $h[n]$ is nonzero for only $n = 0, 1, 2$. As a third method of evaluation, the input signal can be traced through the system in Figure 10.3, as suggested in Example 10.1. Finally, the results can be verified with the MATLAB program

```
n=1:5;
x=[3 4.5 6];
h=[1/3 1/3 1/3];
y=conv(x,h)
stem(n,y,'fill'), grid
```
■

Note that in Examples 10.1 and 10.2, three different descriptions of the system are given:

1. the impulse response in (10.20);
2. the difference equation in (10.19);
3. a block diagram in Figure 10.3.

As a practical matter, usually in technical literature both the difference equation and a block diagram are given as the description of a discrete-time system. The impulse response is seldom given directly; instead the z-transform of the impulse response is specified. This topic is introduced in Section 10.7 and is covered in detail in Chapter 11.

Example 10.2 concerned a system with a finite impulse response. The next example considers a system that has an infinite impulse response.

EXAMPLE 10.3   **Calculation of the impulse response of a discrete system**

Consider the system of Figure 10.5. This system was also considered in Section 9.4. We write the difference equation for this system directly from Figure 10.5:

$$y[n] = ay[n - 1] + x[n].$$

The system unit impulse response is obtained from this equation by applying a unit impulse function, with $x[0] = 1$ and $x[n] = 0, n \neq 0$. The system is causal since the output is a function of the current and past values of input only. Because $x[n]$ is zero for $n < 0$, $y[n]$ is also zero over this range of n. Hence, the value stored in the ideal delay at $n = 0$ in Figure 10.5 is zero. Then,

$$y[0] = h[0] = ay[-1] + x[0] = a(0) + 1 = 1,$$
$$y[1] = h[1] = ay[0] + x[1] = a(1) + 0 = a,$$
$$y[2] = h[2] = ay[1] + x[2] = a(a) + 0 = a^2,$$
$$y[3] = h[3] = ay[2] + x[3] = a(a^2) + 0 = a^3,$$
$$\vdots$$

and the unit impulse response for this system is

$$h[n] = \begin{cases} a^n, & n \geq 0 \\ 0, & n < 0 \end{cases},$$

or $h[n] = a^n u[n]$. The unit impulse response consists of an unbounded number of terms; this system is called an *infinite impulse response* (IIR) system.

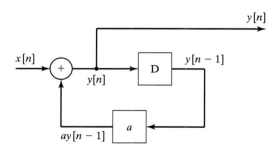

Figure 10.5 Discrete system.

EXAMPLE 10.4   **Step response of a discrete system**

For the system of the last example, let $a = 0.6$; the impulse response is then $h[n] = (0.6)^n u[n]$. We now find the unit step response of this system, with $x[n] = u[n]$. From (10.16), the system output signal is given by

$$y[n] = \sum_{k=-\infty}^{\infty} x[n-k]h[k] = \sum_{k=-\infty}^{\infty} u[n-k](0.6)^k u[k] = \sum_{k=0}^{n} (0.6)^k, \qquad (10.21)$$

because $u[k] = 0$ for $k < 0$ and $u[n-k] = 0$ for $k > n$.

Equation (10.21) gives $y[n]$ as a summation as expected. However, certain summations can be expressed in closed form. Appendix C gives a table of summation formulas. The first formula in this table is given by

$$\sum_{k=0}^{n} a^k = \frac{1 - a^{n+1}}{1 - a}. \qquad (10.22)$$

Note that the right side of this equation is indeterminant for $a = 1$. However, the summation is equal to $(n + 1)$ for $a = 1$, from the left side of (10.22). From the summation (10.22) and the equation (10.21) for $y[n]$, we have

$$y[n] = \sum_{k=0}^{n} (0.6)^k = \frac{1 - (0.6)^{n+1}}{1 - 0.6} = 2.5[1 - (0.6)^{n+1}], \quad n \ge 0.$$

The calculation of values of $y[n]$ yields

$$
\begin{aligned}
y[0] &= 1,   &\quad y[5] &= 2.383, \\
y[1] &= 1.6,   &\quad &\vdots \\
y[2] &= 1.96,   &\quad y[10] &= 2.485, \\
y[3] &= 2.176,   &\quad &\vdots \\
y[4] &= 2.306,   &\quad y[\infty] &= 2.5.
\end{aligned}
$$

Note that the steady-state value of $y[n]$ is 2.5. Figure 10.6 gives a plot of $y[n]$. The exponential nature of this response is evident. (Recall that a signal a^n is exponential, as shown in

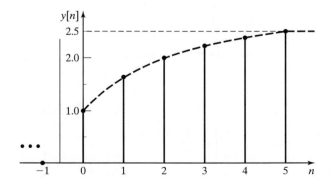

Figure 10.6 Response for Example 10.4.

Section 9.4.) The results of this example can be verified with the following MATLAB program

```
% This MATLAB program performs the convolution operation of
% Example 10.4.
% Establish the input unit-step function.
x=ones (1,11);
% Compute the system impulse response.
for k = 1:10;
h(k) = 0.6^(k-1);
end
% Convolve the input with the impulse response and plot.
c=conv (x,h);
for k=1:11;
    n(k)=k-1;
    y(k)=c(k);
end
 [n' y']
 stem(n, y, 'fill')
```
■

Properties of Convolution

We now discuss three properties of the convolution sum that are related to systems:

1. *Commutative property.* The convolution sum is symmetric with respect to $x[n]$ and $h[n]$:

$$x[n]*h[n] = h[n]*x[n]. \tag{10.23}$$

This property was derived in (10.16) and is illustrated in Figure 10.7. In this figure an LTI system is represented by a block containing the impulse response. The output for each system is identical, from (10.23).

2. *Associative property.* In the convolution of three signals, the result is the same, independent of the order that the convolution is performed. For example,

$$(f[n]*g[n])*h[n] = f[n]*(g[n]*h[n]) = (h[n]*f[n])*g[n]. \tag{10.24}$$

The proof of this property is not given. (See Problem 10.13.)

Figure 10.7 Commutative property.

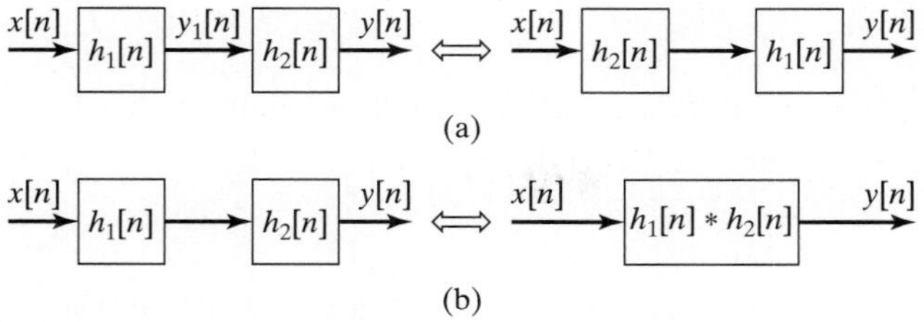

(a)

(b)

Figure 10.8 Associative property.

As an example of this property, consider the output of the system of Figure 10.8(a), which is given by

$$y[n] = y_1[n] * h_2[n] = (x[n] * h_1[n]) * h_2[n].$$

Then, by Property 2,

$$(x[n] * h_1[n]) * h_2[n] = x[n] * (h_1[n] * h_2[n]) = x[n] * (h_2[n] * h_1[n]). \qquad (10.25)$$

Hence, the order of the two systems of Figure 10.8(a) may be reversed without changing the input–output characteristics of the total system, as shown in Figure 10.8(a).

Also, from (10.25), the two cascaded systems of Figure 10.8(a) may be replaced with a single system with the impulse response

$$h[n] = h_1[n] * h_2[n], \qquad (10.26)$$

such that the input–output characteristics are preserved. This property is illustrated in Figure 10.8(b). It follows that for m cascaded LTI systems, the impulse response of the total system is given by

$$h[n] = h_1[n] * h_2[n] * \cdots * h_m(n).$$

3.   *Distributive property*. The convolution sum satisfies the following relationship:

$$x[n] * h_1[n] + x[n] * h_2[n] = x[n] * (h_1[n] + h_2[n]). \qquad (10.27)$$

We prove this relation using the convolution sum of (10.13):

$$x[n] * h_1[n] + x[n] * h_2[n] = \sum_{k=-\infty}^{\infty} x[k] h_1[n-k] + \sum_{k=-\infty}^{\infty} x[k] h_2[n-k]$$

$$= \sum_{k=-\infty}^{\infty} x[k](h_1[n-k] + h_2[n-k])$$

$$= x[n] * (h_1[n] + h_2[n]).$$

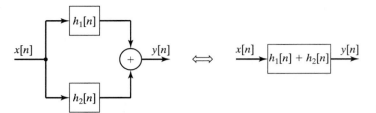

Figure 10.9 Distributive property.

This property is illustrated by two systems in parallel as in Figure 10.9, where the output is given by

$$y[n] = x[n]*h_1[n] + x[n]*h_2[n] = x[n]*(h_1[n] + h_2[n]). \qquad (10.28)$$

Therefore, the total system impulse response is the sum of the impulse responses:

$$h[n] = h_1[n] + h_2[n]. \qquad (10.29)$$

An example of the use of these properties will now be given.

EXAMPLE 10.5   **Impulse response of an interconnected system**

We wish to determine the impulse response of the system of Figure 10.10(a), in terms of the impulse responses in the subsystems. First, the impulse response of the cascaded systems 1 and 2 is given by

$$h_a[n] = h_1[n]*h_2[n],$$

as shown in Figure 10.10(b). The effect of the parallel connection of system a with system 3 is given by

$$h_b[n] = h_a[n] + h_3[n] = h_1[n]*h_2[n] + h_3[n]$$

as shown in Figure 10.10(c). We add the effect of system b cascaded with system 4 to give the total system impulse response, as shown in Figure 10.10(d):

$$h[n] = h_b[n]*h_4[n] = (h_1[n]*h_2[n] + h_3[n])*h_4[n]. \qquad ∎$$

It is shown in this section that the response of a discrete-time LTI system can be calculated using the convolution sum (10.16), provided that the system impulse response is known. Hence, the input–output characteristics of the system are completely specified by its response to an impulse function. Next, a procedure is developed for calculating the impulse response of an LTI system composed of subsystems, where the impulse responses of the subsystems are known. An equivalent and simpler procedure for expressing the input–output characteristics of a system in terms of its subsystem input–output characteristics

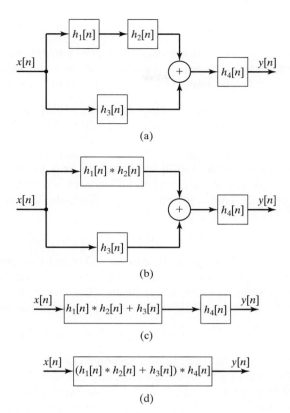

(a)

(b)

(c)

(d)

Figure 10.10 System for Example 10.5.

is the transform approach. This procedure will be presented when the z-transform is covered in Chapter 11.

10.3 PROPERTIES OF DISCRETE-TIME LTI SYSTEMS

In Section 9.6, several properties of discrete-time systems were defined. In this section, we consider these properties as related to LTI systems.

The input–output characteristics of a discrete-time LTI system are completely described by its impulse response $h[n]$. For the input signal $x[n]$, the output signal $y[n]$ is given by the convolution sum in (10.16):

$$y[n] = \sum_{k=-\infty}^{\infty} x[k]h[n-k] = \sum_{k=-\infty}^{\infty} x[n-k]h[k]. \tag{10.30}$$

This equation is now used to derive certain properties for LTI systems. We begin with the memory property.

Memory

Recall that a memoryless (static) system is one whose current value of output depends on only the current value of input. A system with memory is called a dynamic system. Expanding the convolution sum of (10.30), we see that for a memoryless system,

$$y[n] = \cdots + x[n+2]h[-2] + x[n+1]h[-1]$$
$$+ x[n]h[0] + x[n-1]h[1] + \cdots = h[0]x[n],$$

since only $x[n]$ can contribute to the output $y[n]$. Then $h[n]$ must be zero for $n \neq 0$; thus, $h[n] = K\delta[n]$, where $K = h[0]$ is a constant. An LTI system is memoryless if and only if $h[n] = K\delta[n]$. A memoryless LTI system is then a pure gain, described by $y[n] = Kx[n]$. If the gain K is unity ($h[n] = \delta[n]$), the identity system results.

Invertibility

A system is invertible if its input can be determined from its output. An invertible system (impulse response $h[n]$) cascaded with its inverse system (impulse response $h_i[n]$) form the identity system, as shown in Figure 10.11(a). Hence, a discrete-time LTI system with impulse response $h[n]$ is invertible if there exists a function $h_i[n]$ such that

$$h[n]*h_i[n] = \delta[n], \tag{10.31}$$

since the identity system has the impulse response $\delta[n]$.

We do not present a procedure for finding the impulse response $h_i[n]$, given $h[n]$. This problem can be solved using the z-transform of Chapter 11.

A simple example of a noninvertible discrete-time LTI system is given in Figure 10.11(b). The output is zero for n even; hence, the input cannot be determined from the output for n even.

Causality

A discrete-time LTI system is causal if the current value of the output depends on only the current value and past values of the input. This property can be expressed as, for n_1 any integer,

$$y[n_1] = T(x[n]), \quad n \leq n_1$$

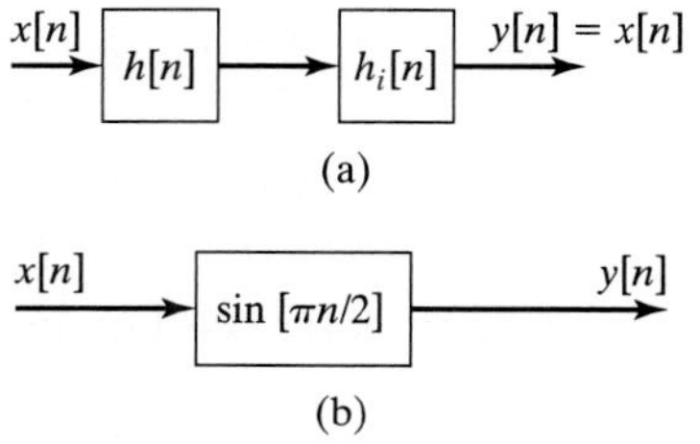

(a)

(b)

Figure 10.11   Illustrations of invertibility.

using the notation of Section 9.5. Recall that a complete input–output description of an LTI system is contained in its impulse response $h[n]$. Because the unit impulse function $\delta[n]$ is nonzero only at $n = 0$, the impulse response $h[n]$ of a causal system must be zero for $n < 0$. The convolution sum for a causal LTI system can then be expressed as

$$y[n] = \sum_{k=-\infty}^{\infty} x[n-k]h[k] = \sum_{k=0}^{\infty} x[n-k]h[k]$$

$$= x[n]h[0] + x[n-1]h[1] + \cdots + x[0]h[n] + x[-1]h[n+1] + \cdots . \tag{10.32}$$

If the impulse response is expressed as $h[n-k]$ for a causal system, this response is zero for $[n-k] < 0$, or for $k > n$. The alternative form of the convolution sum in (10.30) can then be expressed as

$$y[n] = \sum_{k=-\infty}^{n} x[k]h[n-k]$$

$$= x[n]h[0] + x[n-1]h[1] + x[n-2]h[2] + \cdots . \tag{10.33}$$

Note that the expanded sums in (10.32) and (10.33) are identical. Notice that (10.33) makes it explicit that the output $y[n]$ does not depend on future values of the input $x[n]$.

In summary, for a causal discrete-time LTI system, $h[n]$ is zero for $n < 0$ and the convolution sum can be expressed as

$$y[n] = \sum_{k=0}^{\infty} x[n-k]h[k] = \sum_{k=-\infty}^{n} x[k]h[n-k]. \tag{10.34}$$

As an additional point, a signal that is zero for $n < 0$ is called a *causal signal*.

Stability

Recall that a system is bounded-input bounded-output (BIBO) stable if the output remains bounded for *any* bounded input. The boundedness of the input can be expressed as

$$|x[n]| < M,$$

where M is a real constant. Then we can write

$$|y[n]| = \left| \sum_{k=-\infty}^{\infty} x[n-k]h[k] \right| \le \sum_{k=-\infty}^{\infty} |x[n-k]h[k]|$$

$$= \sum_{k=-\infty}^{\infty} |x[n-k]||h[k]| \le \sum_{k=-\infty}^{\infty} M|h[k]| = M \sum_{k=-\infty}^{\infty} |h[k]|. \tag{10.35}$$

Since M is finite, it is *sufficient* that for $y[n]$ to be bounded,

$$\sum_{k=-\infty}^{\infty} |h[k]| < \infty. \tag{10.36}$$

If this relation is satisfied, $h[n]$ is said to be *absolutely summable*.

We now show that (10.36) is *necessary*. Assume that the summation in (10.36) is not bounded. We can choose a bounded input such that $x[n] = (x_+[n] + x_-[n])$, where

$$x_+[n] = \begin{cases} 1, & h[n] \geq 0 \\ 0, & h[n] < 0; \end{cases}$$

$$x_-[n] = \begin{cases} -1 & h[n] < 0 \\ 0, & h[n] \geq 0 \end{cases} . \tag{10.37}$$

Hence, $x[n]$ is either 1 or -1 for each n. The resulting output is

$$y[n] = \sum_{k=-\infty}^{\infty} (x_+[n-k] + x_-[n-k])h[k] = \sum_{k=-\infty}^{\infty} |h[n-k]|. \tag{10.38}$$

This output is unbounded for the bounded input of (10.37), and the system is not stable. Thus, the necessary and sufficient condition for an LTI system to be BIBO stable is that the impulse response be absolutely summable, as given in (10.36). For an LTI causal system, this condition reduces to

$$\sum_{k=0}^{\infty} |h[k]| < \infty. \tag{10.39}$$

EXAMPLE 10.6	**Stability of an LTI discrete system**

We now illustrate the preceding properties with some examples. First, let $h[n] = (\frac{1}{2})^n u[n]$. This system

1. has memory (is dynamic), since $h[n] \neq K\delta[n]$;
2. is causal, because $h[n] = 0$ for $n < 0$;
3. is stable, because, from Appendix C,

$$\sum_{k=-\infty}^{\infty} |h[n]| = \sum_{k=0}^{\infty} \left(\frac{1}{2}\right)^n = \frac{1}{1 - 1/2} = 2.$$

Consider next $h[n] = (2)^n u[n]$. This system has memory and is causal, for the same reasons as for the first system. However, this system is not stable, because

$$\sum_{k=-\infty}^{\infty} |h[n]| = \sum_{k=0}^{\infty} 2^n = 1 + 2 + 4 + 8 + \cdots .$$

This sum is obviously unbounded.

As a final example, consider $h[n] = (\frac{1}{2})^n u[n + 1]$. This system has memory. The system is not causal, since $h[-1] = 2 \neq 0$. The system is stable, because, from earlier,

$$\sum_{k=-\infty}^{\infty} |h[n]| = \sum_{k=-1}^{\infty} \left(\frac{1}{2}\right)^n = 2 + \frac{1}{1 - 1/2} = 4. \qquad \blacksquare$$

Unit Step Response

We now relate the unit step response to the unit impulse response for an LTI system. Suppose that the system input is the unit step function $u[n]$. We denote the unit step response as $s[n]$. Then, from (10.30),

$$s[n] = \sum_{k=-\infty}^{\infty} u[n - k]h[k] = \sum_{k=-\infty}^{n} h[k], \qquad (10.40)$$

since $u[n - k]$ is zero for $(n - k) < 0$, or for $k > n$. Hence, the unit step response can be calculated directly from the unit impulse response.

From (10.40), we form the first difference for $s[n]$:

$$s[n] - s[n - 1] = \sum_{k=-\infty}^{n} h[k] - \sum_{k=-\infty}^{n-1} h[k] = h[n]. \qquad (10.41)$$

Thus, the impulse response can be obtained directly from the unit step response; consequently, the unit step response also completely describes the input–output characteristics of a system.

EXAMPLE 10.7   **Step response from the impulse response**

Consider again the system of Example 10.4, which has the impulse response

$$h[n] = 0.6^n u[n]. \qquad (10.42)$$

This system is dynamic, causal, and stable. The unit step response is then, from (10.40),

$$s[n] = \sum_{k=-\infty}^{n} h[k] = \sum_{k=0}^{n} 0.6^k.$$

From Appendix C, we express this series in the closed form

$$s[n] = \sum_{k=0}^{n} 0.6^k = \frac{1 - 0.6^{n+1}}{1 - 0.6} u[n] = 2.5(1 - 0.6^{n+1})u[n].$$

The factor $u[n]$ is necessary, since $s[n] = 0$ for $n < 0$ (causal system). This result verifies that of Example 10.4, in which the step response was calculated using the convolution summation. Note that the impulse response is obtained from the step response using (10.41); hence, for $n \geq 0$,

$$h[n] = s[n] - s[n - 1]$$
$$= 2.5(1 - 0.6^{n+1})u[n] - 2.5(1 - 0.6^n)u[n - 1].$$

For $n = 0$,

$$h[0] = 2.5(1 - 0.6) = 1.$$

For $n \geq 1$,

$$h[n] = 2.5(1 - 0.6^{n+1} - 1 + 0.6^n)$$
$$= 2.5(0.6^n)(1 - 0.6) = 0.6^n.$$

Hence,

$$h[n] = 0.6^n u[n],$$

which is the given function.   ∎

In this section, the properties of memory, invertibility, causality, and stability were investigated for discrete-time LTI systems. If a system is memoryless, its impulse response is given by $h[n] = K\delta[n]$, with K constant. The impulse response $h[n]$ of a causal system is zero for $n < 0$. A system is stable if its impulse response is absolutely summable, as in (10.36). Invertibility was discussed, but no mathematical test was developed.

10.4   DIFFERENCE-EQUATION MODELS

In Sections 10.1 through 10.3, certain properties of LTI discrete-time systems were developed. We now consider the most common model for systems of this type. LTI discrete-time systems are usually modeled by *linear difference equations with constant coefficients*. We emphasize that *models* of physical systems are being considered, not the physical systems themselves. A common discrete-time LTI physical system is a digital filter. Digital filters are implemented either by digital hardware that is constructed to solve a difference equation, or by a digital computer that is programmed to solve a difference equation. In either case, the difference-equation model is usually accurate, provided that the computer word length is sufficiently long that numerical problems do not occur.

In this section, we consider difference-equation models for LTI discrete-time systems. Then two methods are given for solving linear difference equations with constant coefficients; the first is a classical procedure and the second is an iterative procedure.

Difference-Equation Models

In Example 10.1 and Figure 10.3, we considered a discrete-time system with the difference equation

$$y[n] = (x[n] + x[n - 1] + x[n - 2])/3. \tag{10.43}$$

This is one of the simpler models that we can consider; the current value of the output signal, $y[n]$, is a function of the current value and the last two values of the input signal $x[n]$ only, but is not a function of past output values.

We now consider a first-order discrete-time system model in which the current value of output, $y[n]$, is a function of the last value of output $y[n-1]$ and the current value of the input $x[n]$:

$$y[n] = ay[n-1] + bx[n]. \qquad (10.44)$$

The parameters a and b are constants; Equation (10.44) is a *linear difference equation with constant coefficients*. One of the coefficients is equal to unity, one is a, and one is b.

To show that (10.44) is linear, we use superposition. Suppose that $y_i[n]$ is the solution of (10.44) for the excitation $x_i[n]$, for $i = 1, 2$. By this, we mean that

$$y_i[n] = ay_i[n-1] + bx_i[n], \quad i = 1, 2. \qquad (10.45)$$

We now show that the solution $(a_1y_1[n] + a_2y_2[n])$ satisfies (10.44) for the excitation $(a_1x_1[n] + a_2x_2[n])$, by direct substitution into (10.44):

$$(a_1y_1[n] + a_2y_2[n]) = a(a_1y_1[n-1] + a_2y_2[n-1]) + b(a_1x_1[n] + a_2x_2[n]).$$

This equation is rearranged to yield

$$a_1(y_1[n] - ay_1[n-1] - bx_1[n]) + a_2(y_2[n] - ay_2[n-1] - bx_2[n]) = 0. \quad (10.46)$$

Each term on the left side is equal to zero, from (10.45); hence, the difference equation (10.44) satisfies the principle of superposition and is linear. Also, in (10.44), we replace n with $(n - n_0)$, yielding

$$y[n - n_0] = ay[n - n_0 - 1] + bx[n - n_0]. \qquad (10.47)$$

Thus, an excitation of $x[n - n_0]$ produces a response of $y[n - n_0]$, and (10.44) is also time invariant.

A simple example of a linear difference equation with constant coefficients is the first-order equation

$$y[n] = 0.6y[n-1] + x[n].$$

The equation is first order because the current value of the dependent variable is an explicit function of only the most recent preceding value of the dependent variable, $y[n-1]$.

The general form of an Nth-order linear difference equation with constant coefficients is, with $a_0 \neq 0$,

$$a_0y[n] + a_1y[n-1] + \cdots + a_{N-1}y[n - N + 1] + a_Ny[n - N]$$
$$= b_0x[n] + b_1x[n-1] + \cdots + b_{M-1}x[n - M + 1] + b_Mx[n - M],$$

where $a_0, \ldots, a_N$ and $b_0, \ldots, b_M$ are constants. We consider only the case that these constants are real. The equation is Nth order since the current value of the dependent variable is an explicit function of the last N preceding values of the dependent variable. The general Nth-order equation can be expressed in a more compact form:

$$\sum_{k=0}^{N} a_k y[n-k] = \sum_{k=0}^{M} b_k x[n-k], \quad a_0 \neq 0. \tag{10.48}$$

It can be shown by these procedures that this equation is both linear and time invariant.

Classical Method

Several methods of solution exist for (10.48); in this section, we present one of the classical methods. This method parallels the one given in Appendix E and reviewed for differential equations in Section 3.5. In Chapter 11, the solution by transform methods (the z-transform) is developed.

In practical situations, we use computers to solve both differential and difference equations. If the equations models a system, the solution is called a *system simulation*. Computer simulations are not limited to linear constant-coefficient difference equations; the solution of nonlinear time-varying equations by computer generally presents few additional difficulties as compared to the solution of linear time-invariant equations. However, we have no general analytical techniques for the solution of a nonlinear time-varying difference equation, such that the resulting solution $y[n]$ is expressed as an explicit function of n.

The method of solution of (10.48) to be presented here requires that the general solution $y[n]$ be expressed as the sum of two functions [1]:

$$y[n] = y_c[n] + y_p[n]. \tag{10.49}$$

In this equation, $y_c[n]$ is called the *complementary function* and $y_p[n]$ is a *particular solution*. For the case that the difference equation models a system, the complementary function is often called the *natural response* and the particular solution the *forced response*. We will use this notation. A procedure for finding the natural response is presented first, and the forced response is then considered.

Natural response. To find the natural response, we first write the homogeneous equation, which is (10.48) with the left side set equal to zero—that is,

$$a_0 y[n] + a_1 y[n-1] + \cdots + a_{N-1} y[n-N+1] + a_N[n-N] = 0 \tag{10.50}$$

with $a_0 \neq 0$. The natural response $y_c[n]$ must satisfy this equation. We *assume* that the solution of the homogeneous equation is of the form $y_c[n] = Cz^n$, where C and z are constants to be determined in the solution process.

We now develop the solution procedure. Note that

$$y_c[n] = Cz^n,$$
$$y_c[n-1] = Cz^{n-1} = Cz^{-1}z^n,$$
$$y_c[n-2] = Cz^{n-2} = Cz^{-2}z^n, \tag{10.51}$$
$$\vdots$$
$$y_c[n-N] = Cz^{n-N} = Cz^{-N}z^n.$$

Substitution of these terms into (10.50) yields

$$(a_0 + a_1z^{-1} + \cdots + a_{N-1}z^{-(N-1)} + a_Nz^{-N})Cz^n$$
$$= (a_0z^N + a_1z^{N-1} + \cdots + a_{N-1}z + a_N)Cz^{-N}z^n = 0. \tag{10.52}$$

We assume that our solution $y_c[n] = Cz^n$ is nontrivial ($C \neq 0$); then, from (10.52),

$$a_0z^N + a_1z^{N-1} + \cdots + a_{N-1}z + a_N = 0. \tag{10.53}$$

This equation is called the *characteristic equation*, or the *auxiliary equation*, for the difference equation (10.48). Note that the characteristic equation is a polynomial in z set to zero. The polynomial may be factored as

$$a_0z^N + a_1z^{N-1} + \cdots + a_{N-1}z + a_N = a_0(z - z_1)(z - z_2)\cdots(z - z_N) = 0. \tag{10.54}$$

Hence, N values of z, denoted as z_i, $i = 1, 2, \ldots, N$, satisfy this equation. For the case of no repeated roots, the solution of the homogeneous equation (10.50) may be expressed as

$$y_c[n] = C_1z_1^n + C_2z_2^n + \cdots + C_Nz_N^n, \tag{10.55}$$

since the equation is linear. This solution is called the *natural response* (*complementary function*) of the difference equation (10.48) and contains the N unknown coefficients $C_1, C_2, \ldots, C_N$. These coefficients are evaluated in a later step of the solution procedure. An example is now given.

EXAMPLE 10.8   **Complementary response for an LTI discrete system**

As an example, we consider the first-order difference equation given earlier in the section:

$$y[n] - 0.6y[n-1] = x[n].$$

From (10.53), the characteristic equation is

$$a_0z + a_1 = z - 0.6 = 0 \Rightarrow z = 0.6;$$

thus, the natural response is

$$y_c[n] = C(0.6)^n,$$

where C is yet to be determined.   ■

The unknown constant C cannot be determined until the general (complete) solution has been found. The second part of the general solution is investigated next.

Forced response.   The second part of the general solution of a linear difference equation with constant coefficients,

[eq(10.48)]
$$\sum_{k=0}^{N} a_k y[n - k] = \sum_{k=0}^{M} b_k x[n - k],$$

is called a *forced response (particular solution)* and is denoted by $y_p[n]$. A forced response is any function $y_p[n]$ that satisfies (10.48), that is, that satisfies

$$\sum_{k=0}^{N} a_k y_p[n - k] = \sum_{k=0}^{M} b_k x[n - k]. \tag{10.56}$$

The general solution of (10.48) is then the natural response $y_c[n]$, which is the solution of (10.50), plus a forced response $y_p[n]$, which is the solution of (10.56):

[eq(10.49)]
$$y[n] = y_c[n] + y_p[n].$$

One procedure for evaluating the forced response is to assume that this solution is the sum of functions of the mathematical form of the excitation $x[n]$ and the delayed excitation $x[n - k]$ that differ in form from $x[n]$. This procedure, called the *method of undetermined coefficients*, applies if the forced response as described has a finite number of terms. For example, if

$$x[n] = 3(0.2)^n,$$

we would assume the particular function

$$y_p[n] = P(0.2)^n,$$

where the coefficient P is to be determined. As a second example, if $x[n] = n^3$, then $x[n - k] = (n - k)^3$, which expands into the functional forms n^3, n^2, n^1, and n^0. For this case, the forced response is chosen as

$$y_p[n] = P_1 + P_2 n + P_3 n^2 + P_4 n^3,$$

where the $P_i, i = 1, \ldots, 4$, are unknown. As a third example, if $x[n] = \cos 2n$, the forced response is chosen as

$$y_p[n] = P_1 \cos 2n + P_2 \sin 2n,$$

since a delayed cosine function can be expressed as the sum of a cosine function and a sine function.

The unknown coefficients in $y_p[n]$ are evaluated by direct substitution of the forced response into the difference equation (10.56) and equating coefficients of like mathematical forms that appear on both sides of the equation. An example is given next.

EXAMPLE 10.9   **Total response for an LTI discrete system**

This example is a continuation of Example 10.8. Consider the difference equation

$$y[n] - 0.6y[n-1] = 4u[n].$$

Because the forcing function is constant, the forced response is chosen as

$$y_p[n] = P,$$

where P is constant. Substitution of the forced response into the difference equation yields

$$y_p[n] - 0.6y_p[n-1] = P - 0.6P = 4 \Rightarrow P = 10,$$

or $y_p[n] = P = 10$. From (10.49), the general solution is

$$y[n] = y_c[n] + y_p[n] = C(0.6)^n + 10$$

for $n \geq 0$. Recall that $y_c[n]$ was found in Example 10.8. We now have the problem of evaluating the unknown coefficient C. The procedure for this evaluation is given next; then the solution is completed.   ∎

Once the general solution has been found, as in Example 10.9, the remaining N unknown coefficients $C_1, C_2, \ldots, C_N$ of the natural response (10.55) must be calculated. Thus, we must have N independent conditions to evaluate these unknowns; these N conditions are the initial conditions. Generally, the initial conditions can be expressed as $y[m], y[m+1], \ldots, y[m+N-1]$, where m is given in the specifications of the equation, and we solve for $y[n], n \geq m + N$. For many cases, m is equal to zero, and we solve for $y[n]$ for $n \geq N$. This procedure is illustrated by completing Example 10.9.

EXAMPLE 10.10   **Calculations for Example 10.9**

For the system of Examples 10.8 and 10.9, suppose that the system is initially at rest, that is, the initial conditions are zero. Because the input is $u[n]$ and the system is first order, the required initial condition is $y[0] = 0$. The total solution is, from Example 10.9,

$$y[n] = C(0.6)^n + 10.$$

Thus,

$$y[0] = 0 = C + 10 \Rightarrow C = -10,$$

and the total solution is given by

$$y[n] = 10[1 - (0.6)^n]u[n].$$

As a check of this solution, we solve the difference iteratively for the first three unknown values.

$$y[0] = 0;$$
$$y[1] = 0.6y[0] + 4 = 0.6(0) + 4 = 4;$$
$$y[2] = 0.6y[1] + 4 = 0.6(4) + 4 = 6.4;$$
$$y[3] = 0.6y[2] + 4 = 0.6(6.4) + 4 = 7.84.$$

We now find the first three values from our solution:

$$y[0] = 10[1 - 0.6^0] = 0;$$
$$y[1] = 10[1 - 0.6^1] = 4;$$
$$y[2] = 10[1 - 0.6^2] = 6.4;$$
$$y[3] = 10[1 - 0.6^3] = 7.84.$$

The solution checks for these values. The difference equation is evaluated by the following MATLAB program:

```
n=[0:9];
y(1)=0;
for m=2:10;
   y(m)=0.6*y(m-1)+4;
end
y
stem(n,y,'fill'),grid
```

What is the relationship of m in this program to n in $y[n]$?

The steady-state output $y_{ss}[n]$ can also be verified. From the solution,

$$y[n] = 10[1 - (0.6)^n]u[n] \Rightarrow \lim_{n \to \infty} y[n] = y_{ss}[n] = 10.$$

For the difference equation, as $n \to \infty$, $y_{ss}[n - 1] \to y_{ss}[n]$, and the difference equation is given by

$$y_{ss}[n] - 0.6y_{ss}[n] = 0.4y_{ss}[n] = 4 \Rightarrow y_{ss}[n] = 10.$$

Hence, the steady-state value also checks.   ■

The natural-response part of the general solution of a linear difference equation with constant coefficients,

[eq(10.48)]
$$\sum_{k=0}^{N} a_k y[n - k] = \sum_{k=0}^{M} b_k x[n - k],$$

is independent of the forcing function $x[n]$ and is dependent only on the structure of the system [the left side of (10.48)]; hence, the name *natural response*. It is also called the *unforced response*, or the *zero-input response*. In the preceding example,

$$y_c[n] = C(0.6)^n.$$

The *form* of this component, $(0.6)^n$, is independent of both the input and the initial conditions; it is a function only of the structure of the system. However, the unknown constant C is a function of both the excitation and the initial conditions.

The forced response is also called the *zero-state response*. In this application, the term *zero state* means *zero initial conditions*. The forced response is a function of the system structure *and* of the excitation, but is independent of the initial conditions. For almost all models of physical systems, the natural response goes to zero with increasing time, and then only the forced part of the response remains. (The requirement for this to occur in an LTI system is that the system be BIBO stable.) For this reason, we sometimes refer to the forced response as the *steady-state response*, as in Example 10.10 and the natural response as the *transient response*. When we refer to the steady-state response of a system, we are speaking of the forced response of a difference equation. Of course, the steady-state response of a physical system is found by applying the excitation, allowing the system to settle into steady state, and then measuring the response.

Repeated roots.   The natural response

[eq(10.55)] $$y_c[n] = C_1 z_1^n + C_2 z_2^n + \cdots + C_N z_N^n$$

does not apply if any of the roots of the characteristic equation,

[eq(10.54)]
$$a_0 z^N + a_1 z^{N-1} + \cdots + a_{N-1} z + a_N$$
$$= a_0 (z - z_1)(z - z_2)\cdots(z - z_N) = 0,$$

are repeated. Suppose, for example, that a fourth-order difference equation has the characteristic equation

$$z^4 + a_1 z^3 + a_2 z^2 + a_3 z + a_4 = (z - z_1)^3 (z - z_4). \tag{10.57}$$

The natural response must then be assumed to be of the form

$$y_c[n] = (C_1 + C_2 n + C_3 n^2) z_1^n + C_4 z_4^n. \tag{10.58}$$

The remainder of the procedure for finding the general solution is unchanged. For the general case of an rth-order root z_i in the characteristic equation, the corresponding term in the natural response is

$$\text{term} = (C_1 + C_2 n + C_3 n^2 + \cdots + C_r n^{r-1}) z_i^n. \tag{10.59}$$

Solution by Iteration

A difference equation can *always* be solved by iteration. This procedure was illustrated in Example 10.10 and is now developed further.

The N-order difference equation (10.48) can be expressed as

$$y[n] = -\sum_{k=1}^{N} a_k y[n-k] + \sum_{k=0}^{M} b_k x[n-k]$$
$$= -a_1 y[n-1] - \cdots -a_N y[n-N] \qquad (10.60)$$
$$+ b_0 x[n] + b_1 x[n-1] + \cdots + b_M x[n-M],$$

where we have normalized a_0 to unity, for convenience. As stated before, the initial conditions for this equation are the N values $y[m]$, $y[m+1]$, $\ldots y[m+N-1]$. To illustrate this procedure, we let $m = 0$. From (10.60), we solve for $y[N]$:

$$y[N] = -a_1 y[N-1] - a_2 y[N-2] - \cdots -a_N y[0]$$
$$+ b_0 x[N] + b_1 x[N-1] + \cdots + b_M x[N-M].$$

Note that all terms on the right side are known. Having found $y[N]$, we can now calculate $y[N+1]$:

$$y[N+1] = -a_1 y[N] - a_2 y[N-1] - \cdots -a_N y[1]$$
$$+ b_0 x[N+1] + b_1 x[N] + \cdots + b_M x[N-M+1].$$

Once again, all terms on the right side are known. Knowing $y[N+1]$ allows us to solve for $y[N+2]$:

$$y[N+2] = -a_1 y[N+1] - a_2 y[N] - \cdots -a_N y[2]$$
$$+ b_0 x[N+2] + b_1 x[N+1] + \cdots + b_M x[N-M+2].$$

Using this procedure, we can solve for $y[n]$ for any value of n. This procedure is long if we wish to solve for $y[n]$ for n large; however, the procedure is ideally suited for solution by digital computer. A simulation for a discrete-time system is usually a computer solution of the system difference equation by iteration, as illustrated by the MATLAB program in Example 10.10.

In this section, we consider the modeling of discrete-time LTI systems by linear difference equations with constant coefficients. A classical procedure is given for solving these difference equations, resulting in $y[n]$ as an explicit function of n. As the final topic, the iterative solution of difference equations is developed. This solution does not result in $y[n]$ as an explicit function of n but is easily implemented on a digital computer; this implementation is called a system simulation. In addition, the iterative-solution procedure applies to time-varying nonlinear difference equations.

10.5 TERMS IN THE NATURAL RESPONSE

We now relate the terms of the natural response (complementary function) of a discrete-time LTI system to the signals that were studied in Section 9.4. The mathematical forms of the terms in the natural response are determined by the roots of the characteristic equation

[eq(10.54)]
$$a_0 z^N + a_1 z^{N-1} + \cdots + a_{N-1} z + a_N$$
$$= a_0(z - z_1)(z - z_2)\cdots(z - z_N) = 0.$$

With the roots distinct, the natural response is given by

[eq(10.55)]
$$y_c[n] = C_1 z_1^n + C_2 z_2^n + \cdots + C_N z_N^n.$$

Hence, the general term is given by $C_i z_i^n$, where z_i^n is called a *system mode*. The root z_i of the characteristic equation may be real or complex. However, because the coefficients of the characteristic equation are real, complex roots must occur in complex conjugate pairs. We now consider the different forms of the modes that can appear in the natural response.

z_i Real and Positive

If z_i is real, we let $z_i = e^{\Sigma_i}$ with Σ_i real, and

$$C_i z_i^n = C_i(e^{\Sigma_i})^n = C_i e^{\Sigma_i n} \qquad (10.61)$$

with $\Sigma_i = \ln z_i$. This term is exponential in form.

z_i Complex or Negative

If z_i is complex, we let

$$z_i = |z_i|e^{j\theta_i} = e^{\Sigma_i}e^{j\Omega_i} \qquad (10.62)$$

with $\Sigma_i = \ln|z_i|$ and $\Omega_i = \theta_i$. Since the natural response $y_c[n]$ must be real; two of the terms of $y_c[n]$ can be expressed as, with $C_i = |C_i|e^{j\beta_i}$,

$$\begin{aligned} C_i z_i^n + C_i^*(z_i^*)^n &= |C_i|e^{j\beta_i}e^{\Sigma_i n}e^{j\Omega_i n} + |C_i|e^{-j\beta_i}e^{\Sigma_i n}e^{-j\Omega_i n} \\ &= |C_i|e^{\Sigma_i n}e^{j(\Omega_i n + \beta_i)} + |C_i|e^{\Sigma_i n}e^{-j(\Omega_i n + \beta_i)} \\ &= 2|C_i|e^{\Sigma_i n}\cos(\Omega_i n + \beta_i). \end{aligned} \qquad (10.63)$$

If Σ_i is zero, the term is an undamped sinusoid. If Σ_i is negative, the term is a damped sinusoid, and the term approaches zero as n approaches infinity. If Σ_i is positive, the term becomes unbounded as n approaches infinity.

We see then that the terms that were discussed in detail in Section 9.4 appear in the natural response of an LTI discrete-time system. The natural-response terms are always present, independent of the type of excitation to the system.

Stability

We now relate the stability of a *causal* discrete-time LTI system to the roots of the system characteristic equation. As stated earlier, the general term in the natural response is of the form $C_i z_i^n$, where z_i is a root of the system characteristic equation. The magnitude of this term is given by $|C_i||z_i|^n$. If $|z_i|$ is less than unity, the magnitude of the term approaches zero as n approaches infinity. If $|z_i|$ is greater than unity, the magnitude of the term becomes unbounded as n approaches infinity.

The solution of a constant-coefficient linear difference equation is given by

[eq(10.49)]
$$y[n] = y_c[n] + y_p[n].$$

Recall also that the forced response $y_p[n]$ is of the mathematical form as the system input $x[n]$. Hence, if $x[n]$ is bounded, $y_p[n]$ is also bounded. If all roots of the characteristic equation satisfy the relation $|z_i| < 1$, each term of the natural response $y_c[n]$ is also bounded. Thus, the necessary and sufficient condition that a causal discrete-time LTI system is BIBO stable is that $|z_i| < 1$. We now illustrate the determination of stability with an example.

EXAMPLE 10.11   **Stability of a discrete system**

Suppose that a causal system is described by the difference equation

$$y[n] - 1.25y[n - 1] + 0.375y[n - 2] = x[n].$$

From (10.48) and (10.54), the system characteristic equation is

$$z^2 - 1.25z + 0.375 = (z - 0.75)(z - 0.5) = 0.$$

A MATLAB program that calculates these roots is

```
p=[1 -1.25 .375];
r=roots(p)
results: r=0.75 0.5
```

This system is stable, since the magnitude of each root is less than unity. The natural response is given by

$$y_c[n] = C_1(0.75)^n + C_2(0.5)^n.$$

This function approaches zero as n approaches infinity.

Consider a second causal system described by the difference

$$y[n] - 2.5y[n - 1] + y[n - 2] = x[n].$$

The system characteristic equation is given by

$$z^2 - 2.5z + 1 = (z - 2)(z - 0.5).$$

The system is unstable because the root $z_1 = 2$ is greater than unity. The natural response is given by

$$y_c[n] = C_1(2)^n + C_2(0.5)^n.$$

The instability is evident in the term $C_1(2)^n$.   ■

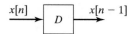

Figure 10.12   Representation of a delay.

10.6   BLOCK DIAGRAMS

In Section 9.5, the representation of discrete-time systems by block diagrams was introduced. In this section, we extend these representations to LTI systems described by difference equations.

The representation of difference equations by block diagrams requires the use of an ideal delay, as discussed earlier. We use the block shown in Figure 10.12 to represent this delay. Recall that if the signal into an ideal delay is $x[n]$, the signal out at that instant is $x[n - 1]$.

We express the operation in Figure 10.12 in the standard form

$$x[n] \rightarrow y[n] = x[n - 1]. \tag{10.64}$$

One implementation of an ideal delay uses a memory location in a digital computer. One digital-computer program segment illustrating this implementation is given by the two statements

```
          ⋮
XNMINUS1 = X
X = XN
          ⋮
```

This segment applies to many high-level languages. In this segment, X is the memory location, XNMINUS1 is the number shifted out, and XN is the number shifted in. Note that the delay is *not* realized if the two statements are reversed in order; if the number XN is shifted in first, the number X stored in the memory location is overwritten and lost. This delay will now be used in an example.

EXAMPLE 10.12   **Simulation diagram for a discrete system**

In Section 10.4, we considered a discrete-time system described by the difference equation

$$y[n] - 0.6y[n - 1] = x[n]. \tag{10.65}$$

We now find a block diagram, constructed of certain specified elements, that satisfies this equation. The difference equation can be written as

$$y[n] = 0.6y[n - 1] + x[n].$$

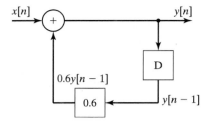

Figure 10.13 System for Example 10.12.

If we assume that $y[n]$ is available, we can realize the right side of this equation using a delay, a multiplication by 0.6, and a summing junction. This is done in the block diagram of Figure 10.13, and the loop is completed to satisfy (10.65). Hence, we have constructed a block diagram that satisfies the given difference equation. ∎

The block diagram of Figure 10.13 can be implemented by a digital computer program; for example, one MATLAB program is given by, with $x[n] = 4$,

```
ynminus1=0;
for n=0:3
    yn=0.6*ynminus1+4
    ynminus1=yn;
end
result: yn=4  6.4  7.84  8.704
```

The first statement sets $y[0]$. The beginning value of n can be zero, since no subscripting is used explicitly in the program. The variable *yn* is the current value of $y[n]$, and *ynminus1* is the current value of $y[n-1]$. The first statement in the *for* loop solves the difference equation, while the second one stores $y[n-1]$ for the next iteration. A semicolon at the end of a statement suppresses printing for that statement. Hence, only *yn* is printed for each iteration of the loop. As an exercise, the reader can relate each statement to the block diagram of Figure 10.13.

Compare this simulation to the one given in Example 10.10. The variables of that simulation were subscripted; these variables are not. Complex simulations are normally written in the manner just described, since subscripting variables requires that the variable be stored for each value of the subscript. In the previous simulation, only two values, *yn* and *ynminus1*, are stored.

A common discrete-time system is a digital filter. If the difference equation of Example 10.12 were a digital-filter equation, the given computer program is a realization of that filter.

As stated earlier, the block diagram of Figure 10.13 is sometimes called a simulation diagram. The given MATLAB program is a machine solution to the difference equation; hence, the program is a digital simulation of the system. One procedure for constructing a simulation of a complex discrete-time system is first to draw a simulation diagram that is based on ideal delays, multiplications by constants, and summing junctions. The computer program is then written directly from the simulation diagram. For the case that the difference equation represents a digital filter, the simulation diagram is also called a *programming form*.

Two Standard Forms

It can be shown that an unbounded number of simulation diagrams can be drawn for a given difference equation. Two standard forms will be given here. We consider some other forms in Chapter 13.

As stated in Section 10.5, with $a_0 \neq 0$, the general form of an Nth-order linear difference equation with constant coefficients is

$$[eq(10.48)] \qquad \sum_{k=0}^{N} a_k y[n - k] = \sum_{k=0}^{M} b_k x[n - k].$$

To introduce the standard forms, a second-order difference equation will be considered. We will then develop the forms for the Nth-order equation of (10.48).

For a second-order system, we can express the difference equation as

$$a_0 y[n] + a_1 y[n - 1] + a_2 y[n - 2] = b_0 x[n] + b_1 x[n - 1] + b_2 x[n - 2]. \quad (10.66)$$

We denote the right side of this equation as

$$w[n] = b_0 x[n] + b_1 x[n - 1] + b_2 x[n - 2].$$

A representation of this equation by a block diagram is shown in Figure 10.14(a). Then (10.66) becomes

$$a_0 y[n] + a_1 y[n - 1] + a_2 y[n - 2] = w[n].$$

Solving for $y[n]$ yields

$$y[n] = \frac{1}{a_0}[w[n] - a_1 y[n - 1] - a_2 y[n - 2]], \qquad (10.67)$$

with $a_0 \neq 0$. This equation can be realized by the system of Figure 10.14(b). The total realization is the cascade (series) connection of the systems in Figure 10.14(a) and (b), with the result given in Figure 10.14(c). This block diagram realizes (10.66) and is called the *direct form I* realization.

A second form for realizing a difference equation is now derived by manipulating form I of Figure 10.14(c). The system of this figure is seen to be two systems in cascade. The first system has the input $x[n]$ and the output $w[n]$, and the second system has the input $w[n]$ and the output $y[n]$. Because the systems are linear, the order of the two systems can be reversed without affecting the input–output characteristics of the total system. The result is shown in Figure 10.15(a).

Note that in Figure 10.15(a), the same signal is delayed by the two sets of cascaded delays; hence, the outputs of the delays labeled "1" are equal, as are the outputs of the delays labeled "2." Therefore, one set of the cascaded delays can be eliminated. The final system is given in Figure 10.15(b), and we see that only two delays are required. This block diagram is called the *direct form II* realization.

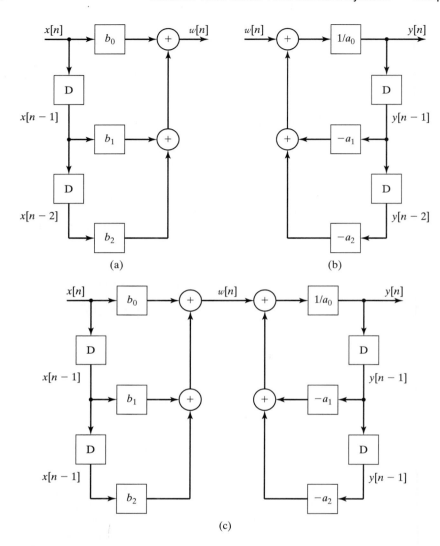

Figure 10.14 Direct form I realization of a second-order system.

Consider now the Nth-order difference equation

[eq(10.48)] $$\sum_{k=0}^{N} a_k y[n-k] = \sum_{k=0}^{M} b_k x[n-k].$$

Solving this equation for $y[n]$ yields

$$y[n] = \frac{1}{a_0}\left[\sum_{k=0}^{M} b_k x[n-k] - \sum_{k=1}^{N} a_k y[n-k]\right] \qquad (10.68)$$

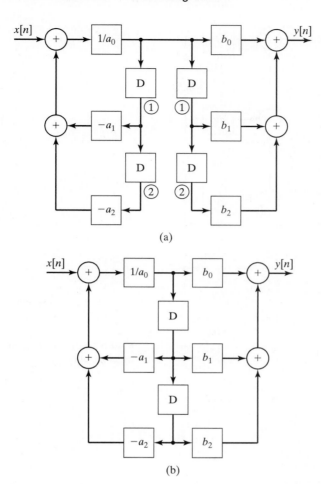

(a)

(b)

Figure 10.15 Direct form II realization of a second-order system.

with $a_0 \neq 0$. Using the procedure illustrated with the second-order systems, we construct the block diagrams for the direct form I and the direct form II. The block diagrams are given in Figures 10.16 and 10.17 for $M = N$. For $M \neq N$, the appropriate coefficients in the figures are set to zero.

EXAMPLE 10.13   **A low-pass digital filter**

We consider again the α-filter of Example 9.8, Section 9.5. Two applications of this filter are in radar-signal processing and in automatic aircraft-landing systems. The purpose of the filter is to remove high-frequency noise from the input signal, while passing the lower-frequency information in that signal. The filter equation is given by

$$y[n] - (1 - \alpha)y[n - 1] = \alpha x[n].$$

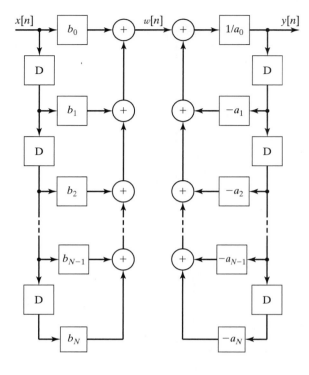

Figure 10.16 Direct form I for an Nth-order system.

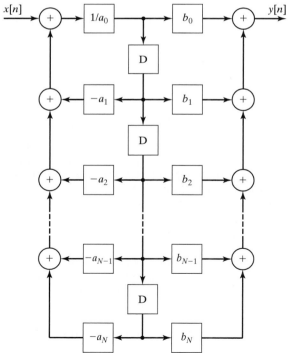

Figure 10.17 Direct form II for an Nth-order system.

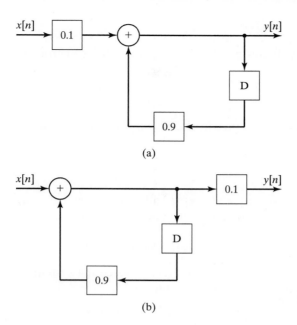

Figure 10.18 Realizations of an α-Filter: (a) form I; (b) form II.

This equation is often solved in real time and is then expressed as

$$y(nT) - (1 - \alpha)y(nT - T) = \alpha x(nT).$$

The parameter α and the sample period T determine the frequency range at which noise is rejected. The mode of the filter is $(1 - \alpha)^n$; thus, $|1 - \alpha|$ must be less than unity for stability. For example, letting $\alpha = 0.1$ results in the filter equation

$$y[n] - 0.9y[n - 1] = 0.1x[n].$$

The form I and form II realizations are given in Figure 10.18. For this filter, the only difference in the two forms is the movement of the gain of 0.1 from the input of the filter to the output. ■

In this section, two standard forms are developed for the realization of difference equations with constant coefficients. These realizations, also called simulation diagrams or programming forms, are constructed of ideal delays, multiplications by constants, and summing junctions. Realizations are discussed in greater detail when state variables are introduced in Chapter 13.

10.7 SYSTEM RESPONSE FOR COMPLEX-EXPONENTIAL INPUTS

First, this section considers further the linearity property for systems. Then the response of discrete-time LTI systems to a certain class of input signals is derived.

Linearity

Consider the discrete-time LTI system depicted in Figure 10.19. This system is denoted by

$$x[n] \rightarrow y[n]. \tag{10.69}$$

From (10.16), this relation can also be expressed as the convolution sum

$$y[n] = \sum_{k=-\infty}^{\infty} x[n - k]h[k]. \tag{10.70}$$

The functions $x[n]$, $y[n]$, and $h[n]$ are all real for physical systems.
  Consider next two real inputs $x_i[n]$, $i = 1, 2$. In (10.69),

$$x_i[n] \rightarrow y_i[n], \quad i = 1, 2, \tag{10.71}$$

and $y_i[n]$, $i = 1, 2$, are real, from (10.70). Since the system of (10.69) is linear, the principle of superposition applies, and

$$a_1 x_1[n] + a_2 x_2[n] \rightarrow a_1 y_1[n] + a_2 y_2[n]. \tag{10.72}$$

  No restrictions exist on the constants a_1 and a_2 in (10.72); hence, these constants may be chosen to be complex. For this development, we choose the constants to be

$$a_1 = 1, \quad a_2 = j = \sqrt{-1}.$$

The superposition property of (10.72) becomes

$$x_1[n] + j x_2[n] \rightarrow y_1[n] + j y_2[n]. \tag{10.73}$$

This result may be stated as follows: For a complex input function to an LTI system, the real part of the input produces the real part of the output, and the imaginary part of the input produces the imaginary part of the output.

Complex Inputs for LTI Systems

The response of an LTI system to the complex-exponential input (see Section 9.4)

$$x[n] = X z^n \tag{10.74}$$

with X and z constant, is now investigated. For the general case, both X and z are complex. We investigate the important case that the system of (10.70) and Figure 10.19 is stable and is modeled by an Nth-order linear difference equation

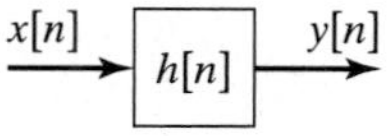

Figure 10.19 LTI system.

with constant coefficients. The exponential input of (10.74) is assumed to exist for all time; hence, the *steady-state system response* will be found. In other words, we will find the forced response for a difference equation with constant coefficients, for a complex-exponential excitation.

The difference-equation model for an Nth-order LTI system is, with $a_N \neq 0$,

$$[eq(10.48)] \qquad \sum_{k=0}^{N} a_k y[n-k] = \sum_{k=0}^{M} b_k x[n-k],$$

where $a_0, \ldots, a_N$ and $b_0, \ldots, b_M$ are real constants. For convenience, we let $M = N$. If this is not the case, certain coefficients must be set to zero.

For the complex-exponential excitation of (10.74), recall from Section 10.4 that the forced response (steady-state response) of (10.48) is of the same mathematical form; hence,

$$y_{ss}[n] = Y z^n, \tag{10.75}$$

where $y_{ss}[n]$ is the steady-state response and Y is a complex constant to be determined. [z is known from (10.74).] We denote the forced response as $y_{ss}[n]$ rather than $y_p[n]$, for clarity. From (10.74) and (10.75), the terms of (10.48) become

$$
\begin{aligned}
a_0 y_{ss}[n] &= a_0 Y z^n; & b_0 x[n] &= b_0 X z^n; \\
a_1 y_{ss}[n-1] &= a_1 Y z^{n-1} & b_1 x[n-1] &= b_1 X z^{n-1} \\
&= a_1 z^{-1} Y z^n; & &= b_1 z^{-1} X z^n; \\
a_2 y_{ss}[n-2] &= a_2 Y z^{n-2} & b_2 x[n-2] &= b_2 X z^{n-2} \\
&= a_2 z^{-2} Y z^n; & &= b_2 z^{-2} X z^n; \\
&\ \ \vdots & &\ \ \vdots \\
a_N y_{ss}[n-N] &= a_N Y z^{n-N} & b_N x[n-N] &= b_N X z^{n-N} \\
&= a_N z^{-N} Y z^n; & &= b_N z^{-N} X z^n. \tag{10.76}
\end{aligned}
$$

These terms are substituted into (10.48), resulting in

$$(a_0 + a_1 z^{-1} + \cdots + a_{N-1} z^{-N+1} + a_N z^{-N}) Y z^n$$

$$= (b_0 + b_1 z^{-1} + \cdots + b_{N-1} z^{-N+1} + b_N z^{-N}) X z^n. \tag{10.77}$$

The only unknown in the steady-state response $y_{ss}[n]$ of (10.75) is Y. In (10.77), the factor z^n cancels, and Y is given by

$$Y = \left[\frac{b_0 + b_1 z^{-1} + \cdots + b_{N-1} z^{-N+1} + b_N z^{-N}}{a_0 + a_1 z^{-1} + \cdots + a_{N-1} z^{-N+1} + a_N z^{-N}} \right] X = H(z) X. \tag{10.78}$$

It is standard practice to denote the ratio of polynomials as

$$H(z) = \frac{b_0 + b_1 z^{-1} + \cdots + b_{N-1} z^{-N+1} + b_N z^{-N}}{a_0 + a_1 z^{-1} + \cdots + a_{N-1} z^{-N+1} + a_N z^{-N}}. \tag{10.79}$$

(We show later that this function is related to the impulse response $h[n]$.) A second common method for specifying the transfer function is given by multiplying the numerator and the denominator of (10.79) by z^N:

$$H(z) = \frac{b_0 z^N + b_1 z^{N-1} + \cdots + b_{N-1} z + b_N}{a_0 z^N + a_1 z^{N-1} + \cdots + a_{N-1} z + a_N}. \tag{10.80}$$

The function $H(z)$ is called a *transfer function* and is said to be Nth order. The order of a transfer function is the same as that of the difference equation upon which the transfer function is based.

We now summarize the foregoing development. Consider an LTI system with the transfer function $H(z)$ as given in (10.79) and (10.80). If the system excitation is the complex exponential $x[n] = X z_1^n$, then, from (10.74), (10.75), and (10.78), the steady-state response is given by

$$x[n] = X z_1^n \rightarrow y_{ss}[n] = X H(z_1) z_1^n. \tag{10.81}$$

We now show that the complex-exponential solution in (10.81) also applies for the case of sinusoidal inputs. Suppose that, in (10.81), $X = |X| e^{j\phi}$ and $z_1 = e^{j\Omega_1}$, where ϕ and Ω_1 are real. Then

$$x[n] = X z_1^n = |X| e^{j\phi} e^{j\Omega_1 n} = |X| e^{j(\Omega_1 n + \phi)}$$
$$= |X| \cos{(\Omega_1 n + \phi)} + j|X| \sin{(\Omega_1 n + \phi)}. \tag{10.82}$$

Because, in general, $H(z_1) = H(e^{j\Omega_1})$ is also complex, we let $H(e^{j\Omega_1}) = |H(e^{j\Omega_1})| e^{j\theta_H}$. The right side of (10.81) can be expressed as

$$y_{ss}[n] = X H(e^{j\Omega_1}) e^{j\Omega_1 n} = |X| |H(e^{j\Omega_1})| e^{j(\Omega_1 n + \phi + \theta_H)}$$
$$= |X| |H(e^{j\Omega_1})| [\cos{(\Omega_1 n + \phi + \theta_H)} + j \sin{(\Omega_1 n + \phi + \theta_H)}].$$

From (10.73) and (10.82), since the real part of the input signal produces the real part of the output signal,

$$|X| \cos{(\Omega_1 n + \phi)} \rightarrow |X| |H(e^{j\Omega_1})| \cos{(\Omega_1 n + \phi + \theta_H)}. \tag{10.83}$$

This result is general for an LTI system and is fundamental to the analysis of LTI systems with periodic inputs.

Suppose that a system is specified by its transfer function $H(z)$. To obtain the system difference equation, we reverse the steps in (10.76) through (10.79). In fact, in $H(z)$ the numerator coefficients b_i are the coefficients of $x[n - i]$, and the denominator coefficients a_i are the coefficients of $y[n - i]$; we can then consider the transfer function to be a shorthand notation for a difference equation. Therefore, the system difference equation can be written directly from the transfer function $H(z)$; consequently, $H(z)$ *is a complete description of the system, regardless of the input function.* For this reason, an LTI system can be represented by the block diagram in Figure 10.20 with the system transfer function given inside the block. It is common engineering practice to specify an LTI system in this manner.

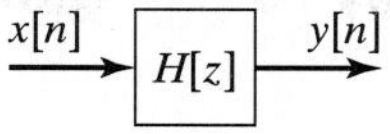

$$\xrightarrow{x[n]} \boxed{H[z]} \xrightarrow{y[n]}$$

Figure 10.20   LTI system.

The form of the transfer function in (10.79) [and (10.80)], which is a ratio of polynomials, is called a *rational function*. The transfer function of a discrete-time LTI system described by a linear difference equation with constant coefficients, as in (10.48), will *always* be a rational function.

We now consider three examples to illustrate the preceding developments.

EXAMPLE 10.14   **Transfer function for a discrete system**

In this example, we illustrate the transfer function using the α-filter of Example 10.13. The difference equation of the α-filter is given by

$$y[n] - (1 - \alpha)y[n - 1] = \alpha x[n].$$

The coefficients, as given in (10.48), are

$$a_0 = 1, \quad a_1 = -(1 - \alpha), \quad b_0 = \alpha.$$

The filter transfer function is normally given in one of two different ways, from (10.79) and (10.80):

$$H(z) = \frac{\alpha}{1 - (1 - \alpha)z^{-1}} = \frac{\alpha z}{z - (1 - \alpha)}.$$

This transfer function is first order. Figure 10.21 shows the α-filter as a block diagram.

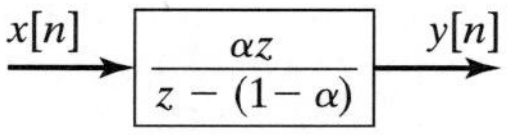

$$\xrightarrow{x[n]} \boxed{\dfrac{\alpha z}{z - (1 - \alpha)}} \xrightarrow{y[n]}$$

Figure 10.21 α-Filter.   ∎

EXAMPLE 10.15   **Sinusoidal response for a discrete system**

In this example, we calculate the system response of an LTI system with a sinusoidal excitation. Consider the α-filter of Example 10.14, with $\alpha = 0.1$. The transfer function is given by

$$H(z) = \frac{0.1z}{z - 0.9}.$$

Suppose that the system is excited by the sinusoidal signal $x[n] = 5 \cos(0.01n + 20°)$. In (10.83), with $e^{j\Omega_1} = e^{j0.01} = 1\underline{/0.573°}$,

$$H(z)|_{z=e^{j0.01}}X = \frac{0.1(e^{j0.01})}{e^{j0.01} - 0.9}(5\underline{/20°})$$

$$= \frac{0.5\underline{/20.573°}}{0.99995 + j0.0100 - 0.9}$$

$$= \frac{0.5 \underline{/20.573°}}{0.1004 \underline{/5.71°}}$$

$$= 4.98 \underline{/14.86°}.$$

A MATLAB program that performs these calculations is as follows:

```
n = [.1 0];
d = [1 -.9];
z = exp(.01*j);
h = polyval(n,z)/polyval(d,z);
ymag = 5*abs(h)
yphase = 20 + angle(h)*180/pi
result: ymag = 4.9777 yphase = 14.8596
```

Thus, from (10.83), the system response is given by

$$y_{ss}[n] = 4.98 \cos(0.01n + 14.86°).$$

Note the calculation required:

$$H(e^{j0.01}) = \frac{0.1(e^{j0.01})}{e^{j0.01} - 0.9} = 0.996 \underline{/-5.14°}.$$

The steady-state response can be written directly from this numerical value for the transfer function, from (10.83).

$$y_{ss}[n] = (0.996)(5)\cos(0.01n + 20° - 5.14°)$$
$$= 4.98 \cos(0.01n + 14.86°). \qquad \blacksquare$$

EXAMPLE 10.16   **Continuation of Example 10.15**

This example is a continuation of Example 10.15; we will demonstrate the low-pass nature of the α-filter. Suppose that the sinusoidal input has the discrete frequency $\Omega = 3$, with the input signal given by $x[n] = \cos(3n)$. Then $z = e^{j3} = 1 \underline{/171.9°}$, and

$$H(e^{j3}) = \frac{0.1(e^{j3})}{e^{j3} - 0.9} = \frac{0.1 \underline{/171.9°}}{1 \underline{/171.9°} - 0.9}$$
$$= \frac{0.1 \underline{/171.9°}}{1.895 \underline{/175.7°}} = 0.0528 \underline{/-3.8°}.$$

Thus,

$$H(e^{j0.01}) = 0.996 \underline{/-5.14°}, \quad H(e^{j3}) = 0.0528 \underline{/-3.8°}.$$

The gain of the filter for the discrete frequency $\Omega = 0.01$ is approximately unity, and for $\Omega = 3$ is approximately $\frac{1}{20}$. Hence, the filter passes frequencies in the vicinity of $\Omega = 0.01$ and rejects frequencies in the vicinity of $\Omega = 3$. A property of discrete-time systems is that the system frequency response is periodic with period $\Omega = 2\pi$. (See Section 9.3.) Hence, we must be careful in drawing general conclusions from this example. The periodic nature of the frequency response is considered in greater detail in Chapter 12. $\blacksquare$

Consider now the case in which the input function is a sum of complex exponentials:

$$x[n] = \sum_{k=1}^{M} X_k z_k^n. \tag{10.84}$$

By superposition and from (10.81), the response of an LTI system with the transfer function $H(z)$ is given by

$$y_{ss}[n] = \sum_{k=1}^{M} X_k H(z_k) z_k^n. \tag{10.85}$$

Stability

It is shown in Section 10.5 that a causal LTI system described by a difference equation is BIBO stable provided that the roots of the characteristic equation, (10.54), all have magnitudes less than unity. The polynomial in (10.54) is identical to that in the denominator in $H(z)$ in (10.80). Hence, the transfer function (10.80) can be expressed as

$$
\begin{aligned}
H(z) &= \frac{b_0 z^N + b_1 z^{N-1} + \cdots + b_{N-1} z + b_N}{a_0 z^N + a_1 z^{N-1} + \cdots + a_{N-1} z + a_N} \\[2mm]
&= \frac{b_0 z^N + b_1 z^{N-1} + \cdots + b_{N-1} z + b_N}{a_0 (z - z_1)(z - z_2) \cdots (z - z_N)}.
\end{aligned} \tag{10.86}
$$

The values z_i (roots of the characteristic equation) are called the *poles* of the transfer function. Hence, a discrete-time system described by the transfer function $H(z)$ is BIBO stable provided that the transfer-function poles all have magnitudes less than one.

Sampled Signals

The foregoing derivations also apply if the discrete-time signals are obtained by sampling continuous-time signals. For this case, the discrete frequency Ω and the continuous frequency ω are related by $\Omega = \omega T$, where T is the sample period. This relation was derived in Section 9.4. Hence, in the preceding equations, for sampled signals all equations apply directly, with Ω replaced with ωT.

Impulse Response

Recall that when the impulse response of an LTI system was introduced, the notation $h[\cdot]$ was reserved for the impulse response. In (10.78), the notation $H(\cdot)$ is used to describe the transfer function of an LTI system. We now show that the transfer function $H(z)$ is directly related to the impulse response $h[n]$, and one can be found if the other is known.

For the complex exponential excitation $x[n] = z^n$, the convolution sum yields the system response:

$$y_{ss}[n] = \sum_{k=-\infty}^{\infty} h[k]x[n-k] = \sum_{k=-\infty}^{\infty} h[k]z^{n-k}$$

$$= z^n \sum_{k=-\infty}^{\infty} h[k]z^{-k}. \tag{10.87}$$

In (10.81), the value of z_1 is not constrained and can be considered to be the variable z. From (10.81) with $X = 1$, and from (10.87),

$$y_{ss}[n] = H(z)z^n = z^n \sum_{k=-\infty}^{\infty} h[k]z^{-k},$$

and we see that the impulse response and the transfer function of a discrete-time LTI system are related by

$$H(z) = \sum_{k=-\infty}^{\infty} h[k]z^{-k}. \tag{10.88}$$

This equation is the desired result. Table 10.1 summarizes the results developed in this section.

Those readers familiar with the bilateral z-transform will recognize $H(z)$ in (10.88) as the z-transform of $h[n]$. Furthermore, with $z = e^{j\Omega}$, $H(e^{j\Omega})$ in (10.88) is the discrete-time Fourier transform of $h[n]$. We see then that both the z-transform (covered in Chapter 11) and the discrete-time Fourier transform (covered in Chapter 12) appear naturally in the study of discrete-time LTI systems.

In practice, it is more common in describing an LTI system to specify the transfer function $H(z)$ rather than the impulse response $h[n]$. However, we can represent an LTI system with either of the block diagrams given in Figure 10.22, with $H(z)$ and $h[n]$ related by (10.88).

TABLE 10.1   Input–Output Functions for an LTI System

$$H(z) = \sum_{k=-\infty}^{\infty} h[k]z^{-k}$$

$$Xz_1^n \to XH(z_1)z_1^n, \quad X = |X|e^{j\phi}$$

$$|X|\cos(\Omega_1 n + \phi) \to |X||H(e^{j\Omega_1})|\cos(\Omega_1 n + \phi + \theta_H)$$

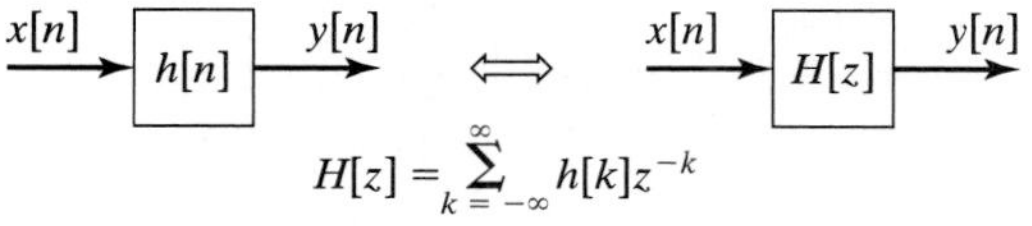

$$H[z] = \sum_{k=-\infty}^{\infty} h[k]z^{-k}$$

Figure 10.22   LTI system.

In this section, we considered the response of LTI systems to complex-exponential inputs, which led us to the concept of transfer functions. Using the transfer-function approach, we can easily find the system steady-state response to sinusoidal inputs. As a final point, the relationship between the transfer function of a system and its impulse response was derived.

▨ SUMMARY

In this chapter, we considered discrete-time linear time-invariant (LTI) systems. First, it was shown that discrete-time signals can be represented as a sum of weighted discrete impulse functions. This representation allows us to model an LTI system in terms of its impulse response.

The modeling of a system by its impulse response is basic to the analysis and design of LTI systems. The impulse response gives a complete input–output description of an LTI system. It was shown that through the convolution summation, the input $x[n]$, the impulse response $h[n]$, and the output $y[n]$ are related by

$$y[n] = \sum_{k=-\infty}^{\infty} x[k]h[n-k] = \sum_{k=-\infty}^{\infty} x[n-k]h[k].$$

The importance of the impulse response of an LTI system cannot be overemphasized. It was also shown that the impulse response of an LTI system can be derived from its step response. Hence, the input–output description of a system is also contained in its step response.

Next some general properties of LTI systems were discussed. These properties included memory, invertibility, causality, and stability.

The most popular method for modeling LTI systems is by ordinary linear difference equations with constant coefficients. This method is used for physical systems that can be modeled accurately by these equations. A linear time-invariant digital filter is an LTI discrete-time system and, in general, is modeled very accurately by a linear difference equation with constant coefficients.

A general procedure was given for solving linear difference equations with constant coefficients. This procedure led to a test that determines stability for causal discrete-time LTI systems.

Next a procedure for representing system models by simulation diagrams was developed. Two simulation diagrams, the direct forms I and II, were given. However, it should be realized that an unbounded number of simulation diagrams exist for a given LTI system. In many applications, the simulation diagrams are called programming forms.

As the final topic, the response of an LTI system to a sinusoidal input signal was derived. This derivation led to the transfer-function description of an LTI system. It is shown in Chapter 11 that the transfer function allows us to find the response of an LTI system to any input signal. Hence, the transfer function is also a complete input–output description of an LTI system.

See Table 10.2.

TABLE 10.2 Key Equations of Chapter 10

Equation Title	Equation Number	Equation
DT unit impulse response	(10.9)	$\delta[n] \rightarrow h[n]$
DT convolution equation	(10.16)	$y[n] = \sum\limits_{k=-\infty}^{\infty} x[k]h[n-k] = x[n]*h[n]$
		$= \sum\limits_{k=-\infty}^{\infty} h[k]x[n-k] = h[n]*x[n]$
Convolution with a DT unit impulse	(10.18)	$\delta[n]*g[n-n_0] = \delta[n-n_0]*g[n] = g[n-n_0]$
Finite sum of DT exponentials	(10.22)	$\sum\limits_{k=0}^{n} a^k = \dfrac{1 - a^{n+1}}{1 - a}$
Convolution sum of a DT inverse system	(10.31)	$h[n]*h_i[n] = \delta[n]$
Convolution sum of a DT causal system	(10.34)	$y[n] = \sum\limits_{k=0}^{n} x[n-k]h[k] = \sum\limits_{k=-\infty}^{n} x[k]h[n-k]$
Condition on DT impulse response for BIBO stability	(10.36)	$\sum\limits_{k=-\infty}^{\infty} |h[k]| < \infty$
Derivation of DT step response from DT impulse response	(10.40)	$s[n] = \sum\limits_{k=-\infty}^{\infty} u[n-k]h[k] = \sum\limits_{k=-\infty}^{n} h[k]$
Derivation of DT impulse response from DT step response	(10.41)	$s[n] - s[n-1] = \sum\limits_{k=-\infty}^{n} h[k] - \sum\limits_{k=-\infty}^{n-1} h[k] = h[n]$
Linear difference equation with constant coefficients	(10.48)	$\sum\limits_{k=0}^{N} a_k y[n-k] = \sum\limits_{k=0}^{M} b_k x[n-k], \quad a_0 \neq 0$
Characteristic equation	(10.53)	$a_0 z^N + a_1 z^{N-1} + \cdots + a_{N-1}z + a_N = 0$
Solution of homogeneous equation	(10.55)	$y_c[n] = C_1 z_1^n + C_2 z_2^n + \cdots + C_N z_N^n$

REFERENCES

L. A. Pipes, *Applied Mathematics for Engineers*. New York: McGraw–Hill, 1946.

PROBLEMS

10.1. Consider the convolution sum

$$y[n] = x[n]*h[n] = \sum_{k=-\infty}^{\infty} x[k]h[n-k].$$

Show that this sum can also be expressed as

$$y[n] = h[n]*x[n] = \sum_{k=-\infty}^{\infty} h[k]x[n-k].$$

(*Hint*: Use a change of variables.)

10.2. Show that, for any function $g[n]$,

$$g[n]*\delta[n] = g[n].$$

10.3. Given the LTI system of Figure P10.3, with the input $x[n]$ and the impulse response $h[n]$, where

$$x[n] = \begin{cases} 2, & 1 \leq n \leq 3 \\ 0, & \text{otherwise} \end{cases} \qquad h[n] = \begin{cases} 3, & -1 \leq n \leq 2 \\ 0, & \text{otherwise} \end{cases}.$$

Parts (a), (b), and (c) are to be solved without finding $y[n]$ for all n.

(a) Solve for the system output at $n = 5$; that is, find $y[5]$.

(b) Find the maximum value for the output $y[n]$.

(c) Find the values of n for which the output is maximum.

(d) Verify the results by solving for $y[n]$ for all n.

(e) Verify the results of this problem using MATLAB.

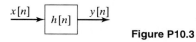

Figure P10.3

10.4. Given the LTI system of Figure P10.3, with the impulse response $h[n] = \alpha^n u[n]$, where α is a constant. This system is excited with the input $x[n] = \beta^n u[n]$, with $\beta \neq \alpha$ and β constant.

(a) Find the system response $y[n]$. Express $y[n]$ in closed form, using the formulas for geometric series in Appendix C.

(b) Evaluate $y[4]$, using the results of Part (a).

(c) Verify the results of Part (b) by expanding the convolution sum for $y[4]$, as in (10.15).

10.5. Consider the LTI system of Figure P10.3, with the input $x[n]$ and the impulse response $h[n]$, where

$$x[n] = \begin{cases} 4, & 1 \leq n \leq 4 \quad \text{and} \quad 8 \leq n \leq 12 \\ 0, & \text{otherwise} \end{cases}$$

$$h[n] = \begin{cases} 3, & 3 \leq n \leq 7 \\ 0, & \text{otherwise.} \end{cases}$$

Parts (a), (b), and (c) are to be solved without finding $y[n]$ for all n.

(a) Solve for the system output at $n = 8$; that is, find $y[8]$.

(b) Find the maximum value for the output $y[n]$.

(c) Find the values of n for which the output is maximum.

(d) Verify the results by solving for $y[n]$ for all n.

(e) Verify the results of this problem using MATLAB.

10.6. Consider the discrete-time LTI system of Figure P10.3. This system has the impulse response

$$h[n] = u[n - 1] - u[n - 5],$$

and the input signal is given by

$$x[n] = \delta[n] + (0.2)^n(u[n-1] - u[n-5]).$$

(a) Sketch $h[n]$.
(b) Find the system output $y[n]$.

10.7. (a) Suppose that the discrete-time LTI system of Figure P10.3 has the impulse response $h[n]$ given in Figure P10.7(a). The system input is the unit step function $x[n] = u[n]$. Find the output $y[n]$.

(b) Repeat Part (a) if the system input is $x[n]$ in Figure P10.7(b).
(c) Verify the results in Part (b) using MATLAB.
(d) Repeat Part (a) if the system input is $x[n]$ in Figure P10.7(c).
(e) Verify the results in Part (d) using MATLAB.
(f) Repeat Part (a) if $h[n]$ is the same function as $x[n]$ in Figure P10.7(b); that is, $h[n] = x[n]$.
(g) Verify the results in Part (f) using MATLAB.

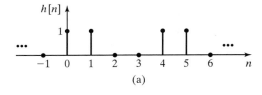
(a)

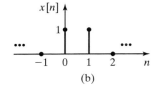
(b)

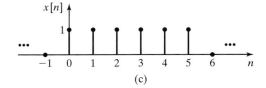
(c)   **Figure P10.7**

10.8. For the LTI system of Figure P10.3, the input signal is $x[n]$, the output signal is $y[n]$, and the impulse response is $h[n]$. For each of the cases that follow, find the output $y[n]$. The referenced signals are given in Figure P10.8.

(a) $x[n]$ in (a), $h[n]$ in (b)
(b) $x[n]$ in (a), $h[n]$ in (c)
(c) $x[n]$ in (a), $h[n]$ in (d)

(d) $x[n]$ in (b), $h[n]$ in (c)

(e) $x[n]$ in (b), $h[n]$ in (f)

(f) $x[n]$ in (α), $h[n]$ in (β), where α and β are assigned by your instructor

(g) Verify the results in each part using MATLAB.

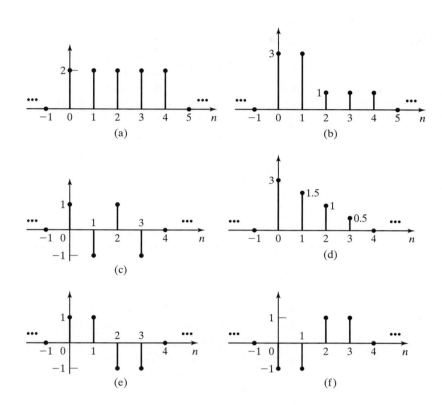

Figure P10.8

10.9. For the system of P10.3, the input signal is $x[n]$, the output signal is $y[n]$, and the impulse response is $h[n]$. For each of the following cases, find the output $y[n]$:

(a) $x[n] = a^{-3n}u[1 - n]$, $h[n] = b^n u[2 - n]$ (where $a \neq b$).

(b) $x[n] = a^n u[-n]$, $h[n] = b^n u[n]$ (where $a \neq b$).

(c) $x[n] = u[n]$, $h[n] = a^n(u[n] - u[n - 100])$.

(d) $x[n] = u[-2 - n]$, $h[n] = b^{-2n}u[n - 1]$ (where $|b| < 1$).

(e) $x[n] = a^{2n}u[n]$, $h[n] = b^n u[n - 2]$, (where $a^2 \neq b$).

(f) $x[n] = u[1 - n]$, $h[n] = (u[n] - u[n - 51])$.

10.10. For the system of Figure P10.3, suppose that $x[n]$ and $h[n]$ are identical and as shown in Figure P10.8(c).

   (a) Find the output $y[n]$ for all n, by sketching $h[k]$ and $x[n - k]$.

   (b) Consider the expanded convolution sum of (10.15). Write out this expansion for each value of n in Part (a), but include only those terms that are nonzero. Evaluate this expansion to verify the results of Part (a).

   (c) Verify the results using MATLAB.

10.11. (a) Consider the two LTI systems cascaded in Figure P10.11. The impulse responses of the two systems are identical, with $h_1[n] = h_2[n] = (0.8)^n u[n]$. Find the impulse response of the total system.

   (b) Repeat Part (a) for $h_1[n] = h_2[n] = \delta[n - 3]$

   (c) Write out the terms in Part (b), as in (10.15), to verify the results.

   (d) Repeat Part (a) for $h_1[n] = h_2[n] = u[n] - [n - 2]$. Express the results in terms of impulse functions.

Figure P10.11

10.12. An LTI discrete-time system has the impulse response

$$h[n] = (2)^n u[n].$$

   (a) Determine whether this system is causal.

   (b) Determine whether this system is stable.

   (c) Find the system response to a unit step input $x[n] = u[n]$.

   (d) Use MATLAB to verify the results in (c) for $n = 0, 1, 2,$ and 3.

   (e) Repeat Parts (a) through (c) for

$$h[n] = (2)^n u[-n].$$

   (f) Repeat Parts (a) and (b) for

$$h[n] = (0.3)^n u[-n].$$

   (g) Repeat Parts (a) and (b) for

$$h[n] = u[-n].$$

10.13. Show that the convolution of three signals can be performed in any order by showing that

$$(f[n]*g[n])*h[n] = f[n]*(g[n]*h[n]).$$

(*Hint*: Form the required summations and use a change of variables.)

10.14. Consider an LTI system with the input and output related by

$$y[n] = 0.8(x[n + 1] + x[n]).$$

(a) Find the system impulse response $h[n]$.
(b) Is this system causal? Why?
(c) Determine the system response $y[n]$ for the input shown in Figure P10.14(a).
(d) Consider the interconnections of the LTI systems given in Figure P10.14(b), where $h[n]$ is the function found in Part (a). Find the impulse response of the total system.
(e) Solve for the response of the system of Part (d) for the input of Part (c).

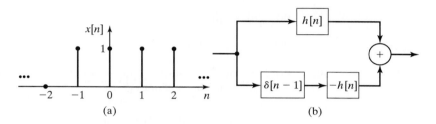

(a)   (b)

Figure P10.14

10.15. Consider a system described by the equation

$$y[n] = \sin\left(\frac{\pi n}{4}\right)x[n].$$

(a) Is this system linear?
(b) Is this system time invariant?
(c) Determine the impulse response $h[n]$.
(d) Determine the response to the input $\delta[n - 1]$.
(e) Can a linear time-varying system be described by its impulse response $h[n]$? Why?

10.16. Determine the causality and the stability for the systems with the following impulse responses:

(a) $h[n] = e^{-n}u[n]$
(b) $h[n] = e^{-n}u[-n]$
(c) $h[n] = e^{n}u[n]$
(d) $h[n] = \cos(n)u[n]$
(e) $h[n] = ne^{-n}u[n]$
(f) $h[n] = e^{-n}\cos(n)u[n]$

10.17. (a) Consider an LTI system with the output given by

$$y[n] = \sum_{k=0}^{\infty} e^{-2k}x[n - k].$$

Find the impulse response of this system by letting $x[n] = \delta[n]$ to obtain $y[n] = h[n]$.

(b) Is this system causal? Why?

(c) Is this system stable? Why?

(d) Repeat Parts (a), (b), and (c) for an LTI system with the output given by

$$y[n] = \sum_{k=-\infty}^{n} e^{-2(n-k)}x[k-1].$$

10.18. Suppose that the system of Figure P10.3 is described by each of the following system equations. Find the impulse response of this system by letting $x[n] = \delta[n]$ to obtain $y[n] = h[n]$.

(a) $y[n] = x[n-3]$

(b) $y[n] = \sum_{k=-\infty}^{n+3} x[k]$

10.19. (a) Find the responses for systems described by the following difference equations with the initial conditions given:

(i) $y[n] - \frac{5}{6}y[n-1] = 2^n u[n],\ y[-1] = 0$

(ii) $y[n] - 0.7y[n-1] = e^{-n}u[n],\ y[-1] = 0$

(iii) $y[n] + 3y[n-1] + 2y[n-2] = 3u[n],\ y[-1] = 0,\ y[-2] = 0$

(iv) $y[n] - 0.7y[n-1] = \cos(n)u[n],\ y[-1] = -1$

(b) Verify that your response satisfies the initial conditions and the difference equation.

(c) Use MATLAB to verify your solutions in Part (a) by finding $y(n)$ for $n = 0, 1, 2,$ and 3.

10.20. Consider a causal system with each of the subsequent system characteristic equations.

(a) Give the modes of the system.

(b) Give the natural response for each of the systems.

(i) $z - 0.5 = 0$

(ii) $z^2 - 1.1z + 0.3 = 0$

(iii) $z^2 + 1 = 0$

(iv) $z^3 - 2z^2 + 1.5z - 0.5 = 0$

(v) $(z - 0.5)^3 = 0$

(vi) $(z - 0.5)(z - 1.5)(z + 0.7) = 0$

10.21. Determine the stability of each of the systems of Problem 10.20.

10.22. Consider a discrete-time LTI system described by the difference equation

$$y[n] - 0.7y[n-1] = 2.5x[n] - x[n-1].$$

(a) Draw the form I realization (block diagram) for this system.

(b) Determine the impulse response $h[n], 0 \leq n \leq 4$, for the system.

(c) Verify the results of Part (b) by tracing the impulse function through the block diagram of Part (a).

(d) Suppose that the system input is given by

$$x[n] = \begin{cases} 1, & n = -2 \\ -3, & n = 0 \\ 2, & n = 1 \end{cases}$$

and $x[n]$ is zero for all other values of n. Express the output $y[n]$ as a function of $h[n]$.

(e) Calculate the output $y[n]$ for $n = -3, -1$, and 1 for $x[n]$ in Part (d) and using the results of Part (b).

10.23. Draw the block diagrams of both the direct forms I and II simulation diagrams for the systems with the following difference equations:

(a) $2y[n] - y[n-1] + 4y[n-2] = 5x[n]$

(b) $y[n] - 5y[n-1] + y[n-2] = 3x[n] - 4x[n-2]$

(c) $y[n] - 2y[n-1] + 3y[n-2] - 5y[n-3]$
$\quad = 4x[n] - 2x[n-1] + 10x[n-2] - 15x[n-3]$

(d) $y[n] = 0.25(x[n] + x[n-1] + x[n-2] + x[n-3])$

(e) $y[n] - 1.8y[n-1] + 0.9y[n-3] = 2x[n] - 3.5x[n-1] + 2.8x[n-2]$

10.24. Determine the stability of each of the systems of Problem 10.23. Use MATLAB as required.

10.25. Consider the system simulation diagram of Figure P10.25. This form is often used in automatic control.

(a) Find the difference equation of the system.

(b) Is this one of the two forms given in Section 10.6? If so, give the form.

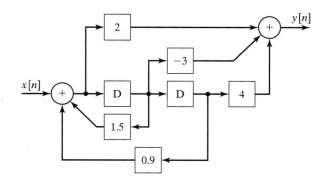

Figure P10.25

10.26. Consider a system described by the difference equation

$$y[n] - 0.9y[n-1] = 2x[n] - 1.9x[n-1].$$

(a) Draw a form I representation for this system.

(b) Draw a form II representation for this system.

(c) Let $x[n] = 0.8^n u[n]$ and $y[-1] = 0$. Solve for $y[n]$ as a function of n.

(d) Write a MATLAB program that solves for $y[n], 0 \leq n \leq 5$, using the form I representation. Run this program to verify the results in Part (c).

10.27. Consider the MATLAB program for the simulation of an LTI system:

```
y(1)=0;
for n=1:6; x(n)=0.7^(n-1); end
for n=2:6
    y(n)=0.9*y(n-1)+x(n)-x(n-1);
end
y
```

(a) Write the system difference equation.

(b) Draw the form I representation for the system.

(c) Draw the form II representation for the system.

(d) Express the input signal $x[n]$ as a function of n.

(e) Solve for $y[n]$ as a function of n.

(f) Verify the solution in part (e) by running the MATLAB program.

10.28. For the system described by the MATLAB program of Problem 10.27,

(a) Find the system difference equation.

(b) Find the particular solution for the difference equation, with the excitation $x[n] = u[n]$.

(c) Find the system transfer function.

(d) Use the transfer function to verify the results of Part (b).

(e) Change the MATLAB program such that $x[n] = u[n]$.

(f) Verify the solution in Parts (b) and (d) by running the program. Recall that only the steady-state response has been calculated.

10.29. (a) Find the transfer function for the difference equation $y[n] = 0.7y[n - 1] = x[n]$

(b) Use the transfer function to find the steady-state response of this system for the input given.

(c) Verify the calculations in Part (b) using MATLAB.

(d) Show that the response satisfies the system difference equation.

10.30. For each of the following problems, state the restriction on the variables a and b (if any) that would be required for any sums to converge. If no restriction is needed, state no restriction.

(a) $\displaystyle\sum_{i=-\infty}^{1} b^i$

(b) $a^n u[n] * b^n u[n + 6]$

(c) $a^n u[n - 2] * b^n u[-n - 4]$

(d) $a^n u[-n + 3] * b^n u[-n - 4]$

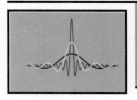

11 THE Z-TRANSFORM

In this chapter, we study the *z-transform*, which is one of several important transforms used in linear-system analysis and design. The *z*-transform offers significant advantages relative to time-domain procedures. When possible, we model discrete-time physical systems with linear difference equations with constant coefficients; one example is a linear time-invariant digital filter. The *z*-transform of a difference equation gives us a good description of the characteristics of the equation (the model) and, hence, of the physical system. In addition, transformed difference equations are algebraic, and therefore easier to manipulate; in particular, the transformed equations are easier to solve.

Using the *z*-transform to solve a difference equation yields the solution as a function of the transform variable *z*. As a consequence, we must have a method for converting functions of the transform variable back to functions of the discrete-time variable; the *inverse z-transform* is used for this purpose.

Several important properties of the *z*-transform are derived in this chapter. These derivations are not mathematically rigorous; such derivations are generally beyond the scope of this book. Thus for some properties, certain constraints apply that are not evident from the derivations. However, these constraints will be stated; see Refs. 1 and 2 for rigorous mathematical derivations related to all aspects of the *z*-transform.

11.1 DEFINITIONS OF Z-TRANSFORMS

We begin by defining the direct *z*-transform and the inverse *z*-transform. We usually omit the term *direct* and call the direct *z*-transform simply the *z*-transform. By definition, the (direct) *z-transform* $F(z)$ of a discrete-time function $f[n]$ is given by the summation

$$\mathcal{Z}_b[f[n]] = F_b(z) = \sum_{n=-\infty}^{\infty} f[n]z^{-n}, \tag{11.1}$$

547

where $\mathscr{Z}_b[\cdot]$ indicates the *z*-transform. Expanding the *z*-transform yields

$$F_b(z) = \cdots + f[-2]z^2 + f[-1]z + f[0] + f[1]z^{-1} + f[2]z^{-2} + \cdots.$$

In general, *z* is complex with $z = \Sigma + j\Omega$. (Recall that the Laplace-transform variable *s* is also complex, with $s = \sigma + j\omega$.)

Definition (11.1) is called the bilateral, or two-sided, *z* transform—hence, the subscript *b*. The inverse *z*-transform is given by

$$\mathscr{Z}_b^{-1}[F_b(z)] = f[n] = \frac{1}{2\pi j}\oint_\Gamma F_b(z)z^{n-1}dz, \quad j = \sqrt{-1}, \tag{11.2}$$

where $\mathscr{Z}_b^{-1}[\cdot]$ indicates the inverse *z*-transform, and Γ is a particular counterclockwise closed path in the *z*-plane. Equation (11.2) is called the *complex inversion integral*. Because of the difficulty of evaluating this integral, we seldom, if ever, use it to find inverse transforms. Instead, we use tables, as is done with other transforms.

Equations (11.1) and (11.2) are called the *bilateral z-transform pair*. We now modify definition (11.1) to obtain a form of the *z*-transform that is useful in many applications. First, we express (11.1) as

$$\mathscr{Z}_b[f[n]] = F_b(z) = \sum_{n=-\infty}^{-1} f[n]z^{-n} + \sum_{n=0}^{\infty} f[n]z^{-n}. \tag{11.3}$$

Next we *define f[n]* to be zero for $n < 0$, such that the first summation in (11.3) is zero. The resulting transform is called the *unilateral*, or *single-sided*, *z*-transform, and is given by the power series

$$\mathscr{Z}[f[n]] = F(z) = \sum_{n=0}^{\infty} f[n]z^{-n}, \tag{11.4}$$

where $\mathscr{Z}[\cdot]$ denotes the unilateral *z*-transform. This transform is usually called simply the *z*-transform, and we will follow this custom. When any confusion can result, we will refer to the transform of (11.1) as the bilateral *z*-transform. We take the approach of making the unilateral transform a special case of the bilateral transform. This approach is not necessary; we could start with (11.4), with $f[n] = 0$ for $n < 0$, as a definition.

The equation for the inverse *z*-transform, (11.2), is the same for both the bilateral and the unilateral *z*-transforms. Hence, (11.2) also gives the inverse unilateral *z*-transform, provided that $F_b(z)$ is replaced with $F(z)$. In addition, the inverse *z*-transform of the unilateral *z*-transform gives the function $f[n]$ for all time and, in particular, gives the value $f[n] = 0, n < 0$ [1].

If $f[n]$ is *z*-transformable [if the summation in (11.4) exists], evaluating (11.4) will yield a function $F(z)$. The evaluation of the inverse transform of $F(z)$ using the complex inversion integral, (11.2), will then yield $f[n]$. We denote this relationship with

$$f[n] \stackrel{\mathscr{Z}}{\longleftrightarrow} F(z). \tag{11.5}$$

Two important properties of the z-transform will now be demonstrated. The (unilateral) z-transform is used in this derivation; however, it is seen that the derivation applies equally well to the bilateral z-transform.

Consider the sum $f[n] = (f_1[n] + f_2[n])$. From (11.4), the z-transform of $f[n]$ is given by

$$\mathcal{Z}[f[n]] = \mathcal{Z}[f_1[n] + f_2[n]] = \sum_{n=0}^{\infty}[f_1[n] + f_2[n]]z^{-n}$$

$$\tag{11.6}$$

$$= \sum_{n=0}^{\infty}f_1[n]z^{-n} + \sum_{n=0}^{\infty}f_2[n]z^{-n} = F_1(z) + F_2(z).$$

Hence, the z-transform of the sum of two functions is equal to the sum of the z-transforms of the two functions. (It is assumed that the involved z-transforms exist.) This property is extended to the sum of any number of functions by replacing $f_2[n]$ in the foregoing derivation with the sum $(f_3[n] + f_4[n])$, and so on.

To derive a second property of the z-transform, we consider the z-transform of $af[n]$, where a is a constant:

$$\mathcal{Z}[af[n]] = \sum_{n=0}^{\infty}af[n]z^{-n} = a\sum_{n=0}^{\infty}f[n]z^{-n} = aF(z). \tag{11.7}$$

Thus, the z-transform of a function multiplied by a constant is equal to the constant multiplied by the z-transform of the function. A transform with the properties (11.6) and (11.7) is said to be a *linear transform*; the z-transform is then a linear transform. These two properties are often stated as a single equation:

$$\mathcal{Z}[a_1f_1[n] + a_2f_2[n]] = a_1F_1(z) + a_2F_2(z). \tag{11.8}$$

Suppose, in (11.7), that the constant a is replaced with the function $g[n]$. Then

$$\mathcal{Z}[f[n]g[n]] = \sum_{n=0}^{\infty}f[n]g[n]z^{-n} \neq \sum_{n=0}^{\infty}f[n]z^{-n}\sum_{n=0}^{\infty}g[n]z^{-n}.$$

Hence,

$$\mathcal{Z}[f[n]g[n]] \neq \mathcal{Z}[f[n]]\mathcal{Z}[g[n]]. \tag{11.9}$$

The z-transform of a product of two functions is *not* equal to the product of the z-transforms of the functions.

In this section, the unilateral and bilateral z-transforms are defined. These transforms are series in the variable z; however, we will see that the series for many useful signals can be expressed in closed form. The complex inversion integral for the inverse z-transform is also given, but we generally use tables for finding inverse transforms. The linearity properties of the z-transform are derived in this section. In the remainder of this chapter, we develop z-transform analysis from the definitions given here.

11.2 EXAMPLES

In this section, we introduce z-transform system analysis with a simple application. First, two examples of derivations of z-transforms are presented. Next, the z-transform is used to find the step response of a first-order digital filter. The z-transform is then developed in more detail in the following sections.

Two *z*-Transforms

Before presenting the first example, we consider the convergent power series from Appendix C:

$$\sum_{n=0}^{\infty} a^n = 1 + a + a^2 + \cdots = \frac{1}{1-a}; \quad |a| < 1. \tag{11.10}$$

Any function in Appendix C can be verified by dividing the numerator by the denominator; for (11.10), this division yields

$$
\begin{array}{r}
1 + a + a^2 + a^3 + \cdots \\
\hline
1 - a \overline{\smash{\big)}\, 1} \\
\underline{1 - a} \\
a \\
\underline{a - a^2} \\
a^2 \\
\underline{a^2 - a^3} \\
a^3 \\
\cdots
\end{array}
\tag{11.11}
$$

The series of (11.10) is useful in expressing certain z-transforms in closed form (not as a series), as we illustrate subsequently. We prefer to express z-transforms in closed form, because of the resulting simplifications in manipulating these transforms.

Also, we recall the unit step function, $u[n - n_0]$:

$$u[n - n_0] = \begin{cases} 1, & n \ge n_0 \\ 0, & n < n_0 \end{cases}. \tag{11.12}$$

This function is plotted in Figure 11.1 and is used in the next example.

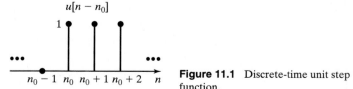

Figure 11.1 Discrete-time unit step function.

EXAMPLE 11.1   ***z*-transform of the unit step function**

The *z*-transform of the unit step function is now derived for the step occurring at $n = 0$. From (11.4) and (11.12),

$$\mathcal{Z}[u[n]] = \sum_{n=0}^{\infty} u[n]z^{-n} = \sum_{n=0}^{\infty} (1)z^{-n} = 1 + z^{-1} + z^{-2} + \cdots.$$

We let $a = z^{-1}$ in the series (11.10), resulting in the *z*-transform in closed form:

$$\sum_{n=0}^{\infty} z^{-n} = \frac{1}{1 - z^{-1}} = \frac{z}{z - 1}.$$

$$S = \frac{a}{1 - r}$$

Hence, the *z*-transform of the unit step function exists *only* for $|z^{-1}| < 1$, that is, for *z* outside the unit circle. We then have the *z*-transform pair:

$$u[n] \xleftrightarrow{\ \mathcal{Z}\ } \frac{1}{1 - z^{-1}}; \quad |z^{-1}| < 1.\qquad\blacksquare$$

The *z*-transform of the unit step function illustrates a problem in notation. The definition of the *z*-transform, (11.4), results in a function of z^{-1}; however, generally our experiences are in working with variables expressed in positive exponents. Some authors leave all *z*-transforms in negative exponents. We choose instead to take the additional step to express $F(z)$ in positive powers of *z*, because this notation is more common and, hence, less prone to error. Then, from Example 11.1, we express the *z*-transform of the unit step function as

$$u[n] \xleftrightarrow{\ \mathcal{Z}\ } \frac{z}{z - 1}; \quad |z| > 1.$$

EXAMPLE 11.2   ***z*-transform of an exponential function**

We now derive the *z*-transform of $f[n] = a^n$. Recall from Section 9.4 that a^n is an exponential function. From (11.4),

$$F(z) = \sum_{n=0}^{\infty} a^n z^{-n} = 1 + az^{-1} + a^2 z^{-2} + \cdots$$
$$= 1 + az^{-1} + (az^{-1})^2 + \cdots.$$

This series is of the form of (11.10) with $|az^{-1}| < 1$; hence, the *z*-transform is given by

$$\mathcal{Z}[a^n] = \sum_{n=0}^{\infty} (az^{-1})^n = \frac{1}{1 - az^{-1}} = \frac{z}{z - a}; \quad |z| > |a|,$$

and we have the *z*-transform pair

$$a^n \xleftrightarrow{\ \mathcal{Z}\ } \frac{z}{z - a}; \quad |z| > |a|.$$

This transform pair is verified with the MATLAB program

```
syms n a f F
f=a^n;
F=ztrans(f);
'f= ', pretty(f)
'F = ', pretty(F)
```

∎

In Example 11.2, we could have expressed $f[n]$ as $a^n u[n]$. However, the unilateral z-transform *requires* that $f[n] = 0$ for $n < 0$. If the factor $u[n]$ is not shown explicitly, it is understood to be present.

As seen from Examples 11.1 and 11.2, the z-transforms of the unit step function $u[n]$ and the exponential function a^n have conditions for existence. The z-transform of any function $f[n]$, denoted as $F(z)$, will generally have similar conditions for existence, such that the infinite series

$$\mathscr{Z}[f[n]] = F(z) = \sum_{n=0}^{\infty} f[n] z^{-n}$$

is convergent. Increasing the magnitude of z increases the likelihood of convergence for any $f[n]$, since $|z^{-n}|$ becomes smaller as n approaches infinity. Hence, the conditions for convergence are of the form $|z| > |r|$, where r is determined by $f[n]$.

The conditions for convergence determine the path Γ in the inversion integral (11.2). Because we do not use (11.2) to find the inverse z-transform, we generally ignore the conditions for convergence. In addition, in the derivations that follow, usually conditions for convergence are not stated; these conditions are evident from the derivation. However, when we introduce the bilateral z-transform in Section 11.7, the conditions for convergence must be considered.

From Examples 11.1 and 11.2, a short table of z-transforms is constructed and given as Table 11.1. Note in this table that the functions $f[n]$ are zero for $n < 0$. Recall that the evaluation of the complex-inversion integral (11.2) yields $f[n] = 0$ for $n < 0$.

Generally, we use a z-transform table to find inverse z-transforms, rather than using the inversion integral of (11.2). In any transform pair

$$f[n] \xleftrightarrow{\ \mathscr{Z}\ } F(z),$$

TABLE 11.1 Two z-Transforms

$f[n],\ n \geqq 0$	$F(z)$
$u[n]$	$\dfrac{z}{z - 1}$
a^n	$\dfrac{z}{z - a}$

given $f[n]$, the transform is $F(z)$; given $F(z)$, the inverse transform is $f[n]$. For example, for the exponential function

$$a^n \overset{\mathscr{Z}}{\longleftrightarrow} \frac{z}{z-a},$$

the z-transform of a^n is $z/(z-a)$; the inverse transform of $z/(z-a)$ is a^n for $n \geqq 0$.

Digital-Filter Example

We now use the z-transform to solve a first-order difference equation. However, first we must derive the real-shifting property of the z-transform. Consider the z-transform of a delayed function $f[n-n_0]u[n-n_0]$ for $n_0 \geqq 0$:

$$\mathscr{Z}[f[n-n_0]u[n-n_0]] = \sum_{n=0}^{\infty} f[n-n_0]u[n-n_0]z^{-n}$$

$$= \sum_{n=n_0}^{\infty} f[n-n_0]z^{-n}$$

$$= f[0]z^{-n_0} + f[1]z^{-n_0-1} + f[2]z^{-n_0-2} + \cdots$$

$$= z^{-n_0}[f[0] + f[1]z^{-1} + f[2]z^{-2} + \cdots] = z^{-n_0}F(z).$$

For $n_0 \geqq 0$, we have the property

$$\mathscr{Z}[f[n-n_0]u[n-n_0]] = z^{-n_0}F(z). \tag{11.13}$$

Of course, n_0 must be an integer. We derive the real-shifting property for $n_0 < 0$ in Section 11.5.

The difference equation for the α-filter is

$$y[n] - (1-\alpha)y[n-1] = \alpha x[n], \tag{11.14}$$

where α is a constant. The value of α is determined by the design specifications for the filter. This equation describes a low-pass digital filter that is used in radar-signal processing. Recall that this filter was used in several examples in Chapter 10.

First, we take the z-transform of the filter equation (11.14), using the linearity property of (11.8) and the real-shifting property of (11.13), we have

$$Y(z) - (1-\alpha)z^{-1}Y(z) = \alpha X(z), \tag{11.15}$$

where $Y(z) = \mathscr{Z}[y[n]]$ and $X(z) = \mathscr{Z}[x[n]]$. Solving for $Y(z)$ yields

$$Y(z) = \frac{\alpha}{1-(1-\alpha)z^{-1}} X(z) = \frac{\alpha z}{z-(1-\alpha)} X(z).$$

In Chapter 10, we derived the ratio of output to input using a different approach, and we called this ratio the *system-transfer function*, denoted as $H(z)$. We use the

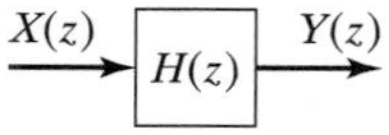

Figure 11.2 Discrete-time LTI system.

same notation here:

$$\frac{Y(z)}{X(z)} = H(z) = \frac{\alpha z}{z - (1 - \alpha)}. \tag{11.16}$$

We represent this linear time-invariant (LTI) discrete-time system by the block diagram of Figure 11.2.

We now find the unit step response of the α-filter. The input signal is $X(z) = \mathcal{Z}[u[n]] = z/(z - 1)$, from Table 11.1. Hence, from (11.16), the output of the α-filter is given by

$$Y(z) = \frac{\alpha z}{z - (1 - \alpha)} X(z) = \frac{\alpha z}{z - (1 - \alpha)} \frac{z}{z - 1}, \tag{11.17}$$

and the output $y[n]$ is the inverse z-transform of this function. For the remainder of this development, we let $\alpha = 0.1$, to simplify the calculations.

We note that Table 11.1 does not contain the function in (11.17). As in Laplace- and Fourier-transform applications, we must use a *partial-fraction expansion* to express $Y(z)$ as the sum of terms that do appear in Table 11.1. A review of partial-fraction expansion procedures is given in Appendix F.

One problem occurs in the use of the partial-fraction expansion procedure of Appendix F. This is illustrated by the z-transform of the exponential function a^n:

$$a^n \xleftrightarrow{\mathcal{Z}} \frac{z}{z - a}.$$

This transform has the variable z in the numerator, while the procedure of Appendix F yields only a constant in the numerator. We solve this problem by finding the partial-fraction expansion of $Y(z)/z$ in (11.17), with $\alpha = 0.1$:

$$\frac{Y(z)}{z} = \frac{0.1z}{(z - 0.9)(z - 1)} = \frac{k1}{z - 0.9} + \frac{k_2}{z - 1}.$$

We now find the constants k_1 and k_2 by partial-fraction expansion:

$$k_1 = (z - 0.9)\left[\frac{0.1z}{(z - 0.9)(z - 1)}\right]_{z=0.9} = \frac{0.1z}{z - 1}\bigg|_{z=0.9} = -0.9,$$

$$k_2 = (z - 1)\left[\frac{0.1z}{(z - 0.9)(z - 1)}\right]_{z=1} = \frac{0.1z}{z - 0.9}\bigg|_{z=1} = 1.$$

Hence,

$$\frac{Y(z)}{z} = \frac{-0.9}{z - 0.9} + \frac{1}{z - 1}.$$

Multiplying by z and rearranging yields

$$Y(z) = \frac{z}{z-1} - \frac{0.9z}{z-0.9},$$

and these terms appear in Table 11.1. The unit step response of the α-filter is then

$$y[n] = 1 - 0.9^{n+1}, \quad n \geq 0.$$

This result can be verified for the first few values of n by the iterative solution of difference equation (11.14), with $\alpha = 0.1$ and $y[-1] = 0$. A MATLAB program that performs the partial-fraction expansion is given by

```
n=[.1 0];d=[1 -1.9 .9];
[r,p,k]=residue(n,d)
```

The inverse z-transform can be verified with the MATLAB program

```
syms Y z
Y=0.1*z^2/ ((z-0.9) * (z-1));
y=iztrans (Y);
'Y(z) = ', pretty(Y)
'y(n) = ', pretty(y)
```

Note the following points from the preceding example:

1. A difference equation with constant coefficients is transformed into the algebraic equation (11.15).
2. The algebraic equation is solved for $\mathcal{Z}[y[n]] = Y(z)$ as a function of the transform variable z and $X(z)$.
3. A table of transforms is used to find the inverse transform rather than using the inversion integral of (11.2).
4. In general, a partial-fraction expansion of $Y(z)/z$ is required to expand complicated functions of z into the simpler functions that appear in tables of z-transforms.

We expand on these conclusions in the developments of the following sections.
In this section, we derived the z-transforms of the unit step function and the discrete-exponential function. In both cases, the transforms are expressed in closed form. Next a first-order difference equation was solved using the z-transform. In the next section, we derive the z-transforms of additional functions.

11.3   z-TRANSFORMS OF FUNCTIONS

The direct unilateral z-transform is defined by

[eq(11.4)] $$\mathcal{Z}[f[n]] = F(z) = \sum_{n=0}^{\infty} f[n]z^{-n}$$

and the inverse z-transform by the inversion integral

$$f[n] = \frac{1}{2\pi j} \oint_\Gamma F(z)z^{n-1}\, dz,$$

from (11.2). The closed path of integration Γ is determined by the region of convergence of the summation in (11.4).

As stated previously, we seldom use the inversion integral to find inverse transforms; hence, the region of convergence is of secondary importance to us. In fact, we seldom state the region of convergence when we give a z-transform. However, the reader should be aware that a particular z-transform does have a region of convergence. In addition, the region of convergence must be known for applications of the bilateral z-transform (11.1).

We now derive several commonly used z-transforms. First, consider the discrete-impulse function, defined in Section 9.1. From (9.13),

$$\delta[n - n_0] = \begin{cases} 1, & n = n_0 \\ 0, & n \neq n_0 \end{cases}. \tag{11.18}$$

For $n_0 \geq 0$, the z-transform of the unit impulse function (unit sample function) is given by

$$\mathcal{Z}[\delta[n - n_0]] = \sum_{n=0}^{\infty} \delta[n - n_0]z^{-n} = z^{-n_0},$$

and we have the z-transform pair

$$\delta[n - n_0] \overset{\mathcal{Z}}{\leftrightarrow} z^{-n_0}. \tag{11.19}$$

For the unit impulse function occurring at $n = 0$ ($n_0 = 0$,),

$$\delta[n] \overset{\mathcal{Z}}{\leftrightarrow} 1. \tag{11.20}$$

Additional transform pairs will now be derived. Consider the z-transform pair from Table 11.1:

$$a^n \overset{\mathcal{Z}}{\leftrightarrow} \frac{z}{z - a}. \tag{11.21}$$

Recall that a^n is exponential and can be expressed as

$$a^n = (e^\beta)^n = e^{\beta n}, \beta = \ln a.$$

Pair (11.21) can then be expressed as

$$a^n = e^{\beta n} \overset{\mathcal{Z}}{\leftrightarrow} \frac{z}{z - a} = \frac{z}{z - e^\beta}. \tag{11.22}$$

Sinusoids

We now consider sinusoidal functions. By Euler's identity,

$$\cos bn = \frac{e^{jbn} + e^{-jbn}}{2}.$$

Hence,

$$\mathscr{Z}[\cos bn] = \tfrac{1}{2}[\mathscr{Z}[e^{jbn}] + \mathscr{Z}[e^{-jbn}]],$$

by the linearity property (11.8). Then, from (11.22), with $\beta = jb$,

$$\mathscr{Z}[\cos bn] = \frac{1}{2}\left[\frac{z}{z - e^{jb}} + \frac{z}{z - e^{-jb}}\right] = \frac{z}{2}\left[\frac{z - e^{-jb} + z - e^{jb}}{(z - e^{jb})(z - e^{-jb})}\right]$$

$$= \frac{z}{2}\left[\frac{2z - (e^{jb} + e^{-jb})}{z^2 - (e^{jb} + e^{-jb})z + 1}\right] = \frac{z(z - \cos b)}{z^2 - 2z \cos b + 1},$$

where Euler's relation was used in the last step. By the same procedure, since $\sin bn = (e^{jbn} - e^{-jbn})/2\,j$,

$$\mathscr{Z}[\sin bn] = \frac{1}{2j}\left[\frac{z}{z - e^{jb}} - \frac{z}{z - e^{-jb}}\right] = \frac{z}{2j}\left[\frac{z - e^{-jb} - z + e^{jb}}{(z - e^{jb})(z - e^{-jb})}\right]$$

$$= \frac{z}{2j}\left[\frac{e^{jb} - e^{-jb}}{z^2 - (e^{jb} + e^{-jb})z + 1}\right] = \frac{z \sin b}{z^2 - 2z \cos b + 1}.$$

The foregoing procedure can also be used for sinusoids with exponentially varying amplitudes. Because

$$e^{-an}\cos bn = e^{-an}\left[\frac{e^{jbn} + e^{-jbn}}{2}\right] = \frac{e^{-an+jbn} + e^{-an-jbn}}{2},$$

it follows that

$$\mathscr{Z}[e^{-an} \cos bn] = \frac{1}{2}\left[\frac{z}{z - e^{-a+jb}} + \frac{z}{z - e^{-a-jb}}\right]$$

$$= \frac{z}{2}\left[\frac{z - e^{-a-jb} + z - e^{-a+jb}}{(z - e^{-a+jb})(z - e^{-a-jb})}\right]$$

$$= \frac{z}{2}\left[\frac{2z - e^{-a}(e^{jb} + e^{-jb})}{z^2 - e^{-a}(e^{jb} + e^{-jb})z + e^{-2a}}\right]$$

$$= \frac{z(z - e^{-a} \cos b)}{z^2 - 2ze^{-a} \cos b + e^{-2a}}.$$

Note the transform pairs:

$$\cos bn \overset{\mathcal{Z}}{\longleftrightarrow} \frac{z(z - \cos b)}{z^2 - 2z \cos b + 1},$$

$$e^{-an}\cos bn \overset{\mathcal{Z}}{\longleftrightarrow} \frac{z(z - e^{-a} \cos b)}{z^2 - 2ze^{-a} \cos b + e^{-2a}}$$

$$= \frac{ze^a(ze^a - \cos b)}{(ze^a)^2 - 2ze^a \cos b + 1}.$$

We see that for these two functions, the effect of multiplying $\cos bn$ by the exponential function e^{-an} is to replace z with ze^a in the z-transform. We now show that this property is general:

$$\mathcal{Z}[e^{-an}f[n]] = \sum_{n=0}^{\infty} e^{-an}f[n]z^{-n} = \sum_{n=0}^{\infty} f[n](ze^a)^{-n}$$

$$= F(z)\Big|_{z \leftarrow ze^a} = F(ze^a). \tag{11.23}$$

Observe that the notation $z \leftarrow ze^a$ is read as "z is replaced with ze^a." Using the z-transform pair for $\sin (bn)$ and (11.23), we see that

$$\sin bn \overset{\mathcal{Z}}{\longleftrightarrow} \frac{z \sin b}{z^2 - 2z \cos b + 1},$$

$$e^{-an} \sin bn \overset{\mathcal{Z}}{\longleftrightarrow} \frac{ze^a \sin b}{(ze^a)^2 - 2ze^a \cos b + 1}$$

$$= \frac{ze^{-a} \sin b}{z^2 - 2ze^{-a}\cos b + e^{-2a}}.$$

We now derive an additional transform as an example.

<table>
<tr><td>**EXAMPLE 11.3**</td><td>**z-transform of the unit ramp function**</td></tr>
</table>

The z-transform of the unit ramp function, $f[n] = n$, is now derived.

$$\mathcal{Z}[n] = \sum_{n=0}^{\infty} nz^{-n} = 0 + z^{-1} + 2z^{-2} + 3z^{-3} + \cdots.$$

We have, from Appendix C, the summation formula

$$\sum_{n=0}^{\infty} na^n = \frac{a}{(1 - a)^2}; \quad |a| < 1.$$

Hence, letting $a = z^{-1}$, we have the *z*-transform

$$\mathcal{Z}[n] = \sum_{n=0}^{\infty} nz^{-n} = \frac{z^{-1}}{(1 - z^{-1})^2} = \frac{z}{(z - 1)^2}$$

and the transform pair

$$n \overset{\mathcal{Z}}{\longleftrightarrow} \frac{z}{(z - 1)^2}. \qquad\blacksquare$$

The following MATLAB program derives the *z*-transform of a unit-ramp function using symbolic math:

```
syms n y Y
y=n;
Y=ztrans(y);
'y(n) = ',pretty(y)
'Y(z) = ', pretty(Y)
```

In this section, we have developed several *z*-transform pairs. These pairs, in addition to several others, are given in Table 11.2. The region of convergence

TABLE 11.2   *z*-Transforms

$f[n], n \geq 0$	$F(z)$	ROC
1. $\delta[n]$	1	All z
2. $\delta[n - n_0]$	z^{-n_0}	$z \neq 0$
3. $u[n]$	$\dfrac{z}{z - 1}$	$|z| > 1$
4. n	$\dfrac{z}{(z - 1)^2}$	$|z| > 1$
5. n^2	$\dfrac{z(z + 1)}{(z - 1)^3}$	$|z| > 1$
6. a^n	$\dfrac{z}{z - a}$	$|z| > |a|$
7. na^n	$\dfrac{az}{(z - a)^2}$	$|z| > |a|$
8. $n^2 a^n$	$\dfrac{az(z + a)}{(z - a)^3}$	$|z| > |a|$
9. $\sin bn$	$\dfrac{z \sin b}{z^2 - 2z \cos b + 1}$	$|z| > 1$
10. $\cos bn$	$\dfrac{z(z - \cos b)}{z^2 - 2z \cos b + 1}$	$|z| > 1$
11. $a^n \sin bn$	$\dfrac{az \sin b}{z^2 - 2az \cos b + a^2}$	$|z| > |a|$
12. $a^n \cos bn$	$\dfrac{z(z - a\cos b)}{z^2 - 2az\cos b + a^2}$	$|z| > |a|$

(ROC) is given for each transform; this information is required later when the bilateral *z*-transform is considered.

In Sections 11.4 and 11.5, we derive several properties for the *z*-transform. It is then shown that these properties easily allow additional transform pairs to be derived. Also, these properties aid us in solving linear difference equations with constant coefficients.

11.4 *z*-TRANSFORM PROPERTIES

In Sections 11.1 through 11.3, three properties were derived for the *z*-transform. These properties are

[eq(11.8)] $\qquad\qquad \mathcal{Z}[a_1 f_1[n] + a_2 f_2[n]] = a_1 F_1(z) + a_2 F_2(z),$

[eq(11.13)] $\qquad\qquad \mathcal{Z}[f[n - n_0]u[n - n_0]] = z^{-n_0}F(z); n_0 \geqq 0,$

and

[eq(11.23)] $\qquad\qquad \mathcal{Z}[e^{-an} f[n]] = F(z)\Big|_{z \leftarrow ze^a} = F(ze^a),$

where $F(z) = \mathcal{Z}[f[n]]$. Property (11.8) is the *linearity property*, and Property (11.13) is the *real-shifting property*. Property (11.23) is sometimes called the *complex-* or *frequency-scaling property*, since multiplication by e^{-an} in the time domain results in a complex scaling in the *z*-plane variable. Property (11.23) is also referred to as the *modulation property*. In general, $z = \Sigma + j\Omega$ is complex, and a may also be complex.

Real Shifting

The shifting property (11.13) applies for a *delay* in time. In this section, the shifting property for an *advance* in time is derived.

First, consider the discrete-time function $f[n]$ of Figure 11.3(a). Note that $f[n] = -1, n < 0; f[0] = f[1] = 1;$ and $f[n] = 1.5, n \geqq 2$. We now consider various ways of shifting this time function.

Figure 11.3(b) is a plot of $f[n]u[n]$, where $u[n]$ is the unit step function. This is the functional form required for the unilateral *z*-transform, even though we usually omit the factor $u[n]$. Figure 11.3(c) shows a plot of the function $f[n - n_0]$, where $n_0 \geqq 0$ is the amount of the delay in time. The function $f[n - n_0]u[n]$ is shown in Figure 11.3(d), and the function $f[n - n_0]u[n - n_0]$ is given in Figure 11.3(e). For this last function,

$$f[n - n_0]u[n - n_0] = \begin{cases} f[n - n_0]; & n \geqq n_0 \\ 0; & n < n_0 \end{cases}. \qquad (11.24)$$

The function in Figure 11.3(f) is considered subsequently.

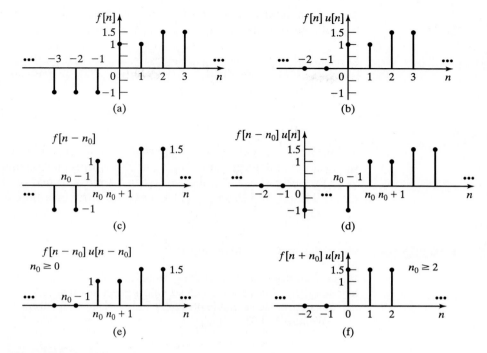

Figure 11.3 Shifted functions.

The reader should carefully note the differences in the functions in Figure 11.3. The function in Figure 11.3(b) is required for the unilateral z-transform of $f[n]$. Figure 11.3(e) illustrates the function for the real-shifting property (11.13).

We now consider the shifted function $f[n + n_0]u[n]$, $n_0 > 0$. This function is illustrated in Figure 11.3(f), for $n_0 \geq 2$; we see that this shift is an advance in time. The z-transform of this function is given by, from (11.4),

$$\mathcal{Z}[f[n + n_0]u[n]] = \sum_{n=0}^{\infty}[f[n + n_0]u[n]z^{-n}$$

$$= f[n_0] + f[n_0 + 1]z^{-1} + f[n_0 + 2]z^{-2} + \cdots.$$

We multiply the last expression by $z^{n_0} z^{-n_0}$, yielding

$$\mathcal{Z}[f[n + n_0]u[n]] = z^{n_0}[f[n_0]z^{-n_0} + f[n_0 + 1]z^{-(n_0+1)} + f[n_0 + 2]z^{-(n_0+2)} + \cdots]$$

$$= z^{n_0} \sum_{n=n_0}^{\infty} f[n]z^{-n}.$$

The last summation is a part of the z-transform of $f[n]$. To complete this transform, we add and subtract $\Sigma f[n]z^{-n}$, $n = 0, 1, \ldots, (n_0 - 1)$. Thus,

$$\mathscr{Z}[f[n + n_0]u[n]] = z^{n_0}[\sum_{n=n_0}^{\infty} f[n]z^{-n} + \sum_{n=0}^{n_0-1} f[n]z^{-n} - \sum_{n=0}^{n_0-1} f[n]z^{-n}]$$

$$= z^{n_0}[F(z) - \sum_{n=0}^{n_0-1} f[n]z^{-n}], \quad n_0 > 0,$$

(11.25)

and we have the real-shifting property for an advance in time. Because we are considering the unilateral z-transform, the factor $u[n]$ may be omitted.

Property (11.25) applies *only* for a time advance of the type illustrated in Figure 11.3(f), and Property (11.13) applies *only* for a time delay of the type illustrated in Figure 11.3(e).

EXAMPLE 11.4   **Illustration of time-shifting properties**

We now illustrate the time-shifting properties of the z-transform. Consider a discrete-time function $f[n]$, which has the first six values given in Table 11.3. Because $f[n]$ is not given in functional form, we cannot express its z-transform in closed form.

We now find $\mathscr{Z}[f[n]]$, $\mathscr{Z}[f[n - 2]u[n - 2]]$, and $\mathscr{Z}[f[n + 2]u[n]]$:

$$F(z) = \mathscr{Z}[f[n]] = \sum_{n=0}^{\infty} f[n]z^{-n}$$

$$= 1 - z^{-1} - 0.5z^{-2} + 1.5z^{-4} + 3.5z^{-5} + \cdots;$$

$$\mathscr{Z}[f[n - 2]u[n - 2]] = \sum_{n=0}^{\infty} f[n - 2]u[n - 2]z^{-n}$$

$$= z^{-2} - z^{-3} - 0.5z^{-4} + 1.5z^{-6} + 3.5z^{-7} + \cdots;$$

$$\mathscr{Z}[f[n + 2]] = \sum_{n=0}^{\infty} f[n + 2]z^{-n}$$

$$= -0.5 + 1.5z^{-2} + 3.5z^{-3} + \cdots.$$

(11.26)

Note that the time advance of two discrete increments results in the loss of the first two samples.

TABLE 11.3 Functions for Example 11.4

n	$f[n]$	$f[n - 2]u[n - 2]$	$f[n + 2]u[n]$
0	1	0	−0.5
1	−1	0	0
2	−0.5	1	1.5
3	0	−1	3.5
4	1.5	−0.5	⋮
5	3.5	0	
⋮	⋮	⋮	

The z-transforms of the two shifted functions, from (11.13) and (11.25), are given by

$$\mathscr{Z}[f[n - 2]u[n - 2]] = z^{-2}F(z),$$

and

$$\mathscr{Z}[f[n + 2]] = z^2[F(z) - 1 + z^{-1}].$$

The results in (11.26) verify these relations.   ■

Initial and Final Values

The initial-value property relates to finding the initial value of a function, $f[0]$, directly from the z-transform of that function. From the definition of the z-transform, (11.4),

$$F(z) = f[0] + f[1]z^{-1} + f[2]z^{-2} + \cdots.$$

We see that $f[0]$ is found from $F(z)$ by taking its limit as z approaches infinity:

$$\lim_{z \to \infty} F(z) = \lim_{z \to \infty} \left[f[0] + \frac{f[1]}{z} + \frac{f[2]}{z^2} + \cdots \right] = f[0].$$

Thus, the initial-value property is given by

$$f[0] = \lim_{z \to \infty} F(z). \tag{11.27}$$

The final-value property relates to finding the final (steady-state) value of a function directly from its z-transform. This derivation is more involved than that for the initial-value property.

We begin the derivation by considering the transform

$$\mathscr{Z}[f[n + 1] - f[n]] = \lim_{k \to \infty} \left[\sum_{n=0}^{k} f[n + 1]z^{-n} - \sum_{n=0}^{k} f[n]z^{-n} \right]$$

$$= \lim_{k \to \infty} [-f[0] + f[1](1 - z^{-1}) + f[2](z^{-1} - z^{-2}) + \cdots$$

$$+ f[k](z^{-k+1} - z^{-k}) + f[k + 1]z^{-k}].$$

We now take the limit of both sides of this equation as z approaches unity; as a result, the terms $(z^{-i+1} - z^i)$ approach zero. Thus,

$$\lim_{z \to 1} \mathscr{Z}[f[n + 1] - f[n]] = \lim_{n \to \infty} [f[n + 1] - f[0]]. \tag{11.28}$$

We have replaced k with n on the right side, for clarity in the remainder of this derivation.

From the shifting property (11.25),

$$\mathcal{Z}[f[n + 1] - f[n]] = z[F(z) - f[0]] - F(z) = (z - 1)F(z) - zf[0],$$

and thus,

$$\lim_{z \to 1} \mathcal{Z}[f[n + 1] - f[n]] = \lim_{z \to 1} [(z - 1)F(z) - zf[0]]. \tag{11.29}$$

Equating the right sides of (11.28) and (11.29) yields

$$\lim_{n \to \infty} [f[n + 1] - f[0]] = \lim_{z \to 1} [(z - 1)F(z) - zf[0]].$$

Because $f[0]$ is a constant, this term cancels and the final-value property is given by

$$\lim_{n \to \infty} f[n] = f[\infty] = \lim_{z \to 1} (z - 1)F(z), \tag{11.30}$$

provided that the limit on the left side exists, that is, provided that $f[n]$ has a final value. [It is shown later that $f[n]$ has a final value provided that all poles of $F(z)$ are *inside* the unit circle, except for possibly a single pole at $z = 1$. In addition, from (11.30), if $f[n]$ has a final value, that value is nonzero only for the case that $F(z)$ has a pole at $z = 1$.]

EXAMPLE 11.5   **Illustrations of initial- and final-value properties**

We illustrate the initial- and final-value properties with an example. Consider the unit step function $u[n]$:

$$\mathcal{Z}[u[n]] = \frac{z}{z - 1}.$$

From the initial-value property (11.27),

$$f[0] = \lim_{z \to \infty} \frac{z}{z - 1} = \lim_{z \to \infty} \frac{1}{1 - 1/z} = 1.$$

We know that the final value of $u[n]$ exists; hence, from the final-value property (11.30),

$$f[\infty] = \lim_{z \to 1} (z - 1) \frac{z}{z - 1} = \lim_{z \to 1} z = 1.$$

Both of these values are seen to be correct.   ∎

EXAMPLE 11.6   **Continuation of Example 11.5**

The sinusoidal function $\sin(\pi n/2)$ is now considered. From Table 11.2,

$$\mathcal{Z}[\sin(\pi n/2)] = \frac{z \sin(\pi/2)}{z^2 - 2z \cos(\pi/2) + 1} = \frac{z}{z^2 + 1}.$$

From the initial-value property (11.27),

$$f[0] = \lim_{z \to \infty} \frac{z}{z^2 + 1} = 0,$$

which is the correct value. From the final-value property (11.30),

$$f[\infty] = \lim_{z \to 1} (z - 1) \frac{z}{z^2 + 1} = 0,$$

which is incorrect, because $\sin(\pi n/2)$ oscillates continually and therefore does not have a final value. ∎

We have now derived several properties of the z-transform. Additional properties are derived in the next section.

11.5 ADDITIONAL PROPERTIES

Two additional properties of the z-transform will now be derived; then a table of properties will be given.

Time Scaling

Independent-variable transformations were introduced in Chapter 9. We now consider the effects of these discrete-time transformations on the z-transform of a function.

Consider first the z-transform of $f[m]$; for convenience, we now denote discrete time by the variable m:

$$F(z) = \mathcal{Z}[f[m]] = \sum_{m=0}^{\infty} f[m]z^{-m} = f[0] + f[1]z^{-1} + f[2]z^{-2} + \cdots.$$

An example of $f[m]$ is plotted in Figure 11.4(a). We now consider the time-scaling transformation $m = n/k$ ($n = mk$), where k is a positive integer, and use the notation

$$f_t[n] = f[m] \Big|_{m=n/k} = f[n/k] \tag{11.31}$$

as in Section 9.2. The n-axis ($n = mk$) is plotted in Figure 11.4(a), and $f[n/k]$ is plotted versus n in Figure 11.4(b).

We now *define* the z-transform of $f[n/k]$ as

$$\mathcal{Z}[f_t[n]] = F_t(z) = \mathcal{Z}[f[n/k]]$$

$$= f[0] + f[1]z^{-k} + f[2]z^{-2k} + \cdots = \sum_{n=0}^{\infty} f[n]z^{-kn}. \tag{11.32}$$

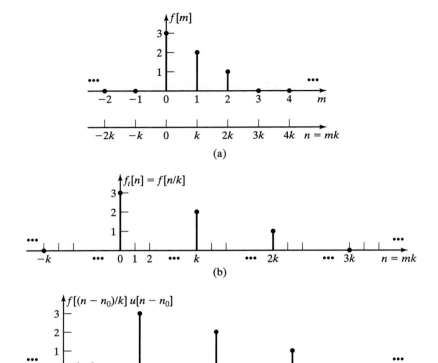

Figure 11.4   Examples of time shifting and scaling.

Note that this *definition* sets the values of $f_t[n]$ to zero for $n \neq mk$, with n a positive integer. We see then that

$$F_t(z) = \mathcal{Z}[f[n/k]] = \sum_{n=0}^{\infty} f[n]z^{-kn}$$

$$= \sum_{n=0}^{\infty} f[n](z^k)^{-n} = F(z)\bigg|_{z \leftarrow z^k} = F(z^k), \qquad (11.33)$$

where $\mathcal{Z}[f[n]] = F(z)$. Hence, we have the time-scaling property (as defined earlier), for k a positive integer:

$$f[n/k] \overset{\mathcal{Z}}{\leftrightarrow} F(z^k). \qquad (11.34)$$

The time scaling $m = n/k$ creates additional sample values, all of which we choose to set to zero. The derivations presented here apply for only this choice; other rules can be used to assign values for the samples created, and then (11.34) does not apply.

We do not consider the time scaling $f[nk]$, where k is a positive integer. Recall from Chapter 9 that this transform results in the loss of sample values. We now illustrate time scaling with an example.

EXAMPLE 11.7   **Illustration of time-scaling property**

Consider the exponential function $f[n] = a^n$. From Table 11.2,

$$f[n] = a^n \overset{\mathscr{Z}}{\leftrightarrow} \frac{z}{z - a} = F(z).$$

We wish to find the z-transform of $f_t[n] = f[n/2]$. We first find $F_t(z)$ from its definition; then we use property (11.33) for verification.
From definition (11.32),

$$\mathscr{Z}[f[n/2]] = 1 + az^{-2} + a^2z^{-4} + \cdots = \sum_{n=0}^{\infty} a^n z^{-2n} = \sum_{n=0}^{\infty} (az^{-2})^n.$$

From Appendix C,

$$\sum_{n=0}^{\infty} b^n = \frac{1}{1 - b}; \quad |b| < 1.$$

Thus, with $b = az^{-2}$,

$$\mathscr{Z}[f[n/2]] = \frac{1}{1 - az^{-2}} = \frac{z^2}{z^2 - a}. \tag{11.35}$$

Direct substitution into the scaling property, (11.33), with $k = 2$ verifies this result.   ∎

Convolution in Time

We now derive the transform for the convolution summation. From (10.13) and the definition of convolution,

$$x[n]*y[n] = \sum_{k=-\infty}^{\infty} x[k]y[n - k] = \sum_{k=0}^{\infty} x[k]y[n - k], \tag{11.36}$$

because $x[k]$ is zero for $k < 0$. Then, from (11.4),

$$\mathscr{Z}[x[n]*y[n]] = \sum_{n=0}^{\infty} \left[\sum_{k=0}^{\infty} x[k]y[n - k] \right] z^{-n}$$

$$= \sum_{k=0}^{\infty} x[k] \left[\sum_{n=0}^{\infty} y[n - k]z^{-n} \right], \tag{11.37}$$

TABLE 11.4 Properties of the *z*-Transform

Name	Property
1. Linearity, (11.8)	$\mathscr{Z}[a_1 f_1[n] + a_2 f_2[n]] = a_1 F_1(z) + a_2 F_2(z)$
2. Real shifting, (11.13)	$\mathscr{Z}[f[n - n_0]u[n - n_0]] = z^{-n_0}F(z), \quad n_0 \geq 0$
3. Real shifting, (11.25)	$\mathscr{Z}[f[n + n_0]u[n]] = z^{n_0}[F(z) - \sum_{n=0}^{n_0-1} f[n]z^{-n}]$
4. Complex shifting, (11.23)	$\mathscr{Z}[a^n f[n]] = F(z/a)$
5. Multiplication by *n*	$\mathscr{Z}[nf[n]] = -z\dfrac{dF(z)}{dz}$
6. Time scaling, (11.33)	$\mathscr{Z}[f[n/k]] = F(z^k),\ k$ a positive integer
7. Convolution, (11.38)	$\mathscr{Z}[x[n]*y[n]] = X(z)Y(z)$
8. Summation	$\mathscr{Z}[\sum_{k=0}^{n} f[k]] = \dfrac{z}{z-1}F(z)$
9. Initial value, (11.27)	$f[0] = \lim\limits_{z \to \infty} F(z)$
10. Final value, (11.30)	$f[\infty] = \lim\limits_{z \to 1}(z-1)F(z),$ if $f[\infty]$ exists

where the order of the summations is reversed in the last step. Next we change variables on the inner summation, letting $m = (n - k)$. Then $n = m + k$ and

$$\mathscr{Z}[x[n]*y[n]] = \sum_{k=0}^{\infty} x[k]\left[\sum_{m=-k}^{\infty} y[m]z^{-m-k}\right]$$

$$= \sum_{k=0}^{\infty} x[k]z^{-k}\sum_{m=0}^{\infty} y[m]z^{-m} = X(z)Y(z). \tag{11.38}$$

The lower limit $m = -k$ is changed to $m = 0$, because $y[m]$ is zero for $m < 0$. Hence, convolution transforms into multiplication in the *z*-domain. Examples of convolution are given later in this chapter when we consider linear systems.

Several properties of the *z*-transform have been developed. These properties are useful in generating tables of *z*-transforms and in applying the *z*-transform to the solutions of linear difference equations with constant coefficients. When possible, we model discrete-time physical systems with linear difference equations with constant coefficients; hence, these properties are useful in both the analysis and design of linear time-invariant physical systems. Table 11.4 gives properties for the *z*-transform and includes some properties in addition to those derived.

11.6 LTI SYSTEM APPLICATIONS

In this section, we illustrate some applications of the *z*-transform to linear time-invariant (LTI) systems. First, we consider transfer functions, and then certain system properties are investigated.

Transfer Functions

When possible, we model discrete-time systems with linear difference equations with constant coefficients; the model is then linear and time invariant. (See Section 10.4.) From (10.48), the general equation for this model is given by

$$\sum_{k=0}^{N} a_k y[n-k] = \sum_{k=0}^{M} b_k x[n-k], \tag{11.39}$$

where $x[n]$ is the input signal, $y[n]$ is the output signal, N is the system order, and the constants a_k, b_k, N, and M are parameters of the system. For convenience, we let $M = N$. (If this is not the case, certain coefficients must be set to zero.) The expansion of (11.39) yields

$$a_0 y[n] + a_1 y[n-1] + \cdots + a_{N-1} y[n-(N-1)] + a_N y[n-N]$$
$$= b_0 x[n] + b_1 x[n-1] + \cdots + b_{N-1} x[n-(N-1)] + b_N x[n-N]. \tag{11.40}$$

From (11.13), for $n_0 \geqq 0$, the time-delay property is given by

$$\mathcal{Z}[f[n-n_0]] = z^{-n_0} F(z), \quad f[n] = 0, \quad n < 0. \tag{11.41}$$

Thus, the z-transform of (11.39), with $M = N$, yields

$$\sum_{k=0}^{N} a_k z^{-k} Y(z) = \sum_{k=0}^{N} b_k z^{-k} X(z), \tag{11.42}$$

where $x[n] = y[n] = 0, n < 0$. Expanding (11.42) gives

$$[a_0 + a_1 z^{-1} + \cdots + a_{N-1} z^{-N+1} + a_N z^{-N}] Y(z)$$
$$= [b_0 + b_1 z^{-1} + \cdots + b_{N-1} z^{-N+1} + b_N z^{-N}] X(z). \tag{11.43}$$

This equation can also be seen directly from (11.40).

By definition, the system transfer function $H(z)$ is the ratio $Y(z)/X(z)$. Therefore, from (11.43),

$$H(z) = \frac{Y(z)}{X(z)} = \frac{b_0 + b_1 z^{-1} + \cdots + b_{N-1} z^{-N+1} + b_N z^{-N}}{a_0 + a_1 z^{-1} + \cdots + a_{N-1} z^{-N+1} + a_N z^{-N}}$$
$$= \frac{b_0 z^N + b_1 z^{N-1} + \cdots + b_{N-1} z + b_N}{a_0 z^N + a_1 z^{N-1} + \cdots + a_{N-1} z + a_N}, \tag{11.44}$$

where the last step is required to express the variable z in positive exponents. Note that this transfer function is identical to that derived in Chapter 10, which applies for only a complex-exponential input signal. This transfer function, (11.44), applies for any input that has a z-transform and, hence, is a generalization of that of Chapter 10. An example is now given.

EXAMPLE 11.8   **Transfer function of a discrete system**

We consider again the α-filter of (11.14), which is depicted in Figure 11.5(a). (See Figure 10.18.) The filter equation is given by

$$y[n] - (1 - \alpha)y[n - 1] = \alpha x[n].$$

For this example, we let $\alpha = 0.1$; then

$$y[n] - 0.9y[n - 1] = 0.1x[n].$$

The z-transform of this equation yields

$$(1 - 0.9z^{-1})Y(z) = 0.1X(z),$$

and the transfer function is

$$H(z) = \frac{Y(z)}{X(z)} = \frac{0.1}{1 - 0.9z^{-1}} = \frac{0.1z}{z - 0.9}. \tag{11.45}$$

Note that we could have written the transfer function directly from (11.39) and (11.44).   ■

It is common when specifying digital filters to give the filter-transfer function $H(z)$ or the difference equation. For example, the block-diagram specification for the α-filter of the last example can be any of the four forms illustrated in Figure 11.5. The representation of the time delay by the transfer function z^{-1} is based on the property

$$\mathscr{Z}[y[n - 1]] = z^{-1}\mathscr{Z}[y[n]] = z^{-1}Y(z).$$

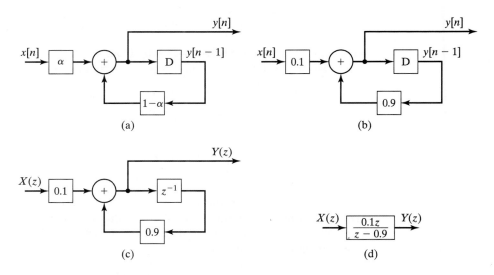

Figure 11.5 α-Filter representation.

Inverse z-Transform

Three procedures for finding the inverse z-transform will now be discussed. The first procedure involves the use of the complex inversion integral, from (11.2):

$$f[n] = \frac{1}{2\pi j} \oint_\Gamma F(z)z^{n-1}\, dz.$$

This integration is in the complex plane and is usually too complicated to be of practical value; hence, we will not use this approach. (See Ref. 3 for applications.)

The second method for finding the inverse z-transform is by partial-fraction expansion, in which a function that does not appear in the z-transform tables is expressed as a sum of functions that do appear in the tables. Partial-fraction expansions are presented in Appendix F; those readers unfamiliar with this topic should read this appendix. We now illustrate the use of partial fractions to find inverse z-transforms.

EXAMPLE 11.9   **Inverse z-transform by partial-fraction expansion**

We solve for the time response of the α-filter of Example 11.8, for the unit step input. From Table 11.2, $X(z) = z/(z - 1)$. From (11.45), the transformed output is given by,

$$Y(z) = H(z)X(z) = \frac{0.1z}{z - 0.9}\frac{z}{z - 1}.$$

We expand $Y(z)/z$ in partial fractions:

$$\frac{Y(z)}{z} = \frac{0.1z}{(z - 0.9)(z - 1)} = \frac{k_1}{z - 0.9} + \frac{k_2}{z - 1}.$$

Then (see Appendix F)

$$k_1 = (z - 0.9)\left[\frac{0.1z}{(z - 0.9)(z - 1)}\right]_{z=0.9} = \frac{0.1z}{z - 1}\bigg|_{z=0.9} = -0.9$$

and

$$k_2 = (z - 1)\left[\frac{0.1z}{(z - 0.9)(z - 1)}\right]_{z=1} = \frac{0.1z}{z - 0.9}\bigg|_{z=1} = 1.$$

Thus, $Y(z)$ is given by

$$Y(z) = \frac{0.1z^2}{(z - 0.9)(z - 1)} = \frac{z}{z - 1} - \frac{0.9z}{z - 0.9},$$

and from Table 11.2, the inverse transform yields

$$y[n] = 1 - 0.9(0.9)^n = 1 - 0.9^{n+1},$$

for $n \geq 0$. The partial-fraction expansion can be verified by modifying the MATLAB program in Section 11.2.

The z-transform solution gives $y[0] = 0.1$; this value is also found from the initial-value property:

$$y[0] = \lim_{z \to \infty} Y(z) = \lim_{z \to \infty} \frac{0.1z^2}{(z - 0.9)(z - 1)} = 0.1.$$

As a final point, we verify the first three values of $y[n]$ by the iterative solution of the filter difference equation:

$$y[n] = 0.9y[n - 1] + 0.1x[n];$$
$$n = 0: \quad y[0] = 0.9(0) + 0.1(1) = 0.1;$$
$$n = 1: \quad y[1] = 0.9(0.1) + 0.1(1) = 0.19;$$
$$n = 2: \quad y[2] = 0.9(0.19) + 0.1(1) = 0.271.$$

[handwritten margin notes: .19 ; 1.42632 ; 1.269 ; n=3: Y[3]=0.9(.271)+.1(1)=.3439]

The solution yields

$$y[n] = 1 - 0.9^{n+1},$$
$$n = 0: \quad y[0] = 1 - 0.9 = 0.1,$$
$$n = 1: \quad y[1] = 1 - 0.81 = 0.19,$$

and

$$n = 2: \quad y[2] = 1 - 0.729 = 0.271,$$

and these values are verified. The values can also be verified with the following MATLAB program:

```
ynminus1 = 0;
xn = 1;
for n = 0:2;
    yn = 0.9*ynminus1 + 0.1*xn
    ynminus1 = yn;
end
result:yn=0.1 0.19 0.271
```

In this program, $yn = y[n]$, $ynminus1 = y[n - 1]$, and $xn = x[n]$. A SIMULINK simulation that also verifies the response is illustrated in Figure 11.6. ∎

The third procedure for finding inverse transforms is the expansion of a transform into a power series of the proper form by long division. This procedure, illustrated earlier in this chapter, involves dividing the numerator of the transform by its denominator. The result of this division is the power series

$$F(z) = \frac{N(z)}{D(z)} = f_0 + f_1 z^{-1} + f_2 z^{-2} + \cdots. \tag{11.46}$$

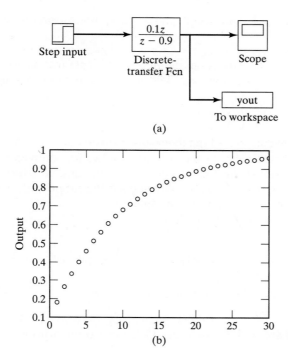

(a)

(b)

Figure 11.6 SIMULINK simulation for Example 11.9.

Comparing this series with the definition of the unilateral z-transform, we see that $f[n]$ in (11.4) is equal to f_n in (11.46). We now illustrate this procedure with an example.

EXAMPLE 11.10   **Inverse z-transform by long division**

The inverse transform in Example 11.9 is now verified by long division. From Example 11.9,

$$Y(z) = \frac{0.1z}{z - 0.9} \frac{z}{z - 1} = \frac{0.1z^2}{z^2 - 1.9z + 0.9}.$$

Dividing the numerator by the denominator yields

$$
\begin{array}{r}
0.1 \quad\;\; + 0.19z^{-1} + 0.271z^{-2} + \cdots \\
z^2 - 1.9z + 0.9{\overline{\smash{\big)}\,0.1z^2 }} \\
\underline{0.1z^2 - 0.19z + 0.09 } \\
0.19z - 0.09 \\
\underline{0.19z - 0.361 + 0.171z^{-1}} \\
0.271 - 0.171z^{-1} \\
\vdots
\end{array}
$$

Hence, $y[0] = 0.1$, $y[1] = 0.19$, and $y[2] = 0.271$, which verifies the values found in Example 11.9.   ∎

We see that the power-series method is practical for evaluating only the first few values of a function unless the long division is implemented on a digital computer.

Complex Poles

Thus far, we have considered the inverse transform only for functions that have real poles. The same partial-fraction procedure applies for complex poles; however, the resulting inverse transform contains complex functions. Of course, the sum of these functions is real. In this section, we develop a different procedure that expresses the inverse transforms as the sum of real functions.

First, consider the real function

$$
y[n] = Ae^{\Sigma n} \cos(\Omega n + \theta) = \frac{Ae^{\Sigma n}}{2} [e^{j\Omega n}e^{j\theta} + e^{-j\Omega n}e^{-j\theta}]
$$

$$
= \frac{A}{2} [e^{(\Sigma+j\Omega)n}e^{j\theta} + e^{(\Sigma-j\Omega)n}e^{-j\theta}],
$$

(11.47)

where Σ and Ω are real. From Table 11.2, the z-transform of this function is given by

$$
Y(z) = \frac{A}{2} \left[\frac{e^{j\theta}z}{z - e^{\Sigma+j\Omega}} + \frac{e^{-j\theta}z}{z - e^{\Sigma-j\Omega}} \right]
$$

$$
= \frac{(Ae^{j\theta}/2)z}{z - e^{\Sigma+j\Omega}} + \frac{(Ae^{-j\theta}/2)z}{z - e^{\Sigma-j\Omega}} = \frac{k_1 z}{z - p_1} + \frac{k_1^* z}{z - p_1^*},
$$

(11.48)

where the asterisk indicates the complex conjugate.

The usual partial-fraction expansion yields terms in the form of (11.48). Hence, given k_1 and p_1 in (11.48), we can solve for the discrete-time function of (11.47), using the following relationship from (11.48):

$$
p_1 = e^{\Sigma}e^{j\Omega} = e^{\Sigma}\angle\Omega \Rightarrow \Sigma = \ln |p_1|; \quad \Omega = \arg p_1
$$

(11.49)

and

$$
k_1 = \frac{Ae^{j\theta}}{2} = \frac{A}{2} \angle\theta \Rightarrow A = 2|k_1|; \quad \theta = \arg k_1.
$$

(11.50)

Thus, we calculate Σ and Ω from the poles, and A and θ from the partial-fraction expansion. We can then express the inverse transform as the sinusoid of (11.47). An illustrative example is given next.

EXAMPLE 11.11   **Inverse z-transform with complex poles**

We find the inverse z-transform of the function

$$Y(z) = \frac{-2.753z}{z^2 - 1.101z + 0.6065} = \frac{-2.753z}{(z - 0.550 - j0.550)(z - 0.550 + j0.550)}$$

$$= \frac{k_1 z}{z - 0.550 - j0.550} + \frac{k_1^* z}{z - 0.550 + j0.550}.$$

Dividing both sides by z, we calculate k_1:

$$k_1 = (z - 0.550 - j0.550)\left[\frac{-2.753}{(z - 0.550 - j0.550)(z - 0.550 + j0.550)}\right]_{z=0.550+j0.550}$$

$$= \frac{-2.753}{2(j0.550)} = 2.50\underline{/90°}.$$

From (11.49) and (11.50),

$$p_1 = 0.550 + j0.550 = 0.7778\underline{/45°},$$

$$\Sigma = \ln|p_1| = \ln(0.7778) = -0.251; \quad \Omega = \arg p_1 = \frac{\pi}{4},$$

and

$$A = 2|k_1| = 2(2.50) = 5; \quad \theta = \arg k_1 = \frac{\pi}{2}.$$

Hence, from (11.47),

$$y[n] = Ae^{\Sigma n}\cos(\Omega n + \theta) = 5e^{-0.251n}\cos\left(\frac{\pi}{4}n + \frac{\pi}{2}\right).$$

This result can be verified by finding the z-transform of this function using Table 11.2. The partial-fraction expansion can be verified with the following MATLAB program:

```
n = [0 0 -2.753];
d = [1 -1.101 0.6065];
[r,p,k]=residue(n,d)
result: r=0+2.5001i 0-2.5001i
        p=0.5505+0.5509i 0.5505-0.5509i
```
■

Causality

We next investigate causal LTI systems. Consider the system of Figure 11.7(a). This system is a unit advance, with the output equal to the input advanced by one

(a)   (b)   **Figure 11.7** Unit advance.

discrete-time increment; that is, $y[n] = x[n + 1]$. For this derivation, we ignore initial conditions. Then the output is given by

$$Y(z) = \mathscr{Z}[x[n + 1]] = z[X(z) - x[0]] = zX(z), \qquad (11.51)$$

from Table 11.4. The unit advance has a transfer function of $H(z) = z$ and can be represented by the block diagram of Figure 11.7(b). In a like manner, it is seen that the transfer function of N cascaded unit advances is $H(z) = z^N$. If we allow N to be negative, this transfer function also applies to $|N|$ unit delays. For example, the transfer function for three cascaded advances is z^3, and the transfer function for three cascaded delays is z^{-3}.

The unit advance is not causal; the system of Figure 11.7 cannot be realized physically. Consider the transfer function given by

$$H(z) = \frac{z^2 + 0.4z + 0.9}{z - 0.6} = z + \frac{z + 0.9}{z - 0.6}, \qquad (11.52)$$

where the last function can be obtained by dividing the numerator of $H(z)$ by its denominator. This system is noncausal, since the system can be represented as a unit advance in parallel with a second system that is physically realizable. This unit advance appears because the numerator of $H(z)$ is of higher order than the denominator.

It is seen from the preceding development that for a *causal system*, the numerator of the transfer function $H(z)$ of (11.44) cannot be of higher order than the denominator, when the exponents are positive. If the transfer function $H(z)$ is expressed in negative exponents as in (11.44), that is, as

$$H(z) = \frac{Y(z)}{X(z)} = \frac{b_0 + b_1 z^{-1} + \cdots + b_{N-1} z^{-N+1} + b_N z^{-N}}{a_0 + a_1 z^{-1} + \cdots + a_{N-1} z^{-N+1} + a_N z^{-N}}, \qquad (11.53)$$

the system is causal provided that $a_0 \neq 0$.

Stability

We now relate bounded-input bounded-output (BIBO) stability of causal systems to the system transfer function. Recall the definition of BIBO stability:

BIBO Stability
A system is stable if the output remains bounded for any bounded input.

For an Nth-order discrete-time LTI causal system, the transfer function can be expressed as

$$[\text{eq}(11.44)] \quad H(z) = \frac{Y(z)}{X(z)} = \frac{b_0 z^N + b_1 z^{N-1} + \cdots + b_{N-1} z + b_N}{a_0 z^N + a_1 z^{N-1} + \cdots + a_{N-1} z + a_N},$$

with $a_0 \neq 0$. The denominator of this transfer function can be factored as

$$a_0 z^N + a_1 z^{N-1} + \cdots + a_{N-1} z + a_N = a_0(z - p_1)(z - p_2) \cdots (z - p_N). \qquad (11.54)$$

The zeros of this polynomial are the *poles* of the transfer function, where, by definition, the poles are those values of z for which $H(z)$ is unbounded.

First, we assume that $H(z)$ has no repeated poles. We can then express the output $Y(z)$ in (11.44) as

$$Y(z) = H(z)\,X(z) = \frac{b_0 z^N + b_1 z^{N-1} + \cdots + b_{N-1}z + b_N}{a_0(z - p_1)(z - p_2)\cdots(z - p_N)}X(z)$$

$$= \frac{k_1 z}{z - p_1} + \frac{k_2 z}{z - p_2} + \cdots + \frac{k_N z}{z - p_N} + Y_x(z),$$

(11.55)

where $Y_x(z)$ is the sum of the terms, in the partial-fraction expansion, that originate in the poles of the input function $X(z)$. Hence, $Y_x(z)$ is the *forced response*.

In the partial fraction expansion of (11.55), it is assumed that the order of the numerator of $H(z)$ is lower than that of the denominator. If the order of the numerator polynomial is equal to or greater than the order of the denominator polynomial, the partial-fraction expansion will include additional terms. [See (F.1).]

The inverse transform of (11.55) yields

$$y[n] = k_1 p_1^n + k_2 p_2^n + \cdots + k_N p_N^n + y_x[n] = y_n[n] + y_x[n]. \qquad (11.56)$$

The terms of $y_n[n]$ originate in the poles of the transfer function, and $y_n[n]$ is the *natural response*. The natural response is always present in the system output, independent of the form of the input signal $x[n]$. The factor p_i^n in each term of the natural response is called a *mode* of the system.

If the input $x[n]$ is bounded, the forced response $y_x[n]$ will remain bounded, since $y_x[n]$ is of the functional form of $x[n]$. [$Y_x(z)$ has the same poles as $X(z)$.] Thus, an unbounded output must be the result of at least one of the natural-response terms, $k_i p_i^n$, becoming unbounded. This unboundedness can occur only if the magnitude of at least one pole, $|p_i|$, is greater than unity.

From the preceding discussion, we see the requirement for BIBO stability:

> An LTI discrete-time causal system is BIBO stable provided that all poles of the system transfer function lie inside the unit circle in the z-plane.

The stable region of the z-plane is illustrated in Figure 11.8. This conclusion was also reached in Chapter 10 by taking a different approach.

The stability criterion can be stated in a different way. If a system is stable, the poles of its transfer function $H(z)$ are restricted to the interior of the unit circle. Because $h[n]$ is a causal function, the region of convergence of $H(z)$ includes the unit circle and the entire region of the finite plane outside the unit circle, as illustrated in Figure 11.9. Hence, the stability criteria can also be stated as follows:

> An LTI discrete-time causal system is BIBO stable provided that the region of convergence of its transfer function includes the unit circle.

The *system characteristic equation* is, by definition, the denominator polynomial of the transfer function set to zero. Hence, the characteristic equation is the

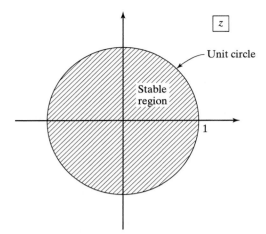

Figure 11.8 Stable region of the z-plane.

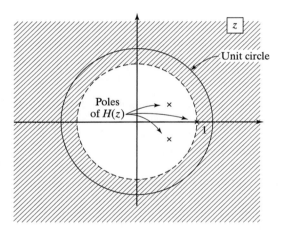

Figure 11.9 Region of convergence of a stable system.

denominator of (11.44) set to zero; that is,

$$a_0 z^N + a_1 z^{N-1} + \cdots + a_{N-1}z + a_N = a_0(z - p_1)(z - p_2)\cdots(z - p_n) = 0 \quad (11.57)$$

is the system characteristic equation. The system is stable provided that the roots of the system characteristic equation are inside the unit circle. A similar development shows that the same requirements apply if $H(z)$ has repeated poles. We now illustrate system stability with an example.

EXAMPLE 11.12   **Stability of a discrete system**

Suppose that the transfer function of an LTI system is given by

$$H(z) = \frac{2z^2 - 1.6z - 0.90}{z^3 - 2.5z^2 + 1.96z - 0.48}.$$

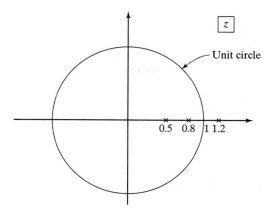

Unit circle

0.5   0.8   1 1.2

Figure 11.10   Pole locations for Example 11.12.

The characteristic equation for this system is seen to be

$$z^3 - 2.5z^2 + 1.96z - 0.48 = (z - 0.5)(z - 0.8)(z - 1.2) = 0.$$

The poles of the transfer function are at 0.5, 0.8, and 1.2, as illustrated in Figure 11.10. Thus, the system is unstable because the pole at $z = 1.2$ is outside the unit circle. The modes of the system are 0.5^n, 0.8^n, and 1.2^n; the system is unstable since $\lim_{n \to \infty} 1.2^n$ is unbounded. The characteristic-equation roots can be calculated with the following MATLAB program:

```
P = [1 -2.5 1.96 -.48];
r = roots(p)
result: r = 1.2 0.8 0.5
```

∎

Invertibility

Recall from Section 9.6 the definition of the inverse of a system:

Inverse of a System
The inverse of a system $H(z)$ is a second system $H_i(z)$ that, when cascaded with $H(z)$, yields the identity system.

Thus, $H_i(z)$ is defined by the equation

$$H(z)H_i(z) = 1 \Rightarrow H_i(z) = \frac{1}{H(z)}. \tag{11.58}$$

These systems are illustrated in Figure 11.11.

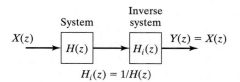

$$H_i(z) = 1/H(z)$$

Figure 11.11   System with its inverse.

We now consider the characteristics of the inverse system. The transfer function $H(z)$ of a causal system can be expressed as

$$H(z) = \frac{b_0 z^M + b_1 z^{M-1} + \cdots + b_M}{a_0 z^N + a_1 z^{N-1} + \cdots + a_N}, \tag{11.59}$$

where the order of the numerator is less than or equal to that of the denominator, or $M \leq N$. Hence, the inverse system has the transfer function

$$H_i(z) = \frac{a_0 z^N + a_1 z^{N-1} + \cdots + a_N}{b_0 z^M + b_1 z^{M-1} + \cdots + b_M}. \tag{11.60}$$

This system is causal if $N \leq M$. Hence, for both the system and its inverse to be causal, $M = N$.

Next we consider stability. For the system of (11.59) to be stable, the poles of the transfer function must lie inside the unit circle. For the inverse system of (11.60) to be stable, the poles of $H_i(z)$ [the zeros of $H(z)$] must lie inside the unit circle. Hence, for the system of Figure 11.11 to be stable, both the poles and zeros of the system must lie inside the unit circle.

In this section, we consider solving for the time response of an LTI system using z-transforms. Generally, we use partial-fraction expansions in the inverse transform; however, long division can also be used. The characteristics of causality, stability, and invertibility are then investigated for LTI systems. It is shown that stability can always be determined for systems modeled by linear difference equations with constant coefficients. This type of system is stable provided that the poles of its transfer function are all inside the unit circle in the z-plane.

11.7 BILATERAL z-TRANSFORM

In Section 11.1, we defined the bilateral z-transform pair

[eq(11.1)] $$\mathcal{Z}_b[f[n]] = F_b(z) = \sum_{n=-\infty}^{\infty} f[n] z^{-n}$$

and

[eq(11.2)] $$\mathcal{Z}_b^{-1}[F_b(z)] = f[n] = \frac{1}{2\pi j} \oint_\Gamma F_b(z) z^{n-1}\, dz, \quad j = \sqrt{-1},$$

where $\mathcal{Z}_b[\cdot]$ denotes the bilateral z-transform. The path of integration Γ in the inverse transform is determined by the region of convergence (ROC) of $F_b(z)$. However, as with the unilateral z-transform, we do not use the inversion integral (11.2) to find inverse transforms; instead, we use tables. Nevertheless, as will be shown, we must know the region of convergence of $F_b(z)$ to determine its inverse transform.

Because the unilateral z-transform is a special case of the bilateral transform, Table 11.2 applies for the bilateral z-transform of functions for which $f[n] = 0, n < 0$, that is, for causal functions. For example, from Table 11.2, the bilateral z-transform pair for the causal function $a^n u[n]$ is given by

$$a^n u[n] \overset{\mathscr{Z}_b}{\longleftrightarrow} \frac{z}{z-a}; \quad |z| > |a|. \tag{11.61}$$

As indicated, we must always include the ROC for a bilateral transform.

The exponential function in (11.61) is sketched in Figure 11.12(a), along with its ROC, for a real. To illustrate the requirement for specifying the ROC, we will derive the bilateral transform of $-a^n u[-n-1]$. This exponential function is plotted in Figure 11.12(b), for a real. From (11.1),

$$\mathscr{Z}_b[-a^n u[-n-1]] = \sum_{n=-\infty}^{\infty} -a^n u[-n-1] z^{-n} = \sum_{n=-\infty}^{-1} -a^n z^{-n}$$

$$= -(a^{-1}z + a^{-2}z^2 + a^{-3}z^3 + \cdots) = \sum_{n=1}^{\infty} -(a^{-1}z)^n, \tag{11.62}$$

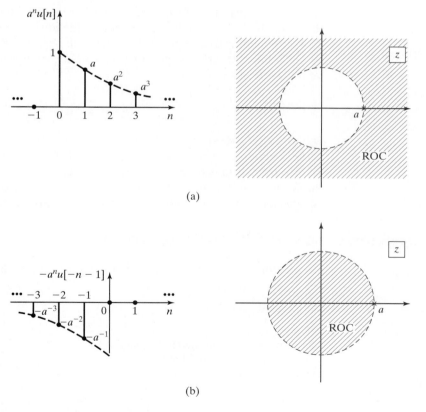

(a)

(b)

Figure 11.12 Exponential functions.

because $u[-n - 1]$ is zero for $n \geq 0$. From Appendix C, we have the convergent power series

$$\sum_{n=k}^{\infty} b^n = \frac{b^k}{1 - b}; \quad |b| < 1. \tag{11.63}$$

We then let $b = a^{-1}z$ and $k = 1$ from (11.62), resulting in the z-transform

$$\mathcal{Z}_b[-a^n u[-n - 1]] = \frac{-a^{-1}z}{1 - a^{-1}z} = \frac{z}{z - a}; \quad |a^{-1}z| < 1. \tag{11.64}$$

The ROC of this transform can also be expressed as $|z| < |a|$ and is also shown in Figure 11.12(b).

We next list the bilateral transform pairs (11.61) and (11.64) together:

$$a^n u[n] \overset{\mathcal{Z}_b}{\longleftrightarrow} \frac{z}{z - a}; \quad |z| > |a|.$$

$$-a^n u[-n - 1] \overset{\mathcal{Z}_b}{\longleftrightarrow} \frac{z}{z - a}; \quad |z| > |a|.$$

Note that the two transforms are identical; only the ROCs are different. Hence, for this case, the ROC must be known before the inverse transform can be determined. This statement is true in general for determining all inverse bilateral z-transforms.

To illustrate the last point with an example, suppose that we are given the bilateral z-transform

$$F_b(z) = \frac{z}{z - 0.5} \tag{11.65}$$

with the ROC not specified. If the ROC is given by $|z| > 0.5$, the inverse transform is the function $f[n] = 0.5^n u[n]$, from (11.61). If the ROC is given by $|z| < 0.5$, the inverse transform is the function $f[n] = -(0.5^n)u[-n - 1]$, from (11.64). Note that the ROC cannot include the pole at $z = 0.5$, since, by definition, $F_b(z)$ is unbounded (the series does not converge) at a pole. We now consider a second example.

EXAMPLE 11.13   **Bilateral z-transform of unit step functions**

Consider the causal unit step function $u[n]$. From (11.61) with $a = 1$, we have the transform pair

$$u[n] \overset{\mathcal{Z}_b}{\longleftrightarrow} \frac{z}{z - 1}; \quad |z| > |1|. \tag{11.66}$$

This step function is plotted in Figure 11.13(a).

Next, consider the noncausal unit step function $-u[-n - 1]$. From (11.64), we have the transform pair

$$\mathcal{Z}_b[-u[-n - 1]] = \frac{z}{z - 1}; \quad |z| < 1. \tag{11.67}$$

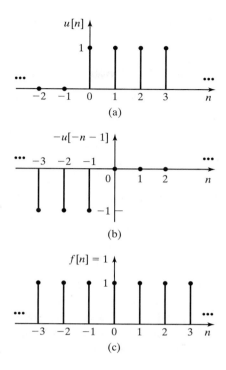

Figure 11.13 Unit step functions.

This step function is plotted in Figure 11.13(b). The z-transforms of both functions are identical, with ROCs different.

As a final step, consider the difference of the two-step function of (11.66) and (11.67):

$$u[n] - [-u[-n - 1]] = u[n] + u[-n - 1] = 1. \tag{11.68}$$

This function is plotted in Figure 11.13(c). However, even though the z-transform is a linear transform, the z-transform of the sum of functions in (11.68) is not equal to the sum of the z-transforms; that is,

$$\mathcal{Z}_b[u[n] + u[-n - 1]] \neq \mathcal{Z}_b[u[n]] + \mathcal{Z}_b[u[-n - 1]]. \tag{11.69}$$

The ROCs of the two transforms on the right side have no regions of the z-plane in common; hence, no values of z exist for which $\mathcal{Z}_b[1]$ is convergent. Consequently, the z-transform of $f[n] = 1$ for all n does not exist. ■

We now give four useful definitions with respect to functions and the bilateral z-transform.

1. A function $f[n]$ is *right sided* if $f[n] = 0$ for $n < n_0$, where n_0 can be a positive or a negative integer. For example, $u[n + 10]$ is right sided, because

$$u[n + 10] = \begin{cases} 0, & n < -10 \\ 1, & n \geq -10 \end{cases}.$$

A second example is given in Figure 11.14(a).

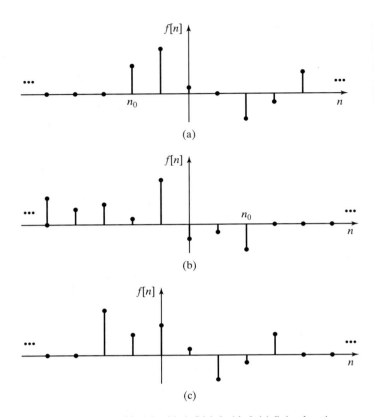

Figure 11.14 Examples for signals: (a) right sided; (b) left sided; (c) finite duration.

   2.   A function $f[n]$ is *left sided* if $f[n] = 0$ for $n > n_0$, where n_0 can be a positive or a negative integer. For example, $u[-n + 10]$ is left sided, because

$$u[-n + 10] = \begin{cases} 1, & n \leq 10 \\ 0, & n > 10 \end{cases}.$$

A second example is given in Figure 11.14(b).

   3.   A function $f[n]$ is *two sided* if it is neither right sided nor left sided. For example, $\cos(n)$ is two sided.

   4.   A function is of *finite duration* if it is both right sided and left sided. For example, $(u[n] - u[n - 10])$ is of finite duration. A second example is given in Figure 11.14(c).

We find these definitions useful when working with bilateral transforms.

Bilateral Transforms

In Chapter 7, a procedure was given for finding bilateral Laplace transforms using unilateral Laplace-transform tables. An equivalent procedure can be developed for finding bilateral *z*-transforms from unilateral *z*-transform tables. However, this procedure is complex and prone to error. Instead, a table of bilateral *z*-transforms is given as Table 11.5. A procedure is now given for using this table.

For a function $f[n]$, we define the two functions $f^+[n]$ and $f^-[n]$:

$$f^+[n] = f[n]u[n]; \quad f^-[n] = f[n]u[-n - 1].$$

Hence, $f^+[n]$ is right sided and $f^-[n]$ is left sided. We express $f[n]$ as

$$f[n] = f^+[n] + f^-[n] \Rightarrow F_b(z) = F_b^+(z) + F_b^-(z), \tag{11.70}$$

provided that $F_b(z)$ exists. Note that each transform in Table 11.5 can be classified as either $F_b^+(z)$ or $F_b^-(z)$, except for the last entry. We now consider an example of the use of this table.

TABLE 11.5 Bilateral *z*-Transform

$f[n]$	$F(z)$	ROC
1. $\delta[n]$	1	All z
2. $\delta[n - n_0]$	z^{-n_0}	$z \neq 0, \quad n_0 \geqq 1$
3. $u[n]$	$\dfrac{z}{z - 1}$	$\lvert z \rvert > 1$
4. $nu[n]$	$\dfrac{z}{(z - 1)^2}$	$\lvert z \rvert > 1$
5. $a^n u[n]$	$\dfrac{z}{z - a}$	$\lvert z \rvert > \lvert a \rvert$
6. $na^n u[n]$	$\dfrac{az}{(z - a)^2}$	$\lvert z \rvert > \lvert a \rvert$
7. $a^n \sin(bn)u[n]$	$\dfrac{az \sin b}{z^2 - 2az \cos b + a^2}$	$\lvert z \rvert > \lvert a \rvert$
8. $a^n \cos(bn)u[n]$	$\dfrac{z(z - a \cos b)}{z^2 - 2az \cos b + a^2}$	$\lvert z \rvert > \lvert a \rvert$
9. $-u[-n - 1]$	$\dfrac{z}{z - 1}$	$\lvert z \rvert < 1$
10. $-a^n u[-n - 1]$	$\dfrac{z}{z - a}$	$\lvert z \rvert < \lvert a \rvert$
11. $-na^n u[-n - 1]$	$\dfrac{az}{(z - a)^2}$	$\lvert z \rvert < \lvert a \rvert$
12. $a^{\lvert n \rvert}, \lvert a \rvert < 1$	$\dfrac{z}{z - a} - \dfrac{z}{z - 1/a}$	$\lvert a \rvert < z < \lvert 1/a \rvert$

EXAMPLE 11.14   **Bilateral *z*-transform of an exponential function**

We now find the bilateral *z*-transform of the two-sided function $f[n] = a^{|n|}$ using (11.70) and Table 11.5. This function is plotted in Figure 11.15 for $0 < a < 1$. It is assumed that a is real for the sketch; however, the following derivation applies for a complex.

For (11.70), we express $f[n]$ as

$$f[n] = a^{|n|} = a^{|n|}[u[n] + u[-n - 1]]$$

$$= a^n u[n] + a^{-n}u[-n - 1] = f^+[n] + f^-[n].$$

From Table 11.5,

$$F_b^+(z) = \mathcal{Z}_b[f^+[n]] = \mathcal{Z}[a^n u[n]] = \frac{z}{z - a}; \quad |z| > |a|,$$

and for the left-sided function,

$$F_b^-(z) = \mathcal{Z}_b[f^-[n]] = \mathcal{Z}_b[a^{-n}u[-n - 1]] = \frac{-z}{z - 1/a}, \quad |z| < 1/|a|.$$

Hence, the *z*-transform of the two-sided function $f[n] = a^{|n|}$ is given by

$$\mathcal{Z}_b[a^{|n|}] = \frac{z}{z - a} - \frac{z}{z - 1/a}; \quad |a| < z < 1/|a|.$$

This transform exists only if $|a| < 1$. This example verifies the last entry in Table 11.5. ∎

Regions of Convergence

Before we consider the inverse bilateral transform, we investigate further the regions of convergence. We first consider a special case. Suppose that the function $f[n]$ is zero except for n equal to zero and unity; hence, $f[n]$ is of finite duration, and

$$F_b(z) = f[0] + f[1]z^{-1}. \tag{11.71}$$

This function exists everywhere except at $z = 0$; hence, the ROC is the entire *z*-plane except for the origin.

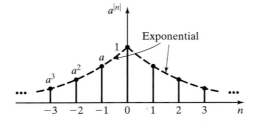

Figure 11.15 Two-sided exponential function.

Now consider the case that $f[n]$ is zero except for $n = -1$ and $n = 0$; once again, $f[n]$ is of finite duration. Then

$$F_b(z) = f[-1]z + f[0]. \qquad (11.72)$$

This function exists everywhere in the finite plane; hence, its ROC is the finite plane.

Consider next the case that $f[n]$ is any sequence of finite duration. In general, $F_b(z)$ can have poles *only* at the origin [as in (11.71)] and at infinity [as in (11.72)]. We conclude then that the bilateral z-transform of a finite sequence exists everywhere in the finite z-plane except possibly at the origin.

From the developments of this section, we see that the ROCs are bounded by circles centered at the origin, and possibly excluding the origin. Hence, the ROC of a bilateral z-transform is of one of four forms:

1. the exterior of a circle centered at the origin (right-sided functions), as illustrated in Figure 11.16(a);

2. the interior of a circle centered at the origin, except for possibly the origin (left-sided functions), as illustrated in Figure 11.16(b);

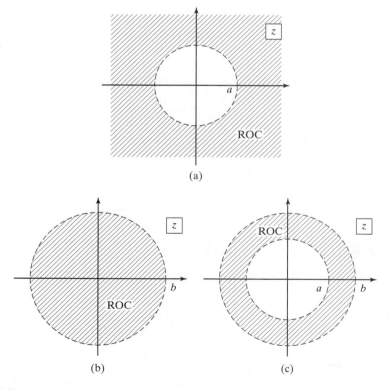

Figure 11.16   Possible regions of convergence.

3. An annular region centered at the origin (two-sided functions), as illustrated in Figure 11.16(c);

4. the entire finite plane, except for possibly the origin (finite-duration functions).

In all cases, the ROCs are open regions; that is, the boundaries of the ROCs are not parts of these regions. Note that all ROCs can be considered to be special cases of Figure 11.16(c).

Finally, we have the property that all poles of transformed functions are exterior to ROCs, since a function is unbounded at a pole of that function. The boundaries of the ROCs will always contain poles, with possibly other poles outside the ROCs.

Inverse Bilateral Transforms

As just stated, the general ROC for a bilateral transform is of the form of Figure 11.16(c). The poles inside the ROC belong to right-sided functions, and the poles outside the ROC belong to left-sided functions.

To determine the inverse bilateral transform, we first find the partial-fraction expansion of $F_b(z)$. Then we express $F_b(z)$ as the sum of two functions, as in (11.70):

$$F_b(Z) = F_b^+(z) + F_b^-(z) \Rightarrow f[n] = f^+[n] + f^-[n], \qquad (11.73)$$

Here, $F_b^+(z)$ contains the terms with poles inside the ROC and $F_b^-(z)$ contains the terms with poles outside the ROC. The inverse transforms are then found directly in Table 11.5.

Three illustrative examples are now given. In these examples, the bilateral z-transforms are identical, with the ROCs different. The bilateral transform used in the examples is given by

$$F_b(z) = \frac{2z^2 - 0.75z}{(z - 0.25)(z - 0.5)} = \frac{z}{z - 0.25} + \frac{z}{z - 0.5}.$$

EXAMPLE 11.15   **An inverse bilateral z-transform**

Consider first the function

$$F_b(z) = \frac{z}{z - 0.25} + \frac{z}{z - 0.5}; \quad |z| > 0.5.$$

The poles and the ROC are plotted in Figure 11.17. Hence, $f[n]$ is right sided, and, from Table 11.5,

$$f[n] = \mathcal{Z}_b^{-1}[F_b(z)] = [0.25^n + 0.5^n]u[n].$$

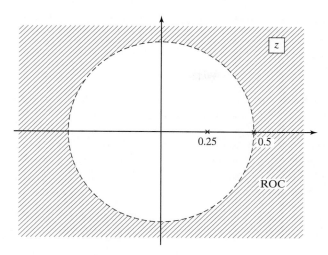

Figure 11.17   ROC for Example 11.15.

∎

EXAMPLE 11.16   **Continuation of Example 11.15**

Consider next the function

$$F_b(z) = \frac{z}{z - 0.25} + \frac{z}{z - 0.5}; \quad |z| < 0.25.$$

The poles and the ROC are plotted in Figure 11.18. Hence, $f[n]$ is left sided. From Table 11.5,

$$f[n] = \mathcal{L}_b^{-1}[F_b(z)] = [-0.25^n - 0.5^n]u[-n - 1].$$

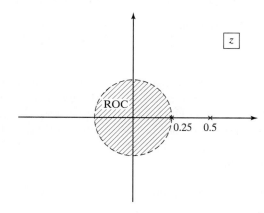

Figure 11.18   ROC for Example 11.16.

∎

EXAMPLE 11.17   **Continuation of Example 11.16**

Finally, consider the function

$$F_b(z) = \frac{z}{z - 0.25} + \frac{z}{z - 0.5}; \quad 0.25 < |z| < 0.5.$$

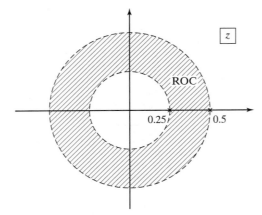

Figure 11.19   ROC for Example 11.17.

The poles and the ROC are plotted in Figure 11.19. The function $z/(z - 0.25)$ is the transform of a right-sided function, and the function $z/(z - 0.5)$ is that of a left-sided function. From Table 11.5,

$$f[n] = \mathscr{Z}_b^{-1}[F_b(z)] = 0.25^n u[n] - 0.5^n u[-n - 1]. \qquad \blacksquare$$

In this section, we have introduced the bilateral z-transform. The importance of the region of convergence is evident. For bilateral transforms, we usually separate a function into the sum of a right-sided function and a left-sided function. Tables are then used for both the bilateral transform and the inverse bilateral transform.

▨ SUMMARY

The unilateral and bilateral z-transforms were introduced in this chapter. The unilateral transform is used in the analysis and design of linear time-invariant (LTI), causal discrete-time systems. The unilateral z-transform is especially useful in the design of causal digital filters and in understanding their characteristics. It is also useful in the analysis of switched LTI discrete-time systems, since this transform allows us to include initial conditions.

The bilateral z-transform is useful in the steady-state analysis of LTI discrete-time systems and in the analysis and design of noncausal systems. Recall that noncausal systems are not realizable in real time. However, noncausal systems are realizable for the digital processing of recorded signals.

The next chapter involves the discrete-time Fourier transform. This transform is the result of applying the Fourier techniques of Chapters 4 through 6 to discrete-time signals, especially to sampled signals.

See Table 11.6.

TABLE 11.6 Key Equations of Chapter 11

Equation Title	Equation Number	Equation
Bilateral z-transform	(11.1)	$\mathcal{Z}_b[f[n]] = F_b(z) = \displaystyle\sum_{n=-\infty}^{\infty} f[n]z^{-n}$
Inverse bilateral z-transform	(11.2)	$\mathcal{Z}_b^{-1}[F_b(z)] = f[n] = \dfrac{1}{2\pi j} \displaystyle\oint_{\Gamma} F_b(z)z^{n-1}\,dz, \quad j = \sqrt{-1}$
Unilateral z-transform	(11.4)	$\mathcal{Z}[f[n]] = F(z) = \displaystyle\sum_{n=0}^{\infty} f[n]z^{-n}$
Transfer function	(11.44)	$H(z) = \dfrac{Y(z)}{X(z)} = \dfrac{b_0 + b_1 z^{-1} + \cdots + b_{N-1}z^{-N+1} + b_N z^{-N}}{a_0 + a_1 z^{-1} + \cdots + a_{N-1}z^{-N+1} + a_N z^{-N}}$ $= \dfrac{b_0 z^N + b_1 z^{N-1} + \cdots + b_{N-1}z + b_N}{a_0 z^N + a_1 z^{N-1} + \cdots + a_{N-1}z + a_N}$
z-transform of right-sided DT exponential	(11.61)	$a^n u[n] \overset{\mathcal{Z}_b}{\longleftrightarrow} \dfrac{z}{z-a}; \quad |z| > |a|$
z-transform of left-sided DT exponential	(11.64)	$\mathcal{Z}_b[-a^n u[-n-1]] = \dfrac{-a^{-1}z}{1-a^{-1}z} = \dfrac{z}{z-a}; \quad |z| < |a|$

REFERENCES

1. G. Doetsch, *Guide to the Applications of the Laplace and z-Transforms*. New York: Van Nostrand Reinhold, 1971.
2. E. I. Jury, *Theory and Application of the z-Transform Method*. New York: Krieger, 1973.
3. C. L. Phillips and H. T. Nagle, *Digital Control System Analysis and Design*, 3 ed. Upper Saddle River, NJ: Prentice Hall, 1996.
4. Martin Vetterli and Jelena Kovačević, *Wavelets and Subband Coding*, Prentice Hall, Englewood Cliffs, NJ, 1995

PROBLEMS

11.1. Express the unilateral z-transforms of the following functions as rational functions. Tables may be used.

    (a) 0.6^n

    (b) $0.7^n + 4(1.5)^n$

    (c) $5e^{-.5n}$

    (d) $5e^{-j.1n}$

    (e) $5\sin 3n$

    (f) $20\cos(2n - \pi/4)$

    (g) $e^{-.5n}\cos 0.3n$

    (h) $e^{-.5n}\cos(0.3n - \pi/4)$

11.2. The signals given are sampled every 0.05 s, beginning at $t = 0$. Find the unilateral z-transforms of the sampled functions, with each transform expressed as a rational function.

  (a) $2e^{-2t}$

  (b) $2e^{-2t} + 2e^{t}$

  (c) $2e^{-0.2t}$

  (d) $5e^{-0.5jt}$

  (e) $5\cos(t)$

  (f) $5e^{-t}\cos(t)$

11.3. (a) The signal e^{-5t} is sampled every 0.2 s, beginning at $t = 0$. Find the z-transform of the sampled signal.

  (b) The signal e^{-t} is sampled every second, beginning at $t = 0$. Find the z-transform of the sampled signal.

  (c) Why are the z-transforms found in Parts (a) and (b) identical?

  (d) A third function e^{at} is sampled every T seconds. Find two different values of (a, T) such that the z-transforms of the sampled function are identical to those of Parts (a) and (b).

11.4. (a) Use the z-transform to evaluate the following series:

  (i) $x = \sum_{n=0}^{\infty} 0.5^n$ **(ii)** $x = \sum_{n=2}^{\infty} 0.5^n$

  (b) Use the z-transform to evaluate the series

$$x = \sum_{n=0}^{\infty} 0.5^n \cos(0.1n).$$

11.5. (a) A function $f[n] = A\cos(\Omega n)$ has the z-transform

$$F(z) = \frac{3z(z - 0.6967)}{z^2 - 1.3934z + 1}.$$

Find A and Ω.

  (b) A function $f(t) = A\cos(\omega t)$ is sampled every $T = 0.0001$ s, beginning at $t = 0$. The z-transform of the sampled function is given in Part (a). Find A and ω.

11.6. The z-transform of a discrete-time function $f[n]$ is given by

$$F(z) = \frac{z^2}{(z - 1)(z + 1)}.$$

  (a) Apply the final-value property to $F(z)$.

  (b) Check your result in Part (a) by finding the inverse z-transform of $F(z)$.

  (c) Why are the results of Parts (a) and (b) different?

11.7. (a) Given $\mathcal{Z}[4^n] = z/(z - 4)$, find the z-transform of $f[n]$ as given, using only the properties of z-transforms:

$$f[n] = n(n - 1)4^n.$$

  (b) Verify the results of Part (a) using the z-transform table.

11.8. A function $y[n]$ has the unilateral z-transform

$$Y(z) = \frac{z^3}{z^3 - 3z^2 + 5z - 9}.$$

(a) Find the z-transform of $y_1[n] = y[n - 3]u[n - 3]$.

(b) Find the z-transform of $y_2[n] = y[n + 3]u[n]$.

(c) Evaluate $y[n]$ for $n = 0$ and 3, $y[n - 3]u[n - 3]$ for $n = 3$, and $y[n + 3]u[n]$ for $n = 0$ by expanding the appropriate z-transform into power series.

(d) Are the values found in Part (c) consistent? Explain why.

11.9. Given $f[n] = a^n u[n]$, find the z-transforms of the following:

(a) $f[n/3]$

(b) $f[n - 3]u[n - 3]$; verify your result by a power series expansion

(c) $f[n + 3]u[n]$; verify your result by a power series expansion

(d) $b^{2n} f[n]$; using two different procedures.

11.10. (a)   Given the following unilateral z-transforms, find the inverse z-transform of each function:

(i) $X(z) = \dfrac{0.4z^2}{(z - 1)(z - 0.6)}$

(ii) $X(z) = \dfrac{0.4z}{(z - 1)(z - 0.6)}$

(iii) $X(z) = \dfrac{0.4}{(z - 1)(z - 0.6)}$

(iv) $X(z) = \dfrac{z}{z^2 - z + 1}$

(b) Verify the partial-fraction expansions in Part (a) using MATLAB.

(c) Evaluate each $x[n]$ in Part (a) for the first three nonzero values.

(d) Verify the results in Part (c) by expanding each $X(z)$ in Part (a) into a power series using long division.

(e) Use the final-value property to evaluate $x[\infty]$ for each function in Part (a).

(f) Check the results of Part (e), using each $x[n]$ found in Part (a).

(g) Use the initial-value property to evaluate $x[0]$ for each function in Part (a).

(h) Check the results of Part (g), using each $x[n]$ found in Part (a).

11.11. Consider the transforms from Problem 11.10.

$$X_1(z) = \frac{0.4z^2}{(z - 1)(z - 0.6)};$$

$$X_2(z) = \frac{0.4z}{(z - 1)(z - 0.6)};$$

$$X_3(z) = \frac{0.4}{(z - 1)(z - 0.6)}.$$

(a) Without calculating the inverse transforms, state how $x_1[n]$, $x_2[n]$, and $x_3[n]$ are related.

(b) Verify the results of Part (a) by finding the inverse transforms.

(c) Verify the partial-fraction expansions in Part (b) using MATLAB.

11.12. Consider the following difference equation and excitation:

$$y[n] - 1.5y[n - 1] + 0.5y[n - 2] = x[n];$$

$$x[n] = \begin{cases} 1, & n = 1 \\ 0, & \text{otherwise} \end{cases}.$$

(a) Find $y[n]$, using the z-transform.

(b) Check the partial-fraction expansions in Part (a) using MATLAB.

(c) Verify the results of Part (a) for $n = 0, 1, 2, 3$, and 4, by solving the difference equation by iteration.

(d) Use MATLAB to check the results in Part (c).

(e) Verify the value of $y[0]$ in Part (a), using the initial-value property.

(f) Will the final-value property give the correct value of $y[\infty]$? If your answer is yes, find the final value. Otherwise, state why the final-value property is not applicable.

11.13. Consider the following difference equation and excitation:

$$y[n] - 0.75y[n - 1] + 0.125y[n - 2] = x[n];$$

$$x[n] = \begin{cases} 1, & n = 0 \\ 0, & \text{otherwise} \end{cases}.$$

(a) Find $y[n]$, using the z-transform.

(b) Check the partial-fraction expansions in Part (a) using MATLAB.

(c) Verify the results of Part (a) for $n = 0, 1, 2, 3$, and 4, by solving the difference equation by iteration.

(d) Use MATLAB to check the results in Part (c).

(e) Verify the value of $y[0]$ in Part (a), using the initial-value property.

(f) Will the final-value property give the correct value of $y[\infty]$? If your answer is yes, find the final value. Otherwise, state why the final-value property is not applicable.

11.14. (a) The difference equation

$$y[n] - 0.75y[n - 1] + 0.125y[n - 2] = x[n].$$

models an LTI system. Find the system transfer function.

(b) Find the unit step response for the system of Part (a).

(c) Use MATLAB to check the partial-fraction expansion in Part (b).

(d) From Part (b), give the values of $y[0]$, $y[1]$, and $y[2]$.

(e) Verify the results of Part (d) by solving the difference equation iteratively.

(f) Verify the results in Part (e) using MATLAB.

11.15. Consider the block diagram of a discrete-time system given in Figure P11.15.

(a) Find the difference-equation model of this system.

(b) Find the system transfer function.

(c) Determine the range of the parameter a for which this system is BIBO stable.

(d) Find the impulse response of this system. Is the answer to Part (c) evident from the impulse response? Why?

(e) Let $a = 0.5$. Find the unit step response for this system.

(f) Let $a = 2.0$. Find the unit step response for this system.

(g) Check the results of Parts (e) and (f) using MATLAB.

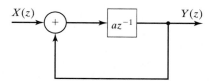

Figure P11.15

11.16. A simple way to smooth data is to just take a weighted average of a number of samples. Consider an LTI system with input and output related by

$$y[n] = ax[n - 1] + (1 - a)x[n].$$

Assume that all initial conditions are 0.

(a) Find the system impulse response $h[n]$.

(b) Find the impulse response of the system needed to recover $x[n]$ from $y[n]$. (*Hint*: consider the inverse system.)

11.17. (a) Use the z-transform tables and the time-scaling property to find the inverse transforms of the following functions:

(a) (i) $F(z) = \dfrac{z^2}{z^2 - 0.7}$

   (ii) $F(z) = \dfrac{z}{z^2 - 0.7}$

(b) Give a sketch of each $f[n]$ in Part (a).

(c) Verify the results in Part (a) by a power-series expansion.

11.18. Find the inverse Z-transform of the following functions:

(a) $F(z) = \dfrac{z^{-9}}{z - a}$

(b) $F(z) = \dfrac{z^{-2}}{z - 3}.$

11.19. Consider a system with the transfer function $H(z)$.

(a) Give any third-order transfer function such that the system is causal, but not stable.

(b) Give any third-order transfer function such that the system is not causal, but stable.

(c) Give any third-order transfer function such that the system is neither causal nor stable.

(d) Give any third-order transfer function such that the system is both causal and stable.

11.20. (a) Determine the stability of the causal systems with the following transfer functions:

$$\text{(i)} \quad H(z) = \frac{3(z - 1.2)}{(z - 1)(z - 0.9)}$$

$$\text{(ii)} \quad H(z) = \frac{3(z + 0.9)}{z(z - 0.9)(z - 1.2)}$$

$$\text{(iii)} \quad H(z) = \frac{3(z - 0.9)}{z(z + 0.9)(z + 1.2)}$$

$$\text{(iv)} \quad H(z) = \frac{3(z - 1)^2}{z^3 - 1.8z^2 + 0.81z}$$

$$\text{(v)} \quad H(z) = \frac{2z - 1.5}{z^3 - 2z^2 + 0.99z}$$

Use MATLAB where required.

(b) For each system that is unstable, give a bounded input for which the output is unbounded.

(c) Verify the results in Part (b) by finding the unbounded term in the response for that input.

11.21. Given the general system transfer function

$$H(z) = \frac{b_0 + b_1 z^{-1} + \cdots + b_M z^{-M}}{a_0 + a_1 z^{-1} + \cdots + a_N z^{-N}}.$$

Show that this system is causal provided that $a_0 \neq 0$. (*Hint*: Consider the impulse response.)

11.22. Given the discrete-time function

$$f[n] = a^n u[n] - b^{2n} u[-n - 1].$$

(a) What condition must hold on a and b for the bilateral z-transform to exist?

(b) Assuming that the preceding condition holds, find the bilateral z-transform and region of convergence of $f[n]$.

11.23. Find the bilateral z-transforms and the regions of convergence for the following functions:

(a) $0.5^n u[n]$

(b) $0.5^n u[n - 5]$

(c) $0.5^n u[n + 5]$

(d) $-0.5^n u[-n - 1]$

(e) $(0.5)^{-n} u[n + 5]$

(f) $0.5^n u[-n]$

11.24. (a) Find the inverse of the bilateral z-transform

$$F_b(z) = \frac{.6z}{(z - 1)(z - .6)}$$

for the following regions of convergence:

(i) $|z| < .6$

(ii) $|z| > 1$

(iii) $.6 < |z| < 1$

(b) Give the final values of the functions of Parts (i) through (iii).

(c) Verify the results of Part (b) using the final-value theorem.

11.25. (a) Given the discrete-time function

$$f[n] = \begin{cases} (\tfrac{1}{2})^n, & -10 \le n \le 20 \\ 0, & \text{otherwise} \end{cases},$$

express the bilateral z-transform of this function in closed form (not as a power series).

(b) Find the region of convergence of the transform of Part (a).

(c) Repeat Parts (a) and (b) for the discrete-time function

$$f[n] = \begin{cases} (\tfrac{1}{2})^n, & -10 \le n \le 10 \\ (\tfrac{1}{4})^n, & n \ge 21 \\ 0, & \text{otherwise} \end{cases}.$$

(d) Repeat Parts (a) and (b) for the discrete-time function

$$f[n] = \begin{cases} (\tfrac{1}{2})^n, & -10 \le n \le 0 \\ (\tfrac{1}{4})^n, & 1 \le n \le 10 \\ 0, & \text{otherwise} \end{cases}.$$

11.26. Consider the bilateral z-transform

$$F(z) = \frac{3z}{z-1} + \frac{z}{z-12} - \frac{z}{z-.6}.$$

(a) Find all possible regions of convergence for this function.

(b) Find the inverse transform for each region of convergence found in Part (a).

11.27. The *wavelet transform* [4] has recently become popular for various signal-processing operations. The *analysis step* involves applying a series of low-pass filters to the input signal. After each application of the filter, the signal is *downsampled* to retain only every other sample point so that the number of wavelet transform coefficients that remains is equal to the number of points in the input function.

In the *synthesis stage*, the wavelet coefficients are repeatedly *upsampled* and filtered to reconstruct the input signal. This problem will demonstrate that the filtering and upsampling steps of the wavelet transform can be interchanged.

To upsample a signal $x[n]$, we form

$$x_M[n] = \begin{cases} x[\tfrac{n}{M}], & \text{if } n \text{ is an integer-multiple of } M \\ 0, & \text{otherwise} \end{cases}.$$

(a) A filter with impulse response $h[n]$ is applied to a signal $x[n]$ to form $y[n]$. The signal $y[n]$ is then upsampled by a factor of M to form $y_M[n]$. Find $Y_M(z)$.

(b) Next the input signal $x[n]$ is first upsampled by M to form $x_M[n]$. The filter $h[n]$ is also upsampled by M to form $h_M[n]$ and is then applied to $x_M[n]$. Find the $\mathcal{Z}$-transform of $x_M[n]*h_M[n]$.

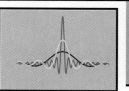
The widespread use of digital computers for data processing has greatly increased the importance of transforms for number sequences. A continuous-time signal cannot be processed by a digital computer. To process a continuous-time signal using a digital computer, it is first necessary to convert the signal into a number sequence by sampling; a digital computer can process only numbers (binary codes). The rate of sampling must be such that the characteristics of the number sequence are essentially those of the continuous-time signal. The choice of sampling rates was discussed in Sections 5.4 and 6.4.

Once a continuous-time signal has been converted to a number sequence, we can determine its characteristics using transforms. In addition, we can determine the characteristics of data-processing algorithms by the use of transforms. One transform for number sequences has already been introduced—the z-transform in Chapter 11. This transform is applied to discrete signals and systems and is approximately the equivalent of the Laplace transform for continuous-time signals and systems.

In this chapter, we define both the *discrete-time Fourier transform* and the *discrete Fourier transform*. (Note the difference in the names of these transforms.) Later, we consider some of the efficient computer algorithms for calculating discrete Fourier transforms. These algorithms fall under the general classification of fast Fourier transforms (FFTs). The discrete-time Fourier transform and the discrete Fourier transform are then applied to discrete-time signals and systems and are equivalent to the Fourier transform for continuous-time signals and systems. The Fourier transform is discussed in Chapters 5 and 6.

In this chapter, we also introduce the discrete cosine transform (DCT). The DCT is closely related to the FFT. The DCT finds wide application as the basis of the image-compression standard issued by the Joint Photographic Experts Group (JPEG). The JPEG algorithm is commonly used on the World Wide Web for compressing images.

12.1 DISCRETE-TIME FOURIER TRANSFORM

In Section 6.4, we saw that the Fourier transform

[eq(6.17)]
$$X_s(\omega) = \sum_{n=-\infty}^{\infty} x(nT)e^{-jn\omega T}$$

appears in modeling the sampling process. If, in this equation, we make the usual substitution $x(nT) = x[n]$ and make the change of variables $\omega T = \Omega$, we have the defining equation of the *discrete-time Fourier transform*:

$$X(\Omega) = \mathscr{DTF}(x[n]) = \sum_{n=-\infty}^{\infty} x[n]e^{-jn\Omega}. \tag{12.1}$$

In this equation, $\mathscr{DTF}(\cdot)$ denotes the discrete-time Fourier transform and Ω is the *discrete-frequency variable*. We see then that the discrete-time Fourier transform is inherent in the Fourier-transform model of the sampling operation in Figure 6.29.

The inverse discrete-time Fourier transform is defined as

$$x[n] = \mathscr{DTF}^{-1}[X(\Omega)] = \frac{1}{2\pi}\int_{\Omega_1}^{\Omega_1+2\pi} X(\Omega)e^{jn\Omega}d\Omega = \frac{1}{2\pi}\int_{2\pi} X(\Omega)e^{jn\Omega}d\Omega, \tag{12.2}$$

where Ω_1 is arbitrary. This is denoted by placing the value 2π directly underneath the integral symbol. We show later that the integrand is periodic with period 2π. This inversion integral can be derived directly from that of the Fourier transform, (5.2) [3]. We denote a discrete-time Fourier transform pair by

$$x[n] \overset{\mathscr{DTF}}{\longleftrightarrow} X(\Omega). \tag{12.3}$$

It is important to note that $x[n]$ is a function of the *discrete* variable n, while the transform $X(\Omega)$ in (12.1) is a function of the *continuous* variable Ω. Hence, $X(\Omega)$ is a continuous function of frequency while $x[n]$ is a discrete function of time.

In general, the discrete sequence $x[n]$ is obtained by sampling a continuous-time signal $x(t)$, or the sequence is interpreted as being the samples of a continuous-time signal. For this case, the discrete-frequency variable Ω is related to the real-frequency variable ω by the equation $\Omega = \omega T$. Hence, we see that discrete frequency Ω is a *scaled* version of real frequency ω. Recall that this scaling also appeared in the study of discrete-time signals and systems in Chapters 9 and 10. The principal application of the discrete-time Fourier transform is in the analysis of sampled signals.

Next, we consider two examples of discrete-time Fourier transforms.

EXAMPLE 12.1   **Discrete-time Fourier transform (DTFT) of a signal**

We now find the discrete-time Fourier transform of the function $x[n] = a^n u[n]$, where

$$a^n u[n] = \begin{cases} a^n, & n \geq 0 \\ 0, & n < 0 \end{cases}.$$

Recall from Section 9.4 that this function is exponential in nature. From (12.1),

$$X(\Omega) = \sum_{n=-\infty}^{\infty} x[n]e^{-jn\Omega} = \sum_{n=0}^{\infty} a^n e^{-jn\Omega}$$

$$= 1 + ae^{-j\Omega} + a^2 e^{-j2\Omega} + \cdots = 1 + ae^{-j\Omega} + (ae^{-j\Omega})^2 + \cdots.$$

We now find $X(\Omega)$ in closed form. From Appendix C, we have the geometric series

$$\sum_{n=0}^{\infty} b^n = 1 + b + b^2 + \cdots = \frac{1}{1-b}; \quad |b| < 1.$$

In $X(\Omega)$, we let $ae^{-j\Omega} = b$, resulting in the transform

$$X(\Omega) = \frac{1}{1 - ae^{-j\Omega}}; \quad |ae^{-j\Omega}| < 1.$$

Because $|e^{-j\Omega}| = |\cos \Omega - j \sin \Omega| = 1$, this transform exists for $|a| < 1$; we then have the transform pair

$$a^n u[n] \xleftrightarrow{\mathcal{DTF}} \frac{1}{1 - ae^{-j\Omega}}, \quad |a| < 1.$$

This transform is valid for either real or complex values of a. The discrete-time Fourier transform of $a^n u[n]$, $|a| > 1$, does not exist. ∎

EXAMPLE 12.2   **The linearity property of the DTFT**

Consider the discrete-time function $x[n] = a^{|n|}$. This function is plotted in Figure 12.1 for a real and $0 < a < 1$ and can be expressed as

$$x[n] = a^n u[n] + a^{-n} u[-n-1] = x_1[n] + x_2[n].$$

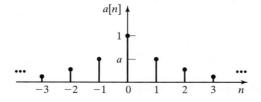

Figure 12.1   A plot of $a^{|n|}$.

The transform of $x_1[n] = a^n u[n]$ was found in Example 12.1. For the second term $x_2[n]$,

$$\mathcal{DTF}[a^{-n}u[-n-1]] = \sum_{n=-\infty}^{\infty} [a^{-n}u[-n-1]]e^{-nj\Omega}$$

$$= \sum_{n=-\infty}^{-1} (ae^{j\Omega})^{-n} = ae^{j\Omega} + (ae^{j\Omega})^2 + (ae^{j\Omega})^3 + \cdots$$

$$= \sum_{n=1}^{\infty} (ae^{j\Omega})^n,$$

since $u[-n-1] = 0, n > -1$. From Appendix C, we have the geometric series

$$\sum_{n=k}^{\infty} b^n = b^k + b^{k+1} + \cdots = \frac{b^k}{1-b}, \quad |b| < 1.$$

In $X(\Omega)$, we let $ae^{j\Omega} = b$, resulting in the transform

$$X(\Omega) = \frac{ae^{j\Omega}}{1 - ae^{j\Omega}}, \quad |ae^{j\Omega}| < 1.$$

Because $|e^{-j\Omega}| = 1$, this transform exists for $|a| < 1$. We now use the linearity property of the discrete-time Fourier transform (proved in the next section), which states that the transform of a sum of functions is equal to the sum of the transforms, provided that the sum of transforms exists:

$$X(\Omega) = X_1(\Omega) + X_2(\Omega) = \frac{1}{1 - ae^{-j\Omega}} + \frac{ae^{j\Omega}}{1 - ae^{j\Omega}}, \quad |a| < 1.$$

The transform of $a^{|n|}, |a| > 1$, does not exist.   ■

z-Transform

We now relate the discrete-time Fourier transform to the z-transform. The bilateral z-transform is defined in Chapter 11 as

[eq(11.1)] $$\mathcal{L}[x[n]] = X(z) = \sum_{n=-\infty}^{\infty} x[n]z^{-n}.$$

Table 11.5 lists the bilateral z-transform of several discrete-time functions. Earlier in this section, the discrete-time Fourier transform was given by

[eq(12.1)] $$X(\Omega) = \mathcal{DTF}[x[n]] = \sum_{n=-\infty}^{\infty} x[n]e^{-jn\Omega}.$$

Comparison of (11.1) and (12.1) yields the following relationship between the z-transform and the discrete-time Fourier transform:

$$\mathcal{DTF}[x[n]] = \mathcal{Z}[x[n]]_{z=e^{j\Omega}}. \tag{12.4}$$

We see from this equation that a table of z-transforms can be used for discrete-time Fourier transforms, provided that the discrete-time Fourier transform exists.

We state without proof that $\mathcal{DTF}[x[n]]$ exists if $x[n]$ is *absolutely summable* [4]—that is, if

$$\sum_{n=-\infty}^{\infty} |x[n]| < \infty. \tag{12.5}$$

This condition is *sufficient*; however, the discrete-time Fourier transforms of some functions that do not satisfy this condition do exist. In general, the transforms for these functions contain impulse functions in the variable Ω; these transforms are discussed later.

The regions of convergence of the z-transforms are given in Table 11.5; these regions are annular and of the form $a < |z| < b$. To use this table for discrete-time Fourier transforms, we make the substitution of (12.4), $z = e^{j\Omega}$. The region of convergence then transforms into

$$a < |z| < b \Rightarrow a < |e^{j\Omega}| < b.$$

Because $|e^{j\Omega}| = 1$, a plot of $|e^{j\Omega}|$ forms the unit circle in the Ω-plane. Hence, Table 11.5 becomes a table of discrete-time Fourier transforms for those functions for which the region of convergence includes the unit circle. This property is illustrated in Figure 12.2. In Figure 12.2(a), the unit circle is in the region of convergence, and the discrete-time Fourier transform does exist. In Figure 12.2(b) the unit circle is outside the region of convergence, and the discrete-time Fourier transform does not exist. An example will now be given.

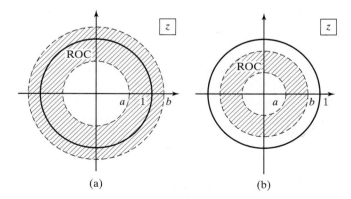

(a)   (b)

Figure 12.2 Regions of convergence in the z-plane.

EXAMPLE 12.3   **Use of the z-transform to find a DTFT**

We wish to find the discrete-time Fourier transform of the function $x[n] = na^n u[n]$. From Table 11.5,

$$\mathscr{L}[na^n u[n]] = \frac{az}{(z-a)^2}; \quad |z| > |a|.$$

Hence, for the discrete-time Fourier transform of $na^n u[n]$ to exist, $|e^{j\Omega}| = 1 > |a|$, and we have the transform

$$\mathscr{DTF}[na^n u[n]] = \frac{ae^{j\Omega}}{(e^{j\Omega} - a)^2}; |a| < 1.$$

Consider next the discrete-time Fourier transform of $\cos(bn)u[n]$. From Table 11.5,

$$\mathscr{L}[\cos(bn)u[n]] = \frac{z^2 - z\cos b}{z^2 - 2z\cos b + 1}; \quad |z| > 1.$$

TABLE 12.1   Discrete-Time Fourier Transforms

$f[n]$	$F(\Omega)$
1. $\delta[n]$	1
2. 1	$2\pi \sum\limits_{k=-\infty}^{\infty} \delta(\Omega - 2\pi k)$
3. $u[n]$	$\dfrac{1}{1 - e^{-j\Omega}} + \sum\limits_{k=-\infty}^{\infty} \pi\delta(\Omega - 2\pi k)$
4. $a^n u[n]; \quad |a| < 1$	$\dfrac{1}{1 - ae^{-j\Omega}}$
5. $na^n u[n]; \quad |a| < 1$	$\dfrac{ae^{j\Omega}}{(e^{j\Omega} - a)^2}$
6. $a^{-n} u[-n-1]$	$\dfrac{ae^{j\Omega}}{1 - ae^{j\Omega}}$
7. $e^{j\Omega_0 n}$	$2\pi \sum\limits_{k=-\infty}^{\infty} \delta(\Omega - \Omega_0 - 2\pi k)$
8. $\cos[\Omega_0 n]$	$\pi \sum\limits_{k=-\infty}^{\infty} [\delta(\Omega - \Omega_0 - 2\pi k) + \delta(\Omega + \Omega_0 - 2\pi k)]$
9. $\sin[\Omega_0 n]$	$\dfrac{\pi}{j} \sum\limits_{k=-\infty}^{\infty} [\delta(\Omega - \Omega_0 - 2\pi k) - \delta(\Omega + \Omega_0 - 2\pi k)]$
10. $\text{rect}[n/N]$	$\dfrac{\sin \Omega(N + \frac{1}{2})}{\sin\Omega/2}$
11. $\dfrac{\sin [\Omega_1 n]}{\pi n}$	$\text{rect}(\Omega/\Omega_1)$
12. $x[n]$ periodic with period N	$2\pi \sum\limits_{k=-\infty}^{\infty} a_k \delta(\Omega - 2\pi k/N)$ $a_k = \dfrac{1}{N} \sum\limits_{n=n_0}^{n_0+N-1} x[n]e^{\frac{-j2\pi kn}{N}}$,

Because the unit circle is not within the given region of convergence, the discrete-time Fourier transform of $\cos(bn)u[n]$ is not the one given by Table 11.5. However, as we will see, this transform does exist.   ∎

Table 12.1 gives several discrete-time Fourier transforms, taken from Table 11.5. Several additional transforms that cannot be obtained from the table of z-transforms are included.

12.2 PROPERTIES OF THE DISCRETE-TIME FOURIER TRANSFORM

In this section, several properties of the discrete-time Fourier transform are given. These properties are based on the defining equations of the transform:

[eq(12.1)]
$$X(\Omega) = \sum_{n=-\infty}^{\infty} x[n]e^{-jn\Omega}$$

and

[eq(12.2)]
$$x[n] = \frac{1}{2\pi}\int_{2\pi} X(\Omega)e^{jn\Omega}d\Omega.$$

First, however, we review the relation of the Fourier transform and the discrete-time Fourier transform. Recall that

$$\mathcal{F}[x_s(t)]_{\omega T=\Omega} = \mathcal{F}\left[\sum_{n=-\infty}^{\infty} x(nT)\delta(t - nT)\right]_{\omega T=\Omega} = \mathcal{DTF}[x(nT) = x[n]], \qquad (12.6)$$

from (5.40). We see that the discrete-time Fourier transform can be interpreted as the Fourier transform of a train of weighted impulse functions. Hence, the properties of the Fourier transform given in Chapter 5 apply directly to the discrete-time Fourier transform, with the change of variables in (12.6). For some properties, we refer to Chapter 5 for the proofs; for other properties, proofs are given.

Before discussing the properties given in Chapter 5, we consider a property that is unique to the discrete-time Fourier transform.

Periodicity

The discrete-time Fourier transform $X(\Omega)$ is periodic in Ω with period 2π; we demonstrate this from the definition of periodicity, $X(\Omega) = X(\Omega + 2\pi)$. From (12.1),

$$X(\Omega + 2\pi) = \sum_{n=-\infty}^{\infty} x[n]e^{-jn(\Omega+2\pi)} = \sum_{n=-\infty}^{\infty} x[n]e^{-jn\Omega}e^{-j2\pi n}$$

$$= \sum_{n=-\infty}^{\infty} x[n]e^{-jn\Omega} = X(\Omega), \qquad (12.7)$$

because $e^{-j2\pi n} = \cos 2\pi n - j \sin 2\pi n = 1$; thus, periodicity is proved. This property is very important, and its implications are covered in detail later in this chapter. We illustrate this property with an example.

EXAMPLE 12.4   **Demonstration of the periodicity of the DTFT**

From Table 12.1, for $x[n] = a^n u[n]$ with $|a| < 1$,

$$X(\Omega) = \frac{1}{1 - ae^{-j\Omega}}.$$

Then

$$X(\Omega + 2\pi) = \frac{1}{1 - ae^{-j(\Omega+2\pi)}} = \frac{1}{1 - ae^{-j\Omega}e^{-j2\pi}}.$$

Because $e^{-j2\pi} = \cos 2\pi - j \sin 2\pi = 1$,

$$X(\Omega + 2\pi) = \frac{1}{1 - ae^{-j\Omega}} = X(\Omega).$$ ∎

Linearity

The linearity property of the Fourier transform states that the Fourier transform of a sum of functions is equal to the sum of the Fourier transforms of the functions, provided that the sum exists. The property applies directly to the discrete-time Fourier transform:

$$\mathscr{DTF}[a_1 x_1[n] + a_2 x_2[n]] = a_1 X_1(\Omega) + a_2 X_2(\Omega).$$

This property was demonstrated in Example 12.2.

Time Shift

It is informative to derive the time-shift property from the definition of the discrete-time Fourier transform. From (12.1),

$$\mathscr{DTF}[x[n - n_0]] = \sum_{n=-\infty}^{\infty} x[n - n_0]e^{-jn\Omega}.$$

We make the change of variables $(n - n_0) = k$ on the right side of this equation. Then, because $n = (k + n_0)$,

$$\mathscr{DTF}[x[n - n_0]] = \sum_{k=-\infty}^{\infty} x[k]e^{-j\Omega(k+n_0)}$$

$$= e^{-j\Omega n_0} \sum_{k=-\infty}^{\infty} x[k]e^{-j\Omega k} = e^{-j\Omega n_0}X(\Omega);$$

thus the time-shift property is given by

$$x[n - n_0] \overset{\mathcal{DTF}}{\longleftrightarrow} e^{-j\Omega n_0} X(\Omega). \qquad (12.8)$$

Note also that this property can be written directly from Table 5.1. We illustrate this property with a numerical example.

EXAMPLE 12.5   **Illustration of the time-shift property of the DTFT**

We find the discrete-time Fourier transform of the sequence shown in Figure 12.3. The sequence is described mathematically by

$$x[n] = \delta[n - 3] + 0.5\delta[n - 4] + 0.25\delta[n - 5] + \cdots = \sum_{k=3}^{\infty} (0.5)^{k-3}\delta[n - k]$$

$$= (0.5)^{n-3} u[n - 3].$$

From Table 12.1, for $|a| < 1$,

$$\mathcal{DTF}[a^n u[n]] = \frac{1}{1 - ae^{-j\Omega}}.$$

Then, from this transform and (12.8),

$$\mathcal{DTF}[(0.5)^{n-3} u[n - 3]] = \frac{e^{-j3\Omega}}{1 - 0.5e^{-j\Omega}}.$$

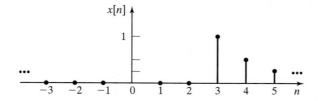

Figure 12.3 A discrete-time sequence.   ■

Frequency Shift

The time-shift property gives the effects in the frequency domain of a shift in the time domain. We now give the time-domain manipulation that results in a shift in the frequency domain:

$$\mathcal{DTF}[e^{j\Omega_0 n} x[n]] = \sum_{n=-\infty}^{\infty} e^{jn\Omega_0} x[n] e^{-jn\Omega}$$

$$= \sum_{n=-\infty}^{\infty} x[n] e^{-jn(\Omega-\Omega_0)} = X(\Omega-\Omega_0).$$

This property is then

$$e^{jn\Omega_0} x[n] \overset{\mathcal{DTF}}{\longleftrightarrow} X(\Omega-\Omega_0). \qquad (12.9)$$

Hence, the multiplication of a time-domain signal by the complex exponential $e^{jn\Omega_0}$ results in a shift of Ω_0 in the frequency domain.

EXAMPLE 12.6   **Application of the frequency-shift property of the DTFT**

From Example 12.1,

$$a^n u[n] \overset{\mathcal{DTF}}{\longleftrightarrow} \frac{1}{1 - ae^{-j\Omega}} , |a| < 1$$

Then, from (12.9),

$$e^{jn\Omega_0} a^n u[n] \overset{\mathcal{DTF}}{\longleftrightarrow} \frac{1}{1 - ae^{-j(\Omega - \Omega_0)}} , |a| < 1 \qquad \blacksquare$$

Symmetry

We state the symmetry properties of the discrete-time Fourier transform without proof. Expressing $X(\Omega)$ in its real and imaginary parts yields

$$X(\Omega) = \text{Re}[X(\Omega)] + j\,\text{Im}[X(\Omega)].$$

For the case that $x[n]$ is a real-valued sequence, we list the symmetry properties:

$$\text{Re}[X(\Omega)] \text{ is even;}$$
$$\text{Im}[X(\Omega)] \text{ is odd;}$$
$$|X(\Omega)| \text{ is even;} \qquad (12.10)$$
$$\arg X(\Omega) \text{ is odd.}$$

Time Reversal

Consider the time-reversed signal $x[-n]$. Then

$$\mathcal{DTF}[x[-n]] = \sum_{n=-\infty}^{\infty} x[-n]e^{-jn\Omega}.$$

Now let $-n = k$. It follows that

$$\mathcal{DTF}[x[-n]] = \sum_{k=-\infty}^{\infty} x[k]e^{jk\Omega} = \sum_{k=-\infty}^{\infty} x[k]e^{-jk(-\Omega)} = X(-\Omega).$$

The time-reversal property is then

$$x[-n] \longleftrightarrow X(-\Omega). \qquad (12.11)$$

The effect of time reversal is frequency reversal.

Convolution in Time

The time-convolution relation is given by

$$\mathcal{DTF}[x[n]*y[n]] = X(\Omega)Y(\Omega).\tag{12.12}$$

We derive this relationship, because of its importance. Consider the definition of convolution, from (10.16):

$$x[n]*y[n] = \sum_{k=-\infty}^{\infty} x[k]y[n-k].$$

Then

$$\mathcal{DTF}[x[n]*y[n]] = \sum_{n=-\infty}^{\infty}\left[\sum_{k=-\infty}^{\infty} x[k]y[n-k]\right]e^{-jn\Omega}.$$

We reverse the order of the summations, yielding

$$\mathcal{DTF}[x[n]*y[n]] = \sum_{k=-\infty}^{\infty} x[k]\left[\sum_{n=-\infty}^{\infty} y[n-k]e^{-jn\Omega}\right].$$

Next we change variables, letting $m = n - k$. Then

$$\mathcal{DTF}[x[n]*y[n]] = \sum_{k=-\infty}^{\infty} x[k]\left[\sum_{m=-\infty}^{\infty} y[m]e^{-j(m+k)\Omega}\right]$$

$$= \sum_{k=-\infty}^{\infty} x[k]e^{-jk\Omega}\sum_{m=-\infty}^{\infty} y[m]e^{-jm\Omega} = X(\Omega)Y(\Omega),$$

and the property is proved.

Convolution in Frequency

The process of multiplying discrete-time signals, as in modulation, results in an operation called *circular convolution* (or *periodic convolution*) for the DTFT of the product:

$$x[n]y[n] \overset{\mathcal{DTF}}{\longleftrightarrow} \sum_{n=-\infty}^{\infty} x[n]y[n]e^{-jn\Omega}.\tag{12.13}$$

We use the inverse discrete-time Fourier transform (12.2) to write

$$x[n] = \frac{1}{2\pi}\int_{2\pi} X(\Omega)e^{jn\Omega}d\Omega.$$

After a change of variables in the integral, $\Omega \to \theta$, we use the result to rewrite (12.13) as

$$x[n]y[n] \overset{\mathscr{DTF}}{\longleftrightarrow} \sum_{n=-\infty}^{\infty} \left[\frac{1}{2\pi} \int_{2\pi} X(\theta)e^{jn\theta}d\theta \right] y[n]e^{-jn\Omega}$$

$$= \frac{1}{2\pi} \sum_{n=-\infty}^{\infty} y[n] \left[\int_{2\pi} X(\theta)e^{-jn(\Omega-\theta)}d\theta \right]$$

$$= \frac{1}{2\pi} \int_{2\pi} X(\theta) \left[\sum_{n=-\infty}^{\infty} y[n]e^{-jn(\Omega-\theta)} \right] d\theta.$$

From the frequency-shift property,

$$\sum_{n=-\infty}^{\infty} y[n]e^{-jn(\Omega-\theta)} = Y(\Omega-\theta),$$

and we can now write the convolution-in-frequency property as

$$x[n]y[n] \overset{\mathscr{DTF}}{\longleftrightarrow} \frac{1}{2\pi} \int_{2\pi} X(\theta)Y(\Omega-\theta)d\theta$$

$$= \frac{1}{2\pi} X(\Omega) \circledast Y(\Omega), \tag{12.14}$$

where the mathematical symbol $\circledast$ is used to denote circular convolution.

We see then that multiplication in time results in circular convolution in frequency. Circular convolution is discussed more thoroughly in Section 12.6.

Multiplication by n

This property is given by

$$\mathscr{DTF}[nx[n]] = j\frac{dX(\Omega)}{d\Omega}. \tag{12.15}$$

The proof of this property is given as a problem.

Parseval's Theorem

Parseval's theorem for discrete-time signals is given by

$$\sum_{n=-\infty}^{\infty} |x[n]|^2 = \frac{1}{2\pi} \int_{2\pi} |X(\Omega)|^2 d\Omega, \tag{12.16}$$

where the integral on the right side is over any range of Ω that is of width 2π. [Recall that $X(\Omega)$ is periodic with period 2π.] The left side is the energy of the signal, and $|X(\Omega)|^2$ is called the energy-density spectrum of the signal.

For convenience, the properties of this section are summarized in Table 12.2.

TABLE 12.2 Properties of the Discrete-Time
Fourier Transform

Signal	Transform
$x[n]$	$X(\Omega) = \sum\limits_{n=-\infty}^{\infty} x[n]e^{-jn\Omega}$
$x[n]$	$X(\Omega) = X(\Omega + 2\pi)$
$a_1 x_1[n] + a_2 x_2[n]$	$a_1 X_1(\Omega) + a_2 X_2(\Omega)$
$x[n - n_0]$	$e^{-j\Omega n_0} X(\Omega)$
$e^{jn\Omega_0} x[n]$	$X(\Omega - \Omega_0)$
$x[n]$ real	$\begin{cases} \mathrm{Re}[X(\Omega)] \text{ is even} \\ \mathrm{Im}[X(\Omega)] \text{ is odd} \\ |X(\Omega)| \text{ is even} \\ \arg X(\Omega) \text{ is odd} \end{cases}$
$x[-n]$	$X(-\Omega)$
$x[n]*y[n]$	$X(\Omega)Y(\Omega)$
$x[n]y[n]$	$\dfrac{1}{2\pi} X(\Omega) \circledast Y(\Omega)$
$nx[n]$	$j \dfrac{dX(\Omega)}{d\Omega}$

Parseval's theorem: $\displaystyle\sum_{n=-\infty}^{\infty} |x(n)|^2 = \frac{1}{2\pi} \int_{2\pi} |X(\Omega)|^2 d\Omega$

12.3 DISCRETE-TIME FOURIER TRANSFORM OF PERIODIC SEQUENCES

In this section, we consider the discrete-time Fourier transform of periodic sequences. The resulting development leads us to the discrete Fourier transform and the fast Fourier transform.

Consider a periodic sequence $x[n]$ with period N, such that $x[n] = x[n + N]$. Of course, N must be an integer. We define $x_0[n]$ to be the values of $x[n]$ over the period beginning at $n = 0$, such that

$$x_0[n] = \begin{cases} x[n], & 0 \le n \le N - 1 \\ 0, & \text{otherwise} \end{cases}. \tag{12.17}$$

An example of a periodic sequence is shown in Figure 12.4, with $N = 3$. In this case, $x_0[n]$ is the sequence composed of $x_0[0] = 0$, $x_0[1] = 1$, and $x_0[2] = 1$, with $x_0[n] = 0$ for all other n.

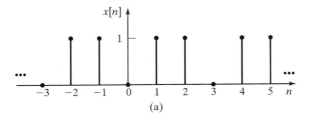

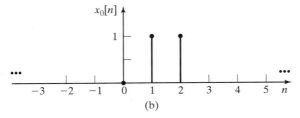

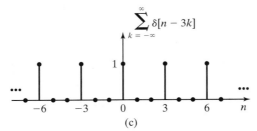

Figure 12.4 A periodic discrete-time sequence.

From (12.1) and (12.17), the discrete-time Fourier transform of $x_0[n]$ is given by

$$X_0(\Omega) = \sum_{n=-\infty}^{\infty} x_0[n]e^{-jn\Omega}$$

$$= x_0[0] + x_0[1]e^{-j\Omega} + x_0[2]e^{-j2\Omega} + \cdots + x_0[N-1]e^{-j(N-1)\Omega}$$

$$= \sum_{n=0}^{N-1} x_0[n]e^{-jn\Omega}. \tag{12.18}$$

Hence, $X_0(\Omega)$ is a finite series in the continuous-frequency variable Ω.

We now find the discrete-time Fourier transform of the periodic sequence $x[n]$ as a function of $X_0(\Omega)$ in (12.18).

We can express the periodic sequence as

$$x[n] = x_0[n] * \sum_{k=-\infty}^{\infty} \delta[n - kN]. \tag{12.19}$$

The train of discrete impulses

$$\sum_{k=-\infty}^{\infty} \delta[n - kN]$$

(illustrated in Figure 12.4(c) for the case that $N = 3$) can be viewed as the sequence generated by sampling a unity amplitude constant signal with the sampling period NT. From Table 5.2,

$$f_s(t) = \sum_{k=-\infty}^{\infty} \delta(t - kNT) \xleftrightarrow{\mathscr{F}} \frac{2\pi}{NT} \sum_{k=-\infty}^{\infty} \delta\left(\omega - \frac{2\pi k}{NT}\right) = F_s(\omega).$$

Making the change of variables $\Omega = \omega T$, we have

$$F(\Omega) = F_s(\omega)\bigg|_{\Omega=\omega T} = \frac{2\pi}{NT} \sum_{k=-\infty}^{\infty} \delta\left(\frac{1}{T}\left[\Omega - \frac{2\pi k}{N}\right]\right).$$

From the definition of the impulse function, we have the relationship

$$\int_{-\infty}^{\infty} \delta(ax)dx = \int_{-\infty}^{\infty} \delta(\lambda)d\frac{\lambda}{a}\bigg|_{\lambda=ax} = \frac{1}{a}\int_{-\infty}^{\infty} \delta(\lambda)d\lambda = \frac{1}{a} \Rightarrow \delta(ax) = \frac{1}{a}\delta(x), a > 0,$$

which allows us to write

$$F(\Omega) = \frac{2\pi}{NT} \sum_{k=-\infty}^{\infty} T\delta\left(\Omega - \frac{2\pi k}{N}\right) = \frac{2\pi}{N} \sum_{k=-\infty}^{\infty} \delta\left(\Omega - \frac{2\pi k}{N}\right).$$

We now have the DTFT pair

$$\sum_{k=-\infty}^{\infty} \delta[n - kN] \xleftrightarrow{\mathscr{DTF}} \frac{2\pi}{N} \sum_{k=-\infty}^{\infty} \delta\left(\Omega - \frac{2\pi k}{N}\right). \qquad (12.20)$$

Note that the function $\delta[\cdot]$ is the *discrete* impulse function, and $\delta(\cdot)$ is the *continuous* impulse function. Hence, the discrete-time Fourier transform of a train of discrete impulse functions is a train of continuous impulse functions.

Recall, from Table 12.2, that convolution in discrete-time n transforms into multiplication in frequency Ω. Then, from (12.19) and (12.20), the discrete-time Fourier transform of $x[n]$ can be written as

$$X(\Omega) = \mathscr{DTF}[x[n]] = \mathscr{DTF}\left[x_0[n]*\left[\sum_{k=-\infty}^{\infty} \delta[n - kN]\right]\right]$$

and

$$\qquad (12.21)$$

$$X(\Omega) = \frac{2\pi}{N} \sum_{k=-\infty}^{\infty} X_0(\Omega)\delta\left(\Omega - \frac{2\pi k}{N}\right)$$

with the term $X_0(\Omega)$ moved inside the summation as shown. Using the property of the continuous impulse function from Table 2.3,

$$f(t)\delta(t - t_0) = f(t_0)\delta(t - t_0), \qquad (12.22)$$

we write (12.21) as

$$X(\Omega) = \frac{2\pi}{N} \sum_{k=-\infty}^{\infty} X_0\left(\frac{2\pi k}{N}\right)\delta\left(\Omega - \frac{2\pi k}{N}\right). \tag{12.23}$$

Recall, from Table 12.2, that a discrete-time Fourier transform is *always* periodic with period 2π; thus, $X_0(\Omega)$ is periodic. Therefore, the Fourier transform of a sequence $x[n]$, which is periodic with period N, results in a function $X(\Omega)$ that is periodic with period 2π. Furthermore, the N distinct values of $x[n]$, $0 \leq n \leq N - 1$, transform into N distinct values of $X_0(2\pi k/N)$, $0 \leq k \leq N - 1$ in frequency.

Figure 12.5 gives an example of $X(\Omega)$ in (12.23). For convenience, we have assumed that $X_0(\Omega)$ is real and triangular. We also assume that $N = 3$ for this example. The discrete-time Fourier transform of $x[n]$, $X(\Omega)$ of (12.23), is then as shown in Figure 12.5(b). In this figure, the lengths of the arrows denote the weights of the impulse functions. The period of $X(\Omega)$ is 2π, with an impulse function occurring every $\Omega = 2\pi/3$. Hence, the values $x[0]$, $x[1]$, and $x[2]$ completely describe the periodic function $x[n]$, and the values $X(0)$, $X(2\pi/3)$, and $X(4\pi/3)$ completely describe the periodic function $X(\Omega)$.

Next we derive the inverse discrete-time Fourier transform of (12.23). From (12.2),

$$\mathcal{DTF}^{-1}[X(\Omega)] = x[n] = \frac{1}{2\pi}\int_0^{2\pi} X(\Omega)e^{j\Omega n}d\Omega$$

$$= \frac{1}{2\pi}\int_0^{2\pi}\left[\frac{2\pi}{N}\sum_{k=-\infty}^{\infty} X_0\left(\frac{2\pi k}{N}\right)\delta\left(\Omega - \frac{2\pi k}{N}\right)\right]e^{j\Omega n}d\Omega \tag{12.24}$$

$$= \frac{1}{N}\sum_{k=0}^{N-1} X_0\left(\frac{2\pi k}{N}\right)e^{j2\pi kn/N}.$$

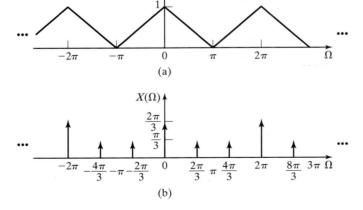

(a)

(b)

Figure 12.5 Discrete-time Fourier transform of a periodic signal.

The result is obtained from the property of the impulse function

$$\int_a^b f(t)\delta(t - t_0)dt = \begin{bmatrix} f(t_0), & a \le t_0 \le b \\ 0, & \text{otherwise} \end{bmatrix}.$$

In (12.24), only the impulse functions $\delta(\Omega)$, $\delta(\Omega - 2\pi/N)$, $\ldots$, $\delta(\Omega - 2\pi[N - 1]/N)$ occur between $0 \le \Omega < 2\pi$.

In summary, for a periodic sequence $x[n] = x[n + N]$,

[eq(12.23)] $$X(\Omega) = \frac{2\pi}{N} \sum_{k=-\infty}^{\infty} X_0\left(\frac{2\pi k}{N}\right)\delta\left(\Omega - \frac{2\pi k}{N}\right)$$

and

[eq(12.24)] $$x[n] = \frac{1}{N}\sum_{k=0}^{N-1} X_0\left(\frac{2\pi k}{N}\right)e^{j2\pi kn/N},$$

where, from (12.18),

$$X_0(\Omega) = \sum_{n=0}^{N-1} x_0[n]e^{-jn\Omega}. \tag{12.25}$$

There are N distinct values of $x[n]$ and N distinct values of $X_0(2\pi k/N)$. An example is now given to illustrate these developments.

EXAMPLE 12.7   **Calculation of the DTFT of a periodic sequence**

Consider again the periodic signal $x[n]$ of Figure 12.4. For this signal, $N = 3$, $x_0[0] = 0$, $x_0[1] = 1$, and $x_0[2] = 1$. From (12.25), the discrete-time Fourier transform of $x_0[n]$ is given by

$$X_0(\Omega) = \sum_{n=0}^{N-1} x_0[n]e^{-jn\Omega} = 0e^{-j0} + (1)e^{-j\Omega} + (1)e^{-j2\Omega}$$

$$= e^{-j\Omega} + e^{-j2\Omega}.$$

This transform is periodic and is plotted in Figure 12.6. From (12.23), the discrete-time Fourier transform of $x[n]$ is then

$$X(\Omega) = \frac{2\pi}{3} \sum_{k=-\infty}^{\infty} X_0\left(\frac{2\pi k}{3}\right)\delta\left(\Omega - \frac{2\pi k}{3}\right),$$

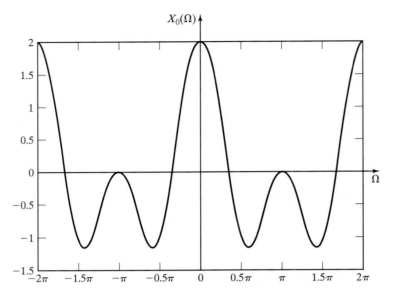

Figure 12.6 $X_0(\Omega)$ for Example 12.7.

where the three distinct values of $X_0(\,\cdot\,)$ are given by

$$X_0(0) = e^{j(0)} + e^{j2(0)} = 2,$$

$$X_0\left(\frac{2\pi}{3}\right) = e^{-j2\pi/3} + e^{-j4\pi/3} = 1\underline{/-120°} + 1\underline{/-240°}$$

$$= -0.5 - j0.866 - 0.5 + j0.866 = -1,$$

and

$$X_0\left(\frac{4\pi}{3}\right) = e^{-j4\pi/3} + e^{-j8\pi/3}$$

$$= -0.5 - j0.866 - 0.5 + j0.866 = -1.$$

The frequency spectrum for $X(\Omega)$ is plotted in Figure 12.7. For this case, $X_0(2\pi k/N)$ are all real.

We now calculate $x[n]$ using the inverse transform. From (12.24),

$$x[n] = \frac{1}{N}\sum_{k=0}^{N-1} X_0\left(\frac{2\pi k}{N}\right)e^{j2\pi kn/N}.$$

Thus,

$$x[0] = \frac{1}{3}\left[X_0(0) + X_0\left(\frac{2\pi}{3}\right) + X_0\left(\frac{4\pi}{3}\right)\right] = \frac{1}{3}[2 - 1 - 1] = 0;$$

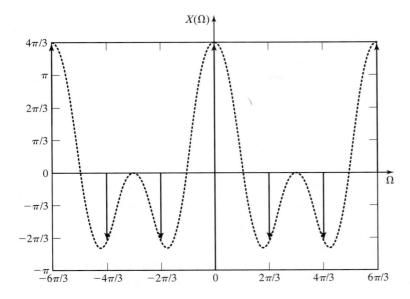

Figure 12.7 $X(\Omega)$ for Example 12.7.

$$x[1] = \frac{1}{3}\left[X_0(0) + X_0\left(\frac{2\pi}{3}\right)e^{j2\pi/3} + X_0\left(\frac{4\pi}{3}\right)e^{j4\pi/3} \right]$$

$$= \frac{1}{3}[2 + (-1)\underline{/120°} + (-1)\underline{/240°}]$$

$$= \frac{1}{3}[2 + 0.5 - j0.866 + 0.5 + j0.866] = 1;$$

$$x[2] = \frac{1}{3}\left[X_0(0) + X_0\left(\frac{2\pi}{3}\right)e^{j4\pi/3} + X_0\left(\frac{4\pi}{3}\right)e^{j8\pi/3} \right]$$

$$= \frac{1}{3}[2 + 0.5 - j0.866 + 0.5 + j0.866] = 1.$$

These values are seen to be correct.   ■

In the next section, a transform, called the discrete Fourier transform, will be defined for the distinct values $x_0[0]$, $x_0[1]$, ..., $x_0[N - 1]$, that is independent of the derivations of this section. However, these derivations give us an interpretation of the discrete Fourier transform for cases in which the values of $x_0[n]$ are samples from a continuous-time signal.

12.4 DISCRETE FOURIER TRANSFORM

In Chapters 5 and 6 and in earlier sections of this chapter, we have presented formulas for calculation of Fourier transforms. However, for both the actual Fourier transform and the discrete-time Fourier transform, the formulas produce continuous functions of frequency. In this section, we develop an approximation of the Fourier transform that can be calculated from a finite set of discrete-time samples of an analog signal and which produces a finite set of discrete-frequency spectrum values. This Fourier transform approximation is well suited for calculation by a digital computer. It is called the *discrete Fourier transform* (DFT).

To use a digital computer to calculate the Fourier transform of a continuous-time signal, we must use sampled values of $x(t)$ in the form of a discrete-time signal $x[n]$. From Section 12.1, we have the discrete-time Fourier Transform (DTFT):

[eq(12.1)]
$$X(\Omega) = \mathcal{DF}(x[n]) = \sum_{n=-\infty}^{\infty} x[n]e^{-jn\Omega}.$$

The DTFT is computed from discrete-time samples, but $X(\Omega)$ is a continuous function of the frequency variable and cannot be represented exactly in a digital computer. However, using digital computations, we can approximate $X(\Omega)$ by calculating discrete-frequency samples of the continuous-frequency function.

To generate the discrete-frequency samples we must, for practical reasons, limit ourselves to a finite set of discrete-time samples. For the purpose of this development, we will let the symbol N represent the number of samples chosen to represent the discrete-time signal. We choose the value of N sufficiently large, so that our set of samples adequately represents all of $x[n]$. We can select our finite set of samples by multiplying the infinite set $x[n]$ by a rectangular windowing function [5]

$$w_R[n] = \begin{cases} 1, n = 0, 1, 2, \dots, N-1 \\ 0, \qquad\qquad \text{otherwise} \end{cases}.$$

Then the set of samples used to calculate the frequency spectrum is

$$x_N[n] = x[n]w_R[n] = \begin{cases} x[n], & n = 0, 1, 2, \dots, N-1 \\ 0, & \text{otherwise} \end{cases}. \qquad (12.26)$$

The frequency spectrum of the signal $x_N[n]$ is given by (12.1)

$$X_N(\Omega) = \mathcal{DF}(x_N[n]) = \sum_{n=-\infty}^{\infty} x_N[n]e^{-jn\Omega} = \sum_{n=0}^{N-1} x[n]e^{-jn\Omega}. \qquad (12.27)$$

For the remainder of the development, we consider the set of N samples of $x_N[n]$ to be the complete signal. We will drop the subscript on $x_N[n]$ and refer to the finite discrete-time sequence as simply $x[n]$. This is justified by the assumption that we have chosen the finite set of samples so that they adequately represent the entire signal.

We now select N samples of $X_N(\Omega)$ to represent the frequency spectrum. We could choose more than N samples, but we must choose at least N to avoid creating errors in the inverse transformation. (See Ref. 3 for a detailed explanation.) It is general practice to compute the same number of samples (N) of the frequency spectrum as were used to represent $x[n]$. Because the DTFT is periodic in Ω with period 2π, the N samples are taken from one period of $X_N(\Omega)$. The ideally sampled frequency spectrum is given by

$$X_S(\Omega) = X_N(\Omega) \sum_{k=0}^{N-1} \delta(\Omega - 2\pi k/N)$$

$$= \sum_{k=0}^{N-1} X_N(2\pi k/N)\delta(\Omega - 2\pi k/N). \tag{12.28}$$

In $X_S(\Omega)$, the value of the frequency spectrum at each sample frequency is shown as the weight of an impulse in frequency. The set of discrete-frequency values $X_N(2\pi k/N)$ is the information that we select to represent the frequency spectrum of the discrete-time sequence $x[n]$. Using the convention applied earlier for discrete-time sequences, we write this discrete-frequency sequence as

$$X[k] = X_N(2\pi k/N), \quad k = 0, 1, 2, \ldots, N-1.$$

Comparison of the equation for $X_S(\Omega)$ with (12.23) shows that the frequency spectrum $X_S(\Omega)$ has the same form as derived in Section 12.3 for periodic signals. (Since $X[k] = 0, k \neq 0, 1, 2, \ldots, N-1$, we could change the limits of summation in our equation for $X_S(\Omega)$ to $-\infty < k < \infty$.) In fact, if we use (12.24) to find the inverse transformation of $X_S(\Omega)$, we will generate a periodic discrete-time sequence, $x_p[n]$, which has N samples per period. However, since we know that our sampled frequency spectrum was generated by a finite, nonperiodic, discrete-time sequence, we can limit our inverse transformation, from (12.24), to the N samples from one period of $x_p[n]$. We determine our time signal to be

$$x[n] = \frac{1}{N} \sum_{k=0}^{N-1} X[k]e^{j2\pi kn/N}, \quad n = 0, 1, 2, \ldots, N-1. \tag{12.29}$$

The transform pair that we have developed for a set of N discrete-frequency samples calculated from a set of N samples of a discrete-time signal is known as the *discrete Fourier transform* (DFT) and the *inverse discrete Fourier transform* (IDFT), respectively:

$$X[k] = \mathcal{DF}[x[n]] = \sum_{n=0}^{N-1} x[n]e^{-j2\pi kn/N}, \quad k = 0, 1, 2, \ldots, N-1;$$

$$x[n] = \mathcal{DF}^{-1}[X[k]] = \frac{1}{N} \sum_{k=0}^{N-1} X[k]e^{j2\pi kn/N}, \quad n = 0, 1, 2, \ldots, N-1. \tag{12.30}$$

The symbol $\mathcal{DF}$ denotes the discrete Fourier transform. We also use the notation

$$x[n] \xleftrightarrow{\mathcal{DF}} X[k]$$

to represent this transform pair.

EXAMPLE 12.8	**Calculation of a discrete Fourier transform (DFT)**

We consider the case that we have three data points. Hence, for $N = 3$, the discrete Fourier transform, (12.30), is given by

$$
\begin{aligned}
X[0] &= x[0] + x[1] + x[2]; \\
X[1] &= x[0] + x[1]e^{-j2\pi/3} + x[2]e^{-j4\pi/3}; \\
X[2] &= x[0] + x[1]e^{-j4\pi/3} + x[2]e^{-j2\pi/3}.
\end{aligned}
\tag{12.31}
$$

In the last term in the third equation, $e^{\frac{-j8\pi}{3}} = e^{\frac{-j6\pi}{3}}e^{\frac{-j2\pi}{3}} = e^{\frac{-j2\pi}{3}}$. The inverse discrete Fourier transform, (12.30), is then

$$
\begin{aligned}
x[0] &= 1/3(X[0] + X[1] + X[2]); \\
x[1] &= 1/3(X[0] + X[1]e^{j2\pi/3} + X[2]e^{j4\pi/3}); \\
x[2] &= 1/3(X[0] + X[1]e^{j4\pi/3} + X[2]e^{j2\pi/3}).
\end{aligned}
\tag{12.32}
$$

Note the symmetries of the two sets of equations. These equations are normally evaluated by computer. A computer program written to evaluate the forward discrete Fourier transform, (12.31), can easily be modified to calculate the inverse discrete Fourier transform, (12.32). A general computer program allows N to be any integer value. ∎

Shorthand Notation for the DFT

A shorthand notation that is commonly used with the discrete Fourier transform will now be given. We define the symbol $W_N = e^{-j2\pi/N}$. Equation (12.30) can then be expressed as

$$X[k] = \mathcal{DF}[x[n]] = \sum_{n=0}^{N-1} x[n]W_N^{kn}, k = 0, 1, \ldots, N-1;$$

$$x[n] = \mathcal{DF}^{-1}[X[k]] = \frac{1}{N}\sum_{k=0}^{N-1} X[k]W_N^{-kn}, n = 0, 1, \ldots, N-1. \tag{12.33}$$

The discrete Fourier transform is normally stated in this form.

Frequency resolution of the DFT

Because the DTFT from which the N frequency samples were taken is periodic with period 2π, the discrete frequency spectrum that we compute using the DFT has a resolution (separation between samples) of

$$\Delta\Omega = 2\pi/N. \tag{12.34}$$

This is illustrated by Figure 12.8 for $N = 8$, where the unit circle represents one (2π) period of the signal's discrete-time Fourier transform. From this, we see that the choice of the number of samples of $x[n]$ used in the calculation determines the resolution of the frequency spectrum, or vice-versa; the resolution required in the frequency spectrum determines the number of samples of $x[n]$ that we must use. In the event that a fixed number, N_1, of time-domain samples is available, but a larger number, N_2, of frequency-domain samples is required to provide adequate resolution, $N_2 - N_1$ zeros can be appended to the time sequence. This process is called *zero padding*. We discuss applications of zero padding in Section 12.6.

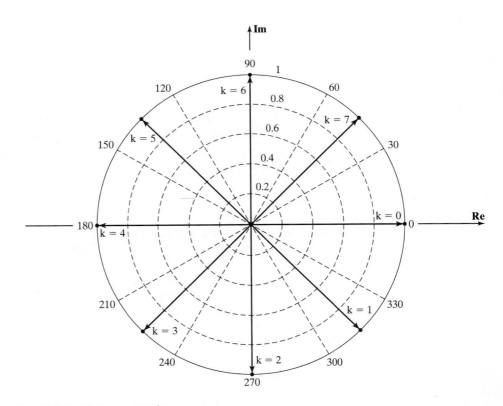

Figure 12.8   Polar plot of W_8^k

Figure 12.8 shows a polar plot of W_8^k, $k = 0, 1, 2, \ldots, 7$, in the complex plane. Three key observations can be made from this plot:

1. Adjacent vectors in the sequence are separated by angles of $2\pi/N$ radians ($N = 8$).

2. Each vector in the plot has an opposite of equal magnitude so that the sum of the two must be zero, and we can extend this finding to state that

$$\sum_{k=0}^{7} W_8^k = 0.$$

3. The vectors are in conjugate pairs, so that each vector has a conjugate mate with an angle of equal value but opposite sign.

Plotted in Figure 12.8, we have the sequence of complex vectors

$$W_8^0 = 1; W_8^1 = 1e^{-j\pi/4}; W_8^2 = 1e^{-j\pi/2}; W_8^3 = 1e^{-j3\pi/4}; \ldots; W_8^7 = 1e^{-j7\pi/4}.$$

If we evaluate $W_8^{kn} = (W_8^k)^n$, $n = 0, 1, 2, \ldots, N - 1$, we have the sequence of vectors

$$(W_8^0)^n = 1; (W_8^1)^n = 1e^{-jn\pi/4}; (W_8^2)^n = 1e^{-jn\pi/2};$$
$$(W_8^3)^n = 1e^{-j3n\pi/4}; \ldots; (W_8^7)^n = 1e^{-j7n\pi/4}.$$

For $n = 0$, all of the vectors have a value of 1. For $n = 1$, the vectors are as shown in Figure 12.8. For $n > 1$, the vectors will repeat the pattern of Figure 12.8. For example, if we examine the set of vectors for n = 5, we have

$$(W_8^0)^5 = 1; (W_8^1)^5 = 1e^{-j5\pi/4}; (W_8^2)^5 = 1e^{-j5\pi/2} = 1e^{-j\pi/2};$$
$$(W_8^3)^5 = 1e^{-j15\pi/4} = 1e^{-j7\pi/4}; (W_8^4)^5 = 1e^{-j20\pi/4} = 1e^{-j\pi};$$
$$(W_8^5)^5 = 1e^{-j25\pi/4} = 1e^{-j\pi/4}; (W_8^6)^5 = 1e^{-j30\pi/4} = 1e^{-j3\pi/2}; (W_8^7)^5 = 1e^{-j3\pi/4}.$$

If we carry this on for all integer values of $k, 0 \leq k \leq N - 1$, and $n, 0 \leq n \leq N - 1$, we find that

$$\sum_{n=0}^{N-1} W_N^{0n} = N \text{ and } \sum_{n=0}^{N-1} W_N^{kn} = 0, k = 1, 2, \ldots, N - 1. \qquad (12.35)$$

Validity of the DFT

It will now be shown that the discrete Fourier transform is valid; that is, given N values of $x[n]$, the forward transform in (12.30) results in N values, $X[k]$. If, then, these N values, $X[k]$, are substituted in the inverse transform in (12.30), the original N values of $x[n]$ are obtained.

We begin by substituting the first equation in (12.30) into the second one. In the first equation, n is the summation variable; we change this summation variable to m, to avoid confusion with variable n in the second equation. Then

$$x[n] = \frac{1}{N}\sum_{k=0}^{N-1} X[k]W_N^{-kn} = \frac{1}{N}\sum_{k=0}^{N-1}\left[\sum_{m=0}^{N-1} x[m]W_N^{km}\right]W_N^{-kn}. \qquad (12.36)$$

Next, the order of the summations is reversed:

$$x[n] = \frac{1}{N}\sum_{m=0}^{N-1} x[m]\sum_{k=0}^{N-1} W_N^{k(m-n)}. \qquad (12.37)$$

From (12.35), we can write

$$\sum_{k=0}^{N-1} W_N^{k(m-n)} = \begin{cases} N, & n = m \\ 0, & n \neq m \end{cases}.$$

Thus, (12.37) becomes

$$x[n] = \frac{1}{N}\sum_{m=0}^{N-1} x[m]N\big|_{n=m} = x[n], \qquad (12.38)$$

and the validity of the discrete Fourier transform is proved.

An example will help us to understand computation of the DFT.

EXAMPLE 12.9   **Calculation of a DFT with a MATLAB program**

We wish to find the discrete Fourier transform of the sequence $x[n]$, given by the data in Table 12.3. For this case, $N = 4$. From (12.30), with $N = 4$ and $W_4 = e^{-j2\pi/4} = 1\underline{/-90°} = -j$,

$$X[0] = x[0] + x[1] + x[2] + x[3] = 1 + 2 + 3 + 4 = 10,$$
$$X[1] = x[0] + x[1](-j) + x[2](-j)^2 + x[3](-j)^3$$
$$= 1 - j2 + -3 + j4 = -2 + j2,$$
$$X[2] = x[0] + x[1](-j)^2 + x[2](-j)^4 + x[3](-j)^6$$
$$= 1 - 2 + 3 - 4 = -2,$$

TABLE 12.3   Values for Example 12.9

n	$x[n]$	k	$X[k]$
0	1	0	10
1	2	1	$-2 + j2$
2	3	2	-2
3	4	3	$-2 - j2$

and

$$X[3] = x[0] + x[1](-j)^3 + x[2](-j)^6 + x[3](-j)^9$$
$$= 1 + j2 - 3 - j4 = -2 - j2.$$

The following MATLAB program can be used to confirm the results of this example:

```
% This MATLAB program computes the DFT using the defining
% equations (12.30) for the transform pair.
N = input...
   ('How many discrete-time samples are in the sequence?')
x = input...
   ('Type the vector of samples, in brackets[...]:')
% Compute the DFT from (12.30)
for k1 = 1:N
   X(k1) = 0;
   k = k1 - 1;
   for n1 = 1:N;
        n = n1 - 1;
        X(k1) = X(k1) + x(n1)*exp(-j*2*pi*k*n/N);
   end
end
x
X
```

The discrete Fourier transform is also listed in Table 12.3. At this time we make no attempt to give meaning to these values. ∎

EXAMPLE 12.10   **Calculation of an inverse DFT, with MATLAB confirmation**

This example is a continuation of the last one. We now find the inverse discrete Fourier transform of $X[k]$ of Table 12.3. From (12.30),

$$x[0] = [X[0] + X[1] + X[2] + X[3]]/4$$
$$= [10 + (-2 + j2) + (-2) + (-2 - j2)]/4 = 1,$$
$$x[1] = [X[0] + X[1](j) + X[2](j)^2 + X[3](j)^3]/4$$
$$= [10 - j2 - 2 + 2 + j2 - 2]/4 = 2,$$
$$x[2] = [X[0] + X[1](j)^2 + X[2](j)^4 + X[3](j)^6]/4$$
$$= [10 + 2 - j2 - 2 + 2 + j2]/4 = 3,$$

and

$$x[3] = [X[0] + X[1](j)^3 + X[2](j)^6 + X[3](j)^9]/4$$
$$= [10 + j2 + 2 + 2 - j2 + 2]/4 = 4,$$

which are the correct values. Note the symmetries of the calculations of the DFT and its inverse. As stated earlier, these symmetries allow the same computer program used to calculate the forward transform to calculate the inverse transform, with slight modification:

```
% This MATLAB program computes the IDFT using the defining
% equations (12.30) for the transform pair.
N = input...
('How many discrete-frequency samples are in the sequence?')
X = input...
('Type the vector of samples, enclosed in brackets[...]:')
% Compute the IDFT from (12.30)
for n1 = 1:N
    x(n1) = 0;
n = n1 - 1;
    for k1 = 1:N;
        k = k1 - 1;
        x(n1) = x(n1) + X(k1)*exp(j*2*pi*k*n/N)/N;
    end
end
x
X
```

An alternative method is often used for computing the inverse discrete Fourier transform (IDFT). The alternative method allows the same algorithm to be used for both the DFT and IDFT. The standard IDFT equation is given by

[eq(12.30)] $$x[n] = \mathcal{DF}^{-1}[X[k]] = \frac{1}{N}\sum_{k=0}^{N-1} X[k]e^{j2\pi kn/N}.$$

The alternative method takes advantage of the properties that the complex conjugate of a sum is equal to the sum of conjugates and the complex conjugate of a product is equal to the product of conjugates. In algebraic form,

$$[X + Y]^* = X^* + Y^* \text{ and } [XY]^* = X^*Y^*.$$

These properties are used on the IDFT equation to write

$$x^*[n] = \frac{1}{N}\sum_{k=0}^{N-1} X^*[k][e^{j2\pi kn/N}]^* = \frac{1}{N}\sum_{k=0}^{N-1} X^*[k]e^{-j2\pi kn/N}.$$

From this, we write the alternative inversion formula

$$x[n] = [x^*[n]]^* = \frac{1}{N}\left[\sum_{k=0}^{N-1} X^*[k]e^{-j2\pi kn/N}\right]^*. \qquad (12.39)$$

The factor in the brackets can be computed using the same algorithm used for computation of the DFT except that values of $X[k]^*$ are used as the input rather than $x[n]$ when we are computing the IDFT. The alternative IDFT procedure is given step by step as follows:

1. Change the sign of the imaginary parts (find the complex conjugate) of $X[k]$.
2. Use the DFT algorithm to find the DFT of $X^*[k]$.
3. Find the complex conjugate of the results of Step 2 for each value of n.
4. Divide the results of Step 3 by N.

EXAMPLE 12.11 **An alternative method of calculating the inverse DFT**

We use the data shown in Table 12.3 to show the alternative procedure for computation of the IDFT. From Table 12.3, we have

$$X[k] = [10, -2 + j2, -2, -2 - j2].$$

From the first step of the inversion procedure, we have

$$X^*[k] = [10, -2 - j2, -2, -2 + j2].$$

The second step is to compute the DFT of $X^*[k]$:

$$4x^*[n] = \sum_{k=0}^{3} X^*[k]e^{-j2\pi kn/4};$$

$$4x^*[0] = 10 + (-2 - j2) + (-2) + (-2 + j2) = 4;$$
$$4x^*[1] = 10 + (-2 - j2)e^{-j\pi/2} + (-2)e^{-j\pi} + (-2 + j2)e^{-j3\pi/2}$$
$$= 10 + (-j)(-2 - j2) + (-1)(-2) + j(-2 + j2) = 8;$$
$$4x^*[2] = 10 + (-2 - j2)e^{-j\pi} + (-2)e^{-j2\pi} + (-2 + j2)e^{-j3\pi}$$
$$= 10 + (-j)(-2 - j2) + (-1)(-2) + (-1)(-2 + j2) = 12;$$
$$4x^*[3] = 10 + (-2 - j2)e^{-j3\pi/2} + (-2)e^{-j3\pi} + (-2 + j2)e^{-j9\pi/2}$$
$$= 10 + (j)(-2 - j2) + (-1)(-2) + (-j)(-2 + j2) = 16.$$

Because the results of Step 2 are all real valued, finding the complex conjugate as specified in Step 3 is not necessary. We divide the results of Step 2 by 4 to complete the evaluation:

$$x[n] = [1, 2, 3, 4].$$

The results are given in Table 12.3.   ■

Summary

In the DFT, we have derived a discrete-frequency approximation of the discrete-time Fourier transform. This transform has wide application for digital signal processing. We have looked at the mathematical roots of the DFT as well as the practical meaning of the development. Methods of computing the DFT and the IDFT have been presented and used in example problems.

In the following section, we study more efficient methods of computing the DFT of discrete-time sequences.

12.5 FAST FOURIER TRANSFORM

As seen in Section 12.4, the discrete Fourier transform pair is given by

$$[eq(12.33)] \qquad X[k] = \sum_{n=0}^{N-1} x[n] W_N^{kn}, \quad k = 0, 1, 2, \ldots, N-1$$

and

$$x[n] = \frac{1}{N} \sum_{k=0}^{N-1} X[k] W_N^{-kn}, \quad W_N = e^{-j2\pi/N}, \quad n = 0, 1, \ldots, N-1,$$

which is readily programmed for calculation by digital computer. From inspection of (12.33), we see that for each value of k, computation of $X[k]$ will require N multiplications. Because $x[n]$, and especially $X[k]$, can have complex values, the computation of an N-point DFT or the inverse DFT generally requires N^2 complex multiplications. The thrust of this section will be to develop an algorithm to compute the discrete Fourier transform more efficiently. The collection of efficient algorithms that are generally used to compute the discrete Fourier transform is known as the *fast Fourier transform (FFT)* [6].

Decomposition-in-Time Fast Fourier Transform Algorithm

We next develop an efficient algorithm for computing the discrete Fourier transform for cases in which the number of samples to be computed is a power of 2 ($N = 2^m$). We start with $N = 2$ and work our way up to $N = 2^3 = 8$ before we try to generalize the process. The result of our efforts is known as the *decomposition-in-time, radix-2 FFT*.

Starting with a two-point DFT, we have

$$X[k] = \sum_{n=0}^{1} x[n] W_2^{nk} = x[0] W_2^{0k} + x[1] W_2^{1k}, k = 0, 1.$$

Because $W_2^{0k} = e^{-j0} = 1$ and $W_2^{1k} = e^{-j\pi k} = (-1)^k$, we write

$$X[0] = x[0] + x[1];$$
$$X[1] = x[0] - x[1].$$

In general, for the two-point DFT, we have

$$X[k] = x[0] + (-1)^k x[1].$$

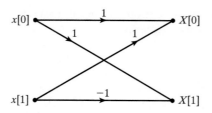

Figure 12.9 Butterfly diagram for a two-point DFT.

The signal flow graph of Figure 12.9 illustrates the process for computing the two-point DFT. This signal flow graph is known as a *butterfly diagram* because of its shape.

We proceed with the four-point DFT:

$$X[k] = \sum_{n=0}^{3} x[n]W_4^{nk} = x[0]W_4^{0k} + x[1]W_4^{1k} + x[2]W_4^{2k} + x[3]W_4^{3k}.$$

As a result of the periodicity of the weighting factor, we can simplify this expression:

$$W_N^{nk} = e^{-j(2\pi/N)nk};$$
$$W_4^{0k} = 1;$$
$$W_4^{1k} = e^{-j(\pi/2)k} = (-j)^k;$$
$$W_4^{2k} = e^{-j\pi k} = (-1)^k;$$
$$W_4^{3k} = W_4^{2k}W_4^{1k} = (-1)^k W_4^{1k}.$$

Using these results, we write

$$X[k] = x[0] + x[1]W_4^{1k} + x[2](-1)^k + x[3](-1)^k W_4^{1k},$$
$$X[k] = [x[0] + x[2](-1)^k] + [x[1] + x[3](-1)^k]W_4^{1k}.$$

To clarify the next step, we define two new variables:

$$x_e[n] = x[2n], \quad n = 0, 1;$$
$$x_o[n] = x[2n + 1], \quad n = 0, 1.$$

Then

$$X[k] = [x_e[0] + x_e[1](-1)^k] + [x_o[0] + x_o[1](-1)^k]W_4^{1k}.$$

The factors in brackets in this equation can be recognized as two-point DFTs:

$$X_e[m] = x_e[0] + x_e[1](-1)^m, \quad m = 0, 1;$$
$$X_o[m] = x_o[0] + x_o[1](-1)^m, \quad m = 0, 1.$$

Note that $X_e[k]$ and $X_o[k]$ are periodic; for example,

$$X_e[2] = x_e[0] + x_e[1](-1)^2 = X_e[0]$$

and

$$X_e[3] = x_e[0] + x_e[1](-1)^3 = X_e[1].$$

The four-point DFT then is

$$
\begin{aligned}
X[0] &= X_e[0] + X_o[0]W_4^{1(0)} = X_e[0] + X_o[0]; \\
X[1] &= X_e[1] + X_o[1]W_4^{1(1)} = X_e[1] + X_o[1]W_4^1; \\
X[2] &= X_e[0] - X_o[0]W_4^{1(2)} = X_e[0] - X_o[0]; \\
X[3] &= X_e[1] - X_o[1]W_4^{1(3)} = X_e[1] - X_o[1]W_4^1.
\end{aligned}
\tag{12.40}
$$

where the changed term is $W_4^{1(k)}$ for the chosen value k.

We see that the four-point DFT can be computed by the generation of two two-point DFTs followed by a *recomposition* of terms as shown in the signal flow graph of Figure 12.10. In other words,

$$[\text{4-point DFT of } x[n]] = [\text{2-point DFT of } x_e[n]] + W_4^{1k}[\text{2-point DFT of } x_0[n]].$$

The Equations (12.40) are known as the *recomposition equations* of the four-point DFT.

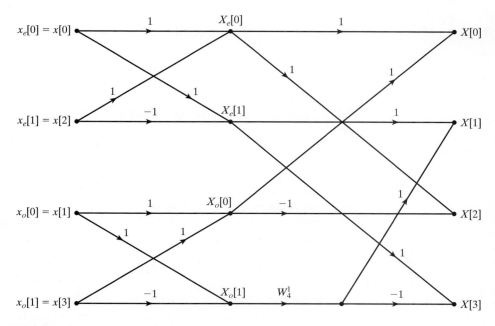

Figure 12.10 Signal flow graph for a four-point DFT.

We now proceed with the eight-point DFT:

$$X[k] = \sum_{n=0}^{7} x[n]W_8^{nk}.$$

The weighting factors for the eight-point DFT are

$$W_8^{0k} = 1;$$
$$W_8^{1k} = e^{-j(\pi/4)k};$$
$$W_8^{2k} = e^{-j(\pi/4)2k} = e^{-j(\pi/2)k} = W_4^{1k};$$
$$W_8^{3k} = e^{-j(\pi/4)3k} = [e^{-j(\pi/4)2k}]e^{-j(\pi/4)k} = W_8^{1k}W_4^{1k};$$
$$W_8^{4k} = e^{-j(\pi/4)4k} = e^{-j\pi k} = W_4^{2k};$$
$$W_8^{5k} = e^{-j(\pi/4)5k} = e^{-j(\pi/4)4k}e^{-j(\pi/4)k} = W_8^{1k}W_4^{2k};$$
$$W_8^{6k} = e^{-j(\pi/4)6k} = W_4^{3k};$$
$$W_8^{7k} = e^{-j(\pi/4)7k} = e^{-j(\pi/4)k}e^{-j(\pi/4)6k} = W_8^{1k}W_4^{3k}.$$

Using the weighting factors in the forms given previously, we write the eight-point DFT as

$$X[k] = x[0] + x[1]W_8^{1k} + x[2]W_4^{1k} + x[3]W_8^{1k}W_4^{1k} + x[4]W_4^{2k}$$
$$+ x[5]W_8^{1k}W_4^{2k} + x[6]W_4^{3k} + x[7]W_8^{1k}W_4^{3k}. \tag{12.41}$$

Much as we did in the derivation of the four-point FFT, we define

$$x_e[n] = x[2n], \quad n = 0, 1, 2, 3$$

and

$$x_o[n] = x[2n + 1], \quad n = 0, 1, 2, 3.$$

Using these newly defined variables, we write (12.41) as

$$X[k] = [x_e[0] + x_e[1]W_4^{1k} + x_e[2]W_4^{2k} + x_e[3]W_4^{3k}]$$
$$+ W_8^{1k}[x_o[0] + x_o[1]W_4^{1k} + x_o[2]W_4^{2k} + x_o[3]W_4^{3k}].$$

In this form, we recognize the factors in brackets as the four-point DFTs of $x_e[n]$ and $x_o[n]$, respectively. Therefore, it is seen that the eight-point FFT is found by the recomposition of two, four-point FFTs. Figure 12.11 illustrates the procedure for computing the eight-point FFT.

In general, the N-point, radix-2 FFT is computed by the recomposition of two $(N/2)$-point FFTs. The generalized procedure is illustrated in Figure 12.12.

As we said at the start of this discussion, our reason for deriving the FFT algorithm is for computational efficiency in calculating the DFT. Table 12.4 shows the number of complex multiplications required for both the DFT and the FFT for several values of N. We see that the increased efficiency of the radix-2 FFT algorithm becomes more significant as the number of points in the DFT becomes larger.

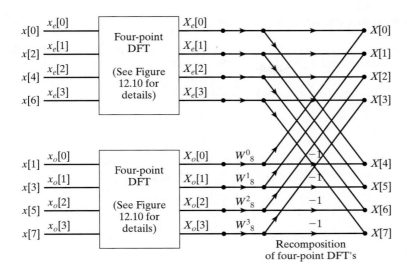

Figure 12.11   Decomposition-in-time fast Fourier transform.

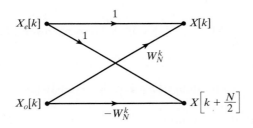

Figure 12.12   Butterfly diagram for an N-point FFT.

TABLE 12.4   DFT and FFT Comparison (Number of Complex Multiplications Required)

N	Standard DFT	FFT
2	4	1
4	16	4
8	64	12
16	256	32
32	1,024	80
64	4,096	192
128	16,384	448
256	65,536	1,024
512	262,144	2,304
1,024	1,048,576	5,120
N (a power of 2)	N^2	$\dfrac{N}{2}\log_2 N$ [6]

EXAMPLE 12.12   **The decomposition-in-time method of calculating the FFT**

The decomposition-in-time (DIT) method of the FFT will be used to compute the DFT of the discrete sequence listed in Table 12.3:

$$x[n] = [1, 2, 3, 4].$$

Referring to Figure 12.10 for the four-point FFT, we find the following:

$$X_e[0] = x[0] + x[2] = 1 + 3 = 4;$$
$$X_e[1] = x[0] - x[2] = 1 - 3 = -2;$$
$$X_o[0] = x[1] + x[3] = 2 + 4 = 6;$$
$$X_o[1] = x[1] - x[3] = 2 - 4 = -2.$$

Thus,

$$X[0] = X_e[0] + X_o[0] = 4 + 6 = 10,$$
$$X[1] = X_e[1] + W_4^1 X_o[1] = -2 + (-j)(-2) = -2 + j2,$$
$$X[2] = X_e[0] - X_o[0] = 4 - 6 = -2,$$

and

$$X[3] = X_e[1] - W_4^1 X_o[1] = -2 - (j)(-2) = -2 - j2,$$

which is in agreement with the results found in Example 12.9 and listed in Table 12.3.   ■

Decomposition-in-Frequency Fast Fourier Transform

The idea behind the decomposition-in-frequency (DIF) FFT algorithm is similar to that of the decomposition-in-time (DIT) FFT presented previously. The DIT FFT and the DIF FFT require the same number of complex multiplications to compute. We begin the derivation of the DIF FFT with the equation of the standard DFT, namely,

$$X[k] = \sum_{n=0}^{N-1} x[n] W_N^{kn}, \quad k = 0, 1, 2, \ldots, N - 1,$$

and divide the summation in half, so that

$$X[k] = \sum_{n=0}^{(N/2)-1} x[n] W_N^{kn} + \sum_{n=(N/2)}^{N-1} x[n] W_N^{kn}, \quad k = 0, 1, 2, \ldots, N - 1.$$

Next, we rewrite the second summation as

$$\sum_{n=(N/2)}^{N-1} x[n] W_N^{kn} = \sum_{n=0}^{(N/2)-1} x[n + N/2] W_N^{k(n+N/2)} = \sum_{n=0}^{(N/2)-1} x[n + N/2] W_N^{kn} W_N^{kN/2},$$

where

$$W_N^{kN/2} = e^{-j(2\pi k/N)N/2} = e^{-j\pi k} = (-1)^k = \begin{cases} 1, & k \text{ even} \\ -1, & k \text{ odd} \end{cases}.$$

We now have

$$X[k] = \sum_{n=0}^{(N/2)-1} [x[n] + (-1)^k x[n + N/2]]W_N^{kn}, \quad k = 0, 1, 2, \ldots, N - 1,$$

which we decompose into the two frequency sequences

$$X[2k] = \sum_{n=0}^{(N/2)-1} [x[n] + x[n + N/2]]W_N^{2kn}, \quad k = 0, 1, 2, \ldots, \frac{N}{2} - 1$$

and

$$X[2k + 1] = \sum_{n=0}^{(N/2)-1} [x[n] - x[n + N/2]]W_N^{(2k+1)n}, \quad k = 0, 1, \ldots, \frac{N}{2} - 1.$$

Now we consider the weighting factors of the two sequences:

$$W_N^{2kn} = W_{N/2}^{kn} \text{ and } W_N^{(2k+1)n} = W_{N/2}^{kn}W_N^n.$$

Then

$$X[2k] = \sum_{n=0}^{(N/2)-1} [x[n] + x[n + N/2]]W_{N/2}^{kn}, \qquad k = 0, 1, 2, \ldots, \frac{N}{2} - 1;$$

$$X[2k + 1] = \sum_{n=0}^{(N/2)-1} [[x[n] - x[n + N/2]]W_N^n]W_{N/2}^{kn}, \quad k = 0, 1, \ldots, \frac{N}{2} - 1.$$

$$(12.42)$$

Each of the frequency sequences of (12.42) can be recognized as an $(N/2)$-point DFT:

$$X[2k] = \frac{N}{2}\text{-point DFT of } [x[n] + x[n + N/2]];$$

$$X[2k + 1] = \frac{N}{2}\text{-point DFT of } [x[n] - x[n + N/2]]W_N^n.$$

Figure 12.13 shows the butterfly diagram for the DIF FFT algorithm. Figure 12.14 illustrates the DIF FFT process for a four-point DIF FFT.

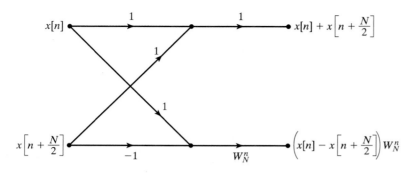

Figure 12.13 A decomposition-in-frequency FFT flow diagram.

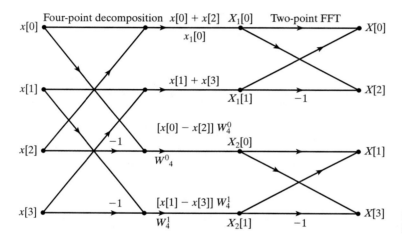

Figure 12.14 A four-point decomposition-in-frequency FFT.

EXAMPLE 12.13   **The decomposition-in-frequency method of calculating the FFT**

The decomposition-in-frequency (DIF) method of the FFT will be used to compute the DFT of the discrete sequence

$$x[n] = [1, 2, 3, 4]$$

listed in Table 12.3. Referring to Figure 12.14 for the four-point DIF FFT, we find the following:

$$X_1[0] = x[0] + x[2] = 1 + 3 = 4;$$
$$X_1[1] = W_4^0[x[1] + x[3]] = 2 + 4 = 6;$$
$$X_2[0] = x[0] - x[2] = 1 - 3 = -2;$$
$$X_2[1] = W_4^1[x[1] - x[3]] = -j[2 - 4] = j2.$$

Hence,

$$X[0] = X_1[0] + X_1[1] = 4 + 6 = 10,$$
$$X[1] = X_2[0] + X_2[1] = -2 + j2 = -2 + j2,$$
$$X[2] = X_1[0] - X_1[1] = 4 - 6 = -2,$$

and

$$X[3] = X_2[0] - X_2[1] = -2 - j2 = -2 - j2$$

which is in agreement with the previous results found in Examples 12.9 and 12.12 and listed in Table 12.3.   ∎

Summary

In this section, we have presented some efficient algorithms for computation of the discrete Fourier transform (DFT). These algorithms are collectively known as the *fast Fourier transform* (FFT). In most practical applications, such as those considered in the following section, the DFT is computed using an FFT algorithm.

12.6   APPLICATIONS OF THE DISCRETE FOURIER TRANSFORM

In this section, we look at several typical applications of the discrete Fourier transform. The primary application is to approximate the Fourier transform of signals. The other applications we consider—convolution, filtering, correlation, and energy spectral density estimation—are all based on the DFT being an approximation of the Fourier transform.

Calculation of Fourier Transforms

In Section 12.4, we introduced the DFT as a discrete-frequency approximation of the discrete-time Fourier transform (DTFT). The DFT is also used to calculate approximations of the Fourier transforms of analog signals.

The steps required for the calculation of the DFT as an approximation of the Fourier transform follow. (Steps 1 and 2 can be done in reverse order.)

1. Determine the resolution required for the DFT to be useful for its intended purpose. The discrete frequency resolution is determined by

$$\Delta\Omega = \frac{2\pi}{N}$$

and establishes a lower limit on N, the number of sample values of the signal required for the DFT computation.

2. Determine the sampling frequency required to sample the analog signal so as to avoid aliasing. Shannon's sampling theorem establishes the requirement

$$\omega_s > 2\omega_M,$$

where ω_M is the highest significant frequency component of the analog signal.

3. Accumulate N samples of the analog signal over a period of NT seconds ($T = 2\pi/\omega_s$). If the DFT calculation of Step 4 is to be computed using a radix-2 DFT, then N must be an integer power of 2. (In some cases, the set of samples will be fixed. In that case, zero padding can be used to create a total of N elements.)

4. Calculate the DFT. This can be done using (12.33) directly, but more often an FFT algorithm, as described in Section 12.5, is used to execute the calculation.

If these steps are executed properly, the resulting DFT should be a good approximation of the Fourier transform of the analog signal. We now look at some examples of the process and point out some details that must be taken into consideration.

EXAMPLE 12.14   **Using the FFT to approximate the Fourier transform**

The Fourier transform of the rectangular pulse shown in Figure 12.15(a) will be computed. From our previous study of the Fourier transform, we see that this waveform can be described as

$$f(t) = \text{rect}[(t - 1)/2].$$

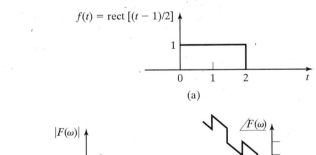

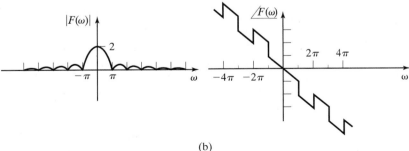

(b)

Figure 12.15 A rectangular pulse and its Fourier transform.

From Tables 5.1 and 5.2, we determine that

$$F(\omega) = 2 \text{ sinc } (\omega)e^{-j\omega}.$$

$F(\omega)$ is sketched in Figure 12.15(b). From Figure 12.15(b) we see that the frequency spectrum has its first null at $\omega = \pi$ rad/s. We also see that although the frequency spectrum does not have an absolute bandwidth, the magnitude of the sidelobes decreases with frequency. We choose a sampling frequency of $\omega_s = 10\pi$ rad/s. This requires a sampling period of $T = 0.2$ s. We choose to use 16 samples for the computation. Therefore, the discrete-frequency resolution of our calculated frequency spectrum will be $\frac{2\pi}{16}$. The discrete-time sequence of samples $f[n]$ is shown in Figure 12.16(a). The 16-point, radix-2 FFT is used to calculate the DFT. The magnitude of the discrete-frequency sequence is shown in Figure 12.16(b).

Because the DFT is calculated from discrete-time samples, we must multiply $|F[k]|$ by the sampling period T to cancel out the factor of $1/T$ inherent to the Fourier transform of sampled signals (5.42).

We now want to see how good an approximation of $F(\omega)$ we have in the $T \cdot F[k]$ that we compute from the 16 samples of $f(t)$. This will be determined by calculating the value of $F(\omega)$ at a few frequencies ($\omega < \omega_s/2$) that correspond to a value of k in $F[k]$. Table 12.5 gives a comparison of the values.

From Table 12.5 and Figure 12.17, we see that the Fourier-transform approximation found by computing the DFT is not perfect. Although the approximation is reasonably accurate for samples where $\omega < \omega_s/2 = 15.708$ rad/s, there are some errors in both magnitude and phase.

The results shown in Table 12.5 and Figure 12.17 can be reproduced by the following MATLAB program:

```
% This MATLAB program reproduces the results of
% Example 12.14.
Ts=0.2
fn=[1 1 1 1 1 1 1 1 1 1 1 0 0 0 0 0];
n=[0:length(fn)-1];
figure(1),stem(n,fn,'fill'),grid
Fk=fft(fn,length(fn));
w=2*pi/length(fn)/Ts*n;
Ft=2*sin(w).*exp(j*w)./w;
Ft(1)=2;
' k omega (w) |F(w)| Ts*|F[k]| phase(F(w)) phase(F[k])'
[n', w', abs(Ft)', Ts*abs(Fk)', angle(Ft)', angle(Fk)']
figure(2),stem(w,Ts*abs(Fk),'filled'),hold,plot(w,abs(Ft),'r+'),
title('Fast Fourier Transform')
grid, xlabel('omega'), ylabel('magnitude'),hold
```  ■

One source of error such as seen in Example 12.14 is the "windowing" or truncation of the periodic extension of the discrete-time sequence that is implied in the DFT development. This windowing has the effect of multiplying the periodic extension of the sequence $x[n]$ by a sequence that represents samples of a rectangular pulse with duration NT. The effect we have just described is illustrated in Figure 12.18. In Figure 12.18(a) we see a nonperiodic, ideally sampled signal $x(nT)$. Figure 12.18(b) shows the periodic extension of the signal that is implicit in the DFT

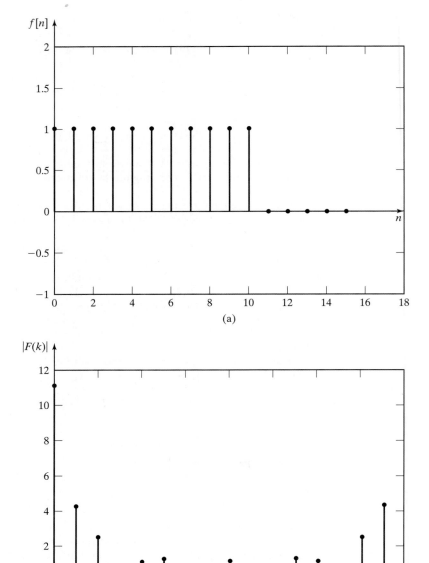

Figure 12.16 A sampled rectangular pulse and its 16-point DFT.

TABLE 12.5 Fourier Transform and DFT Approximation

k	ω	$|F(\omega)|$	$T|F[k]|$	% Error	$\angle F(\omega)$	$\angle F[k]$
0	0	2.000	2.2000	10.0	0	0
1	1.9635	0.9411	0.8524	9.4	−1.9635	−1.9635
2	3.9270	0.3601	0.4828	34.1	−0.7854	−0.7854
3	5.8905	0.1299	0.0702	45.9	−2.7489	0.3927
4	7.8540	0.2546	0.2000	21.5	−1.5708	−1.5708
5	9.8175	0.0780	0.2359	202.6	−0.3927	−0.3927
6	11.7810	0.1200	0.0828	31.0	−2.3562	0.7854
7	13.7445	0.1344	0.1133	15.7	−1.1781	−1.1781

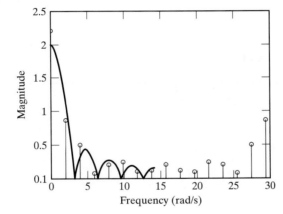

Figure 12.17 Comparison of the DFT with the Fourier transform.

implementation. The rectangular windowing function is shown in Figure 12.18(c). The product of the windowing function and the periodic extension results in the discrete-time sequence $x[n]$. The sequence $x[n]$ contains the values of $x_p(nT)$ for $0 \leq n \leq N - 1$. $x[n]$ represents one period of the periodic extension of the original signal. These are the values used to calculate $X[k]$ from (12.33). From this presentation, we see that the DFT we calculate is not only based on sampled values of the analog signal, but also involves a convolution with the Fourier transform of the windowing function.

This implied multiplication of the periodic extension of the sampled signal by the windowing function results in the phenomenon called *spectrum-leakage distortion*, which arises from the spectrum spreading that develops from truncating a signal. This phenomenon can be illustrated by considering the truncated cosine shown in Figure 12.19(a). This signal is the product of a cosine wave and a rectangular pulse as shown in Figure 12.19(b). From our previous study of the Fourier transform and Table 5.2, we know that

$$\cos(\omega_1 t) \xleftrightarrow{\mathcal{F}} \pi[\delta(\omega - \omega_1) + \delta(\omega + \omega_1)]$$

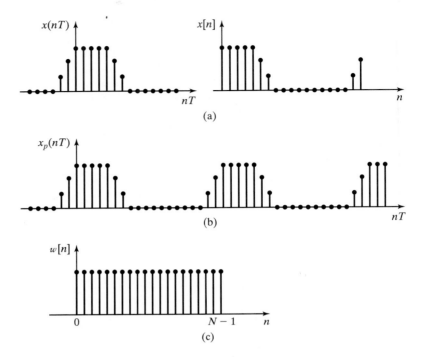

Figure 12.18 Periodic extension and windowing of a sequence.

and

$$\text{rect}(t/T) \xleftrightarrow{\;\mathscr{F}\;} T\,\text{sinc}(T\omega/2)$$

as shown in Figure 12.19(c). From the Fourier transform properties (Table 5.1),

$$f_1(t)f_2(t) \xleftrightarrow{\;\mathscr{F}\;} \frac{1}{2\pi} F_1(\omega)*F_2(\omega).$$

The frequency spectrum of the truncated cosine is shown in Figure 12.19(d). From this figure, we see that the frequency spectrum of the truncated cosine is spread across all frequencies. It is this spreading, caused by the truncation of the signal, that causes spectrum-leakage distortion in the DFT.

Several alternative window shapes, other than the rectangular window we have used, have been developed to alleviate spectrum-leakage distortion. The two best known are the Hamming window, proposed by Richard W. Hamming, and the Hanning window, named for Julius von Hann [5].

The Hamming window is given by the equation

$$w[n] = 0.54 - 0.46\cos\left(\frac{2\pi n}{N-1}\right), \quad 0 \leq n \leq N-1 \qquad (12.43)$$

$$\frac{2\pi n}{255}$$

$$N = 256$$

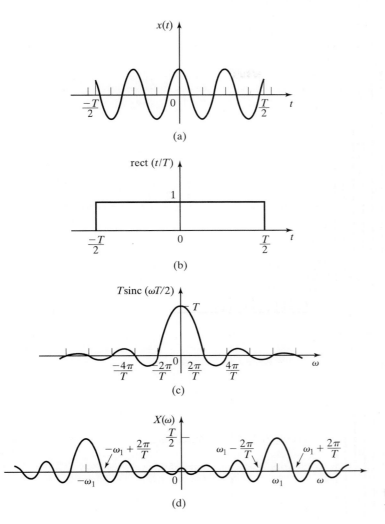

Figure 12.19   Spectrum spreading effect of windowing.

and the Hanning window by

$$w[n] = 0.50 - 0.50\cos\left(\frac{2\pi n}{N-1}\right), \quad 0 \leqq n \leqq N-1. \qquad (12.44)$$

Figure 12.20 shows an example of the Hamming window and the Hanning window for a 32-point sequence. It can be seen that both of these windowing functions gradually approach zero at the limits of the sequence rather than cut off abruptly as the rectangular window does. This gradual change in the time domain results in a narrower bandwidth in the frequency domain and, therefore, less spectrum-leakage distortion in the DFT. The following example demonstrates the effect of spectrum-leakage distortion and the difference in spectrum-leakage distortion that is made by changing the shape of the window.

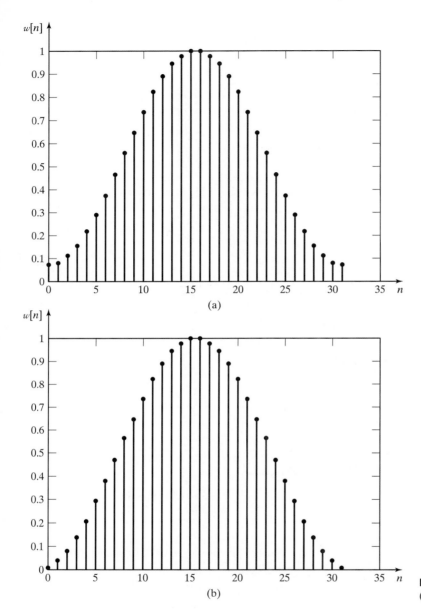

Figure 12.20   (a) Hamming window; (b) Hanning window.

EXAMPLE 12.15   **Comparison of windowing functions on the accuracy of the FFT**

The Fourier transform of the signal $x(t) = \cos(10\pi t)$ is to be approximated by calculating the DFT.

1. It has been determined that a discrete-frequency resolution of $\Delta\Omega \leq 0.1$ is required for our purpose. This is equivalent to $\Delta\Omega = 2\pi/62.83$. Because we plan to use a radix-2 FFT

for the DFT calculation we choose a 64-point FFT. This will result in a discrete-frequency resolution of $\Delta\Omega = \pi/32$. To calculate a 64-point FFT, we must use 64 samples of the signal ($N = 64$).

2. The signal is absolutely bandlimited. Actually, it contains only one frequency component at $\omega = 10\pi$ rad/s. The Nyquist frequency for this signal is 20π rad/s. We arbitrarily choose to sample the signal at $\omega_s = 100\pi$ rad/s ($T = 0.02$ s), which is five times the minimum sampling frequency.

3. Figure 12.21(a) shows the sampled data sequence. We are now assuming a rectangular window of $NT = (64)(0.02) = 1.28$ s duration.

4. The DFT is calculated using a 64-point FFT algorithm. Figure 12.21(b) shows the discrete-frequency sequence. We see in Figure 12.21(b) the evidence of spectrum spreading. We know that the frequency spectrum of the original analog signal has discrete-frequency components at $\omega = 10\pi$ only. The DFT shows nonzero frequency components throughout the spectrum.

A Hanning window will be used to reduce the errors in the DFT caused by spectrum spreading. Figure 12.21(c) shows the discrete-time sequence $x[n]$ produced using the Hanning window. Figure 12.21(d) shows the new discrete-frequency spectrum. It is seen that the spectrum-leakage distortion is reduced by using a Hanning window instead of a rectangular window.

The results shown in this example can be reproduced using the following MATLAB program:

```
% This MATLAB program reproduces the results of
% Example 12.15.
win = input...
('Is window to be:(1) rectangular or (2) Hanning?')
Ts=1/50;
t = Ts*(0:63);
for n = 1:64
    xn(n) = cos(10*pi*t(n));
end
if win == 2
xw = hanning(64)'.*xn;
else
    xw = xn;
end
figure(1), stem(t,xw,'filled'), title('Windowed Time Sequence')
grid, xlabel('seconds'), ylabel('magnitude')
X = fft(xw,64);
w = 2/Ts*(-32:31)/64;
figure(2),stem(w, Ts*abs(fftshift(X)),'filled'),title('Magnitude of
T*F[k]')
grid,xlabel('omega/pi'),ylabel('magnitude')
```                                                                    ■

We see that the DFT can give a good approximation of the Fourier transform. The errors in the Fourier transform approximation can be reduced by choosing a windowing function that causes less spectrum-leakage distortion. Increasing the sampling rate and increasing the number of samples used in the calculation also

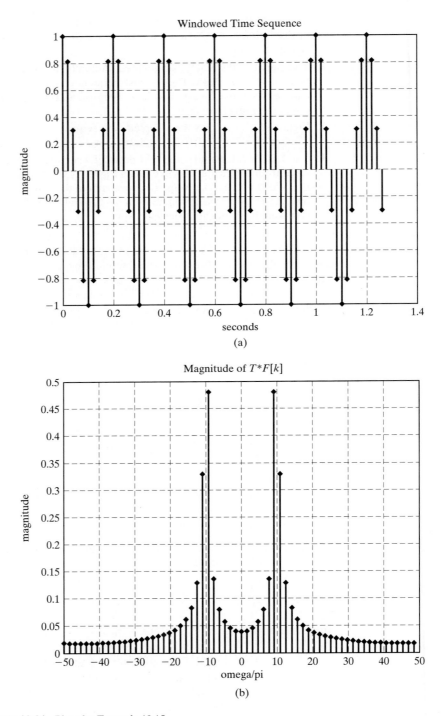

Figure 12.21 Plots for Example 12.15.

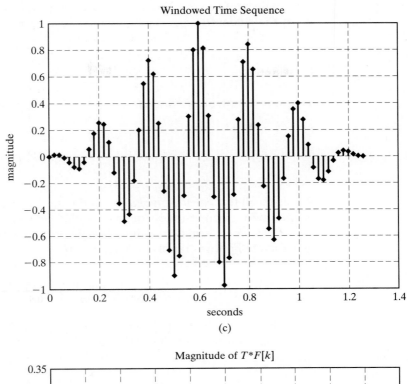

(c)

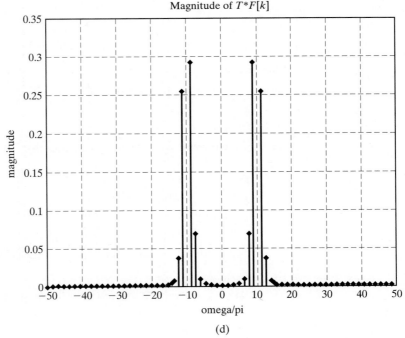

(d)

Figure 12.21   (cont.)

tend to decrease errors. Problems at the end of this chapter provide the opportunity for the student to investigate various ways of decreasing errors when using the DFT to approximate the Fourier transform.

Convolution

The convolution of two discrete-time signals is described by the equation

[eq(10.16)]
$$x[n]*h[n] = \sum_{m=-\infty}^{\infty} x[m]h[n - m].$$

We shall call this *linear convolution*. In this section, we discuss another convolution operation for discrete-time sequences, called *circular convolution*.

First, we determine the time-domain process corresponding to the product of two discrete Fourier-transform functions. If

$$Y[k] = X[k]H[k],$$

then what is the relationship among $y[n]$, $x[n]$, and $h[n]$? We approach the answer to this question by beginning with the definition of the DFT:

[eq(12.30)]
$$X[k] = \mathcal{DF}[x[n]] = \sum_{n=0}^{N-1} x[n]e^{-j2\pi kn/N};$$

$$x[n] = \mathcal{DF}^{-1}[X[k]] = \frac{1}{N}\sum_{k=0}^{N-1} X[k]e^{j2\pi kn/N}.$$

Using (12.30), we write the transform equation for $H[k]$. We have

$$H[k] = \sum_{m=0}^{N-1} h[m]e^{-j2\pi km/N}, \quad m = 0, 1, 2, \ldots, N - 1;$$

$$Y[k] = \left[\sum_{n=0}^{N-1} x[n]e^{-j2\pi kn/N}\right]\left[\sum_{m=0}^{N-1} h[m]e^{-j2\pi km/N}\right];$$

$$y[n] = \frac{1}{N}\sum_{k=0}^{N-1}\left[\sum_{l=0}^{N-1} x[l]e^{-j2\pi kl/N}\right]\left[\sum_{m=0}^{N-1} h[m]e^{-j2\pi km/N}\right]e^{j2\pi kn/N};$$

$$y[n] = \frac{1}{N}\sum_{l=0}^{N-1} x[l]\sum_{m=0}^{N-1} h[m]\left[\sum_{k=0}^{N-1} e^{j2\pi k(n-l-m)/N}\right].$$

From (12.35), we see that the term in brackets can be evaluated as

$$\sum_{k=0}^{N-1} e^{j2\pi k(n-l-m)/N} = \begin{cases} N, & n-l-m=0 \\ 0, & \text{otherwise} \end{cases}$$

and, therefore,

$$y[n] = \frac{1}{N}\sum_{l=0}^{N-1} x[l]\sum_{m=0}^{N-1} h[m]N\delta[n-l-m].$$

Because the impulse function is zero except when $m = n - l$, we can rewrite the equation as

$$y[n] = \sum_{l=0}^{N-1} x[l]h[n-l], \tag{12.45}$$

which is clearly related to the equation for linear convolution (10.16). However, the summation is over only one period rather than for all time. This equation represents the process called *periodic convolution* or *circular convolution*. In this book, we usually use the latter title.

The symbol ⊛ is used to signify the operation of circular convolution as described by (12.45):

$$y[n] = x[n] \circledast h[n] = \sum_{l=0}^{N-1} x[l]h[n-l]. \tag{12.46}$$

In this development, we have established the discrete Fourier transform pair:

$$x[n] \circledast h[n] \overset{\mathcal{DF}}{\longleftrightarrow} X[k]H[k]. \tag{12.47}$$

The process of (12.46) is called circular convolution because it is easily (if sometimes tediously) evaluated using two concentric circles, as shown in Figure 12.22. The circular convolution of (12.46) can be evaluated by writing the N values of $x[n]$

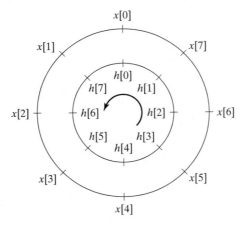

Figure 12.22 Circular convolution.

equally spaced at angles of $2\pi/N$ around the outer circle in a counterclockwise direction. The N values of $h[n]$ are then written equally spaced at angles of $2\pi/N$ in a clockwise direction on the inner circle. $y[0]$ is calculated by multiplying the corresponding values on each radial line and then adding the products. Succeeding values of $y[n]$ are then found in the same way after rotating the inner circle counterclockwise through the angle $2\pi n/N$ and finding the sum of products of the corresponding values. The circular convolution process is demonstrated in the following example.

EXAMPLE 12.16   **Circular convolution of two discrete sequences**

We wish to evaluate the circular convolution of the sequences

$$x_1[n] = [1, 2, 3, 4]; \quad x_2[n] = [0, 1, 2, 3];$$
$$y[n] = x_1[n] \circledast x_2[n].$$

We will do this by two methods, first using the circular convolution process as described above and second by using the DFT and IDFT.
The first value in the convolution sequence, $y[0]$, is calculated from Figure 12.23(a):

$$y[0] = (1)(0) + (2)(3) + (3)(2) + (1)(4) = 16.$$

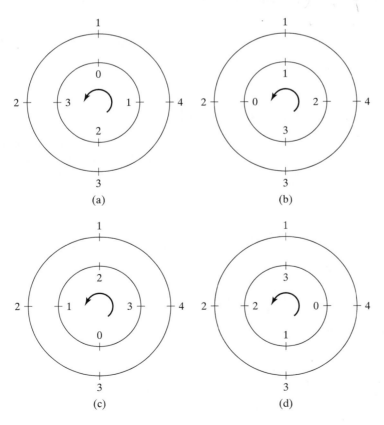

(a)

(b)

(c)

(d)

Figure 12.23   Circular convolution for Example 12.16.

The inner circle is rotated counterclockwise through $2\pi/4 = \pi/2$ radians to form Figure 12.23(b) for calculation of the second term:

$$y[1] = (1)(1) + (2)(4) + (3)(3) + (0)(2) = 18.$$

The inner circle is rotated counterclockwise an additional $\pi/2$ radians to form Figure 12.23(c) for calculation of the third term:

$$y[2] = (2)(1) + (3)(4) + (0)(3) + (1)(2) = 16.$$

Repeating the process for the fourth term, we have

$$y[3] = (3)(1) + (0)(4) + (1)(3) + (2)(2) = 10.$$

Note that the sequence

$$y[n] = [16, 18, 16, 10]$$

is not the same as would result from linear convolution of the two sequences. Linear convolution would give

$$x_1[n]*x_2[n] = [0, 1, 4, 10, 16, 17, 12].$$

It is seen that circular convolution yields a four-element sequence, whereas linear convolution yields seven elements.

The DFT and IDFT can be used to compute the circular convolution of the two sequences. From (12.47),

$$Y[k] = X_1[k]X_2[k],$$

where

$$X_1[k] = \mathcal{DF}\{x_1[n]\} = [10, -2 + j2, -2, -2 - j2],$$

as found in Example 12.9, and

$$X_2[k] = \mathcal{DF}\{x_2[n]\} = [6, -2 + j2, -2, -2 - j2].$$

The two discrete-frequency sequences are multiplied term by term to find

$$\begin{aligned} Y[k] &= [(10)(6), (-2 + j2)(-2 + j2), (-2)(-2), (-2 - j2)(-2 - j2)] \\ &= [60, -j8, 4, j8]. \end{aligned}$$

Then

$$y[n] = \mathcal{IDF}\{Y[k]\} = [16, 18, 16, 10],$$

as was found from circular convolution of the discrete-time sequence.   ∎

In Example 12.16 we saw that circular convolution did not yield the same sequence as linear convolution. This is generally the case. Circular convolution of two N-sample sequences yields an N-sample sequence. Linear convolution of an N_1-sample sequence with an N_2-sample sequence yields an $(N_1 + N_2 - 1)$-sample sequence. (These last two statements were illustrated in Example 12.16.)

The circular convolution of two sequences, of lengths N_1 and N_2, respectively, can be made equal to the linear convolution of the two sequences by zero padding both sequences so that they both consist of $N_1 + N_2 - 1$ samples.

EXAMPLE 12.17   Circular convolution with zero padding

We shall make the circular convolution of the two sequences given in Example 12.16, namely,

$$x_1[n] = [1, 2, 3, 4] \quad \text{and} \quad x_2[n] = [0, 1, 2, 3],$$

equal to the linear convolution by zero padding both sequences so that each has $4 + 4 - 1 = 7$ samples:

$$x_1'[n] = [1, 2, 3, 4, 0, 0, 0] \quad \text{and} \quad x_2'[n] = [0, 1, 2, 3, 0, 0, 0].$$

The circular convolution process is illustrated in Figure 12.24 and results in

$$y'[n] = [0, 1, 4, 10, 16, 17, 12].$$

This is the result shown in Example 12.16 for linear convolution of the two four-sample sequences. ∎

The procedure of zero-padding sequences so that the circular convolution result is the same as the linear convolution of the original sequences is important in signal processing applications. This procedure allows us to use DFTs of two sequences to compute their linear convolution.

EXAMPLE 12.18   Linear convolution using the DFT

We shall compute the linear convolution of the two sequences given in Example 12.16,

$$x_1[n] = [1, 2, 3, 4] \quad \text{and} \quad x_2[n] = [0, 1, 2, 3],$$

using the DFT. We begin by zero padding so that each sequence consists of $N_1 + N_2 - 1 = 7$ samples:

$$x_1'[n] = [1, 2, 3, 4, 0, 0, 0] \quad \text{and} \quad x_2'[n] = [0, 1, 2, 3, 0, 0, 0].$$

The DFTs of the two sequences are

$$X_1'[k] = [10, -2.0245 - j6.2240, 0.3460 + j2.4791, 0.1784 - j2.4220,$$
$$0.1784 + j2.4220, 0.3460 - j2.4791, -2.0245 - j6.2240]$$

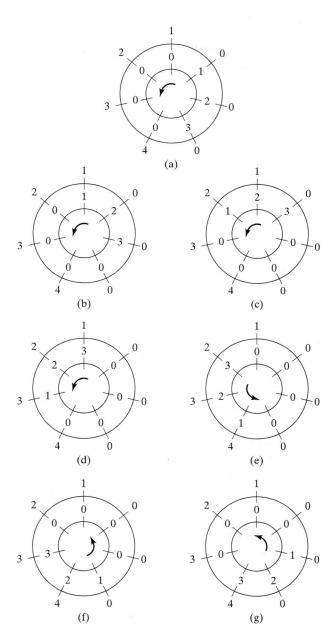

Figure 12.24 Circular convolution for Example 12.17.

and

$$X'_2[k] = [6, -2.5245 - j4.0333, -0.1540 + j2.2383, -0.3216 - j1.7950,$$
$$-0.3216 + j1.7950, -0.1540 - j2.2383, -2.5245 + j4.0333].$$

We find $Y'[k]$ by multiplying the corresponding elements of the two discrete-frequency sequences:

$$Y'[k] = [60, -19.9928 + j23.8775, -5.6024 + j0.3927, -5.8342 - j0.8644,$$
$$-4.4049 + j0.4585, -5.6024 - j0.3927, -19.9928 + j23.8775].$$

Now we find the IDFT to complete the convolution calculation:

$$\mathscr{IDF}\{Y'[k]\} = y'[n] = [0, 1, 4, 10, 16, 17, 12].$$

The following MATLAB program be used to reproduce the results of this example:

```
% This MATLAB program reproduces the results of
% Example 12.18.
x1 = [1 2 3 4];
x2 = [0 1 2 3];
y = conv(x1,x2);
% Zero-pad the vectors to 7 elements each.
x1p = zeros(1:7);, x1p(1:4) = x1(1:4);
x2p = zeros(1:7);, x2p(1:4) = x2(1:4);
X1 = fft(x1p,7);
X2 = fft(x2p,7);
Y = X1.*X2;
yp = ifft(Y);
y
yp
```

■

The discovery that we can use the DFTs of two sequences to compute their convolution is an important result for signal processing, as shown in examples to follow. We should now consider the computational efficiency of calculating a convolution using the DFT rather than the direct method. Figure 12.25 shows a block-diagram representation of the convolution process using the DFT method. In this system, both DFTs and the IDFT are computed using a radix-2 FFT algorithm. The two N-element sequences to be convolved are extended by zero padding so that

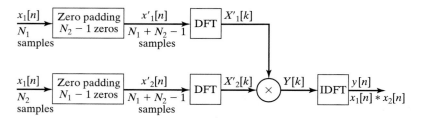

Figure 12.25 Block diagram of convolution using DFTs.

each of the extended sequences contains $2N - 1$ elements. To simplify the algebra without significant loss of accuracy, we consider that each sequence has $2N$ elements when we calculate the number of multiplications required to complete the convolution calculation. If we assume that the discrete-time samples can take on complex values, computing the FFT of each extended sequence requires $N \log_2(2N)$ complex multiplications. The IFFT calculation will also require $N \log_2(2N)$ complex multiplications. Multiplication of the two discrete-frequency sequences, $X_1[k] \times X_2[k]$, requires $2N$ complex multiplications. This adds up to a total of

$$3N \log_2(2N) + 2N$$

complex multiplications in computing the convolution of two N-element sequences using the DFT method.

Direct convolution of the two sequences requires N^2 complex multiplications. If we assume that N is chosen as a power of 2, so that a radix-2 FFT can be used, the DFT method is more efficient for $N \geq 32$.

For long-duration discrete-time sequences, a block of arbitrary length may be selected for the purpose of estimating the Fourier transform. In other words, we can choose to use any N consecutive values of a long sequence to compute the DFT. This concept is discussed in more detail in the following section.

Filtering

An important application of the DFT is the filtering of signals. If we review the convolution process, we see that if $x[n]$ represents the signal input to a filter with discrete-time impulse response $h[n]$, then the output signal can be calculated by linear convolution:

$$[eq(10.16)] \qquad y[n] = x[n]*h[n] = \sum_{m=-\infty}^{\infty} x[m]h[n - m].$$

If the input signal, $x[n]$, has N_1 samples and the impulse-response sequence has N_2 samples, the convolution can be computed using the DFT if we first extend both sequences by zero padding so that they each contain $N_1 + N_2 - 1$ elements. This was discussed previously in the section on convolution. If we let $x'[n]$ and $h'[n]$ represent the extended sequences of the input signal and the impulse response, respectively, we can use the convolution property of the DFT to find

$$x[n]*h[n] = x'[n] \circledast h'[n] \xleftrightarrow{\mathscr{DF}} X'[k]H'[k].$$

Therefore,

$$Y[k] = X'[k]H'[k]$$

and

$$y[n] = \mathscr{IDF}\{X'[k]H'[k]\},$$

as shown in Figure 12.26.

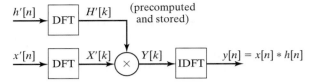

Figure 12.26　Block diagram of filtering using DFTs.

EXAMPLE 12.19　**Application of the DFT in filtering**

A filter has the discrete impulse response shown in Figure 12.27(a). The input signal sequence is shown in Figure 12.27(b). In this case, the output sequence can be calculated more efficiently by direct convolution, but we will use the DFT method for demonstration purposes (see Figure 12.26). Before calculating the DFT, we extend the two sequences by zero padding so that they have eight elements each. This provides more than the required $N_1 + N_2 - 1 = 7$ elements and allows a radix-2 FFT algorithm to be used to compute the DFTs and the IDFT:

$$x'[n] = [1, 2, 3, 4, 0, 0, 0, 0];$$
$$h'[n] = [1, 1, 0, 0, 0, 0, 0, 0].$$

From the FFT computation, we get

$$X'[k] = [10, -0.4142 - j7.2426, -2 + j2, 2.4142 - j1.2426, -2, 2.4142 + j1.2426,$$
$$-2 - j2, -0.4142 + j7.2426]$$

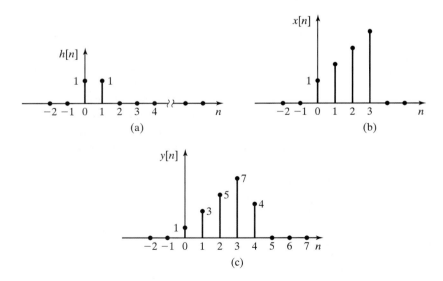

Figure 12.27　Discrete-time sequence for Example 12.19.

and

$$H'[k] = [2, 1.7071 - j0.7071, 1 - j1, 0.2929 - j0.7071,$$
$$0, 0.2929 + j0.7071, 1 + j1, 1.7071 + j0.7071].$$

We then multiply corresponding elements of the two discrete-frequency sequences to form $Y'[k]$:

$$Y[k] = X'[k]H'[k];$$
$$Y[k] = [20, -5.8284 - j12.0711, 0 + j4, -0.1716 - j2.0711,$$
$$0, -0.1716 + j2.0711, 0 - j4, -5.8284 + j12.0711].$$

Finally, the output discrete-time sequence is computed by finding the inverse discrete Fourier transform:

$$y[k] = \mathscr{IDF}\{Y[k]\} = [1, 3, 5, 7, 4, 0, 0, 0].$$

The output sequence is shown in Figure 12.27(c).

The following MATLAB program can be used to confirm the results of this example:

```
% This MATLAB program reproduces and confirms the
% results of Example 12.19.
x = [1 2 3 4 0 0 0 0];
h = [1 1 0 0 0 0 0 0];
% The filter function is a MATLAB primitive.
% It will be used to confirm the DFT method.
yf = filter(h,1,x);
X = fft(x,8);
H = fft(h,8);
Y = X.*H;
y = ifft(Y);
y    yf
```
■

The illustrated use of the DFT in filtering is limited to relatively short signal sequences. Because all sample values of the signal must be accumulated before the process begins, the system would require a great deal of memory for the storage of samples from long-duration signals. Also, long time delays would be encountered in generating the output signal, because there is no output until the entire input sequence is accumulated and processed. One method of alleviating these limitations is called *block filtering*.

In block filtering, the long input sequence is divided into blocks of appropriate length for DFT calculation. Figure 12.28 shows how the blocks of data are prepared for block filtering by the *overlap-add technique*. The blocks of input data are chosen so that they do not overlap in time. Each block of the input sequence contains N samples.

As shown in Figure 12.28, the unit impulse response of the filter is a sequence of M samples. The convolution of $h[n]$ and $x_b[n]$ produces a sequence of

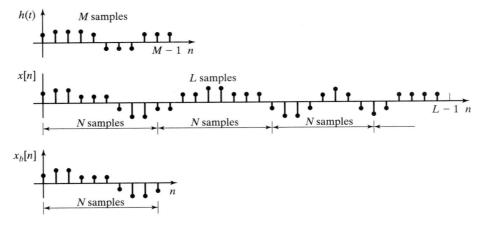

Figure 12.28 Signal preparation for block filtering.

$M + N - 1$ samples. Therefore, both sequences must be zero padded so that each contains $M + N - 1$ elements so that circular convolution (using the DFTs of the two sequences) can be used to make the calculation. The output sequences $y_b[n]$ are $M + N - 1$ in length and therefore overlap when they are fit together in the NT-duration time periods to which they are restricted. The elements in the intervals

$$\begin{aligned} n &= N & to & \quad N + M - 2, \\ n &= 2N & to & \quad 2N + M - 2, \\ & \vdots & & \\ n &= kN & to & \quad kN + M - 2 \end{aligned}$$

overlap in time and must be added together to form the output sequence $y[n]$.

EXAMPLE 12.20   **Block filtering**

For the filter with the unit impulse response,

$$h[n] = [1, 1],$$

block filtering is to be used to compute the output sequence. The block filter system is shown in Figure 12.29, where the inputs to the FFT operations are seen to be the zero-padded impulse response and the zero-padded block of signal data.

The input signal sequence $x[n] = [3, 1, 2, -1, -2, 1, 1.5, 2, 0.5, 2]$ is shown in Figure 12.30.

We will perform block filtering using a four-point DFT. The four-point DFT is chosen for this example so that the calculations will remain relatively simple and so that a radix-2 FFT can be used for the computations.

Because the unit impulse response has $M = 2$ elements, we break the input sequence into blocks of $N = 3$ elements. Then the convolution has four elements and can be computed

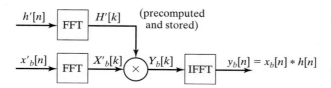

Figure 12.29 A block filtering system.

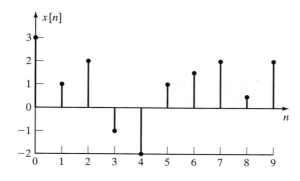

Figure 12.30 A discrete-time sequence to be block filtered.

using the four-point FFT. To compute the convolution result using the DFT, we first extend the two sequences so that each has

$$N + M - 1 = 3 + 2 - 1 = 4$$

elements. This means that the circular convolution of the extended sequences is equal to the linear convolution of the original sequences. In mathematical terms, if

$$h'[n] \overset{\mathscr{DF}}{\longleftrightarrow} H'[k] \quad \text{and} \quad x_b'[n] \overset{\mathscr{DF}}{\longleftrightarrow} X'_b[k],$$

then the output sequence for each block of input is given by

$$y_b[n] = h[n]*x_b[n] = h'[n] \circledast x_b'[n] \overset{\mathscr{DF}}{\longleftrightarrow} H'[k]X'_b[k].$$

The block filter system is implemented by first extending the unit impulse response of the filter to four elements by zero padding. We then compute the four-point DFT to find—which is stored for use with each block of the input signal.

We proceed by extending each three-element block of the input sequence as it is received to four elements by zero padding. The DFT of each block of input data is then computed. Finally, the DFT of the output sequence for each block of the input is computed as

$$Y_b[k] = H'[k]X_b'[k],$$

and the output sequence for each block of input data is

$$y_b[n] = \mathscr{IDF}\,\{Y_b[k]\}.$$

Because $y_b[k]$ is a four-point DFT, $y_b[n]$ is a four-element sequence. Therefore, as we fit the block outputs together to find the total output sequence, we must fit a series of four-element

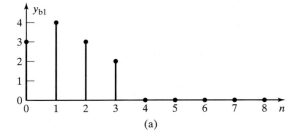

(a)

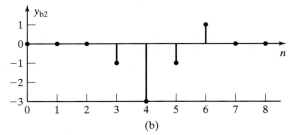

(b)

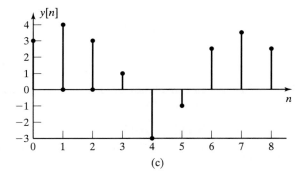

(c)

Figure 12.31 Block filter output sequences.

sequences into three-element time slots. There is an overlap of data elements that results in the fourth element of each sequence being summed with the first element of the following sequence. Figure 12.31 illustrates the process just described. Figure 12.31(a) and (b) show the block-filtered output sequences for the first two blocks. Figure 12.31(c) shows the total output signal computed from the block outputs. The sequence shown for $y[n]$ in Figure 12.31(c) is identical to the sequence that would be calculated by convolving the original input sequence $x[n]$ with the filter's unit impulse response $h[n]$.

The following MATLAB program performs the calculations described in this example:

```
% Enter the input sequence.
xn=[3 1 2 -1 -2 1 1.5 2 0.5 2]
% Enter the impulse response.
hn=[1 1]
% Select the 3-element blocks of input data.
xb1=[xn(1) xn(2) xn(3)]
```

```
xb2=[xn(4) xn(5) xn(6)]
xb3=[xn(7) xn(8) xn(9)]
% zero pad the blocks to 4 elements
xb1p=[xb1 0]
xb2p=[xb2 0]
xb3p=[xb3 0]
% zero pad hn to 4 elements
hp=[hn 0 0]
% Compute the DFT of each zero-padded block of input data.
X1=fft(xb1p);
X2=fft(xb2p);
X3=fft(xb3p);
% Compute the DFT of the zero-padded impulse response.
Hp=fft(hp)
% Multiply the H[k] and X[k]to calculate Y[k] for each block.
Y1=Hp.*X1;
Y2=Hp.*X2;
Y3=Hp.*X3;
% Compute IDFT[Y[k]] = y[n], for each block of output data.
y1=ifft(Y1);
y2=ifft(Y2);
y3=ifft(Y3);
% Shift each block of output into the proper time slot.
y1s=[y1 0 0 0 0 0];
y2s=[0 0 0 y2 0 0 0];
y3s=[0 0 0 0 0 0 y3];
% Add the shifted, zero-padded output blocks to form the output signal.
y=y1s+y2s+y3s;
% Display the first 9 elements of the output signal.
yp=y(1,1:9)
% Calculate the output signal using convolution to compare results.
yn=conv(hn,xn)
% Prepare data for plotting.
y1sp=y1s(1,1:9);
y2sp=y2s(1,1:9);
n=0:8;
stem(n,y1sp,'filled')
figure(2)
stem(n,y2sp, 'filled')
figure(3)
stem(n,yp,'filled')
```

■

We see that the DFT is useful for filtering discrete-time sequences. For cases where the input sequence is too long to allow the convolution operation to be computed directly, we have discussed the overlap-add method of block filtering. Block filtering allows us to break the input sequence into manageable-length sequences for convolution with the filter's unit impulse response.

Correlation

The *correlation* of signals is a signal-processing technique often used in estimating the frequency content of a noisy signal. Correlation is also used for detection of targets in a radar or sonar signal. Correlation is divided into two cases: *cross-correlation*, which is the correlation of two different signals, and *autocorrelation*, which is the correlation of a signal with itself.

The cross-correlation operation is described by the equation

$$R_{xy}[p] = \sum_{m=-\infty}^{\infty} x[m]y[p+m]. \tag{12.48}$$

We see that (12.48) is much like (10.16), the equation we used to define discrete-time convolution. In performing graphical correlation, we do essentially the same operations as in convolution except that $y[m]$ is not reversed before we begin shifting the sequence in time.

EXAMPLE 12.21   **Cross-correlation using a graphical technique**

We will compute the cross-correlation of the two sequences

$$x_1[n] = [1, 2, 3, 4] \quad \text{and} \quad x_2[n] = [0, 1, 2, 3]$$

using graphical techniques. Figure 12.32 illustrates the process. We can see that

$$R_{x_1x_2}[p] = 0, \; |p| > 3.$$

The nonzero values are found by shifting $x_2[p+m]$ along the scale for $-3 \leq p \leq 3$ and calculating the sum of products of concurrent values:

$$R_{x_1x_2}[0] = (3)(4) + (2)(3) + (1)(2) + (0)(1) = 20;$$
$$R_{x_1x_2}[4] = R_{x_1x_2}[-3] = (3)(0) + (2)(0) + (1)(0) + (0)(0) = 0;$$
$$R_{x_1x_2}[3] = (3)(1) + (2)(0) + (1)(0) + (0)(0) = 3;$$
$$R_{x_1x_2}[2] = (3)(2) + (2)(1) + (1)(0) + (0)(0) = 8;$$
$$R_{x_1x_2}[1] = (3)(3) + (2)(2) + (1)(1) + (0)(1) = 14;$$
$$R_{x_1x_2}[-1] = (3)(0) + (2)(4) + (1)(3) + (0)(2) = 11;$$
$$R_{x_1x_2}[-2] = (3)(0) + (2)(0) + (1)(4) + (0)(3) = 4. \quad \blacksquare$$

We see that the correlation process is much like convolution. Like circular convolution, we can use *circular correlation* of extended sequences to find the correlation of two sequences. If the two sequences each have N samples, they must be extended to $2N - 1$ samples by zero padding. The circular correlation process is explained in detail by the following example.

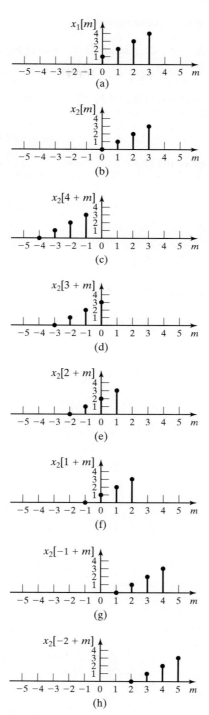

Figure 12.32 Illustration of the cross correlation process.

EXAMPLE 12.22   **Cross-correlation using circular correlation**

We will compute the cross-correlation of the two sequences

$$x_1[n] = [1, 2, 3, 4] \quad \text{and} \quad x_2[n] = [0, 1, 2, 3]$$

using the circular correlation technique. The first step is to extend the two sequences using zero padding so that each has $N_1 + N_2 - 1$ elements:

$$x_1'[n] = [1, 2, 3, 4, 0, 0, 0] \quad \text{and} \quad x_2'[n] = [0, 1, 2, 3, 0, 0, 0].$$

Figure 12.33 illustrates the process of circular correlation. The two concentric circles are divided into arcs of $2\pi/(N_1 + N_2 - 1)$ radians. The two extended discrete-time sequences are

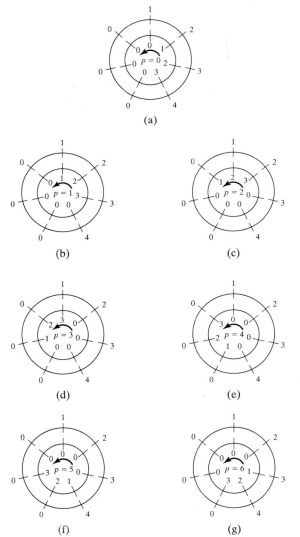

Figure 12.33   Illustration of circular correlation.

then arranged on the two concentric circles as shown in Figure 12.33(a). Both sequences are written in the clockwise direction with $x'_1[n]$ on the outer circle and $x'_2[n]$ on the inner. From Figure 12.33(a), we compute the sum of products of corresponding elements for $p = 0$. The inner circle is then rotated counterclockwise one step at a time, and the sums of products are calculated for succeeding values of p as shown in Figure 12.33(b) through (g):

$$R_{x_1x_2}[0] = (0)(1) + (1)(2) + (2)(3) + (3)(4) + (0)(0) + (0)(0) + (0)(0) = 20;$$
$$R_{x_1x_2}[1] = (1)(1) + (2)(2) + (3)(3) + (0)(4) + (0)(0) + (0)(0) + (0)(0) = 14;$$
$$R_{x_1x_2}[2] = (2)(1) + (3)(2) + (0)(3) + (0)(4) + (0)(0) + (0)(0) + (1)(0) = 8;$$
$$R_{x_1x_2}[3] = (3)(1) + (0)(2) + (0)(3) + (0)(4) + (0)(0) + (1)(0) + (2)(0) = 3;$$
$$R_{x_1x_2}[4] = (0)(1) + (0)(2) + (0)(3) + (0)(4) + (1)(0) + (2)(0) + (3)(0) = 0;$$
$$R_{x_1x_2}[5] = (0)(1) + (0)(2) + (0)(3) + (1)(4) + (2)(0) + (3)(0) + (0)(0) = 4;$$
$$R_{x_1x_2}[6] = (0)(1) + (0)(2) + (1)(3) + (2)(4) + (3)(0) + (0)(0) + (0)(0) = 11.$$

The cross-correlation sequence computed by circular correlation appears to be different from the result of linear correlation calculated in Example 12.21. However, we recognize that, like circular convolution, circular correlation results in a periodic sequence. In this case, the period of the correlation sequence is $N_1 + N_2 - 1 = 7$. Therefore, $R_{x_1x_2}[p - 7] = R_{x_1x_2}[p]$ and the sequence can be rewritten for $R_{x_1x_2}[p - 7]$ beginning with $p = 4$ to match the linear correlation results.   ∎

Equation (12.48) can be rewritten as

$$R_{xy}[p] = \sum_{n=-\infty}^{\infty} x[n]y[p - (-n)] = x[-n]*y[n]. \tag{12.49}$$

In working with discrete-time sequences, we usually use a finite number, N, of samples. In doing this, we implicitly assume that the signal is periodic with period NT, as discussed previously. If the signal is periodic, then

$$x[-n] = x[N - n], \quad 0 \leq n \leq N - 1.$$

From (12.49), we see that the algorithm that we have used for discrete convolution can be used to compute the correlation if we reorder the appropriate sequence before we begin.

We can make use of the DFT in the computation of correlation just as we used it for convolution calculations. Recall that for the DFT method of computing the convolution to be valid, the discrete-time sequences were zero padded so that each contained $2N - 1$ elements. The same zero padding is necessary to allow the DFT to be used to calculate the correlation. Taking the DFT of both sides of (12.49) after zero padding the two sequences, we have

$$\mathcal{DF}\{R_{xy}[p]\} = \mathcal{DF}\{x[-n]*y[n]\} = \mathcal{DF}\{x'[-n]\} \times \mathcal{DF}\{y'[n]\}.$$

Let $m = -n$; then

$$\mathscr{DF}\{x'[-n]\} = \sum_{k=0}^{N-1} x'[m]e^{-j2\pi k(-m)/N} = X'[-k]$$

and

$$\mathscr{DF}\{R_{xy}[p]\} = X'[-k]Y'[k].$$

If the discrete-time sequence $x[n]$ is real valued, as sequences consisting of sample values of physical signals always are, then it can be shown that

$$X'[-k] = X'*[k].$$

Therefore, the correlation function can be derived from

$$R_{xy}[p] = \mathscr{IDF}\{X'[-k]Y'[k]\}$$

$$= \mathscr{IDF}\{X'*[k]Y'[k]\}. \tag{12.50}$$

We must take note of the fact that $R_{yx} \neq R_{xy}$. Reversing the order of the signals in the correlation equation gives

$$R_{yx}[p] = \sum_{n=-\infty}^{\infty} y[n]x[p - (-n)] = x[n]*y[-n]. \tag{12.51}$$

If the calculation is by the DFT method, then

$$R_{yx}[p] = \mathscr{IDF}\{X'[k]Y'[-k]\} = \mathscr{IDF}\{X'[k]Y'*[k]\}. \tag{12.52}$$

| EXAMPLE 12.23 | **Cross-correlation using the DFT** |

The cross-correlation of the two discrete-time sequences given in Example 12.21 will be calculated using the DFT method.
Zero padding is used to extend the two sequences so that

$$x'_1[n] = [1, 2, 3, 4, 0, 0, 0] \quad \text{and} \quad x'_2[n] = [0, 1, 2, 3, 0, 0, 0].$$

We compute the DFT of each sequence using an FFT algorithm to get

$$X'_1[k] = [10, -2.0245 - j6.2240, 0.3460 + j2.4791, 0.1784 - j2.4220,$$
$$0.1784 - j2.4220, 0.3460 - j2.4791, -2.0245 + j6.2240];$$
$$X'_2[k] = [6, -2.5245 - j4.0333, -0.1540 + j2.2383, -0.3216 - j1.7950,$$
$$-0.3216 + j1.7950, -0.1540 - j2.2383, -2.5245 + j4.0333].$$

We write the conjugates of the two discrete-frequency sequences for use in implementing (12.51) and (12.52):

$$X_1'^*[k] = [10, -2.0245 + j6.2240, 0.3460 - j2.4791, 0.1784 + j2.4220,$$
$$0.1784 - j2.4220, 0.3460 + j2.4791, -2.0245 - j6.2240];$$
$$X_2'^*[k] = [6, -2.5245 + j4.0333, -0.1540 - j2.2383, -0.3216 + j1.7950,$$
$$-0.3216 - j1.7950, -0.1540 + j2.2383, -2.5245 - j4.0333].$$

To compute $R_{x_1 x_2}[p]$, we calculate $X_1'^*[k]X_2'[k]$ by multiplying the corresponding elements of the two sequences to get

$$\mathcal{DF}\{R_{x_1 x_2}[p]\} = [60, 30.2141 - j7.5469, 5.458 + j1.1562, 4.2901 - j1.0991,$$
$$4.201 + j1.0991, 5.458 - j1.1562, 30.2141 + j7.5469].$$

We use the IFFT algorithm to find the inverse discrete Fourier transform:

$$R_{x_1 x_2}[p] = [20, 14, 8, 3, 0, 4, 11].$$

We see that this is the same result found by circular convolution in Example 12.22. If we recall that $R_{x_1 x_2}[p - 7] = R_{x_1 x_2}[p]$, we see that this can be rewritten to give the same results as found by linear correlation in Example 12.21.

The results of the computations just described can be confirmed by the following MATLAB program:

```
% This MATLAB program confirms the results of
% Example 12.23.
% Type in the vectors end zero-pad them.
x1 = [1 2 3 4];
x1p = [1 2 3 4 0 0 0];
x2 = [0 1 2 3];
x2p = [0 1 2 3 0 0 0];
% Form x2(-n)
N = length(x1);
for n = 1:N
      x1r(n) = x1(N + 1 - n);
end
% Use the convolution method to compute
% the correlation.
ypc = conv(x1r,x2);
% Use the DFT method to compute the correlation.
X2 = fft(x2p,7);
X1 = fft(x1p,7);
X1c = conj(X1);
Y = X1c.*X2;
y = ifft(Y);
% Compare the results of the two solutions.
ypc
y
```

TABLE 12.6 Properties of Correlation

$$R_{xy}[p] = R_{yx}[-p]$$
$$R_{xx}[p] = R_{xx}[-p]$$
$$R_{xy}[p] = x[-n]*y[n]$$
$$R_{yx}[p] = x[n]*y[-n]$$

Next we calculate $R_{x_2 x_1}[p]$. From (12.52), we obtain

$$\mathcal{DF}\{R_{x_2 x_1}[p]\} = X'_1[k]X'_2{}^*[k];$$

$$R_{x_2 x_1}[p] = \mathcal{IDF}\{[60, 30.2141 + j7.5469, 5.458 - j1.1562, 4.2901 + j1.0991,$$
$$4.201 - j1.0991, 5.458 + j1.562, 3.2141 - j7.5469]\}$$
$$= [20, 11, 4, 0, 3, 8, 14].$$

We see that this is not the same sequence as $R_{x_1 x_2}[p]$. ∎

Some properties of correlation have been discussed earlier. We list a few of the more important properties in Table 12.6.

Energy Spectral Density Estimation

An important application of autocorrelation is in the estimation of the energy spectral density of signals. From the energy spectral density, we learn the various frequency components that make up the signal and their relative strengths. The process of autocorrelation helps to eliminate the random components of noisy signals while accentuating the valid components of the signal's frequency spectrum.

For this application, we define another function, the *sample estimate of the autocorrelation* as

$$r_N[p] = \frac{1}{N}\sum_{n=0}^{N-1} x[n]x[p+n], \quad p = 0, 1, \ldots, N-1. \tag{12.53}$$

We see that this is similar to the correlation function (12.48), except for the division by N and that the range of n is restricted to the range $0 \le n \le N - 1$.

The *energy spectral density estimate* is given by

$$S_x[k] = \mathcal{DF}\{r_N[p]\} = \frac{1}{N}X[k]X^*[k]. \tag{12.54}$$

The function defined by (12.54) is also called the *periodogram spectrum estimate*.

In practice, the sequence $x[n]$ of (12.53) is often selected from a long sequence of sampled values of a continuous-time signal. The long sequence is broken into a set of shorter sequences, $x_m[n]$, using a windowing function. This can be done in a manner similar to that discussed in the section on block filtering. Usually, however, the short sequences, $x_m[n]$, are chosen so that they overlap in time as shown in Figure 12.34. The autocorrelation of the short sequences is then feasible and the sample estimate of the autocorrelation is calculated for each $x_m[n]$ from (12.53):

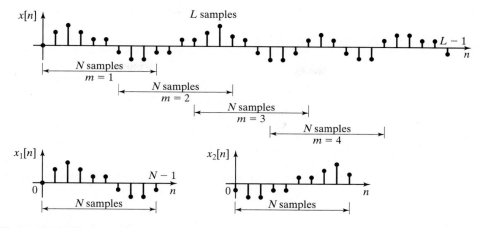

Figure 12.34   Blocks of data for a periodogram spectrum estimate.

$$r_{Nm}[p] = \frac{1}{N} \sum_{n=0}^{N-1} x_m[n] x_m[p + n], \, p = 0, 1, \ldots, N - 1.$$

An FFT algorithm is then used to compute the energy spectral density estimate (12.54) for that sample sequence

$$S_{xm}[k] = \mathcal{DF}\{r_{Nm}[p]\}.$$

In the *averaging periodogram method*, the energy spectral density sequences, $S_{xm}[k]$, for several sample sequences are averaged together to give the periodogram estimate

$$\overline{S}_x = \frac{1}{M} \sum_{m=0}^{\infty} S_{xm}[k]. \tag{12.55}$$

Summary

In this section, we have looked at applications of the DFT. We saw ways that the DFT is used for digital signal processing in addition to providing an estimate of the Fourier transform of a signal.

12.7 THE DISCRETE COSINE TRANSFORM

Closely related to the DFT is the discrete cosine transform (DCT). Instead of the exponential kernel

$$e^{-j2\pi kn/N}$$

of the DFT, the DCT uses a cosine kernel

$$\cos(2\pi kn/N).$$

In two dimensions, the DCT is the basis of the image-compression standard issued by the Joint Photographic Experts Group (JPEG). The JPEG algorithm is commonly used on the World Wide Web for compressing images. Any image that you download that has a *.jpg* extension has been compressed with JPEG. The DCT is also used in the MPEG standard for video compression and in many streaming video players.

Let $f[i, j]$ be a pixel block of size M $\times$ N. Its two-dimensional DCT, $F[u, v]$, is

$$F[u, v] = \frac{2}{\sqrt{MN}} C(u)C(v) \sum_{i=0}^{M-1} \sum_{j=0}^{N-1} f[i, j]$$

$$\times \cos((2i + 1)u\pi/2M)\cos((2j + 1)v\pi/2N), \qquad (12.56)$$

where

$$C(x) = \begin{cases} \dfrac{1}{\sqrt{2}}, & x = 0 \\ 1, & x \neq 0 \end{cases}.$$

The Inverse MXN DCT $f[i, j]$ is recovered as

$$f[i, j] = \frac{2}{\sqrt{MN}} \sum_{u=0}^{M-1} \sum_{v=0}^{N-1} C(u)C(v)F[u, v]$$

$$\times \cos((2i + 1)u\pi/2M)\cos((2j + 1)v\pi/2N), \qquad (12.57)$$

For the JPEG standard, $f[i, j]$ is a pixel block of size 8 $\times$ 8. Then, from (12.56), its two-dimensional DCT, $F[u, v]$, is

$$F[u, v] = \frac{1}{4} C(u)C(v) \sum_{i=0}^{7} \sum_{j=0}^{7} f[i, j]\cos((2i + 1)u\pi/16)$$

$$\times \cos((2j + 1)v\pi/16). \qquad (12.58)$$

The Inverse 8×8 DCT $f(i, j)$ is recovered as

$$f[i, j] = \frac{1}{4} \sum_{u=0}^{7} \sum_{v=0}^{7} C(u)C(v)F[u, v]$$
$$\times \cos((2i + 1)u\pi/16)\cos((2j + 1)v\pi/16). \quad (12.59)$$

In the JPEG standard, first the input image is level shifted. (For an eight-bit image, the value 128 is subtracted from each pixel.) Then the image is divided into 8×8 pixel blocks. The 64 DCT coefficients $F[u, v]$ are then uniformly scalar quantized. The lowest frequency DC term, $F[0, 0]$, is coded predictively based on the DC terms from previous 8×8 image blocks, and the remaining 63 DCT coefficients are coded in a zigzag pattern using lossless compression.

The DCT works well for image compression because it provides *energy compaction*. This means that the energy in the spatial or image domain is typically concentrated in a smaller number of DCT coefficients, hence, the low-magnitude DCT coefficients can be coded at very low bit rates without degrading the image quality significantly.

For more information on the JPEG standard, see *www.jpeg.org*. For comprehensive presentation of the DCT, see Ref. 7.

▨ SUMMARY

In this chapter, we have considered the frequency spectrum of discrete-time signals. First, we looked at the discrete-time Fourier transform (DTFT). We saw that the DTFT is a scaled version of the Fourier transform of a discrete-time signal. We also saw that the DTFT is a continuous function of frequency and, therefore, is not useful for digital computer calculations. Because of this limitation, a discrete-frequency approximation of the DTFT is more often used.

The discrete Fourier transform (DFT) is a discrete-frequency approximation of the DTFT. It consists of computed samples of the DTFT. The DFT is computed from a sequence of N discrete-time values and results in N discrete frequency values. The discrete-frequency values of the DFT are separated in frequency by $2\pi/NT$ (rad/s), where T is the sampling period of the discrete-time sequence.

There exist several well-known efficient algorithms for computing the DFT on a digital computer. These are collectively known as the fast Fourier transform (FFT).

A few practical applications of the DFT for signal processing and analysis were considered, including approximation of the Fourier transform, convolution, filtering, and spectrum estimation.

In the final section of the chapter, the discrete cosine transform was introduced. See Table 12.7.

TABLE 12.7 Key Equations of Chapter 12

Equation Title	Equation Number	Equation
Discrete-time Fourier transform	(12.1)	$X(\Omega) = \mathcal{DTF}(x[n]) = \displaystyle\sum_{n=-\infty}^{\infty} x[n]e^{-jn\Omega}$
Inverse discrete-time Fourier transform	(12.2)	$x[n] = \mathcal{DTF}^{-1}[X(\Omega)] = \dfrac{1}{2\pi}\displaystyle\int_{\Omega_1}^{\Omega_1+2\pi} X(\Omega)e^{jn\Omega}d\Omega = \dfrac{1}{2\pi}\displaystyle\int_{2\pi} X(\Omega)e^{jn\Omega}d\Omega$
Relation of Fourier transform of sampled signal to DTFT	(12.6)	$\mathcal{F}[x_s(t)]_{\omega T=\Omega} = \mathcal{F}\left[\displaystyle\sum_{n=-\infty}^{\infty} x(nT)\delta(t-nT)\right]_{\omega T=\Omega} = \mathcal{DTF}[x(nT) = x[n]]$
Periodicity of DTFT	(12.7)	$X(\Omega + 2\pi) = X(\Omega)$
One period of a periodic signal ($x_0[n]$)	(12.17)	$x_0[n] = \begin{cases} x[n], & 0 \le n \le N-1 \\ 0, & \text{otherwise} \end{cases}$
Periodic signal in terms of $x_0[n]$	(12.19)	$x[n] = x_0[n] * \displaystyle\sum_{k=-\infty}^{\infty} \delta[n-kN]$
DTFT of DT impulse train	(12.20)	$\displaystyle\sum_{k=-\infty}^{\infty} \delta[n-kN] \overset{\mathcal{DTF}}{\longleftrightarrow} \dfrac{2\pi}{N}\displaystyle\sum_{k=-\infty}^{\infty}\delta\left(\Omega - \dfrac{2\pi k}{N}\right)$
DTFT of periodic signal	(12.23)	$X(\Omega) = \dfrac{2\pi}{N}\displaystyle\sum_{k=-\infty}^{\infty} X_0\left(\dfrac{2\pi k}{N}\right)\delta\left(\Omega - \dfrac{2\pi k}{N}\right)$
DFT and IDFT	(12.30)	$X[k] = \mathcal{DF}[x[n]] = \displaystyle\sum_{n=0}^{N-1} x[n]e^{-j2\pi kn/N}, \quad k = 0,1,2,\ldots,N-1$ $x[n] = \mathcal{DF}^{-1}[X[k]] = \dfrac{1}{N}\displaystyle\sum_{k=0}^{N-1} X[k]e^{j2\pi kn/N}, \quad n = 0,1,2,\ldots,N-1$
DFT and IDFT with shorthand notation	(12.33)	$X[k] = \mathcal{DF}[x[n]] = \displaystyle\sum_{n=0}^{N-1} x[n]W_N^{kn}, \quad k = 0,1,\ldots,N-1$ $x[n] = \mathcal{DF}^{-1}[X[k]] = \dfrac{1}{N}\displaystyle\sum_{k=0}^{N-1} X[k]W_N^{-kn}, \quad n = 0,1,\ldots,N-1$
Orthogonality of DT exponentials	(12.35)	$\displaystyle\sum_{n=0}^{N-1} W_N^{0n} = N \text{ and } \displaystyle\sum_{n=0}^{N-1} W_N^{kn} = 0, \quad k = 1,2,\ldots,N-1$
Circular convolution	(12.46)	$y[n] = x[n]\circledast h[n] = \displaystyle\sum_{l=0}^{N-1} x[l]h[n-l]$
DFT of circular convolution	(12.47)	$x[n]\circledast h[n] \overset{\mathcal{DF}}{\longleftrightarrow} X[k]H[k]$

REFERENCES

1. H. Nyquist, "Certain Topics in Telegraph Transmission Theory," *Transactions of AIEE*, vol. 47, April 1928.
2. C. E. Shannon, "Communication in the Presence of Noise," *Proceedings of the IRE*, vol. 37, January 1949.
3. G. E. Carlson, *Signal and Linear System Analysis*, 2nd ed. New York: John S. Wiley & Sons, 1998.
4. S. S. Soliman and M. D. Srinath, *Continuous and Discrete Signals and Systems*, 2nd ed. Upper Saddle River, NJ: Prentice Hall, 1997.
5. L. B. Jackson, *Signals, Systems and Transforms*, Reading, MA: Addison–Wesley, 1991.
6. R. D. Strum and D. E. Kirk, *First Principles of Discrete Systems and Digital Signal Processing*. Reading, MA: Addison–Wesley, 1988.
7. K. Sayood, *Introduction to Data Compression*, 2nd ed. San Francisco: Morgan Kaufmann Publishers, 2000.

PROBLEMS

12.1. (a) Use Discrete-Time Fourier Transform Tables 12.1 and 12.2 to find the frequency spectra of the signals listed subsequently.

    (b) Plot the magnitude and phase frequency spectra of each of the signal listed over the frequency range $|\omega| \leqq 2\omega_s$:

        (i) $f(t) = 8\cos(2\pi t) + 4\sin(4\pi t)$ sampled with $T_s = 0.1$ s.

        (ii) $g[n] = 4\cos[0.5\pi n]u[n]$.

12.2. Find the discrete-time Fourier transform (DTFT) of each signal shown in Figure P12.2.

12.3. Prove the linearity property of the DFT.

12.4. Prove that $\mathcal{DFT}\{nx[n]\} = j\dfrac{dX(\Omega)}{d\Omega}$.

12.5. Given a filter with impulse response

$$h[n] = \delta[n] + 2\delta[n-1] + \delta[n-2],$$

    (a) Find $H(\Omega)$, the DTFT of $h[n]$.

    (b) Find the phase of $H(\Omega)$ and simplify as much as possible. (Because $h[n]$ is a symmetric, finite impulse response filter, $H(\Omega)$ has *linear phase*.)

12.6. A discrete-time function $x[n] = a^n u[n]$, $|a| < 1$ is convolved with a second discrete-time signal $h[n]$ to form an output signal

$$y[n] = \frac{a}{a-b}a^n u[n] + \frac{b}{b-a}b^n u[n],$$

where $|b| < 1$. Find $h[n]$.

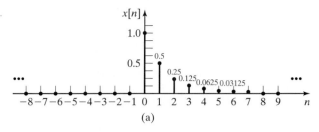

(a)

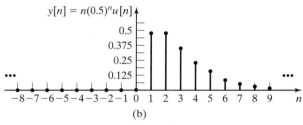

(b)

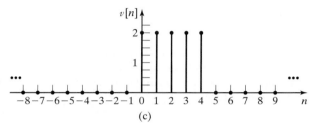

(c)

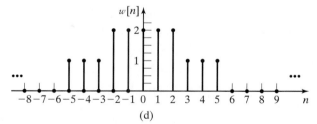

(d)

Figure P12.2

12.7. Given $x_0[n] = \delta[n] + \delta[n-2] + \delta[n-4]$, find $X_0(\Omega)$ and $X(\Omega)$ (the DTFT of the periodic version of $x_0[n]$). Assume $N = 5$.

12.8. Given a discrete-time function $y[n] = x[\frac{n}{3}]$, where $x[n]$ has DTFT $X(\Omega)$, find $Y(\Omega)$, the DTFT of $y[n]$, in terms of $X(\Omega)$.

12.9. Consider a discrete-time periodic function $x[n]$ with DTFT

$$X(\Omega) = \frac{2\pi}{4} \sum_{-\infty}^{\infty} X_0\left(\frac{2\pi k}{4}\right)\delta\left(\Omega - \frac{2\pi k}{4}\right).$$

The values of $X_0(\frac{2\pi k}{4})$ are

$$X_0\left(\frac{2\pi k}{4}\right) = \begin{cases} 4, & k = 0 \\ 0, & k = 1 \\ 4, & k = 2 \\ 0, & k = 3. \end{cases}$$

Find $x_0[n]$, where $x_0[n]$ is one period of $x[n]$, that is,

$$x_0[n] = \begin{cases} x[n], & 0 \le n \le N - 1 \\ 0, & \text{otherwise.} \end{cases}$$

12.10. We wish to design a finite impulse response (FIR) filter $h[n]$. We have the following constraints on the DTFT $H(\Omega)$ of the filter:

$$H(\Omega) = \begin{cases} 0, & \Omega = 0 \\ 1, & \Omega = \dfrac{\pi}{2} \\ 0, & \Omega = \pi \\ 1, & \Omega = \dfrac{-\pi}{2} \end{cases}.$$

Find an $h[n]$ that satisfies these constraints. This is known as *frequency sampling*.

12.11. **(a)** Compute the eight-point DFT of the sequence shown in Figure P12.2(a)
(b) Use MATLAB to confirm the results of Part (a).

12.12. **(a)** Compute the eight-point DFT of the sequence shown in Figure P12.2(b)
(b) Use MATLAB to confirm the results of Part (a).

12.13. The signal $x(t) = \text{rect}[(t - 2)/4]$ is shown in Figure P12.13.

(a) Compute the four-point DFT of the signal when it is sampled with $T_s = 2$ ms. Plot the magnitude and phase spectra.

(b) Use MATLAB to compute the eight-point DFT of the signal when it is sampled with $T_s = 1$ ms. Plot the magnitude and phase spectra.

(c) Use MATLAB to compute the 16-point DFT of the signal when it is sampled with $T_s = 0.5$ ms. Plot the magnitude and phase spectra.

(d) Compare the results of Parts (a), (b), and (c). Comment on their relationship.

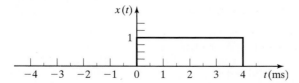

Figure P12.13

12.14. Find and sketch $X[k]$, the Discrete Fourier Transform (DFT) of

$$x[n] = e^{\frac{j6\pi n}{8}},$$

where $n = 0, 1, \ldots, 7$.

12.15. The signal $x(t) = 5\cos(8\pi t)$ is sampled eight times starting at $t = 0$ with $T = 0.1$ s.

(a) Compute the DFT of this sequence.
(b) Use MATLAB to confirm the results of Part (a).
(c) Determine the Fourier transform of $x(t)$ and compare it with the results of Part (a) and (b). Explain the differences.

12.16. Repeat Problem 12.15(a) and (b) after multiplying the sequence $x[n]$ by an eight-point Hanning window. Discuss the differences between this DFT and that found in Problem 12.15.

12.17. Find

$$A[k] = \sum_{n=0}^{N-1} \cos\left[\frac{2\pi kn}{N}\right] \cos\left[\frac{2\pi pn}{N}\right],$$

for $k = 0, 1, \ldots, N - 1$ and $p \in [0, N - 1]$.

12.18. Given that the DFT of the eight-point sequence $[1, .5, 0, 0, 0, 0, 0, 0]$ is $Y[k]$, express the DFT of the finite-length sequence $x[n] = [.5, 1, .5, 1, .5, 1, .5, 1]$ in terms of $Y[k]$.

12.19. Given the two four-point sequences

$$x[n] = [1, .75, .5, .25]$$

and

$$y[n] = [.75, .5, .25, 1],$$

express the DFT $Y[k]$ in terms of the DFT $X[k]$.

12.20. The DFT of the analog signal $f(t) = 7\cos(100\,t)\cos(40\,t)$ is to be computed.

(a) What is the minimum sampling frequency to avoid aliasing?
(b) If a sampling frequency of $\omega_s = 300$ rad/s is used, how many samples must be taken to give a frequency resolution of 1 rad/s?

12.21. An analog signal is sampled at 1024 equally spaced times, and its DFT is computed.

(a) What is the separation in rad/s between successive frequency components?
(b) What is the highest frequency that can be allowed in the analog signal if aliasing is to be prevented?

12.22. Given the two four-point sequences

$$x[n] = [-2, -1, 0, 2] \quad \text{and} \quad y[n] = [-1, -2, -1, -3],$$

find the following:

(a) $x[n] * y[n]$, the linear convolution;

(b) $x[n] \circledast y[n]$, the circular convolution;

(c) $R_{xy}[p]$, the cross-correlation of $x[n]$ and $y[n]$;

(d) $R_{yx}[p]$, the cross-correlation of $y[n]$ and $x[n]$;

(e) $R_{xx}[p]$, the autocorrelation of $x[n]$;

(f) Use MATLAB to confirm the results of Parts (a) through (e).

12.23. Extend the sequences given in Problem 12.22 by zero padding, and perform circular convolution on the extended sequences so that the result equals that obtained by linear convolution of the original sequences.

12.24. The four-point DFTs of $x[n]$ and $h[n]$ are

$$X[k] = [12, -2 - j2, 0, -2 + j2]$$

and

$$H[k] = [2.3, 0.51 - j0.81, 0.68, 0.51 + j0.81].$$

Find the value of $y[2]$, where $y[n] = x[n] \circledast h[n]$.

12.25. The four-point DFTs of two discrete-time signals are

$$X[k] = [22, -4 + j2, -6, -4 - j2]$$

and

$$Y[k] = [8, -2, -j2, 0, -2 + j2].$$

(a) If $v[n] = x[n] * y[n]$, find $v[2]$.

(b) If $w[n] = x[n] \circledast y[n]$, find $w[2]$.

(c) Find $R_{xy}[2]$.

(d) Find $R_{yx}[2]$.

(e) Find $R_{xx}[2]$.

(f) Calculate the periodogram spectral estimate of the signal $x[n]$.

12.26. (a) Draw the four-point FFT signal-flow diagram, and use it to solve for the DFT of the sequence shown in Figure P12.26.

(b) Use MATLAB to confirm the results of Part (a).

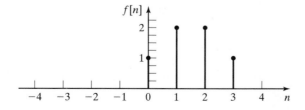

Figure P12.26

12.27. (a) Draw an eight-point DIT FFT signal-flow diagram, and use it to solve for the DFT of the sequence shown in Figure P12.2(a).

   (b) Use MATLAB to confirm the results of Part (a).

12.28. (a) Draw an eight-point DIF FFT signal-flow diagram and use it to solve for the DFT of the sequence shown in Figure P12.2(a).

   (b) Use MATLAB to confirm the results of Part (a).

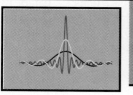

13 STATE VARIABLES FOR DISCRETE-TIME SYSTEMS

$\mathbf{W}$e introduced state-variable models for continuous-time systems in Chapter 8. State-variable models for discrete-time systems are introduced in this chapter. Although the development here closely parallels that of Chapter 8, that material is not a prerequisite for this chapter.

In Chapter 10, the modeling of discrete-time linear time-invariant (LTI) systems by linear difference equations with constant coefficients was discussed. We considered the representation of these systems by block diagrams using the form I and form II simulation diagrams. The form I representation required $2N$ delays for an Nth-order system, while the form II representation required N delays. In this chapter, we expand this representation considerably. This will lead us to a system model that is set of N first-order coupled difference equations to represent an Nth-order system. These models are called *state-variable models*, or simply *state models*.

State-variable models have the following advantages:

1. An internal structure of the system is given, in addition to the input–output model. Thus a state-variable model is more general than an input–output model, such as a transfer function.

2. The modeling of nonlinear time-varying systems using state variables is a relatively simple extension of state-variable modeling of LTI systems.

3. State-variable models are required for certain analysis and design procedures. These models have wide applications in the design of computer-controlled continuous-time systems [1, 2].

4. State-variable theory leads us to many different implementations of a given digital filter. For a given filter, certain implementations may have advantages over other implementations, in terms of numerical accuracy, random-noise generation, and so on.

Analysis and design via state-variable models require the use of matrix mathematics. The required mathematics are reviewed in Appendix G; the terms relating to matrices used in this chapter are defined in that appendix.

13.1 STATE-VARIABLE MODELING

In Chapters 9 and 10, we introduced the modeling of discrete-time systems by difference equations. If the discrete-time systems are linear and time invariant (LTI), we can represent these systems by transfer functions. Let the notation $\mathcal{Z}[\,\cdot\,]$ denote the z-transform. For an LTI system with an input of $U(z) = \mathcal{Z}[u[n]]$ and output of $Y(z) = \mathcal{Z}[y[n]]$, we can write, from Chapter 10,

$$Y(z) = H(z)U(z), \tag{13.1}$$

where $H(z)$ is the system transfer function. This discrete-time system can be represented by either of the two block diagrams of Figure 13.1. The impulse response $h[n]$ and the transfer function $H(z)$ are related by $H(z) = \mathcal{Z}[h[n]]$. The variable $u[n]$ is used to denote the input in (13.1), since we use the variable $x[n]$ to denote state variables. The use of $u[n]$ as the symbol for the general input function can lead to confusion, because that symbol is also used for the discrete-time unit step function. However, this is the notation commonly used for the input signal in state-variable models.

We now introduce state variables using an example; then a general development will be given. Consider a system modeled by the second-order linear difference equation with constant coefficients:

$$y[n + 2] - 0.7y[n + 1] + 0.9y[n] = 2u[n]. \tag{13.2}$$

In this equation, $u[n]$ is the input and $y[n]$ is the output. We commonly write difference equations for state models in terms of advances rather than delays. Of course, (13.2) can also be written as

$$y[n] - 0.7y[n - 1] + 0.9y[n - 2] = 2u[n - 2] \tag{13.3}$$

by replacing n with $(n - 2)$ in (13.2).

Ignoring the initial conditions, from Table 11.4, we find that the z-transform of (13.2) yields

$$(z^2 - 0.7z + 0.9)Y(z) = 2U(z).$$

Hence, the transfer function is given by

$$H(z) = \frac{Y(z)}{U(z)} = \frac{2}{z^2 - 0.7z + 0.9}. \tag{13.4}$$

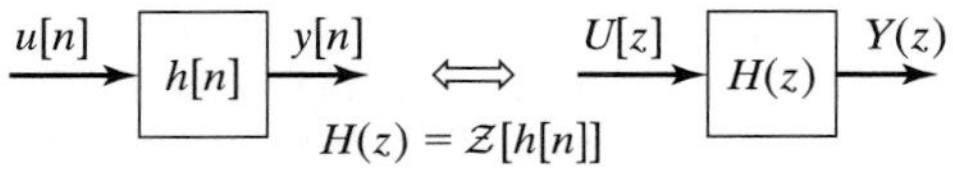

$$H(z) = \mathcal{Z}[h[n]]$$

Figure 13.1 LTI system.

Equations (13.2) and (13.4) give a difference-equation model and the transfer-function model, respectively, of the system. We will now derive a third model.

For (13.2), we define two state variables

$$x_1[n] = y[n]$$

and   (13.5)

$$x_2[n] = y[n + 1] = x_1[n + 1].$$

We now write (13.2) as

$$y[n + 2] = 0.7y[n + 1] - 0.9y[n] + 2u[n].$$

Using (13.5), we express this equation as

$$\begin{aligned} x_2[n + 1] = y[n + 2] &= 0.7y[n + 1] - 0.9y[n] + 2u[n] \\ &= 0.7x_2[n] - 0.9x_1[n] + 2u[n]. \end{aligned}$$ (13.6)

Hence, we have the state equations from (13.5) and (13.6):

$$\begin{aligned} x_1[n + 1] &= x_2[n]; \\ x_2[n + 1] &= -0.9x_1[n] + 0.7x_2[n] + 2u[n]; \\ y[n] &= x_1[n]. \end{aligned}$$ (13.7)

We can derive the difference equation (13.2) from (13.7) by reversing the steps above. Hence, (13.7) is a model of the system that has the same input–output characteristics as (13.2).

The first two equations in (13.7) are coupled first-order difference equations, and the third equation relates the state variables to the output variable. The equations are normally written in vector-matrix form:

$$\begin{bmatrix} x_1[n + 1] \\ x_2[n + 1] \end{bmatrix} = \begin{bmatrix} 0 & 1 \\ -0.9 & 0.7 \end{bmatrix} \begin{bmatrix} x_1[n] \\ x_2[n] \end{bmatrix} + \begin{bmatrix} 0 \\ 2 \end{bmatrix} u[n];$$

$$y[n] = \begin{bmatrix} 1 & 0 \end{bmatrix} \begin{bmatrix} x_1[n] \\ x_2[n] \end{bmatrix}.$$ (13.8)

These equations form a complete set of state equations for the system of (13.2). As we discuss in Section 13.6, this set is not unique; that is, we can define other variables to be the states of the system.

The standard form of the state equations of a discrete-time LTI system is given by

$$\mathbf{x}[n + 1] = \mathbf{A}\mathbf{x}[n] + \mathbf{B}u[n]$$

and   (13.9)

$$\mathbf{y}[n] = \mathbf{C}\mathbf{x}[n] + \mathbf{D}u[n],$$

where boldface type denotes vectors and matrices. In these equations,

$\mathbf{x}[n] = (N \times 1)$ state vector for an Nth-order system;

$\mathbf{u}[n] = (r \times 1)$ vector composed of the system input signals;

$\mathbf{y}[n] = (p \times 1)$ vector composed of the defined output signals;

$\mathbf{A} = (N \times N)$ system matrix;

$\mathbf{B} = (N \times r)$ input matrix;

$\mathbf{C} = (p \times N)$ output matrix;

$\mathbf{D} = (p \times r)$ matrix that represents the direct coupling between the input and the output.

Expanding the vectors in (13.9) yields

$$\mathbf{x}[n + 1] = \begin{bmatrix} x_1[n+1] \\ x_2[n+1] \\ \vdots \\ x_N[n+1] \end{bmatrix}, \qquad \mathbf{x}[n] = \begin{bmatrix} x_1[n] \\ x_2[n] \\ \vdots \\ x_N[n] \end{bmatrix},$$

$$\mathbf{u}[n] = \begin{bmatrix} u_1[n] \\ u_2[n] \\ \vdots \\ u_r[n] \end{bmatrix}, \quad \text{and} \quad \mathbf{y}[n] = \begin{bmatrix} y_1[n] \\ y_2[n] \\ \vdots \\ y_p[n] \end{bmatrix}.$$

(13.10)

As stated earlier, it is standard notation to denote the input functions as $u_i[n]$. We illustrate the ith state equation in (13.9):

$$x_i[n + 1] = a_{i1}x_1[n] + a_{i2}x_2[n] + \cdots + a_{iN}x_N[n]$$
$$+ b_{i1}u_1[n] + \cdots + b_{ir}u_r[n].$$

(13.11)

The ith output equation is

$$y_i[n] = c_{i1}x_1[n] + c_{i2}x_2[n] + \cdots + c_{iN}x_N[n]$$
$$+ d_{i1}u_1[n] + \cdots + d_{ir}u_r[n].$$

(13.12)

We now define the *state* of a system:

The state of a system at any time n_0 is the information that together with all inputs for $n \geq n_0$, determines the behavior of the system for $n \geq n_0$.

It will be shown that the state vector $\mathbf{x}[n]$ of the standard form of the state-variable equations, (13.9), satisfies this definition.

We refer to the two matrix equations of (13.9) as the *state equations* of the system. The first equation, a difference equation, is called the *state equation*, and the second one, an algebraic equation, is called the *output equation*. The state equation is a first-order matrix difference equation, and the state vector $\mathbf{x}[n]$ is its solution. Given knowledge of $\mathbf{x}[n]$ and the input vector $\mathbf{u}[n]$, the output equation, which is algebraic, yields the output $\mathbf{y}[n]$.

In the state equation in (13.9), the only variables that may appear on the left side of the equation are $x_i[n + 1]$, and the only variables that may appear on the right side are $x_i[n]$ and $u_i[n]$. Only $y_i[n]$, $x_i[n]$, and $u_i[n]$ may appear in the output equation (no $x_i[n + 1]$ or $u_i[n + 1]$). Valid equations that model an LTI system can be written without following these rules; however, those equations will not be in the standard form.

The standard form of the state equations, (13.9), allows for more than one input and more than one output. Systems with more than one input or more than one output are called *multivariable systems*. For a single-input system, the matrix $\mathbf{B}$ is an $(N \times 1)$ column matrix and the input is the scalar $u[n]$. For a single-output system, the matrix $\mathbf{C}$ is a $(1 \times N)$ row matrix and the output is the scalar $y[n]$. An example is now given to illustrate a multivariable system.

EXAMPLE 13.1	**State variables for a third-order discrete system**

Consider the system described by the coupled difference equations

$$y_1[n + 2] + 2y_1[n] + 3y_2[n] = u_1[n] + 9u_2[n]$$

and

$$y_2[n + 1] + 4y_2[n] - 6y_1[n + 1] = 5u_1[n],$$

where $u_1[n]$ and $u_2[n]$ are the input signals and $y_1[n]$ and $y_2[n]$ are the output signals. We define the states as the outputs, and, where necessary, the advanced outputs. Thus,

$$x_1[n] = y_1[n]; \quad x_2[n] = y_1[n + 1] = x_1[n + 1]; \quad x_3[n] = y_2[n].$$

From the system difference equations, we write

$$
\begin{aligned}
y_1[n + 2] = x_2[n + 1] &= -2y_1[n] - 3y_2[n] + u_1[n] + 9u_2[n] \\
&= -2x_1[n] - 3x_3[n] + u_1[n] + 9u_2[n]; \\
y_2[n + 1] = x_3[n + 1] &= -4y_2[n] + 6y_1[n + 1] + 5u_1[n] \\
&= 6x_2[n] - 4x_3[n] + 5u_1[n].
\end{aligned}
$$

We rewrite the state equations in the following order:

$$
\begin{aligned}
x_1[n + 1] &= x_2[n]; \\
x_2[n + 1] &= -2x_1[n] - 3x_3[n] + u_1[n] + 9u_2[n]; \\
x_3[n + 1] &= 6x_2[n] - 4x_3[n] + 5u_1[n].
\end{aligned}
$$

The output equations are

$$y_1[n] = x_1[n]$$

and

$$y_2[n] = x_3[n].$$

These equations may be written in vector-matrix form as

$$\mathbf{x}[n + 1] = \begin{bmatrix} 0 & 1 & 0 \\ -2 & 0 & -3 \\ 0 & 6 & -4 \end{bmatrix} \mathbf{x}[n] + \begin{bmatrix} 0 & 0 \\ 1 & 9 \\ 5 & 0 \end{bmatrix} \mathbf{u}[n]$$

and

$$\mathbf{y}[n] = \begin{bmatrix} 1 & 0 & 0 \\ 0 & 0 & 1 \end{bmatrix} \mathbf{x}[n].$$

Thus, we have derived a set of state equations for the system given.   ■

We have introduced in this section the standard form of the state model of an LTI discrete-time system. We illustrated the state model with two systems described by difference equations. Next we develop a procedure for obtaining the state model directly from the system transfer function $H(z)$.

13.2 SIMULATION DIAGRAMS

In Section 13.1, we presented two examples of finding the state model of a system directly from the system difference equations. The procedure in those examples is useful. In this section, we extend that procedure, resulting in a state model that is obtained directly from the transfer function of a system.

The procedure is based on the form II representation, or simulation diagram, of a system, as given in Section 10.6. The simulation diagram, repeated in Figure 13.2, represents an Nth-order system with the transfer function

$$H(z) = \frac{Y(z)}{U(z)} = \frac{b_0 z^N + b_1 z^{N-1} + \cdots + b_{N-1} z + b_N}{z^N + a_1 z^{N-1} + \cdots + a_{N-1} z + a_N}, \qquad (13.13)$$

where $U(z)$ is the (transform of the) input and $Y(z)$ is the output. Note that the denominator coefficient a_0 has been normalized to unity. If this coefficient is not unity for a given transfer function, we divide both numerator and denominator by a_0 to obtain (13.13). Hence, (13.13) is general.

Equation (13.13) can be expressed as

$$(z^N + a_1 z^{N-1} + \cdots + a_{N-1} z + a_N) Y(z)$$
$$= (b_0 z^N + b_1 z^{N-1} + \cdots + b_{N-1} z + b_N) U(z). \qquad (13.14)$$

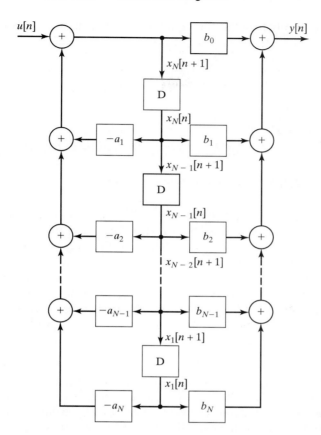

Figure 13.2 Direct form II for an Nth-order system.

Using either the procedures of Chapter 10 or the inverse z-transform, we write the system difference equation as

$$y[n + N] + a_1 y[n + N - 1] + \cdots + a_{N-1} y[n + 1] + a_N y[n]$$
$$= b_0 u[n + N] + b_1 u[n + N - 1] + \cdots + b_{N-1} u[n + 1] + b_N u[n].$$

This equation can be written in the compact notation

$$\sum_{k=0}^{N} a_k y[n + N - k] = \sum_{k=0}^{N} b_k u[n + N - k], \qquad (13.15)$$

with $a_0 = 1$. Thus, the system modeled by the transfer function (13.13) has the difference equation of (13.15). Recall that initial conditions are ignored in deriving the transfer function (13.13); hence, the initial conditions must be ignored when finding its inverse z-transform.

We now give a procedure for writing the state model for the system of (13.13) and Figure 13.2. First, we label the output of each delay in Figure 13.2 as a state variable, as shown in Figure 13.2. If the output of a delay is $x_i[n]$, its input must be $x_i[n + 1]$. Next, the signals $x_i[n + 1]$ are added to the simulation diagram, as shown in Figure 13.2. We then write the equations for the input signals to the delays of Figure 13.2:

$$x_1[n + 1] = x_2[n];$$
$$x_2[n + 1] = x_3[n];$$
$$\vdots \qquad\qquad\qquad (13.16)$$
$$x_{N-1}[n + 1] = x_N[n];$$
$$x_N[n + 1] = -a_N x_1[n] - a_{N-1} x_2[n] - \cdots - a_2 x_{N-1}[n] - a_1 x_N[n] + u[n].$$

From Figure 13.2, the equation for the output signal is

$$y[n] = (b_N - a_N b_0)x_1[n] + (b_{N-1} - a_{N-1}b_0)x_2[n]$$
$$+ \cdots + (b_1 - a_1 b_0)x_N[n] + b_0 u[n]. \qquad (13.17)$$

We now write (13.16) and (13.17) as matrix equations:

$$\mathbf{x}[n + 1] = \begin{bmatrix} 0 & 1 & 0 & \cdots & 0 & 0 \\ 0 & 0 & 1 & \cdots & 0 & 0 \\ \vdots & \vdots & \vdots & & \vdots & \vdots \\ 0 & 0 & 0 & \cdots & 0 & 1 \\ -a_N & -a_{N-1} & -a_{N-2} & \cdots & -a_2 & -a_1 \end{bmatrix} \mathbf{x}[n] + \begin{bmatrix} 0 \\ 0 \\ \vdots \\ 0 \\ 1 \end{bmatrix} u[n]; \quad (13.18)$$

$$\mathbf{y}[n] = [(b_N - a_N b_0)(b_{N-1} - a_{N-1}b_0) \cdots (b_1 - a_1 b_0)]\mathbf{x}[n] + b_0 u[n].$$

Note that the state equations can be written directly from the transfer function (13.13) or from the difference equation (13.15), since the coefficients a_i and b_i are given in these two equations. The intermediate step of drawing the simulation diagram is not necessary. An example is now given.

EXAMPLE 13.2   **State equations from a transfer function**

Suppose that we have a system with the transfer function

$$H(z) = \frac{2z^2 + 3z + 1.5}{z^2 - 1.1z + 0.8} = \frac{b_0 z^2 + b_1 z + b_2}{z^2 + a_1 z + a_2}.$$

We write the state equations directly from (13.18):

$$\mathbf{x}[n + 1] = \begin{bmatrix} 0 & 1 \\ -0.8 & 1.1 \end{bmatrix} \mathbf{x}[n] + \begin{bmatrix} 0 \\ 1 \end{bmatrix} u[n];$$

$$y[n] = [(1.5 - 0.8 \times 2)(3 + 1.1 \times 2)]\mathbf{x}[n] + 2u[n]$$
$$= [-0.1 \quad 5.2]x[n] + 2u[n]. \qquad (13.19)$$

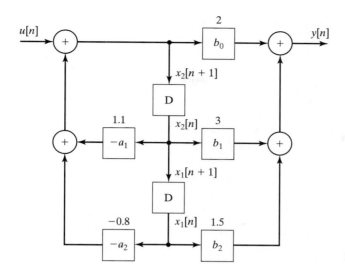

Figure 13.3 Second-order system.

The form II simulation diagram is given in Figure 13.3, with the numerical parameters and the state variables as shown. The state model (13.19) is verified directly from this diagram. ■

The procedure for writing (13.19) from Figure 13.2 can be used to write the state equations, given any form of a simulation diagram for that system; that is, the form II realization is not required. In this procedure, the delay outputs are chosen as the states; hence, the delay inputs are the states advanced by one discrete-time increment. The equations are then written for the delay inputs $x_i[n + 1]$ and the system output $y[n]$ as functions of the system input $u[n]$, the delay outputs $x_i[n]$, and the system parameters. We now discuss this procedure further.

In a simulation diagram, a signal is altered when transmitted through a delay, and hence, its designation is changed. We denote $x_i[n + 1]$ as the input to a delay, and $x_i[n]$ is then the delay output. The state-equation procedure is simplified by omitting the delays from the diagram; the equations are written directly from the remaining diagram. This omission is especially useful in complex simulation diagrams. The system of Figure 13.3, with the delays omitted, is shown in Figure 13.4. Note that all signals are shown in this figure. In addition, the effects of the delays are included, by the designations $x_i[n + 1]$ and $x_i[n]$.

The final step of the procedure is to write the equations for the delay inputs and the system outputs on the diagram as functions of the system inputs, the delay outputs, and the system parameters. The results for Figure 13.4 are given by

$$x_1[n + 1] = x_2[n];$$
$$x_2[n + 1] = -0.8x_1[n] + 1.1x_2[n] + u[n];$$

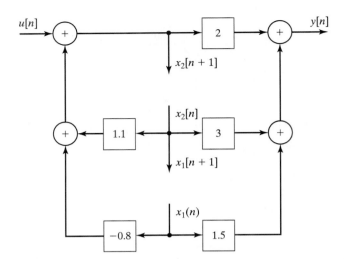

Figure 13.4 Second-order system.

$$y[n] = (1.5 - 1.6)x_1[n] + (3 + 2.2)x_2[n] + 2u[n]$$
$$= -0.1x_1[n] + 5.2x_2[n] + 2u[n].$$

These equations check those obtained in Example 13.2 using (13.18). A second example of this procedure will now be given.

EXAMPLE 13.3   **State equations from a simulation diagram**

We now write the state equations for the simulation diagram of Figure 13.5(a). Note that this diagram is not one of the two standard forms developed in Chapter 10. This system has two delays, and the system is second order. First, we redraw the simulation diagram with the state variables shown and the delays removed. The result is given in Figure 13.5(b). From this diagram, we write the state equations in the matrix format:

$$\mathbf{x}[n+1] = \begin{bmatrix} 0.9 & 0.7 \\ 0 & 0.96 \end{bmatrix} \mathbf{x}[n] + \begin{bmatrix} 1 \\ 2 \end{bmatrix} u[n];$$

$$y[n] = [1.5 \quad 2.5]\mathbf{x}[n]. \qquad \blacksquare$$

In Section 13.1, a procedure was given for writing state equations from difference equations. In this section, a procedure is developed for writing state equations directly from a transfer function. This procedure is then extended to writing state equations directly from a simulation diagram.

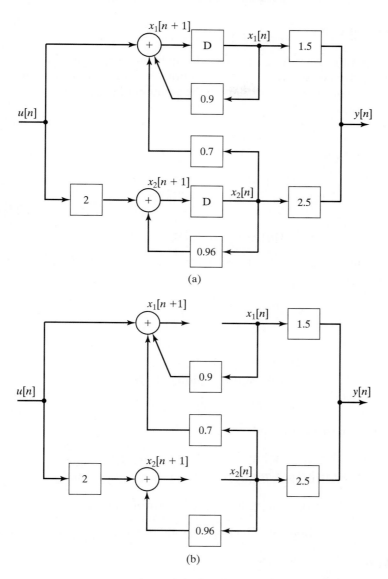

Figure 13.5 System for Example 13.3.

A model of a discrete-time physical system can be specified by

1.   difference equations,
2.   a transfer function,
3.   a simulation diagram, and
4.   state equations.

We have illustrated all four models in this section and Section 13.1. In Section 13.5, we give a procedure for obtaining the transfer function from the state equations. However, first we consider the solution of state equations.

13.3   SOLUTION OF STATE EQUATIONS

We have developed procedures for writing the state equations for a system, given the system difference equations, a transfer function, or a simulation diagram. In this section, we present two methods for finding the solution of state equations.

Recursive Solution

Consider the state equations

$$\mathbf{x}[n + 1] = \mathbf{Ax}[n] + \mathbf{Bu}[n]$$

and   (13.20)

$$\mathbf{y}[n] = \mathbf{Cx}[n] + \mathbf{Du}[n].$$

We assume that the initial state vector $\mathbf{x}[0]$ is known and that the input vector $\mathbf{u}[n]$ is known for $n \geq 0$. In a recursive manner, we write, for $n = 1$,

$$\mathbf{x}[1] = \mathbf{Ax}[0] + \mathbf{Bu}[0]$$

and for $n = 2$,

$$\begin{aligned} \mathbf{x}[2] &= \mathbf{Ax}[1] + \mathbf{Bu}[1] \\ &= \mathbf{A}(\mathbf{Ax}[0] + \mathbf{Bu}[0]) + \mathbf{Bu}[1] \\ &= \mathbf{A}^2\mathbf{x}[0] + \mathbf{ABu}[0] + \mathbf{Bu}[1]. \end{aligned}$$

In a like manner, we can show that

$$\begin{aligned} \mathbf{x}[3] &= \mathbf{A}^3\mathbf{x}[0] + \mathbf{A}^2\mathbf{Bu}[0] + \mathbf{ABu}[1] + \mathbf{Bu}[2]; \\ \mathbf{x}[4] &= \mathbf{A}^4\mathbf{x}[0] + \mathbf{A}^3\mathbf{Bu}[0] + \mathbf{A}^2\mathbf{Bu}[1] + \mathbf{ABu}[2] + \mathbf{Bu}[3]. \end{aligned}$$

We see from this pattern that the general solution is given by

$$\begin{aligned} \mathbf{x}[n] &= \mathbf{A}^n\mathbf{x}[0] + \mathbf{A}^{n-1}\mathbf{Bu}[0] + \mathbf{A}^{n-2}\mathbf{Bu}[1] + \cdots + \mathbf{ABu}[n - 2] + \mathbf{Bu}[n - 1] \\ &= \mathbf{A}^n\mathbf{x}[0] + \sum_{k=0}^{n-1} \mathbf{A}^{(n-1-k)}\mathbf{Bu}[k], \end{aligned} \qquad (13.21)$$

where $\mathbf{A}^0 = \mathbf{I}$. We define the *state-transition matrix* $\mathbf{\Phi}[n]$ from this solution:

$$\mathbf{\Phi}[n] = \mathbf{A}^n. \qquad (13.22)$$

This matrix is also called the *fundamental matrix*. From (13.21), the solution can then be expressed as

$$\mathbf{x}[n] = \mathbf{\Phi}[n]\mathbf{x}[0] + \sum_{k=0}^{n-1} \mathbf{\Phi}[n - 1 - k]\mathbf{Bu}[k], \qquad (13.23)$$

and from (13.20), the output vector is given by

$$\mathbf{y}[n] = \mathbf{C}\mathbf{x}[n] + \mathbf{D}\mathbf{u}[n]$$

$$= \mathbf{C}\boldsymbol{\Phi}[n]\mathbf{x}[0] + \sum_{k=0}^{n-1}\mathbf{C}\boldsymbol{\Phi}[n-1-k]\mathbf{B}\mathbf{u}[k] + \mathbf{D}\mathbf{u}[n]. \qquad (13.24)$$

Note that the summation in (13.24) is a convolution sum. This is not surprising, because the output signal $y[n]$ for an LTI system is expressed as a convolution sum in (10.13).

The complete solution of the state equations is given in Equation (13.24). However, in practice, we generally do not solve for $\boldsymbol{\Phi}[n]$ as a function of n. Instead, the solution of state equations is calculated recursively on a digital computer using the state equations. An example is now given to illustrate the recursive nature of the solution.

EXAMPLE 13.4   **Recursive solution of state equations**

Consider the discrete-time system with the transfer function

$$H(z) = \frac{z+3}{z^2 - 5z + 6} = \frac{b_1 z + b_2}{z^2 + a_1 z + a_2}.$$

From (13.18), we write the state equations

$$\mathbf{x}[n+1] = \begin{bmatrix} 0 & 1 \\ -6 & 5 \end{bmatrix}\mathbf{x}[n] + \begin{bmatrix} 0 \\ 1 \end{bmatrix}u[n];$$

$$y[n] = [3 \quad 1]\mathbf{x}[n].$$

Assume that the initial state is $x[0] = [2 \quad 2]^T$ and that the input signal is a unit step function, such that $u[n] = 1$ for $n \geq 0$. The recursive solution is obtained by evaluating the state equations first for $n = 1$, next for $n = 2$, then for $n = 3$, and so on:

$$\mathbf{x}[1] = \begin{bmatrix} 0 & 1 \\ -6 & 5 \end{bmatrix}\mathbf{x}[0] + \begin{bmatrix} 0 \\ 1 \end{bmatrix}u[0] = \begin{bmatrix} 0 & 1 \\ -6 & 5 \end{bmatrix}\begin{bmatrix} 2 \\ 2 \end{bmatrix} + \begin{bmatrix} 0 \\ 1 \end{bmatrix}(1) = \begin{bmatrix} 2 \\ -1 \end{bmatrix};$$

$$y[1] = [3 \quad 1]\begin{bmatrix} 2 \\ -1 \end{bmatrix} = 5$$

and

$$\mathbf{x}[2] = \begin{bmatrix} 0 & 1 \\ -6 & 5 \end{bmatrix}\begin{bmatrix} 2 \\ -1 \end{bmatrix} + \begin{bmatrix} 0 \\ 1 \end{bmatrix} = \begin{bmatrix} -1 \\ -16 \end{bmatrix};$$

$$y[2] = [3 \quad 1]\begin{bmatrix} -1 \\ -16 \end{bmatrix} = -19.$$

Also,

$$\mathbf{x}[3] = \begin{bmatrix} 0 & 1 \\ -6 & 5 \end{bmatrix} \begin{bmatrix} -1 \\ -16 \end{bmatrix} + \begin{bmatrix} 0 \\ 1 \end{bmatrix} = \begin{bmatrix} -16 \\ -73 \end{bmatrix};$$

$$y[3] = \begin{bmatrix} 3 & 1 \end{bmatrix} \begin{bmatrix} -16 \\ -73 \end{bmatrix} = -121.$$

Hence, one can recursively determine the states and the output at successive time instants. This procedure is well suited to digital computer implementation. ■

In Example 13.4, the output appears to be diverging, which indicates an unstable system. Recall from Section 10.5 that the system characteristic equation is the denominator of the transfer function set to zero; that is,

$$z^2 - 5z + 6 = (z - 2)(z - 3) = 0.$$

We can see that the transfer-function poles occur at $z = 2$ and $z = 3$, which are outside the unit circle; hence, the system is unstable. The modes of the system are given by $(2)^n$ and $(3)^n$, and the instability is evident.

A MATLAB program that performs this recursive solution in the last example is as follows:

```
A = [0 1;-6 5] ;B = [0;1] ;C = [3 1] ;
xn = [2;2] ;
for n = 0:3
    yn = C*xn
    xnplus1 = A*xn+B*1;
    xn = xnplus1;
end
result: y = 8 5 -19 -121
```

z-Transform Solution

The general solution of the state equations

$$\mathbf{x}[n + 1] = \mathbf{A}\mathbf{x}[n] + \mathbf{B}u[n], \tag{13.25}$$

developed before, is given by

[eq(13.23)] $$\mathbf{x}[n] = \mathbf{\Phi}[n]\mathbf{x}[0] + \sum_{k=0}^{n-1} \mathbf{\Phi}[n - 1 - k]\mathbf{B}u[k],$$

where the state-transition matrix is $\mathbf{\Phi}[n] = \mathbf{A}^n$. We now show that this solution can also be found by a z-transform approach.

Recall the z-transform property of Table 11.4:

$$\mathcal{Z}(x_i[n + 1]) = zX_i(z) - zx_i[0].$$

The first scalar equation of (13.25) is given by

$$x_1[n + 1] = a_{11}x_1[n] + a_{12}x_2[n] + \cdots + a_{1N}x_N[n]$$
$$+ b_{11}u_1[n] + \cdots + b_{1r}u_r[n]. \qquad (13.26)$$

The z-transform of this equation yields

$$zX_1(z) - zx_1[0] = a_{11}X_1(z) + a_{12}X_2(z) + \cdots + a_{1N}X_N(z)$$
$$+ b_{11}U_1(z) + \cdots + b_{1r}U_r(z). \qquad (13.27)$$

We will find the complete solution; hence, the initial condition $x_1[0]$ is included. The second equation in (13.25) is given by

$$x_2[n + 1] = a_{21}x_1[n] + a_{22}x_2[n] + \cdots + a_{2N}x_N[n]$$
$$+ b_{21}u_1[n] + \cdots + b_{2r}u_r[n], \qquad (13.28)$$

which has the z-transform

$$zX_2(z) - zx_2[0] = a_{21}X_1(z) + a_{22}X_2(z) + \cdots + a_{2N}X_N(z)$$
$$+ b_{21}U_1(z) + \cdots + b_{2r}U_r(z). \qquad (13.29)$$

The z-transform of the remaining $(n - 2)$ equations in (13.25) yield equations of the same form. We see, then, that these transformed equations may be written in matrix form as

$$z\mathbf{X}(z) - z\mathbf{x}[0] = \mathbf{A}\mathbf{X}(z) + \mathbf{B}\mathbf{U}(z).$$

We wish to solve this equation for $\mathbf{X}(z)$; to do this, we collect all terms containing $\mathbf{X}(z)$ on the left side of the equation:

$$z\mathbf{X}(z) - \mathbf{A}\mathbf{X}(z) = z\mathbf{x}[0] + \mathbf{B}\mathbf{U}(z). \qquad (13.30)$$

It is necessary to factor $\mathbf{X}(z)$ in the left side to solve this equation. First, the term $z\mathbf{X}(z)$ is written as $z\mathbf{I}\mathbf{X}(z)$, where $\mathbf{I}$ is the identity matrix (see Appendix G):

$$z\mathbf{I}\mathbf{X}(z) - \mathbf{A}\mathbf{X}(z) = (z\mathbf{I} - \mathbf{A})\mathbf{X}(z) = z\mathbf{x}[0] + \mathbf{B}\mathbf{U}(z). \qquad (13.31)$$

This additional step is necessary since the subtraction of the matrix $\mathbf{A}$ from the scalar z is not defined; we cannot factor $\mathbf{X}(z)$ directly in (13.30). Equation (13.31) may now be solved for $\mathbf{X}(z)$:

$$\mathbf{X}(z) = z(z\mathbf{I} - \mathbf{A})^{-1}\mathbf{x}[0] + (z\mathbf{I} - \mathbf{A})^{-1}\mathbf{B}\mathbf{U}(z). \qquad (13.32)$$

The solution $\mathbf{x}[n]$ is the inverse z-transform of this equation.

Comparing (13.32) and (13.23), we see that the state transition matrix $\mathbf{\Phi}[n]$ is given by

$$\mathbf{\Phi}[n] = \mathscr{Z}^{-1}[\mathbf{\Phi}(z)] = \mathscr{Z}^{-1}[z(z\mathbf{I} - \mathbf{A})^{-1}]. \qquad (13.33)$$

The matrix $\Phi(z) = z(z\mathbf{I} - \mathbf{A})^{-1}$ is called the *resolvant* of $\mathbf{A}$ [3]. Note that for an Nth-order system, the state transition matrix is an $N \times N$ matrix. The inverse z-transform of a matrix, as in (13.32), is defined as the inverse transform of the elements of the matrix. A computer algorithm is available for calculating $z(z\mathbf{I} - \mathbf{A})^{-1}$ [3].

Finding the inverse z-transforms indicated in (13.32) in general is difficult, time consuming, and prone to error. A more practical procedure for calculating the state vector $\mathbf{x}[n]$ is a recursive computer solution, as described earlier. An example is now presented to illustrate the calculation of a state transition matrix.

EXAMPLE 13.5 **Transition matrix for a second-order system**

We use the system of Example 13.4. From this example, the state equations are given by

$$\mathbf{x}[n + 1] = \begin{bmatrix} 0 & 1 \\ -6 & 5 \end{bmatrix}\mathbf{x}[n] + \begin{bmatrix} 0 \\ 1 \end{bmatrix}u[n]$$

and

$$y[n] = \begin{bmatrix} 3 & 1 \end{bmatrix}\mathbf{x}[n].$$

To find the state transition matrix, we first calculate the matrix $(z\mathbf{I} - \mathbf{A})$:

$$z\mathbf{I} - \mathbf{A} = z\begin{bmatrix} 1 & 0 \\ 0 & 1 \end{bmatrix} - \begin{bmatrix} 0 & 1 \\ -6 & 5 \end{bmatrix} = \begin{bmatrix} z & -1 \\ 6 & z - 5 \end{bmatrix}.$$

To find the inverse of this matrix, we calculate its adjoint matrix (see Appendix G):

$$\text{Adj}(z\mathbf{I} - \mathbf{A}) = \begin{bmatrix} z - 5 & 1 \\ -6 & z \end{bmatrix}.$$

The determinant of $(z\mathbf{I} - \mathbf{A})$ is given by

$$\det(z\mathbf{I} - \mathbf{A}) = z(z - 5) - (-1)(6)$$
$$= z^2 - 5z + 6 = (z - 2)(z - 3).$$

As we show later, this determinant is always equal to the denominator of the transfer function. The inverse of a matrix is the adjoint matrix divided by the determinant:

$$z(z\mathbf{I} - \mathbf{A})^{-1} = \begin{bmatrix} \dfrac{z(z - 5)}{(z - 2)(z - 3)} & \dfrac{z}{(z - 2)(z - 3)} \\ \dfrac{-6z}{(z - 2)(z - 3)} & \dfrac{z^2}{(z - 2)(z - 3)} \end{bmatrix}$$

$$= \begin{bmatrix} \dfrac{3z}{z - 2} + \dfrac{-2z}{z - 3} & \dfrac{-z}{z - 2} + \dfrac{z}{z - 3} \\ \dfrac{6z}{z - 2} + \dfrac{-6z}{z - 3} & \dfrac{-2z}{z - 2} + \dfrac{3z}{z - 3} \end{bmatrix}.$$

The state transition matrix is the inverse z-transform of this matrix. Thus, from Table 11.2,

$$\Phi[n] = \begin{bmatrix} 3(2)^n - 2(3)^n & -(2)^n + (3)^n \\ 6(2)^n - 6(3)^n & -2(2)^n + 3(3)^n \end{bmatrix}.$$

We see that the state transition matrix for a second-order system is a 2×2 matrix. In a like manner, the state transition matrix for an Nth-order system is $N \times N$.

This MATLAB program performs the computation of $\Phi[n]$:

```
syms z n A I2 Phiz phin
A=[0 1;-6 5]
I2=[1 0;0 1];
Phiz=z*inv(z*I2-A)
phin=iztrans(Phiz)
```
∎

Note that in Example 13.5, the modes of the system are evident in the state-transition matrix. This will always be the case, because the denominator of the elements of the resolvent matrix $z(z\mathbf{I} - \mathbf{A})^{-1}$ is always the denominator of the transfer function. This property of the state transition matrix will be proved in Section 13.4.

The complete solution of the state equations is given by

[eq(13.32)]
$$\mathbf{X}(z) = z(z\mathbf{I} - \mathbf{A})^{-1}\mathbf{x}[0] + (z\mathbf{I} - \mathbf{A})^{-1}\mathbf{B}U(z).$$

We now illustrate the complete solution with an example.

EXAMPLE 13.6   *z*-transform solution of state equations

Consider the same system as described in Examples 13.4 and 13.5. The state equations are given by

$$\mathbf{x}[n + 1] = \begin{bmatrix} 0 & 1 \\ -6 & 5 \end{bmatrix}\mathbf{x}[n] + \begin{bmatrix} 0 \\ 1 \end{bmatrix}u[n];$$

$$y[n] = [3 \quad 1]\mathbf{x}[n],$$

with

$$(z\mathbf{I} - \mathbf{A})^{-1} = \begin{bmatrix} \dfrac{z-5}{(z-2)(z-3)} & \dfrac{1}{(z-2)(z-3)} \\ \dfrac{-6}{(z-2)(z-3)} & \dfrac{z}{(z-2)(z-3)} \end{bmatrix}.$$

Suppose that a unit step function is applied as the input. Then $U(z) = z/(z - 1)$ from Table 11.2, and the second term in (13.32) becomes

$$(z\mathbf{I} - \mathbf{A})^{-1}\mathbf{B}U(z) = \begin{bmatrix} \dfrac{z-5}{(z-2)(z-3)} & \dfrac{1}{(z-2)(z-3)} \\ \dfrac{-6}{(z-2)(z-3)} & \dfrac{z}{(z-2)(z-3)} \end{bmatrix}\begin{bmatrix} 0 \\ 1 \end{bmatrix}\dfrac{z}{z-1}$$

$$= \begin{bmatrix} \dfrac{z}{(z-1)(z-2)(z-3)} \\ \dfrac{z^2}{(z-1)(z-2)(z-3)} \end{bmatrix} = \begin{bmatrix} \dfrac{z/2}{z-1} + \dfrac{-z}{z-2} + \dfrac{z/2}{z-3} \\ \dfrac{z/2}{z-1} + \dfrac{-2z}{z-2} + \dfrac{3z/2}{z-3} \end{bmatrix}.$$

The inverse z-transform of this matrix is, from Table 11.2, for $n \geq 0$,

$$\mathscr{Z}^{-1}[(z\mathbf{I} - \mathbf{A})^{-1}\mathbf{B}U(z)] = \begin{bmatrix} \dfrac{1}{2} - (2)^n + \dfrac{1}{2}(3)^n \\[2mm] \dfrac{1}{2} - 2(2)^n + \dfrac{3}{2}(3)^n \end{bmatrix}.$$

The state transition matrix was derived in Example 13.4. Hence, from (13.32), the complete solution of the state equations is

$$\mathbf{x}[n] = \mathscr{Z}^{-1}[z(z\mathbf{I} - \mathbf{A})^{-1}\mathbf{x}[0] + (z\mathbf{I} - \mathbf{A})^{-1}\mathbf{B}U(z)]$$

$$= \begin{bmatrix} 3(2)^n - 2(3)^n & -(2)^n + (3)^n \\ 6(2)^n - 6(3)^n & -2(2)^n + (3)^n \end{bmatrix}\begin{bmatrix} x_1[0] \\ x_2[0] \end{bmatrix} + \begin{bmatrix} \dfrac{1}{2} - (2)^n + \dfrac{1}{2}(3)^n \\[2mm] \dfrac{1}{2} - 2(2)^n + \dfrac{3}{2}(3)^n \end{bmatrix},$$

and the state variables are given by

$$x_1[n] = [3(2)^n - 2(3)^n]x_1[0] + [-(2)^n + (3)^n]x_2[0] + \frac{1}{2} - (2)^n + \frac{1}{2}(3)^n$$

$$= \frac{1}{2} + (3x_1[0] - x_2[0] - 1)(2)^n + (-2x_1[0] + x_2[0] + \frac{1}{2})(3)^n$$

and

$$x_2[n] = [6(2)^n - 6(3)^n]x_1[0] + [-2(2)^n + 3(3)^n]x_2[0] + \frac{1}{2} - 2(2)^n + \frac{3}{2}(3)^n$$

$$= \frac{1}{2} + (6x_1[0] - 2x_2[0] - 2)(2)^n$$

$$+ \left(-6x_1[0] + 3x_2[0] + \frac{3}{2}\right)(3)^n.$$

For example, for $\mathbf{x}[0] = [2 \quad 2]^T$ as in Example 13.4, and for $n = 3$,

$$x_1[3] = \frac{1}{2} + (3(2) - 2 - 1)(2)^3 + \left(-2(2) + 2 + \frac{1}{2}\right)(3)^3$$

$$= \frac{1}{2} + 24 - \frac{81}{2} = -16;$$

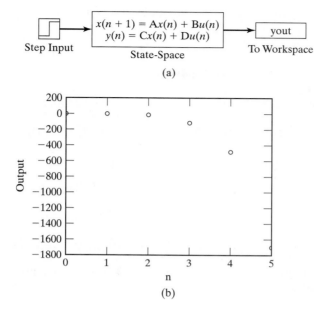

Figure 13.6 SIMULINK simulation for Example 13.6.

$$x_2[3] = \frac{1}{2} + (6(2) - 2(2) - 2)(2)^3 + \left(-6(2) + 3(2) + \frac{3}{2}\right)(3)^3$$

$$= \frac{1}{2} + 48 - \frac{243}{2} = -73;$$

$$y[3] = 3x_1[3] + x_2[3] = 3(-16) + (-73) = -121.$$

These values check those calculated in Example 13.4. It is seen that calculating the solution of the state equations by the z-transform is long and involved, even for a second-order system. The necessity for reliable digital computer solutions is evident. A SIMULINK simulation is illustrated in Figure 13.6. Note the large value of the magnitude of the output; this results from the system being unstable.   ■

In this section, two expressions for the solution of state equations are derived. The first, (13.23), expresses the solution in the discrete-time domain, while the second, (13.32), expresses the solution as a z-transform. The state-transition matrix is found either by matrix multiplication, (13.22), or the z-transform, (13.33).

13.4 PROPERTIES OF THE STATE TRANSITION MATRIX

Three properties of the state transition matrix are derived in this section. First, for an unforced system, from (13.23),

$$\mathbf{x}[n] = \mathbf{\Phi}[n]\mathbf{x}[0] \Rightarrow \mathbf{x}[0] = \mathbf{\Phi}[0]\mathbf{x}[0],$$

and, hence,

$$\Phi[0] = \mathbf{I}, \tag{13.34}$$

where $\mathbf{I}$ is the identity matrix. This property can be used in the verification of the calculation of $\Phi[n]$.

Next, from (13.22), $\Phi[n] = \mathbf{A}^n$. Thus, the second property is seen to be

$$\Phi[n_1 + n_2] = \mathbf{A}^{n_1+n_2} = \mathbf{A}^{n_1}\mathbf{A}^{n_2} = \Phi[n_1]\Phi[n_2]. \tag{13.35}$$

The third property is derived from the relationships

$$\Phi[-n] = \mathbf{A}^{-n} = [\mathbf{A}^n]^{-1} = \Phi^{-1}[n].$$

Consequently,

$$\Phi^{-1}[n] = \Phi[-n]. \tag{13.36}$$

In summary, the three properties of the state transition matrix are given by the following equations:

[eq(13.34)] $\qquad\qquad\qquad \Phi[0] = \mathbf{I};$

[eq(13.35)] $\qquad\qquad \Phi[n_1 + n_2] = \Phi[n_1]\Phi[n_2];$

[eq(13.36)] $\qquad\qquad\qquad \Phi^{-1}[n] = \Phi[-n].$

An example illustrating these properties will now be given.

EXAMPLE 13.7   **Illustration of properties of a state transition matrix**

We use the state transition matrix from Example 13.6 to illustrate the three properties:

$$\Phi[n] = \begin{bmatrix} 3(2)^n - 2(3)^n & -(2)^n + (3)^n \\ 6(2)^n - 6(3)^n & -2(2)^n + 3(3)^n \end{bmatrix}.$$

From (13.34), the first property is satisfied:

$$\Phi[0] = \begin{bmatrix} 3(2)^0 - 2(3)^0 & -(2)^0 + (3)^0 \\ 6(2)^0 - 6(3)^0 & -2(2)^0 + 3(3)^0 \end{bmatrix} = \begin{bmatrix} 1 & 0 \\ 0 & 1 \end{bmatrix} = \mathbf{I}.$$

The second property, (13.35), yields

$$\Phi[n_1]\Phi[n_2] = \begin{bmatrix} 3(2)^{n_1} - 2(3)^{n_1} & -(2)^{n_1} + (3)^{n_1} \\ 6(2)^{n_1} - 6(3)^{n_1} & -2(2)^{n_1} + 3(3)^{n_1} \end{bmatrix}$$

$$\times \begin{bmatrix} 3(2)^{n_2} - 2(3)^{n_2} & -(2)^{n_2} + (3)^{n_2} \\ 6(2)^{n_2} - 6(3)^{n_2} & -2(2)^{n_2} + 3(3)^{n_2} \end{bmatrix}.$$

The $(1, 1)$ element of the product is given by

$$(1, 1) \text{ element} = [3(2)^{n_1} - 2(3)^{n_1}][3(2)^{n_2} - 2(3)^{n_2}]$$
$$+ [-(2)^{n_1} + (3)^{n_1}][6(2)^{n_2} - 6(3)^{n_2}] = 9(2)^{n_1+n_2}$$
$$- 6(2)^{n_1}(3)^{n_2} - 6(2)^{n_2}(3)^{n_1} + 4(3)^{n_1+n_2} - 6(2)^{n_1+n_2}$$
$$+ 6(2)^{n_1}(3)^{n_2} + 6(2)^{n_2}(3)^{n_1} - 6(3)^{n_1+n_2}.$$

Combining these terms yields

$$(1, 1) \text{ element} = 3(2)^{n_1+n_2} - 2(3)^{n_1+n_2}, \tag{13.37}$$

which is the $(1, 1)$ element of $\Phi[n_1 + n_2]$. In a like manner, the other three elements of the product matrices can be verified.

To illustrate the third property, (13.36), we assume that the property is true for this example. Hence,

$$\Phi[n]\Phi[-n] = \begin{bmatrix} 3(2)^n - 2(3)^n & -(2)^n + (3)^n \\ 6(2)^n - 6(3)^n & -2(2)^n + 3(3)^n \end{bmatrix}$$
$$\times \begin{bmatrix} 3(2)^{-n} - 2(3)^{-n} & -(2)^{-n} + (3)^{-n} \\ 6(2)^{-n} - 6(3)^{-n} & -2(2)^{-n} + 3(3)^{-n} \end{bmatrix} = \mathbf{I}.$$

As with the last property, we test only the $(1, 1)$ element of the product. From (13.37), with $n_1 = n$ and $n_2 = -n$,

$$(1, 1) \text{ element} = 3(2)^{n-n} - 2(3)^{n-n}$$
$$= 3 - 2 = 1.$$

In a like manner, the other three elements of the product matrix can be verified.   ■

In this section, three properties of the state transition matrix are developed. The first property, $\Phi[0] = \mathbf{I}$, is easily applied as a check of the calculation of a state-transition matrix.

13.5 TRANSFER FUNCTIONS

A procedure was given in Section 13.2 for writing state equations of a system from the transfer function. In this section, we investigate the calculation of transfer functions from state equations.

The standard form of the state equations is given by

$$\mathbf{x}[n + 1] = \mathbf{Ax}[n] + \mathbf{B}u[n]$$

and

$$y[n] = \mathbf{Cx}[n] + Du[n] \tag{13.38}$$

for a single-input single-output system. The z-transform of the first equation in (13.38) yields [see (13.30)]

$$z\mathbf{X}(z) = \mathbf{A}\mathbf{X}(z) + \mathbf{B}U(z). \tag{13.39}$$

Because we are interested in the transfer function, the initial conditions are ignored. Collecting terms for $\mathbf{X}(z)$ yields

$$(z\mathbf{I} - \mathbf{A})\mathbf{X}(z) = \mathbf{B}U(z), \tag{13.40}$$

and thus, $\mathbf{X}(z)$ is given by

$$\mathbf{X}(z) = (z\mathbf{I} - \mathbf{A})^{-1}\mathbf{B}U(z). \tag{13.41}$$

The z-transform of the output equation in (13.38) yields

$$Y(z) = \mathbf{C}\mathbf{X}(z) + DU(z). \tag{13.42}$$

From (13.41) and (13.42), the input–output relationship for the system is given by

$$Y(z) = [\mathbf{C}(z\mathbf{I} - \mathbf{A})^{-1}\mathbf{B} + D]U(z). \tag{13.43}$$

Since the system transfer function is defined by the equation $Y(z) = H(z)U(z)$, we see that the transfer function is given by

$$H(z) = \frac{Y(z)}{U(z)} = \mathbf{C}(z\mathbf{I} - \mathbf{A})^{-1}\mathbf{B} + D = \mathbf{C}\Phi(z)\mathbf{B} + D. \tag{13.44}$$

Because $\mathbf{C}$ is $1 \times N$, $(z\mathbf{I} - \mathbf{A})^{-1}$ is $N \times N$, and $\mathbf{B}$ is $N \times 1$, the product $\mathbf{C}(z\mathbf{I} - \mathbf{A})^{-1}\mathbf{B}$ is 1×1, or a scalar, as required. An example is given to illustrate this result.

EXAMPLE 13.8 **Transfer function from state equations**

Consider the system of Examples 13.4 and 13.5, which has the transfer function

$$H(z) = \frac{Y(z)}{U(z)} = \frac{z + 3}{z^2 - 5z + 6}.$$

The state equations were found to be

$$\mathbf{x}[n + 1] = \begin{bmatrix} 0 & 1 \\ -6 & 5 \end{bmatrix}\mathbf{x}[n] + \begin{bmatrix} 0 \\ 1 \end{bmatrix}u[n];$$

$$y[n] = [3 \quad 1]\mathbf{x}[n].$$

The resolvant matrix $(z\mathbf{I} - \mathbf{A})^{-1}$ was calculated in Example 13.5. Then, from (13.44) and Example 13.5, with $D = 0$,

$$H(z) = \mathbf{C}(z\mathbf{I} - \mathbf{A})^{-1}\mathbf{B}$$

$$= \begin{bmatrix} 3 & 1 \end{bmatrix} \begin{bmatrix} \dfrac{z - 5}{z^2 - 5z + 6} & \dfrac{1}{z^2 - 5z + 6} \\ \dfrac{-6}{z^2 - 5z + 6} & \dfrac{z}{z^2 - 5z + 6} \end{bmatrix} \begin{bmatrix} 0 \\ 1 \end{bmatrix}$$

$$= \begin{bmatrix} 3 & 1 \end{bmatrix} \begin{bmatrix} \dfrac{1}{z^2 - 5z + 6} \\ \dfrac{z}{z^2 - 5z + 6} \end{bmatrix} = \dfrac{z + 3}{z^2 - 5z + 6}.$$

This transfer function checks the one given.   ∎

Although (13.44) does not appear to be useful in calculating the transfer function for higher-order systems, relatively simple computer algorithms exist for evaluating the resolvant matrix $(z\mathbf{I} - \mathbf{A})^{-1}$[3]. A MATLAB program that performs the calculations of Example 13.8 is given by

```
A=[0 1;-6 5],B=[0;1],C=[3 1],D=0
[num,den]=ss2tf(A,B,C,D)
Hz=tf(num,den,-1)
```

Stability

We saw in Section 11.6 that bounded-input bounded-output (BIBO) stability can be determined from the transfer function of an LTI system. From (13.13), the transfer function of (13.44) can be expressed as a rational function:

$$H(z) = \mathbf{C}(z\mathbf{I} - \mathbf{A})^{-1}\mathbf{B} + D = \frac{b_0 z^N + \cdots + b_{N-1}z + b_0}{z^N + \cdots + a_{N-1}z + a_0}. \qquad (13.45)$$

From Section 11.6, this system is BIBO stable provided that all poles of $H(z)$ are inside the unit circle, where the poles of the transfer function are the zeros of the denominator polynomial in (13.45).

The transfer function $H(z)$ can be expressed as

$$\mathbf{C}(z\mathbf{I} - \mathbf{A})^{-1}\mathbf{B} + D = \mathbf{C}\left[\frac{\text{adj}(z\mathbf{I} - \mathbf{A})}{\det(z\mathbf{I} - \mathbf{A})} \right]\mathbf{B} + D. \qquad (13.46)$$

Hence, the denominator polynomial of $H(z)$ is the determinant of $(z\mathbf{I} - \mathbf{A})$; the poles of the transfer function are the roots of

$$\det(z\mathbf{I} - \mathbf{A}) = 0. \qquad (13.47)$$

This equation is then the system characteristic equation. Note that stability is a function only of the system matrix $\mathbf{A}$ and is not affected by $\mathbf{B}$, $\mathbf{C}$, or D. In fact,

(13.47) is also the characteristic equation of a multivariable system, which has more than one input or more than one output. We now consider an example illustrating stability.

EXAMPLE 13.9   **Stability from state equations**

We consider the system of Example 13.8. The state equation is given by

$$\mathbf{x}[n + 1] = \begin{bmatrix} 0 & 1 \\ -6 & 5 \end{bmatrix}\mathbf{x}[n].$$

We have ignored the input and output terms, since stability is independent of these terms. From (13.47), the characteristic equation is given by

$$\det(z\mathbf{I} - \mathbf{A}) = \det\begin{bmatrix} z & -1 \\ 6 & z - 5 \end{bmatrix}$$

$$= z^2 - 5z + 6 = (z - 2)(z - 3) = 0.$$

The roots are at $z = 2$ and $z = 3$. Both poles are outside the unit circle; hence, the system is unstable, as is noted following Example 13.4. A MATLAB program that calculates the system characteristic equation is given by

```
A=[0,1;-6 5]
charpoly=poly(A)
charroots=roots (charpoly)
```
■

In this section, a procedure is developed for calculating the transfer function of a system from its state equations. The procedure can be implemented on a computer and is used extensively in the practice of engineering for calculating transfer functions of high-order systems. As a final point, it is shown that the characteristic polynomial of a system is equal to $\det(z\mathbf{I} - \mathbf{A})$; hence, this determinant determines the modes of a system. These modes determine the stability of a system and the characteristics of the transient response of a stable system.

13.6 SIMILARITY TRANSFORMATIONS

In this chapter, procedures have been presented for finding a state-variable model from the system difference equations, the system-transfer function, or a system-simulation diagram. In this section, a procedure is given for finding different state models from a given state model. It is seen that a system has an unlimited number of state models. The state models have the same input–output characteristics (same transfer function), whereas the internal characteristics are different.

The procedure is identical to that developed in Section 8.6 for continuous-time system; thus, only the results are reviewed. The state model for a discrete-time single-input, single-output system is given by

$$\mathbf{x}[n + 1] = \mathbf{A}\mathbf{x}[n] + \mathbf{B}u[n];$$
$$y[n] = \mathbf{C}\mathbf{x}[n] + Du[n], \tag{13.48}$$

and the transfer function for this model is, from (13.44),

$$H(z) = \frac{Y(z)}{U(z)} = \mathbf{C}(z\mathbf{I} - \mathbf{A})^{-1}\mathbf{B} + D. \tag{13.49}$$

The similarity transformation is defined by

$$\mathbf{v}[n] = \mathbf{Q}\mathbf{x}[n] = \mathbf{P}^{-1}\mathbf{x}[n] \Rightarrow \mathbf{x}[n] = \mathbf{P}\mathbf{v}[n], \tag{13.50}$$

where $\mathbf{v}(n)$ are the new state variables. The transformed state equations are then

$$\mathbf{v}[n + 1] = \mathbf{A}_v\mathbf{v}[n] + \mathbf{B}_v u[n]$$

and

$$y[n] = \mathbf{C}_v\mathbf{v}[n] + D_v u[n], \tag{13.51}$$

where the transformed matrices for the $\mathbf{v}(n)$-states are

$$\mathbf{A}_v = \mathbf{P}^{-1}\mathbf{A}\mathbf{P}, \quad \mathbf{B}_\mathbf{v} = \mathbf{P}^{-1}\mathbf{B},$$
$$\mathbf{C}_\mathbf{v} = \mathbf{C}\mathbf{P}, \text{ and } D_v = D. \tag{13.52}$$

Three examples illustrating similarity transformations for discrete-time systems are now given. A MATLAB program that performs the calculations in these examples is given in Example 13.12.

EXAMPLE 13.10   **State-variable transformation for a second-order system**

Consider the system of Example 13.4, which has the transfer function

$$H(z) = \frac{Y(z)}{U(z)} = \frac{z + 3}{z^2 - 5z + 6}.$$

From Example 13.4, the state equations are given by

$$\mathbf{x}[n + 1] = \begin{bmatrix} 0 & 1 \\ -6 & 5 \end{bmatrix}\mathbf{x}[n] + \begin{bmatrix} 0 \\ 1 \end{bmatrix}u[n]$$

and

$$y[n] = [3 \quad 1]\mathbf{x}[n].$$

We arbitrarily define the elements of $\mathbf{v}[n]$ as follows:

$$v_1[n] = x_1[n];$$
$$v_2[n] = x_1[n] + x_2[n].$$

Thus, from (13.50),

$$\mathbf{v}[n] = \mathbf{Q}\mathbf{x}[n] = \begin{bmatrix} 1 & 0 \\ 1 & 1 \end{bmatrix} \mathbf{x}[n]$$

and

$$\mathbf{P}^{-1} = \mathbf{Q} = \begin{bmatrix} 1 & 0 \\ 1 & 1 \end{bmatrix} \Rightarrow \mathbf{P} = \begin{bmatrix} 1 & 0 \\ -1 & 1 \end{bmatrix}.$$

Hence, the components of $\mathbf{x}[n] = \mathbf{P}\mathbf{v}[n]$ can be expressed as

$$x_1[n] = v_1[n];$$
$$x_2[n] = -v_1[n] + v_2[n].$$

It is seen from this example that, given the vector $\mathbf{v}[n]$ and the transformation $\mathbf{P} = \mathbf{Q}^{-1}$, we can solve for the vector $\mathbf{x}[n]$. Or, given the vector $\mathbf{x}[n]$ and the transformation $\mathbf{Q}$, we can solve for the vector $\mathbf{v}[n]$. ∎

EXAMPLE 13.11   **Similarity transformation for a second-order system**

This example is a continuation of the last example. From (13.52), the system matrices for the transformed matrices become

$$\mathbf{A}_v = \mathbf{P}^{-1}\mathbf{A}\mathbf{P} = \begin{bmatrix} 1 & 0 \\ 1 & 1 \end{bmatrix}\begin{bmatrix} 0 & 1 \\ -6 & 5 \end{bmatrix}\begin{bmatrix} 1 & 0 \\ -1 & 1 \end{bmatrix}$$

$$= \begin{bmatrix} 0 & 1 \\ -6 & 6 \end{bmatrix}\begin{bmatrix} 1 & 0 \\ -1 & 1 \end{bmatrix} = \begin{bmatrix} -1 & 1 \\ -12 & 6 \end{bmatrix},$$

$$\mathbf{B}_v = \mathbf{P}^{-1}\mathbf{B} = \begin{bmatrix} 1 & 0 \\ 1 & 1 \end{bmatrix}\begin{bmatrix} 0 \\ 1 \end{bmatrix} = \begin{bmatrix} 0 \\ 1 \end{bmatrix},$$

and

$$\mathbf{C}_v = \mathbf{C}\mathbf{P} = \begin{bmatrix} 3 & 1 \end{bmatrix}\begin{bmatrix} 1 & 0 \\ -1 & 1 \end{bmatrix} = \begin{bmatrix} 2 & 1 \end{bmatrix}.$$

The transformed state equations are then

$$\mathbf{v}[n + 1] = \mathbf{A}_v\mathbf{v}[n] + \mathbf{B}_v u[n] = \begin{bmatrix} -1 & 1 \\ -12 & 6 \end{bmatrix}\mathbf{v}[n] + \begin{bmatrix} 0 \\ 1 \end{bmatrix} u[n]$$

and

$$y[n] = \mathbf{C}_v\mathbf{v}[n] = [2 \quad 1]\mathbf{v}[n].$$ ■

EXAMPLE 13.12   **Transfer function for system of Example 13.11**

To verify the state model of Example 13.11, we now derive the transfer function. From (13.44),

$$H_v(z) = \mathbf{C}_v(z\mathbf{I} - \mathbf{A}_v)^{-1}\mathbf{B}_v.$$

First, we calculate $(z\mathbf{I} - \mathbf{A}_v)^{-1}$. Hence,

$$z\mathbf{I} - \mathbf{A}_v = z\begin{bmatrix} 1 & 0 \\ 0 & 1 \end{bmatrix} - \begin{bmatrix} -1 & 1 \\ -12 & 6 \end{bmatrix} = \begin{bmatrix} z+1 & -1 \\ 12 & z-6 \end{bmatrix}.$$

Therefore,

$$\det(z\mathbf{I} - \mathbf{A}) = z^2 - 5z - 6 + 12 = z^2 - 5z + 6.$$

Then, letting $\det(z\mathbf{I} - \mathbf{A}) = \Delta(z)$, we obtain

$$(z\mathbf{I} - \mathbf{A})^{-1} = \frac{\text{adj}(z\mathbf{I} - \mathbf{A})}{\det(z\mathbf{I} - \mathbf{A})} = \begin{bmatrix} \dfrac{z-6}{\Delta(z)} & \dfrac{1}{\Delta(z)} \\ \dfrac{-12}{\Delta(z)} & \dfrac{z+1}{\Delta(z)} \end{bmatrix}.$$

Thus, the transfer function is given by

$$H_v(z) = \mathbf{C}_v(z\mathbf{I} - \mathbf{A}_v)^{-1}\mathbf{B}_v$$

$$= [2 \quad 1]\begin{bmatrix} \dfrac{z-6}{\Delta(z)} & \dfrac{1}{\Delta(z)} \\ \dfrac{-12}{\Delta(z)} & \dfrac{z+1}{\Delta(z)} \end{bmatrix}\begin{bmatrix} 0 \\ 1 \end{bmatrix}$$

$$= \begin{bmatrix} \dfrac{2z-24}{\Delta(z)} & \dfrac{z+3}{\Delta(z)} \end{bmatrix}\begin{bmatrix} 0 \\ 1 \end{bmatrix} = \frac{z+3}{z^2 - 5z + 6},$$

and the transfer function is as given in Example 13.10. The following MATLAB program performs the calculations of Examples 13.11 and 13.12:

```
A=[0 1;-6 5];B=[0;1];C=[3 1];D=0;P=[1 0;1 1];
sys1=ss(A,B,C,D,-1)
sys2=ss2ss(sys1,P)
[num,den]=ss2tf(A,B,C,D,1);
'H(z)=',Hz=tf(num,den,-1)
```
■

Properties

Similarity transformations have been demonstrated through examples. Certain important properties of these transformations are derived next. Consider first the determinant of $(z\mathbf{I} - \mathbf{A}_v)$. From (13.52),

$$\det(z\mathbf{I} - \mathbf{A}_v) = \det(z\mathbf{I} - \mathbf{P}^{-1}\mathbf{AP}) = \det(z\mathbf{P}^{-1}\mathbf{IP} - \mathbf{P}^{-1}\mathbf{AP})$$

$$= \det[\mathbf{P}^{-1}(z\mathbf{I} - \mathbf{A})\mathbf{P}]. \tag{13.53}$$

For two square matrices,

$$\det \mathbf{R}_1\mathbf{R}_2 = \det \mathbf{R}_1 \det \mathbf{R}_2. \tag{13.54}$$

Then we can express (13.53) as

$$\det(z\mathbf{I} - \mathbf{A}_v) = \det \mathbf{P}^{-1} \det(z\mathbf{I} - \mathbf{A}) \det \mathbf{P}, \tag{13.55}$$

because, for a matrix $\mathbf{R}$, $\mathbf{R}^{-1}\mathbf{R} = \mathbf{I}$. Thus,

$$\det \mathbf{R}^{-1}\mathbf{R} = \det \mathbf{R}^{-1} \det \mathbf{R} = \det \mathbf{I} = 1. \tag{13.56}$$

Hence, (13.55) yields the first property:

$$\det(z\mathbf{I} - \mathbf{A}_v) = \det(z\mathbf{I} - \mathbf{A}). \tag{13.57}$$

The roots of $\det(z\mathbf{I} - \mathbf{A})$ are the *characteristic values*, or the *eigenvalues*, of $\mathbf{A}$. (See Appendix G.) Thus, the eigenvalues of $\mathbf{A}_v$ are equal to those of $\mathbf{A}$, from (13.57). Since the transfer function is unchanged under a similarity transformation, and since the eigenvalues are the poles of the transfer function, we are not surprised that they are unchanged.

A second property is now derived. From (13.57) with $z = 0$,

$$\det \mathbf{A}_v = \det \mathbf{A}. \tag{13.58}$$

The determinant of $\mathbf{A}_v$ is equal to the determinant of $\mathbf{A}$. This property can also be seen from the fact that the determinant of a matrix is equal to the product of its eigenvalues. (See Appendix G.)

The third property of a similarity transformation can also be seen from the fact that the eigenvalues of $\mathbf{A}_v$ and of $\mathbf{A}$ are equal. The trace (sum of the diagonal elements) of a matrix is equal to the sum of the eigenvalues; hence,

$$\operatorname{tr} \mathbf{A}_v = \operatorname{tr} \mathbf{A}. \tag{13.59}$$

A fourth property was demonstrated in Example 13.12. Since the transfer function is unchanged under a similarity transformation,

$$\mathbf{C}_v(z\mathbf{I} - \mathbf{A}_v)^{-1}\mathbf{B}_v + D_v = \mathbf{C}(z\mathbf{I} - \mathbf{A})^{-1}\mathbf{B} + D. \tag{13.60}$$

The proof of this property is left as an exercise.

To summarize the properties of similarity transforms, we first let $\lambda_1, \lambda_2, \ldots, \lambda_N$ denote the eigenvalues of the matrix $\mathbf{A}$. Then, for the similarity transformation of (13.51) and (13.52), the following are true:

1.   The eigenvalues of $\mathbf{A}$ and $\mathbf{A}_v$ are equal:

$$\det(z\mathbf{I} - \mathbf{A}) = \det(z\mathbf{I} - \mathbf{A}_v)$$

$$= (z - \lambda_1)(z - \lambda_2)\cdots(z - \lambda_N). \tag{13.61}$$

2.   The determinants of $\mathbf{A}$ and $\mathbf{A}_v$ are equal:

$$\det \mathbf{A} = \det \mathbf{A}_v = \lambda_1\lambda_2\cdots\lambda_N. \tag{13.62}$$

3.   The traces of $\mathbf{A}$ and $\mathbf{A}_v$ are equal:

$$\text{tr } \mathbf{A} = \text{tr } \mathbf{A}_v = \lambda_1 + \lambda_2 + \cdots + \lambda_N. \tag{13.63}$$

4.   The following transfer functions are equal:

$$\mathbf{C}_v(z\mathbf{I} - \mathbf{A}_v)^{-1}\mathbf{B}_v + D_v = \mathbf{C}(z\mathbf{I} - \mathbf{A})^{-1}\mathbf{B} + D. \tag{13.64}$$

SUMMARY

In earlier chapters, we specified the model of a discrete-time LTI system by a difference equation or a transfer function. In both cases, the system input–output characteristics are given. In this chapter, a third model, the state-variable model, is developed. This model is a set of coupled first-order difference equations. The state model can be specified either by state equations or by a simulation diagram.

The state model gives an internal model of a system, in addition to the input–output characteristics. Methods are presented in this chapter to derive any one of the three models from any other one.

It is also demonstrated in this chapter that a state model for a given system is not unique. Similarity transformations may be used to develop any number of different state models from a given state model. The different state models have the same input–output characteristics, while the internal models are all different.

A state model of a discrete-time LTI system is required for applying certain types of analysis and design procedures [1, 2]. State models are especially useful in computer-aided analysis and design.

See Table 13.1.

TABLE 13.1   Key Equations of Chapter 13

Equation Title	Equation Number	Equation
State equations of DT LTI system	(13.9)	$\mathbf{x}[n+1] = \mathbf{Ax}[n] + \mathbf{Bu}[n]$ and $\mathbf{y}[n] = \mathbf{Cx}[n] + \mathbf{Du}[n]$
State equations in matrix form	(13.18)	$\mathbf{x}[n+1] = \begin{bmatrix} 0 & 1 & 0 & \cdots & 0 & 0 \\ 0 & 0 & 1 & \cdots & 0 & 0 \\ \vdots & \vdots & \vdots & & \vdots & \vdots \\ 0 & 0 & 0 & \cdots & 0 & 1 \\ -a_N & -a_{N-1} & -a_{N-2} & \cdots & -a_2 & -a_1 \end{bmatrix} \mathbf{x}[n] + \begin{bmatrix} 0 \\ 0 \\ \vdots \\ 0 \\ 1 \end{bmatrix} u[n]$ $\mathbf{y}[n] = [(b_N - a_N b_0)(b_{N-1} - a_{N-1}b_0) \cdots (b_1 - a_1 b_0)]\mathbf{x}[n] + b_0 u[n]$
State-transition matrix	(13.22)	$\mathbf{\Phi}[n] = \mathbf{A}^n$
Solution of state equation	(13.23)	$\mathbf{x}[n] = \mathbf{\Phi}[n]\mathbf{x}[0] + \sum_{k=0}^{n-1} \mathbf{\Phi}[n-1-k]\mathbf{Bu}[k]$
Convolution solution of state equation	(13.24)	$\mathbf{y}[n] = \mathbf{Cx}[n] + \mathbf{Du}[n]$ $= \mathbf{C\Phi}[n]\mathbf{x}[0] + \sum_{k=0}^{n-1} \mathbf{C\Phi}[n-1-k]\mathbf{Bu}[k] + \mathbf{Du}[n]$
State-transition matrix	(13.33)	$\mathbf{\Phi}[n] = \mathscr{Z}^{-1}[\mathbf{\Phi}(z)] = \mathscr{Z}^{-1}[z(z\mathbf{I} - \mathbf{A})^{-1}]$
Transfer function	(13.44)	$H(z) = \dfrac{Y(z)}{U(z)} = \mathbf{C}(z\mathbf{I} - \mathbf{A})^{-1}\mathbf{B} + D = \mathbf{C\Phi}(z)\mathbf{B} + D$
Singularity transformation	(13.50)	$\mathbf{v}[n] = \mathbf{Qx}[n] = \mathbf{P}^{-1}\mathbf{x}[n] \Rightarrow \mathbf{x}[n] = \mathbf{Pv}[n]$
Transformed state equations	(13.51)	$\mathbf{v}[n+1] = \mathbf{A}_v \mathbf{v}[n] + \mathbf{B}_v u[n]$ and $y[n] = \mathbf{C}_v \mathbf{v}[n] + D_v u[n]$
Transformed matrices	(13.52)	$\mathbf{A}_v = \mathbf{P}^{-1}\mathbf{AP}, \qquad \mathbf{B}_v = \mathbf{P}^{-1}\mathbf{B}$ $\mathbf{C}_v = \mathbf{CP}, \quad \text{and} \quad D_v = D$

REFERENCES

1. C. L. Phillips and H. T. Nagle, *Digital Control Systems*, 3d ed. Upper Saddle River, NJ: Prentice Hall, 1996.
2. G. F. Franklin and J. D. Powell, *Digital Control of Dynamic Systems*, 3d ed. Reading, MA: Addison–Wesley, 1997.
3. B. Friedlander, *Control System Design*. New York: McGraw–Hill, 1986.

PROBLEMS

13.1. Find a set of state equations for each of the systems described by the following difference equations:

    (a) $y[n + 1] - 0.8y[n] = 1.9u[n]$
    (b) $y[n] + 0.8y[n - 2] = u[n - 2]$
    (c) $y[n] = u[n - 1]$

13.2. **(a)** Find a state model for the α-filter described by the equation

$$y[n + 1] - (1 - \alpha)y[n] = \alpha x[n + 1],$$

    where $x[n]$ is the input signal. *Hint:* Draw a simulation diagram first.
    (b) Verify the results of (a) by (i) finding the transfer function from the describing equation and (ii) finding the transfer function from the state equations.

13.3. The simulation diagram for the $\alpha - \beta$ filter is given in Figure P13.3. This filter is second order and is used in radar-signal processing. The input $u[n]$ is the unfiltered target-position data, the output $y[n]$ is the filtered position data, and the output $v[n]$ is an estimate of the target velocity. The parameter T is the sample period. The parameters α and β are constants and depend on the design specifications for the filter.

    (a) Write the state equations for the filter, with the state variables equal to the outputs of the delays and the system outputs equal to $y[n]$ and $v[n]$.
    (b) Let $\beta = 0$ in Part (a). Show that the resulting equations are equivalent to those of the α-filter of Problem 13.2.

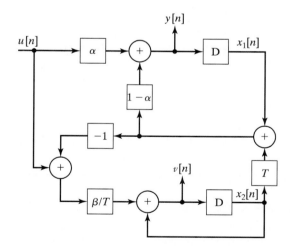

Figure P13.3

13.4. (a) Draw a simulation diagram for the system described by the difference equation

$$2y[n + 1] - 1.8y[n] = 3u[n + 1].$$

(b) Write the state equations for the simulation diagram of Part (a).

(c) Give the system transfer equation for the simulation diagram of Part (a).

(d) Use the results in Part (b) and MATLAB to verify Part (c).

(e) Repeat Parts (a) through (d) for the difference equation

$$y[n + 2] - 1.5y[n + 1] + 0.9y[n] = 2u[n].$$

(f) Repeat Parts (a) through (d) for the difference equation

$$y[n + 3] - 2.9y[n + 2] + 3.4y[n + 1] - 0.72y[n] = 2u[n].$$

13.5. (a)   Draw a simulation diagram for the system described by the transfer function

$$\frac{Y(z)}{U(z)} = \frac{4(z - 0.9)}{z - 0.8}.$$

(b) Write the state equations for the simulation diagram of Part (a).

(c) Give the system difference equation.

(d) The MATLAB program

```
n = [4 -3.6] ;
d = [1 -0.8] ;
[A,B,C,D] = tf2ss(n,d)
```

generates a set of state equations for Part (a).

(i)   Run this program.

(ii) Draw a simulation diagram for these state equations.

(e) Repeat Parts (a) through (d) for the transfer function

$$\frac{Y(z)}{U(z)} = H(z) = \frac{2z^2 + 3}{z^2 - 1.96z + 0.8}.$$

(f) Repeat Parts (a) through (d) for the transfer function

$$\frac{Y(z)}{U(z)} = H(z) = \frac{2z^2 - 3.1z + 1.73}{z^3 + 2.3z^2 - 2.1z + 0.65}.$$

13.6. (a) Write the state equations for the system modeled by the simulation diagram of Figure P13.6.

(b) Use the results of Part (a) to find the system transfer function.

(c) Use MATLAB to check the results in Part (b).

(d) Use the system transfer function to draw a simulation diagram that is different from that of Figure P13.6.

(e) Write the state equations for the simulation diagram of Part (d).

(f) Use the results of Part (e) to calculate the system transfer function, which will verify the results of Parts (d) and (e).

(g) Use MATLAB to verify the results in Part (f).

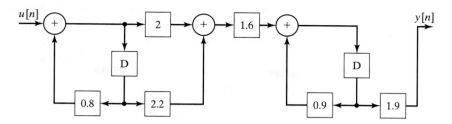

Figure P13.6

13.7. (a) Write the state equations for the system modeled by the simulation diagram of Figure P13.7.
   (b) Use the results of Part (a) to find the system-transfer function.
   (c) Use MATLAB to verify the results in Part (b).
   (d) Use the system transfer function to draw a simulation diagram that is different from that of Figure P13.7.
   (e) Write the state equations for the simulation diagram of Part (d).
   (f) Use MATLAB to check the results in Parts (d) and (e).

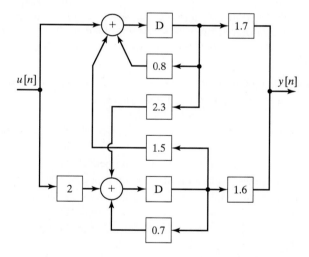

Figure P13.7

13.8. Consider a system described by the state equations

$$x[n + 1] = \begin{bmatrix} 1.9 & 0.8 \\ -1 & 0 \end{bmatrix} x[n] + \begin{bmatrix} 0 \\ 0.95 \end{bmatrix} u[n];$$

$$y[n] = [1.5 \quad -1.3]x[n] + 2u[n].$$

   (a) Draw a simulation diagram of this system.
   (b) Find the transfer function directly from the state equations.

(c) Use MATLAB to verify the results in Part (b).

(d) Draw a different simulation diagram of the system.

(e) Write the state equations of the simulation diagram of Part (d).

(f) Verify the simulation diagram of Part (d) by showing that its transfer function is that of Part (b).

(g) Use MATLAB to verify the results in Part (f).

(h) Repeat Parts (a) through (g) for the state equations

$$x[n + 1] = 0.82x[n] + 3.2u[n];$$

$$y[n] = x[n].$$

(i) Repeat Parts (a) through (g) for the state equations

$$\mathbf{x}[n + 1] = \begin{bmatrix} 0 & 1 & 0 \\ 0 & 0 & 1 \\ 1 & 0 & 1 \end{bmatrix} \mathbf{x}[n] + \begin{bmatrix} 2 \\ 0 \\ 0 \end{bmatrix} u[n];$$

$$y[n] = \begin{bmatrix} 0 & 0 & 1 \end{bmatrix} \mathbf{x}[n].$$

13.9. Figure P13.9 gives the simulation diagram of an automatic control system. The plant is the system that is controlled and is a discrete model of a continuous-time system. The compensator, a digital filter, is a system added to give the closed-loop system certain specified characteristics. It can be shown that a system of this type is modeled as shown [1].

(a) Write the state equations for the plant only, with the input $m[n]$ and the output $y[n]$.

(b) Give the transfer function $H_p(z)$ for the plant.

(c) Write the difference equation for the plant.

(d) Write the state equations for the compensator only, with the input $e[n]$ and the output $m[n]$.

(e) Give the transfer function $H_c(z)$ for the compensator.

(f) Write the difference equation for the compensator.

(g) Write the state equations for the closed-loop system, with the input $u[n]$ and the output $y[n]$. Choose as states those of Parts (a) and (d).

(h) Give the transfer function $H(z)$ for the closed-loop system.

(i) Use MATLAB to verify the results in Part (h).

(j) Write the difference equation for the closed-loop system.

(k) It can be shown that the closed-loop transfer function is given by

$$\frac{Y(z)}{U(z)} = H(z) = \frac{H_c(z)H_p(z)}{1 + H_c(z)H_p(z)}.$$

Verify your results in Part (h) by showing that this equation is satisfied by the derived transfer functions in Parts (b) and (e).

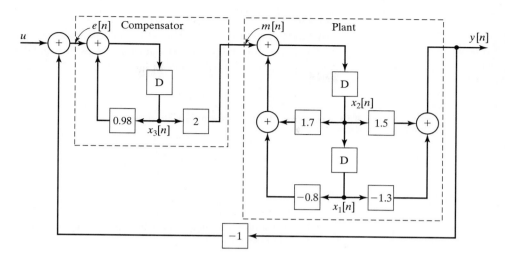

Figure P13.9

13.10. Consider the α-filter of Problem 13.2. Parts of this problem are repeated from Problem 13.2. Use those results if available.

    (a) Write the state equations of the filter, with the state variable equal to the output $y[n]$.
    (b) Use the results of Part (a) to find the filter transfer function.

13.11. Consider the $\alpha-\beta$ filter of Figure P13.3. Parts of this problem are repeated from Problem 13.2. Use those results if available.

    (a) Write the state equations of the filter, with the state variables equal to the delay outputs and the output equal to $y[n]$.
    (b) Use the results of Part (a) to find the filter-transfer function $H(z) = Y(z)/U(z)$.
    (c) Show that, with $\beta = 0$, the transfer function is that of the α-filter in Problem 13.10(b).

13.12. Consider the system of Figure P13.6.

    (a) Write the state equations, with the outputs of the delays as the states.
    (b) Find the state-transition matrix.
    (c) Find the system output for $u[n] = 0$ and the initial states given by $\mathbf{x}(0) = [1 \quad 2]^T$.
    (d) Find the system unit step response, with $\mathbf{x}(0) = \mathbf{0}$, using (13.32).
    (e) Verify the results of Part (d), using the z-transform and the system-transfer function.
    (f) Find the system response, with the initial conditions given in Part (c) and the input in Part (d).
    (g) Verify the results in Part (f) using SIMULINK.

13.13. Consider the system of Figure P13.7. Replace the gains of 1.5 and 2.3 with gains of zero for this problem.

    (a) Write the state equations, with the outputs of the delays as the states.
    (b) Find the state-transition matrix.

(c) Find the system output for $u[n] = 0$ and the initial states given by $x(0) = [1 \quad 9]^T$.

(d) Find the system unit step response, with $x(0) = 0$, using the z-transform as in (13.32).

(e) Verify the results of Part (d), using the z-transform and the system-transfer function.

(f) Find the system response, with the initial conditions given in Part (c) and the input in Part (d).

(g) Verify the results in Part (f) using SIMULINK.

13.14. (a) Consider the system described by

$$\mathbf{x}[n+1] = \begin{bmatrix} 0 & 1 \\ 0 & 0 \end{bmatrix}\mathbf{x}[n] + \begin{bmatrix} 0 \\ 1 \end{bmatrix}u[n];$$

$$y[n] = [1 \quad 0]\mathbf{x}[n].$$

Find the state transition matrix by two different procedures.

(b) Draw a simulation diagram for the system, and describe how the system can be realized physically.

13.15. Consider the system described by the state equations

$$\mathbf{x}[n+1] = \begin{bmatrix} 0 & 0 \\ 1 & 0 \end{bmatrix}\mathbf{x}[n] + \begin{bmatrix} 1 \\ 1 \end{bmatrix}u[n];$$

$$y[n] = [0 \quad 1]\mathbf{x}[n].$$

(a) Find the state transition matrix.

(b) Verify the results of Part (a), using a different procedure.

(c) Find the initial-condition response for $x(0) = [1 \quad 2]^T$.

(d) Verify the calculation of the state vector $x[n]$ in Part (c), by substitution in the equation $\mathbf{x}[n+1] = \mathbf{Ax}[n]$.

(e) Calculate the system unit step response, with $x(0) = 0$, using iteration.

(f) Calculate the system unit step response, with $x(0) = 0$, using (13.32).

(g) Verify the results of Parts (e) and (f), using the system transfer function and the z-transform.

(h) Verify the results in Part (g) using MATLAB.

13.16. Consider the system described by the state equations

$$x[n+1] = 0.95x[n] + u[n];$$

$$y[n] = 3x[n].$$

(a) Find the state transition matrix.

(b) Find the initial-condition response for $x(0) = 1$.

(c) Verify the calculation of the state $x[n]$ in Part (b), by substitution in the equation $x[n+1] = Ax[n]$.

(d) Calculate the system unit step response, with $x(0) = 0$, using (13.32).

(e) Verify the results of Part (d), using the system-transfer function and the z-transform.

(f) Verify the results in Part (e) using MATLAB.

13.17. Consider the system of Problem 13.15, given by

$$\mathbf{x}[n + 1] = \begin{bmatrix} 0 & 0 \\ 1 & 0 \end{bmatrix} \mathbf{x}[n] + \begin{bmatrix} 1 \\ 1 \end{bmatrix} u[n];$$

$$y[n] = [0 \quad 1]\mathbf{x}[n].$$

(a) Find the transfer function for this system.
(b) Use a similarity transformation to find a different state model for this system.
(c) Use MATLAB to verify the results of Parts (a) and (b).
(d) Calculate the transfer function of Part (b). This function should equal that of Part (a).
(e) Verify the results in Part (d) using MATLAB.
(f) You have just verified Property 4, (13.64), of similarity transformations. Verify the other three properties in (13.61), (13.62), and (13.63).

13.18. Consider the system of Problem 13.9, given by

$$\mathbf{x}[n + 1] = \begin{bmatrix} 1.9 & 0.8 \\ -1 & 0 \end{bmatrix} \mathbf{x}[n] + \begin{bmatrix} 0 \\ 0.95 \end{bmatrix} u[n];$$

$$y[n] = [1.5 \quad -1.3]\mathbf{x}[n] + 2u[n].$$

(a) Find the transfer function for this system.
(b) Use a similarity transformation to find a different state model for this system.
(c) Use MATLAB to verify the results of Parts (a) and (b).
(d) Calculate the transfer function of Part (b). This function should equal that of Part (a).
(e) Verify the results in Part (d) using MATLAB.
(f) You have just verified Property 4, (13.64), of similarity transformations. Verify the other three properties in (13.61), (13.62), and (13.63).

13.19. Consider the system of Problem 13.18.

(a) Determine if this system is stable.
(b) Give the system modes.
(c) Use MATLAB to check the results in Part (a).

13.20. Consider the system of Problem 13.8(i).

(a) Use MATLAB to determine if this system is stable.
(b) Give the system modes.

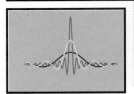

INTEGRALS

1. $\displaystyle\int u\,dv = uv - \int v\,du$

2. $\displaystyle\int e^u\,du = e^u + C$

3. $\displaystyle\int \cos u\,du = \sin u + C$

4. $\displaystyle\int \sin u\,du = -\cos u + C$

5. $\displaystyle\int u e^u\,du = e^u(u - 1) + C$

6. $\displaystyle\int e^{au}\cos bu\,du = \frac{e^{au}(a\cos bu + b\sin bu)}{a^2 + b^2} + C$

7. $\displaystyle\int e^{au}\sin bu\,du = \frac{e^{au}(a\sin bu - b\cos bu)}{a^2 + b^2} + C$

8. $\displaystyle\int u\cos u\,du = \cos u + u\sin u + C$

9. $\displaystyle\int u\sin u\,du = \sin u - u\cos u + C$

TRIGONOMETRIC IDENTITIES

1. $\cos(a \pm b) = \cos a \cos b \mp \sin a \sin b$

2. $\sin (a \pm b) = \sin a \cos b \pm \cos a \sin b$

3. $\cos a \cos b = \frac{1}{2}[\cos(a + b) + \cos(a - b)]$

4. $\sin a \sin b = \frac{1}{2}[\cos(a - b) - \cos(a + b)]$

5. $\sin a \cos b = \frac{1}{2}[\sin (a + b) + \sin (a - b)]$

6. $\cos 2a = \cos^2 a - \sin^2 a = 2 \cos^2 a - 1 = 1 - 2 \sin^2 a$

7. $\sin 2a = 2 \sin a \cos a$

8. $\cos^2 a = \frac{1}{2}(1 + \cos 2a)$

9. $\sin^2 a = \frac{1}{2}(1 - \cos 2a)$

$\sin C.$

$\frac{1}{2}(\sin (a+b) + \sin a - b$

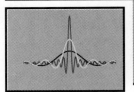

B

LEIBNITZ'S AND L'HÔPITAL'S RULES

LEIBNITZ'S RULE

Given the integral

$$g(t) = \int_{a(t)}^{b(t)} f(t, x) \, dx,$$

where $a(t)$ and $b(t)$ are differentiable in t and $f(t, x)$ and $\partial f(t, x)/\partial t$ are continuous in both t and x, it follows that

$$\frac{dg(t)}{dt} = \int_{a(t)}^{b(t)} \frac{\partial f(t, x)}{\partial t} \, dx + f[b(t), t] \frac{db(t)}{dt} - f[a(t), t] \frac{da(t)}{dt}.$$

EXAMPLE B.1

Let

$$s(t) = \int_{-\infty}^{t} h(\tau) \, d\tau$$

Then

$$\frac{ds(t)}{dt} = \int_{-\infty}^{t} \frac{\partial h(\tau)}{\partial t} \, d\tau + h(t) \frac{dt}{dt} - h(\alpha) \frac{d\alpha}{dt} \bigg|_{\alpha \to -\infty}.$$

The first and third terms on the right are zero, with the result

$$\frac{ds(t)}{dt} = h(t).$$

$\blacksquare$

L'Hôpital's Rule

If the functions $f(x)$ and $g(x)$ both vanish for $x = a$, such that $f(x)/g(x)$ assumes the indeterminant form 0/0, then

$$\lim_{x \to a} \frac{f(x)}{g(x)} = \lim_{x \to a} \frac{f'(x)}{g'(x)},$$

provided that the limit exists as x approaches a from one or both sides. This rule also applies for the indeterminant form ∞/∞.

1. $\displaystyle\sum_{k=0}^{n} a^k = \frac{1 - a^{n+1}}{1 - a}$

2. $\displaystyle\sum_{k=0}^{\infty} a^k = \frac{1}{1 - a}; \quad |a| < 1$

3. $\displaystyle\sum_{k=n}^{\infty} a^k = \frac{a^n}{1 - a}; \quad |a| < 1$

4. $\displaystyle\sum_{k=n_1}^{n_2} a^k = \frac{a^{n_1} - a^{n_2+1}}{1 - a}, \; n_2 > n_1$

5. $\displaystyle\sum_{k=0}^{\infty} k a^k = \frac{a}{(1 - a)^2}; \quad a < 1$

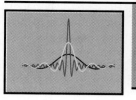

D

COMPLEX NUMBERS AND EULER'S RELATION

A review of complex numbers, complex functions, and Euler's relation is presented in this appendix. It is assumed that the reader has a background in this area.

We begin the review of complex numbers by defining the complex plane [1, 2]. The complex plane is shown in Figure D.1. The abscissa in this plane is the real axis, denoted by Re; the ordinate is the imaginary axis, denoted by Im. An example of a complex number is the number

$$s = 3 + j2, \quad j = \sqrt{-1}. \tag{D.1}$$

We will use the engineering notation j for $\sqrt{-1}$, rather than i, as used by mathematicians. The complex number s, which is plotted in Figure D.1, can also be expressed as

$$s = 3 + j2 = 3 + \sqrt{-4}.$$

A third method of expressing s is $s = (3, 2)$.

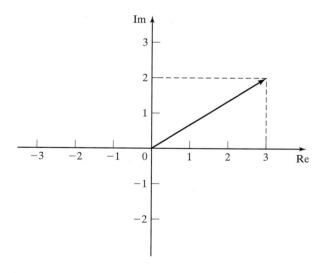

Figure D.1 Complex plane.

In general, a complex number s can be expressed as

$$s = a + jb, \tag{D.2}$$

where both a and b are real. The real number a is known as the real part of s, which is denoted as $a = \text{Re}(s)$. The real number b is known as the imaginary part of s, which is denoted as $b = \text{Im}(s)$. Note that the imaginary part of s is b, not jb; the imaginary part of a complex number is real. A complex number expressed as in (D.2) is said to be in the *rectangular form*. We will define other forms for a complex number later.

The following relationships are seen from the definition of j:

$$\begin{array}{ll}
j = \sqrt{-1}; & j^5 = j(j^4) = j; \\
j^2 = -1; & j^6 = j^2(j^4) = -1; \\
j^3 = j(j^2) = -j; & j^7 = -j; \\
j^4 = (j^2)^2 = 1; & j^8 = 1; \\
 & j^9 = j \\
 & \vdots
\end{array} \tag{D.3}$$

Also, the reciprocal of j is $-j$; that is,

$$\frac{1}{j} = \frac{1}{j}\frac{j}{j} = \frac{j}{-1} = -j. \tag{D.4}$$

COMPLEX-NUMBER ARITHMETIC

All real arithmetic operations apply to complex-number arithmetic, but we must remember the definition of j. First, the complex numbers $s_1 = a + jb$ and $s_2 = c + jd$ are equal,

$$s_1 = s_2 \quad \therefore a + jb = c + jd \tag{D.5}$$

if and only if $a = c$ and $b = d$. Hence, an equation relating complex numbers is in fact two equations relating real numbers. The real parts of the numbers must be equal, and the imaginary parts of the numbers must also be equal.

Let the complex number s_3 be the sum of s_1 and s_2, where $s_1 = 2 + j2$ and $s_2 = 3 - j1$. Then

$$s_3 = s_1 + s_2 = 2 + j2 + 3 - j1 = 5 + j1.$$

This addition can be represented in the complex plane as shown in Figure D.2(a). In the summation of complex numbers, the real part of the sum is equal to the sum of the real parts, and the imaginary part of the sum is equal to the sum of the imaginary parts.

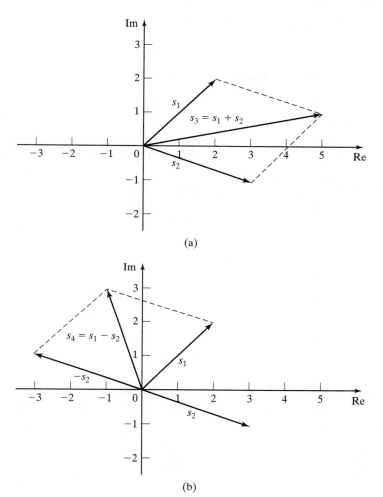

(a)

(b)

Figure D.2 Complex number addition and subtraction.

The difference of the two complex numbers is illustrated by

$$s_4 = s_1 - s_2 = (2 + j2) - (3 - j1) = -1 + j3.$$

In subtraction, the real part of the difference is equal to the difference of the real parts, and the imaginary part of the difference is equal to the difference of the imaginary parts. This subtraction in the complex plane is illustrated in Figure D.2(b) and can be considered to be the addition of s_1 and $-s_2$. Note that the negative of a complex number in the complex plane is that complex number rotated by 180°.

The rules of real multiplication apply to complex multiplication, with the values of powers of j given by (D.3). For example, let s_5 be the product of s_1 and s_2 as just defined:

$$s_5 = s_1 s_2 = (2 + j2)(3 - j1) = 6 - j2 + j6 - (j)^2 2$$
$$= (6 + 2) + j(6 - 2) = 8 + j4.$$

This multiplication is performed in MATLAB by

```
s1 = 2 + j*2; s2 = 3 - j ;
s5 = s1*s2
result: s5 = 8 + 4i
```

The rules of real division also apply to complex division; however, the presence of j in the denominator complicates the operation. We often wish to express the quotient of two complex numbers as a complex number in rectangular form. First, we define the *conjugate* of a complex number. The conjugate of $s = a + jb$, denoted by s^*, is defined as

$$s^* = (a + jb)^* = a - jb.$$

The conjugate of a complex number is obtained by changing the sign of its imaginary part. A property of complex numbers is that the product of a number with its conjugate is real; that is,

$$ss^* = (a + jb)(a - jb) = a^2 + b^2. \tag{D.6}$$

For division of two complex numbers, we multiply both the numerator and the denominator by the conjugate of the denominator to express the quotient in rectangular form; that is,

$$\frac{a + jb}{c + jd} = \frac{a + jb}{c + jd}\frac{c - jd}{c - jd} = \frac{ac + bd}{c^2 + d^2} + j\frac{bc - ad}{c^2 + d^2}. \tag{D.7}$$

As an example, for s_1 and s_2 as defined earlier,

$$\frac{s_1}{s_2} = \frac{2 + j2}{3 - j1} = \frac{2 + j2}{3 - j1}\frac{3 + j1}{3 + j1} = \frac{6 - 2}{9 + 1} + j\frac{6 + 2}{9 + 1} = 0.4 + j0.8.$$

The following MATLAB statement, performs this division:

```
s1 = 2+j*2; s2 = 3-j;
s6 = s1/s2
result: s6 = 0.4 + 0.8i
```

In general, the quotient of two complex numbers can be expressed as

$$\frac{s_1}{s_2} = \frac{s_1 s_2^*}{s_2 s_2^*} = \frac{\text{Re}(s_1 s_2^*)}{s_2 s_2^*} + j\frac{\text{Im}(s_1 s_2^*)}{s_2 s_2^*}. \tag{D.8}$$

Euler's Relation

Before continuing with the review of complex numbers, it is necessary to develop an important relation. Recall, from calculus, the power series [1]

$$e^x = 1 + x + \frac{x^2}{2!} + \frac{x^3}{3!} + \cdots;$$

$$\cos x = 1 - \frac{x^2}{2!} + \frac{x^4}{4!} - \frac{x^6}{6!} + \cdots;$$

and

$$\sin x = x - \frac{x^3}{3!} + \frac{x^5}{5!} - \frac{x^7}{7!} + \cdots. \qquad (D.9)$$

In each of these functions, x is unitless. For the trigonometric functions, we assign the units of radians to x, but recall that radians are unitless. The series of (D.9) are valid for x real or complex. Hence, trigonometric functions of complex arguments are defined. The functions of (D.9) are complex functions of complex arguments.

Consider now the complex exponential e^{jy}. From (D.9),

$$e^{jy} = 1 + jy + \frac{(jy)^2}{2!} + \frac{(jy)^3}{3!} + \frac{(jy)^4}{4!} + \frac{(jy)^5}{5!} + \cdots.$$

From (D.3), we can express this complex exponential as

$$e^{jy} = \left[1 - \frac{y^2}{2!} + \frac{y^4}{4!} - \cdots \right] + j\left[y - \frac{y^3}{3!} + \frac{y^5}{5!} - \cdots \right]$$

and from (D.9),

$$e^{jy} = \cos y + j \sin y. \qquad (D.10)$$

This equation is known as *Euler's* relation. If in (D.10), y is replaced with $-y$, we obtain the relation

$$e^{-jy} = \cos y - j \sin y, \qquad (D.11)$$

since $\cos y$ is even and $\sin y$ is odd. Adding (D.10) and (D.11) yields

$$\cos y = \frac{e^{jy} + e^{-jy}}{2}, \qquad (D.12)$$

and subtracting (D.11) from (D.10) yields

$$\sin y = \frac{e^{jy} - e^{-jy}}{2j}. \qquad (D.13)$$

The four expressions (D.10) through (D.13) are of such importance that they should be memorized. In the engineering use of these four expressions, y is usually real; however, these expressions are valid for y complex.

We will again suppose that s is complex. Then

$$e^s = e^{a+jb} = e^a e^{jb} = e^a(\cos b + j \sin b)$$

$$= e^a \cos b + je^a \sin b. \tag{D.14}$$

Thus, the exponential raised to a complex power is itself complex, with the real and imaginary parts as given in (D.14). For example, for the value of s_1 given previously,

$$e^{s_1} = e^{2+j2} = e^2(\cos 2 + j \sin 2)$$

$$= e^2 \cos 114.6° + je^2 \sin 114.6° = -3.076 + j6.718,$$

because 1 rad = 57.30°. This evaluation is performed in **MATLAB** by

```
exp (2+2*j)
result: -3.0749 + 6.7188i
```

Conversion between Forms

We will now consider expressing a complex number in a form other than the rectangular form, based on the foregoing developments. Euler's relation is given by

$$e^{j\theta} = \cos \theta + j \sin \theta.$$

Letting A and θ be real numbers, we see that

$$Ae^{j\theta} = A(\cos \theta + j \sin \theta) = A \cos \theta + jA \sin \theta$$

$$= a + jb = s. \tag{D.15}$$

Thus, a complex number s can be expressed as a real number multiplied by a complex exponential; this form is called the *exponential form*. For example,

$$5e^{j\pi/6} = 5 \cos 30° + j5 \sin 30° = 4.33 + j2.50.$$

This evaluation is performed in MATLAB by

```
5*exp(j*pi/6)
result: 4.3301 + 2.5i
```

Equation (D.15) gives a procedure for conversion from the exponential form of a complex number to the rectangular form. We now develop a procedure for converting from the rectangular form to the exponential form. From (D.15), with $s = a + ib$,

$$a^2 + b^2 = A^2 \cos^2\theta + A^2 \sin^2\theta = A^2,$$

and thus,

$$A = (a^2 + b^2)^{1/2} = ([\text{Re}(s)]^2 + [\text{Im}(s)]^2)^{1/2}. \tag{D.16}$$

In addition,

$$\frac{b}{a} = \frac{A \sin \theta}{A \cos \theta} = \tan \theta,$$

or

$$\theta = \tan^{-1} \frac{b}{a} = \tan^{-1} \frac{\text{Im}(s)}{\text{Re}(s)}. \tag{D.17}$$

The relationship of the rectangular form to the exponential form is illustrated in the complex plane in Figure D.3 and is quadrant dependent.

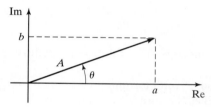

Figure D.3 Rectangular form and exponential form.

REFERENCES

1. R. V. Churchill, J. W. Brown, and R. F. Verkey, *Complex Variables and Applications*, 4th ed. New York: McGraw–Hill, 1989.

2. R. E. Larson and R. P. Hostetler, *Algebra and Trigonometry*, 2d ed. Lexington, MA: D.C. Heath, 1993.

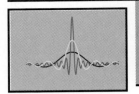

This appendix presents a procedure for solving linear differential equations with constant coefficients. These equations can be expressed as

$$\sum_{k=0}^{n} a_k \frac{d^k y(t)}{dt^k} = \sum_{k=0}^{m} b_k \frac{d^k x(t)}{dt^k},$$ (E.1)

with a_k and b_k constants. The solution procedure is called the *method of undetermined coefficients*.

The method of undetermined coefficients requires that the general solution $y(t)$ be expressed as the sum of two functions [1]:

$$y(t) = y_c(t) + y_p(t).$$ (E.2)

In this equation, $y_c(t)$ is called the *complementary function* and $y_p(t)$ is a *particular solution*. A procedure for finding the complementary function is presented now, and the particular solution is then considered.

Complementary Function

To find the complementary function, we first write the homogeneous equation, which is (E.1) with the left side set equal to zero; that is,

$$a_n \frac{d^n y(t)}{dt^n} + a_{n-1} \frac{d^{n-1} y(t)}{dt^{n-1}} + \cdots + a_1 \frac{dy(t)}{dt} + a_0 y(t) = 0,$$ (E.3)

with $a_n \neq 0$. The complementary function $y_c(t)$ must satisfy this equation. We assume that the solution of the homogeneous equation is of the form $y(t) = Ce^{st}$, where C and s are constants to be determined. Then

$$y(t) = Ce^{st};$$

$$\frac{dy(t)}{dt} = Cse^{st};$$

$$\frac{d^2 y(t)}{dt^2} = Cs^2 e^{st};$$ (E.4)

$$\vdots$$

$$\frac{d^n y(t)}{dt^n} = Cs^n e^{st}.$$

Substitution of these terms into (E.3) yields

$$(a_n s^n + a_{n-1} s^{n-1} + \cdots + a_1 s + a_0) C e^{st} = 0.$$ (E.5)

If we assume that our solution $y(t) = C e^{st}$ is nontrivial ($C \neq 0$), then, from (E.5),

$$a_n s^n + a_{n-1} s^{n-1} + \cdots + a_1 s + a_0 = 0.$$ (E.6)

The equation is called the *characteristic equation*, or the *auxiliary equation*, for the differential equation (E.1). The polynomial may be factored as

$$a_n s^n + a_{n-1} s^{n-1} + \cdots + a_1 s + a_0 = a_n (s - s_1)(s - s_2) \cdots (s - s_n) = 0.$$ (E.7)

Hence, n values of s, denoted as $s_i, 1 \leq i \leq n$, satisfy this equation; that is, $y_i(t) = C_i e^{s_i t}$ for the n values of s_i in (E.7) satisfies the homogeneous equation (E.3), with C_i constant. Because the differential equation is linear, the sum of these solutions is also a solution. For the case of no repeated roots, the solution of the homogeneous equation (E.3) may be expressed as

$$y_c(t) = C_1 e^{s_1 t} + C_2 e^{s_2 t} + \cdots + C_n e^{s_n t}.$$ (E.8)

This solution is called the *complementary function* of the differential equation (E.1) and contains the n unknown coefficients $C_1, C_2, \ldots, C_n$. These coefficients are evaluated in a later step.

Particular Solution

The second part of the general solution of a linear differential equation with constant coefficients,

[eq(E.1)] $$\sum_{k=0}^{n} a_k \frac{d^k y(t)}{dt^k} = \sum_{k=0}^{m} b_k \frac{d^k x(t)}{dt^k},$$

is called a *particular solution*, or a *particular integral*, and is denoted by $y_p(t)$ in (E.2). A particular solution is any function $y_p(t)$ that satisfies (E.1); that is, y_p satisfies

$$\sum_{k=0}^{n} a_k \frac{d^k y_p(t)}{dt^k} = \sum_{k=0}^{m} b_k \frac{d^k x(t)}{dt^k}.$$ (E.9)

The general solution of (E.1) is then the sum of the complementary function, (E.8), and the particular solution, (E.9), as given in (E.2):

[eq(E.2)] $$y(t) = y_c(t) + y_p(t).$$

One procedure for evaluating the particular solution is to assume that the particular solution is the sum of functions of the mathematical form of the excitation $x(t)$ and all derivatives of $x(t)$ that differ in form from $x(t)$. This procedure is called the *method of undetermined coefficients* and applies if the particular solution as described has a finite number of terms. For example, if

$$x(t) = 5e^{-7t},$$

we assume the particular function to be

$$y_p(t) = Pe^{-7t},$$

where the (constant) coefficient P is to be determined. As another example, if $x(t) = 170 \cos 377t$, we assume the particular function

$$y_p(t) = P_1 \cos 377t + P_2 \sin 377t,$$

where the (constant) coefficients P_1 and P_2 are to be determined. The unknown coefficients in $y_p(t)$ are evaluated by direct substitution of the particular solution into the differential equation, as in (E.9), and equating coefficients of like mathematical forms that appear on either side of the equation.

GENERAL SOLUTION

Once the general solution has been formed, the remaining n unknowns $C_1, C_1, \ldots C_n$ of the complementary function (E.8) must be calculated. Thus, we must have n independent conditions to evaluate these unknowns, and these are the n initial conditions

$$y(0), \left. \frac{dy(t)}{dt} \right|_{t=0}, \ldots, \left. \frac{dy^{n-1}(t)}{dt^{n-1}} \right|_{t=0}.$$

REPEATED ROOTS

The complementary function

[eq(E.8)] $$y_c(t) = C_1 e^{s_1 t} + C_2 e^{s_2 t} + \cdots + C_n e^{s_n t}$$

does not apply if any of the roots of the characteristic equation,

$$[eq(E.7)] \quad a_n s^n + a_{n-1} s^{n-1} + \cdots + a_1 s + a_0 = a_n(s - s_1)(s - s_2)\cdots(s - s_n) = 0,$$

are repeated. Suppose, for example, that a fourth-order differential equation has the characteristic equation

$$s^4 + a_3 s^3 + a_2 s^2 + a_1 s + a_0 = (s - s_1)^3(s - s_4). \tag{E.10}$$

The complementary function must then be assumed to be of the form

$$y_c(t) = (C_1 + C_2 t + C_3 t^2)e^{s_1 t} + C_4 e^{s_4 t}. \tag{E.11}$$

The remainder of the procedure for finding the general solution is unchanged. For the general case of an rth-order root s_i in the characteristic equation, the corresponding term in the complementary function is

$$\text{term} = (C_1 + C_2 t + C_3 t^2 + \cdots + C_r t^{r-1})e^{s_i t}. \tag{E.12}$$

REFERENCE

1. W. E. Boyce and R. C. DePrima, *Elementary Differential Equations and Boundary Value Problems*, 2d ed. New York: Wiley, 1992.

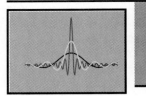

F

PARTIAL-FRACTION EXPANSIONS

The solution of a problem using transforms often results in transformed functions that do not appear in the table of transforms. In most cases, the function is a ratio of polynomials in the transform variable. The ratio of two polynomials is called a *rational function*. We need a procedure for expressing a rational function as a sum of lower-order rational functions, such that these lower-order functions do appear in transform tables. We now present a method, *partial-fraction expansions*, to accomplish this.

An example of a partial-fraction expansion is the relationship

$$\frac{c}{(s + a)(s + b)} = \frac{k_1}{s + a} + \frac{k_2}{s + b}.$$

Given the constants a, b, and c, we wish to find the constants k_1 and k_2. The right side of this equation is called a partial-fraction expansion. A general procedure for calculating partial-fraction expansions will now be presented. In this development, we use the Laplace transform variable s; however, the procedure is independent of the specification of the variable and can be used with any appropriate transform.

The general form of the rational functions that we consider is given by

$$\begin{aligned}
F(s) &= \frac{\beta_m s^m + \beta_{m-1} s^{m-1} + \cdots + \beta_1 s + \beta_0}{s^n + a_{n-1} s^{n-1} + \cdots + a_1 s + a_0} \\
&= \alpha_{m-n} s^{m-n} + \alpha_{m-n-1} s^{m-n-1} + \cdots + \alpha_1 s + \alpha_0 \qquad \text{(F.1)} \\
&\quad + \frac{b_{n-1} s^{n-1} + \cdots + b_1 s + b_0}{s^n + a_{n-1} s^{n-1} + \cdots + a_1 s + a_0}
\end{aligned}$$

for the case that the order of the numerator is higher than that of the denominator. If the numerator order is less than that of the denominator, the parameters α_i are all zero. The expansion shown in (F.1) can be performed using long division.

Partial-fraction expansions are used in finding the inverse transform of the rational function on the right side of (F.1). We will now present a procedure for expanding a rational function as partial fractions. For the general case,

$$F(s) = \frac{b_m s^m + \cdots + b_1 s + b_0}{s^n + a_{n-1} s^{n-1} + \cdots + a_1 s + a_0} = \frac{N(s)}{D(s)}, \quad m < n, \tag{F.2}$$

where $N(s)$ is the numerator polynomial and $D(s)$ is the denominator polynomial. To perform a partial-fraction expansion, first the roots of the denominator polynomial, $D(s)$, must be found. Then we can express the denominator polynomial as

$$D(s) = (s - p_1)(s - p_2) \cdots (s - p_n) = \prod_{i=1}^{n} (s - p_i), \tag{F.3}$$

where $\prod$ indicates the product of terms, and the p_i are called the *poles* of $F(s)$ [the values of s for which $F(s)$ is unbounded].

We first consider the case that the roots of $D(s)$ are distinct (there are no repeated roots). The function $F(s)$ in (F.2) can then be expressed as

$$F(s) = \frac{N(s)}{D(s)} = \frac{N(s)}{\displaystyle\prod_{i=1}^{n} (s - p_i)} = \frac{k_1}{s - p_1} + \frac{k_2}{s - p_2} + \cdots + \frac{k_n}{s - p_n}. \tag{F.4}$$

This partial-fraction expansion is completed by calculating the constants $k_1, k_2, \ldots, k_n$. To calculate $k_j, 1 \leq j \leq n$, first we multiply $F(s)$ by the term $(s - p_j)$:

$$(s - p_j)F(s) = \frac{k_1(s - p_j)}{s - p_1} + \cdots + k_j + \cdots + \frac{k_n(s - p_j)}{s - p_n}. \tag{F.5}$$

If we evaluate this equation at $s = p_j$, all terms on the right are zero except the term k_j. Therefore,

$$k_j = (s - p_j)F(s)|_{s=p_j}, \quad j = 1, 2, \ldots, n, \tag{F.6}$$

which is the desired result.

Next, we consider the case that the denominator has repeated roots. For example, suppose that the rational function is given by

$$F(s) = \frac{N(s)}{(s - p_1)(s - p_2)^r}$$

$$= \frac{k_1}{s - p_1} + \frac{k_{21}}{s - p_2} + \frac{k_{22}}{(s - p_2)^2} + \cdots + \frac{k_{2r}}{(s - p_2)^r}. \tag{F.7}$$

All terms on the right side of this equation must be included, because combining terms of the right side yields the left side. The coefficient for the simple-root term is calculated from (F.6), and the coefficients of the repeated-root terms are calculated from the equation

$$
k_{2j} = \frac{1}{(r-j)!} \frac{d^{r-j}}{ds^{r-j}}[(s-p_2)^r F(s)]\Big|_{s=p_2} \tag{F.8}
$$

with $0! = 1$, and for any function $G(s)$, $d^0 G(s)/ds^0 = G(s)$. This equation is given without proof [1].

The preceding developments apply to complex poles as well as real poles. Suppose that $F(s)$ has a single pair of complex poles at $s = a \pm jb$. If we let $p_1 = a - jb$ and $p_2 = a + jb$, then, with the numerator order of $F(s)$ less than that of the denominator, (F.4) can be written as

$$
F(s) = \frac{k_1}{s-a+jb} + \frac{k_2}{s-a-jb} + \frac{k_3}{s-p_3} + \cdots + \frac{k_n}{s-p_n}. \tag{F.9}
$$

The coefficients k_1 and k_2 can be evaluated using (F.6) as before. It is seen, however, that k_1 and k_2 are complex valued and that k_2 is the conjugate of k_1. All of the remaining coefficients are real.

REFERENCE

1. R. V. Churchill, *Operations Mathematics*, 2d ed. New York: McGraw–Hill, 1972.

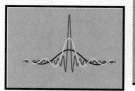

G

REVIEW OF MATRICES

This appendix presents a brief review of definitions, properties, and the algebra of matrices. It is assumed that the reader has a background in this area. Those readers interested in more depth are referred to Refs. 1 through 3. MATLAB statements are given for performing the mathematical operations, where appropriate.

The study of matrices originated in linear algebraic equations. As an example, consider the equations

$$x_1 + x_2 + x_3 = 3;$$
$$x_1 + x_2 - x_3 = 1;$$
$$2x_1 + x_2 + 3x_3 = 6.$$

(G.1)

In a *vector-matrix* format, we write these equations as

$$\begin{bmatrix} 1 & 1 & 1 \\ 1 & 1 & -1 \\ 2 & 1 & 3 \end{bmatrix} \begin{bmatrix} x_1 \\ x_2 \\ x_3 \end{bmatrix} = \begin{bmatrix} 3 \\ 1 \\ 6 \end{bmatrix}.$$

(G.2)

We define the following:

$$\mathbf{A} = \begin{bmatrix} 1 & 1 & 1 \\ 1 & 1 & -1 \\ 2 & 1 & 3 \end{bmatrix}; \quad \mathbf{x} = \begin{bmatrix} x_1 \\ x_2 \\ x_3 \end{bmatrix}; \quad \mathbf{u} = \begin{bmatrix} 3 \\ 1 \\ 6 \end{bmatrix}.$$

(G.3)

Then (G.2) can be expressed as

$$\mathbf{Ax} = \mathbf{u}.$$

(G.4)

In this equation, $\mathbf{A}$ is a 3×3 (3 rows, 3 columns) *matrix*, $\mathbf{x}$ is a 3×1 *matrix*, and $\mathbf{u}$ is a 3×1 *matrix*. Usually, matrices that contain only one row or only one column are called *vectors*. A matrix of only one row and one column is a scalar. In (G.1), for example, x_1 is a scalar. One statement for entering the matrix $\mathbf{A}$ into MATLAB is

```
|A = [1 1 1; 1 1 -1; 2 1 3];
```

733

The general matrix $\mathbf{A}$ is written as

$$\mathbf{A} = \begin{bmatrix} a_{11} & a_{12} & \cdots & a_{1n} \\ a_{21} & a_{22} & \cdots & a_{2n} \\ \vdots & \vdots & \vdots & \vdots \\ a_{m1} & a_{m2} & \cdots & a_{mn} \end{bmatrix} = [a_{ij}], \tag{G.5}$$

where $[a_{ij}]$ is a convenient notation for the matrix $\mathbf{A}$. This matrix has m rows and n columns and, thus, is an $m \times n$ matrix. The element a_{ij} is the element common to the ith row and the jth column.

Some definitions will now be given.

Identity Matrix
The identity matrix is an $n \times n$ (square) matrix with all main diagonal elements a_{ii} equal to 1 and all off-diagonal elements a_{ij} equal to 0, $i \neq j$. For example, the 3×3 identity matrix is

$$\mathbf{I} = \begin{bmatrix} 1 & 0 & 0 \\ 0 & 1 & 0 \\ 0 & 0 & 1 \end{bmatrix}. \tag{G.6}$$

The MATLAB statement for generating the 3×3 identity matrix is

```
I3 = eye(3);
```

If the matrix $\mathbf{A}$ is also $n \times n$, then

$$\mathbf{AI} = \mathbf{IA} = \mathbf{A}. \tag{G.7}$$

Diagonal Matrix
A diagonal matrix is an $n \times n$ matrix with all off-diagonal elements equal to zero:

$$\mathbf{D} = \begin{bmatrix} d_{11} & 0 & 0 \\ 0 & d_{22} & 0 \\ 0 & 0 & d_{33} \end{bmatrix}. \tag{G.8}$$

Symmetric Matrix
The square matrix $\mathbf{A}$ is symmetric if $a_{ij} = a_{ji}$ for all i and j.

Transpose of a Matrix
To take the transpose of a matrix, interchange the rows and the columns. For example,

$$\mathbf{A} = \begin{bmatrix} a_{11} & a_{12} & a_{13} \\ a_{21} & a_{22} & a_{23} \\ a_{31} & a_{32} & a_{33} \end{bmatrix} \text{ and } \mathbf{A}^T = \begin{bmatrix} a_{11} & a_{21} & a_{31} \\ a_{12} & a_{22} & a_{32} \\ a_{13} & a_{23} & a_{33} \end{bmatrix}, \tag{G.9}$$

where $\mathbf{A}^T$ denotes the transpose of $\mathbf{A}$. A property of the transpose is

$$(\mathbf{AB})^T = \mathbf{B}^T \mathbf{A}^T. \tag{G.10}$$

The transpose in MATLAB is denoted with the apostrophe; that is, $\mathbf{A}'$ is the transpose of the matrix $\mathbf{A}$.

Trace
The trace of a square matrix is equal to the sum of its diagonal elements. Given an $n \times n$ matrix $\mathbf{A}$,

$$\text{trace } \mathbf{A} = \text{tr } \mathbf{A} = a_{11} + a_{22} + \cdots + a_{nn}. \tag{G.11}$$

In MATLAB, the trace is found with

```
t = trace(A)
```

Eigenvalues
The *eigenvalues* (*characteristic values*) of a square matrix $\mathbf{A}$ are the roots of the polynomial equation

$$|\lambda\mathbf{I} - \mathbf{A}| = 0, \tag{G.12}$$

where $|\cdot|$ denotes the determinant. Equation (G.12) is the *characteristic equation* of the square matrix $\mathbf{A}$.

Eigenvectors
The *eigenvectors* (*characteristic vectors*) of a square matrix $\mathbf{A}$ are the vectors $\mathbf{x}_i$ that satisfy the equation

$$\lambda_i\mathbf{x}_i = \mathbf{A}\mathbf{x}_i, \tag{G.13}$$

where λ_i are the eigenvalues of $\mathbf{A}$.

The eigenvalues and the eigenvectors are calculated in MATLAB with the statement

```
[v,d] = eig(A)
```

where d denotes a diagonal matrix with the eigenvalues as the diagonal elements and v denotes corresponding eigenvectors.

Properties
Two properties of an $n \times n$ matrix $\mathbf{A}$ are

$$|\mathbf{A}| = \prod_{i=1}^{n} \lambda_i$$

and

$$\text{tr } \mathbf{A} = \sum_{i=1}^{n} \lambda_i, \tag{G.14}$$

where Π denotes the product of factors, and Σ denotes the sum of terms.

Determinants
With both $\mathbf{A}$ and $\mathbf{B}$ $n \times n$,

$$|\mathbf{AB}| = |\mathbf{A}||\mathbf{B}|. \tag{G.15}$$

The MATLAB statement for the determinant is

```
d = det(A)
```

Minor
The minor m_{ij} of element a_{ij} of an $n \times n$ matrix $\mathbf{A}$ is the determinant of the $(n-1) \times (n-1)$ matrix remaining when the ith row and jth column are deleted from $\mathbf{A}$. For example, m_{21} of $\mathbf{A}$ in (G.9) is

$$m_{21} = \begin{vmatrix} a_{12} & a_{13} \\ a_{32} & a_{33} \end{vmatrix} = a_{12}a_{33} - a_{13}a_{32}. \tag{G.16}$$

Cofactor
The cofactor c_{ij} of the element a_{ij} of the square matrix $\mathbf{A}$ is given by

$$c_{ij} = (-1)^{i+j} m_{ij}. \tag{G.17}$$

For (G.16),

$$c_{21} = (-1)^{2+1}(a_{12}a_{33} - a_{13}a_{32}) = -a_{12}a_{33} + a_{13}a_{32}. \tag{G.18}$$

Adjoint
The matrix of cofactors of the matrix $\mathbf{A}$, when transposed, is called the adjoint of $\mathbf{A}$ (adj $\mathbf{A}$). For $\mathbf{A}$ of (G.3),

$$\text{adj } \mathbf{A} = \begin{bmatrix} c_{11} & c_{12} & c_{13} \\ c_{21} & c_{21} & c_{23} \\ c_{32} & c_{32} & c_{33} \end{bmatrix}^T = \begin{bmatrix} 4 & -5 & -1 \\ -2 & 1 & 1 \\ -2 & 2 & 0 \end{bmatrix}^T. \tag{G.19}$$

Inverse
The inverse of a square matrix $\mathbf{A}$ is given by

$$\mathbf{A}^{-1} = \frac{adj A}{|A|}, \tag{G.20}$$

where $\mathbf{A}^{-1}$ denotes the inverse of $\mathbf{A}$ and $|\mathbf{A}|$ denotes the determinant of $\mathbf{A}$. For $\mathbf{A}$ of (G.3) and (G.19), $|\mathbf{A}| = -2$ and

$$\mathbf{A}^{-1} = \begin{bmatrix} -2 & 1 & 1 \\ \frac{5}{2} & -\frac{1}{2} & -1 \\ \frac{1}{2} & -\frac{1}{2} & 0 \end{bmatrix}. \tag{G.21}$$

The MATLAB statement for the inverse is

```
ainv = inv(A)
```

Two properties of the inverse matrix are

$$\mathbf{A}^{-1}\mathbf{A} = \mathbf{A}\mathbf{A}^{-1} = \mathbf{I}$$

and

$$(\mathbf{AB})^{-1} = \mathbf{B}^{-1}\mathbf{A}^{-1}. \tag{G.22}$$

where both $\mathbf{A}$ and $\mathbf{B}$ are $n \times n$. Note that the matrix inverse is defined only for a square matrix and exists only if the determinant of the matrix is nonzero. If $\mathbf{A}$ has an inverse, so does $\mathbf{A}^{-1}$, with $(\mathbf{A}^{-1})^{-1} = \mathbf{A}$. For $\mathbf{A}$ square and $|\mathbf{A}| \neq 0$,

$$(\mathbf{A}^{-1})^T = (\mathbf{A}^T)^{-1} = \mathbf{A}^{-T} \tag{G.23}$$

and

$$|\mathbf{A}^{-1}| = \frac{1}{|\mathbf{A}|}, \tag{G.24}$$

where the notation $\mathbf{A}^{-T}$ is defined by (G.23).

Algebra of Matrices

The algebra of matrices must be defined such that the operations indicated in (G.2), and any additional operation we may wish to perform, lead us back to (G.1).

Addition
To form the sum of matrices $\mathbf{A}$ and $\mathbf{B}$ of equal order, we add corresponding elements a_{ij} and b_{ij}, for each ij. For example,

$$\begin{bmatrix} 1 & 2 \\ 3 & 4 \end{bmatrix} + \begin{bmatrix} 5 & 6 \\ 7 & 8 \end{bmatrix} = \begin{bmatrix} 6 & 8 \\ 10 & 12 \end{bmatrix}. \tag{G.25}$$

Multiplication by a Scalar
To multiply a matrix $\mathbf{A}$ by a scalar k, multiply each element of $\mathbf{A}$ by k, that is, $k\mathbf{A} = [ka_{ij}]$.

Multiplication of Vectors
The multiplication of the $1 \times n$ (row) vector $\mathbf{x}$ with an $n \times 1$ (column) vector $\mathbf{y}$ is defined as

$$[x_1 x_2 \cdots x_n] \begin{bmatrix} y_1 \\ y_2 \\ \vdots \\ y_n \end{bmatrix} = x_1 y_1 + x_2 y_2 + \cdots + x_n y_n. \tag{G.26}$$

Multiplication of Matrices
An $m \times p$ matrix $\mathbf{A}$ may be multiplied by only a $p \times n$ matrix $\mathbf{B}$; that is, the number of columns of $\mathbf{A}$ must equal the number of rows of $\mathbf{B}$. Let

$$\mathbf{AB} = \mathbf{C},$$

where $\mathbf{C}$ is of order $m \times n$. Then the ijth element of $\mathbf{C}$ is equal to the multiplication (as vectors) of the ith row of $\mathbf{A}$ with the jth column of $\mathbf{B}$.
Multiplication in MATLAB is performed by the statement

| C = A*B

As an example, consider the product $\mathbf{AA}^{-1}$ from (G.3) and (G.21):

$$\mathbf{AA}^{-1} = \begin{bmatrix} 1 & 1 & 1 \\ 1 & 1 & -1 \\ 2 & 1 & 3 \end{bmatrix} \begin{bmatrix} -2 & 1 & 1 \\ \frac{5}{2} & -\frac{1}{2} & -1 \\ \frac{1}{2} & -\frac{1}{2} & 0 \end{bmatrix} = \begin{bmatrix} 1 & 0 & 0 \\ 0 & 1 & 0 \\ 0 & 0 & 1 \end{bmatrix}. \qquad (G.27)$$

OTHER RELATIONSHIPS

Other important matrix relationships will now be given.

Solution of Linear Algebraic Equations
Given the linear equations (G.1) expressed in vector-matrix format (G.4)

$$\mathbf{Ax} = \mathbf{u},$$

the solution is

$$\mathbf{x} = \mathbf{A}^{-1}\mathbf{u}. \qquad (G.28)$$

MATLAB performs this operation with the statement

x = inv(A)*u

For example, from (G.21), for the equations (G.1),

$$\mathbf{x} = \mathbf{A}^{-1}\mathbf{u} = \begin{bmatrix} -2 & 1 & 1 \\ \frac{5}{2} & -\frac{1}{2} & -1 \\ \frac{1}{2} & -\frac{1}{2} & 0 \end{bmatrix} \begin{bmatrix} 3 \\ 1 \\ 6 \end{bmatrix} = \begin{bmatrix} 1 \\ 1 \\ 1 \end{bmatrix}. \qquad (G.29)$$

This solution is easily verified by substitution back into the original Equations (G.1).

Cramer's Rule
Given the n linear algebraic equations in vector-matrix form,

$$\mathbf{Ax} = \mathbf{u},$$

where $\mathbf{A}$ is $n \times n$, $\mathbf{x}$ is $n \times 1$, and $\mathbf{u}$ is $n \times 1$, the ith component of $\mathbf{x}$ is given by

$$x_i = \frac{|\mathbf{A}_i|}{|\mathbf{A}|}. \qquad (G.30)$$

In this equation, $|\mathbf{A}|$ is the determinant of the matrix $\mathbf{A}$, and $|\mathbf{A}_i|$ is the determinant of the matrix formed by replacing the ith column in $\mathbf{A}$ with the vector $\mathbf{u}$. Equation (G.30) is called Cramer's rule.

Differentiation

The derivative of a matrix is obtained by differentiating the matrix element by element. For example, let

$$\mathbf{x}(t) = \begin{bmatrix} x_1(t) \\ x_2(t) \end{bmatrix}. \tag{G.31}$$

Then, by definition,

$$\frac{dx(t)}{dt} = \dot{\mathbf{x}}(t) = \begin{bmatrix} \dfrac{dx_1(t)}{dt} \\ \dfrac{dx_2(t)}{dt} \end{bmatrix}. \tag{G.32}$$

Integration

The integral of a matrix is obtained by integrating the matrix element by element. For (G.31),

$$\int \mathbf{x}(t)dt = \begin{bmatrix} \int x_1(t)\,dt \\ \int x_2(t)\,dt \end{bmatrix}. \tag{G.33}$$

REFERENCES

1. F. R. Gantmacher, *Theory of Matrices*, Vols. I and II. New York: Chelsea, 1959.

2. G. Strang, *Linear Algebra and Its Applications*, 2d ed. New York: Academic Press, 1988.

3. G. H. Golub and C. F. Van Loan, *Matrix Computations*, 2d ed. Baltimore, MD: Johns Hopkins University Press, 1996.

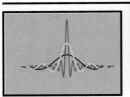

H

ANSWERS TO SELECTED PROBLEMS

CHAPTER 2

CHAPTER 2

2.3. **(a)** $x_2(t) = -2x_1(-t/4) + 2$

2.4. **(a)** $x_2(t) = -(\frac{1}{2}x_1(t/4) - 1)$

   (c) $x_1(t) = -2x_2(4t) + 2$

2.8. **(a)** odd

   (b) even

   (c) even

   (d) even

   (e) neither

2.12. **(a)** $T_0 = \dfrac{2\pi}{3}, \omega_0 = 3$

   (b) $T_0 = \dfrac{\pi}{4}, \omega_0 = 8$

   (c) $T_0 = \pi, \omega_0 = 2$

   (d) $T_0 = 2\pi, \omega_0 = 1$

   (e) $T_0 = \dfrac{2\pi}{5}, \omega_0 = 5$

   (f) $T_0 = \dfrac{2\pi}{5}, \omega_0 = 5$

2.15. **(a)** $T_0 = 2\pi, \omega_0 = 1$

   (b) $T_0 = \pi, \omega_0 = 2$

       (c) aperiodic

       (d) $T_0 = 12, \omega_0 = \dfrac{\pi}{6}$

2.16. $\dfrac{1}{a} \sin^2\left(\dfrac{b}{a} - 4\right)$

2.18. $\frac{1}{2}[x(2) + x(-2)]$

2.22. **(a)** $1 - u(t)$

       (b) $1 - u(t - 3)$

       (c) $t[1 - u(t)]$

       (d) $(t - 3)[1 - u(t - 3)]$

2.26. **(a)** (i) Has memory; (ii) Invertible; (iii) Stable; (iv) Time invariant; (v) Linear;

       (b) Causal for $\alpha \geq 1$

2.28. $2y_1(t + 1) + y_1(t)$

2.30. (i) Has memory; (ii) Invertible; (iii) Causal; (iv) Stable; (v) Time invariant; (vi) Linear

2.31. $h(t) = 2tu(t) - 2(t - 1)u(t - 1)$

2.32. **(a)** (i) Memoryless; (ii) Not invertible; (iii) Causal; (iv) Stable; (v) Time invariant; (vi) Not linear

       (b) (i) Memoryless; (ii) Not invertible; (iii) Causal; (iv) Stable; (v) Time invariant; (vi) Not linear

       (c) (i) Memoryless; (ii) Not invertible; (iii) Causal; (iv) Stable; (v) Time invariant; (vi) Not linear

       (d) (i) Memoryless; (ii) Not invertible; (iii) Causal; (iv) Stable; (v) Time invariant; (vi) Not linear

CHAPTER 3

3.1. **(a)** $(t - 2)u(t - 2)$

       (b) $\frac{1}{5}(e^{5t} - 1)u(t)$

       (c) $tu(t)$

       (d) $\left(\dfrac{t^2}{2} + t + \dfrac{1}{2}\right) \times u(t + 1)$

3.3. $(t - t_0 - t_1)u(t - t_0 - t_1)$

3.7. **(a)** $(1 - e^{t-2})[u(t - 1) - u(t - 2)] + (2 - e^{t-1} - e^{t-2})$

             $\times [u(t) - u(t - 1)] + (2e^t - e^{t-1} - e^{t-2}) \times u(-t)$

(b) $(1 - e^{-(t-1)})[u(t - 1) - u(t - 3)] + (e^{-(t-3)} - e^{-(t-1)})u(t - 3)$

(c) $e^{-1}u(2 - t) + e^{-(t-1)}u(t - 2)$

(d) $\frac{1}{a}(1 - e^{-at})[u(t) - u(t - 2)] + \frac{1}{a}(1 - e^{-2a})u(t - 2)$

(e) $(1 - e^{-400})u(-t) + (e^{-t} - e^{-400})[u(t) - u(t - 400)]$

(f) $2(1 - e^{-(1-t)}) \times u(1 - t)$

3.10. $(t - 5)u(t - 5)$

3.12. **(a)** $te^{-t}u(t)$

(b) $\delta(t)$

(c) $\delta(t - 2)$

(d) $(t - 4)u(t - 4) - 2(t - 6)u(t - 6) + (t - 8)u(t - 8)$

3.13. $z(t) = y(t + a)$

3.14. **(a)** $\delta(t - 7)$

(b) $u(t - 7)$

(c) $(t - 7)u(t - 7)$

3.16. **(a)** $y(t) = [h_1(t)*h_2(t) + h_1(t)*h_3(t)*h_4(t) + h_4(t)*h_5(t)]*x(t)$

(b) $h(t) = 5u(t) + 5tu(t) + \frac{1}{2}(1 - e^{-2t})u(t)$

3.19. **(a)** Not causal

(b) Stable

(c) $y(t) = e^t u(-t) + u(t)$

(d) Causal, unstable, $y(t) = (e^t - 1)u(t)$

3.23. **(a)** **(i)** $h(t) = e^{-2(t-1)}u(t - 1)$

(ii) Causal

(iii) Stable

(b) **(i)** $h(t) = e^{-2(t-1)}$

(ii) Not causal

(iii) Unstable

3.25. **(a)** Causal

(b) Stable

(c) Not causal and not stable

3.26. **(i)** $y(t) = \frac{3}{2} - \frac{5}{2}e^{-2t}, t \geq 0$

(ii) $y(t) = 3e^{-t} - 2e^{-2t}, t \geq 0$

3.27. **(a)** Stable

(b) Stable

(c) Unstable

3.28. Modes are at $s = 2$ and $s = \frac{1}{2}$. Not stable.

3.32. **(a)** $a = 4$ and $K = 14.14$

(b) $H(s) = \dfrac{14.14}{s + 4}$

3.36. **(a)** $H(s) = H_1(s)H_2(s) + H_1(s)H_3(s)H_4(s) + H_4(s)H_5(s)$

(b) $H(s) = H_1(s)H_2(s)H_3(s) + H_1(s)H_2(s)H_4(s) + H_1(s)H_5(s)$

(c) $H(s) = \dfrac{H_1(s)}{1 + H_1(s)H_2(s)}$

▨ CHAPTER 4

4.3. **(i)** $C_3 = \dfrac{1}{2}, C_{-3} = \dfrac{1}{2}, C_5 = \dfrac{1}{2j}, C_{-5} = -\dfrac{1}{2j}, C_k = 0$, all other k

(ii) $C_1 = 1, C_3 = \dfrac{1}{2}, C_{-3} = \dfrac{1}{2}, C_4 = \dfrac{1}{2j}, C_{-4} = -\dfrac{1}{2j}, C_k = 0$, all other k

(iii) Aperiodic

(iv) $C_1 = \dfrac{1}{2j}, C_{-1} = -\dfrac{1}{2j}, C_2 = \dfrac{1}{2j}, C_{-2} = -\dfrac{1}{2j}, C_k = 0$, all other k

4.6. **(a)** π

(b) π

(c) 0

4.10. **(a)** $C_k = \dfrac{3j}{k\pi}\left[1 - \cos\left(\dfrac{k\pi}{2}\right)\right]$ and $C_0 = 0$

(b) $C_k = \dfrac{j}{2\pi k}[2e^{-j2\pi k} - e^{-jk\pi} - e^{-j\frac{3}{2}k\pi}]$ and $C_0 = \dfrac{3}{4}$

(c) $C_k = \dfrac{-1}{k^2\pi^2}[e^{-jk\pi}(-jk\pi - 1) + 1]$ and $C_0 = \dfrac{1}{2}$

(d) $C_k = \dfrac{e^{-jk\pi} - 1}{-jk\pi} - \dfrac{1}{k^2\pi^2}[e^{-jk\pi}(-jk\pi-1) + 1]$ and $C_0 = \frac{1}{2}$

(e) $C_k = \dfrac{1}{\pi(1 - k^2)}[e^{j\frac{\pi}{2}k} - jk]$

4.11. **(a)** $C_k = \dfrac{-8}{(\pi k)^2}$, k odd, $C_k = 0$, k even

(b) $C_0 = 3.6$, $C_k = 1.6\,\text{sinc}(0.2\pi k)$, $k \neq 0$

(c) $C_0 = 0$, $C_k = \dfrac{4j}{\pi k}$, $k \neq 0$

(d) $C_0 = 10$, $C_k = \dfrac{40}{(\pi k)^2}$, k odd, $C_k = 0$, k even

(e) $C_0 = \dfrac{12}{\pi}$, $C_k = \dfrac{-12}{\pi(4k^2 - 1)}$, $k \neq 0$

4.14. **(a)** $C_k = 10$, k odd, $C_k = 0$, k even

(b) $C_k = 5[1 - e^{-jk\pi}]$

4.17. **(a)** $m = 2$

(b) $m = 1$

(c) $m = 1$

(d) $m = 2$

(e) $m = 2$

4.21. $2 + e^{j\pi t} + \cos\left(3\pi t + \dfrac{\pi}{4}\right)$

4.23. **(a)** $\omega_0 = \dfrac{\pi}{3}$

(b) $C_0 = \dfrac{1}{6}$, $C_k = \dfrac{1}{j2\pi k}(1 - e^{-jk\omega_0})$, $k \neq 0$

4.24. $C_{yk} = H(jk\omega_0)\,C_{xk} = \dfrac{10}{jk\omega_0 + 5}C_{xk}$

4.31. **(a)** $y(t) = \dfrac{1}{2} - \dfrac{1}{4}\dfrac{\alpha}{\alpha + j\omega_0}e^{j4t} - \dfrac{1}{4}\dfrac{\alpha}{\alpha - j\omega_0}e^{-j4t}$

(b) $y(t) = 1 + \dfrac{1}{2}\dfrac{\alpha}{\alpha + j\omega_0}e^{jt} + \dfrac{1}{2}\dfrac{\alpha}{\alpha - j\omega_0}e^{-jt}$

$$+ \dfrac{1}{2}\dfrac{\alpha}{\alpha + 8j\omega_0}e^{j8t} + \dfrac{1}{2}\dfrac{\alpha}{\alpha - 8j\omega_0}e^{-j8t}$$

CHAPTER 5

5.2. **(a)** $\dfrac{-b\omega^2 + j\omega b^2}{\omega^4 + \omega^2 b}$

(b) $F(\omega) = A\pi e^{j\phi}\delta(\omega - \omega_0) + A\pi e^{-j\phi}\delta(\omega + \omega_0)$

(c) $\dfrac{1}{a - j\omega}$

(d) $Ce^{j\omega t_0}$

5.6. **(a)** $F(\omega) = \dfrac{\frac{A}{2}}{\beta + j(\omega - \omega_0)} + \dfrac{\frac{A}{2}}{\beta - j(\omega + \omega_0)}$

(b) $F(\omega) = \dfrac{A\pi}{j}[\delta(\omega - \omega_1) - \delta(\omega + \omega_1)] + \beta\pi[\delta(\omega - \omega_2) + \delta(\omega + \omega_2)]$

(c) $F(\omega) = 12\pi \, \text{rect}\,(\omega)$

(d) $18\,\text{sinc}\left(\dfrac{3\omega}{2}\right)e^{-j4\omega}$

5.8. **(a)** $\dfrac{j2\omega}{\omega^2 + 1}$

(b) $\frac{1}{2}e^{-|\omega|}$

(c) $2\pi[e^{-|\omega - 2|} + e^{-|\omega + 2|}]$

5.10. $F(\omega) = aA\text{sinc}^2\left(\dfrac{a\omega}{2}\right)$

5.11. $y(t) = \dfrac{\pi}{2}\cos(t)$

5.15. **(a)** $g(t) = \dfrac{50}{\pi}\,\text{sinc}(10t)$

(b) $g(t) = \dfrac{25}{\pi}[\text{sinc}(10t + 5\pi) + \text{sinc}\,(10t - 5\pi)]$

5.18. $G_p(\omega) = \displaystyle\sum_{n=-\infty}^{\infty} 20\pi\,\text{sinc}\left(\dfrac{n\pi}{2}\right)\delta\left(\omega - \dfrac{n\pi}{2}\right)$

5.20. (a) $\pi \, \text{sinc}(t)$

(b) $\dfrac{\pi}{2}\text{sinc}\,(t) + \dfrac{\pi}{4}\,\text{sinc}^2\left(\dfrac{t}{2}\right)$

(c) 0

5.21. $-4 < A < 4$

5.27. 91%

5.28. $P_y = \dfrac{30}{\pi} = 9.55 \text{ W}$

CHAPTER 6

6.5. $H(\omega) = \dfrac{1}{1 + j\left(\frac{\omega L}{R} - \frac{1}{\omega RC}\right)}$. This is a bandpass filter.

6.7. $H(\omega) = \dfrac{1}{\sqrt{1 + [(\frac{\omega}{\omega_1})^2]^2}}$

6.12. (a) The first null of $f_1(t)$ is at $\frac{2\pi}{\tau}$, the first null of $f_2(t)$ is at $\frac{2\pi}{\tau}$, and the first null of $f_3(t)$ is at $\frac{4\pi}{\tau}$.

(b) Shorter time duration results in wider bandwidth.

6.22. (a) High-pass filter

(b) Low-pass filter

6.26. $Y(\omega) = \frac{1}{4}[2M(\omega) + M(\omega - 2\omega_c) + M(\omega + 2\omega_c)]$ and $Z(\omega) = \frac{1}{2}M(\omega)$

6.28. (a) $g_1(t) = \frac{1}{2}f_1(t) + \frac{1}{2}f_1(t)\cos(2\omega_c t) + \frac{1}{2}f_2(t)\sin(2\omega_c t)$

(b) $g_2(t) = \frac{1}{2}f_2(t) + \frac{1}{2}f_1(t)\sin(2\omega_c t) - \frac{1}{2}f_2(t)\cos(2\omega_c t)$

(c) $e_1(t) = \frac{1}{2}f_1(t)$ and $e_2(t) = \frac{1}{2}f_2(t)$

6.32. (a) 19 signals can be multiplexed.

(b) 785.4 (k-rad/sec)

6.33. (a) $x(t)$ must be band limited.

(b) $H(\omega) = 0, |\omega| > \dfrac{\omega_s}{2}$

(c) $200\pi, 300\pi, 700\pi, 800\pi, 1200\pi, 1300\pi, -200\pi, -300\pi,$

$-700\pi, -800\pi, -1200\pi, -1300\pi$

(d) $x(t) = \cos(300\pi t)$

CHAPTER 7

7.3. **(a)** $f(t) = 5tu(t) - 5(t - 2)u(t - 2) - 15u(t - 2) + 5u(t - 4)$

(b) $F(s) = \dfrac{5}{s^2} - \dfrac{5}{s^2}e^{-2s} - \dfrac{15}{s}e^{-2s} + \dfrac{5}{s}e^{-4s}$

7.5. **(a)** $F(s) = \dfrac{s}{s^2 - a^2}$

7.8. **(a)** $\dfrac{5}{2}tu(t) - 5(t - 2)u(t - 2) + \dfrac{5}{2}(t - 4)u(t - 4)$

(b) $\dfrac{\frac{5}{2}}{s^2} - \dfrac{5e^{-2s}}{s^2} + \dfrac{\frac{5}{2}e^{-4s}}{s^2}$

(c) $\dfrac{5}{2}u(t) - 5u(t - 2) + \dfrac{5}{2}u(t - 4)$

(d) $\dfrac{1}{s}\left(\dfrac{5}{2} - 5e^{-2s} + \dfrac{5}{2}e^{-4s}\right)$

(e) $\dfrac{1}{s^2}\left(\dfrac{5}{2} - 5e^{-2s} + \dfrac{5}{2}e^{-4s}\right)$

(f) $\dfrac{1}{s}\left(\dfrac{5}{2} - 5e^{-2s} + \dfrac{5}{2}e^{-4s}\right)$

7.10. **(a)** 2

(b) The final value theorem gives 0, which is an error. There is no final value.

7.11. **(a)** $\dfrac{1}{s^2}$

(b) $\dfrac{s^2 - b^2}{(s^2 + b^2)^2}$

(c) $\dfrac{n!}{s^{n+1}}$

7.13. **(a)** $f(t) = 2.5(1 - e^{-2t})u(t)$

(b) $f(t) = (1.5 - 2e^{-t} + 0.5e^{-2t})u(t)$

7.15.   **(a)** $f(t) = (1 - e^{-3(t-2)})u(t - 2)$

   (b) $f(t) = (1 - e^{-3t})u(t) - (1 - e^{-3(t-2)})u(t - 2)$

7.21.   **(a)** $\dfrac{1}{s + 2}$, $\text{Re}(s) > -2$

   (b) $\dfrac{e^{-8}e^{-4s}}{s + 2}$, $\text{Re}(s) > -2$

   (c) $\dfrac{1}{s - 2}$, $\text{Re}(s) < 2$

   (d) $\dfrac{e^{-8}e^{4s}}{s - 2}$, $\text{Re}(s) < 2$

   (e) $\dfrac{e^{8}e^{4s}}{s + 2}$, $\text{Re}(s) > -2$

   (f) $\dfrac{e^{8}e^{-4s}}{s - 2}$, $\text{Re}(s) < 2$

7.24.   **(a)** $f(t) = -9u(-t) + 8e^{-t}u(-t)$

   (b) $f(t) = 9u(t) - 8e^{-t}u(t)$

   (c) $f(t) = -9u(-t) - 8e^{-t}u(t)$

   (d)
       (a) $f(\infty) = 0$
       (b) $f(\infty) = 9$
       (c) $f(\infty) = 0$

7.26.   **(a)** $\text{Re}(a) > 0$ and $\text{Re}(b) > 0$
   (b) Either $\text{Re}(a) > 0$ and $\text{Re}(b) < 0$ or $\text{Re}(a) < 0$ and $\text{Re}(b) > 0$
   (c) $\text{Re}(a) < 0$ and $\text{Re}(b) < 0$

7.27. $\frac{3}{2}e^{-4t}u(t) + \frac{1}{2}e^{-2t}u(-t)$

7.28.   The poles at -10 and -5 are right-sided and the pole at 3 is left-sided.

CHAPTER 8

8.1. **(a)** $x_1 = i, v_i = u, y = v_R$

$$\dot{x} = -\frac{R}{L}x + \frac{1}{L}u$$

$$y = Rx$$

(b) $x = v_R = Ri, i = \frac{1}{R}x$

$$\dot{x} = -\frac{R}{L}x + \frac{R}{L}u$$

$$y = x$$

8.3. **(b)** $x_1 = y; x_2 = \dot{y};$

$$\underline{\dot{x}} = \begin{bmatrix} 0 & 1 \\ -2 & -3 \end{bmatrix}\underline{x} + \begin{bmatrix} 0 \\ 4 \end{bmatrix}u$$

$$y = [1 \quad 0]\,\underline{x}$$

8.4. **(d)**

(b) $\underline{\dot{x}} = \begin{bmatrix} 0 & 1 \\ -1 & -3 \end{bmatrix}\underline{x} + \begin{bmatrix} 0 \\ 1 \end{bmatrix}u$

$$y = [0 \quad 50]\,\underline{x}$$

(c) $\ddot{y} + 3\dot{y} + y = 50\dot{u}$

8.5. **(b)** $\dot{x} = -2x + u$

$$y = 4x$$

(c) $\dfrac{Y(s)}{U(s)} = \dfrac{4}{s + 2}$

8.6. **(a)** $\dot{x} = -3x + 6u$

$$y = 4x$$

(b) $sI - A = s + 3$

8.10. **(b)** $H(s) = \dfrac{2s^2 + 7s + 1}{s^2 + 3s - 4}$

8.12. **(a)** $\dot{x} = -\dfrac{R}{L}x + \dfrac{1}{L}u$

$$y = Rx$$

(b) $H(s) = \dfrac{\frac{R}{L}}{s + \frac{R}{L}}$

(c) $\dfrac{V_R(s)}{V_i(s)} = \dfrac{\frac{R}{L}}{s + \frac{R}{L}}$

8.15. **(a)** $\dot{x} = -3x + 6u$

$y = 4x$

(b) $\Phi(t) = e^{-3t}$

(c) $y_c(t) = 8e^{-3t},\ t > 0$

(d) $y_p(t) = 8(1 - e^{-3t}),\ t > 0$

(f) $y(t) = 8,\ t > 0$

8.21. **(a)** $H(s) = \dfrac{2s^2 + 7s + 1}{s^2 + 3s - 4}$

8.24. **(a)** $H(s) = \dfrac{2s}{s^3 - s^2 - 1}$

(b) $\underline{\dot{v}} = \begin{bmatrix} 2 & 1 & 0 \\ -2 & -1 & 1 \\ 1 & 0 & 0 \end{bmatrix} \underline{v} + \begin{bmatrix} 0 \\ 2 \\ 0 \end{bmatrix} u$

$y = [1 \quad 0 \quad 0]\,\underline{v}$

8.26. **(a)** Not stable

(b) e^{-4t}, e^{-t}

(c) $A = [-4 \quad 5; 0 \quad 1];\ \mathrm{eig}(A)$

CHAPTER 9

9.2. **(a)** **(i)** $2\delta[n] - 4\delta[n - 1]$

(ii) $-4\delta[n + 4] - 4\delta[n + 2] + 2\delta[n] + 2\delta[n - 2]$

(iii) $-4\delta[n + 2] - 4\delta[n + 1] + 2\delta[n] + 2\delta[n - 1]$

(iv) $-4\delta[n] - 4\delta[n - 1] + 2\delta[n - 2] + 2\delta[n - 3]$

(v) $2\delta[n - 1] + 2\delta[n - 2] - 4\delta[n - 3] - 4\delta[n - 4]$

(vi) $-4\delta[n + 4] - 4\delta[n + 3] + 2\delta[n + 2] + 2\delta[n + 1]$

9.5. (a) $a_1 = 3, b_1 = 0$

(b) $a_2 = 2, b_2 = 1$

(c) $a_3 = 3, b_3 = -1$ or $a_3 = -3, b_3 = 11$

9.7. (a) $x_e[n] = -\delta[n+2] - \delta[n-2]$ and $x_o[n]$
$$= \delta[n+2] + 2\delta[n+1] - 2\delta[n-1] - \delta[n-2]$$

(b) $x_e[n] = \frac{1}{2}\delta[n+4] + \delta[n+3] + \frac{3}{2}\delta[n+2] + \delta[n+1]$
$$+ \delta[n-1] + \frac{3}{2}\delta[n-2] + \delta[n-3] + \frac{1}{2}\delta[n-4]$$
$$\text{and } x_o[n] = \frac{1}{2}\delta[n+4] + \delta[n+3]$$
$$+ \frac{1}{2}\delta[n+2] - \frac{1}{2}\delta[n-2] - \delta[n-3] - \frac{1}{2}\delta[n-4]$$

9.9. (i) Neither

(ii) Odd

(iii) Even

(iv) Even

(v) Even

(vi) Neither

9.13. (a) Periodic, $N_0 = 10$

(b) Periodic, $N_0 = 16$

(c) Periodic, $N_0 = 5$

(d) Periodic, $N_0 = 80$

9.14. (a) Periodic, $N_0 = 14$

(b) Not periodic

(d) Periodic, $N_0 = 1$

(d) Not periodic

(e) Periodic, $N_0 = 14$

(f) Not periodic

(g) Periodic, $N_0 = 14$

(h) Periodic, $N_0 = 14$

(i) Not periodic

9.17. (i) Periodic, $N_0 = 2$

(ii) Periodic, $N_0 = 20$

(iii) Periodic, $N_0 = 4$

(iv) Not periodic

(v) Periodic, $N_0 = 4$

(vi) Periodic, $N_0 = 40$

9.19. (a) $x_a[n] = 2\delta[n + 1] + 2\delta[n] - 4\delta[n - 1] - 4\delta[n - 2]$

(b) $x_b[n] = -2\delta[n + 1] + 2\delta[n] + 2\delta[n - 1]$

(c) $x_c[n] = -2\delta[n + 1] + 4\delta[n] - 4\delta[n - 1] + 4\delta[n - 3]$

(d) $x_d[n] = 2\delta[n + 1] + 2\delta[n - 1] + 4\delta[n - 3]$

9.21. (a) $y[n] = T_1(x[n]) + T_3[T_2\{x[n] - T_4(y[n])\}]$

9.23. (a) Has memory, not invertible, not causal, stable, time invariant, not linear

(b) Has memory, invertible, not causal, stable, time varying, linear

(c) Memoryless, not invertible, causal, stable, time invariant, not linear

(d) Memoryless, invertible, causal, stable, time invariant, not linear

(e) Memoryless, not invertible, causal, not stable, time varying, not linear

9.28. $h[n] = 4\delta[n - 1] + 8u[n - 2]$

CHAPTER 10

10.3. (a) $y[5] = 6$

(b) $y[3] = 18$

(c) $n = 2, 3$

(d) $y[n] = 6\delta[n] + 12\delta[n - 1] + 18\delta[n - 2] + 18\delta[n - 3]$
$+ 12\delta[n - 4] + 6\delta[n - 5]$

10.4. (a) $y[n] = \dfrac{\beta^{n+1} - \alpha^{n+1}}{\beta - \alpha}$

(b) $\beta^4 + \alpha\beta^3 + \alpha^2\beta^2 + \alpha^3\beta + \alpha^4$

10.9. (a) $b^2\left(\dfrac{1}{a^3}\right)^{n-2}\left(\dfrac{a^3b - (a^3b)^{n-3}}{a^3b - 1}\right)u[3 - n]$

(b) $\dfrac{a^{n+1}}{a - b}u[-1 - n] + \dfrac{ab^n}{a - b}u[n]$

(c) $\left(\dfrac{1 - a^{n+1}}{1 - a}\right)(u[n] - u[n - 100]) + \dfrac{1 - a^{100}}{1 - a}u[n - 100]$

(d) $\dfrac{1}{b^2 - 1}\left(\dfrac{1}{b^2}\right)^{n+1}u[n + 1] + \dfrac{1}{b^2 - 1}u[-2 - n]$

$$(e) \left[\frac{b^{n+1} - b^2(a^2)^{n-1}}{b - a^2} \right] u[n - 2]$$

(f) $51u[-n] + (52 - n)(u[n - 1] - u[n - 52])$

10.11. (a) $h[n] = (n + 1)(.8)^n u[n]$

(b) $h[n] = \delta[n - 6]$

(c) $h[n] = \delta[n - 6]$

(d) $h[n] = \delta[n] + 2\delta[n - 1] + \delta[n - 2]$

10.12. (a) Causal

(b) Not stable

(c) $(2^{n+1} - 1)u[n]$

(e) Noncausal, stable, $y[n] = 2u[n] + 2^{n+1}u[-n - 1]$

(f) Noncausal, not stable

(g) Noncausal, not stable

10.14. (a) $h[n] = .8\delta[n + 1] + .8\delta[n]$

(b) Noncausal

(c) $y[n] = .8u[n + 2] + .8u[n + 1]$

10.16. (a) Causal, stable

(b) Noncausal, not stable

(c) Causal, not stable

(d) Causal, not stable

(e) Causal, stable

(f) Causal, stable

10.18. (a) $h[n] = \delta[n - 3]$

(b) $h[n] = u[n + 3]$

10.19. (i) $y[n] = -\frac{5}{7}(\frac{5}{6})^n + \frac{12}{7}(2)^n, n \geq 0$

(ii) $y[n] = 2.108(0.7)^n - 1.108e^{-n}, n \geq -1$

10.21. (i) Stable

(ii) Stable

(iii) Not stable

(iv) Not stable

(v) Stable

(vi) Not stable

10.27.   **(a)** $y[n] - 0.9y[n-1] = x[n] - x[n-1]$

       (d) $x[n] = (0.7)^n u[n]$

       (e) $y[n] = 1.5[(0.7)^n - (0.9)^n]$

10.29.   **(a)** $H(z) = \dfrac{z}{z - 0.7}$

       (b) $y_{ss}[n] = 1.168 \cos(n - 43.5°)$

       (d) $y_{ss}[n] = 0.847 \cos n + 0.804 \sin n$

$$+ \, 0.153 \cos n - 0.803 \sin n \approx \cos n$$

10.30.   **(a)** $|b| > 1$

       (b) No restriction

       (c) $|a| < |b|$

       (d) No restriction

CHAPTER 11

11.1.   **(a)** $\dfrac{z}{z - 0.6}$

       (b) $\dfrac{5z^2 - 4.3z}{z^2 - 2.2z + 1.05}$

       (c) $\dfrac{5z}{z - 0.6065}$

       (d) $\dfrac{5z}{z - e^{-j.1}}$

       (e) $\dfrac{5z \sin 3}{z^2 - 2z \cos 3 + 1}$

       (f) $\dfrac{14.14z^2 + 18.75z}{z^2 + 0.832z + 1}$

11.8.   **(a)** $\dfrac{1}{z^3 - 3z^2 + 5z - 9}$

       (b) $\dfrac{6z^3 + 7z^2 + 36z}{z^3 - 3z^2 + 5z - 9}$

11.10. **(iii)**

    (a) $x[n] = \frac{2}{3}\delta[n] + u[n] - \frac{5}{3}(0.6)^n$

    (c) $x[2] = 0.4, x[3] = 0.64, x[4] = 0.784$

    (d) $X(z) = 0.4z^{-2} + 0.64z^{-3} + 0.784z^{-4} + \cdots$

    (e) $x[\infty] = 1$

    (g) $x[0] = 0$

11.13. **(a)** $(2(0.5)^n - (0.25)^n)u[n]$

    (c) $y[0] = 1, y[1] = 0.75, y[2] = 0.4375, y[3] = 0.2344, y[4] = 0.1211$

    (e) $y[0] = 1$

11.17. **(a)** **(i)** $(0.7)^{n/2}, n = 0, 2, 4, \ldots$

        (ii) $(0.7)^{\frac{n-1}{2}}, n = 1, 3, 5, \ldots$

11.20. **(ii)**

    (a) Not stable

    (b) Unit impulse function

    (c) $14.6(1.2)^n$

11.22. **(a)** $|a| < |b^2|$

    (b) $\dfrac{z}{z - a} + \dfrac{z}{z - b^2}, |a| < |z| < |b^2|$

11.23. **(a)** $\dfrac{z}{z - 0.5}, |z| > 0.5$

    (b) $(0.5)^5 \dfrac{z^{-4}}{z - 0.5}, |z| > 0.5$

    (c) $32z^5 + 16z^4 + 8z^3 + 4z^2 + 2z + \dfrac{z}{z - 0.5}, |z| > 0.5$

    (d) $\dfrac{z}{z - 0.5}, |z| < 0.5$

    (e) $\dfrac{z^5}{32} + \dfrac{z^4}{16} + \dfrac{z^3}{8} + \dfrac{z^2}{4} + \dfrac{z}{2} + \dfrac{z}{(z - 2)}, |z| > 2$

    (f) $1 - \dfrac{z}{z - 0.5}, |z| < 0.5$

11.24. **(a)** $-\frac{3}{2}u[-n - 1] + 3/2 (0.6)^n u[-n - 1]$

    (b) $\frac{3}{2}u[n] - 3/2 (0.6)^n u[n]$

    (c) $-\frac{3}{2}u[-n - 1] - 3/2 (0.6)^n u[n]$

11.26.   **(a)** $|z| < 0.6$

$0.6 < |z| < 1$

$1 < |z| < 12$

$|z| > 12$

(b) $|z| < 0.6, (0.6)^n u[-n-1] - 3u[-n-1] - 12^n u[-n-1]$

$0.6 < |z| < 1, -(0.6)^n u[n] - 3u[-n-1] - 12^n u[-n-1]$

$1 < |z| < 12, -(0.6)^n u[n] + 3u[n] - 12^n u[-n-1]$

$|z| > 12, -(0.6)^n u[n] + 3u[n] + 12^n u[n]$

CHAPTER 12

12.2.   **(a)** $\dfrac{1}{1 - 0.5e^{-j\Omega}}$

(b) $\dfrac{0.5e^{j\Omega}}{(e^{j\Omega} - 0.5)^2}$

(c) $2e^{-2j\Omega}(1 + 2\cos(\Omega) + 2\cos(2\Omega))$

(d) $2\cos 5\Omega + 2\cos 4\Omega + 2\cos 3\Omega + 4\cos 2\Omega + 4\cos \Omega + 2$

12.5.   **(a)** $1 + 2e^{-j\Omega} + e^{-2j\Omega}$

(b) $-\Omega$

12.9. $[2, 0, 2, 0]$

12.10. $[\frac{1}{2}, 0, -\frac{1}{2}, 0]$

12.13.   **(a)** $[3, -j, 1, j]$

12.17. $\dfrac{N}{2}(\delta[k - p] + \delta[k + p])$

12.19. $W^{3k} X[k]$ or $W^{-k} X[k]$

12.22.   **(a)** $[2, -3, 0, 5, 7, -2, -6]$

(b) $[9, -5, -6, 5]$

(c) $[-6, -3, 5, 6, -2, 4, -1]$

(d) $[-6, -1, 4, -2, 6, 5, -3]$

(e) $[9, 2, -2, -4, -4, -2, 2]$

12.25. **(a)** 30

(b) 38

(c) 12

(d) 30

(e) 60

(f) [121, 5, 9, 5]

CHAPTER 13

13.1. **(b)** $\underline{x}[n + 1] = \begin{bmatrix} 0 & 1 \\ -0.8 & 0 \end{bmatrix} \underline{x}[n] + \begin{bmatrix} 0 \\ 1 \end{bmatrix} u[n]$

$y[n] = \begin{bmatrix} 1 & 0 \end{bmatrix} \underline{x}[n]$

13.4. **(b)** $x[n + 1] = 0.9x[n] + u[n]$

$y[n] = 1.35x[n] + 1.5u[n]$

(c) $H(z) = \dfrac{1.5z}{z - 0.9}$

13.5. **(b)** $x[n + 1] = 0.8x[n] + u[n]$

$y[n] = -0.4x[n] + 4u[n]$

(c) $y[n] = 0.8y[n - 1] + 4u[n] - 3.6u[n - 1]$

13.6. **(a)** $\underline{x}[n + 1] = \begin{bmatrix} 0.8 & 0 \\ 6.08 & 0.9 \end{bmatrix} \underline{x}[n] + \begin{bmatrix} 1 \\ 3.2 \end{bmatrix} u[n]$

$y[n] = \begin{bmatrix} 0 & 1.9 \end{bmatrix} \underline{x}[n]$

(b) $H(z) = \dfrac{6.08z + 6.69}{(z - 0.8)(z - 0.9)}$

13.8. **(b)** $H(z) = \dfrac{2z^2 - 5.035z + 5.0865}{z^2 - 1.9z + 0.8}$

13.10. **(a)** $x_1[n + 1] = (1 - \alpha)x_1[n] + \alpha u[n]$

$y[n] = (1 - \alpha)x_1[n] + \alpha u[n]$

(b) $H(z) = \dfrac{\alpha z}{z - (1 - \alpha)}$

13.12.   **(a)** $\underline{x}[n + 1] = \begin{bmatrix} 0.8 & 0 \\ 6.08 & 0.9 \end{bmatrix} \underline{x}[n] + \begin{bmatrix} 1 \\ 3.2 \end{bmatrix} u[n]$

$y[n] = \begin{bmatrix} 0 & 1.9 \end{bmatrix} \underline{x}[n]$

(b) $\Phi[n] = \begin{bmatrix} 0.8^n & 0 \\ 6.08[0.9^n - 0.8^n] & 0.9^n \end{bmatrix}$

(c) $\underline{x}[n] = \begin{bmatrix} 0.8^n \\ 62.8(0.9)^n - 60.8(0.8)^n \end{bmatrix}$

$y[n] = 119.3(0.9)^n - 115.5(0.8)^n, n \geq 0$

(d) $y[n] = 638.4 + 577.6(0.8)^n - 1216(0.9)^n, n \geq 0$

13.19.   **(a)** Not stable

(b) $(1.27)^n, (0.63)^n$

(c) $a = \begin{bmatrix} 1.9 & 0.8; -1 & 0 \end{bmatrix}$; eig(a)